Name	Melting point, °C	Boiling point, °C	Refractive index	Density	Solubility in water, g/100 mL	Dipole moment, μ	Dielectric constant, ϵ
				Dipolar aprotics			
Acetone **(FL)**	−94	56	1.3585	0.791	misc	2.9	20.7
Acetonitrile **(FL, TOX)**	−48	81	1.3440	0.786	misc	3.94	36.2
Nitromethane	−29	101	1.3820	1.137	9.1	3.46	38.6
Dimethylformamide (DMF)	−61	153	1.4305	0.944	misc	3.7	36.7
Dimethyl sulfoxide (DMSO)	18	189	1.4780	1.101	misc	3.96	47
N,N-Dimethylacetamide	−20	165	1.4375	0.937	misc	3.8	37.8
Formamide	2	210	1.4440	1.134	misc	3.7	110
Hexamethylphosphoramide (HMPA, HMPT) **(TOX)**	7	230	1.4579	1.030	misc	—	—
Tetramethylene sulfone	27	285	1.4840	1.261	misc	4.7	44
				Miscellaneous			
Carbon disulfide **(FL, TOX)**	−112	46	1.6270	1.266	0.3	0	2.64
Ethyl acetate **(FL)**	−84	76	1.3720	0.902	10	1.8	6
Methyl ethyl ketone (MEK)	−86	80	1.3780	0.805	27.5	2.5	18.5
Water	0.00	100	1.330	1.000	—	1.8	81.5
Formic acid **(Irritant)**	8.5	101	1.3721	1.220	misc	1.41	58
Pyridine	−42	115	1.5090	0.978	misc	2.19	12.3
Acetic acid	16	117	1.3720	1.049	misc	1.7	6.2
Nitrobenzene **(TOX)**	5	210	1.5513	1.204	0.2	4.01	35

Notes: (1) All compounds in a pure state are completely colorless except nitrobenzene which is light yellow. (2) For azeotrope information, see Table 4.1 on pp. 146 and 147.

Safety abbreviations: **FL**, flammable; **TOX**, toxic.

Other abbreviations: insol, insoluble; v.sl., very slightly; misc, miscible in all proportions; aq, aqueous.

EXPERIMENTAL ORGANIC CHEMISTRY

McGRAW-HILL SERIES IN CHEMISTRY

David L. Adams, Educational Consultant

Burgoyne: A Short Course in Organic Chemistry
Companion: Chemical Bonding
Compton: Inside Chemistry
Durst and Gokel: Experimental Organic Chemistry
Pine, Hendrickson, Cram, and Hammond: Organic Chemistry
Roach and Leddy: Basic College Chemistry
Russell: General Chemistry
Sienko and Plane: Chemistry: Principles and Applications
Waser, Trueblood, and Knobler: Chem One

EXPERIMENTAL ORGANIC CHEMISTRY

H. DUPONT DURST
Professor of Chemistry
University of Puerto Rico, Rio Piedras

GEORGE W. GOKEL
Professor of Chemistry
University of Maryland, College Park

McGRAW-HILL BOOK COMPANY
New York St. Louis San Francisco Auckland Bogotá Hamburg Johannesburg
London Madrid Mexico Montreal New Delhi Panama
Paris São Paulo Singapore Sydney Tokyo Toronto

EXPERIMENTAL ORGANIC CHEMISTRY

1234567890 DODO 89876543210

This book was set in Times Roman by Progressive Typographers. The editors were Jay Ricci and Sibyl Golden; the designer was Anne Canevari Green; the production supervisor was Phil Galea.
The drawings were done by James Stamos.
R. R. Donnelley & Sons Company was printer and binder.

Cover photograph of Cinchonine (quinine) in polarized light by Eric V. Gravé.

Library of Congress Cataloging in Publication Data

Durst, Horatio Dupont
Experimental organic chemistry.

(McGraw-Hill series in chemistry)
Includes index.
1. Chemistry, Organic—Laboratory manuals.
I. Gokel, George W., date joint author.
II. Title.
QD261.D87 547'.0028 79-18061
ISBN 0-07-018393-7

To
Margaret and Kathy,
Our Mothers, and
the Memory of Our Fathers

Better to light up than merely to shine . . .
Thomas d'Aquino

CONTENTS

Preface xix

General Information **1**

Safety **1**

Maintaining Records **4**

Calculation of Yield **8**

The Chemical Literature **9**

References Cited in Text **13**

Additional Reference Materials **14**

PART I EXPERIMENTAL TECHNIQUES

1 Physical Measurements **19**

1.1 Melting Points **19**
Experimental Procedures for Determining Melting Points

1.2 Boiling Points **32**
Experimental Procedures for Determining Boiling Points

1.3 Refractive Index **37**

1.4 Density and Specific Gravity **39**
Experimental Procedures for Determining Density and Specific Gravity

1.5 Polarimetry **41**

2 Basic Laboratory Techniques **45**

2.1 Distillation **45**
Fractional Distillation Experiment

2.2 Steam Distillation **62**
Isolation of Carvone from Caraway Seeds
2,4-DNP Derivative of Carvone

2.3 Sublimation **67**
Purification of Camphor

2.4 Crystallization **70**
Experimental Procedure for Recrystallization

2.5 Extraction **81**
Determination of Partition Coefficient
Separation of Benzoic Acid and Fluorenone

2.6 Chromatography **92**
Thin Layer Chromatography Experiment
Column Chromatography Experiment (Separation of Fluorene and Fluorenone)

2.7 Methods of Heating **114**

2.8 Removing Noxious Gases **122**

3 Qualitative Characterizations **125**

3.1 Color **126**

3.2 Odor **128**

3.3 The Flame Test and Beilstein Test **129**
Procedure for Flame Test
Procedure for Beilstein's Test for Halogens

3.4 Elemental Analysis **132**
Sodium Alloy Fusion Tests for Nitrogen, Sulfur, and Halogens

4 Solubility and Reactivity **135**

4.1 Introduction **135**

4.2 Solvation **135**

4.3 Hydrogen Bonding **136**

4.4 Lewis Base Properties **139**

4.5 London Forces **140**

4.6 Dipolar Aprotic Solvents **143**

4.7 Phase-Transfer Processes: The "Standard Catalyst" Solution 145

References 151

5 Spectroscopic Identification of Organic Compounds **152**

5.1 Ultraviolet (uv) Spectroscopy 153

5.2 Infrared (ir) Spectroscopy 161

5.3 Nuclear Magnetic Resonance (nmr) 173

5.4 Mass Spectrometry 185

Questions and Exercises 187

References 190

PART II THE EXPERIMENTS

6 Alkanes **195**

6.1 Introduction 195

6.2 Preparation 197

6.3 Solubility 198

Solubility of Alkanes

Solubility of Alkanes and Alkenes in Sulfuric Acid

Questions and Exercises 199

7 Alkenes **200**

Introduction 200

Preparation of Bromine Solution

Bromine Addition to Alkenes

7.1 Dehydration of Cyclohexanol to Cyclohexene 205

7.2 Bromination of *trans*-Stilbene 208

7.3 Isomerization of Maleic Acid to Fumaric Acid 210

7.4 Dichlorocarbene Addition of Cyclohexene by Phase-Transfer Catalysis 212

Questions and Exercises 215

8 Synthesis of Diphenylacetylene (Tolan) from Stilbene Dibromide **217**

Introduction 217

Synthesis of Diphenylacetylene (Tolan)

Questions and Exercises 222

9 The Diels-Alder Reaction **223**

9.1 Reaction of Sulfolene and Maleic Anhydride **225**
Synthesis of 4-Cyclohexene-1,2-dicarboxylic Anhydride

9.2 Reaction of Cyclopentadiene with Maleic Anhydride **229**
Synthesis of cis-Norbornene-5,6-endo-dicarboxylic Anhydride

Questions and Exercises **231**

10 Alkyl Halides **232**

10.1 Synthesis of *n*-Butyl Bromide by an S_N2 Reaction **234**

10.2 Synthesis of *tert*-Butyl Chloride by an S_N1 Reaction **237**

Questions and Exercises **238**

11 Esters and Amides **240**

Introduction **240**

11.1 Synthesis of *n*-Butyl Benzoate **243**

11.2 Synthesis of Esters by the Fischer Esterification **246**
A Synthesis of Methyl Benzoate
B Synthesis of Methyl 4-Chlorobenzoate
C Synthesis of Isoamyl Acetate (Pear Oil)

11.3 Aspirin **253**
Synthesis of Aspirin

11.4 Synthesis of *N,N*-Diethyl-*m*-toluamide: Formation of an Amide from an Acid Chloride **258**

11.5 Synthesis of Acetanilide and Phenacetin: Anhydride Acylation of Amines **261**
A Synthesis of Acetanilide
B Synthesis of Phenacetin (4-Ethoxyacetanilide)

Questions and Exercises **266**

12 Reactions of the Grignard Reagent **268**

12.1 Synthesis of Phenylmagnesium Bromide **271**

12.2 Grignard Synthesis of Alcohols **275**
A Synthesis of Triphenylcarbinol from Benzophenone
B Synthesis of Triphenylcarbinol from Methyl or n-Butyl Benzoate
C Synthesis of Benzhydrol from Benzaldehyde
D Synthesis of 4-Chlorobenzhydrol from 4-Chlorobenzaldehyde

12.3 Carbonation of Grignard Reagents **286**
A Synthesis of Benzoic Acid
B Synthesis of 3,5-Dimethylbenzoic Acid

12.4 Synthesis of Insect Pheromones by Grignard Reactions **292**
A Synthesis of Valeric Acid by Carbonation of a Grignard Reagent
B Synthesis of 4-Methyl-3-heptanol by a Grignard Reaction

Questions and Exercises **300**

13 Nucleophilic Substitution at Saturated Carbon **302**

13.1 Synthesis and Hydrolysis of Phenylacetonitrile **303**
Synthesis of Phenylacetonitrile from Benzyl Chloride and Its Hydrolysis to Phenylacetic Acid

13.2 Synthesis and Hydrolysis of 4-Chlorobenzyl Acetate **308**
Synthesis of 4-Chlorobenzyl Acetate from 4-Chlorobenzyl Chloride and Its Hydrolysis to 4-Chlorobenzyl Alcohol

13.3 Synthesis of 4-Methylphenoxyacetic Acid and 2,4-Dichlorophenoxyacetic Acid **312**
A Preparation of 4-Methylphenoxyacetic Acid from 4-Methylphenol
B Synthesis of 2,4-Dichlorophenoxyacetic Acid from 2,4-Dichlorophenol

13.4 Preparation of bis-4-Chlorobenzyl Ether by the Williamson Ether Synthesis **318**

13.5 The Malonic Ester and Acetoacetic Ester Condensations **320**
A Synthesis of Ethyl n-Butylacetoacetate by the Acetoacetic Ester Condensation
B Synthesis of Ethyl n-Butylmalonate by the Malonic Ester Condensation

13.6 Synthesis of Benzyltriethylammonium Chloride: A Phase-Transfer Catalyst **331**

Questions and Exercises **333**

14 Oxidation and Reduction **334**

14.1 Air Oxidation of Fluorene to Fluorenone **335**
A Air Oxidation of Fluorene to Fluorenone
B Partial Oxidation of Fluorene

14.2 Chromium Trioxide Oxidation of Benzhydrol and Isoborneol **342**
A Chromium Trioxide Oxidation of Benzhydrol to Benzophenone
B Chromium Trioxide Oxidation of Isoborneol to Camphor

14.3 Hypochlorite Oxidation of Benzhydrol and 4-Chlorobenzhydrol **352**
A Hypochlorite Oxidation of Benzhydrol to Benzophenone

B Hypochlorite Oxidation of 4-Chlorobenzhydrol to 4-Chlorobenzophenone

14.4 Sodium Borohydride Reduction of 4-Chlorobenzaldehyde and Fluorenone **358**

A Reduction of 4-Chlorobenzaldehyde to 4-Chlorobenzyl Alcohol

B Reduction of Fluorenone to Fluorenol

14.5 Reduction of Nitrobenzene to Aniline **364**

Questions and Exercises **370**

15 Condensations of Aldehydes and Ketones **371**

15.1 The Aldol Condensation **373**

A Synthesis of Dibenzalacetone by the Aldol Condensation

B Synthesis of Benzalacetophenone (Chalcone) by the Aldol Condensation

15.2 Reaction of Activated Hydrocarbons **379**

A Synthesis of 9-Benzalfluorene from Fluorene and Benzaldehyde

B Reduction of 9-Benzalfluorene to 9-Benzylfluorene by Hydride Transfer

C Direct Synthesis of 9-Benzylfluorene from Fluorene

15.3 Synthesis of 3,4-Methylenedioxycinnamonitrile by Acetonitrile Condensation **385**

15.4 The Benzoin Condensation **389**

Synthesis of Benzoin

15.5 The Cannizzaro Reaction **394**

Cannizzaro Reaction of 4-Chlorobenzaldehyde

Questions and Exercises **400**

16 The Friedel-Crafts Reaction **403**

16.1 The Friedel-Crafts Acylation Reaction **404**

A Synthesis of 4-Chlorobenzophenone

B Synthesis of 4-Bromoacetophenone

16.2 The Synthesis of Acetylferrocene **411**

A The Phosphoric Acid–Catalyzed Method

B The Aluminum Chloride–Catalyzed Method

16.3 The Friedel-Crafts Alkylation **416**

Synthesis of 1,4-Di-tert-butyl-2,5-dimethoxybenzene

Questions and Exercises **419**

17 Enol Bromination **420**

17.1 Synthesis of 4-Bromophenacyl Bromide **421**

A Synthesis of 4-Bromophenacyl Bromide from 4-Bromoacetophenone

B Synthesis of 4-Bromophenacyl Bromide from 4-Bromoacetophenone Using Pyridinium Bromide Perbromide

Questions and Exercises **427**

18 Electrophilic Aromatic Substitution **428**

18.1 Electrophilic Aromatic Nitration **430**

A Nitration of Chlorobenzene

B Alternate Procedure for the Mononitration of Chlorobenzene

C Nitration of Bromobenzene

D Synthesis of 1-Chloro-2,4-dinitrobenzene

E Synthesis of 1-Bromo-2,4-dinitrobenzene

F Synthesis of 3-Nitrobenzoic Acid by Nitration and Hydrolysis

18.2 Electrophilic Aromatic Bromination **448**

A Bromination of p-Xylene

B Synthesis of p-Bromoacetanilide Using Molecular Bromine

C Alternate Bromination of Acetanilide Using a Bromine Complex

Questions and Exercises **456**

19 Nucleophilic Aromatic Substitution **458**

19.1 Synthesis of 2,4-Dinitrophenylhydrazine by Nucleophilic Aromatic Substitution **459**

Questions and Exercises **460**

20 The Chemistry of Natural Products **462**

20.1 Isolation of the Naturally Occurring Stimulant Caffeine **463**

Isolation of Caffeine from Tea Leaves

20.2 Isolation of an Essential Oil from the Spice Clove **466**

Isolation of Eugenol from Cloves

20.3 Optical Resolution Using Naturally Occurring, Optically Active Tartaric Acid **469**

Resolution of Racemic Phenethylamine Using Tartaric Acid

Questions and Exercises **471**

21 The Wittig Reaction **473**

21.1 Synthesis of Diethyl Benzylphosphonate by the Arbuzov Reaction **475**

21.2 Synthesis of *trans*-Stilbene by the Wittig Reaction **478**

21.3 Synthesis of 1,4-Diphenyl-1,3-butadiene **479**

Questions and Exercises **481**

PART III QUALITATIVE ORGANIC ANALYSIS **483**

22 Tactics of Investigation **485**

22.1 Introduction **486**

22.2 Preliminary Examination **488**

22.3 Purification **490**

22.4 Boiling Points **491**
Microreflux and Capillary Methods for Boiling-Point Determination

22.5 Distillation **493**

22.6 Melting Behavior **493**

22.7 Flame Test **494**

22.8 Beilstein Test **494**
Beilstein's Test for Halogens

22.9 Specific Gravity **496**
Approximate and Precise Methods for Determination of Specific Gravity

22.10 Refractive Index **497**
Determination of the Refractive Index

22.11 Solubility **498**
Determination of Solubility in 5% Aqueous Base
Determination of Solubility in 5% Aqueous Hydrochloric Acid
Determination of Solubility in Concentrated Sulfuric Acid

22.12 Carrying On **504**

Appendix: Specific Instructions for Index of Refraction **506**

23 Carboxylic Acids and Phenols **508**

23.1 Introduction **509**

23.2 Historical **509**

23.3 Traditional Acids **510**

23.4 Operational Distinctions **511**

23.5 Typical Acids **512**

23.6 Derivatization and Reactivity **513**

Neutralization Equivalents of Acids
Amide Derivatives of Carboxylic Acids
Anilides and p-Toluidides of Carboxylic Acids
Formation of Methyl and Ethyl Esters of Carboxylic Acids
Phenacyl Ester Formation
Quaternary Ion–Mediated Formation of Phenacyl Esters

23.7 Phenols: The Other Acidic Class **525**
Ferric Chloride Enol Test
Aryloxyacetic Acid Derivatives
Bromination of Phenols
Schotten-Baumann Benzoylation of Phenols
Urethane Derivatives of Hydrocarbon-Soluble and -Insoluble Phenols

23.8 Spectroscopic Confirmation of Structure **530**

24 Amines **531**

24.1 Introduction **532**

24.2 Historical **532**

24.3 Classes of Amines **533**

24.4 Acidity and Basicity **534**

24.5 Operational Distinctions **538**
Hinsberg's Test: Classification and Derivatization
Diazotization of Primary Amines
The PTC-Hofmann Carbylamine Test for Primary Amines

24.6 Reactivity **544**

24.7 Derivatives of Primary and Secondary Amines **545**
Schotten-Baumann Benzoylation of Amines
Hydrochloride Salts of Amines
Formation of Phenylthiourea Derivatives

24.8 Derivatives of Tertiary Amines **548**
Formation of Picrate Derivatives
Formation of Tertiary Amine Methiodide Salts
Formation of p-Toluenesulfonate Salts

25 The Carbonyl Group **552**

25.1 General Tendencies **553**

25.2 Odor **554**

25.3 Structural Variety **555**

25.4 Other Structural Variations **558**

25.5 Classification 558
2,4-Dinitrophenylhydrazine Classification Test for Aldehydes and Ketones
Classification Test for Aldehydes: The Tollens Test
Baeyer Test for Unsaturation (Phase-Transfer Method)
Purpald Classification Test for Aldehydes
Schiff's Test
The Iodoform Test

25.6 Spectroscopic Confirmation of Structure 569

25.7 Derivatives of Aldehydes and Ketones 571
2,4-Dinitrophenylhydrazones of Ketones and Aldehydes: Diethylene Glycol and Ethanol Procedures
Semicarbazone Derivatives
Oxime Derivatives
Oxidation of Aldehydes to the Corresponding Acids by the Potassium Permanganate Method and the Cannizzaro Reaction
Borohydride Reduction

26 Alcohols **579**

26.1 Historical and General 580

26.2 Classes of Alcohols 580

26.3 Properties of Alcohols 582

26.4 Operational Distinctions 584
Preliminary Classification of Alcohols: The Baeyer Test for Unsaturation
Classification of Alcohols: Pyridinium Chlorochromate,Chromic Anhydride, and Ceric Ammonium Nitrate Reagents and the Lucas Alcohol Test

26.5 Spectroscopic Confirmation of Structure 592

26.6 Derivatives of Alcohols 593
Phenylurethanes and α-Naphthylurethanes
Benzoate Esters from the Acid Chloride and the Acid

27 Esters, Amides, Nitriles, and Ureas **600**

27.1 General and Historical 601

27.2 Characterization of the Classes 601

27.3 Operational Distinctions 602

27.4 Classification Tests 604
Ferric Chloride Test for Esters
Diagnostic Test for Nitriles and Amides

27.5 General Classification Scheme **607**

27.6 Spectroscopic Confirmation of Structure **607**

27.7 Derivative Formation Reactions **608**
Saponification Equivalent of Esters and Amides
Ester Saponification and Fragment Isolation
3,5-Dinitrobenzoate Derivatives of Esters
Saponification of Amides
Phenylthiourea Derivatives
Saponification of Amides and Nitriles

28 Derivative Tables **618**

28.1 Liquid Carboxylic Acids **621**

28.2 Solid Carboxylic Acids **623**

28.3 Liquid Alcohols **626**

28.4 Solid Alcohols **629**

28.5 Liquid Aldehydes **630**

28.6 Solid Aldehydes **632**

28.7 Amides **633**

28.8 Liquid Primary and Secondary Amines **639**

28.9 Liquid Tertiary Amines **642**

28.10 Solid Primary and Secondary Amines **643**

28.11 Solid Tertiary Amines **645**

28.12 Liquid Esters **646**

28.13 Solid Esters **651**

28.14 Liquid Ketones **654**

28.15 Solid Ketones **656**

28.16 Liquid Nitriles **658**

28.17 Solid Nitriles **659**

28.18 Liquid Phenols **661**

28.19 Solid Phenols **662**

Index **665**

PREFACE

The organic chemist best equipped to cope with the complexities of research is the one who is best informed and whose background is broadest. It is the major and overriding goal of this book to provide a broad, basic, coverage of experimental organic chemistry and to provide such coverage as safely as possible using the *research* or *investigative approach*.

This book is intended for the laboratory which accompanies a year-long organic chemistry course. There are typically two sorts of students who take this course, those majoring in chemistry and those going on to professional school (medicine, dentistry, etc.) or graduate school in a biological discipline. One tends to view the goals of chemical education in a slightly different way for chemistry majors and nonmajors. For the nonmajor student it is the goal of this book to provide a perspective on chemical techniques and the differences in conditions, reaction rates, work-up procedures, and so on, which are often not clear to the student after reading the lecture text. The experimental differences which can be observed in the laboratory often lead to a much clearer understanding of reactivity differences among molecules in the same class and among classes of molecules. For the student likely to continue in science, the above-stated goal is important. In addition to that, there is the goal of grounding the student in the proper approach to research.

The *research* or *investigative approach* involves several facets. The first and most important of these is recognition of the problem. After the scope of an investigation has been defined, background information must be accumulated and experiments must be done to extend the background information and address the problem at hand. The results obtained in the investigation must be analyzed, sorted through, and examined. If any unambiguous conclusions emerge, the problem is solved. It is common, however, for results to appear ambiguous. In such cases, additional experiments are often required. Naturally, the undergraduate student will not spend nearly as much time solving the problem as a researcher might, and the scope of the problems will not be nearly so broad. Nevertheless, considering the student's limited chemical background, the problems presented in this book should prove interesting and challenging.

The reader will find that throughout this book, strict attention has been paid to a *research-like organization*. It is our belief that the best researcher is the most informed one, so potential hazards and safety problems have been clearly identified. Note that many chemicals pose virtually no threat if handled properly; if handled carelessly, they can be quite dangerous.

In order to accomplish the overall goals of this book, the following organization has been used. First there is a *general information* section which begins with safety. *Safety* is absolutely crucial to the conduct of all experiments whether in student or research laboratories. Although most experiments that are chosen for undergraduate laboratories are especially safe, the fact that many students are conducting operations simultaneously increases the possibility of an accident. Furthermore, the general public has recently become cognizant of potential problems associated with chemicals. Some chemicals have been restricted from use in the chemical industry. We have been extremely careful in the design of this book to *identify potential hazards* and *exclude hazardous or restricted chemicals*. Although one always fears an oversight, we hope that this manual is as safe and up-to-date as possible.

The general information discussion continues with how to maintain a *notebook*. This is discussed in considerable detail because, after all, if there is no record of the experiment one really need not have done it. *Calculation of yield* is discussed next because this relatively simple concept is crucial to all experiments and students often have unnecessary difficulty with it. Finally, acknowledging that this volume is but a brief entry in the literature of organic chemistry, we include some *additional references* so that students, who we hope will be stimulated by this book, may search other sources and broaden their knowledge independently.

The first experimental section of this book, Part I, is titled "Experimental Techniques." A perennial problem in the undergraduate organic laboratory is determining what experiment to do in laboratory while the student is naming alkanes. We have attempted to solve this problem by introducing the student to fundamental experimental techniques and providing the basic information required throughout his or her studies.

Since all chemical compounds must be purified in one way or another, we have begun with the problems of *characterization*. The first experiments, in Chapter 1, deal with melting- and boiling-point determination, application of refractive index, and characterization by density, specific gravity and, where applicable, optical rotation. The first few weeks of organic laboratory teach manipulations required in all future chemical work. This approach continues into the second chapter, "Basic Laboratory Techniques." With purification and physical measurement techniques in hand, the student will be able to determine quite readily whether or not the desired compound has been obtained in the desired purity.

At the end of Chapter 2 (Sections 2.7 and 2.8) there is a discussion of methods of heating and the removal of noxious gases. In these brief sections we present simple and economical methods for heating reaction mixtures and for coping with certain problems of heating and ventilation.

In Chapter 4, we introduce *solubility* and *reactivity*. Most students taking this course will have learned about physical properties and solution dynamics in introductory chemistry. Nevertheless, most students have trouble conceptualizing the relationships of solutes and solvents. Since the two are intimately related, and since an understanding of reactivity derives from these concepts, we introduce these concepts together and at an early stage.

We have made an effort to introduce modern experimental methods into this lab manual wherever possible. Naturally this has included the important *phase-transfer catalysis technique*. Since the student will be learning this technique, and since it is not discussed in many organic chemistry texts, we introduce the concepts here so that the student can understand why the technique is applied. In general, phase-transfer catalysis is used throughout this laboratory manual not only so that the student may be introduced to a new synthetic method, but also because it allows experiments to be conducted less expensively and in greater safety than many traditional methods. It is hoped that this general solubility and reactivity chapter will start students thinking the way practicing organic chemists do.

In Chapter 5, *spectroscopy* is introduced. Detailed discussions of spectroscopy are found in lecture texts both in a separate chapter and where new classes of compounds are introduced. We have included here a relatively abbreviated discussion of spectroscopy which we believe covers most of the major points of theory and practice and which is a ready reference for students working in the laboratory. While not a full text on spectroscopy, this section will suffice for problems encountered in the sophomore laboratory.

Throughout this book, we have presented infrared and nuclear magnetic resonance spectra of starting materials and products wherever significant information can be obtained from them. We have focused very heavily on differences in starting material and products which are reflected in the spectra. If the student examines the spectra for each preparation and then refers back to the general discussion, we hope that the value of spectroscopy as an analytical tool will be obvious.

In Part II, "The Experiments," we present what we believe to be a representative and broad range of experiments. Since it would take a volume many, many times the size of this one to cover all classes of reactions in organic chemistry, we have necessarily had to choose carefully from among those which we considered to be the most important ones. We have attempted to integrate techniques so that two experiments similar in appearance use quite different reagents or quite different experimental approaches to do conceptually similar things. Within each chapter there may be five, six, or more experiments which are conceptually related but which involve either different principles or different techniques. For example, instructors might wish to do two different Grignard reactions: one for the synthesis of a tertiary alcohol and one for the synthesis of a carboxylic acid. Although both of these involve a Grignard nucleophile, distinctly different experimental techniques are required.

We have attempted to present the experiments much as one might organize them in a research notebook. We have presented the equation by which one hopes to prepare the product, the starting materials required, any potential safety problems, and then the detailed procedure.

The procedures presented here have been chosen not only to represent safe approaches involving a wide variety of techniques and a broad range of compounds, but also because the products are relatively easy to isolate and purify. We feel that a demonstration of reactivity that is overly difficult is not a pedagogically sound approach to instruction at this level, even though this very often occurs in the research laboratory. The products also have another characteristic: They have in general been chosen so they may be used as starting materials in other reactions, thereby allowing the instructor to assign multireaction sequences. Alternatively, these products are commercially available at relatively low prices so that the instructor who wishes to conduct only one portion of a sequence can do so economically.

The *questions and exercises* presented at the end of the chapters are designed to reinforce this research format. Many of them present hypothetical problems which might arise if an incorrect quantity or starting material was used accidentally. Such mistakes occur more commonly than one might wish in the research laboratory, and both students and instructors need to be able to cope with such problems.

It is appropriate at this point to add a note about *safety*. Because of our concern for the safety of the students and instructors who conduct these experiments, we have made some rather hard choices. We have eliminated some difficult or dangerous experiments. Two such experiments are the chlorination of cyclohexane using sulfuryl chloride and the nitration of benzene. Sulfuryl chloride is a difficult reagent to handle. This does not mean that it has not been and cannot be handled safely under close supervision. However, this book is designed to be useful in large laboratories involving many students and we have therefore deleted such "traditional" experiments that we feel are difficult or cumbersome to supervise and manage.

For a different reason we have eliminated the nitration of benzene. Although this reaction is an excellent example of an electrophilic aromatic substitution, benzene is now believed to be dangerous. *No benzene is used anywhere in this laboratory manual,* and we believe that experiments involving benzene should not be conducted in the undergraduate laboratory. We feel that safety issues will become more, not less, crucial in the years to come and that such experiments will eventually be deleted from all laboratory manuals.

Every laboratory instructor has a special feeling about which sort of experiments should be conducted as "special projects." The choice of these experiments often reflects the instructor's training and interest. It would be a feat if a laboratory manual could be printed in less than 1000 pages and include photochemistry, kinetics, heterocycles, electrochemistry, and so on, in addition to all the basics. We believe that in most cases such experiments can be supplied by the instructor.

Rather than include a range of special topics in this book, we have taken a somewhat different approach. We have included a major section, Part III, on *qualitative organic analysis.* We have done so for several reasons. Most important is that we believe that the qualitative organic approach is unique in organic pedagogy for its ability to teach a student how to reason, investigate, and eventually do organic chemical research. In doing qualitative organic chemistry, the student must cope with a broad range of compounds, new names, new reagents, and examples of reactions that might otherwise never be encountered.

We also recognize that in many undergraduate curricula qualitative organic analysis still plays an important role. The most up-to-date books available for qualitative organic analysis tend to be very large and expensive; too much so to be useful supplements in most cases. The less expensive books are either too brief or out-of-date and do not involve the use of dipolar aprotic solvents, phase-transfer catalysis, or any of the other modern methods which can improve the applicability of qualitative analysis.

To be sure, spectroscopic identification of compounds has supplanted the use of qualitative organic analysis for structure determination in many research laboratories. Nevertheless, papers appearing in the literature frequently describe cases in which simple spectroscopic identification was not practical and degradation or derivative formation was required. Sometimes derivative formation is required just so that a useful spectrum can actually be obtained. We have tried in this part to present qualitative organic analysis in such a way that the student can easily follow from step to step, find the useful procedures, and execute the various operations with a minimum of confusion. We also have included a great variety of procedures and classification tests, and have left out those which are much less reliable and primarily of historic interest. These are generally covered in the lecture text, so the student will not miss them because of their absence here.

We have picked derivatives to represent those classes of compounds which it is safe and practical to examine. Included among these are carboxylic acids, phenols, amines, alcohols, carbonyl compounds, esters, and so on. We have left out such compounds as alkanes and thiols which are vile smelling, dangerous, or relatively uninformative. We have therefore introduced in this section general reactions of manageable compounds and have tried to avoid safety problems. Many potential problem compounds are listed in the tables (Chapter 28) to make the selection process more informative for the student; not all of these will be used in this context.

We believe that our broad coverage, emphasis on safety, and research orientation continuing into a detailed discussion of qualitative organic analysis make this laboratory manual unique among those currently available. We sincerely hope that it will meet the needs of many instructors, especially those who wish to instill in their students an appreciation of experimental techniques as well as of reactivity and spectral concepts.

We would like to thank our colleagues who have cooperated with us in the class testing of this manual and who have offered numerous and generally helpful suggestions. Notable among these are Professors Joseph A. Dixon, Herman G. Richey, Maurice Shamma, Robert Minard, and R. A. Olofson; Drs. Dennis Hoskin, Blanche Garcia, Stephen A. DiBiase, and Stephen H. Korzeniowski; and Messers. David Forrest, Arthur Shores, and Craig Diamond. We would like to extend our particular thanks to Professors David L. Adams of North Shore Community College, Joseph Casanova of California State University at Los Angeles, Mary Chisholm of the Behrend College, William W. Epstein of the University of Utah, James L. Jensen of California State University at Long Beach, Michael M. King of The George Washington University, C. Peter Lillya of the University of Massachusetts, R. Daniel Little of the University of California at Santa Barbara, Harry A. Morrison of Purdue University, Jack Timberlake of the University of New Orleans, and William P. Weber of the University of Southern California, all of whom read the manuscript with a great deal more care than could have reasonably been expected. We are in their debt.

We wish to express our sincerest thanks to Drs. Irwin Klundt and Charles J. Pouchert for permission to use physical constants from the *Aldrich Catalog-Handbook,* and to reproduce infrared spectra from C. J. Pouchert, *The Aldrich Library of Infrared Spectra,* 2d edition (Aldrich Chemical Company, Milwaukee, 1975), and proton magnetic resonance spectra from C. J. Pouchert and J. R. Campbell, *The Aldrich Library of NMR Spectra* (Aldrich Chemical Company, Milwaukee, 1974).

There is no sufficient way to thank our wives and our research students for their help and understanding during this project. Finally, we regret any errors either of omission or commission which have survived the proofreading process. We would appreciate having these called to our attention and apologize in advance for any inconvenience they may cause.

H. Dupont Durst
George W. Gokel

GENERAL INFORMATION

SAFETY

Accidents of almost any sort are rare in undergraduate organic chemical laboratories. This is the fortunate consequence of careful design of the experiments and also of careful attention of instructors and most students. Nevertheless, the hazard of accidents exists at all times. We call attention to the possibility here so that you will be aware of how important safety is in *any* laboratory at *any* time.

In this book special potential difficulties or hazards are listed at the beginning of every experiment. Very detailed directions are given, and if these directions are followed, no problem should be encountered. However, one can never be certain that everyone in the laboratory is equally alert, so it is a good idea to know all the regulations and safety procedures in case someone close to you has an accident and cannot cope effectively with it.

The best general advice regarding safety that we can give here is threefold:

1 *Read the directions given in every experiment very carefully and in advance*. Know precisely what you are going to do before you come into the laboratory.
2 *After reading the experiment and any ancillary material needed, think about the directions*. Try to visualize in your mind how the apparatus is going to be set up, what operations will be required, and how you are going to carry out the experiment that is assigned to you. Know what hazards are associated with certain things that are mentioned in the experiment. Think about what procedures might be hazardous. Consider when a flame or an open vessel might be dangerous. Think about when a hood might be required.

3 *Use common sense.* If something looks dangerous, it probably is. If someone near you appears not to be using good judgment, help him or her out. Point out the hazard. If you keep someone from having an accident, you have performed a very important and valuable service, not only for that person but for everyone in the laboratory whose time would be wasted.

The following are rules which are of a more or less general nature. Every laboratory has its own safety rules, regulations, and procedures. Each instructor will have his or her own preference for how certain things should be done in an emergency or at a time of difficulty. If you are issued a separate set of safety regulations, read them carefully. Use the regulations and procedures about which you have been informed. If no specific directions are given, the rules which we have recorded below will serve as general safety regulations and should assist you if any difficulty arises.

1. Wear safety glasses. Wearing appropriate safety glasses in undergraduate laboratories is required by law in most, if not all, states. Wearing them is extremely important. The loss of sight in even one eye can be very debilitating indeed, and sight can be lost very quickly if certain chemicals get into your eyes. Contact lenses are not a substitute for safety glasses. Contact lenses in many ways are more dangerous than ordinary glasses or no glasses at all because chemicals can get under the contact lens and be held in proximity to the tissue they are damaging. **Should any chemical reach your eyes, immediately flush your eyes with lots of water.** Rinse them thoroughly. Inform your laboratory instructor as soon as your eyes have been washed. In any situation, if there is evidence of a hazard, inform someone in authority. Have your eyes checked at the student health center or at some nearby hospital. Be certain *on medical authority* that no danger persists.

Safety glasses (with side shields if available) should also be worn even if you feel you are not carrying out a hazardous operation. The person at the bench next to or across from you might be doing something potentially dangerous, and you could be in danger even if you are doing nothing more than reading your laboratory manual. **Wear your safety glasses at all times when you are in a chemical laboratory. There is no exception to this rule.**

2. Any accident which occurs should be reported immediately to the laboratory instructor. This includes cuts, burns, spills, or any potential hazard which might occur. The laboratory instructor may administer first aid or may arrange for medical attention. It is very important that your laboratory instructor know if anything has occurred, because the instructor will be much more knowledgeable about what hazards specific incidents may pose to your health.

3. If you spill anything on you, wash it off immediately. Most chemicals are dangerous only if they linger. Even concentrated sulfuric acid will not be very injurious if it is washed off immediately. If it is allowed to remain on the skin for any period of time, very severe damage can be done. Remember, whenever there appears to be any kind of danger, you should save yourself and your laboratory partners from injury. Laboratory equipment, even buildings, can be replaced, but life and limb are irreplaceable. If there is any danger, save yourself; keep yourself from being hurt.

4. Know the location of all fire extinguishers, eye-wash stations, and safety showers. In the event of an emergency you can move to the appropriate emergency station immediately or you can direct others to do so.

5. Flames. There are many sophisticated devices in use for heating flasks or reactions vessels, but the bunsen burner, or open flame, remains common. It poses a danger, however, because any flammable vapors which come in contact with it can easily ignite. If you find it necessary to distill using a flame, be certain that no flammable liquid is near. Ask anyone who happens to be using a flammable substance to wait until your heating operation is complete, or else move to another area. Remember, especially when working with a laboratory partner, that the two of you are to help keep each other safe.

6. In the event of a fire move quickly away from the burning area. Remember again that equipment and experiments can be replaced and that you should try to save yourself immediately. Move away from the flame quickly. Inform the instructor and other people so that they can move rapidly away from the danger. A fast and effective way to do this is to shout "Fire!"

7. If someone's clothing is burning, immediately help the person to the ground and smother the flames with a fire blanket (this will be located in a strategic position in the laboratory). If no blanket is available roll the person over the floor while spraying the flames with a fire extinguisher.

8. Use gloves, glasses, and inexpensive and protective clothing when you come to the laboratory. The breathing and handling of small amounts of noxious substances probably do not pose an immediate danger to a chemical worker. Nevertheless, experienced chemists always, under all circumstances, try to avoid contacting any potentially noxious chemical in any way. Even mild dishwashing detergents can cause chapped hands. Organic solvents are much more potent than dishwashing liquids and should be treated with appropriate respect. When pouring liquids or when cleaning glassware which may contain harsh materials, use rubber gloves. If you are handling a material which expels noxious vapors, confine it to a well-ventilated hood. Remember, if you can smell a substance, you are breathing it. If you are breathing it, it is going into your lungs and there is the potential for some sort of tissue damage.

9. The use of benzene. In April 1977, the Occupational Safety and Health Administration, often referred to as OSHA, imposed new standards regarding workers exposed to benzene. The new standard reduces worker exposure from 10 parts per million (ppm) to 1 ppm, averaged over 8 h. This action was taken because benzene is believed to be responsible for an abnormally high incidence of leukemia in workers exposed to it. Although no use of benzene is required by this laboratory manual, if you encounter the use of benzene in any other preparative situation, the following safety regulations should always apply:

a When using benzene always work in a well-ventilated hood (hood flow: 150 linear ft/s with the hood door open).

b Never breathe benzene vapors.

c Avoid any situation which could lead to benzene spillage on skin or clothing. If benzene spills on your clothing, wash it off, remove the clothing, and clean yourself.

d If any benzene is spilled on the lab bench, wash the area with water and if possible confine the spill to a hood area.

10. A final note regarding clothing. Since there is the possibility of destroying clothing in a laboratory accident, inexpensive clothing or a lab coat or apron should be worn. Expensive clothing is destroyed as easily as inexpensive clothing and it would be good to wear jeans and an old shirt to laboratory. In addition, sandals or thong shoes should be avoided and are prohibited in most laboratories. Shorts, open shirts, midriff blouses, and any other clothes which leave large areas of skin unprotected are extremely undesirable. Whatever you wear, remember that clothing, like equipment, can be replaced, whereas tissue, life, and limb cannot be.

MAINTAINING RECORDS

The object of exercises and experiments in the organic chemistry laboratory is to learn techniques and to obtain a general understanding of how compounds react. In the research laboratory, the chemist usually has a background in basic techniques and known reactions. The aim, therefore, is to gain new knowledge or to prepare new compounds. Since an experiment is usually done by a single research worker, the knowledge gained from it will be only that worker's unless some mechanism exists for passing it on to others. Since the human memory is fallible, a detailed record of each experiment must be kept. It must be possible for some other reasonably skilled worker to reproduce exactly what the first researcher did for the experiment to be of value. The information developed in an experiment is not worth having if it cannot be reproduced. In order to assure the accurate recording and transmission of knowledge, each research chemist maintains a notebook in which records are kept.

Several general rules concerning the keeping of laboratory records will be mentioned here. Each laboratory instructor and research chemist has a preference regarding the exact details of keeping a notebook. You should check with your laboratory instructor and find out what preferences are enforced in your laboratory. Gradually, as you gain more experience and especially if you go on in chemistry, you will develop preferences of your own. These preferences will be a synthesis of background information, such as that discussed here, and your own experience. Some guidelines are presented below.

1. *Use a hardbound permanent notebook.* Although spiral-bound notebooks are fine for taking class notes, if a permanent record is needed, a bound notebook is preferable. The notebook itself need not be expensive; it simply needs to be permanently bound. Hardbound composition notebooks, approximately 8 by 10 in, are usually satisfactory for most laboratories. Occasionally, special laboratory notebooks with carbon-copy pages are required, and these may be specified for your course.

2. *If the pages of the notebook are not numbered, you should number them yourself.* The pages should be numbered consecutively starting with the first page. Generally, the upper right-hand corner of the right-hand page and the upper left-hand corner of the left-hand page are numbered. Often the first four or five pages are left blank (although the numbering is begun on the first page) so that a table of contents may gradually be added as the notebook is filled. Since each experiment will usually not consume more than two lines in the table of contents, the total number of pages in the notebook may be divided by the number of lines per page and then multiplied by two to determine how many pages should be reserved for the contents. A table of contents is optional and may not be required in your laboratory section. Again, check with your instructor.

3. *Use a new page for each experiment.* Date each page as you begin taking notes. Before you begin an experiment, you should have read the procedure carefully. You will know how long the experiment is and have some idea about how much information will have to be recorded. If several pages will be required to take all the necessary notes, leave two or three extra pages after you begin the entry before beginning the next one. This should be done even if you are doing two experiments on the same day. This is largely a matter of convenience, for when you wish to reproduce the experiment you can simply flip from page to page and not have to search around your notebook. On the other hand there is no rule which says you cannot refer to a different page in your notebook if it becomes necessary. Some instructors suggest that results be recorded on the right-hand page and that the left-hand page be left for making notes.

4. *Write in ink in your notebook so that the record will be permanent.* If you err, simply line through the error and add the correction. The choice of what kind of pen you use is largely your own. There are a couple of considerations however which are worth keeping in mind. A ball-point pen is often more convenient than a fountain pen, marker, or "roller-pen." A ball-point pen is especially convenient because the ink does not smear as readily

when water happens to contact it. However, it has the disadvantage that often it does not give very good photocopies. A second disadvantage of a ball-point pen is that the ink diffuses slowly into the fiber of the paper. As a result, after several years the ink will tend to smear throughout the page, making a carefully written notebook page almost illegible. Usually, however, the undergraduate chemical laboratory record needs only to survive for two academic terms or a year.

If the notebook is maintained for the research laboratory, be it academic or industrial, a fountain pen or a marker is usually preferable. The ink from a fountain pen must diffuse quickly into the paper and dry quickly so that no further diffusion is possible. This is so because if the ink remained wet for a long time it would be extremely inconvenient to use. The dyes which are used in fountain-pen ink usually can be photocopied much more readily and provide a much more legible copy than do ball-point-pen inks. If you are considering maintaining a long-term record of an experiment, you may wish to consider these problems.

5. *The general format of a notebook page is approximately as set forth in each of the experiments described in this book.* In the upper right- or left-hand corner should be a page number, and on the upper left-hand side should be the date on which the experiment is commenced. The title of the intended experiment should follow. After that, an equation describing the intended transformation should be written. Next, in a left-hand column, the various reagents which will be used should be listed. To the right, next to each reagent, should be the quantity of the reagent and any important information concerning its purity or source. The procedure should then be written down exactly as it is carried out. If the procedure is exactly as described in another source, such as this laboratory manual, it is usually not necessary to write down the entire procedure. If the procedure is followed exactly, it should be noted and the reference cited. On the other hand, if anything unusual happens during the procedure or any observation is made which differs from that suggested by the reference source, it should be carefully noted. After the procedure has been carried out, the weight of the product should be recorded, its physical properties noted, and a yield calculation carried out. All notes regarding the weights of flasks and the yield calculations should be written on the notebook page. If there is an error in calculation, it can often be retraced by the fact that the notes have been made in the book. If an error is made, simply line through it and continue writing. The notebook is intended to be a permanent record, not necessarily a work of art. It should, of course, be kept carefully, and it should be as legible as possible in view of the frantic circumstances often surrounding the carrying out of an experiment.

A sample notebook page is shown here. It refers directly to an experiment described in this book (Exp. 14.1). Look at the experiment even though

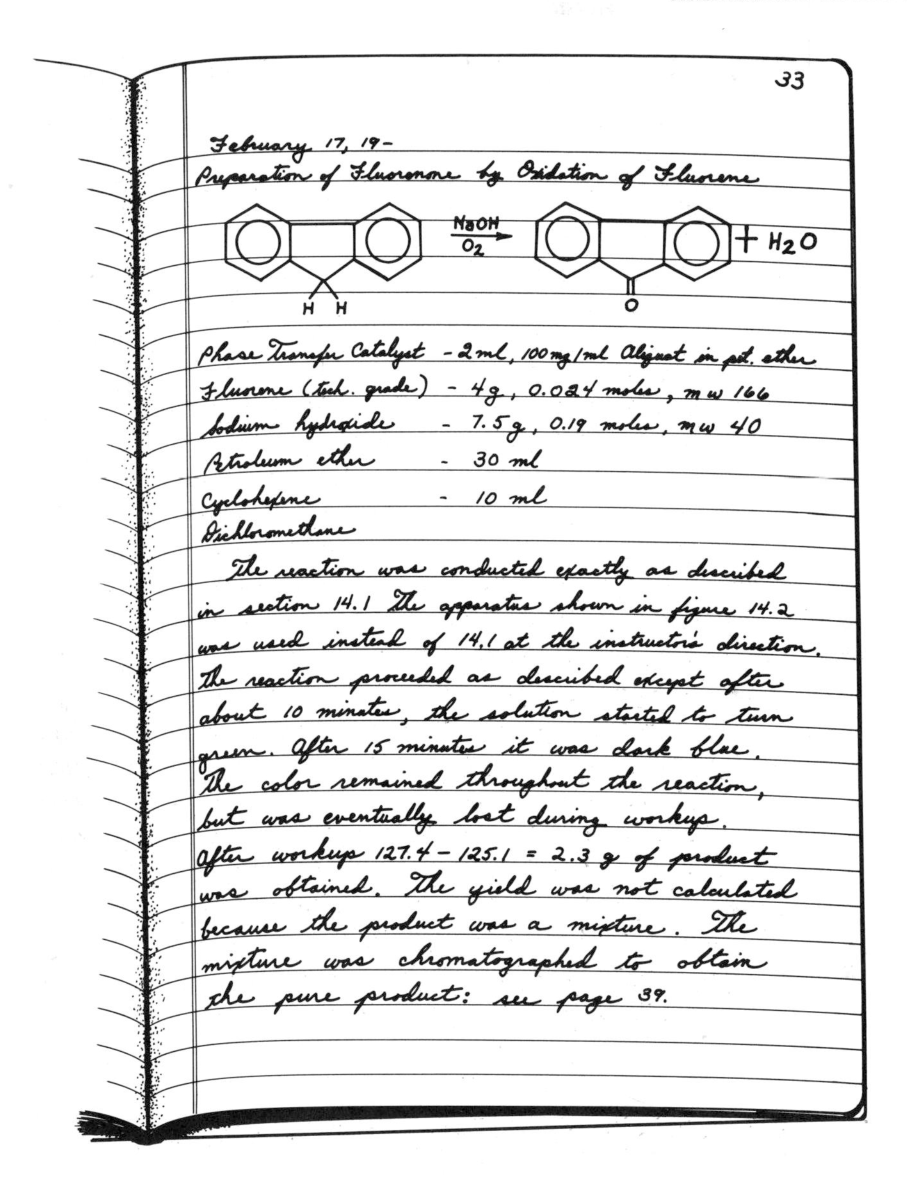

33

February 17, 19-

Preparation of Fluorenone by Oxidation of Fluorene

Phase Transfer Catalyst - 2 ml, 100 mg/ml Aliquat in pet. ether
Fluorene (tech. grade) - 4 g, 0.024 moles, mw 166
Sodium hydroxide - 7.5 g, 0.19 moles, mw 40
Petroleum ether - 30 ml
Cyclohexene - 10 ml
Dichloromethane

The reaction was conducted exactly as described in section 14.1 The apparatus shown in figure 14.2 was used instead of 14.1 at the instructor's direction. The reaction proceeded as described except after about 10 minutes, the solution started to turn green. After 15 minutes it was dark blue. The color remained throughout the reaction, but was eventually lost during workup. After workup 127.4 − 125.1 = 2.3 g of product was obtained. The yield was not calculated because the product was a mixture. The mixture was chromatographed to obtain the pure product: see page 39.

you may not yet be familiar with all the techniques. Note how differences are recorded during the procedure and the fact that the entire procedure is not recorded. In this latter connection some instuctors may feel that it is of value to you to write down a procedure in your own notebook in detail so that you will gain practice in recording procedures. This is a perfectly valid principle and should be followed if your instructor so directs you. In this case, the entire procedure may be recorded word for word from your manual or other reference, but it is usually advisable to rephrase it in your own words so that you get practice, not only in writing your own procedures, but also in recording them exactly as they occur. Read over the sample notebook page and note exactly how the record of this experiment has been kept. Note also that this is not the only acceptable format for a notebook, but is one which is used in many laboratories. Your instructor will advise you if a different format is desired, and you should follow those instructions.

CALCULATION OF YIELD

One of the exercises crucial to any experimental work in organic chemistry is the calculation of yield. This calculation, which can be accomplished quite simply in most cases, often causes students a great deal of unnecessary difficulty. The concept underlying yield is a very simple one: How much material was obtained in relation to the greatest amount that could possibly have been obtained?

In order to calculate the yield it is necessary to write the balanced equation for converting reactants to products. After this first step, the limiting reagent must be identified, i.e., that reactant among those which form a portion of the product or from which the product derives that is present in the smallest molar amount. The conversion of starting material into product must be conceptualized on a molar basis, as illustrated by the acid-catalyzed esterification of benzoic acid with methyl alcohol.

$$C_6H_5\text{—}COOH + CH_3OH \underset{\Delta}{\overset{H^+}{\rightleftharpoons}} C_6H_5\text{—}COOCH_3 + H_2O$$

Esterification reactions are typically conducted by heating the carboxylic acid in the presence of a very large excess of the alcohol. A trace of acid is also present as a catalyst. For example, benzoic acid is refluxed with 10 molar equivalents of methyl alcohol in the presence of a drop of sulfuric acid. The product, methyl benzoate, forms from the acid and the alcohol, and one molecule of water is produced for each molecule of ester. The reagent present in the smallest molar amount is sulfuric acid. However, this is *not* the limiting reagent because it does not appear in the product; its function is purely catalytic. Since there is 10 times as much methanol as benzoic acid, the latter is the limiting reagent.

How much ester can possibly be produced if the esterification reaction is conducted with 12.2 g benzoic acid? First determine how many moles of benzoic acid correspond to 12.2 g. The molecular weight of benzoic acid is 122; therefore 12.2 g represents 0.10 mol. No matter how vigorous the conditions or how devoutly we hope, no more than 0.10 mol of methyl benzoate can possibly be obtained. One-tenth of a mole of methyl benzoate corresponds to 13.6 g because 1 mol methyl benzoate weighs 136 g. If our reaction yields 6.8 g of product, we will have obtained half the amount that we possibly could have obtained, i.e., the percent yield is 50%. This figure is obtained by dividing the number of grams of product (actual yield) by the number of grams which could have been produced (theoretical yield) and multiplying by 100.

$$\frac{\text{Grams of product (produced)}}{\text{Grams of product (possible)}} \times 100 = \%\ \text{yield}$$

In this case the equation would be

$$\frac{6.8\ \text{g}}{13.6\ \text{g}} \times 100 = 50\%$$

The calculation can be done by comparing either the number of moles or the number of grams produced with the number which could have been produced. The equation for the mole calculation is shown below:

$$\frac{\text{Moles of product (obtained)}}{\text{Moles of product (possible)}} \times 100 = \%\ \text{yield}$$

The same would be true of a milligram calculation. The important consideration is the amount of material obtained in relation to the amount theoretically possible. The yield calculation is carried out in this way regardless of the reagents which are used or circumstances of the reaction.

The preparations presented in many laboratory manuals specify "typical yield." This is the yield a skilled worker can expect to obtain after by-products or other impurities have been removed. It should not be confused with the theoretical yield. The theoretical yield of any reaction is 100%; a typical yield may be only 10%.

THE CHEMICAL LITERATURE

The experiments found in this laboratory manual have been selected to represent many classes of organic reactions. The standard preparations presented in the first part of this book, together with the many reactions presented in the qualitative organic analysis section, exemplify most of the reactions that you will encounter in your undergraduate organic chemistry course. There are, however, many preparations which are not represented in this book and many reactions in which you may have an interest which you cannot find in this book. In addition, further pursuit of organic chemistry will almost certainly lead you to other classes of compounds and to preparations which are not simple extensions of those presented here. When difficulty arises in locating preparations for new compounds or in preparing a completely new class of compounds, the chemical literature should be examined to find what information it reveals.

Locating Preparations

If the synthesis of a relatively simple compound is required, a simple extension of a preparation given in this book may suffice. For example, if reduction of *p*-methoxybenzaldehyde to *p*-methoxybenzyl alcohol is desired, the preparation for *p*-chlorobenzyl alcohol (Sec. 14.4) can be simply adapted. If no extension from a procedure in this book is obvious, it is usually a good idea to look for a specific preparation of the desired compound. Other organic laboratory manuals may be consulted if the compound is a relatively simple one.

There are two sets of books worth mentioning in this connection. The five-volume set entitled *Organic Syntheses*[1] represents collections of annual volumes which have been published for over 50 years. In each of these 10-year compilations you will find many preparations of useful compounds. These preparations have the special advantage that each procedure has been checked independently by a member of the editorial board. Although these procedures are not foolproof (no procedure really is), they are as tried and true as is practically possible. Each volume is well indexed according to compound type, reaction type, name of compound, etc., and it is relatively easy to locate a particular compound. In addition, both an up-to-date cumulative index[2] and a reaction index[3] are available.

Occasionally, general information about the reactions or the preparation of a certain type of compound is desired. In such a case, it is often useful to consult the series of books called *Organic Reactions*.[4] The series now numbers almost 30 volumes, and each volume contains several lengthy review chapters dealing with a specific class of compounds. It is a characteristic of each review article that general information is given, along with an extensive discussion of the particular compounds which have been studied or prepared, according to the scope of the chapter. Not only can the *Organic Reactions* review chapters give you general information, but they can often point you to specific references available in the original literature.

In addition to these major sources of information, there are several organic chemistry laboratory manuals which have become more or less standard. Among these is a book called *Reactions of Organic Compounds*, by W. J. Hickinbottom.[5] Although this book is dated, detailed preparations for many classes of compounds are given in it. Most of the compounds that you would encounter in the undergraduate organic laboratory are probably dealt with in this book. A second major source of information and preparations is *Preparative Organic Chemistry*, by G. Hilgetag and A. Martini.[6] This is a massive volume and, although it is found in few private collections, it is available in almost

[1] Full references to the books discussed in this and the next section are listed in References Cited in Text, pp. 13 and 14.

every library. A more common volume of considerable utility is the book by A. I. Vogel entitled *Experimental Organic Chemistry*.[7] It also contains a large number of preparations and may be useful in locating a particular compound.

Locating Compounds and Derivatives

If specific information concerning a compound is desired, it may have to be sought in sources other than general texts. For example, if one wishes to locate the melting point, the color, or even the melting point of a derivative of a particular compound, special sources must be consulted. The source of greatest scope in this connection is Beilstein's *Handbuch der Organischen Chemie*,[8] which is published in Germany. This set consists of many, many volumes and describes an enormous number of compounds. The information which is provided in it includes melting points of derivatives, properties such as solubility and color, and methods of preparation. This is undoubtedly the best possible source to consult. Unfortunately, it is printed only in the German language, making it difficult for many American undergraduate students to use. There are several excellent books describing how Beilstein's text is indexed, and with one of these books most students with only a rudimentary knowledge of German can use this important reference work.

It will probably be more convenient for most students to consult somewhat simpler reference sources. One of these is the *Dictionary of Organic Compounds*.[9] This multivolume work lists a great number of compounds, indexed alphabetically. For each compound as much information as is practically possible is included. Typical information is melting point, color, method of preparation, refractive index if applicable, solvent from which the compound may be recrystallized, and what derivatives are easily formed from it. This source thus constitutes a major and important reference for students undertaking qualitative organic analysis.

Two other convenient reference books are the *Handbook of Chemistry and Physics*[10] and the *Handbook of Tables for Organic Compound Identification*.[11] The *Handbook of Chemistry and Physics* is a general information handbook which contains much more than the melting points of organic compounds. Major sections of this book are devoted to the chemical and physical properties of organic compounds. It is therefore an important reference work for qualitative analysis and chemistry in general.

For qualitative organic analysis, where specific reference to derivative melting points is desired, the best available source is the *Handbook of Tables for Organic Compound Identification*. Rather than being arranged alphabetically, it is arranged by compound class. Within a class of compounds, e.g., aldehydes, all individual compounds are classified according to whether they are liquids or solids and further categorized in order of increasing boiling or melting point. As a result, if one has minimal information concerning the compound, such as the fact that it is a liquid aldehyde, one can locate the refractive index, melting point of the dinitrophenylhydrazone derivative, and other properties of most of the reasonable possibilities by referring to these tables.

Using *Chemical Abstracts*

The Chemical Abstracts Service under the auspices of the American Chemical Society indexes all chemical literature as it appears. At various intervals all compounds, subjects, and authors are indexed. This is done most frequently on a semiannual basis, and at 5- to 10-year intervals major cumulative indexes are published.

The author index lists contributions to the chemical literature alphabetically by the last name of each author of each paper. If there are several authors, the name of each will be found in the author index of *Chemical Abstracts*. An author's name will be of relatively little value if you are searching for a particular compound or preparation unless you happen to know who did the work.

If you are interested in a general class of compounds or even a particular compound which may be listed as a subject in *Chemical Abstracts,* you should avail yourself of the general subject index. Classes and individual compounds are listed in this by their *Chemical Abstracts* names (there are often several possible and you should try all). Subheadings will usually indicate whether or not the particular paper cited will be of interest to you. It is sometimes necessary to peruse a number of papers in order to find the specific information desired.

To locate references to the preparation of a particular compound, it is usually most expeditious to consult the molecular formula index. Compounds are listed by molecular formula in alphabetical order. For example, acetone would be listed under C_3H_6O. Other elements which are present in compounds containing carbon and hydrogen will be listed in alphabetical order after those two elements; for example, trichloroacetone would be listed under $C_3H_3Cl_3O$. Abstracts of publications in which the desired compound is mentioned may be located with relative ease by this approach.

It is often the case that several compounds have the same molecular formula. You must therefore scan the list for that formula which corresponds to the structure in which you are interested. When you locate the particular compound you may once again find several entries. Many of these abstracts will be of little value; several papers may have to be checked. The use of the compound index has one particular advantage: it allows you to survey abstracts and papers which mention your compound in the fastest possible way and removes the ambiguity of nomenclature.

Learning about a Subject

Access to the original literature may easily be gained from a perusal of the *Chemical Abstracts* subject index. This is usually not the best way of surveying the literature because most papers in the original literature deal with relatively narrow subject matter. In order to learn about a particular chemical subject, the best thing to do is look in the card file in the library. It is an astounding fact that there are often complete volumes on what one might consider to be the most obscure subjects. If you can find a complete book on the subject, it will be of great help because it will contain a survey of the literature and usually an extensive discussion of principles.

If there is no monograph listed in the card file on the particular subject in which you are interested, it is usually a good idea to consult the *Index of Reviews in Organic Chemistry.*[12] If you cannot locate this index, check with the librarian who will almost always know the location of this important reference tool. Check through the *Index of Reviews* to discover what review literature has been published concerning your particular subject. This will refer you to original monographs, which you may or may not have in your library, or to such volumes as *Organic Reactions, Organic Syntheses,* and many other special review series and journals. Checking the index issues of such review journals as *Quarterly Reviews, Accounts of Chemical Research,* and *Chemical Reviews* will often turn up useful references. In principle, any information yielded by the review journal indexes will also be found in the *Index of Reviews.* As a result, checking these may in some cases be redundant.

The Original Literature

The original reports of chemical discoveries usually appear in specialized chemical journals. One of the most important of these for the organic chemist is the *Journal of Organic Chemistry.* In addition to this journal, devoted entirely to organic chemistry, several other journals are important in this connection. The journals *Tetrahedron* and *Tetrahedron Letters* are also devoted exclusively to organic chemistry. A number of British journals published by the Chemical Society of London also contain much useful information, as does the *Journal of the American Chemical Society.* In Germany, *Justus Liebigs Annalen der Chemie* and *Chemische Berichte* are also important chemical journals. Most national chemical societies publish a journal, part of which may be devoted to organic chemistry. In general, it is not worthwhile to begin leafing through a journal. It is far more practical to consult a monograph, a review, or *Chemical Abstracts* before jumping into the original literature. The original journals are mentioned here only so that you will know some of the names in the event that you have to consult this literature. A list of additional reference materials is found after the reference citation for this section.

REFERENCES CITED IN TEXT

1 *Organic Syntheses,* vols. 1–5, Wiley, New York, 1932–.

2 L. Shriner and H. Shriner (eds.), *Organic Syntheses, Collective Volumes I, II, III, IV, V, Cumulative Indices,* Wiley, New York, 1976.

3 S. Sugasawa and S. Nakai, *Reaction Index of Organic Synthesis,* Hirokawa Publishing Company, Tokyo, and Wiley, New York, 1967.
4 *Organic Reactions,* Wiley, New York, 1942– (30 volumes).
5 W. J. Hickinbottom, *Reactions of Organic Compounds,* Wiley, New York, 1962.
6 G. Hilgetag and A. Martini, *Weygand/Hilgetag Preparative Organic Chemistry,* Wiley, New York, 1972.
7 A. I. Vogel, *A Textbook of Practical Organic Chemistry,* 3d ed., Wiley, New York, 1966.
8 *Beilsteins Handbuch der Organischen Chemie,* Springer-Verlag, K. G., Berlin, 1918–.
9 J. R. A. Pollock and R. Stevens (eds.), *Dictionary of Organic Compounds,* 4th ed., Oxford University Press, New York, 1965.
10 R. C. Weast (ed.), *Handbook of Chemistry and Physics,* 56th ed., CRC Press, Cleveland, 1975.
11 Z. Rappoport, *Handbook of Tables for Organic Compound Identification,* CRC Press, Cleveland, 1967.
12 D. A. Lewis, *Index of Reviews in Organic Chemistry,* Cumulative Issue, The Chemical Society, London, 1971.

ADDITIONAL REFERENCE MATERIALS

I. General Reference

Carey, F. A. and R. J. Sundberg: *Advanced Organic Chemistry, Part B, Reactions and Synthesis,* Plenum, New York, 1977.

House, H. O.: *Modern Synthetic Reactions,* 2d ed., W. A. Benjamin, Menlo Park, Calif., 1972.

Norman, R. O. C.: *Principles of Organic Synthesis,* Methuen, London, 1968.

Wagner, R. B. and H. D. Zook: *Synthetic Organic Chemistry,* Wiley, New York, 1953.

II. General Synthetic Chemistry

Fuson, R. C.: *Reactions of Organic Compounds,* Wiley, New York, 1962.

Hickinbottom, W. J.: *Reactions of Organic Compounds,* 3d ed., Wiley, New York, 1957.

Ireland, R. E.: *Organic Synthesis,* Prentice-Hall, Englewood Cliffs, N.J., 1969.

Parham, W. E.: *Synthesis and Reactions in Organic Chemistry,* Wiley, New York, 1970.

Turner, S.: *The Design of Organic Synthesis,* Elsevier, New York, 1976.

Vogel, A. I.: *Textbook of Practical Organic Chemistry,* 3d ed., Wiley, New York, 1956.

III. Reagents and Methods

Buehler, C. A. and D. E. Pearson: *Survey of Organic Synthesis* (2 vols.), Wiley-Interscience, New York, 1970.

Fieser, M. and L. F. Fieser: *Reagents for Organic Synthesis* (7 vols.), Wiley-Interscience, New York, 1968–.

Harrison, I. T. and S. Harrison (vols. 1 and 2); L. S. Hegedus and L. Wade (vol. 3): *Compendium of Organic Synthetic Methods,* Wiley-Interscience, New York, 1971.

Pizey, J. S.: *Synthetic Reagents* (3 vols.), Halstead Press, Wiley, New York, 1974.

Sandler, S. R. and W. Karo: *Organic Functional Group Preparations,* Academic Press, New York, 1968.

Organic Reactions, Wiley, New York, 1942–, 30 vols.

Organic Syntheses, vols. 1–5, Wiley, New York, 1932–. (See above for indexes.)

Theilheimer, W.: *Synthetic Methods of Organic Chemistry,* S. KArger, New York, 1948–.

Annual Reports in Organic Synthesis, Academic Press, New York, 1970—.

IV. Reviews of Syntheses

Akhrem, A. A. and Y. A. Titov: *Total Steroid Synthesis,* Plenum, New York, 1970.

Anand, N., J. S. Bindra, and S. Ranganathan: *Art in Organic Synthesis,* Holden Day, San Francisco, 1970.

ApSimon, J.: *The Total Synthesis of Natural Products* (3 vols.), Wiley-Interscience, 1973.

Djerassi, C.: *Steroid Reactions,* Holden Day, San Francisco, 1963.

Danishefsky, S. E. and S. Danishefsky: *Progress in Total Synthesis,* Appleton-Century-Crofts, New York, 1971.

Fleming, I.: *Selected Organic Syntheses,* Wiley, New York, 1973.

Nakanishi, K., T. Goto, S. Ito, S. Natori, and S. Nozoe: *Natural Product Chemistry,* vols. 1 and 2, Academic Press, New York, 1975.

V. Indexes and Journals Containing Reviews

Index of Reviews in Organic Chemistry

Accounts of Chemical Research

Angewandte Chemie, International Edition in English

Chemical Reviews

Chemical Society Reviews (formerly *Quarterly Reviews*)

Journal of Chemical Education

Organic Preparations and Procedures

Synthesis

Tetrahedron

VI. Reference Books on Qualitative Organic Analysis

Cheronis, N. D. and J. B. Entriken: *Identification of Organic Compounds,* Wiley-Interscience, New York, 1963.

Cheronis, N. D., J. B. Entriken, and E. M. Hodnett: *Semimicro Qualitative Organic Analysis,* 3d ed., Wiley-Interscience, New York, 1965.

Criddle, W. J. and G. P. Ellis: *Spectral and Chemical Characterization of Organic Compounds,* Wiley, New York, 1976.

Pasto, D. and C. Johnson: *Laboratory Text for Organic Chemistry,* Prentice-Hall, Englewood Cliffs, N.J., 1979.

Shriner, R. L., R. C. Fuson, and D. Y. Curtin: *The Systematic Identification of Organic Compounds,* 5th ed., Wiley, New York, 1964.

EXPERIMENTAL TECHNIQUES

PHYSICAL MEASUREMENTS

1.1 MELTING POINTS

Background and Principles

Many organic compounds are solids at room temperature as a result of intermolecular forces which hold together the individual molecules of a crystal lattice. The strength and nature of these intermolecular forces are responsible for differences in melting points. The shape of the molecule is also important, as is its ability to orient itself in space with other molecules of the same substance. In general, if the crystal packing forces are very strong, the melting point will tend to be high. If the intermolecular forces are relatively weak, the melting point will be lower. Clearly, a compound will be a liquid if its melting point is below room temperature.

Almost all pure compounds undergo a distinct transition from the solid to the liquid phase. This transition is usually sharper (occurring over a narrower temperature range) than that from liquid to solid. The point at which a crystal undergoes the transition from solid to liquid is called the *melting point* and is characteristic of the particular substance.

The true melting point of a pure compound is usually defined as that temperature at which the solid and the liquid phase are in equilibrium.

Although the solid-liquid transition is referred to as the melting point, the observed melting of a pure solid usually occurs over a range of 1 to 2°C. In general, the purer the substance, the narrower will be the melting range. For now it is important only to know that there are certain factors which will reduce the sharpness of a melting point. The most important of these, the presence of an impurity, would usually cause the melting point to be reduced. Common impurities are: (1) the solvent from which the substance was recrystallized; (2) starting material still contaminating the product; (3) by-products produced in the formation of the product; and (4) water from either the solvent or the atmosphere. A solvate or a hydrate is formed when molecules of solvent or water are present and penetrate the crystal lattice. If an impurity cannot align itself satisfactorily in the lattice, it lowers the melting point to an extent which depends on how the structure of the lattice is altered and how the crystal forces related to the lattice are altered by the presence of the impurity. For example, the anhydrous compound known as citric acid has a melting point of 153°C, but the monohydrate melts sharply at 100°C. Substances which melt at a higher temperature when hydrated than when anhydrous are also known.

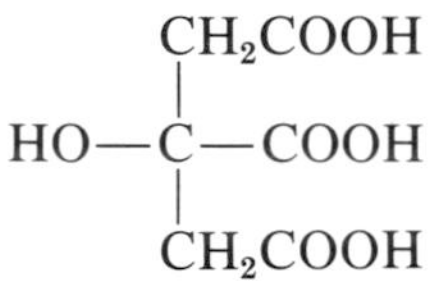

$$\begin{array}{c} CH_2COOH \\ | \\ HO-C-COOH \\ | \\ CH_2COOH \end{array} \qquad \left[\begin{array}{c} CH_2COOH \\ | \\ HO-C-COOH \\ | \\ CH_2COOH \end{array}\right] \cdot H_2O$$

Citric acid (mp 153°C) 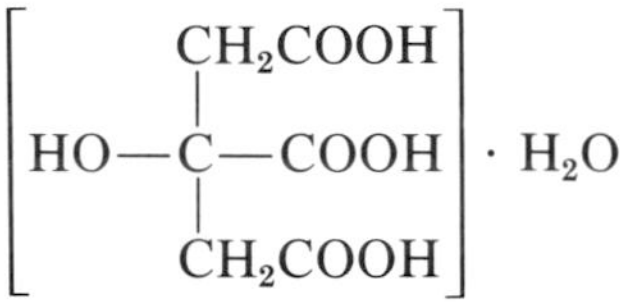 Citric acid monohydrate (mp 100°C)

Students often err when determining a melting point by looking only for the beginning of the melting phenomenon. The entire range over which the compound melts should be observed and recorded. Although impurities usually lower a melting point, sometimes the presence of an impurity will not only broaden but will raise the melting range. A common example of this is the melting behavior of pure versus impure *p*-toluenesulfonyl chloride, which is a reactive sulfonylating agent. This compound is readily hydrolyzed by atmospheric water, even during storage. *p*-Toluenesulfonic acid is the resulting impurity. The chloride melts at 68°C, the acid at 103°C. In this case, the impure material melts at a higher rather than a lower temperature because the amount of chloride is reduced.

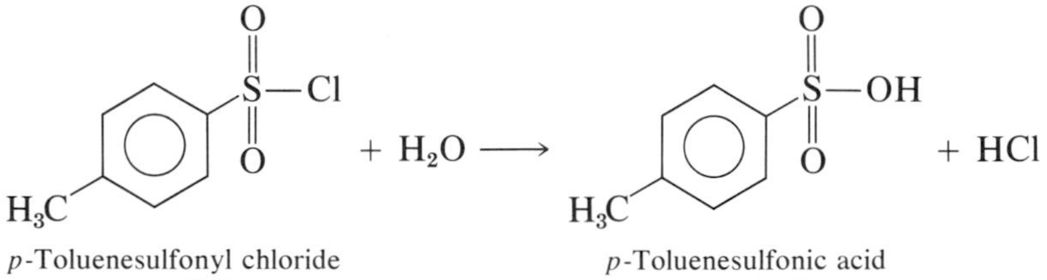

$$H_3C-C_6H_4-SO_2-Cl + H_2O \longrightarrow H_3C-C_6H_4-SO_2-OH + HCl$$

p-Toluenesulfonyl chloride (mp 68°C) *p*-Toluenesulfonic acid (mp 103°C)

Several possible problems should be kept in mind. For example, some substances decompose at their melting points instead of melting sharply. This is often due to a chemical reaction which takes place when the compound is heated or to the fact that the material is unstable relative to its various components at elevated temperature. Characteristically, the material begins to melt, then changes color, loses gas, or simply disappears. When this occurs, the melting point should be recorded and (dec) should be added after the temperature. If decomposition occurred at 141 to 142°C, the melting point would be recorded as 141–142°C (dec).

Mixture Melting Point

While it is important to recognize that these possibilities exist, the most important thing to keep in mind is that no matter what the melting point is, if the melting range is narrow (the melting point is sharp), the substance is probably pure. If the material melts sharply (and at the same temperature) alone and when mixed with a sample of material believed to be the same substance, then both materials are the same and both are pure. Testing a material in this way is called determining a "mixed" or "mixture" melting point.

Glasses

Another special problem which may be encountered is the formation of a glass. Window glass, as you may recall from introductory chemistry, is not crystalline but is a supercooled solution. The same situation is occasionally observed with organic compounds. A pure substance may have what appears to be a distinct melting point, but when it is melted and resolidified, the apparent melting point may change. This behavior is because the orientation of the molecules in the glass has changed and the temperature at which the material undergoes the solid-liquid transition also changes. This alteration will continue to occur as many times as the material is melted, and a single, characteristic melting point will not be achieved (see Fig. 1.1).

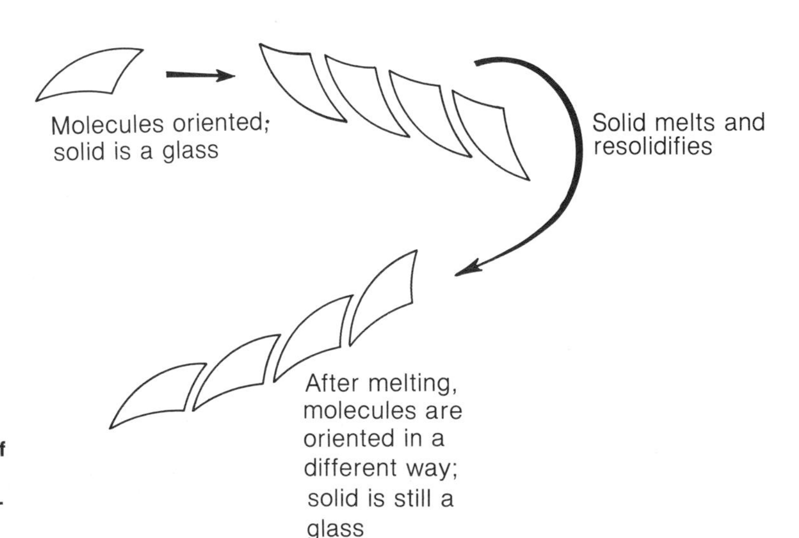

Figure 1.1
The crystals of a glass can orient in many ways, none of which are likely to yield a truly crystalline material.

Isomorphs

A third special problem is that in which molecules are isomorphous. Isomorphous molecules are those which tend to pack into the same lattice because they are approximately the same size and shape. An example is the ditosylate of diethylene glycol (mp 85°C), which is isomorphous with the ditosylate of 1,5-pentanediol (mp 75°C). These substances, which differ in melting point by 10°C, pack in prismatic needles when recrystallized from 95% ethanol, and when mixed they melt sharply at 80°C, halfway between the melting point of each compound. Although such situations are quite rare, they are sometimes encountered in the chemical laboratory.

$$CH_3-C_6H_4-\overset{O}{\underset{O}{\overset{\|}{\underset{\|}{S}}}}-O-CH_2-CH_2-CH_2-CH_2-CH_2-O-\overset{O}{\underset{O}{\overset{\|}{\underset{\|}{S}}}}-C_6H_4-CH_3$$

1,5-Pentanediol ditosylate

$$CH_3-C_6H_4-\overset{O}{\underset{O}{\overset{\|}{\underset{\|}{S}}}}-O-CH_2-CH_2-O-CH_2-CH_2-O-\overset{O}{\underset{O}{\overset{\|}{\underset{\|}{S}}}}-C_6H_4-CH_3$$

Diethylene glycol ditosylate

A related phenomenon is observed when a pure compound can exist in two or more different molecular arrangements in the crystal (different crystal habits). If we think about a crystal being approximately the shape of an oblong, twice as long as it is wide, we can imagine it packing in two different ways: (1) it could pack side by side by side by side by side; or (2) it could pack two long, one side, two long, one side, two long, and so on. These are illustrated below.

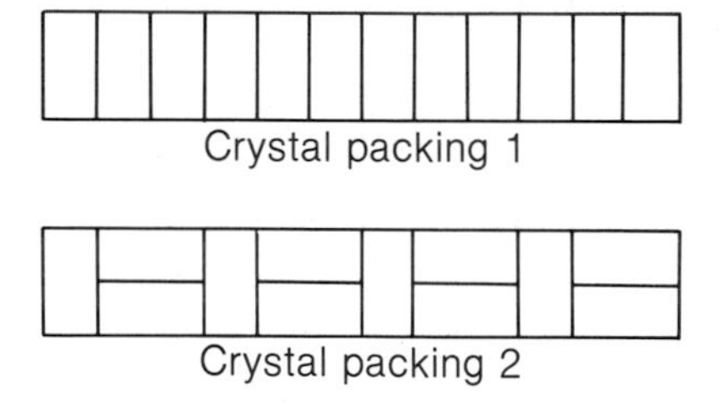
Crystal packing 1

Crystal packing 2

It should be clear that the two different crystal habits will involve different forces, different lattice energies, and therefore different melting points. Pure benzophenone is known to have two crystal habits with melting points that differ by more than 20°C.

The Apparatus

An apparatus is necessary for determining the melting point of a substance. The simplest, of course, consists of something to contain a sample, a medium in which to heat it, and some device for recording temperature. It is possible to use a beaker filled with water or some other heat-transfer fluid and a thermometer to measure the temperature. The sample is then introduced as a solid and, if it is insoluble in the medium, its melting point can be observed. Because introduction of the substance directly into the medium is relatively inconvenient, the sample is usually encased in a glass capillary attached to the thermometer.

An apparatus which appears primitive but which functions very effectively consists of a beaker filled with mineral oil, a thermometer suspended from a clamp on a ring stand, a glass stirring rod, and a device for heating the oil bath (see Fig. 1.2*a*). The sample, which is contained in the capillary, may be attached to the thermometer by a rubber band or piece of rubber tubing so that it will be very close to the thermometer bulb. The solution is agitated so that there will be uniform heat transfer as the oil is heated. Because the sample is in contact with the bulb, its temperature and that of the thermometer will be the same. A small stirring plate, if available, is convenient for heating and stirring the oil at the same time, but this can also be accomplished with a bunsen burner.

An apparatus somewhat more sophisticated, but still inexpensive, is the Thiele tube used in many laboratories. This consists of a test tube with a U-shaped side arm attached to it so that the oil can circulate. It can be seen from Fig. 1.3 that the thermometer, with sample attached, is suspended in the test-tube side of the apparatus and the side arm is heated. As the side arm is heated, the oil is also heated by convection and conduction. The side arm allows for the circulation of oil into the test tube, and the heating and circulation process continues as heating continues. This is rather an efficient means for achieving uniform heating without the necessity of introducing a stirring apparatus. When suspending the thermometer in this apparatus, care should be taken to ensure that the test-tube end is not sealed, as that would cause substantial pressure buildup.

The Thomas-Hoover apparatus, which is still more sophisticated, is illustrated in Fig. 1.4. Essentially the same apparatus described above, it also consists of a beaker, a heating medium, and a thermometer. As can be seen from the picture, however, there is also a good deal of automation which makes this kind of device far more convenient to use than the simple beaker. On the other hand, as these devices frequently cost \$300 to \$600 and, because they are only more convenient, not more accurate, the Thiele tube apparatus is more commonly used in undergraduate laboratories. This is true because the initial expense and replacement cost are low.

The final type of apparatus commonly used for determining melting points

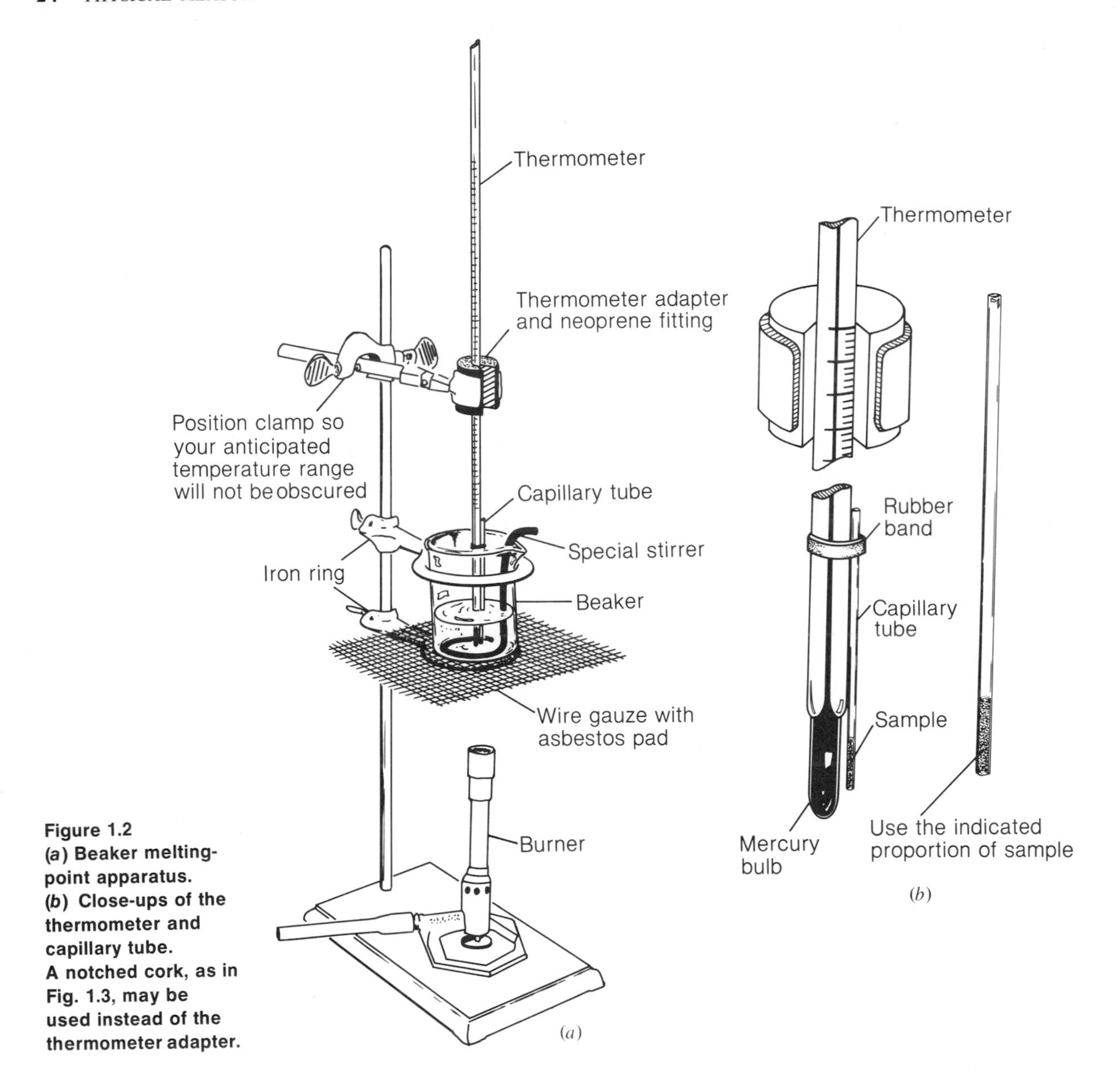

Figure 1.2
(*a*) Beaker melting-point apparatus.
(*b*) Close-ups of the thermometer and capillary tube.
A notched cork, as in Fig. 1.3, may be used instead of the thermometer adapter.

is the so-called "hot stage" (Figs. 1.5 and 1.6). While there are various designs, all basically utilize a solid heating block (aluminum or stainless steel) on which the sample is placed and a thermometer inserted into the heating block to record the temperature. The advantage of a device of this kind is that it contains no oil to heat up and decompose. Whereas these devices can frequently attain a temperature in the vicinity of 350°C, no oil should be heated beyond 250°C, at least not for a sustained period, because the oil may decompose and/or catch

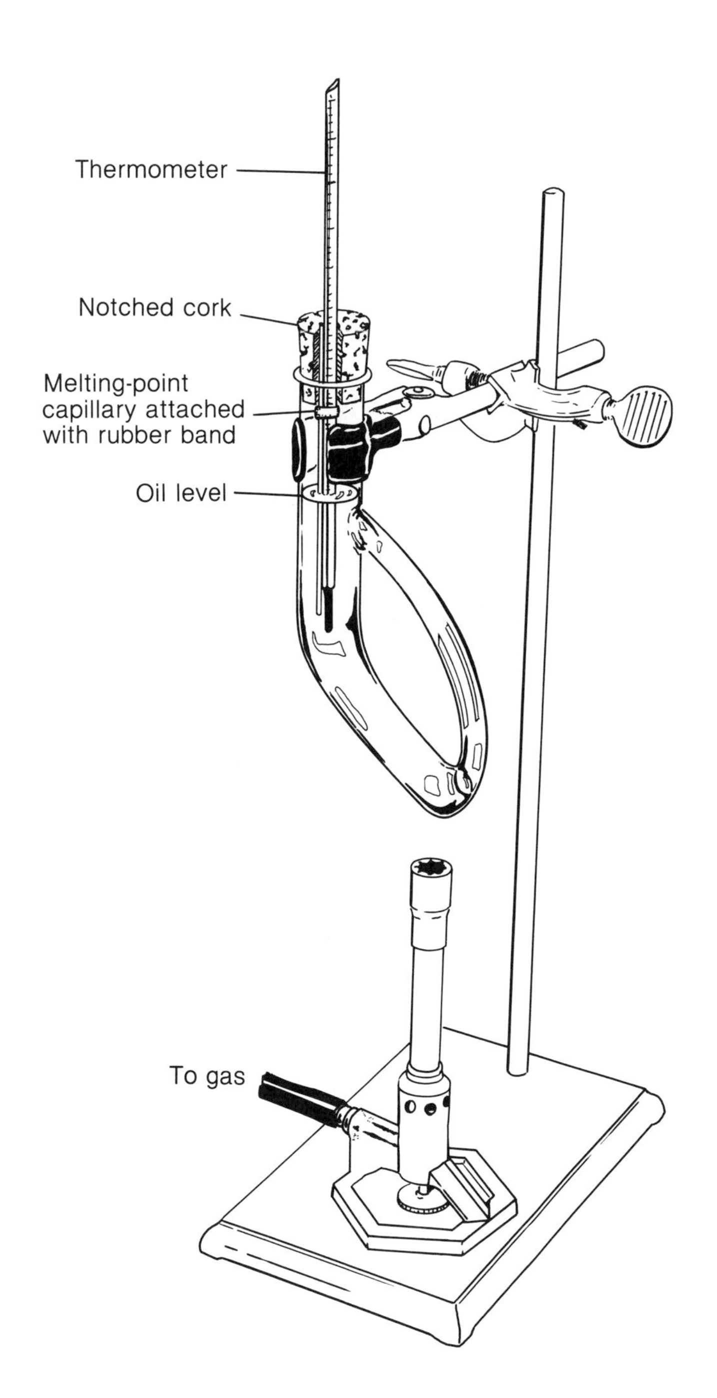

Figure 1.3
Thiele tube, shown with a bunsen burner.

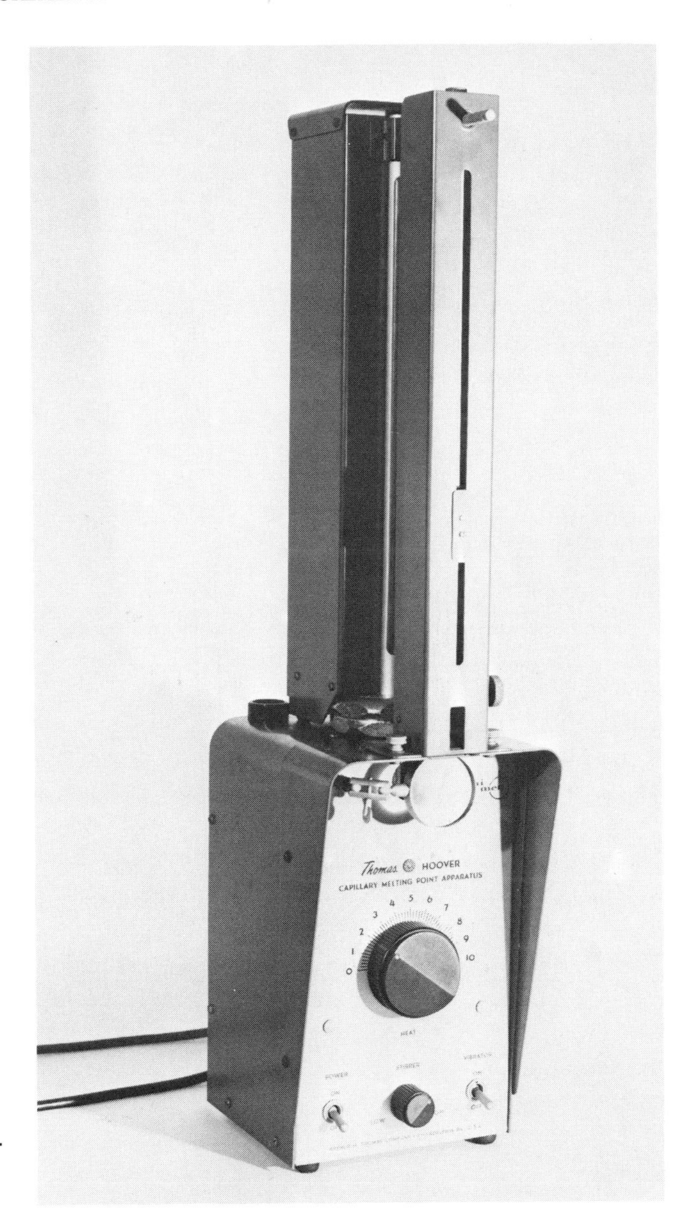

Figure 1.4 Thomas-Hoover melting-point determination device.

fire. In the Fisher-Johns apparatus (Fig. 1.5) a thermometer is inserted horizontally into the block; the sample, between two microscope cover slips, is placed on the block; the melting process is then observed from above through a magnifying glass. In the Mel-Temp apparatus (Fig. 1.6) the thermometer is inserted vertically into a block which, except for the lack of a heating medium, is essentially like a Thomas-Hoover apparatus. In another apparatus, the hot stage is viewed through a microscope. Many other variations exist but the principle is

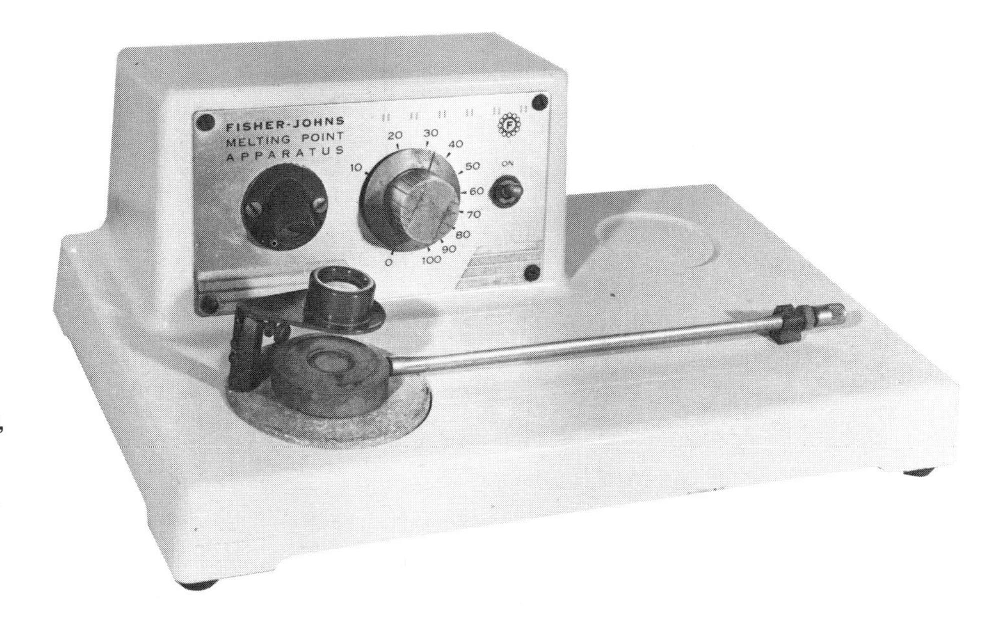

Figure 1.5 Fisher-Johns melting-point apparatus, an example of the so-called hot-stage melting-point apparatus. See text for additional discussion.

the same for all: A method must be available for the gradual heating of the sample, and it must be possible to observe the melting process as it occurs.

Regardless of the kind of device used, there are two principal sources of error. The first occurs when the sample is heated too quickly. If the medium is heated very rapidly, the thermometer may record a temperature which is different from that actually influencing the sample. It is therefore advantageous to have the thermometer as close to the sample as possible.

The second error commonly encountered is a mechanical one related to the thermometer. The thermometer may be misread. Parallax error occurs when the thermometer is viewed from the wrong angle. Your eyes should always be on the same horizontal plane as the top of the mercury column when you read the thermometer.

Thermometer Calibration

A small, systematic error may arise because of the thermometer itself. Some thermometers record temperatures a degree or two higher or lower than the true value. The remedy for this error is to prepare a calibration graph (Fig. 1.7), from which the correct temperature may be determined. The details of preparing such a curve are given in the experimental procedure below.

Sample Preparation

The availability of a sample is the first concern. Once the sample has been prepared, a capillary must be obtained. Ordinarily the capillary tubes used are made of soft glass and have a diameter of approximately 1 mm. Sealed capillar-

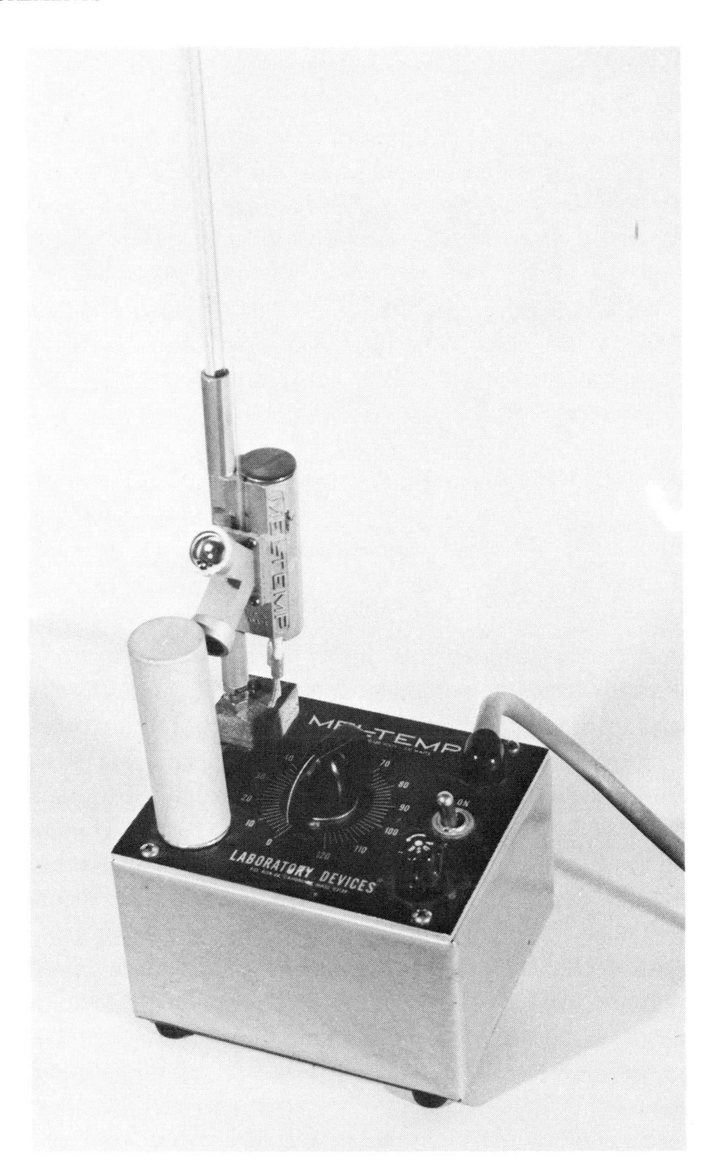

Figure 1.6
Mel-Temp apparatus, a hot-stage melting-point apparatus for student use.

ies cost about a penny apiece and are available in the storeroom. Even though these should be routinely used, one can seal a capillary tube by the following simple technique. A length of capillary tubing about 10 cm long and about 1 mm in diameter is held at one end while the other end is continuously twirled in the flame of a bunsen burner. It will soon be apparent that the end has sealed. Continuous twirling of the tube during this process will prevent the hot end from bending and appearing limp. This is important because the sealed capillary should be completely sealed and straight.

The simplest technique for charging a capillary is to take a small amount of the substance, which has been ground into a fine powder, place it on a spatula, and tip the open end of the capillary against it. The friction between the glass and the sample will ordinarily be enough to hold about 1 mm depth of the sample in the top of the capillary tube. The problem now is to get the sample from the open to the closed end. A number of methods are in common use. In the authors' experience the most effective way is to use a piece of glass tubing about 4 mm in diameter (5 mm is acceptable) and 30, 40, or 50 cm long. This piece of glass is then held up straight on top of the laboratory bench and the capillary, sealed end down, is simply dropped into it. The force of the capillary striking the bench will usually cause the sample to go to the bottom. There is no need to fear that the capillary will be shattered by this process as long as it is dropped straight and not thrown or angled.

Occasionally a very fluffy substance will not go to the bottom of the capillary without a great deal of difficulty. In these cases, it is best to use a piece of fine wire to force the sample down by repeated prodding. The wire should be considerably narrower than the capillary itself, as use of a large piece which just barely fits almost invariably causes the capillary to split.

Another kind of sample often encountered, which may at first seem to present an insurmountable problem, is a highly crystalline compound. In these cases, it is best to grind a minute sample of the crystals on a piece of glass. They can be pulverized by placing a spatula on top of the crystals, your thumb on top of the spatula, and rotating with very small circular motions. Once the crystals are sufficiently pulverized, they can be placed in the capillary without further difficulty.

Now that the capillary has been prepared and filled with the sample, the melting point should be determined. The melting-point capillary is either attached to the thermometer or simply immersed in the heating medium. The medium is then quickly heated above the melting point of the substance. This rapid heating allows you to determine the approximate melting point of the substance in a short time. Once the approximate melting point of the substance is known, a second sample can be placed in the melting-point apparatus and the temperature of the surrounding medium heated to within 15 to 20°C of the previously observed approximate melting point. Slow heating, about 1 to 2°C/min, should be used from this point until the actual melting process is complete. It is important that the same sample not be used both times when a melting point is determined in this way. Certain difficulties arise if an attempt is made to reuse a sample because chemical reactions can often be caused by this simple heating process. When heated, *o*-dicarboxybenzene, commonly known as phthalic acid, loses water to form phthalic anhydride, a different substance with a different melting point. If the substance is remelted in the same capillary, an entirely different melting point is observed. This would be a grave source of error indeed.

When conducting a melting-point determination, it is very important to record all the changes a sample undergoes during the melting process. The entire temperature range over which a process known as *shrinking* occurs should be recorded. This is a condition in which the sample contracts and pulls away from the edges of the capillary tube. Color changes should also be recorded, as should the temperature at which any effervescence or loss of gas occurs. Any residual solid which did not melt during the melting process should be carefully noted. If this material decomposes by falling apart, evaporating, turning black, etc., the melting point at which this occurs is recorded with a parenthetical "dec" (for decomposition). Any future observers will thereby know that this compound undergoes some kind of transformation at the melting point.

PROCEDURE

MELTING-POINT DETERMINATION

A. Thermometer calibration

This experiment is intended to familiarize you with the actual conduct of a melting-point determination. You will have in hand a melting-point apparatus (Fig. 1.4 or 1.5) and capillary tubes. You will be provided with four compounds: benzophenone, mp 48°C; benzil, mp 95°C; benzoic acid, mp 121°C; and adipic acid, mp 153°C. Beginning with the lowest-melting compound, determine the melting point or range of each of these materials (it is not necessary to first determine a crude melting point) in order to gather the necessary data to calibrate your thermometer. Note that a melting point may not actually be a "point" but a range of one to two degrees. In such cases, the range should be reported. Plot (use graph paper) your observed melting point versus the literature value for each of these compounds, as in Fig. 1.7. In the future, use this calibration curve to correct your melting points for variations due to your thermometer.

With a calibration graph and your firsthand experience in determining melting points, you are now ready for your first organic investigation. Although relatively simple and straightforward, this experiment embodies all the principles which will be followed throughout this book.

B. Identifying a compound by melting point

You will be given an unknown sample, which could be one of the compounds for which you have already determined the melting point or one of the following:

Compound	mp, °C
Thymol	50
p-Dichlorobenzene	53
Myristic acid	54
p-Nitrobenzyl alcohol	93
α-Naphthol	95
Glutaric acid	98
(*E*)-Stilbene	125
o-Benzoylbenzoic acid	126
Diphenylacetic acid	148
Benzilic acid	151
Adipic acid	153

From the table and from the compounds you have already encountered, you will notice there are four groups of compounds which have essentially the same melting point. This experiment is to be performed in the following two stages.

Determination of melting point You will be given an unknown sample. First, by rapid heating, determine a rough melting point. Care must be taken not only to see whether the melting behavior yields essentially the same melting point as that previously observed, but also whether there is any char-

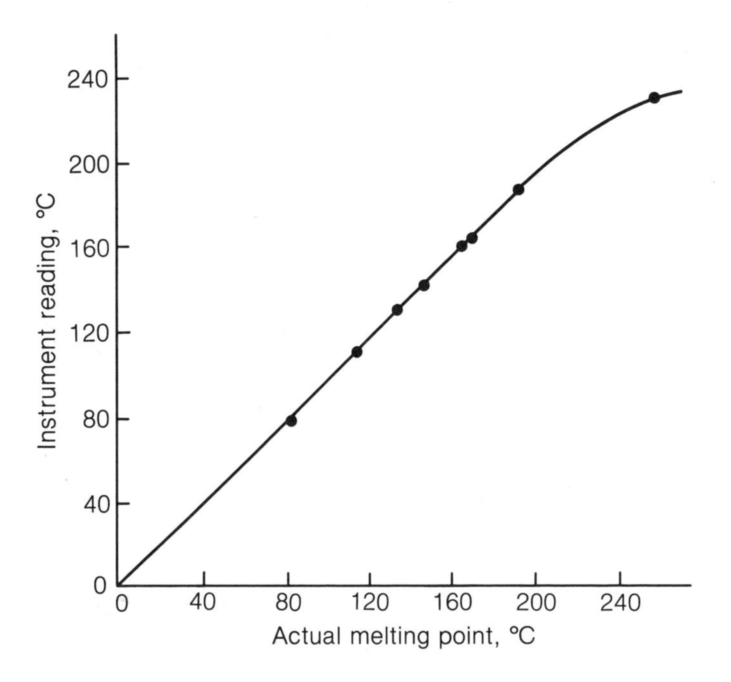

Figure 1.7
A typical calibration curve.

acteristic change (e.g., shrinking or decomposition) similar to that already observed for another compound in the same melting-point range. You should always be on the lookout for clues of this kind with which you might already be familiar.

Identity and mixed melting point The second and more difficult part is to determine the identity of the substance by a mixture melting point. The melting point you have just observed can apply to three or four possible compounds. To determine which compound you have, grind and mix your compound separately with equal amounts of each of the three or four possibilities. Only a few milligrams of each substance is required. As long as there are roughly equal amounts of each substance, the mixing can be approximate. After the mixtures are ground on a glass plate and intimately mixed, transfer each sample to a capillary. Place all three or four capillaries in the melting-point apparatus, and determine the melting points of all the mixtures simultaneously. Only the substance which is a 1 : 1 mixture with itself will give a sharp melting point. Because the other systems will consist of the unknown sample contaminated by about 50% impurity, their melting points will be much lower and broader. Many substances may have the same melting point, but it is very unlikely that two substances which are not identical will still have the same melting point when they are mixed. The mixed melting point is an important criterion for determining purity and is still used in the modern laboratory. The sharpness of a melting point is a key guide to purity, but identity must be confirmed by other evidence.

1.2 BOILING POINTS

When a liquid is heated, thermal energy is transferred to it. The molecules of the liquid acquire additional kinetic energy, and eventually some of them are ejected from the surface of the liquid. Molecules are continually escaping from the solution and are also continuously returning by the process known as *condensation*. The rate at which this occurs at a given pressure depends upon the volatility of the liquid, i.e., how much energy must be added to overcome the intermolecular forces restraining it to the liquid phase. The boiling point characteristic of a substance is the point at which the partial pressure of the vapor above the substance is equal to atmospheric pressure. At this point there is an equilibrium between molecules which are being ejected from the liquid and molecules which are returning to it (condensing). In general, the higher the temperature, the greater will be the number of molecules entering the vapor phase. The boiling point of a substance is usually referred to atmospheric pressure (760 mmHg). If the pressure of the surroundings is lowered (i.e., if a vacuum is applied) the substance will boil at a correspondingly lower temperature. This is illustrated by the pressure-temperature diagram shown in Fig. 1.8.

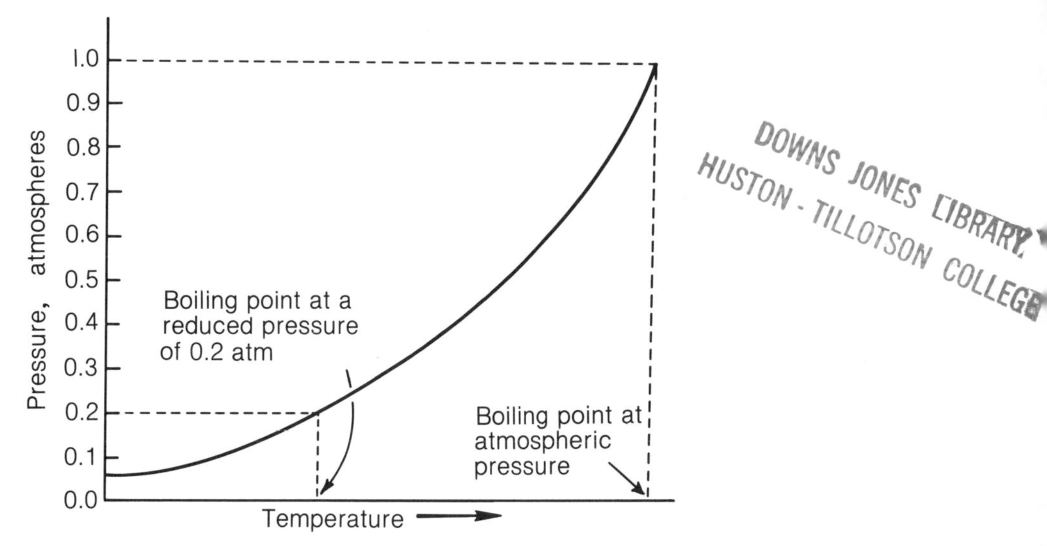

Figure 1.8 Relationship between boiling point and pressure for a pure substance.

The boiling point of a liquid is affected by a variety of factors. Among these are certain properties of the molecules themselves. All things being equal, a higher-molecular-weight material will have a higher boiling point than will a corresponding lower-molecular-weight material. Likewise, the boiling point of a spherically shaped molecule will generally be lower than that of a molecule which has relatively more surface area in proportion to its weight. This is because the spherical compound is capable of fewer intermolecular interactions in solution (see Fig. 1.9). In addition to shape, the polarity of functional groups attached to the molecule will also influence the boiling point. Carboxylic acid groups, for example, can dimerize in the liquid phase, raising the effective molecular weight and the boiling point. A similar, but somewhat less dramatic, effect is encountered with alcohols, which can form hydrogen bonds (see above and Sec. 4.3).

Compounds having polar functional groups, such as ketones, will tend to pair their dipoles, and this is another force which needs to be overcome in order to achieve boiling. These and other related phenomena are discussed in Chap. 4.

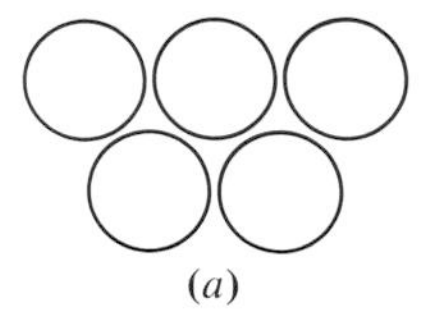

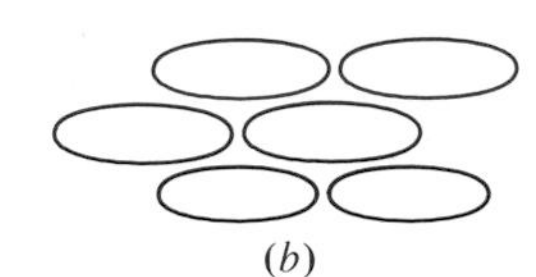

Figure 1.9
(*a*) Spherical molecules have relatively small contact points. (*b*) Elongated molecules are capable of greater contact.

PROCEDURE

METHODS FOR DETERMINING THE BOILING POINT OF WATER

In the experiment described here, we use the well-known boiling point of water to assess various experimental methods for determining it. The boiling point of pure distilled water is 100°C at atmospheric pressure (760 mmHg, 760 torr, 1 atm). We will attempt to determine this boiling point by three different methods. In so doing, we will check the accuracy of the thermometer in use and we will also learn something about the reliability of these various methods.

Method A

Fill each of two 250-mL beakers with 100 mL pure distilled water. Place one of the beakers on a steam bath. Place the second beaker on an iron ring with a piece of iron screen and heat it with a free flame. (The second beaker may be placed on a hot plate, if available.) Figure 1.10 illustrates the correct setup. When the beaker of water which is heated with a flame or a hot plate begins to boil, immerse a thermometer in it and measure the temperature. As soon as that measurement is obtained, remove the thermometer from the beaker of boiling water and transfer it to the beaker on the steam bath. Be sure to record the two temperatures. Now, if an experiment were to call for use of a hot-water bath or if one wished to distill a material which boiled at 97°C, would it be possible to use a steam bath?

Method B

Assemble a simple distillation apparatus (Fig. 1.11) consisting of a 50-mL round-bottom flask, a still head fitted with a thermometer and adapter, a reflux condenser, a takeoff adapter, and a 25-mL receiving flask. All the ground glass joints should be lightly greased to ensure good contact and for easy

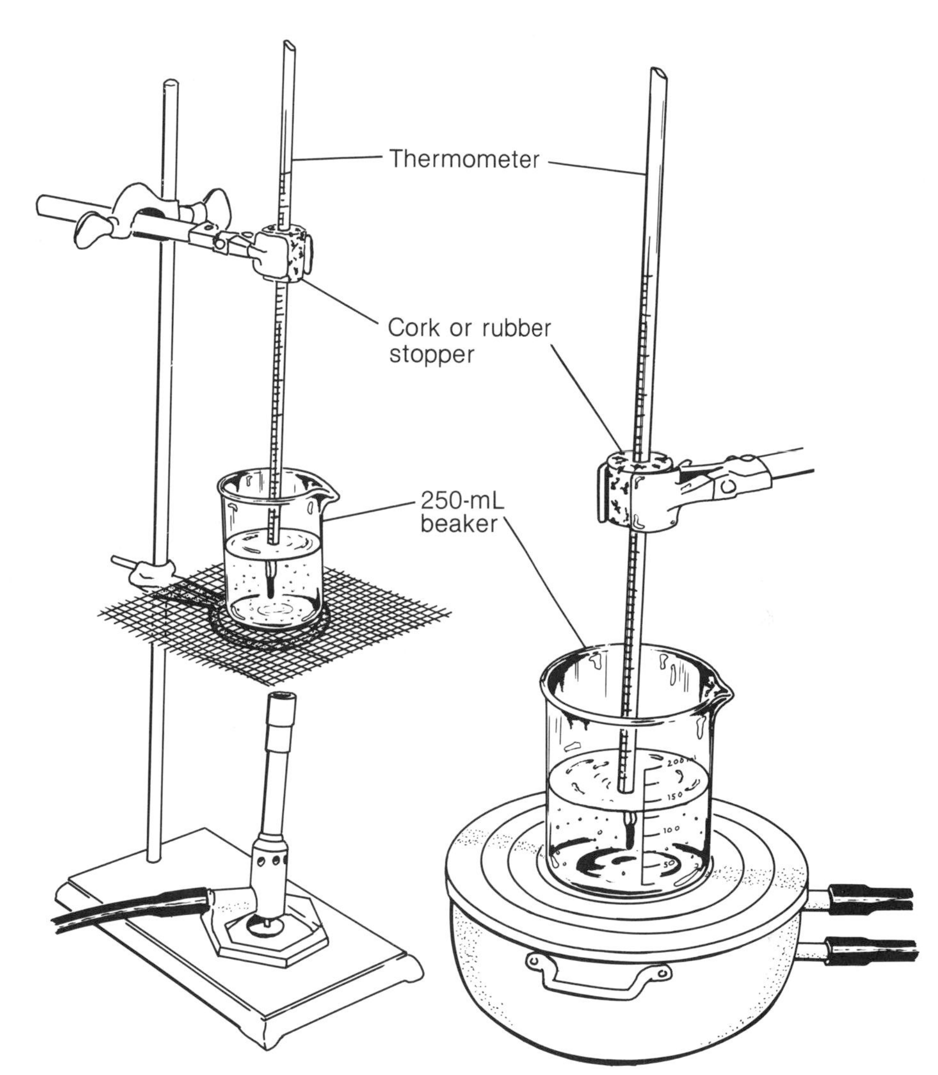

Figure 1.10 Apparatus for measuring the temperature of hot water.

disassembly of the apparatus after the experiment. In setting up the apparatus, be sure to clamp the neck of the distilling flask and the condenser. The takeoff adapter should also be secured, either by clamping or by using a rubber band. The water hose going into the condenser should be at the lower end of the condenser so that it will fill completely. Be sure the outlet hose is placed in a drain. The distilling flask should be placed at a height appropriate for heating with a free flame. The thermometer should be placed below the orifice in the T-head so that it is well bathed by the boiling water vapor. Fill the 50-mL round-bottom flask half full with water (about 25 mL), add a boiling chip, and

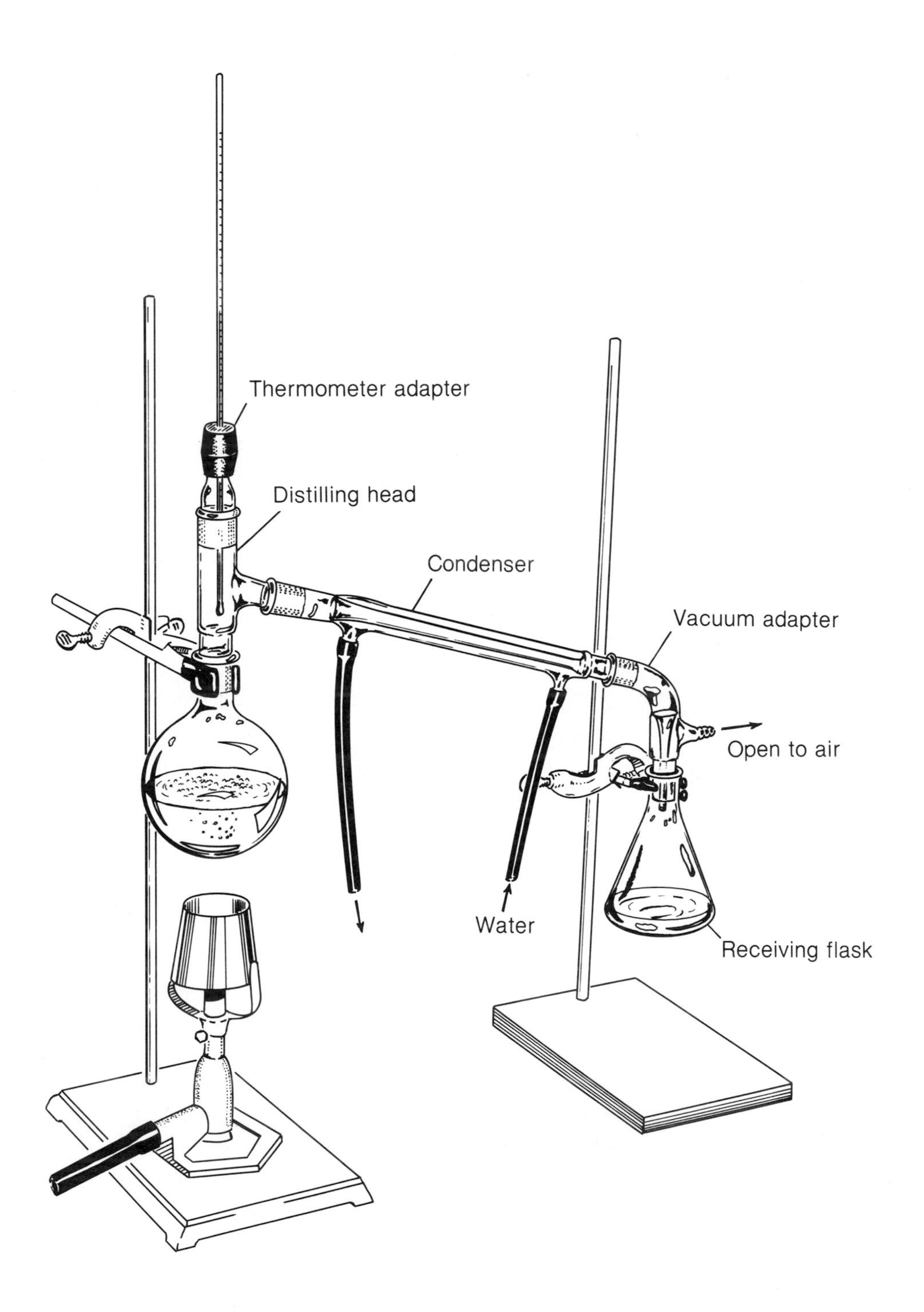

Figure 1.11
Simple distillation apparatus.

heat with the free flame until the water begins to distill. When a continuous distillation of a couple of drops per minute is obtained, observe the boiling point on the thermometer. Does it read 100.0°C? Is it higher? Is it lower? Why?

Method C Microreflux

A technique which is often used to determine the boiling point of a small sample of material involves use of the microreflux apparatus. This is described in Sec. 22.4 and is especially valuable when only small amounts of material are available. Clamp a 13 × 100 mm test tube containing 1 mL distilled water and a boiling chip to a ring stand and suspend or clamp a thermometer so that the bulb of the thermometer is suspended about 2 cm above the top of the liquid (Fig. 1.12). Heat with a free flame until the vapors bathe the bulb of the thermometer. Note the temperature. Note how much additional heating is required to observe a boiling point of 100°C. You should know from previous experiments whether or not your thermometer is completely accurate. You now know how to accomplish a microreflux and you also know the relative accuracy of this technique.

1.3 REFRACTIVE INDEX

The refractive index of a substance is an intrinsic property just as is the boiling point, but a somewhat more sophisticated instrument is required to determine it. The refractive index is defined as the ratio of the speed of light through a vacuum to the speed of light through the sample. The velocity of light through a medium is determined by the interaction of the light waves with the electrons in bonding orbitals and nonbonding orbitals of the substance. The speed of light through the medium will therefore be related to the structure of the molecule and in particular to what functional groups are present.

The lowest common refractive index of a liquid is that of water, 1.33. In general, organic materials have refractive indexes that fall in the range of 1.3 to 1.6. At the lower end are found the alcohols (related structurally to water) and ketones, and at the higher end are found such compounds as chloroform, benzene, nitrobenzene, and aniline.

There are two kinds of refractometers in common use. One is the Fisher refractometer (Fig. 1.13*a*), and the other is the Abbe-Spencer refractometer (Fig. 1.13*b*). Both of these refractometers are useful for observing values between 1.3 and 1.7.

Each of these consists of a monochromatic light source such as a sodium lamp, a constant temperature bath so that the temperature of the material may be held constant, and an optical piece, whose use is described below. The determination of the refractive index is really quite simple. First the refractometer prism is cleaned, usually with 95% ethanol, and allowed to dry thoroughly

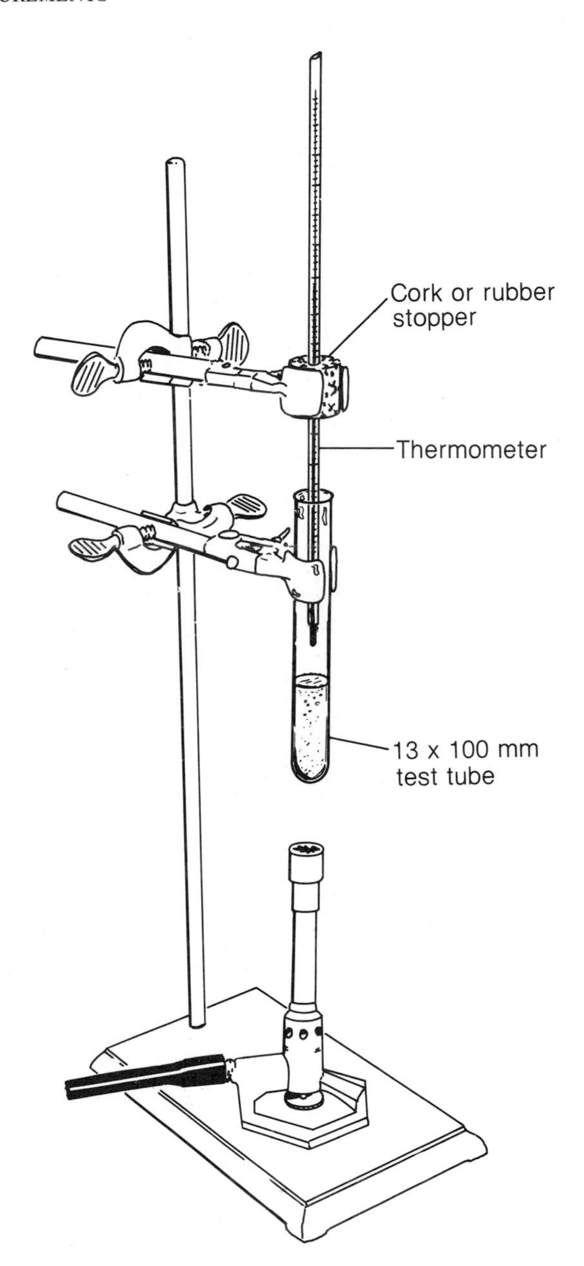

Figure 1.12
Microreflux device.

before the sample is applied. Next the sample is placed on the lower prism so that the entire width of the prism plate is covered. (*Note:* An eyedropper is best used for this and care should be exercised that the glass from the eyedropper does not contact the prism, as it may scratch it.) After the sample has been applied, the upper prism is then brought down into contact with the lower prism. The liquid should now form an unbroken layer between the two prisms.

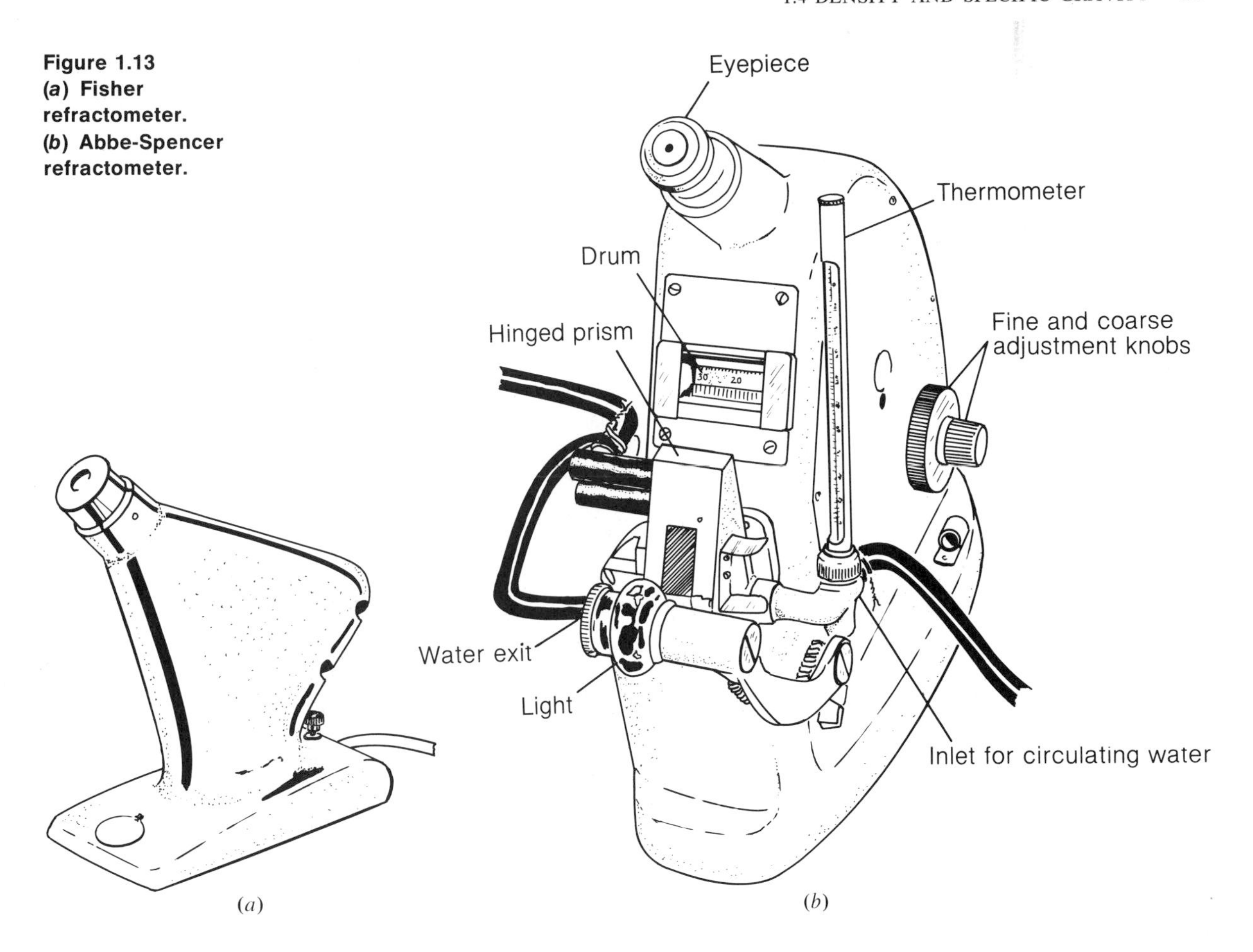

Figure 1.13
(*a*) Fisher refractometer.
(*b*) Abbe-Spencer refractometer.

Manipulation of the controls will now permit determination of the exact refractive index. If more detailed directions are required, refer to Sec. 22.10.

1.4 DENSITY AND SPECIFIC GRAVITY

Density is defined as the total quantity of material per unit space (or mass per unit volume). Specific gravity is defined as the ratio of the density of a material to the density of some standard material, usually water at 4°C. A density of 1, for chemical purposes, is 1 g/cm³ (or 1 g/mL) of material. A specific gravity of 1 means that the material has the same weight as an equal volume of water at 4°C. In this connection, note that water is the standard, having a density at 4°C of 1.00 g/mL.

Most nonhalogenated compounds have densities between about 0.6 and 1.4. Materials having higher densities are generally halogenated or polyhalogenated substances. From Table 1.1 it can be seen that pentane has a density of about 0.6, acetone and methanol about 0.8, ethyl acetate 0.9, and halogen-substituted materials have densities that increase with increasing halogen substitution. For

TABLE 1.1
Selected Densities

Compound	Density, g/mL
Pentane	0.626
Acetonitrile	0.786
Acetone	0.791
Methanol	0.791
Toluene	0.865
Ethyl acetate	0.902
Water	1.000
Chlorobenzene	1.060
Dichloromethane	1.325
Chloroform	1.492
Carbon tetrachloride	1.594
Bromoform	2.890

example, chlorobenzene has a density of 1.1, dichloromethane 1.3, chloroform 1.5, carbon tetrachloride 1.6, and finally bromoform ($HCBr_3$) the extremely high density of 2.9.

An accurate way of measuring a substance, especially one which is noxious and would be difficult to weigh on a balance, is to divide the required number of grams by the density of the substance (see tables in Chap. 28 for density data). The exact volume of material required can now be calculated. The liquid can then be transferred safely and accurately by pipet **(no mouth pipetting)** or graduated cylinder.

There are several methods in common use for determining the density, or specific gravity, of a material. It should be kept in mind that specific gravity is approximately equal to density and, except for the most refined measurements, is usually adequate. Three common procedures are presented below, roughly in order of increasing accuracy. The first is a fairly rough method for determining density and is particularly useful for large quantities. The third method is useful for small quantities and is relatively accurate.

PROCEDURE

DETERMINATION OF DENSITY AND SPECIFIC GRAVITY

A. Large-scale determination of density

A volumetric flask or graduated cylinder of volume 10 to 100 mL is placed on a balance and its weight determined. The appropriate amount of liquid is poured in and the graduated cylinder again weighed. The total number of grams is divided by the total number of milliliters to obtain an approximate density for the material. (*Note:* The accuracy of this method depends on the amount of liquid available. The larger the volume measured by this method, the lower the error.)

B. Smaller-scale determination of density

A clean, *dry* 10-mL volumetric flask or graduated cylinder is weighed on a balance and then filled to the mark with the liquid whose density is to be measured. A second weight is obtained, and the total number of grams is divided by 10 to obtain the density of the material.

C. Determination of specific gravity

A clean, dry 1-mL volumetric flask is accurately weighed. The flask is then filled to the mark with the appropriate liquid and reweighed. The material is then removed and the flask cleaned and allowed to dry. This can be accomplished easily by rinsing with ethyl alcohol and allowing to stand briefly. The clean, *dry* flask is again accurately weighed, filled to the mark with distilled water, and weighed again. Since the volumes of the flask are the same in both cases, the number of grams of material is divided by the number of grams of water to obtain the specific gravity of the substance.

1.5 POLARIMETRY

Light, like other electromagnetic radiation, is a wave/particle phenomenon. Ordinary light is polychromatic: It consists of many colors and of a variety of wavelengths. A propagated wave, i.e., one which is traveling forward, may have electric and magnetic vectors in all possible directions. One of the components of light is circularly polarized light. Right-hand circularly polarized light is illustrated in the diagram below.

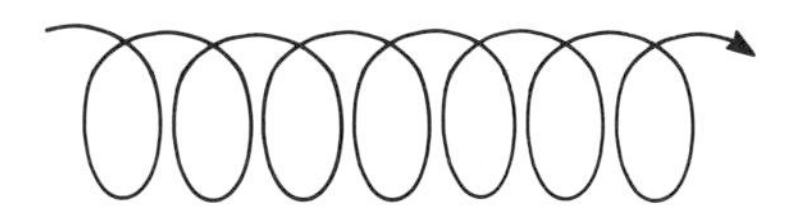

Propagation of circularly polarized light

The sum of right-hand circularly polarized light and left-hand circularly polarized light yields a form of radiation known as plane-polarized light. This effectively means that all of the vectors in any direction outside of the given plane have been excluded. Light of this sort can be obtained by passing nominally monochromatic light through a polaroid filter.

Right- and left-hand circularly polarized light, i.e., helically propagated light, interacts with an organic compound in such a way that the velocity, and therefore also the refractive index, of the left- versus the right-hand components is different. As the light waves travel through a solution containing the substance, the radiation interacts with the bonding and nonbonding electrons present, and the radiation is slowed in accordance with the intensity of the interaction. For a non-optically active material interactions occurring in one direction will be counterbalanced by interactions in every other direction. The extent to which the left-hand light is slowed down is equaled by the extent to which the right-hand light is slowed down, and there will be no change in the angle of the plane.

If an optically active material is present in the solution, an asymmetric interaction is possible. Assume for the sake of argument that the left-hand polarized light interacts more strongly with an *S* optical isomer and is slowed more than the right-hand polarized light. The net effect of this will be a tilt in the angle of the plane. A polarizing filter not angled as is the incoming light would not allow the light to pass through because its orientation would be nonidentical with that of the transmitted light. In order to observe the transmitted radiation, the filter must be at the same angle the light has turned. That angle is the angle of rotation, or the *optical rotation.* An optically active material will not rotate the plane of polarized light if the solution contains equal amounts of isomers having opposite configurations and rotations. In that case, all the interactions causing a leftward turn will be just equaled by those causing a rightward turn, and the net effect will be no change in the plane of the polarized light. This is the definition of a *racemic mixture*.

The change in optical rotation is measured by a polarimeter. A polarimeter is a device consisting of a monochromatic light source, a polarizing filter, a sample cell, and an analyzer. The light coming from the lamp passes through a polarizing filter. Only the electromagnetic waves oriented in the plane transmitted by the filter will emerge. As these pass through the sample, asymmetric interactions cause a tilt in the plane. If the second polarizing filter (analyzer) is held in the same orientation as the first, no light will pass. If the analyzer is tilted through an angle α equal to the tilt of the light plane, the light can emerge. The angle is measured in degrees and is referred to as the angle of rotation or optical rotation. A schematic diagram and an illustration of the actual device are shown in Figs. 1.14*a* and 1.14*b*, respectively.

The equation describing the rotation is

$$[\alpha]_\lambda^t = \frac{\alpha}{\ell c}$$

where $[\alpha]_\lambda^t$ = the specific rotation characteristic of a compound at the temperature t and the wavelength λ

α = the observed rotation

ℓc = the path length, dm

c = the concentration, g/100 mL

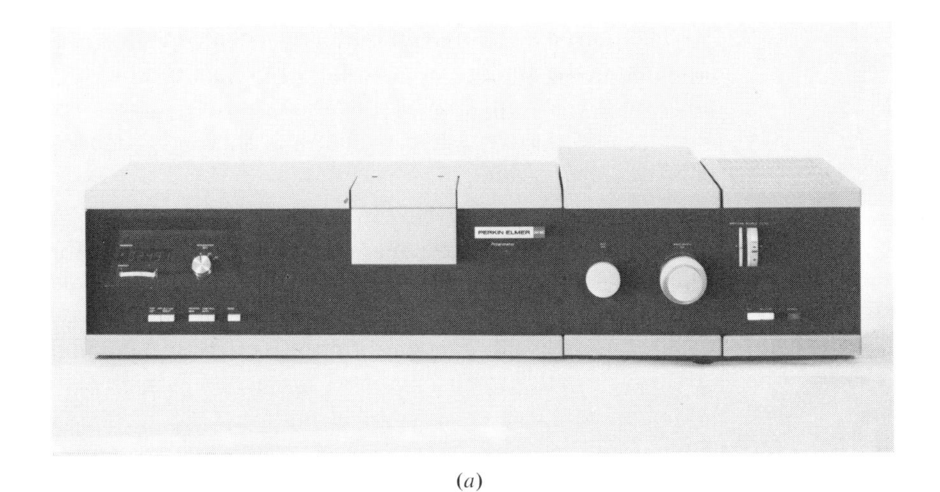

(*a*)

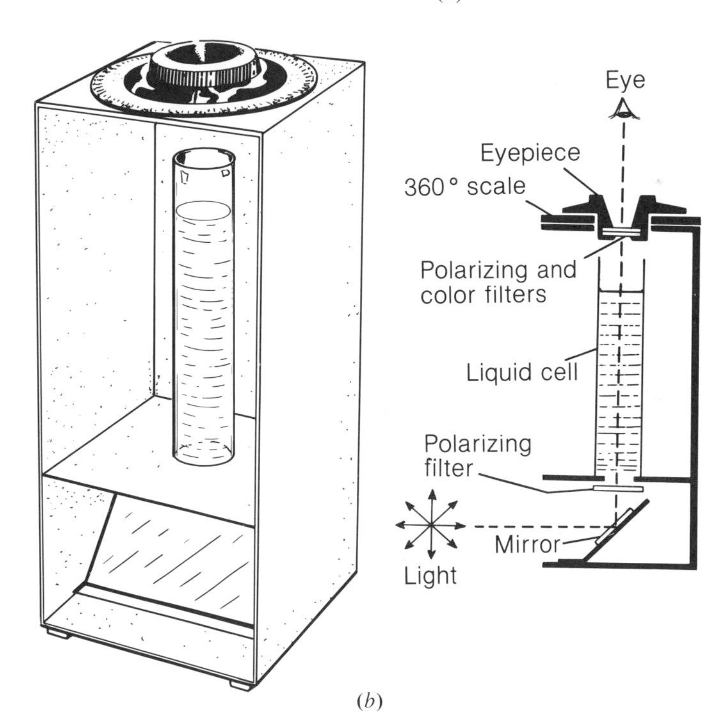

(*b*)

Figure 1.14
(*a*) Perkin-Elmer digital polarimeter. (*b*) A hand-operated polarimeter.

The molecule (*S*)-1-phenylethyl alcohol has a specific rotation at the D line of sodium (589 nm) and 27°C of +49.5°: $[\alpha]_{\mathrm{D}}^{27} = 49.5°$ (c = 1 ethanol). In order to obtain this reading 20 mg of the alcohol would have to be dissolved in 2 mL of some solvent, in this case ethyl alcohol. The concentration, c, would then be

$$\frac{0.02 \text{ g}}{2 \text{ mL}} \times \frac{100 \text{ mL}}{\text{g}} = 1$$

Therefore, $c = 1$ in the specified solvent.

The reason so many variables are specified is that the rotation of a material depends on the amount of material (its concentration), but it also depends very much on the solvent in which it was recorded. The solvent used and the temperature and wavelength at which the rotation is recorded should all be noted with the observation.

The magnitude of a specific rotation is not confined to the range 0 to 360°. It seems unreasonable to assume that a plane could go around more than once, and indeed it cannot. The observed rotation, α, must be less than 360° or it cannot be recorded. However, the specific rotation can easily be well beyond that because of the concentration effect. Some common compounds which have low rotations are camphor, about −40°, and cholesterol, about −10°. Some compounds have very high rotations: the peculiar helical system, heptahelicene, has an optical rotation of 6200°. The structures of these three compounds are shown below.

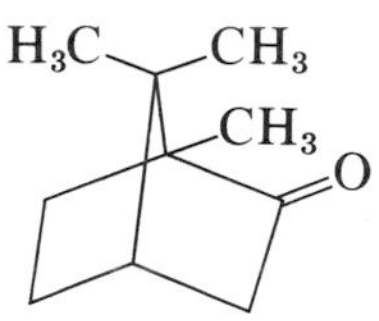

H_3C CH_3 CH_3 O

Camphor

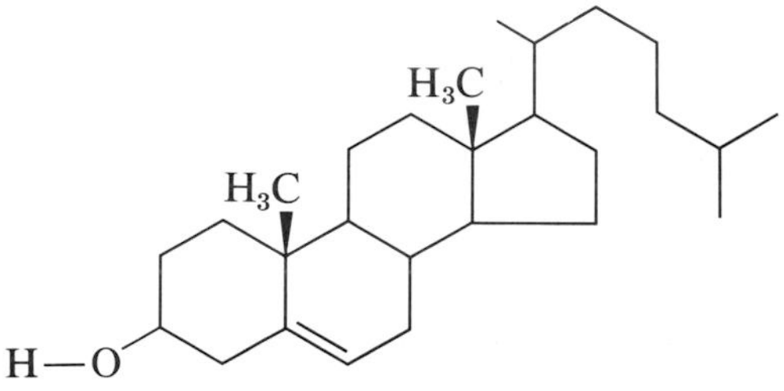

H_3C H_3C H—O

Cholesterol

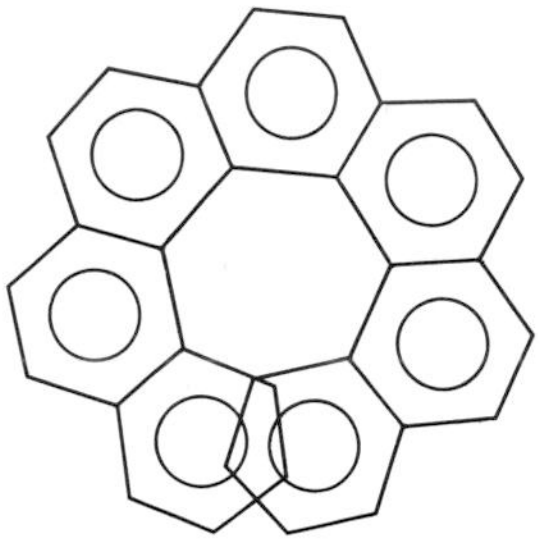

Heptahelicene

The use of a polarimeter consists of obtaining an appropriately concentrated solution, placing it in the polarimeter, and observing the angle of rotation. A calculation will then give the specific rotation. A hand operated polarimeter is illustrated in Fig. 1.14*b*. Instructions for using a polarimeter will depend on the particular model and should be available in the laboratory for the specific polarimeter at hand.

II

BASIC LABORATORY TECHNIQUES

2.1 DISTILLATION

The process of distillation is an excellent method for the purification of a liquid which is stable at its boiling point. The technique may also be adapted to materials which are unstable near their boiling points and this modification (vacuum distillation) is discussed later in this section. Distillation is an especially valuable method for purification because it can be applied with relative ease to large liquid samples. Moreover, the only additional "reagent" involved in a distillation is heat. Heat can be removed from the reaction mixture much more conveniently than can a solvent, so that contamination of the product is less of a problem. Of course, none of this applies if the material is unstable at its boiling point and decomposes.

A liquid is a fluid containing closely packed atoms or molecules of varying energy. When a molecule of the liquid approaches the vapor-liquid phase boundary, it may pass from the liquid phase into the gas phase if it has sufficient energy. The molecule must be energetic enough to overcome the forces which hold it in the liquid phase. The only molecules which can escape from the liquid phase into the vapor phase are those which have sufficient energy to overcome these forces (see Secs. 4.3 to 4.5).

A certain number of molecules are present in the vapor phase above the liquid. As these molecules approach the surface of the liquid, they may enter the liquid phase and thus become part of the condensed phase. In so doing, the molecule will relinquish some of its kinetic energy (i.e., its motion will be slowed). During the process of vaporization, energetic molecules are lost to the vapor phase, but the system gains that energy during condensation. Heating the liquid will cause more molecules to enter the vapor phase, and cooling the vapor reverses this process.

When the system is in equilibrium, as many molecules are escaping into the vapor from the liquid phase as are returning from the vapor to the liquid. The extent of this equilibrium is measured as the vapor pressure. If the energy of the system is increased but equilibrium is maintained, more molecules in the liquid phase will have energy sufficient to escape into the vapor phase. Although more molecules are also returning from the vapor phase, the number of molecules in the vapor phase increases and so does the vapor pressure. The exact number of molecules in the vapor phase depends mainly on the temperature, the pressure, and the strength of the intermolecular forces exerted in the liquid phase.

If two different components are present in the liquid phase, the vapor above the liquid will contain some molecules of each component. If for convenience we designate the two components of the liquid as A and B, then the number of A molecules in the liquid phase will be determined by the volatility of A and by the mole fraction of A in the mixture. In other words, the relative amounts of the components A and B in the vapor phase will be related to the vapor pressure of each pure liquid, and the total vapor pressure of the mixture above the liquid is the sum of the two partial pressures. This relationship can be expressed mathematically as Raoult's law:

$$P_{\text{total}} = P_{\text{A}} + P_{\text{B}} \qquad \text{where } P_{\text{A}} = P^{\circ}_{\text{A}} N_{\text{A}} \text{ and } P_{\text{B}} = P^{\circ}_{\text{B}} N_{\text{B}}$$

when P_{A} = partial pressure of A
P_{B} = partial pressure of B
P°_{A} = vapor pressure of pure A
P°_{B} = vapor pressure of pure B
N_{A} = mole fraction of A
N_{B} = mole fraction of B

At atmospheric pressure (1 atm = 760 mmHg), the sum of the partial pressures equals 1 atm. If the liquid composition and partial pressure of one component are known, the other partial pressure may be determined. Likewise, if the vapor composition and the mole fraction of one of the components are known, the liquid composition can be determined.

For a multicomponent case, the vapor composition may be determined from the partial pressure exerted by each component individually and the mole fraction of the respective component in the mixture. It is an unusual case if more than two or three components are present to any significant extent in a mixture which is being separated by distillation.

Simple Distillation

When a liquid substance is heated and its vapors allowed to condense in a vessel different from that used for heating, a distillation has been carried out. (See

Sec. 2.3 to distinguish this from the process known as sublimation.) When a pure substance is distilled, a simple distillation has been effected. What actually happens in this process is that the liquid is heated in a vessel (a distilling flask) until it vaporizes. The vapor passes into a condenser and is reconverted to the liquid, which is then collected in a receiver flask.

A simple distillation is often considered to be any distillation which does not involve a fractionating column (see below) or in which an essentially pure material is separated either from a nonvolatile or a very minor component. Neither of these definitions is precisely true. If the process of distillation involves separation of one substance from another, regardless of how different their boiling points or vapor pressures may be, a fractional distillation has been carried out. A simple distillation apparatus (Fig. 1.11), i.e., one which does not use a fractionating column, may be used for either a fractional or a simple distillation. Regardless of apparatus, a simple distillation may be carried out only if a pure substance is being distilled.

Fractional Distillation

Fractional distillation is the most common of all distillation procedures. In the most favorable situation for a distillation, the contaminant boils at a temperature which is very different from the boiling point of the major component. It should be borne in mind that even the highest-boiling substance will contribute to the vapor phase in proportion to its individual vapor pressure and the mole fraction of the component present. The only way a vapor above a liquid can consist of a pure component is for the liquid to be a pure sample. The situation is not hopeless, however, because often the contribution of the vapor from the second component to that of the major component is small and for practical purposes may be neglected.

In general, if the boiling points of two liquids differ by less than about 70°C, a simple distillation apparatus will be insufficient to separate them. Use of a distillation column will usually facilitate the separation. The longer the distillation column, the better will be the separation effected. Generally, packing the column with an inert material will also improve the separation. Nevertheless, if two components differ in boiling point by less than about 10°C, they can be separated only with the use of special equipment.

A vapor composition diagram for a two-component system (components A and B) is shown in Fig. 2.1. The vertical axes on both sides of the diagram indicate temperature and the horizontal axis indicates percent composition. Note that the left extremity of the horizontal axis corresponds to 100% of component B and the right extremity to 100% of component A. Component B in this mixture has a lower boiling point than component A. Component A has a higher boiling point and a lower partial pressure. The boiling point difference between

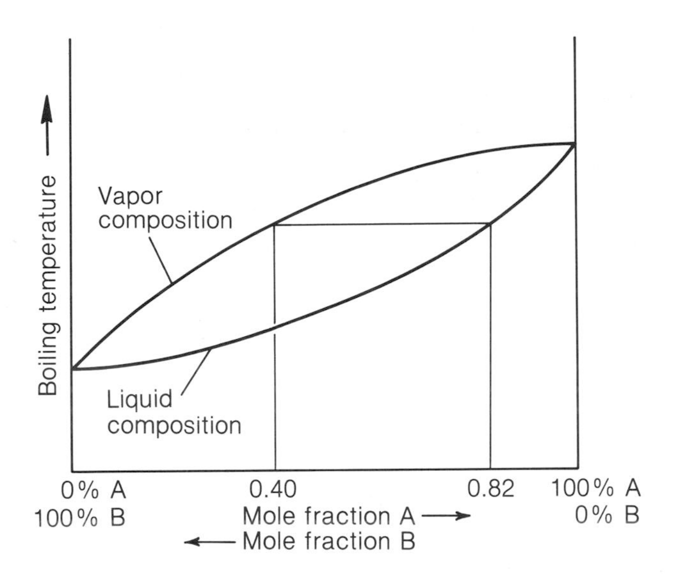

Figure 2.1 Liquid-vapor composition curve.

these two compounds is about 100°C, and separation can probably be effected in a single vaporization-condensation cycle (simple distillation apparatus).

Separation of A and B is possible, even though A is present to a substantial extent in the liquid phase, if the partial pressure of A is so low that its contribution to the total pressure above the liquid is very small. As a result, the vapor above the liquid is almost exclusively component B. If the vapor above the liquid is condensed in a separate flask, it will be almost pure B. As more and more B is removed from the liquid phase and condensed, component A will make a greater and greater contribution. If B is removed completely, the total pressure above the liquid will be contributed by A. The process just described is the essence of the fractional distillation process.

In most cases, the vapor above the liquid will contain predominantly one component but be contaminated by the other. If the vapor above the liquid were separated and condensed, the liquid obtained (the condensate) would be enriched in B in accordance with the difference in the boiling points between the two components. The condensate, which is enriched in B, can now be allowed to evaporate. The vapor above this new liquid will be very much richer in B than it was in the original mixture because it has already been through one vaporization-condensation cycle. If we were to take the vapor above this second liquid and condense it, we would have even more B in the condensate than there was after the first condensation. If this process is repeated often enough, pure B will eventually be obtained.

If we consider each of these evaporation, or vaporization-condensation, cycles to be a distinct operation, we can define this cycle as a theoretical plate. In three vaporization-condensation cycles, we will have effected the equivalent of three simple distillations. If we are able to achieve the same separation as would have been obtained from three simple distillations in a single step, the distillation is said to have three theoretical plates. As the number of theoretical plates increases, the ease of separating components also increases. As the ease of separation increases, the likelihood of separating components whose boiling points are more similar also increases.

A distillation or fractionating column (see Fig. 2.2) provides a lengthened pathway between the distilling flask and the condenser which leads to the receiver. Along this extended column the process of vaporization and condensation (the equivalent of several small distillations) occurs. Each vaporization-condensation cycle is the equivalent of a simple distillation. As more of these

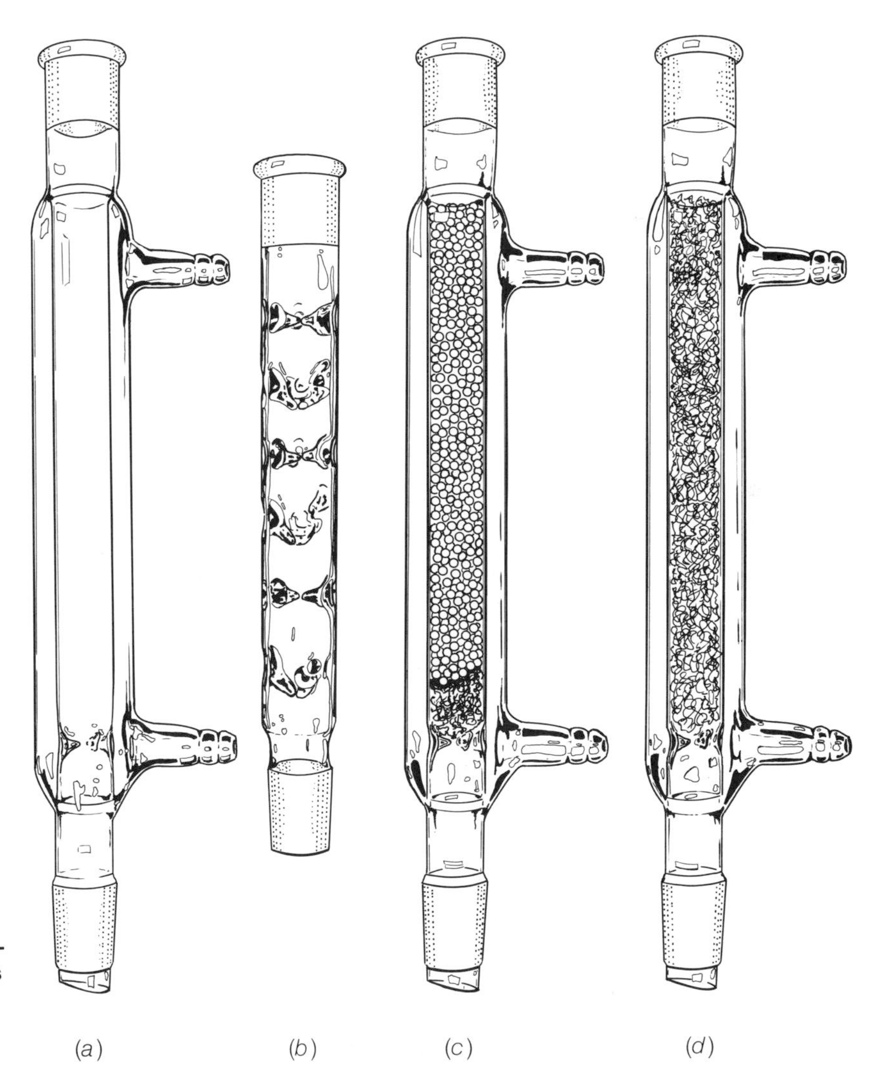

Figure 2.2
Distillation columns. (*a*) Air-cooled condenser used as distilling column. (*b*) Vigreux column. (*c*) Air-cooled condenser packed with glass beads. (*d*) Air-cooled condenser packed with metal sponge.

cycles occur, the vapor will become enriched in the more volatile component and the condensate will be enriched in the less volatile component. This leads to a more effective separation. The more efficient the column is in effecting this condensation-vaporization cycle, the more efficient will be the distillation.

As a practical matter, most simple distillations afford one theoretical plate. The simple distillation is therefore useful for separating components whose boiling points differ by more than 70°C. If a fractionating column is used, two to four theoretical plates can usually be achieved, and components whose boiling points differ by only 20 or 30°C may usually be separated. The experiment which you will do in this section is designed to demonstrate the difference between simple (Fig. 2.3) and fractional (Fig. 2.4) distillation apparatus.

Column efficiency

Much study and invention have been devoted to the question of increasing column efficiency. We mention here only the barest essentials of this work.

Since each theoretical plate corresponds to a single vaporization-condensation cycle, anything which increases the possibility of condensation without adversely affecting the vaporization stage will increase the efficiency of the column. In a practical sense, anything which increases the surface area of the column will increase condensation and increase the overall efficiency of the column.

A simple expedient for increasing the surface area of a column is to construct it with a series of indentations in the side. These small indentations are illustrated in Fig. 2.2*b*. A column of the sort illustrated is called a *Vigreux column,* and the indentations enhance the vapor separation. As the vapor travels up the column, it encounters more glass surface than it would if the column had smooth sides. As a consequence, the vapor can exchange heat with the glass, condense, and return to the distillation pot. Vigreux columns are convenient distilling columns and afford several theoretical plates.

Increasing the glass surface in a one-piece column, either by the introduction of indentations or by extending the length of the column, suffers from practical limitations. First, a column which has a high efficiency but is 1 km long would be extremely difficult to use in most laboratories. Likewise, one cannot introduce as many indentations as one might want, simply because the column would become so restricted that it would merely be a narrower column than it was before rather than an efficient distilling column.

The most practical way of increasing the surface area within a distillation column without making its length unwieldy is to fill the column with a so-called packing material (Fig. 2.2 *c*, *d*). The packing will offer to the vapor a large surface area, so that heat exchange and condensation can readily occur. The loose or open nature of the packing allows an unobstructed path for vapor. Materials which are usually used as column packings are glass chips or spheres, stainless-steel turnings, metal saddles, or copper wool (e.g., Chore Boy). A simple frac-

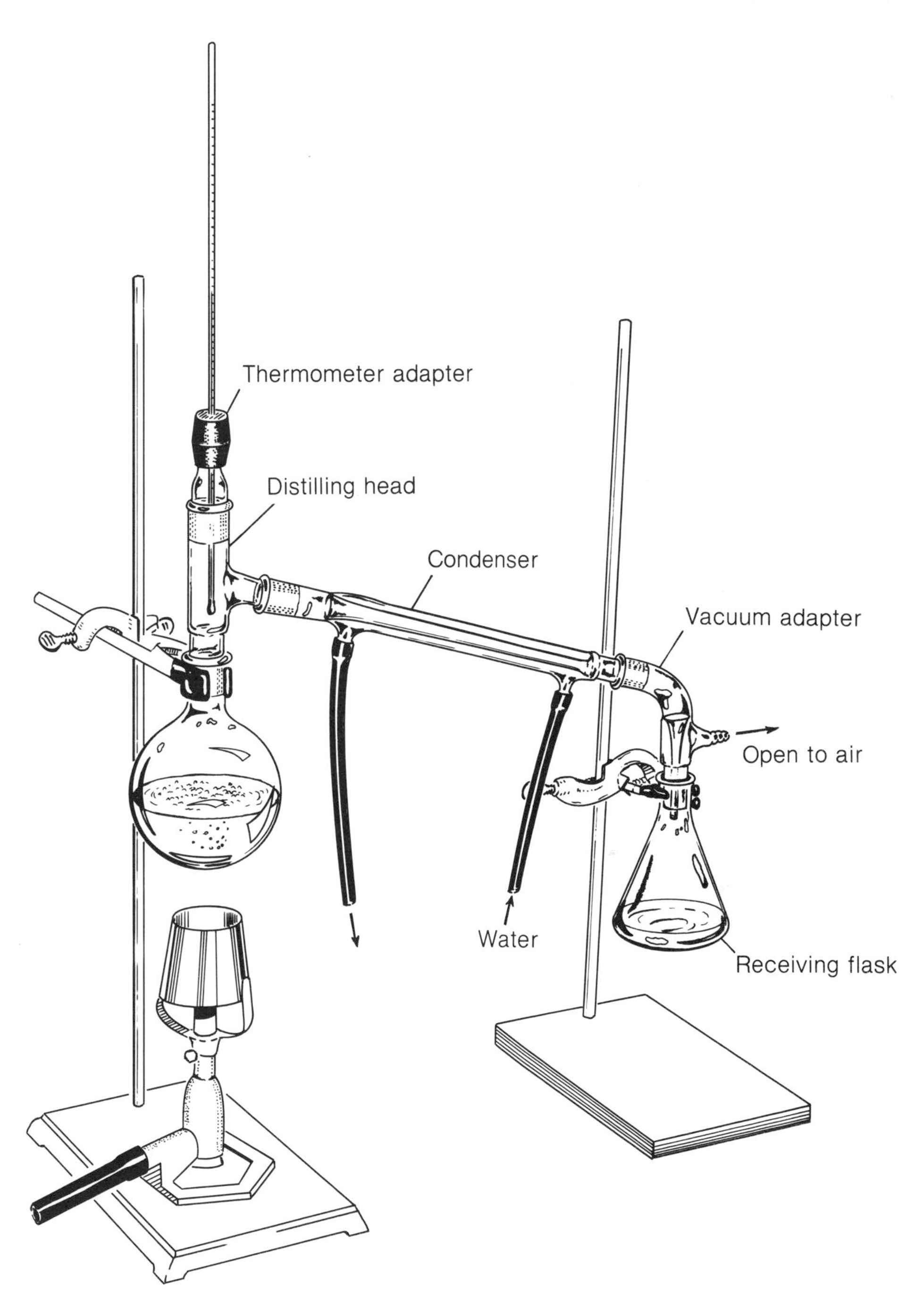

Figure 2.3
Simple distillation apparatus.

tionating column and a packed fractionating column are shown in Figs. 2.4 and 2.5, respectively.

The choice of packing material will be determined to some extent by the chemical reactivity of the compound to be distilled. Stainless steel and glass are both generally useful column packings. Magnesium turnings would afford the same type of surface that stainless steel does but would be inappropriate for the distillation of an alkyl halide. Heating an alkyl halide in a magnesium-packed column would probably lead to Grignard reagent formation. Other undesired reactions can be imagined for certain other column packings.

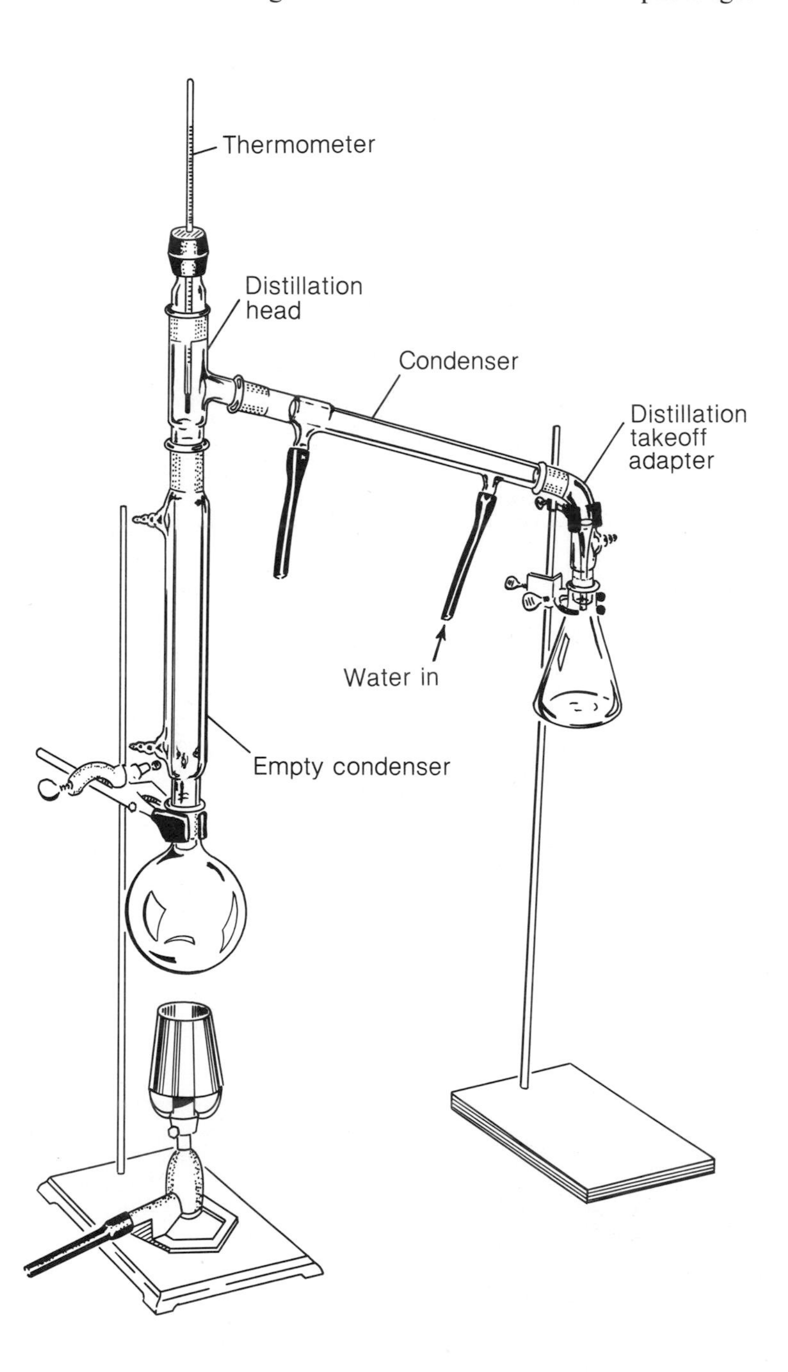

Figure 2.4
Simple fractionating column.

More sophisticated devices

Other and more sophisticated devices have been designed to achieve effective separation. A bubble-cap column is illustrated in Fig. 2.6*a*. As can be surmised from the illustration, bubble-cap columns are made with great difficulty and are quite expensive. They permit very efficient separations but have the drawback that their design prohibits large volumes of liquid from being distilled per unit time.

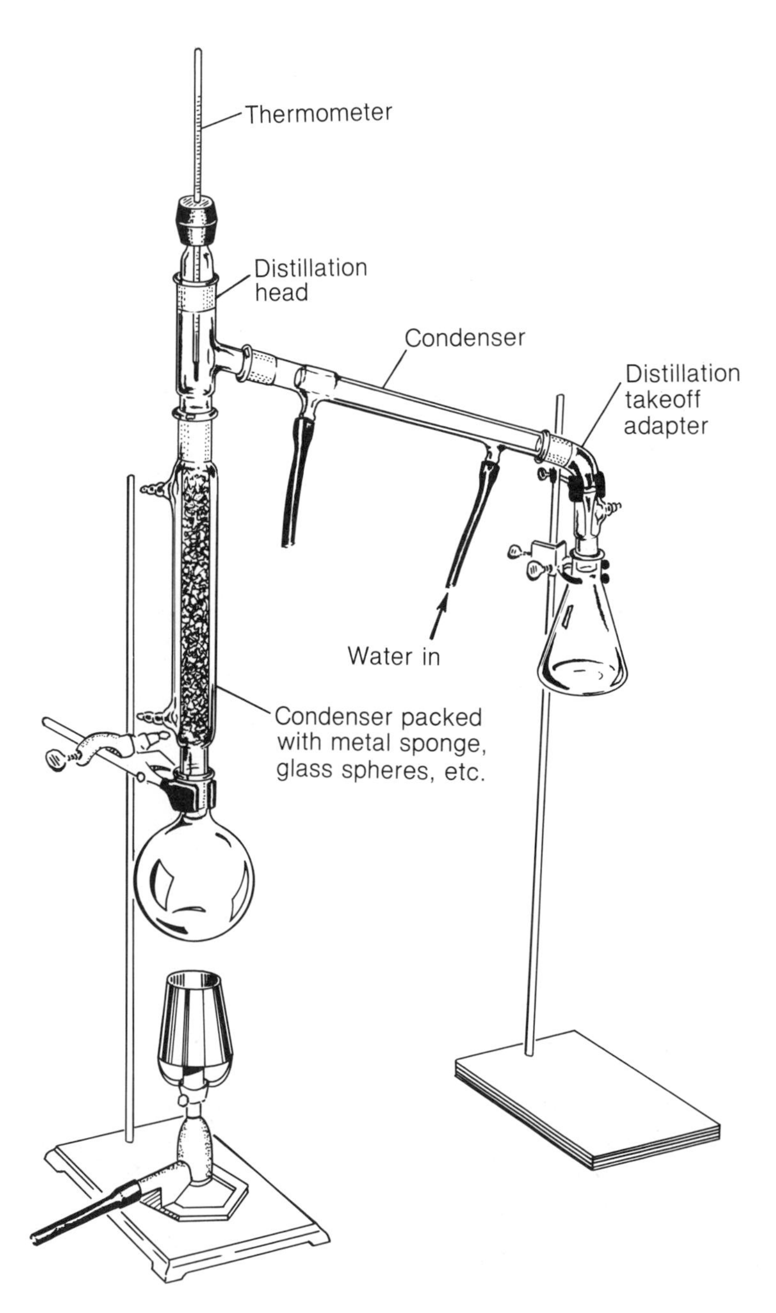

Figure 2.5
Packed fractionating column.

Figure 2.6
(*a*) Close-up of part of a bubble cap distillation column. (*b*) A spinning-band distillation column.

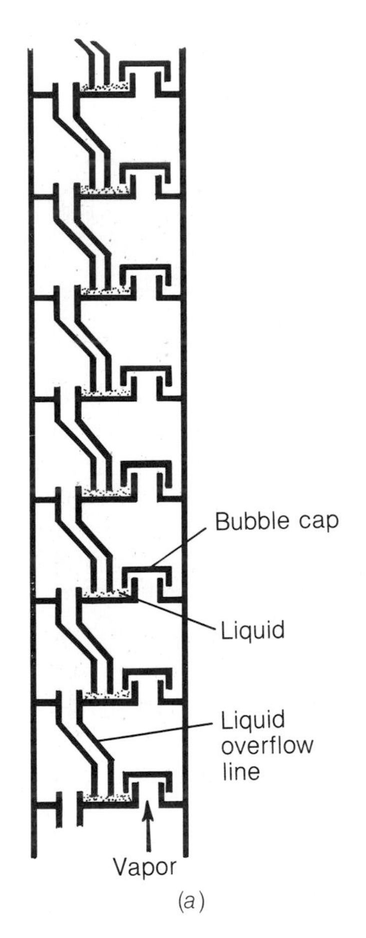

(*a*)

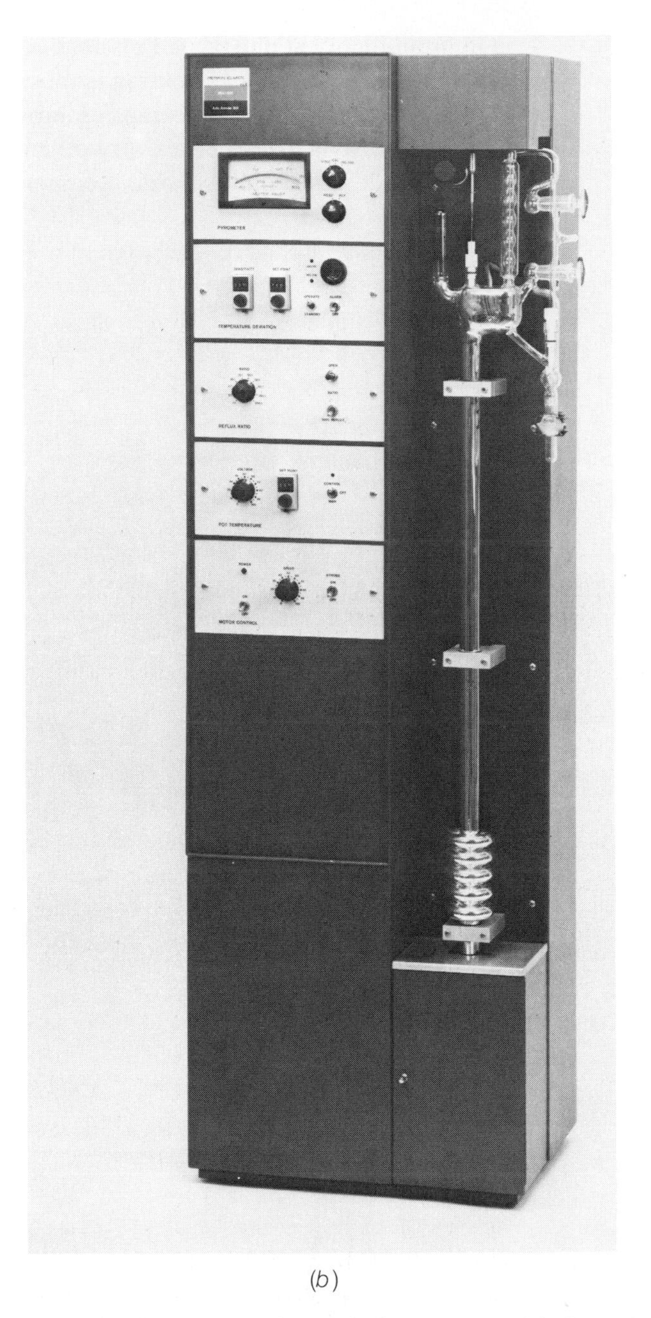

(*b*)

The spinning-band column (see Fig. 2.6*b*) is a very sophisticated electronic and mechanical device which affords many theoretical plates. This device utilizes a rapidly turning inner core of some inert material (stainless steel or Teflon), which throws the vapor and liquid against the wall of the distillation column. As the column spins, temperature gradients are decreased and condensation is increased. The spinning-band column is usually affixed to a temperature-control center, which allows the temperature of the distillation pot, the column, and the distillation head all to be monitored and maintained separately. These columns have very high numbers of theoretical plates and have the advantage that they do not retard the distillation process as the bubble-cap columns do.

In thinking about the operation of a distillation column, one point should be clear: The greater the number of condensation steps which occur in the column, the smaller must be the amount of vapor which reaches the condenser. In a simple distillation very little of the material condenses in the column and returns to the distillation pot. Another way of expressing this phenomenon is to say that the reflux ratio is low. In a very efficient column (one in which much vapor is condensed and returned to the pot), the reflux ratio is high. If the reflux ratio is high, then relatively little material will be distilled per unit time.

In considering what conditions are appropriate for any separation, one must always balance column efficiency against distillation rate. If a compound can be purified by simple distillation, it would be a waste of time and resources to attempt a fractional distillation. The ideal situation is the one in which the greatest possible throughput is achieved but the separation is complete.

Vacuum Distillation

Thus far, we have discussed distillation only of those compounds which do not decompose at their boiling points. Stability at the boiling point is a requirement for all distillations conducted at atmospheric pressure. Recall from Raoult's law that the contribution of each component to the vapor pressure depends not only on the mole fraction and partial pressures of the individual components but also on the total pressure. In the discussions above, the total pressure has always been 1 atm.

If the total pressure above the solution is lower than atmospheric pressure, then the contributions of vapor pressures at a lower temperature will be great enough to permit distillation. This results simply from the fact that the liquid or mixture of liquids boils at a temperature much lower than that required at atmospheric pressure. Chefs who work in high-altitude regions must frequently cope with this problem. If a hard-boiled egg requires 15 min in boiling water at sea level, it will have to be boiled longer at a high altitude because the pressure at this altitude will be lower and hence the water will be boiling at a lower temperature. At a height of 1000 ft above sea level the boiling point difference is only 2 to 3°C. In a city as high as Denver (the Mile High City) the boiling point difference is more substantial. The increase in volatility of a liquid because of reduced pressure can be a great advantage in purifying it. Distillation at reduced pressure is also attended by some complications. The most important problem is that the boiling point of an impurity is also reduced by the decrease in pressure. As a result, the boiling points of the two materials are closer at reduced pressure than they would be at atmospheric pressure. The efficiency of a distillation column must be higher if the difference in boiling points is lower. Despite this potential problem, vacuum distillation is a very powerful separation technique and is used routinely in the purification of liquid materials. A simple apparatus for vacuum distillation is shown in Fig. 2.7.

Azeotrope Formation

Occasionally, a mixture of two or more liquids affords a vapor which is in equilibrium with the liquid phase and both vapor and liquid have the same composition. When liquid and vapor phases of the same composition are in equilibrium, an *azeotrope* is said to have formed. The mixture of these components will distill without any change in composition until one of the components has been

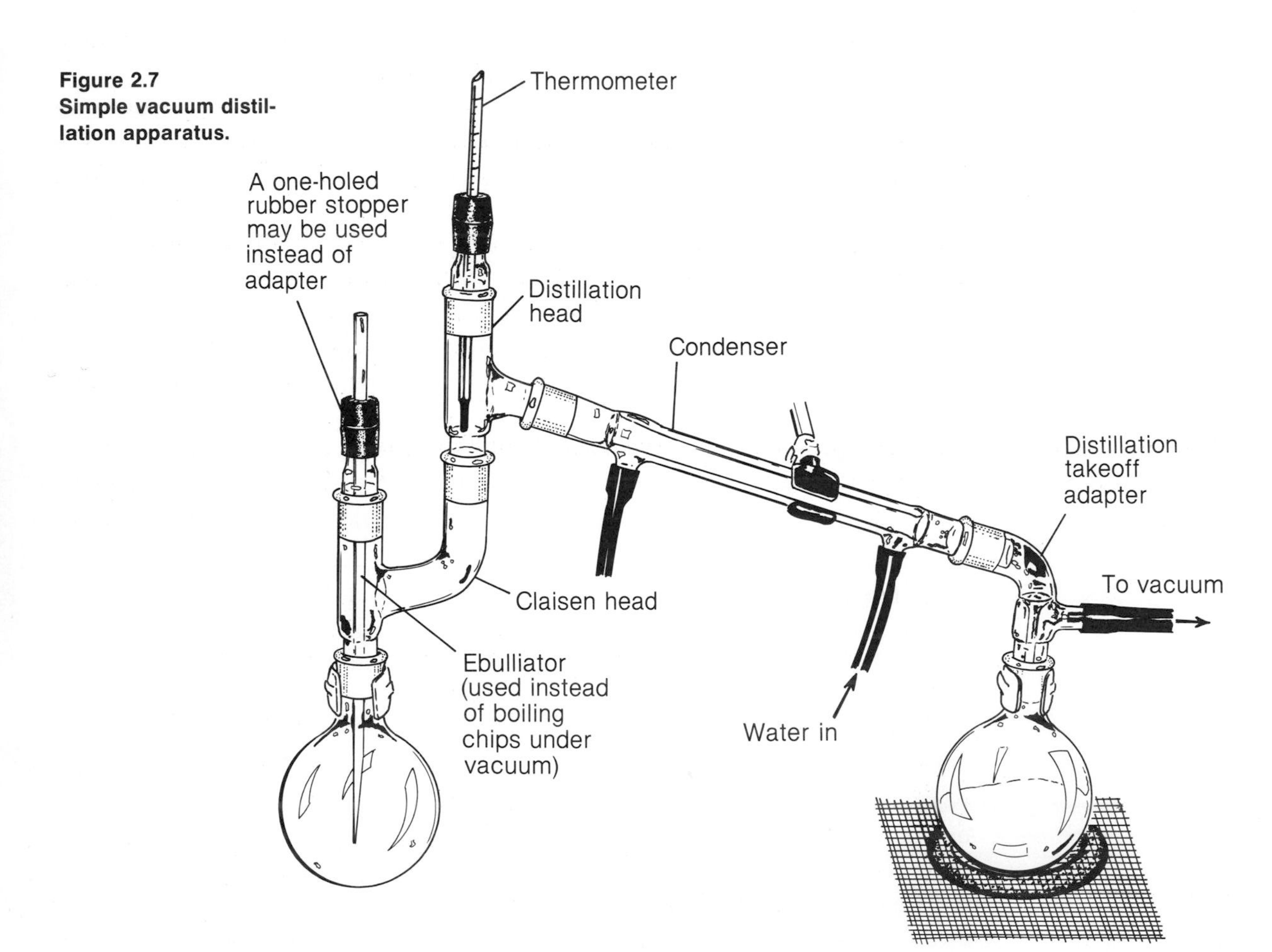

Figure 2.7 Simple vacuum distillation apparatus.

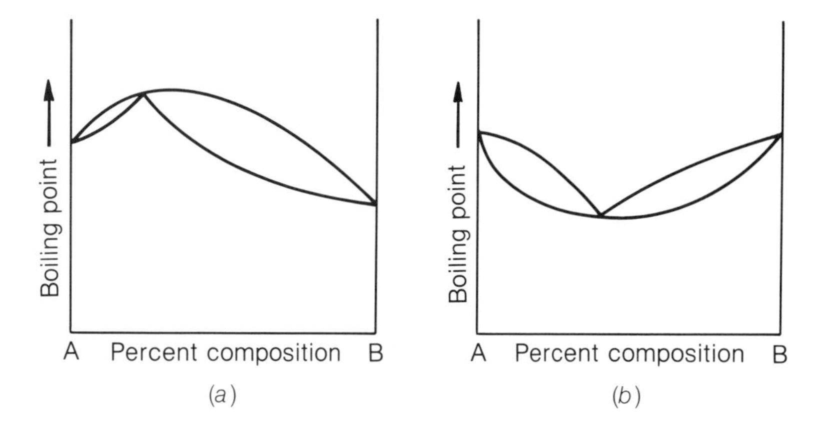

Figure 2.8 (*a*) Maximum-boiling azeotrope. (*b*) Minimum-boiling azeotrope.

consumed. Before one of the components is completely removed from the solution, no separation can be achieved no matter how efficient the column is.

Both maximum- and minimum-boiling azeotropes are known, but the minimum-boiling azeotropes are by far the more common. Phase diagrams for the two types of azeotropes are shown in Fig. 2.8*a* and *b*. The first diagram is for a minimum-boiling azeotrope and the second is for a maximum-boiling azeotrope. The minimum or maximum point in each diagram corresponds to the azeotrope and is the point at which liquid and vapor of identical composition are in equilibrium. The nature of the difficulty associated with azeotrope formation can be gleaned from these diagrams. Although no separation can be achieved until one component is exhausted, the other component may often be obtained in a pure state.

Ethanol and water form an azeotrope consisting of 95% ethanol and 5% water. Since the concentration of water is small in this azeotrope, it would be very inefficient to distill the ethanol until all water had been removed. As a re-

TABLE 2.1
Common azeotropes

Mixture	Boiling point of azeotropes, °C	Composition
Benzene-water	69.4	8.9% water
Toluene-water	85.0	20.2% water
Ethanol-water	78.0	5% water
Ethanol-benzene	67.8	32.4% ethanol
Methanol-benzene	58.3	39.5% methanol
Methanol–carbon tetrachloride	55.7	20.6% methanol
Hexane-water	61.6	12.9% water
Chlorobenzene-water	90.0	20% water

sult, ethanol from most commercial sources is so-called 95% ethanol. In fact, High Proof liquor is usually 190 proof (95% ethanol) rather than 200 proof or absolute. The last 5% of water in ethanol can be removed but often only at the expense of adding an additional contaminant (such as benzene).

Azeotropes do not form only between two compounds as similar as ethanol and water. The hydrocarbon toluene, which bears no similarity to water, forms an azeotrope with it. The percentage of water in this azeotrope is much greater than it is in the ethanol azeotrope. Distilling toluene causes the azeotrope to be removed first, leaving behind pure, dry toluene. The drying of many hydrocarbons which form aqueous azeotropes is routinely effected by *azeotropic distillation*. Note again that no increase in column efficiency will allow for the separation of an azeotrope. Common azeotropes are listed in Table 2.1.

PROCEDURE

FRACTIONAL DISTILLATION

The apparatus used for this experiment is illustrated in Fig. 2.9. It consists of a round-bottom boiling flask containing the mixture of liquids to be separated and two or three boiling chips (to aid even boiling). The round-bottom flask should be clamped to a ring stand at a height appropriate for the insertion either of a steam bath or other heating apparatus. (Check with the instructor for directions.) In the neck of the distillation flask fix a reflux condenser. Attach a water-cooled condenser to the side arm of the distillation head and clamp it to a second ring stand as illustrated in Fig. 2.9. Attach the hoses connected to the condenser so that water comes in at the lower level and leaves through a tube at the higher level, so that the condenser is always filled with water. Be certain that the drain tube is fixed securely in a drain. Be sure the water pressure through the tubing does not cause the tubing to jump from the drain. Attach a distillation takeoff adapter (if available) to the end of the condensing column. Place a graduated cylinder on the bench top in such a location that condensate dripping from either the end of the condenser column or the adapter (if used) drips into it. There should be as small a gap as possible between the end of the distillation equipment and the top of the graduated cylinder. This is especially important if a free flame is used to heat the distillation flask. In an experiment of this sort, it is always wise to conduct the entire operation in an efficient hood. In so doing, potentially flammable vapors will be drawn to the rear of the fume cupboard and not into the laboratory where they may either be breathed or ignited.

The empty reflux condenser between the distilling flask and the distillation head will serve as the fractionating column. The air space between the

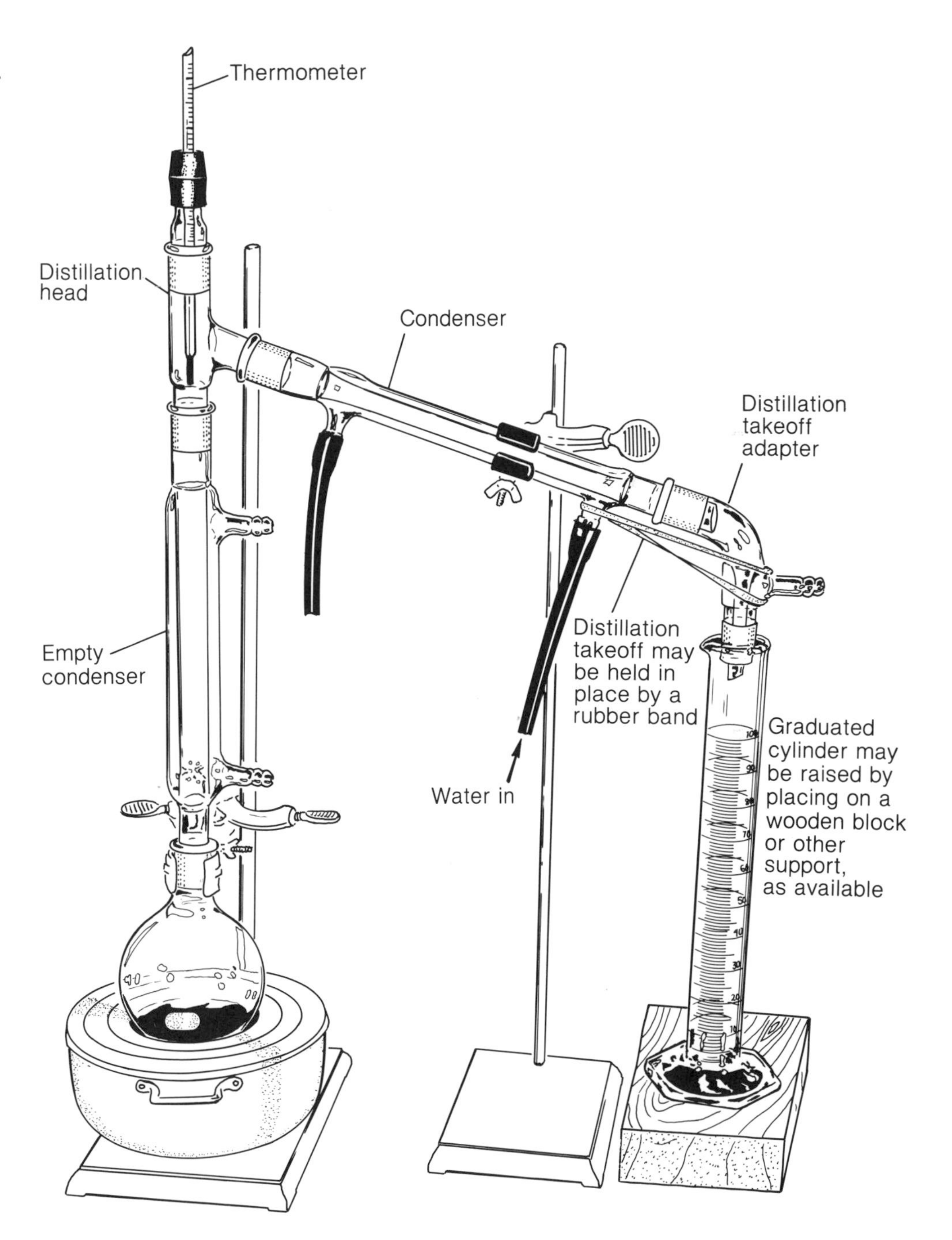

Figure 2.9 Distillation apparatus.

outer jacket and the inner surface of the condenser is a reasonably good insulator. As a consequence, one to two theoretical plates will be realized. In the second part of this experiment, the same column will be packed with an inert material, and the difference in efficiency may be assessed. The round-bottom flask which is chosen should be of sufficient size that the entire liquid sample does not fill it more than half full. This prevents overflowing when the liquid is heated.

The apparatus should be set up in such a way that all the joints, which have been lightly greased, fit snugly into the appropriate opposite joints. If this is done properly there will be no undue strain on any part of the apparatus. In this context, a common mistake is to introduce stress by tightening a clamp too much while the apparatus is at an unfortunate angle. Note that the distillation column should be as nearly vertical as possible. When the mixture is warmed, heat should be applied at such a rate that boiling and condensation (reflux) occur. One can assess the reflux ratio (see p. 55) by counting the number of drops that fall into the graduated cylinder compared with the number of drops which fall back into the pot during the same period of time. If the drop size is similar in both places, the reflux ratio is the number of drops returning to the pot divided by the number of drops of condensate. Ordinarily, the higher the reflux ratio, the better will be the separation, although this generalization is limited somewhat by column efficiency.

Part 1

In the first part of this experiment, two students should work as partners. Each partner should set up the apparatus described above and should distill a mixture consisting of dichloromethane (bp 41°C) and ethyl acetate (bp 76°C). One partner should distill the mixture through an empty distillation column, and the other partner should use a column containing a packing material designated by the instructor.

Place 30 mL of a 1:1 (v/v, i.e., equal volumes) dichloromethane–ethyl acetate solution in a 50-mL round-bottom flask. (The flask will be more than half full, but it is safe in this case.) Distill the mixture using a steam bath as heat source. Use a 10-mL graduated cylinder as a receiver. Adjust the steam pressure so that 1 to 2 mL/min is collected in the receiver without vigorous boiling. Record the temperature of the thermometer in the distillation head after each full milliliter of distillate is collected. This should afford approximately 25 data points by the end of the experiment. Using graph paper, each student should plot the temperature observed versus the volume of distillate collected. Compare the two sets of results and determine which column was more efficient. Estimate from the graphs the relative efficiencies of the two columns.

Part 2

Working individually now, each student should use the packed-column apparatus. Obtain a 30-mL sample from your instructor. The mixture will contain two components from the list shown in Table 2.2. The components are not necessarily present in equal amounts.

Carefully distill the sample (packed column) using a steam bath or other heating apparatus designated by your instructor. If you use a flame, be certain that the distillation apparatus is in the hood and away from any other flammable materials. As the distillation progresses, monitor the temperature on the thermometer at the still head. Attempt to collect two fractions of approximately constant boiling point. Between these constant-boiling fractions there will often be a small fraction which has a wide boiling range. As you monitor the boiling point, be careful also to note the approximate volume of the distillate collected. After you have distilled the entire mixture (always leave a small amount of liquid in the bottom of the round-bottom flask, i.e., never distill to dryness, because it is sometimes dangerous to do so), weigh each of the samples. Determine the percent composition of the original mixture. Determine the refractive index of each component and, assuming a linear relationship between refractive index and composition, use the data in Table 2.2 to identify the components of the mixture. This is a simple unknown identification procedure and one which you will use many times in later experiments in this and other laboratories.

TABLE 2.2
Boiling points and refractive indexes of selected compounds[1]

Liquid	Boiling point, °C	Refractive index; n_D^{25}
Dichloromethane	41	1.4216
Acetone	56	1.3561
Chloroform	61	1.4439
Methanol	65	1.3290
Hexane	69	1.3723
Carbon tetrachloride	77	1.4570
Ethyl acetate	77	1.3720
Benzene	80	1.4979
Cyclohexane	81	1.4260
Water	100	1.3330
Methylcyclohexane	101	1.4222
Toluene	111	1.4941

[1] *To the student:* For safety and practical reasons, not all these compounds may be used in each set of mixtures.

After you have obtained the two fractions and identified each of the two liquids, carefully label the two samples, record their identities in your notebook, and submit the samples and results as directed by your instructor.

2.2 STEAM DISTILLATION

The purification of a liquid by distillation takes advantage of the fact that, in a mixture of two or more miscible liquids, the vapor above the solution will be richer in the more volatile component and can therefore be separated by careful fractionation. But what situation results if the components of a mixture are insoluble in or immiscible with each other? This is a very common occurrence in organic chemistry, since most hydrocarbon solvents are insoluble in aqueous solutions and vice versa. Can a purification be effected by distillation of two immiscible liquids?

When two solutions coexist in contact with the atmosphere, both solutions contribute to the partial pressure above the surface of the liquids. As the temperature is raised, the vapor pressure above the surface of the liquid will increase. Remember that in any boiling process it is the vapor pressure of the solution reaching the atmospheric pressure exerted upon it that is observed, and it is only at this point that boiling and distillation commence. Therefore when water is boiling at sea level, the temperature at which boiling and distillation occur (100°C) indicates that the vapor pressure of water at that temperature is 760 mmHg.

The question is asked above: What will happen if two immiscible liquids are raised in temperature? It is not unreasonable to assume that each liquid of the immiscible pair will exert its vapor pressure independently of the other. While the vapor pressures are not strictly independent, in many cases this is a good approximation. Thus as the temperature is raised, a point will be reached at which the combined vapor pressure of the two liquids will equal atmospheric pressure, so that distillation will commence. Condensation of the vapor phase will yield a two-phase mixture of the aqueous and organic component.

It follows from Raoult's law that the ratio of the vapor pressures of two liquids is in direct proportion to the molar concentrations of the two substances in the gas phase. Thus, if each of two components has a vapor pressure of 380 mmHg at some temperature, their molar ratio is formulated as follows:

$$\frac{P_A}{P_B} = \left(\frac{M_A}{M_B}\right)_{vap}$$

where P_A = vapor pressure of pure A
P_B = vapor pressure of pure B
M_A = moles A
M_B = moles B

The consequence of this in practical organic chemistry is that a high-boiling component with a rather low vapor pressure may be obtained when it is distilled together with an immiscible liquid. Thus, high-boiling materials can be isolated and purified by combining them in a distillative process with some immiscible liquid of lower boiling point.

The reason for the name *steam distillation* should be clear. Many commonly encountered organic compounds are immiscible with water. Water also has several characteristics which make it a favorable choice: it is available, in-

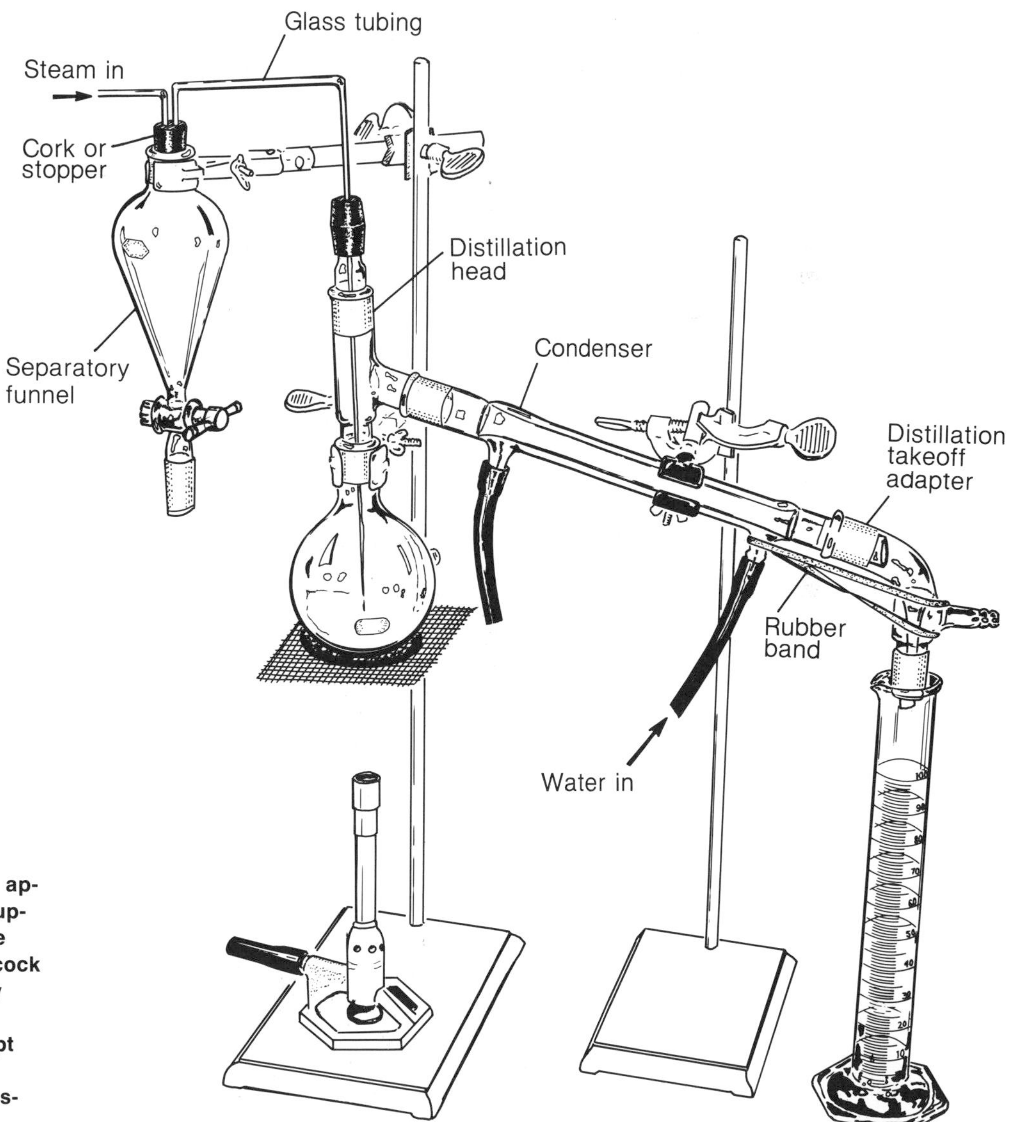

Figure 2.10
Steam distillation apparatus; steam supplied from outside source. The stopcock on the separatory funnel should be kept closed except to drain excess water from the system.

expensive, and of low molecular weight. Because of its low molecular weight, a large number of moles of water can be distilled over even though the volume of water will not be great. The above equation shows that, even though the ratio of component A to water (B) is not very favorable, if enough water is codistilled with A, a reasonable amount of A will be obtained.

A major advantage of this technique is that high-boiling compounds which decompose at or near their boiling point can be steam distilled at a temperature low enough to prevent decomposition. Because steam distillation is such an efficient and inexpensive process (only water and heat are needed), this procedure is frequently used to isolate and purify natural oils from their biological sources. In the specific example below, caraway seeds are steam distilled to obtain two principal volatile oils, (+)-carvone and limonene. The steam distillation selectively removes these two components from the caraway seeds. Steam distillation may also be used to purify reaction products (an example of this is found in Exp. 18.2A) and also to remove high-boiling solvents such as chlorobenzene (as in Exp. 16.1A).

Steam distillations are efficiently performed on large amounts of material by generating steam from an outside source and admitting it to an enclosed container of the material (Fig. 2.10). When steam distilling small amounts of material, it is nearly as efficient to suspend the source of the oil in water in a round-bottom distillation flask. When this mixture is heated, the steam which is internally generated initiates the steam-distillation process. As the volume of water is dissipated by distillation, more is added through an addition funnel in order to maintain the level of the water in the distillation flask. This apparatus setup is shown in Fig. 2.11. In virtually every other way, the steam-distillation process is equivalent to normal distillations in that the vapors are condensed in the usual fashion and collected as a mixture in a receiving flask. Normally, the volatile oils can then be separated from the aqueous solution by saturation with salt (salting out) and extraction of the oils with a solvent such as methylene chloride or ether. Removal of the extraction solvent yields the pure oil.

PROCEDURE

STEAM DISTILLATION

Isolation of carvone from caraway seeds

Place 50 g whole or ground caraway (*Carum carvi*) seeds and 150 mL water in a 500-mL round-bottom flask and assemble the apparatus as shown in either Fig. 2.10 or Fig. 2.11 (check with your instructor). After assembling the apparatus (be sure to fire polish all cut glass ends and to lightly grease all

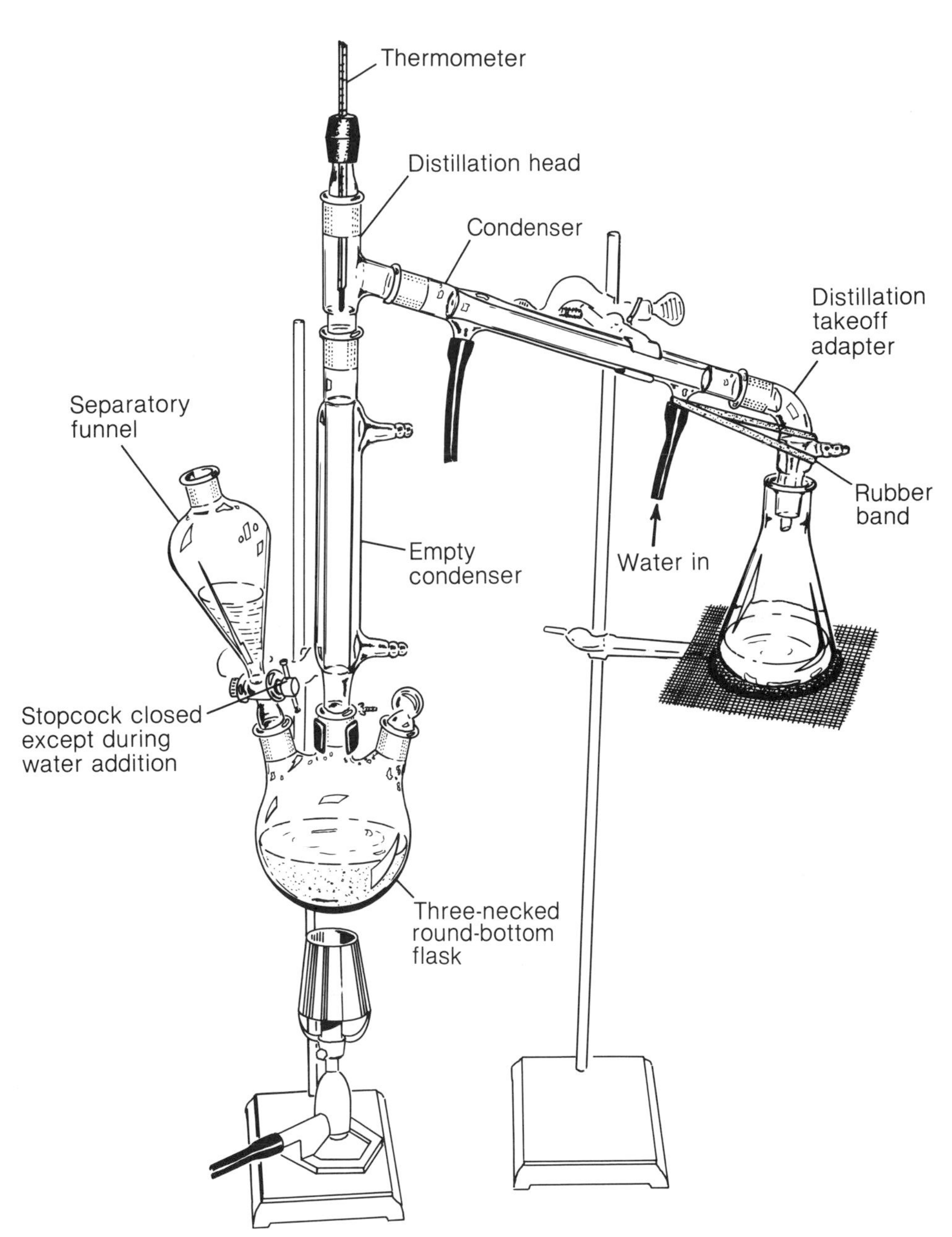

Figure 2.11 Steam distillation apparatus; steam generated internally.

ground glass joints), run the steam at a moderate rate into an unused steam bath for 1 to 2 min in order to flush condensed water and metal fragments out of the steam line; then connect it to the apparatus.

Do not shut the steam off without first withdrawing the piece of cut glass tubing from the distilling flask (Fig. 2.10, only). (Why?)

Regulate the steam so that distillation occurs as rapidly as possible (as fast as the distillate will condense, or as quickly as possible without excessive bumping). Use a bunsen burner as needed to ensure that the distillation flask remains a little less than half full.

Collect the milky white distillate in a 250-mL Erlenmeyer flask cooled in ice. After collecting 200 mL, the distillate should be almost clear (if not, collect an additional 50 mL in a 125-mL Erlenmeyer flask), and the steam distillation of the oil is essentially over.

Dissolve approximately 15 g sodium chloride in the distillate and agitate to dissolve the salt. Because you have a 250-mL separatory funnel, divide the distillate into two 100-mL portions. Extract each portion once with 15 mL diethyl ether. Combine the ether extracts (approximately 30 mL) and concentrate the combined extract on a steam bath. You should obtain approximately 1 g of a clear oil which has the same odor as caraway seeds. Save your sample for future analysis. Remember that diethyl ether is very volatile and flammable, so that no flame should be close to you while you are working with this solvent.

2,4-Dinitrophenylhydrazine (2,4-DNP) derivative of carvone

Before doing this experiment, read Sec. 25.5A carefully.

The 2,4-DNP derivative may be formed from the crude product mixture (obtained above) or from purified (+)-carvone. Set aside a few drops of material; weigh the rest and adjust the scale of the following reaction appropriately.

Place 0.5 g 2,4-dinitrophenylhydrazine and 35 mL diethylene glycol in a 125-mL Erlenmeyer flask. The solution is swirled and heated briefly on a steam bath to dissolve the hydrazine. To this warm, red-orange solution (ignore small amounts of precipitated solid) add 0.5 g of the carvone-limonene mixture while swirling to dissolve the material. Now add approximately 1.5 mL concentrated HCl and swirl the solution to thoroughly mix in the acid. The color of the solution should change from red-orange to light yellow. Allow the mixture to stand at room temperature for 15 to 30 min to permit crystallization of the derivative.

At the end of the reaction period, 10 mL water is added in small portions to the solution while swirling. The residue is suction filtered and the collected solid washed with 10 to 15 mL 50% aqueous ethanol.

The residue is recrystallized from ethanol (steam bath). Report the melting point of your 2,4-DNP derivative, as shown below.

	Observed mp	Literature mp
2,4-Dinitrophenylhydrazone	187 to 189°C	191°C

2.3 SUBLIMATION

Most organic chemists tend to think of extraction, distillation, crystallization, or some form of chromatography as the major purification methods available to them. Although often overlooked, sublimation has historically been one of the more important methods for purifying solids, especially if a large quantity of material is involved. Today, sublimation continues to be important for the purification of solid materials on an industrial scale.

The Principle

When the vapor pressure of a solid which is being heated reaches the external pressure on the system before the temperature reaches the melting point of the solid, the substance will undergo a phase transition directly from solid to vapor. In a distillation, the phase transition is from liquid to vapor. The situation here is similar but the phase transition is different.

Condensation of the vapors constitutes a reversal of this process. The vapor condenses directly to a solid on a cool surface. This overall process is called *sublimation*.

It should be obvious that a solid which can be purified by sublimation may be freed of ionic or nonvolatile impurities with relative ease. Industrial-grade iodine, for example, contains a large number of nonvolatile contaminants. The forces holding iodine molecules together are relatively weak because the iodine molecule is cylindrically shaped and there is relatively little polar contact to hold the molecules together. As it sublimes, iodine passes directly from the solid to the vapor phase. This is often observed as a purple vapor above the dark purple solid. The vapor can easily be condensed on a cold surface. What remains behind in the sublimation vessel consists of impurities which cannot readily vaporize.

Sublimation is usually easiest for those materials which do not possess strong intermolecular forces. Cylindrical and spherical molecules are the most common examples of molecules which sublime readily. Ball-shaped camphor and cylindrical ferrocene (see below and Chaps. 14 and 16) both sublime readily.

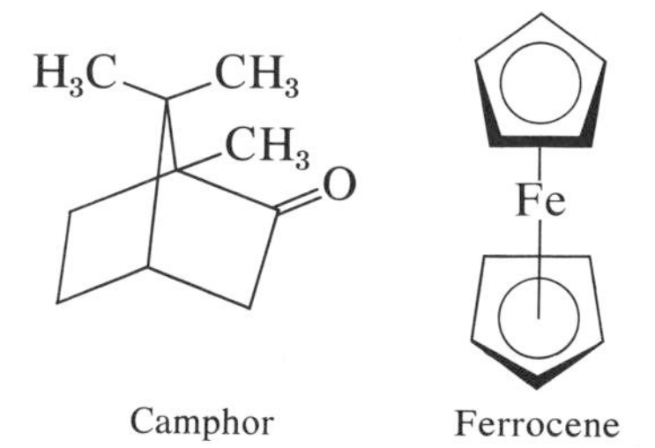

Camphor Ferrocene

In your everyday experience, you have probably encountered the sublimation of cylindrical carbon dioxide (CO_2, dry ice), which vaporizes without ever becoming a liquid. The fact that naphthalene and 1,4-dichlorobenzene sublime readily makes them useful mothballs.

An important advantage of sublimation as a purification technique is the same as that noted for distillation: the only "reagent" necessary is the application of heat. After sublimation, the heating apparatus is simply turned off, and there is no solvent or other reagent which must be removed.

The major difficulty with the sublimation technique is that it is not a very selective process. Despite this shortcoming, many compounds which are nearly pure may be sublimed to purity very easily. One must always be aware that if a contaminant in a desirable compound sublimes easily, the technique of sublimation will not be very efficient.

A few compounds sublime readily at atmospheric pressure and can be recovered by allowing them access to a cool surface. In general, however, most materials sublime only when heated below their melting points and at reduced pressure. Generally, a pressure near 0.1 torr is used, primarily because it is easy to attain this reduced pressure with an inexpensive mechanical vacuum pump. If a higher pressure is acceptable, a pressure of about 20 torr, readily attained with a water aspirator, is often chosen.

PROCEDURE

PURIFICATION OF CAMPHOR

This discussion describes the purification of camphor contaminated by sodium chloride (table salt). Camphor is an inexpensive organic material which can be obtained commercially. It is also an oxidation product from many natural sources.

Set up the sublimation apparatus as shown in Fig. 2.12 or 2.13. Place the contaminated camphor in the bottom of the sublimation apparatus and insert an ice-filled test-tube cold finger. Attach the vessel to a water aspirator and evacuate to about 20 torr. After evacuation, heat the bottom of the sublimation apparatus using a hot-water bath or an oil bath. You should observe a very rapid disappearance of solid as the bath temperature approaches 70°C. At the same time, white camphor should appear on the cold finger. After 20 to 30 min most of the desired material will have sublimed.

Remove the sublimation flask from the heating bath. Break the vacuum carefully and admit air into the chamber slowly. Remove the cold finger bearing the purified material, scrape off the camphor (spatula) onto a preweighed watch glass, and weigh it. Calculate the yield and assess the purity of the sample by melting point. Pure camphor melts at 178°C. Its melting point is depressed in proportion to the molal concentration of impurities according to the equation

$$\Delta t = \frac{100Kw}{W(MW)}$$

where K = 39.17°C (for camphor)
w = the weight of camphor
W = the weight of the impurity
MW = the molecular weight of the impurity

A reasonable value to assume here for the molecular weight of an impurity is 100, reducing the overall equation to

$$\Delta t = \frac{Kw}{W}$$

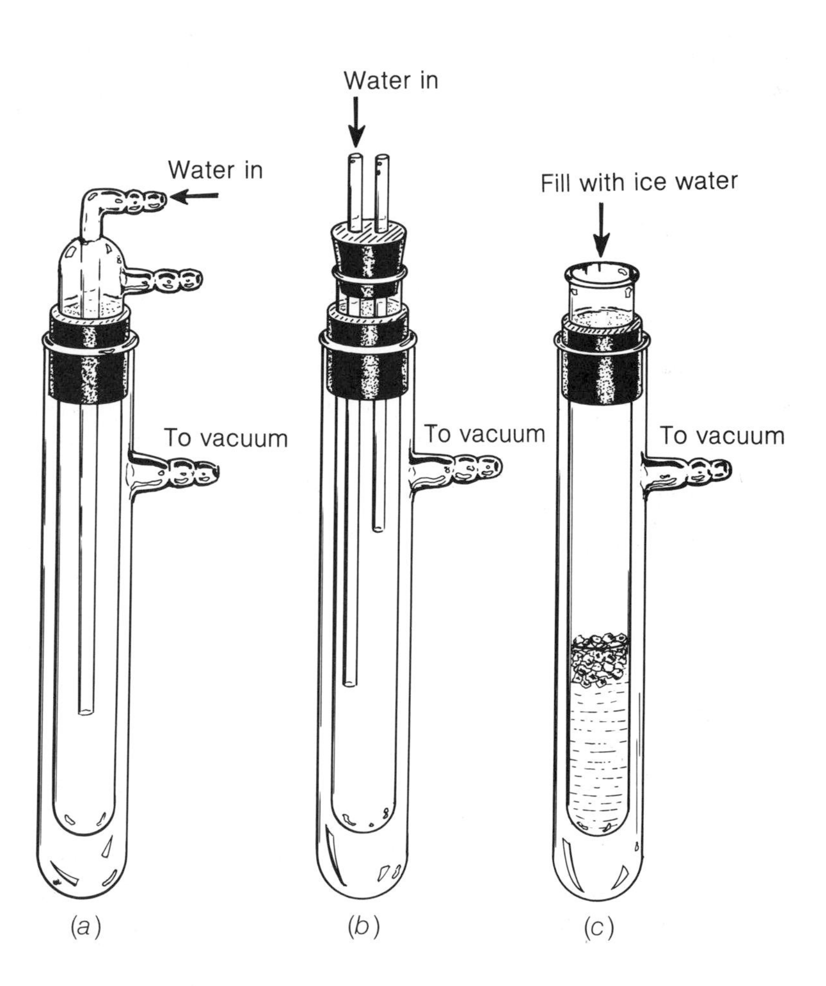

Figure 2.12 Sublimation apparatus. (*a*) and (*b*) Water-cooled. (*c*) Ice-cooled.

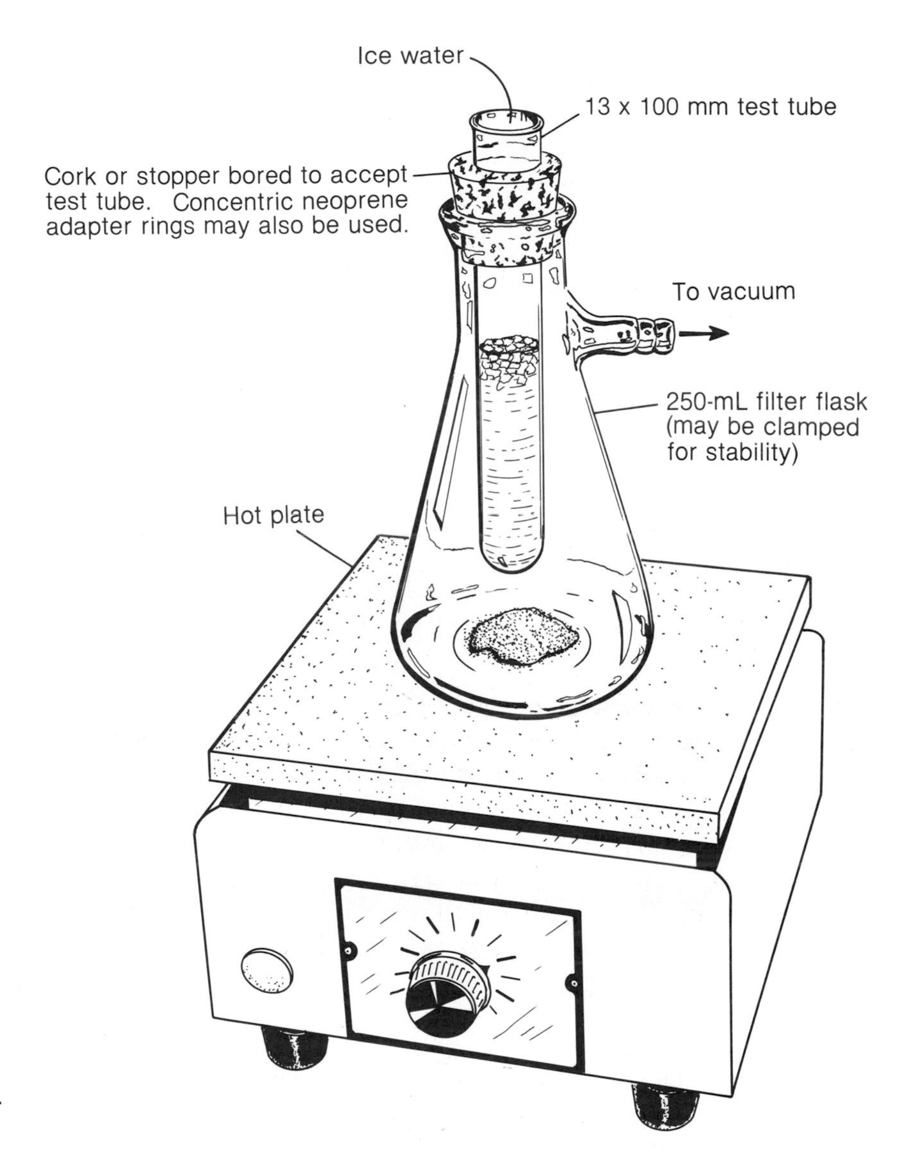

Figure 2.13 Apparatus for the sublimation of camphor.

2.4 CRYSTALLIZATION

In nature, molecules can exist as solids, liquids, or gases. The majority of the known compounds are either liquids or solids, although the state depends on the temperature of the surrounding medium. Water, for example, is a vapor above 100°C, a liquid between 0° and 100°C, and a solid below 0°C.

Probably the best technique for the purification of a liquid is distillation (see Sec. 2.1). Crystallization is usually the best method for purifying a solid. In distillation, volatility differences allow for separation; in crystallization, solubility differences allow molecules to be separated from each other or from contaminants.

In the crystallization process, molecules gradually deposit from solution and attach to each other in an orderly array (the lattice). As the aggregates of molecules become large enough to be visible, they appear as plates, needles, etc. The high symmetry of these macroscopic aggregates suggests how orderly the crystal lattice is. As the molecules deposit one after another in an orderly fashion, molecules of a different shape or size are excluded. The forces which hold molecules together are subtle ones. Molecules which do not have precisely the same kinds and arrangement of forces will not be held in the lattice. Likewise, molecules which are similar in structure but either smaller or larger will be excluded. These notions are illustrated in Fig. 2.14.

The melting point of a pure substance (see Sec. 1.1) will depend on the strength of the intermolecular forces. If an impurity is present, the melting point will usually be lowered. Common impurities are the solvent from which the substance was crystallized, water from either the solvent or the atmosphere, by-products from the reaction used to form the product, and unreacted starting material. It is these substances which recrystallization helps to eliminate.

Advantages of the Technique

One advantage of recrystallization is that it may be used on a wide range of scales. It is not uncommon for 5 to 10 mg of material to be crystallized from a few drops of solvent and thus be obtained in a pure form. Likewise, such materials as sugar (sucrose) may be crystallized on a ton scale before being shipped to the grocery store. In the crystallization process, the solvent may be reused and thus recycled. There is an obvious economic advantage to this process. In the drug industry, crystallization is the most common technique for producing the ultrapure materials required for clinical use.

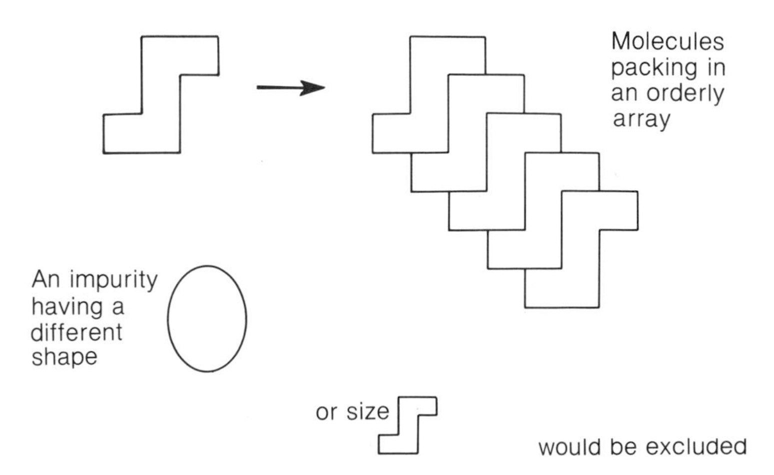

Figure 2.14 The crystallization process.

Solvent Selection

The most crucial aspect of the recrystallization process is selecting the appropriate solvent. The best solvent for recrystallization is one in which the material is insoluble at room temperature but completely soluble at elevated temperature. In discussing crystallization, *cold* usually refers to the temperature of an ice-water bath and *hot* refers to steam-bath temperature. The use of high-boiling solvents and/or dry ice will expand the temperature range for many different cases. As you gain experience, you will notice that, as a practical matter, the range of crystallization temperatures varies between 0 and 100°C.

In considering what solvent to use (see Chap. 4), the rule "like dissolves like" should always be kept in mind. A solvent in which the substance is very soluble will be a poor one for recrystallization. Likewise, a solvent in which the compound is almost totally insoluble, even at elevated temperatures, will also be a poor solvent. The best compromise is usually a solvent in which a compound is relatively insoluble at low temperatures but soluble at high temperatures.

It is readily observed that benzoic acid is relatively insoluble in cold water. This is so because the molecule has appreciable hydrocarbon character despite the presence of a polar carboxylic acid function. At high temperatures (in boiling water), however, benzoic acid is reasonably soluble. At low temperatures the water structure is too stable and orderly to be disrupted by benzoic acid. At higher temperatures the water structure is thermally randomized and it is not so energetically expensive for benzoic acid to enter it. When benzoic acid is dissolved in boiling water and the solution allowed to cool, crystals of benzoic acid are deposited from the solution as it again becomes orderly. Likewise, energy is gained in the crystal packing of benzoic acid.

The choice of solvent should also be governed by another factor. Ease of solvent removal facilitates purification by crystallization. Water may be removed quite readily from benzoic acid, so it is an excellent solvent for this crystallization. Sometimes solvents cannot be removed from the sample and themselves afford significant contamination of the solid material.

A third consideration in choosing a solvent is the temperature at which the crystals will be deposited from solution. In the days before the possible dangers of using benzene were widely recognized, this solvent was often used to crystallize organic materials. On cooling in an ice bath, crystals would invariably appear, but these crystals were more often than not solidified solvent (benzene solidifies at 6°C). Benzene is rarely used now in undergraduate laboratories due to a virtual ban from routine use by the Occupational Safety and Health Administration (OSHA), but cyclohexane is a safe and common solvent which exhibits the same behavior. If cyclohexane is chosen as a recrystallization solvent, be careful to keep the internal temperature above 10°C. If a cyclohexane solution is filtered, and the product is collected but disappears, there is usually a good chance that the solid was frozen solvent.

A final consideration is reactivity: A substance should not be crystallized from a solvent with which it reacts. For example, the common solvent pyridine (a base) would be a poor choice for the crystallization of benzoic acid. Likewise a carboxylic acid anhydride should not be crystallized from a nucleophilic solvent such as water or an alcohol (see Exps. 11.3, 11.5A, and 11.5B).

Mixed-Solvent Systems

It sometimes happens that no single solvent will meet all the requirements of a good crystallization solvent. In such cases, one must resort to so-called mixed solvent systems.

The choice of a mixed solvent system is usually predicated on the mutual solubility of two (or more) solvents in each other and the high affinity of the compound to be purified for one of the two solvents. It is also necessary that the other solvent in the pair have a low affinity for the compound to be purified. Other requirements to be kept in mind are (1) the boiling points of the two solvents should be relatively close to one another, so that one of the solvents is not completely boiled off during the heating and addition process; and (2) the two solvents should be completely miscible so that no third phase appears during the crystallization.

The solvent pairs which are used most often for recrystallization in organic chemistry are those listed below.

Methanol-water
Ethanol-water
Acetone–petroleum ether (hexane)
Ethyl acetate–petroleum ether (hexane, cyclohexane)
Ether–petroleum ether (pentane)
Dichloromethane–petroleum ether (pentane)
Toluene–petroleum ether (heptane)
Acetone-water

Benzoic acid crystallizes readily from water, but if a mixed solvent system were required it could be chosen on the basis of certain rational considerations. First, benzoic acid is a substituted benzene, so its solubility in toluene might have been predicted. Benzoic acid is so soluble in toluene that it does not crystallize from it. Toluene by itself would be an unacceptable solvent for recrystallization of this compound. Benzoic acid is almost insoluble in cyclohexane. A mixture of toluene and cyclohexane would have just the properties that are needed to recrystallize benzoic acid.

Procedure for Choosing a Solvent

For most of the recrystallizations described in this manual, a solvent is suggested. When no solvent is specified, an appropriate one may be determined by a combination of rational analysis and experimentation. Keeping in mind the "like dissolves like" rule, the structure of the compound should be considered. The solvent should dissolve the compound when hot and deposit it when cold.

After a certain amount of rationalizing has been done, a 50-mg sample of the compound should be placed in each of five to ten 10 × 75 mm test tubes. A few drops of a different solvent should be added to each test tube. If the sample dissolves immediately, that solvent will not be of use for recrystallization.

Heat the tubes containing samples which did not dissolve immediately. If solution occurs, set the tubes aside and see if crystals form. If your first choices of solvent are not successful, try other solvents or solvent combinations. If you have exhausted all the possibilities you considered rational, try some irrational ones. This approach may seem unscientific, but remember that crystallization is controlled by a subtle balance of forces, not all of which are understood. It may be of little solace, but the procedure just described is the one a research scientist would use to determine the correct solvent for a crystallization.

If a mixed solvent system appears to be required, a small quantity of the material to be crystallized is dissolved in the smallest amount of hot solvent which has good solubility properties for the substance. To the warm solution is added a small amount of the poor solvent, a drop at a time, while the test tube is heated. Eventually, a cloudiness (or cloud point) will be observed. The poor solvent is now just beginning to precipitate solid. As soon as the cloud point is observed, one or two drops of the good solvent is added to help the crystallizing material back into solution. The cloudiness should disappear. The solution is set aside and allowed to cool *slowly*.

The Recrystallization Process

If a single solvent has been selected for recrystallization, proceed as follows. Place the solid sample in an Erlenmeyer flask about twice the size needed to hold the anticipated volume of solvent. This volume may be approximated from the test tube experiments described earlier.

First, place a small amount of solvent and a boiling chip in the Erlenmeyer flask and set it on a steam bath (or other safe heating surface). Add solvent a little at a time while warming until the solid material completely dissolves. The sample should be observed carefully as it dissolves, because there are sometimes small amounts of impurities present which do not dissolve in the solvent. If the bulk of the material seems to dissolve and a residue remains, *do not* add a large amount of solvent in the hope that the residue will eventually dissolve. Nature may actually be helping out. If this situation arises, decant the solution (containing the dissolved sample) from the residue or gravity filter the hot liquid. After decantation, the residue can be characterized as sample or impurity.

When all the sample has dissolved in the solvent, remove the heating source and lightly stopper the mouth of the flask. (Instead of being stoppered or corked, the mouth of the flask may be covered with tissue paper. This will prevent dust from entering the flask.) As the solution gradually cools, crystals usually deposit. If no crystals are apparent by the time the solution reaches room temperature, cool the solution in an ice bath. If further incentive is required, carefully scratch the sides and bottom of the Erlenmeyer flask or add a seed crystal. Allow the solution to cool in the ice bath for as long as necessary to complete the crystallization process.

After crystallization occurs, the solid must be separated from the solvent which deposited it. This is usually accomplished by filtration, most commonly by so-called suction filtration. A Buchner funnel and filter flask setup for this purpose are illustrated in Fig. 2.15. A piece of filter paper just large enough to cover the bottom of the Buchner funnel is the barrier which will collect the crystalline mass but will allow the solvent to pass through. The filter funnel is

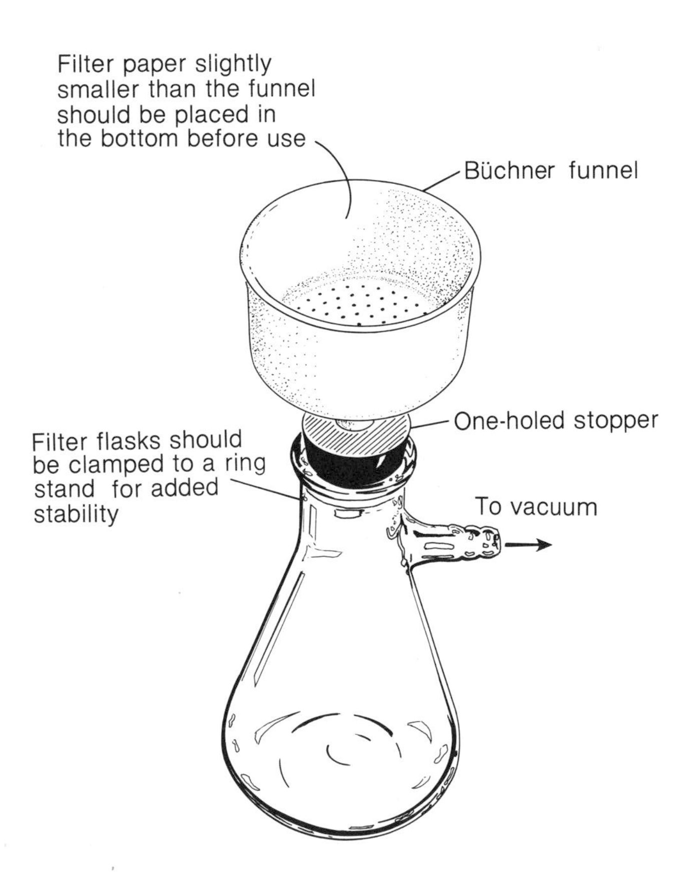

Figure 2.15 Apparatus for suction filtration.

inserted into the neck of the flask and vacuum is applied through the side arm. When a suspension of crystals in a solvent is poured onto the filter paper, the negative pressure in the flask draws the solvent through the filter paper. (It is useful to wet the filter paper with solvent before initiating the filtration so that the paper can lie flat on the Buchner funnel.)

After the mass has been collected, it is good to disperse the crystals as widely on the filter paper as possible. The dispersed mass will dry more readily as air is pulled through. After the crystals have been air dried, they are usually transferred to either a watch glass or another piece of filter paper and allowed to stand in the air. Further drying, if needed, may be accomplished as discussed below.

Purification during the crystallization process

Occasionally a sample is so impure that several recrystallizations are required to purify it. A technique which is used often in such cases is treatment with activated charcoal. The sample is dissolved in a small amount of solvent and charcoal is added. It is often the case that impurities in a sample are more polar than the sample itself and the polar impurities are adsorbed by the charcoal. The hot solution is then gravity filtered from the charcoal, which retains the unwanted impurities. Treatment and filtration of a sample by this technique requires some additional discussion.

Treatment with activated charcoal

If appropriate, water or an alcohol is the best solvent. Dissolve the sample in the solvent as described above. For about 10 g of sample, 1 to 3 g of charcoal is added to the cold solvent, and the solution is heated on a steam bath with constant swirling to raise the temperature of the solution to the boiling point of the solvent. The large surface area of charcoal will foment bumping if vigorous swirling does not accompany the heating process. It is usually a good idea to add the charcoal to the cold solvent; this is because the charcoal can act as a boiling surface and the spontaneous effervescence could cause material loss.

A real problem in working with charcoal is how to transfer this fine powder without spraying charcoal dust everywhere. However it is done, it is wise to wear gloves when transferring charcoal. One of the best methods for transferring and stirring charcoal is to fill with charcoal a small wash bottle the outlet tube of which has been cut down to a length of about 2 cm. The wash bottle can then be inverted and the charcoal squirted into the reaction mixture in the appropriate amount. This alleviates the need for any auxiliary transfer devices for the charcoal. An illustration of this simple device is shown in Fig. 2.16.

After charcoal has been added, the solution should be warmed and swirled for about 30 s. After brief, vigorous swirling and heating near the boiling point of the solvent, the solution should be removed from the heating source. The charcoal must now be removed from the mixture.

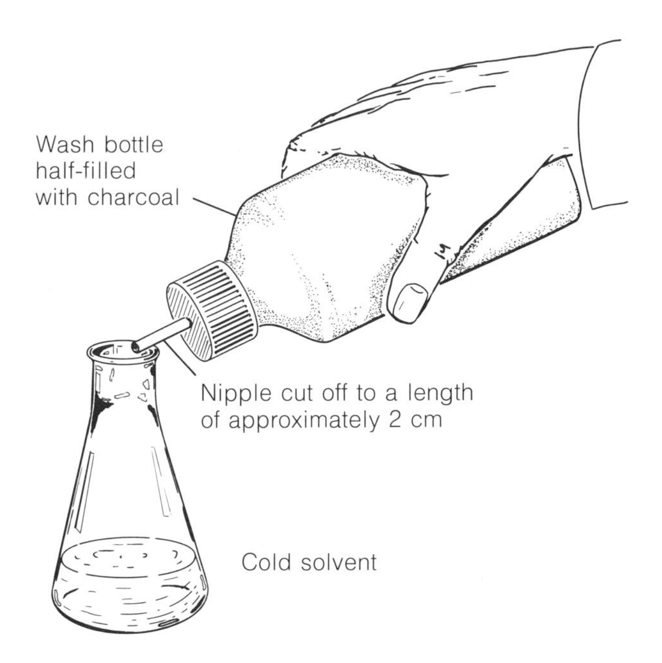

Figure 2.16
Using a wash bottle to transfer activated charcoal.

Gravity filtration

Vacuum filtration using a Buchner funnel is usually ineffective for removing charcoal from a reaction mixture, especially when the reaction mixture is warm. As the warm liquid is drawn through the funnel into the suction flask, some evaporation occurs. This evaporation usually results in rapid cooling of the solvent and premature crystallization of the sample; often the filter paper becomes clogged and the filtration must be terminated. A means for circumventing this difficulty is to use hot gravity filtration.

Hot gravity filtration is carried out as follows. Set up an Erlenmeyer flask of the appropriate size fitted with a conical funnel. A paper clip or short piece of wire inserted in the top of the Erlenmeyer flask provides an air gap between the funnel and the flask. Fold a piece of filter paper into quarters and insert it in the funnel. An alternative is to flute the filter paper (Fig. 2.17) and place it in the funnel as shown in Fig. 2.18. Wet the filter paper with a small amount of the recrystallization solvent and allow it to drip through into the flask. Place the flask on the steam bath and heat it until the solvent begins to reflux. The hot vapors will warm the flask and allow filtration at elevated temperature. Pour the hot solution containing the suspended charcoal into the funnel. The charcoal should be removed as it falls through the filter into the flask. Most of or all the charcoal should be trapped on the filter paper in the funnel. Sometimes the entire procedure must be repeated.

After charcoal treatment, the sample may be recrystallized in the usual way. If the treatment was carried out in the recrystallization solvent, the pure sample should crystallize on cooling.

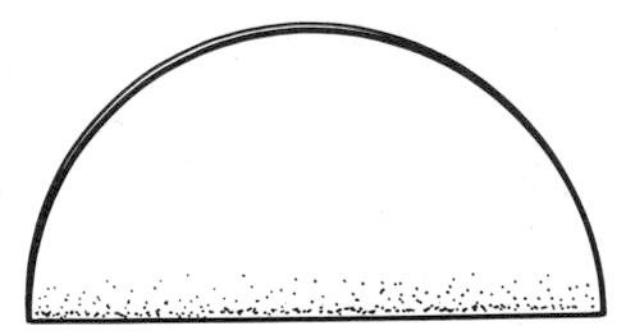

1. Fold the paper in half.

2. Fold the paper in quarters.

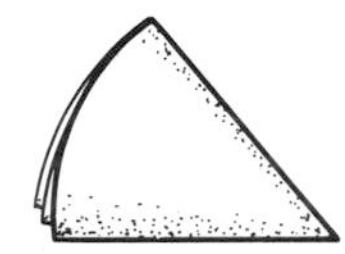

3. Fold the paper in eighths.

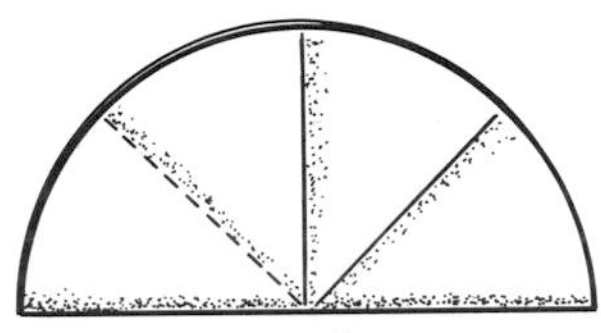

4. Open the paper to half-folded.

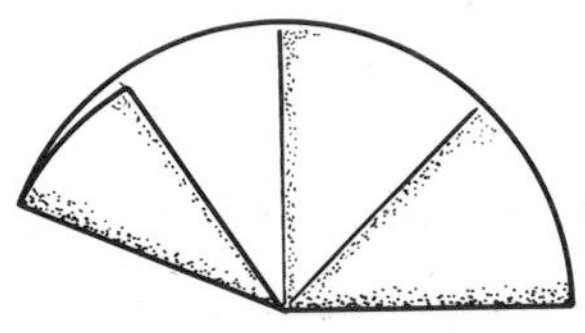

5. Using thumb and forefingers, fold over to eighth line.

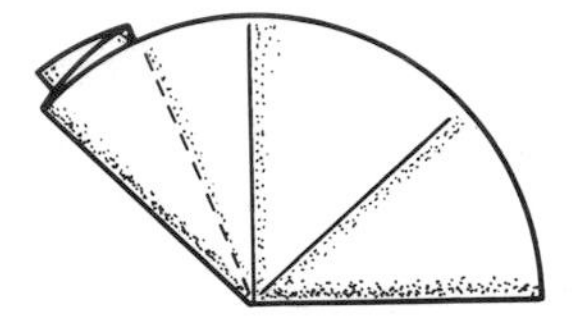

6. Fold back to the quarter line.

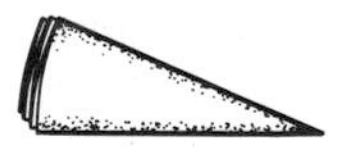

7. Continue the process of alternate folding until the paper is completely folded to one-eighth of its original size.

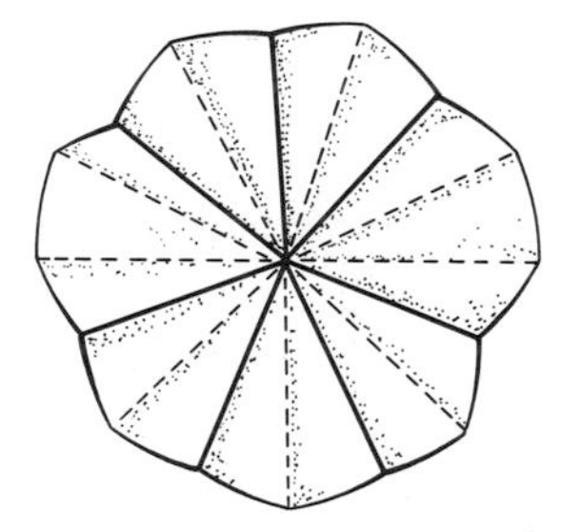

8. Fully fluted paper will be obtained when the paper is opened.

Figure 2.17
Fluting filter paper.

Use of Filter Aid (Celite)

Celite, a filter aid, is sometimes referred to as diatomaceous earth and consists essentially of finely ground silica. It has a very high surface area and may be of considerable use in certain filtering operations.

If there is a very finely divided substance in a solvent, it may be very difficult to remove it. In such cases, Celite may be slurried with water and then filtered through a paper-lined funnel on a filter flask. When the water is drawn through the Buchner funnel, a pad of Celite between 0.125 and 0.25 in should

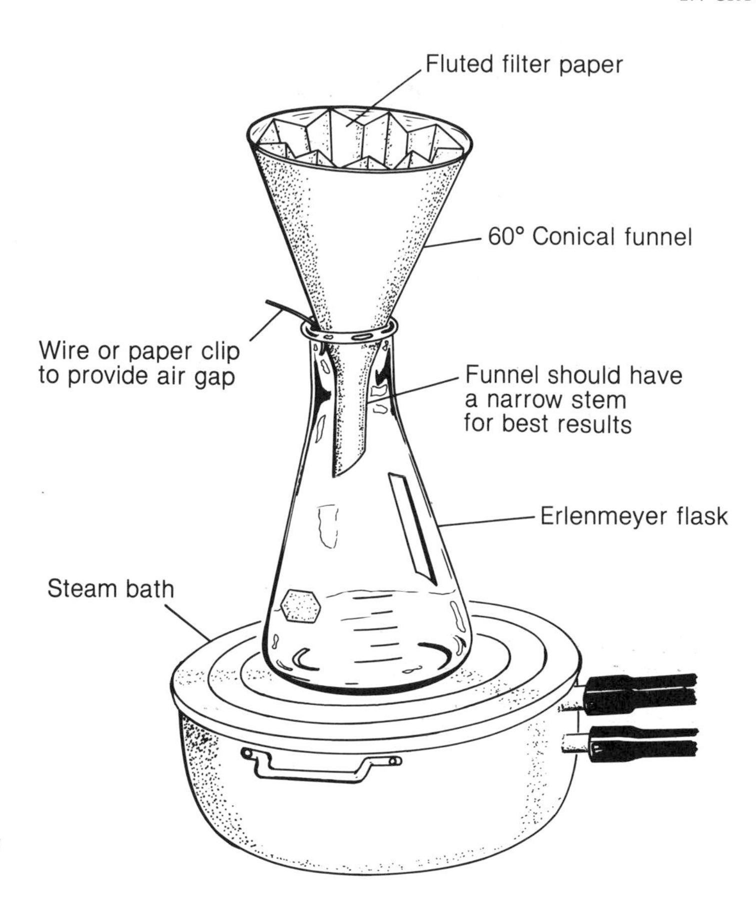

Figure 2.18
Apparatus for gravity filtration.

remain. This filter pad may then be used to remove finely divided particles. Before filtering the solution, remember to remove the water in the filter flask if the mother liquor is desired.

During a charcoal treatment (see above) it is sometimes useful to add a little bit of filter aid to assist the gravity filtration. In addition, a pad of Celite may be used to break up emulsions when these present a problem during extraction (see Sec. 2.5). Simply run the emulsified solutions through a pad of Celite and then return the solutions to the separatory funnel. The interface will often be clearly defined after doing this.

Seed Crystals

The best advice that anyone can be given about seed crystals is: If you get one, don't ever let it go. Often a material will not crystallize because the nucleation process will not start or has not started. By nucleation is meant providing a nucleus on which the crystal may start to grow. Deposition of crystals from the solution may often be induced by scratching the side of the vessel. Sometimes, addition of a small amount of foreign material will initiate crystallization. The most reliable method is to add a crystal of the product (seed crystal) and thereby induce nucleation. In all three of these procedures, one is affording the material a surface on which to grow and thus catalyzing the crystallization process.

A genuine seed crystal is really the best nucleation surface. In fact it is not entirely clear why foreign matter should serve as a nucleation point at all. However, in some cases even dust in the air or lint from a beard has been credited with seeding a crystallization.

Most of the crystallization problems encountered in a student laboratory situation involve materials which crystallize with relative ease. In fact, many of the exercises in this manual were chosen because the products can be purified by recrystallization. In most of the experiments in this book, scratching the inner surface of the crystallization vessel will initiate crystallization. In a research situation, however, a seed crystal can be invaluable.

In order to obtain a seed crystal, it is sometimes useful to place a drop or two of the solution on a watch glass and allow the solvent to evaporate. The minute amount of crystal formed by this process may be scraped off the plate and added to the crystallization mixture. Sometimes blowing a fine stream of air onto the surface of a crystallization solution will induce solids to form. The formation of crystals on the solution surface may induce crystallization in the bulk solution. Alas, there are cases where nothing works and one is faced with the prospect of scratching and stirring for extended periods of time (certain incantations are alleged to help at these times). A good knowledge of crystallization technique coupled with patience and a sense of humor will usually be rewarded.

PROCEDURE

RECRYSTALLIZATION

Part 1

Samples of benzoic acid, cinnamic acid, iodoform, and urea should be available from your instructor. Examine each of these with the solvents hexane, toluene, ethanol, acetone, and water to find which solvents are suitable for each recrystallization. Save the tube with the best recrystallization for each substance. Show the tubes to your instructor.

Part 2

A 2-g sample of one of the above solids should be recrystallized twice. The melting point of a sample should be determined before recrystallization and after each recrystallization. After the first recrystallization, set aside a small sample of the material so that a melting point can be obtained on it when it dries and so that it can be used for seed crystals if necessary. When you obtain a pure sample, put it in a labeled container and show it to your instructor.

Part 3

Take 2 g of a 50:50 urea–cinnamic acid mixture and choose a solvent for separation of pure cinnamic acid by recrystallization. Recrystallize it as described in part 2, noting the weight of recrystallized material recovered and the melting point at each stage.

With the knowledge of procedure you have acquired, you should be able to work out the details of this separation (choice of solvent, when and how to filter, etc.)

Show your best sample of cinnamic acid to your instructor along with the melting-point, recrystallization solvent, and weight data.

2.5 EXTRACTION

Solvent Affinity

The saying most often repeated in the chemical laboratory is probably: Like dissolves like. Think for a moment why this should be so: those compounds which have similar structures should have an affinity for each other. For example, an alcohol such as methanol or ethanol, which contains a hydroxyl group and only a small organic residue, should be quite soluble in water. On a molecular basis this is because the functional group (the hydroxyl group) of each can interact in a similar way. One might also anticipate that a carboxylic acid such as acetic acid (vinegar) would be quite soluble in water because both water and acetic acid have polar functional groups.

On the other hand, a hydrocarbon such as cyclohexane would probably not be soluble in water because it is more organic (or gasolinelike) than waterlike. We would certainly expect cyclohexane to be soluble in pentane because both molecules are hydrocarbons.

The principle underlying the extraction technique is actually a simple extension of the above discussion. A compound which is exposed to two different solvents will dissolve in that solvent which is most similar to it in its molecular properties. For example, if 100 mL of water and 100 mL of pentane were placed in a flask and 10 g of sodium chloride added to this mixture, we would find that the sodium chloride would dissolve only in the aqueous layer. If 10 mL of cyclohexene were added instead of sodium chloride, we would expect that the cyclohexene would dissolve almost exclusively in the pentane layer. In the latter case, no hydrocarbon would be found in the water layer. Thus, like dissolves like.

If methanol is added to the water-pentane mixture, it will dissolve in the aqueous phase because the hydroxyl group is very waterlike. Very little methanol would dissolve in the pentane layer because the polar hydroxyl group dominates the solubility behavior of this small molecule.

It should be clear that extraction can be a very powerful method for the purification of organic compounds. This is especially so if ionic materials are used in the preparation of organic products. For example, in the reaction shown below the product ester has a hydrocarbon chain and does not have a strongly polar functional group. The by-product of the reaction is a salt (NaBr).

$$CH_3CH_2CH_2CH_2CH_2Br + NaOCOCH_3 \rightarrow CH_3CH_2CH_2CH_2CH_2OCOCH_3 + NaBr$$

If the reaction mixture is shaken with a suspension of pentane and water, the salt will dissolve in the water layer and the nonpolar ester group will dissolve in the hydrocarbon. Separation of the layers followed by evaporation of each would afford pure organic product and pure salt.

Partition Coefficients

The solubility of a compound in a solvent is characteristic of the compound and the solvent at any given temperature. For example, at 25°C benzoic acid has the following solubilities in the listed solvents:

Water, 3.4 g/L
Ethanol, 450 g/L
Chloroform, 222 g/L
Carbon tetrachloride, 33.3 g/L

Notice that benzoic acid is slightly soluble in polar (water) and nonpolar (CCl_4) solvents, but very soluble in solvents of intermediate polarity. This is once again a demonstration of the saw that like dissolves like.

The partition coefficient is defined as the ratio of the solubility of a compound in one of two mutually immiscible phases to its solubility in the other. In order to determine the partition coefficient for benzoic acid between water and hexane, the simple procedure given on p. 87 can be used. In the table below are given some representative partition coefficients between water and an organic solvent.

Compound	Solvent pair (equal volumes)	Partition coefficient
Benzoic acid	Carbon tetrachloride–water	3.8
Aniline	Dichloromethane-water	3.3
Nitrobenzene	Dichloromethane-water	51.5
1,2-Dihydroxybenzene	Dichloromethane-water	0.2

Extraction Volumes

The various elements of an extraction should be apparent from the procedure given above. There are several additional things which should also be kept in mind. One of these is that extraction with two small volumes of solvent is generally better than extraction with a single large volume. In other words, if one wished to extract cyclohexene from water into pentane and were going to use 100 mL of each solvent, a more complete extraction could be effected if the 100 mL of water solution were extracted twice with 50-mL portions of pentane than once with 100 mL. This is because cyclohexene will distribute itself primarily in the pentane phase regardless of the exact volume ratios. After equilibration, more cyclohexene will dissolve in the second "empty" pentane phase than if this phase already contained cyclohexene.

Funnel Size

A practical consideration in carrying out an extraction is the size of the apparatus. The device traditionally used is the separatory funnel (see Fig. 2.19). The separatory funnel should be 30 to 50% larger than the total volume of solvent to be used at one time so that room is left for shaking the solvent. For example, if 100 mL of water is to be extracted with 50 mL of pentane, a 250-mL separatory funnel should be used. This generalization is limited by the fact that separatory funnels are manufactured only in certain sizes. The common sizes are 60, 125, 250, 500, and 1000 mL, 2, 4, and 6 L, and larger.

Performing the Extraction

The extraction cannot be performed until the two solvents have both been poured into the separatory funnel. The order of addition will generally be dictated by convenience. For example, if an aqueous solution is extracted with two portions of dichloromethane, the organic phase will be on the bottom. When the dichloromethane is drawn off, the aqueous phase will already be present in the funnel when more organic solvent is added. As you become more experienced, you will note that certain solvent combinations (aqueous sodium hydroxide and chloroform, for example) lead to emulsions (suspensions of globules of one liquid in the other). As emulsification cannot always be anticipated, it is usually a good idea to pour the second solvent into the funnel as gently as possible.

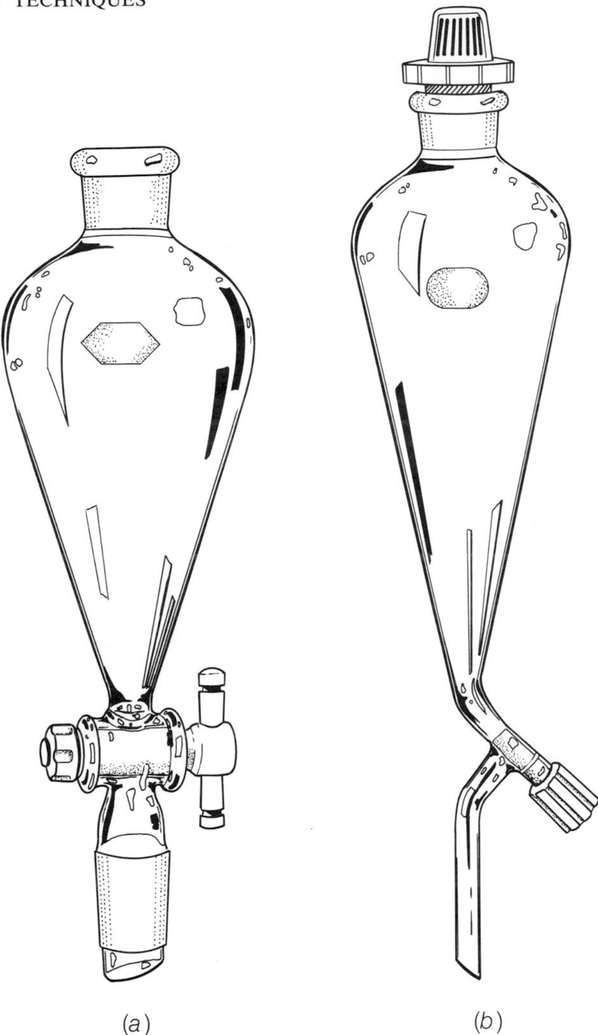

Figure 2.19 Separatory funnels. (a) Pear-shaped funnel with Teflon stopcock. (b) With Rotaflo, stopcock.

Shaking the funnel

After both phases have been transferred to the separatory funnel, stopper the funnel and grasp the stoppered end with one hand and the stopcock with the other (see Fig. 2.20). Shake the separatory funnel using a swirling motion, gently but for at least 30 s. Very gentle shaking will usually not give complete equilibration unless continued for a minute or so. More vigorous shaking will allow equilibrium to be reached more rapidly, but the risk of forming an emulsion increases.

A note of caution

Often during an extraction the liquids warm as the solutions mix. If a significant temperature rise occurs during shaking some of the solvent or product may be forced into the vapor phase. The pressure in the vapor phase will therefore be increased. This pressure must be relieved safely to prevent injury (see Fig. 2.21).

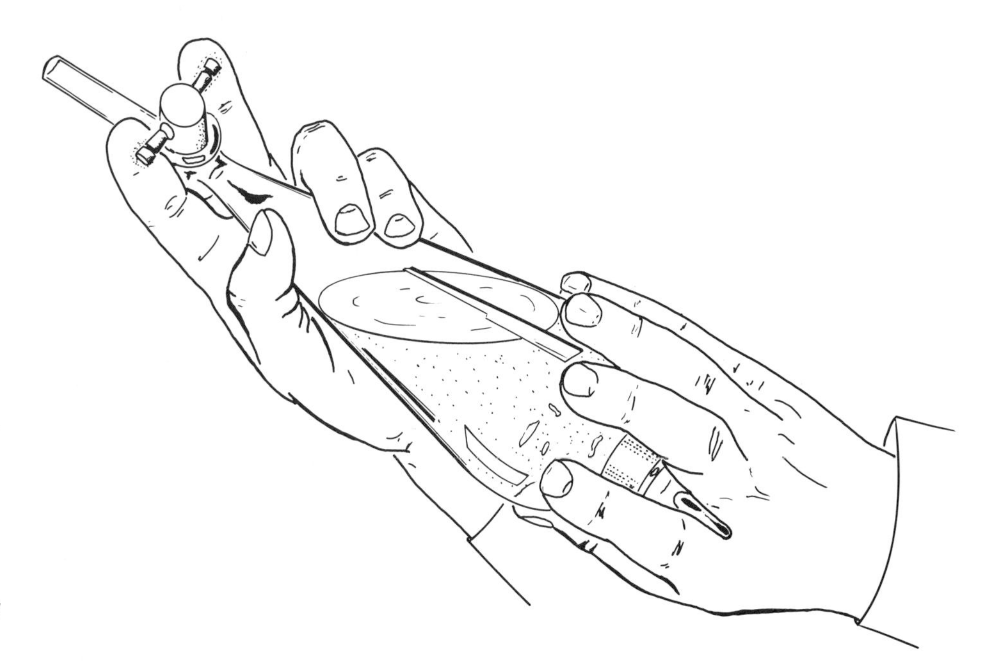

Figure 2.20
Correct manner of shaking a separatory funnel.

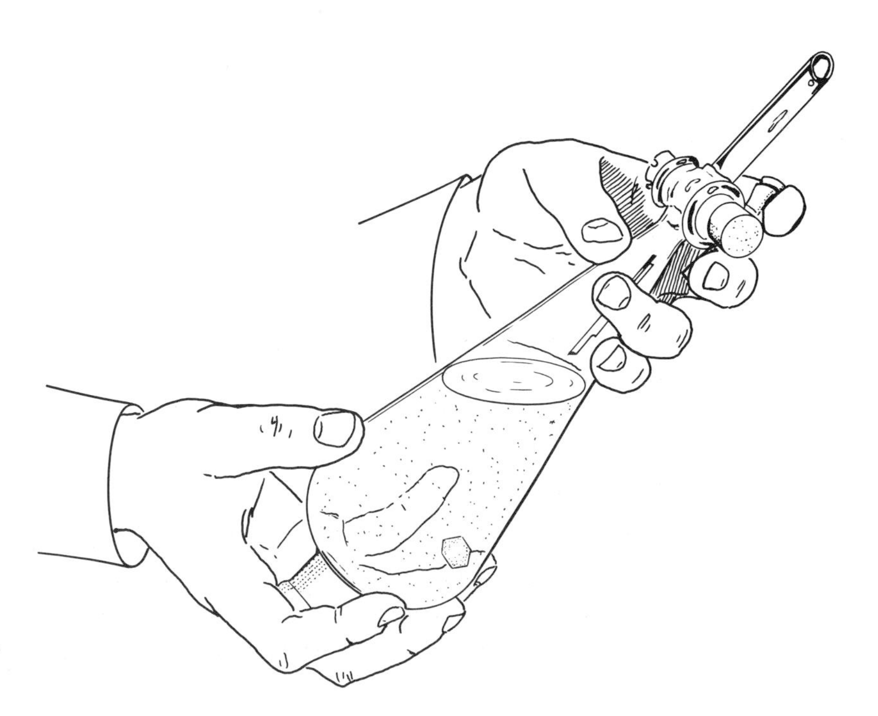

Figure 2.21
Correct method of venting a full separatory funnel.

After shaking, rest the funnel in one hand, grasping the stopper. Tilt the funnel so that the stopcock end is pointed up and away from you (preferably into a hood) and rotate the stopcock to the open position. Be certain that the level of the liquid is *below* the stopcock opening or the liquid will be forced out when the stopcock is opened. This process is referred to as venting. Experienced laboratory workers open the stopcock as described after every shaking, whether or not they anticipate a pressure buildup.

Draining the funnel

After the shaking is completed, slip the separatory funnel into an iron ring (for support, Fig. 2.22) and allow the phase boundary to become a sharp line. Sometimes, during extraction of a very dark mixture, the phase boundary is very difficult to discern. When this happens, hold the separatory funnel up to the window and view the mixture illuminated from behind. Alternatively, a flashlight may be shined through the mixture.

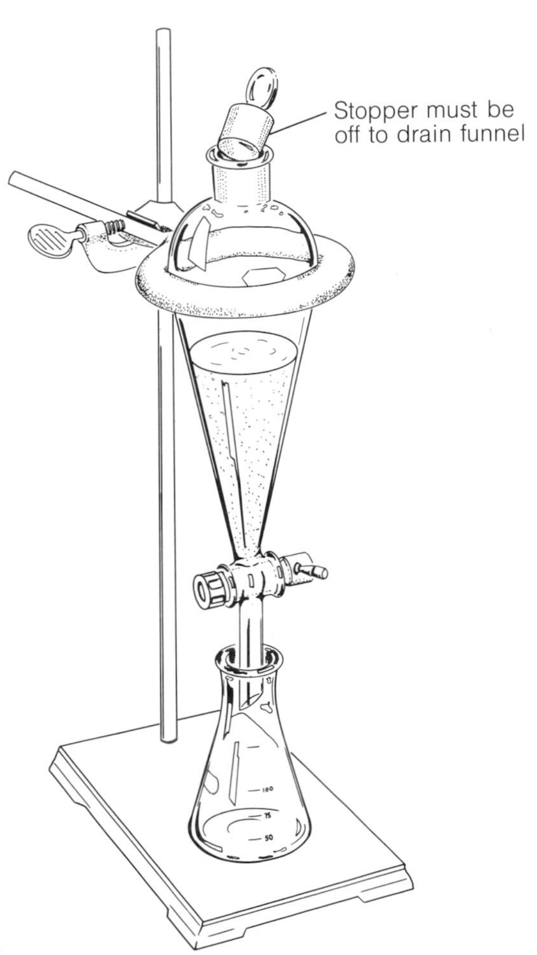

Figure 2.22
Separatory funnel positioned for draining.

In order to drain the lower phase, remove the glass stopper at the top of the separatory funnel and turn the stopcock to the open position. (**Be certain that a receiver flask is in place.**) The stopcock position (either fully or partially open) should be determined by how close the phase boundary is to the top of the stopcock. Start with a fully open stopcock and then gradually reduce the opening as the phase boundary nears the bottom. This procedure is recommended to prevent the upper phase from escaping from the separatory funnel with the lower phase. When the phase boundary is just above the stopcock, turn it to the off position. The last few drops of lower phase may be collected by twisting the stopcock rapidly through the open position. If a second extraction with the same solvent is to be performed, it is usually not necessary to completely remove the last few drops of the lower phase. It is only the last extraction step in which the complete separation between the two phases should be attempted.

PROCEDURE

DETERMINATION OF A PARTITION COEFFICIENT

Place 75 mL water and 75 mL dichloromethane in a 250-mL separatory funnel (see Fig. 2.19). Add about 1 g benzoic acid (determine its weight precisely) and shake the mixture vigorously. Remove the lower dichloromethane layer and dry it briefly over sodium or magnesium sulfate. Decant the dichloromethane solution into a preweighed 250-mL Erlenmeyer flask. Evaporate the dichloromethane by heating on a steam bath. After all the solvent has evaporated, a white solid should remain. Weigh the Erlenmeyer flask. Its new weight should be the tare weight plus the weight of the benzoic acid. The amount of benzoic acid which dissolved in water can be determined by difference. For example, if 1.09 g benzoic acid was added and 0.72 g was found in the organic layer, 0.37 g must have dissolved in water.

The partition coefficient can be determined as follows. Since the volume of the layers is equal (each 75 mL), the partition coefficient is the ratio of the weight of benzoic acid in the organic phase to that in the aqueous phase. Thus, if 0.72 g benzoic acid was found in the dichloromethane layer and 0.37 g in water, the partition coefficient would be $0.72 \div 0.37 = 1.95$. Thus, in this hypothetical experiment the partition coefficient of benzoic acid between water and dichloromethane is nearly 2.

There is no rule that says 75 mL of each solvent and 1 g of material must be used. These are convenient quantities to manipulate under organic laboratory conditions, although much smaller quantities could be used in an analytical lab. In general, it is best to use the smallest amount of compound commensurate with accuracy. This is because the partition coefficient does not always remain linear at high concentrations of sample.

Acid-Base Extractions

One of the most important extraction techniques is that used to separate substances which are acidic or basic. Acid-base extraction involves a chemical reaction rather than a simple physical partitioning as described above. For example, if benzoic acid is partitioned between water and dichloromethane, some of the benzoic acid will be found in each layer. The amount present in each layer is characterized by the partition coefficient.

If we attempted to partition sodium chloride between water and dichloromethane, we could be sure that virtually all the sodium chloride would be found in the aqueous phase and none would be found in the organic phase. If we could convert benzoic acid into a saltlike material, we would dramatically increase its aqueous-phase solubility. We could therefore remove it from a reaction mixture as a water-soluble salt.

As an example, consider the formation of methyl benzoate from benzoic acid and methanol. It seems likely that some benzoic acid will remain as a byproduct after the reaction is terminated. The equation for this reaction is shown below.

$$C_6H_5C(=O)OH + CH_3OH \overset{H^+}{\rightleftharpoons} C_6H_5C(=O)OCH_3 + H_2O$$

Assuming that some preliminary purification has been completed to remove excess methanol and water, the product should consist largely of methyl benzoate contaminated by benzoic acid. Simple solvent partitioning might separate the ester from the acid. It is unlikely, however, that this approach would lead to a good separation. It would be much more efficient if a simple process could be used which would remove only one component of the mixture. In this example, if benzoic acid could be converted to a salt (treatment with $NaHCO_3$) without changing the methyl benzoate, a simple aqueous extraction would remove all the unreacted benzoic acid (as sodium benzoate). Such a process would drastically change the extraction characteristics of one component and allow purification.

It turns out that benzoic acid reacts readily with sodium bicarbonate (baking soda) to form sodium benzoate, carbon dioxide, and water.

$$\underset{\text{(Slightly soluble in water)}}{C_6H_5CO_2H} + NaHCO_3 \longrightarrow \underset{\text{(Very soluble in water)}}{C_6H_5CO_2Na} + H_2CO_3$$

$$H_2CO_3 \rightleftharpoons H_2O + CO_2$$

If the above reaction mixture is dissolved in ether solution and then shaken with aqueous sodium bicarbonate solution, all the benzoic acid present will be converted to the sodium salt and extracted into the aqueous layer. In this simple and inexpensive operation a complete separation is effected. The same strategy may be used to separate acidic phenols from neutral or basic contaminants. In the case of the weakly acidic phenols, a stronger base (e.g., sodium hydroxide) must be used. The principle is exactly the same.

The separation of an amine from neutral or acidic compounds may be accomplished by using the same principles. Naturally, the solvent required would be aqueous acid rather than aqueous base. In this case, the acid would protonate the amine to form a water-soluble ammonium salt.

Salting Out

Since the ability of a sample to partition between two phases is a characteristic of the substance and the two phases, altering one of the liquids will alter the partition coefficient. For example, an organic material which has a low solubility in water will have an even lower solubility if the aqueous phase contains a large amount of sodium chloride (salt). This is so because the functional groups of water will be involved in stabilizing and solvating the sodium ions and the chloride ions from the sodium chloride and will therefore be relatively unavailable for solvating those functional groups which are contained in the organic material.

If an organic material had a partition coefficient of 1 between ether and water, its partition coefficient would be changed in favor of the ether phase if the water phase were saturated with salt. The effect of forcing an organic material out of an aqueous layer by the addition of salt (many salts will do) is termed *salting out*. This technique is used commonly in two different ways.

Salting out to separate a mixture

Most extractions which are carried out in the organic laboratory involve an organic solvent and water. Sometimes the organic compound is soluble in both water and the other solvent and the extraction fails. In this situation, it is often beneficial to add salt to the aqueous solution. The ionic compound interacts strongly with water, keeping the water from solvating the organic substrate. Since the organic compound is no longer soluble in the ionic salt water solution, it migrates readily into the organic phase.

2-Propanol (rubbing alcohol, isopropyl alcohol) is soluble in both water and ether. In fact, in the presence of enough 2-propanol a mixture of water and ether becomes homogeneous. If NaCl is added, it interacts strongly with the water, reducing the solubility of the alcohol. The 2-propanol now prefers the ether, and two phases become apparent. The lower phase is salt water (brine), and the upper phase is a solution of organics.

Washing with saturated salt solution

Just as an organic material is less soluble in a concentrated aqueous salt solution, water enters the salt layer to help solvate the ions which it contains. For example, if ether is shaken with water, a small amount of ether enters the aqueous phase and a small amount of water enters the ether phase. On the other hand, if the ether phase is washed with saturated sodium chloride solution (brine), the solvation demands of the salt cause almost all the water to be retained in the aqueous phase. Likewise, any ether which has entered the aqueous phase will be released from the aqueous phase and seek the organic phase. An extension of this simple property leads to the conclusion that saturated sodium chloride solution may be used as a preliminary *drying agent* for common organic solvents.

If an aqueous phase is extracted with several portions of organic solvent, the organic phase will almost always contain small portions of water. If the organic phase is shaken with saturated sodium chloride solution, virtually all the water will be removed in the aqueous phase and the organic solution will contain much less water. As a result, most chemists routinely wash organic phases with saturated sodium chloride solution immediately before further drying with the solid drying agents.

Drying Agents

For inorganic chemical reactions, water is sometimes called the universal solvent. In organic reactions, water is more often regarded as a reactant or contaminant. After reaction and extractive workup, almost all organic solutions contain small amounts of water. Even washing with saturated brine solution does not remove all the water. As a consequence, organic solutions must be dried with certain "inert" drying agents to remove the last traces of water.

Quite a number of inorganic drying agents are known and have been used to dry organic liquids. In general, the object is to remove the water from an organic sample without transforming the product. Although many drying agents are known, not every drying agent can be used in every case. One must consider the capacity of the drying agent (that is, its total ability to absorb water), the rate at which it dries the solution, its cost, and also the possibility of reactions with the material being dried. The seven most common drying agents and the relative merits and risks associated with each are discussed below.

Sodium sulfate

Sodium sulfate is the most common general-purpose drying agent. It is an inexpensive drying agent and has a very large capacity because it forms a heptahydrate. It is also relatively inert and, in general, does not react with organic materials. The disadvantages of sodium sulfate are that it acts slowly and above about 30°C the heptahydrate breaks down and the drying capacity is reduced. At room temperature, sodium sulfate is usually the drying agent of choice.

Magnesium sulfate

Magnesium sulfate is in almost as common use as sodium sulfate. It enjoys the advantages of high capacity and low cost. The principal differences between sodium and magnesium sulfate are that magnesium sulfate will occasionally react with substances in the Lewis acid sense and also (and this is an advantage) that the rate of drying with magnesium sulfate is faster than with sodium sulfate.

Calcium chloride

Calcium chloride is one of the most inexpensive of all drying agents. It also has a very high capacity and effects drying quite rapidly. It is the reagent of choice when drying hydrocarbons or alkyl halides. Unfortunately, it is much more reactive than is sodium or magnesium sulfate and should never be used to dry amines or alcohols. This applies not only to alcohols as solvents but to alcohols dissolved in another solvent.

Calcium sulfate

Calcium sulfate (often sold under the trade name Drierite) is a good general, high-capacity, and fast drying agent. It is somewhat more expensive than the three reagents named above but is very often used for drying in the research laboratory, where expense is somewhat less important than it is in the undergraduate laboratory.

Potassium carbonate

Potassium carbonate is also an effective drying reagent which has a high capacity and is relatively inexpensive. It is, however, a basic reagent and cannot be used to dry acidic materials. Its primary use as a drying agent is with inert or basic materials, with which it will not react chemically.

Sodium or potassium hydroxide

Sodium and potassium hydroxide are very basic and very reactive materials but they absorb water readily. They are good drying agents for very basic liquids such as amines. For most other purposes, however, their reactivity precludes their routine use as drying agents.

Molecular sieves

Molecular sieves are various aluminum silicates (zeolites) which have pores and channels of varying sizes in their structures. When these substances contact water or other small molecules, the latter diffuse into the channels and are trapped. Molecular sieves are excellent drying agents which have high capacity and dry liquids completely. In certain reactions molecular sieves have exhibited Lewis-acid properties, but this is rare. The only significant disadvantages associated with their use is the fact that they dry slowly and they are moderately expensive compared with other drying agents.

PROCEDURE

SEPARATION OF BENZOIC ACID AND FLUORENONE

You will be given a 2-g mixture containing benzoic acid and fluorenone. Dissolve the entire sample in 50 mL dichloromethane. Extract the organic phase with two 25-mL portions of 5% sodium hydroxide solution. After each extraction, separate the phases. After the second extraction place the organic phase, which should be yellow in color, in a 125-mL Erlenmeyer flask clearly marked "organic phase." After separating the aqueous phase from the organic phase, pour the water solution back into the separatory funnel and wash it with 10 mL dichloromethane (this procedure is called *back washing* the aqueous phase). The dichloromethane wash solvent should be added to the organic phase. Return the aqueous phase to the appropriate flask. Acidify the aqueous phase by dropwise addition of 6 *N* HCl while swirling the flask. Monitor the change in acidity by using pH paper. The pH will change slowly until a pH of approximately 7 is reached. The solution will then rapidly become acidic. The transition from neutrality to acidity will also be accompanied by precipitation of benzoic acid, which is relatively insoluble in cold water. If the neutralization process has caused the aqueous layer to warm up, cool the flask in an ice bath. Filter the mixture using a Buchner funnel and filter paper and collect the white solid. Allow the benzoic acid to air dry on the filter by pulling air through the solid for several minutes. Weigh the air-dried benzoic acid to determine how much benzoic acid was contained in the original 2-g mixture.

Dry the organic phase with granular anhydrous sodium sulfate to remove water, gravity filter the organic solution, and remove the solvent on a steam bath (**hood**). After all the organic solvent has been removed, allow the yellow oil that remains to solidify, scrape the solid yellow material out of the flask, and weigh it. Determine the percentage of fluorenone in the original mixture. The total amount of material from the two determinations should be approximately 2 g. Calculate the number of moles of each component in your original mixture from the two values you have just determined.

Confirm your results by some other method (which would be most appropriate?) to ensure that the separation is quantitative. Submit your results to your instructor.

2.6 CHROMATOGRAPHY

It is a characteristic of all chemical systems that, if there is a possibility of existing in two or more states or environments, given enough time the system will invariably come to equilibrium. This means that all the forces operating in the system in every possible way will eventually come into balance. This process occurs in distillation, for example, where at equilibrium the amount of vapor escaping from the surface of the liquid is balanced by the amount of vapor returning to the liquid phase (see Sec. 2.1). Equilibration will also occur in the partitioning of a substance between two immiscible phases (see Sec. 2.5). The third component will partition itself between the two immiscible phases in accordance with its relative solubilities in each of the other two substances. At equilibrium, the third component will be rapidly exchanging between the two immiscible phases; however, the amount of the third component going into immiscible phase 1 from immiscible phase 2 will be exactly equal to the amount of material returning.

In the second example above, although a three-component system is involved, it is relatively easy to understand what occurs. As a specific example, a system might consist of two layers such as water and ether, and the third component might be a small amount of toluene (methylbenzene). As one might imagine, the toluene will be very soluble in ether and not very soluble in water. If we shake the three-component mixture and allow enough time for all the forces to come into equilibrium (and for the phases to completely separate) we will find that virtually all the toluene is in the ether phase and almost none is in the aqueous phase. The commonly encountered statement that like dissolves like is just a shorthand way of saying that the intermolecular forces acting between toluene and ether are much greater than those operating between toluene and water.

In the various techniques of chromatography the separation of mixtures (sometimes very complex multicomponent mixtures) is achieved by exposing the mixture to a two-phase system and then allowing the system to come to equilibrium. The two phases involved in separation may be two immiscible liquids (extraction, liquid-liquid chromatography), a gas and a liquid phase (gas-liquid chromatography), a gas and a solid phase (gas-solid chromatography), or a liquid and a solid phase (liquid-solid chromatography).

In chromatography one phase of the two-phase system is almost always a stationary phase with respect to the other. Therefore we often speak of a stationary and a mobile phase in discussions of chromatography.

Although it is sometimes difficult to establish the role of each phase, we can say that liquid-liquid chromatography corresponds roughly to multiple extractions, whereas gas-solid and liquid-solid chromatography involve adsorption of the sample on a surface. As we will see in later discussion, paper chromatography is primarily a liquid-liquid extraction phenomenon. Thin-layer and column chromatography are essentially similar and utilize adsorption as the principal means for purification, but there is some element of liquid-liquid extraction involved. The extent to which the latter plays a role depends on the extent to which water is present on the adsorbent surface. The less water present, the more *activated* the adsorbent is said to be. In gas-liquid (glc) or vapor-phase (vpc) chromatography, partitioning is due to extraction, distillation, and a small element of adsorption.

The Kinds of Chromatography

Column and thin-layer chromatography

Column and thin-layer chromatography (tlc) differ only by the means of supporting the adsorbent. Both techniques usually involve either alumina (Al_2O_3) or silica gel (silicic acid, $SiO_2 \cdot xH_2O$) as the stationary phase and a mobile organic phase. The principles which apply to one variation of the chromatographic technique in general apply to all other variations as well. The principal difference between column and thin-layer chromatography is that in column chromatography the adsorbent is supported in a tube or column (usually glass), whereas in thin-layer chromatography the adsorbent is deposited in a thin layer on a glass or plastic plate. It should also be noted in passing that the common use of glass as a support medium is dictated by its mechanical properties and low cost and that any other inert support might be substituted for it.

Another difference between column and thin-layer chromatography is that in the former the mobile phase is run down the column, i.e., the driving force is gravity, whereas in tlc the capillary action of the adsorbent draws the solvent up from a reservoir at the bottom of the plate. Notwithstanding this difference, the fundamental process of separation remains the same in both cases: the partitioning of an unknown compound between the mobile liquid phase and an immiscible phase.

It should be stressed that in any kind of chromatographic procedure, the separation of two or more components is ultimately achieved by a relatively subtle balance of all the forces involved. In column chromatography, for example, different components of the mixture will adsorb onto the surface of the solid with different affinities (binding powers). The mobile phase will displace and dissolve these components in accordance with its affinity for the solid surface and the solubility of the substance in it. The most important single factor is that equilibrium be established.

The two most common adsorbents, silica gel and alumina, are different but work in essentially the same way. Both effect separation primarily by holding components on the surface by a combination of Lewis-acid and Lewis-base forces; these forces act on the samples to different extents. The varying amounts of water on the adsorbent surface also play a role in the separation. The selection of either silica or alumina for a particular separation will depend on numerous factors. Ultimately, however, the decision as to which adsorbent to use will be dictated by which of the two substances gives the best and most efficient separation of the mixture.

On the practical level, there are many types of aluminas and silica gels and these different types are sold in many sizes. In general, the coarser the adsorbent is, the faster the solvent will percolate through it. However, there is a penalty for this because the surface area is lower in coarse adsorbents than in fine ones, so the number of contact sites is reduced. In addition, solvent moves through the coarse adsorbents so rapidly that there is insufficient time to allow all forces to come to equilibrium. This results in a decrease in the effectiveness of some separations as the particle size increases. When choosing the particle size, it is best to select the smallest particle size and highest flow rate which will still permit separation of the mixture's components.

Another obvious conclusion from the description above is that the greater the ratio of adsorbent to the total weight of the unknown mixture, i.e., the longer the column length or plate length, the better the separation will be. This is because the forces which effect the separation will have more chance to act on each component. The penalty paid for increased amounts of adsorbent is an increase in the time necessary to effect separation.

High-pressure liquid chromatography (hplc)

The alumina and silica gel used in the thin-layer technique are generally much finer (smaller particle size) than column-chromatography-grade silica gel and alumina. The greater amount of surface area on thin-layer-grade adsorbent generally results in a much better separation than can usually be attained by column chromatography. This fine grade of adsorbent is not generally used for column chromatography because the flow rate is usually too slow. One means of effecting excellent separations and still maintaining reasonable solvent flow is to force the solvent through the column of fine adsorbent under pressure.

Medium- to high-pressure liquid chromatography has been used for a number of years in many laboratories, but specially designed equipment has only recently become available for this purpose. The equipment consists basically of a chromatography column and a pump to force solvent through it. Figure 2.23 shows a thin-layer plate, a typical column chromatography apparatus, and a high-pressure liquid chromatographic device. With the advent of commercial hplc equipment has come a host of new column packing materials which have been substituted for silica gel and alumina in many cases. The expense of the three techniques increases substantially from *a* to *c* in Fig. 2.23.

Paper chromatography

Paper chromatography is related to thin-layer and column chromatography. The principal difference is that instead of an adsorption process a sort of continuous extraction process is used to effect separation. In paper chromatography the cellulose of the paper has been impregnated with an immobile phase, usually water. As the mobile organic phase passes over this stationary aqueous phase, the unknown compound partitions itself between the two phases. A component of the mixture which has a significant affinity for water will spend more time in the immobile aqueous layer and will not migrate as rapidly as a component with lower water affinity. Two advantages of paper chromatogra-

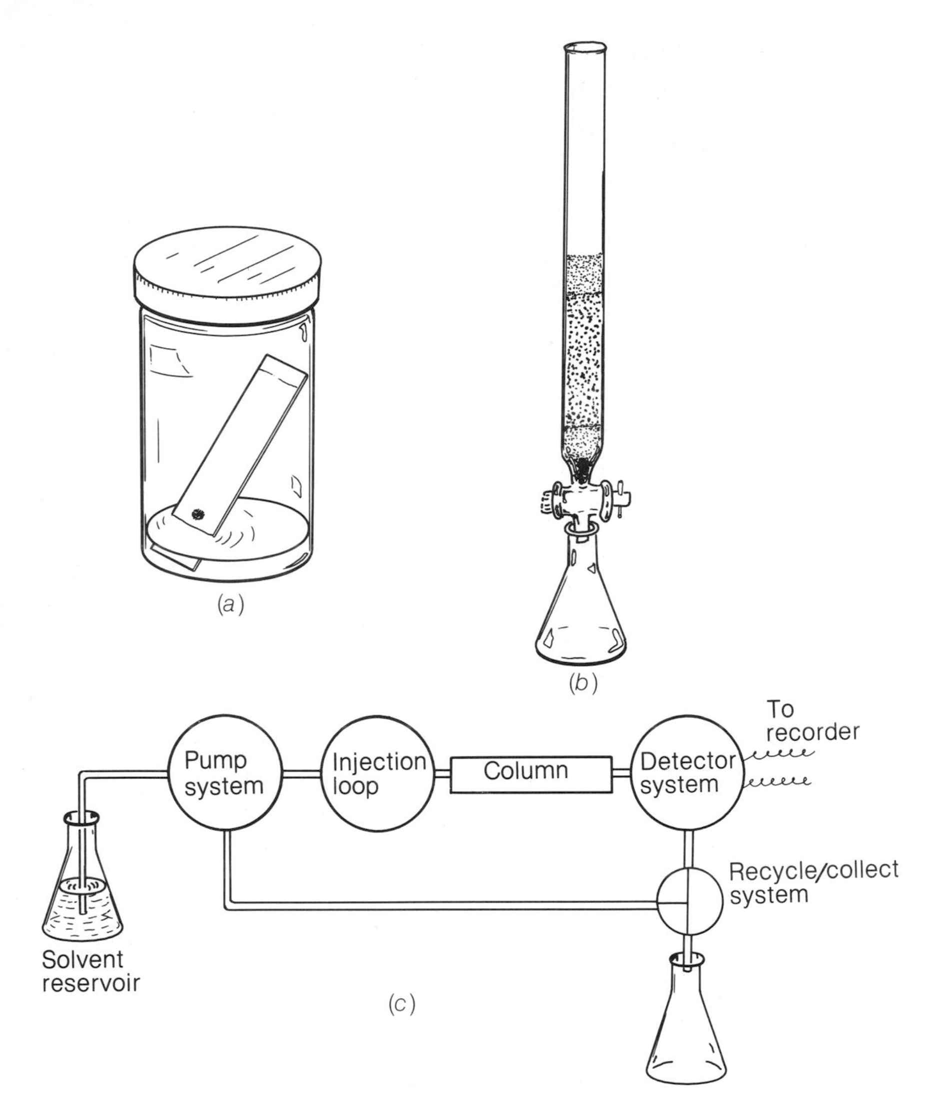

Figure 2.23 Chromatography equipment. (*a*) Thin-layer plate. (*b*) Chromatography column. (c) Schematic diagram of a high-pressure liquid chromatography (hplc) system.

phy are that it is inexpensive (it is often carried out on filter paper) and that it is useful for highly polar substances.

The technique is powerful for separating small amounts of water-soluble materials such as amino acids, nucleic acids, and sugars. Almost all the routine applications of paper chromatography involve biologically derived or related materials. Paper chromatography is used extensively for biochemical and medical analysis. A major drawback of paper chromatography is that paper chromatograms are often somewhat slow. Some require 16 to 24 h per chromatogram, which makes them less satisfactory than some other modern analytical techniques.

Vapor-phase (vpc) or gas-liquid (glc) chromatography

Gas-liquid chromatography (glc) is related to paper chromatography in an important sense. As in paper chromatography, an inert solid support is involved. In paper chromatography paper is the medium which supports the immobile phase. In gas chromatography the immobile phase is usually supported on clay or firebrick. These supports are generally stable at elevated temperatures. The immobile phase may be a wax or an oil. The stationary phases in common use include hydrocarbons, polyesters, polyethers, polyamides, and silicone polymers.

In the glc technique, both stationary and mobile phases are required, the mobile phase being a gas. Thermal equilibrium is maintained by heating the chromatography column in an oven. The mixture to be analyzed is injected into the gas stream flowing through the column and partitions itself between the immobile liquid and mobile gas phases. If a compound spends relatively more time in the liquid phase during the chromatography procedure, then its movement through the column will be retarded with respect to a compound which spends more time in the gas phase. If a mixture contains one component which spends most of its time in the gas phase and another component which spends most of its time in the liquid phase, the two will separate. The component which has a greater affinity for the gas phase will be eluted first from the chromatographic column.

A block diagram of a glc apparatus is shown in Fig. 2.24. The glc is a simple apparatus consisting of an oven containing the chromatography column, an injection device to facilitate introduction and vaporization of the mixture, and an exit port which leads to a detector. Each part of the apparatus is temperature controlled because the partition coefficient, and therefore the separation, is temperature dependent.

One requisite of gas chromatography is that the sample must be vaporized before it reaches the column and throughout the chromatography procedure. The oven, injection port, and detector are usually maintained at a temperature

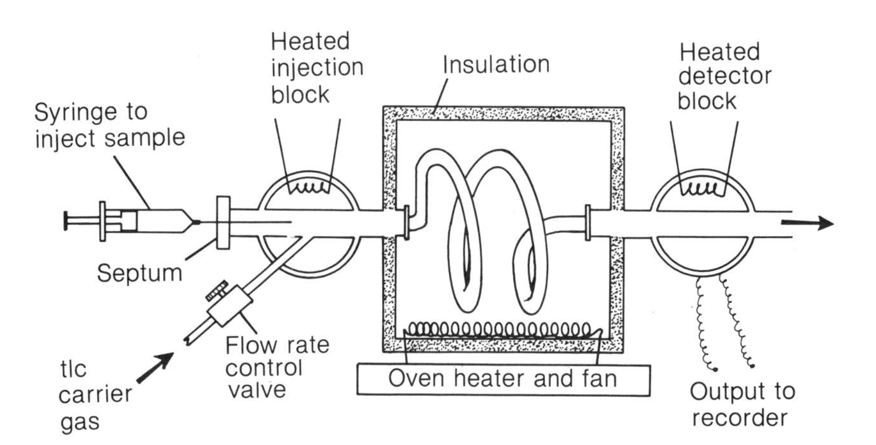

Figure 2.24
Block diagram of gas-liquid chromatographic apparatus.

which is above the boiling point of the sample. The detector and injector are usually kept hotter than the oven. In principle any unreactive gas may be used as the mobile phase. As a practical matter, helium is the mobile phase of choice, although nitrogen is used where helium is unavailable.

The Mobile Phase

The mobile phases which are used to effect separation in all the chromatographic techniques fall into two broad categories, those that are useful for gas chromatography and those that are useful for everything else. As indicated above, helium is the prevalent gas used as the mobile phase. In parts of the world where helium is expensive, nitrogen is used instead. In paper, column, and thin-layer chromatography, the mobile phases are almost always organic solvents, occasionally mixed with water or ammonia.

The effectiveness of a mobile phase in eluting, or moving, a compound along a stationary phase depends on the adsorbent, the sample, and the solvent (as well as on other variables such as temperature, surface area, and flow rate). It is the polarity of the sample, solvent, and phases which in aggregate is the important factor.

The more polar the sample compound, the more tightly it will be bound to the stationary phase. Conversely, the more polar the solid phase, the more tightly it will bind a compound in the mobile phase. Therefore, the greater the polarity of the solvent, the greater its ability to dislodge and displace a polar compound from the surface. The net effect of these forces is a competition for the compound between the immobile phase or surface and the mobile phase. One would expect a sample to spend most of its time on that surface or in that medium which is most like it in polarity.

The order in which compounds will be eluted from silica or alumina is the reverse of the compound's ability to bind to the adsorbent. Binding strength decreases as shown below.

Binding strength: Salts > organic acids > amines > alcohols > carbonyl compounds > arenes > alkylhalides > ethers > alkenes > alkanes

The eluting power of solvents parallels this order. In other words, the weakest solvents are those of the alkane class and the most powerful solvents are organic acids or aqueous salts. The eluotropic strength of common organic solvents used in chromatography decreases as shown below.

Eluting power: Acetic acid > water > ethyl alcohol > acetone > ethyl acetate > diethyl ether > CH_2Cl_2 > $CHCl_3$[1] > toluene > benzene[1] > CCl_4[1] > hexane > pentane

[1] Solvents less often used today because of possible health hazard.

There is an obvious parallel between the binding of a compound to a support and that compound's eluting power. Note that the only inversion in the two series is that diethyl ether is a good solvent but ethers are poorly bound to most adsorbents. The inversion is due to the fact that diethyl ether is polar when compared with most other ethers and the polarity scale for compounds is a general one based on the polarity characteristics of all ethers.

Thin-Layer Chromatographic Techniques

Advantages

There are several advantages which recommend thin-layer chromatography (tlc) over paper and column chromatography for routine use. Thin-layer chromatography may be used analytically to determine the separation characteristics of a mixture before attempting a large-scale chromatographic separation. Its advantage over paper chromatography is that the same type of adsorbent is used in both tlc and column chromatography. The information obtained in a small tlc experiment can often be applied to a column chromatogram, whereas the conditions used in paper chromatography do not usually translate directly. A second advantage of tlc is that the time required to develop a chromatogram is short (10 to 20 min) compared with that required for a similar separation in column chromatography; a third is that better separations of components in a mixture (resolution) are often obtained with tlc than with the column technique.

Choice and preparation of plates

The first consideration in choosing a tlc adsorbent is whether the adsorbent should be silica gel or alumina. Silica gel is SiO_2 and alumina is Al_2O_3. Silica gel is probably the more widely used of the two adsorbents and is generally preferred for less polar or acidic compounds. Alumina is preferred for more polar, or basic substances.

Once the desired adsorbent is determined, the type of plate must be chosen. For routine analytical work, plates approximately 1 × 3 in (microscope slides) are used, although much larger plates are used in certain applications. The 1 × 3 in plates may be purchased commercially or prepared in the laboratory. If purchased commercially, these plates may use either glass or plastic backing for the adsorbent. The plates, which are prepared immediately before use, generally are made from glass. In most of the experiments you will do in the organic laboratory, the adsorbent will be specified. You should consult with your instructor to determine which sort of plate will be used in your laboratory.

If you will be using commercially prepared plates, cut or break them to obtain the size required for your purpose. If you are to prepare them yourself, you may use one of the two procedures given below.

Dipping microscope slides

Begin the procedure by obtaining a 4- to 8-oz bottle or a 150-mL tall-form beaker. Weigh the vessel you have chosen, add about one-fourth of its volume of either silica gel or alumina, and weigh again. The difference in the two weights is the weight of the adsorbent. Now add three to four times the weight of dichloromethane and stir until a slurry with a consistency similar to that of pancake batter is obtained. In order to achieve this consistency a little more solvent or adsorbent may be required.

As soon as the slurry is prepared, place two microscope slides back to back and, grasping them at the top between your thumb and forefinger, dip them into the slurry. Make a quick circular motion and then withdraw the coated slides. Run the thumb and forefinger of your other hand down the side of the slides to remove excess adsorbent slurry. Now separate the slides and set them aside (a hood is preferable) to dry. After about 10 min of air drying, the slides should be ready for use.

Coating microscope slides

If you have chosen to prepare the slides by the dropper procedure of coating, the following procedure may be employed. Place 2 g of adsorbent (silica gel G) in a 10-mL beaker and add 4 mL of water. Stir the material with a glass rod or with an eyedropper. (If the latter is used, be careful not to splinter the tip.) When a smooth slurry is obtained, place six to seven microscope slides side by side, draw the slurry into the eyedropper, and cover the slides as evenly as possible. Place the slides containing the wet adsorbent on a hot plate maintained at about 110°C or under a heat lamp for 2 to 3 min or until the plates are visibly dry. By this method, plates ready for use in about 15 min may be prepared.

It is interesting to note that even in modern research laboratories much of the thin-layer chromatographic analysis which is carried out is done with microscope slides prepared by dipping.

The capillary applicator

The standard method of applying sample to a tlc plate is to use a small-bore capillary tube. Many chemical supply houses supply capillary tubes for tlc use. A disadvantage is that commercial tubes are expensive. A very good capillary may be made from a melting-point tube or a disposable pipet. Heat the tube with a bunsen burner until the glass is soft and pliable. Remove the glass from the bunsen burner and pull firmly on either end of the tube. A fine capillary should form and this can be broken in the middle and used directly (see Fig. 2.25*a*). If the capillary becomes plugged, it can be discarded and another capillary prepared. Most practicing chemists prepare several such tubes at a time. The exact capillary which is used in any particular course or laboratory is somewhat dependent upon the laboratory supervisor. Even though the same general technique may be applied to different types and sizes of glass, each laboratory usually has its own customs. Check with your instructor before initiating an analysis.

Application

Once a supply of usable capillaries is in hand, the next step is to make up a series of sample and standard solutions. The usual procedure is to dissolve a few milligrams (usually no more than 10 mg) in a 1-dram glass vial and add sev-

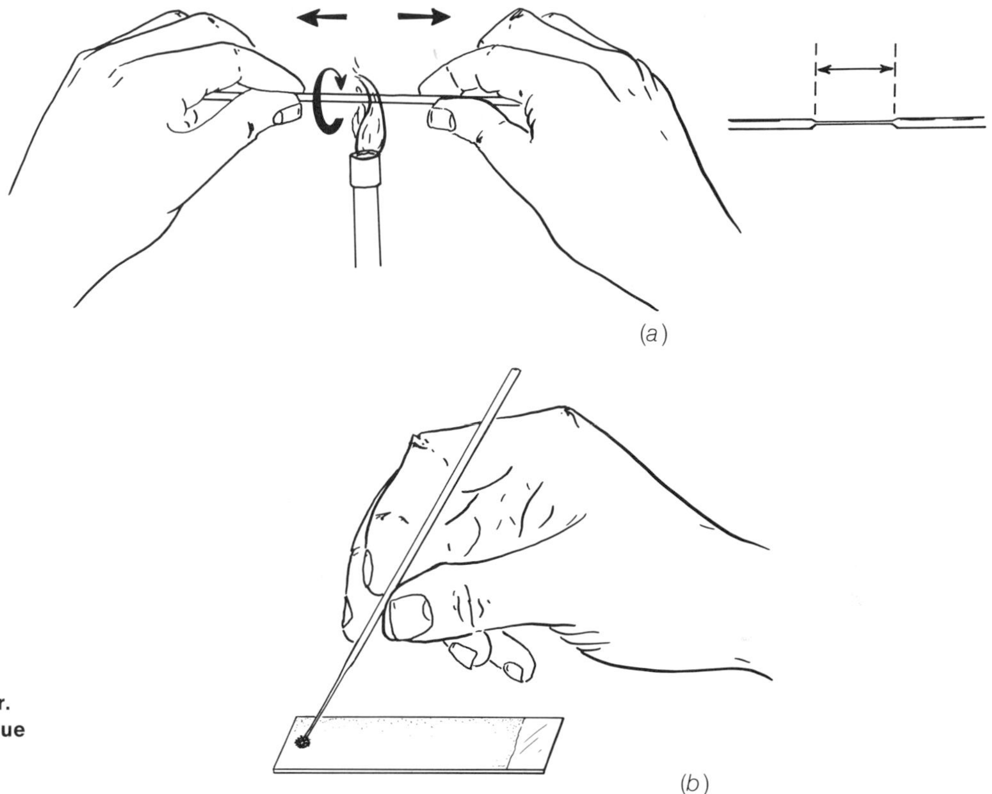

Figure 2.25
(*a*) Preparing a tlc capillary applicator. (*b*) Correct technique for spotting the plate.

eral drops of solvent. The same ratio of weight to solvent should be used for both sample and standard solutions. After the solutions have been prepared, the capillary tube is dipped into the liquid. Because of capillary action, a substantial amount of liquid rises into the tube. The sample solution can be applied to the plate by lightly touching the capillary to the adsorbent (Fig. 2.25*b*), removing the tube, allowing the spot to dry, and then touching the plate again. It is useful to rinse the capillary several times with solvent and then use it to apply the next solution. By doing this, one should achieve an approximately equivalent application of the sample and standard side by side on the plate.

Since most analytical applications of tlc are performed on either 1 × 3 in glass slides of 1 × 4 in plastic plates, we describe the spotting procedure for them here. Using either type of plate, spot the sample in the middle of the width of the plate about 0.25 in from the bottom of the plate. If two spots are to be run on one plate, they should be positioned 0.25 in from the bottom of the plate so that the two spots trisect the width of the plate. It is usually not a good idea to spot more than two samples on each plate (see Fig. 2.26*a*).

You will notice after spotting several plates that, in order to keep the spots as small as possible during application, it is advantageous if the solvent used to dissolve the sample evaporates quickly after the solution is applied to the plate.

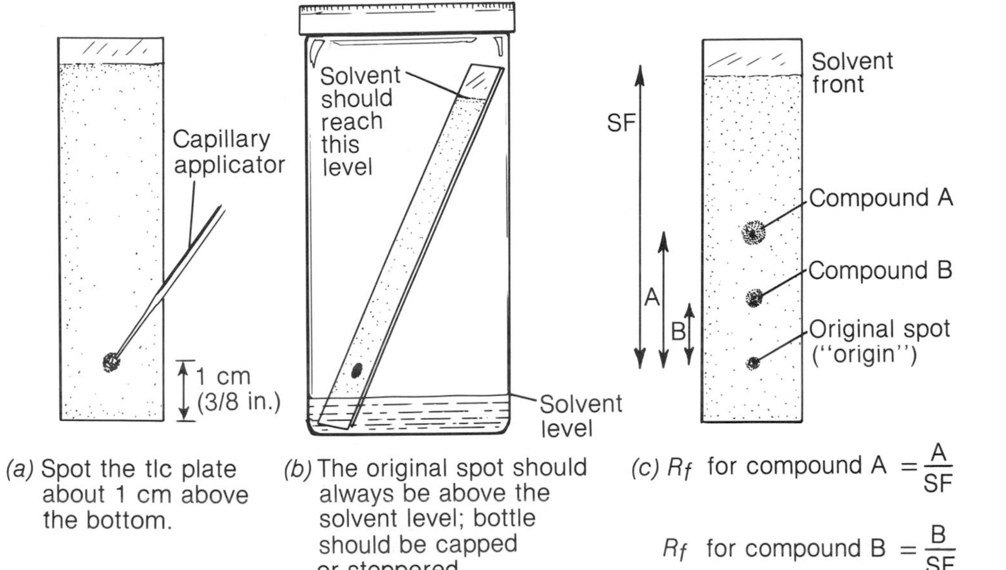

Figure 2.26 Steps in using the tlc technique. (*a*) Applying the solutions to the plate. (*b*) Developing the plate. (*c*) Determining the R_f.

You can encourage this process by spotting the material on the plate and gently blowing over the spot to evaporate most of the solvent. Again, it should be emphasized that the accuracy and reliability of a particular chromatographic separation will be enhanced by using as small a spot as is practically possible. Great attention and care should be given to this point because it will ultimately save time.

Solvents

Although the adsorbent (alumina or silica gel) will usually be specified in a laboratory exercise, it is often left to the investigator to determine which solvents are best for the compounds at hand. Recall from the discussion above that the eluting power of solvents parallels solvent polarity. Because of this relationship, almost any solvent will rapidly elute an alkane hydrocarbon. If the compound under study runs with the solvent front, the solvent is probably too polar for the sample under study. The best solvent or combination of solvents which will separate the components of the mixture can usually be determined by trial and error. A good solvent choice is one which moves the sample halfway up the plate. The ideal separation for a two-component mixture is one-third and two-thirds of the way up the plate. Examination of several different solvent combinations may be necessary before the right experimental mix of solvents is found.

In almost every case, it is more advantageous to choose a single solvent to separate the components of a mixture than it is to use a solvent mixture. This is because small differences in the proportions of the two components can have a dramatic effect on the separation. Unfortunately, a single solvent works really well in only a small percentage of separations. It is much more common to use a mixture of two (or more) solvents to maximize the solvent's separating power. When a mixed solvent system is used, the solvent mixture should be chosen carefully, prepared carefully, and used carefully. For example, if it is determined that a 1:1 mixture of dichloromethane and pentane is appropriate, the following procedure should be used to mix the solvent.

Mixing a tlc solvent

For the mixing, measure 5 mL pentane into a clean, dry 10-mL graduated cylinder. Transfer the 5 mL pentane to a 25-mL Erlenmeyer flask and cork or stopper the flask to prevent evaporation. Now measure 5 mL dichloromethane using the same graduated cylinder. Transfer this solvent to the 25-mL Erlenmeyer flask, cork the flask, and swirl several times to intimately mix the two solvents. (*Note:* Ten milliliters of solvent is more than is routinely required for analytical thin-layer chromatograms on a microscope slide.) Since both these solvents are volatile, they should be stored in a tightly corked Erlenmeyer flask and they should not be allowed to remain open to the atmosphere for any significant length of time. Exposure to the atmosphere will alter the relative concentrations of the two solvents (in this case not a serious problem), so care should be taken that the original ratio of solvents is maintained until the thin-layer chromatographic analysis is completed.

Development of the plate

The most useful developing chamber (see Fig. 2.27) for a 1 × 3 in tlc plate is a 4-oz or 150-mL round-mouth bottle or a 200-mL Berzelius beaker (a beaker that is made without a lip). (The 1 × 4 in prepared plates require a slightly larger bottle, but the 200-mL Berzelius beaker will usually suffice for either. Place enough development solvent in the vessel to cover the bottom to a depth of approximately 0.125 in (in other words, just cover the bottom with solvent). The amount of solvent required depends on the size of the jar used, so no definite volume can be specified. It is very unusual, however, for more than 5 to 10 mL of solvent to be used in an analysis of this type.

It is usually a good idea to cap the jar and swirl it for several seconds to ensure that the solvent has saturated the atmosphere above the liquid. (Sometimes the chamber is lined with filter paper to facilitate equilibration.) After the solvent and atmosphere have equilibrated, remove the cover and place the tlc slide in the container, making sure that the sample spot lies *above* the solvent level. When the slide is in place, close or stopper the chamber. The progress of the chromatogram may be assessed by following the migration of the solvent on the plate. When the solvent front has reached a position approximately 0.25 in from the top of the adsorbent, the plate should be removed.

After development, the solvent front is marked with a pin, pen, or pencil and the solvent allowed to evaporate in the hood. Record in your notebook not

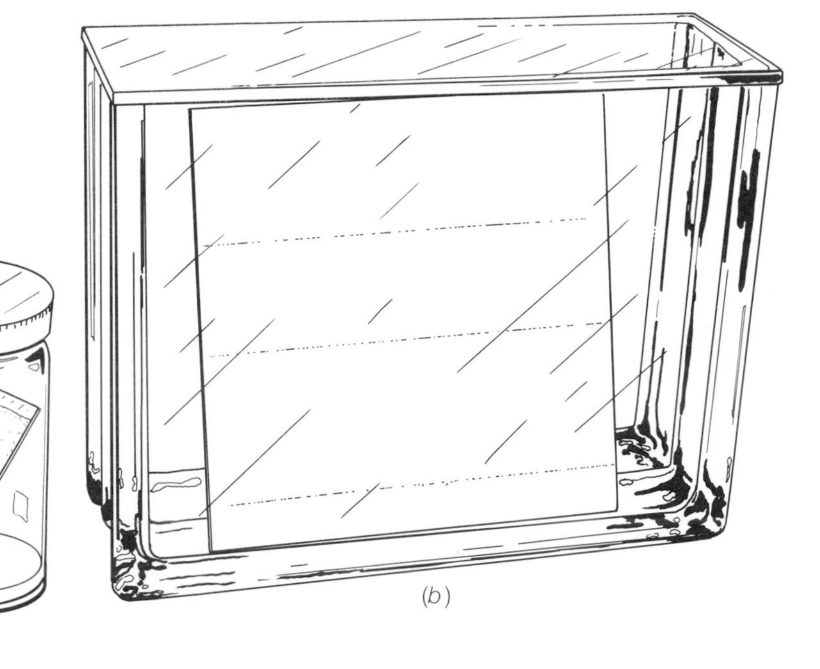

Figure 2.27 Developing chambers for thin-layer chromatogram. (*a*) A 4-oz bottle used as a developing chamber for a microscope-slide tlc plate. (*b*) A large glass chamber for developing preparative thin-layer plates.

only the adsorbent that was used but the identity of solvent or the identities and the ratio of the two (or more) solvents.

It is usually a good strategy to begin the examination of an unknown mixture with the least polar solvent available. If no sample development is observed with a solvent such as hexane, then a more polar solvent (ether, for example) should be tried. If the sample moves with the ether solvent front, a solvent of polarity intermediate between hexane and ether should be used. The intermediate solvent could be a single substance (in this case dichloromethane or toluene), or the solvent polarity could be adjusted by mixing ether and hexane. Figure 2.28 shows this process schematically.

Visualization

After marking the solvent front, the plate is placed in a hood and allowed to dry. After the solvent has evaporated, the location of the sample on the plate must be determined. This process is called *visualization*. Visualization of the developed plate is usually accomplished by exposing the dry plate to iodine vapors or to an ultraviolet (uv) light source.

Ultraviolet visualization When a uv lamp is used (see Fig. 2.29), any organic material which contains a uv chromophore (a part of the molecule which absorbs uv radiation) will appear on the plate as a dark spot on a light background. Most tlc adsorbents contain a fluorescent compound, which, when exposed to uv radiation, appears as a light background. Since the organic material is deposited on the surface of the adsorbent, it will, if it has a uv chromophore,

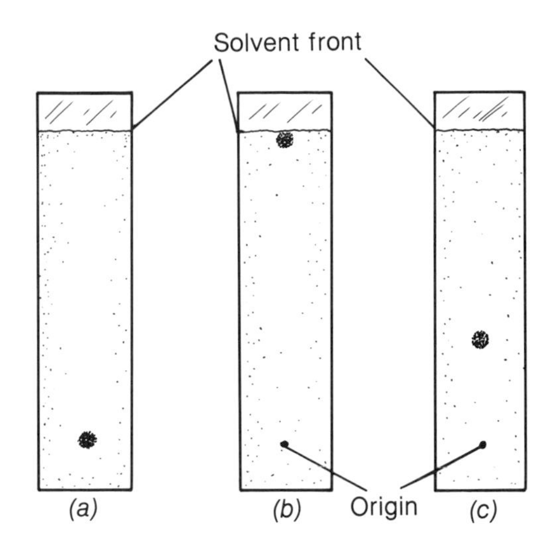

Figure 2.28 Choosing a solvent for thin-layer chromatography. (*a*) Solvent not polar enough. (*b*) Solvent too polar. (c) Correct solvent polarity.

act as a shade (or curtain) to prevent the uv radiation from reaching the surface of the adsorbent. The location of the spot should be marked as before.

If uv visualization is to be used, place the dry plate under a uv light source, preferably in a darkened area (see Fig. 2.29). There are numerous uv light sources, but the two most common ones are a hand-held uv lamp which can be positioned directly over the plate, and a uv cabinet with a black drape arrangement in front and a viewing port at the top. If a hand lamp is used, hold it to the

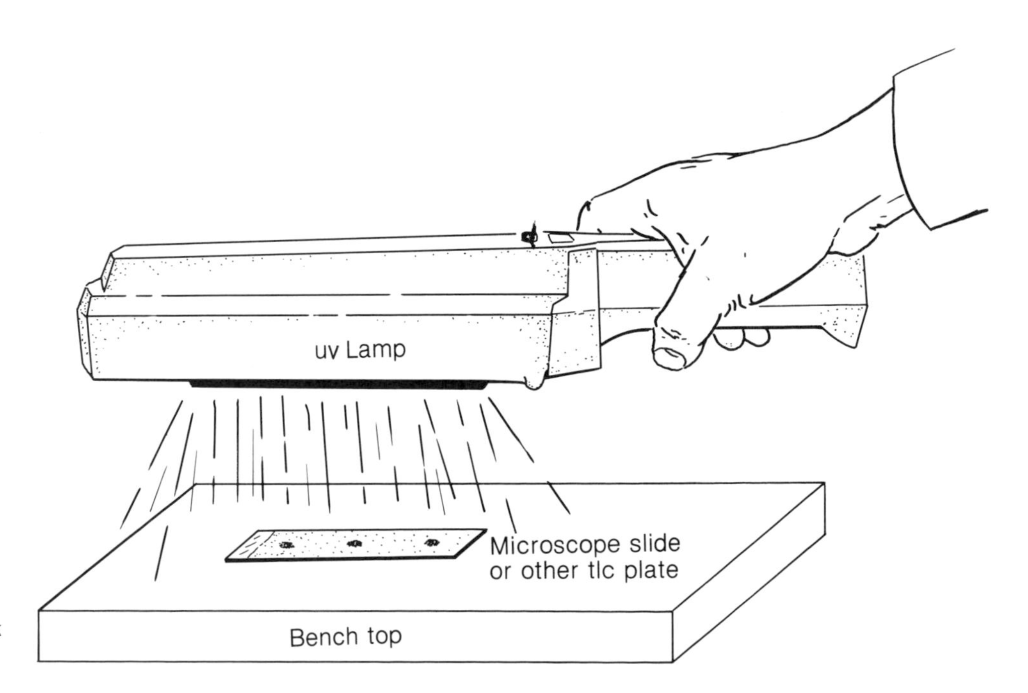

Figure 2.29 Ultraviolet visualization of a tlc plate. (Caution: Do not look into light source.)

side of the plate at a height of about 4 in. Look only at the plate, *never look directly at the lamp.*

(Danger: uv radiation is damaging to the cornea of the eye)

The sample spots will appear dark on a light background. Mark the spot(s) with a pin, pen, or pencil.

Iodine visualization If iodine visualization is desired, begin by placing a few crystals of iodine in a jar. Close the jar and wait until purple vapor is visible above the crystals. When the solvent has evaporated from the tlc plate, carefully place it in the iodine chamber and recap as soon as possible. If the plate was not solvent-free, iodine vapor will be adsorbed at all points on the plate. Otherwise, only those spots where the sample is present will darken. When you see distinct dark spots on the light background, remove the plate from the chamber and mark the spots. The iodine vapor is volatile and the plate will gradually lighten until the spots are no longer visible. Mark the sample spots with pin, pen, or pencil. It is good notebook technique to draw a small diagram of the plate indicating the location and shape of the spots.

It is useful to check the plate first with uv radiation and then subject it to iodine vapors. Record in your notebook which spots are visible in both cases and which spots are visible using only one of the two techniques.

Determining the R_f.

Once the organic materials have been located on the plate, the distance from the origin (that point where they were spotted on the plate) to the position that they now occupy must be measured. The origin of the chromatogram is usually indicated by a small depression which remains at the origin after application of the sample. A second distance now needs to be measured, i.e., the distance from the origin to the solvent front. Again, this is just a linear measurement.

The usual method for determining the relative movement of a compound is to compare the distance traveled by the sample with the distance traveled by the solvent. This simple ratio is called the *R_f value,* which stands for "ratio to front." It is also occasionally referred to as *retardation factor*. It is determined by the formula below.

$$R_f = \frac{\text{distance traveled by unknown (from origin)}}{\text{distance traveled by solvent front (from origin)}}$$

The R_f value that is obtained depends on the solvent used, the thickness of the adsorbent, the type of layer (silica gel, alumina, etc.), and, to a minor extent, the temperature. All these variables must be specified when an R_f value is reported. Notice that if a compound moves with the solvent front, its R_f value is 1.0. If the compound has moved half the distance that the solvent front moves, its R_f is 0.5. An R_f value cannot be greater than 1.0 or less than 0.

Because the R_f value for a particular compound depends on several variables, the value reported for a given compound is not absolute, as is a melting point. On the other hand, if one compares a sample and a standard on the same tlc plate, all the variables which affect the R_f will apply equally to both samples, and the R_f of the sample can be related directly to the R_f of the standard.

In a practical sense, it is often advantageous to use the smallest possible amount of sample and standard for tlc analysis. This is because the tlc adsorbent may easily be overloaded, which affects both the R_f value of the sample and the separating power of the tlc plate. Overloading is usually manifested by elongation of spots or by streaking. An overloaded plate is useless; it should be discarded and the results obtained from it ignored. The sample mixture should be diluted and the chromatogram repeated on a second plate. A successful thin-layer chromatogram should show relatively symmetrical spots and no streaks (Fig. 2.26*c*).

It is always desirable to compare a sample directly with a standard on the same tlc plate. If the same R_f value is obtained for both compounds with two or more solvent systems, this is good presumptive evidence that the two compounds have the same structure. It is this comparison of R_f values under standard conditions which is the basis for identifying an unknown compound in a mixture. The power of the separation technique, combined with its utility on a small scale, makes its application in organic chemistry extraordinarily important.

PROCEDURE

Five stock solvent systems will be available to you in the laboratory. These will be:

1 100% toluene
2 50% toluene:50% dichloromethane
3 5% toluene:95% dichloromethane
4 50% ether:50% dichloromethane
5 75% ether:25% toluene

(*Note:* The solvents above are listed approximately in increasing order of polarity.)

Obtain from your instructor a dichloromethane solution containing two compounds. Only a very small amount of this solution will be required for this exercise. By trial and error, determine which one of the five solvents or solvent systems listed above is best for separating the two compounds. Use both uv and iodine visualization to locate the samples on each plate. Obviously, the best solvent is the one which gives the best separation of the components. The ideal solvent would be the one giving R_f values of 0.3 and 0.7 for the two materials. Use each tlc plate only once. Note the R_f of all spots on the plate and record this information in your notebook. (*Note:* If you have used commercial plastic thin-layer plates, it is sometimes advantageous to affix the plate to the notebook page (transparent tape) after marking the appropriate spots. This is more difficult to do with glass plates. However, a technique sometimes used with glass plates is to place the plate on the lab bench and cover it with about 4 in of transparent tape. Press down on the tape so that the adsorbent adheres to the tape. Lift the tape and place it in the notebook with the appropriate comments.

After determining which solvent system is best for the mixture, obtain a solution of each of eight known (standard) compounds. Each standard will be in dichloromethane solution. By spotting the unknown mixture on the left-hand side of each chromatographic plate and one of the eight known compounds on the right-hand side, determine the identity of each component of the mixture. The mixture should contain two of the eight standard compounds, which are:

Acetophenone
Benzamide
Benzoic acid
p-Benzoquinone
Biphenyl
Diethyl phthalate
β-Naphthol
p-Nitroanisole

Submit the results of this chromatographic study to your instructor.

Column Chromatographic Techniques

The principles which apply to column chromatography are discussed in the introduction to this section. It is noted there that column chromatography and thin-layer chromatography are very similar. The major difference between the two lies in the manner of execution.

In column chromatography, the adsorbent is supported in a column of glass, metal, or plastic. Because of this mechanical support, a variety of materials may be used for separation. The most common adsorbents are still alumina and silica gel, just as in thin-layer chromatography, but such materials as charcoal, clay, diatomaceous earth, cellulose, starch, and even sugar have been used as well. If alumina or silica gel is to be used for the column chromatography, a preliminary tlc examination of the mixture using either alumina or silica will be of value.

In most laboratory exercises and indeed for many analytical applications, the appropriate adsorbent will already have been determined. In a research situation it is usually wise to obtain preliminary thin-layer chromatograms on both alumina and silica plates with use of a variety of solvents. In this way, the appropriate adsorbent and solvent combination may be determined rapidly and inexpensively. This information can then be applied more or less directly to the column separation.

Once the adsorbent and solvent have been selected for a particular determination (usually by a combination of intuition and experiment) the column should be packed with the adsorbent. The size of the column used will depend on the total amount of adsorbent needed to effect a particular separation. The amount of adsorbent required will depend roughly on the amount of sample to be separated. Although there are no hard and fast rules, or ratios, for determining how much adsorbent will be needed, it is generally assumed that 20 g silica gel or 30 g alumina will be required for each gram of sample. In many cases, this amount will be insufficient and more adsorbent will be required. When necessary, a ratio of 60:1 or even 100:1 may be used. Unless contraindicated by economic considerations, the 20:1 and 30:1 ratios can be regarded as minimum values.

Packing the column

Clamp the column securely (in two places) in a vertical position, using a ring stand or a bench rack for support. Once the column is clamped in place, examine it from the front and then from the side to make sure that the column is vertical (see Fig. 2.30). Solvent will not percolate evenly through the adsorbent and the separation will almost certainly be poorer if the apparatus is not vertical.

Once the column is secure in a vertical position, gently press a small piece of glass wool down the inside of the column, all the way to the very bottom (see Fig. 2.31). A glass rod may be used for this purpose. Next, pour a small amount of sand (enough to make a layer 1 cm deep) onto the glass wool plug. The sand will not participate in the separation; it is used as a mechanical support for the adsorbent. The glass wool is to prevent the sand and small particles of adsorbent from escaping.

When the glass wool and sand are in place, close the stopcock (or pinchclamp) and half fill the column with the solvent to be used for the separation. If more than one solvent or a gradually increasing percentage of polar solvent (a solvent gradient) is to be used, start with the least polar solvent. Next add the appropriate amount of adsorbent in small portions, so that as it settles to the bottom of the column the solvent wets it evenly and no lumps are formed. As the adsorbent is added, the chromatographic tube should be agitated by gently

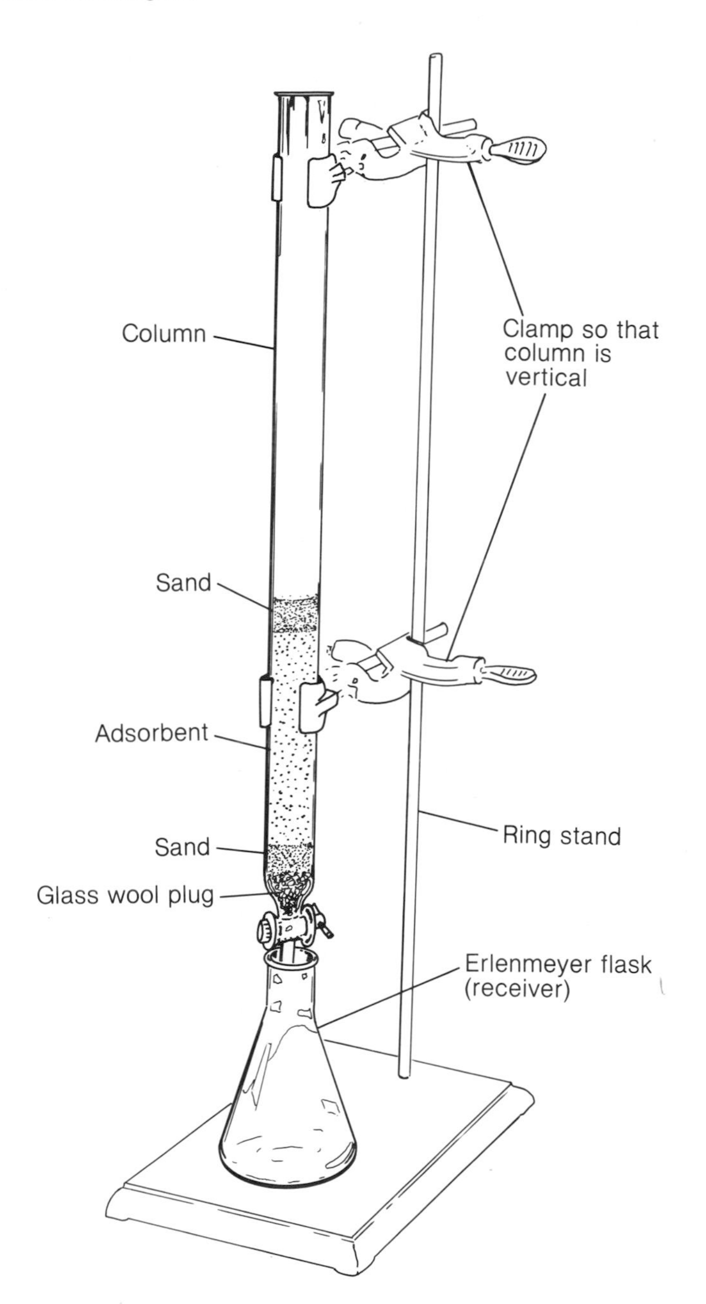

Figure 2.30 A chromatographic column.

tapping with a rubber stopper so that the adsorbent settles evenly (**be careful of channels or cracks**). It is also useful to open the stopcock to allow a slow dripping of solvent as the adsorbent is added through the top of the tube. Again, gentle vibration will help prevent channeling of the adsorbent. Continue tapping the side of the column even after all the adsorbent has been added; allow a

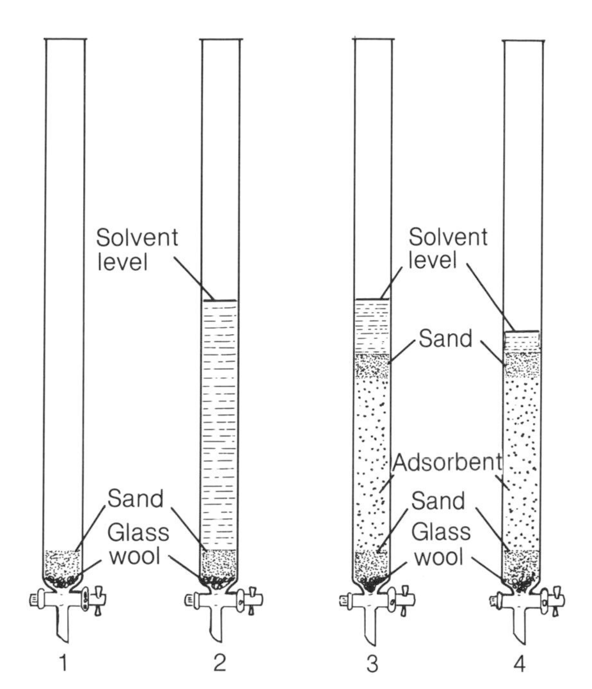

Figure 2.31 Steps in packing a chromatographic column. See text for further details.

few milliliters of solvent to pass through and pack the adsorbent; and add another 1-cm sand layer to the top of the column. The top layer of sand protects the adsorbent from spattering when fresh solvent is added. A piece of filter paper cut just slightly smaller than the inside diameter of the column and placed on top of the sand serves the same purpose. It should be added after the sample. The solvent should be drained to slightly above the level of the sand.

In any column chromatography, the level of the solvent should never be allowed to fall below the level of the packing. Never allow any portion of the column to become dry.

When either alumina or silica gel (especially silica gel) contacts solvent, a significant amount of heat may be released. In the procedure described above, the adsorbent is allowed to trickle through a layer of solvent before it settles in the bottom of the tube. The heat produced is thereby dissipated. If too much heat is generated, the solvent will bubble, and cracks will appear in the adsorbent. Any crack or channel in a column of adsorbent will adversely affect the overall separating power of the column. One way to avoid overheating is to premix the adsorbent in an Erlenmeyer flask or beaker (usually necessary only when silica gel is used) with a portion of the solvent and swirl the mixture for several minutes to dissipate the heat of adsorption. The wet adsorbent can then be poured in small portions into the column as described above.

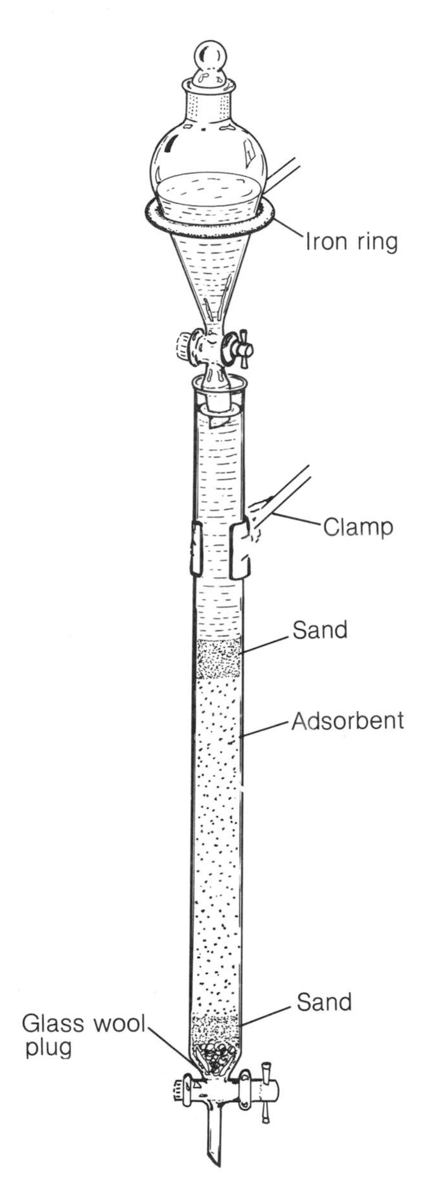

Figure 2.32
A constant-flow solvent delivery system.

Addition of the sample

After the column has been packed, the sample mixture may be applied to it. This is usually done by carefully adding a concentrated solution of the sample (in the column solvent) directly to the top of the column. The column solvent should be no more than 1 or 2 mm above the sand layer when the sample is added. After sample application, the stopcock is opened and the sample is drawn onto the adsorbent. Again, care should be taken that the column does not run dry.

When the sample has been drawn onto the column, close the stopcock (no solvent flow through the column) and add a small amount of solvent. Open the stopcock long enough for the solvent to reach the top of the sand. Repeat this process to ensure that the sample is completely drawn onto the adsorbent. This procedure is intended to ensure that the sample is adsorbed onto the column in the narrowest possible band. The efficiency of the separation (the resolution) is critically dependent on achieving the narrowest sample band which is practically possible.

Elution of the sample

Solvent should flow over the adsorbent steadily, evenly, and slowly enough for equilibrium to be reached. There are several techniques used to ensure a constant flow of solvent during the elution process. The most obvious way of doing this is to have a flask of the solvent in reserve and, as the layer of the solvent falls, to add more solvent. This process has the disadvantage that the column must be continually monitored by the investigator.

One common method for delivering a constant flow of solvent is to suspend a separatory funnel full of solvent in a ring stand above the column (Fig. 2.32). The separatory funnel is stoppered and placed so that the delivery tube at the bottom of the separatory funnel is inside the column. The stopcock is opened and solvent is allowed to trickle into the column until the level of solvent in the column is above the lowest point of the delivery tube. Now solvent will run from the separatory funnel only when the solvent level in the column falls and an air gap develops between the top of the column solvent and the lowest point of the delivery tube. This technique ensures continuous delivery of solvent to the chromatography column and eliminates the need for continuous monitoring.

PROCEDURE

SEPARATION OF FLUORENE AND FLUORENONE

Prepare a chromatography column about 1 cm in diameter and approximately 50 cm long (a 50-mL buret will do) as described above; using about 10 g activated alumina and approximately 20 mL petroleum ether. Obtain an approximately 1 : 1 mixture of fluorene and fluorenone from your instructor or from Exp. 14.1B. Dissolve 0.5 g of this mixture in 1 mL cyclohexane (slight warming on a steam bath may be necessary). Add the solution directly to the top of the chromatography column. Add a few milliliters (usually about 3) of petroleum ether after the sample has been drawn onto the top of the column. Allow the 3 mL to slowly percolate onto the column. Begin normal elution with petroleum ether and collect the eluate in a 100-mL beaker.

As you collect the solvent, from time to time allow a drop or two of eluate to fall onto a watch glass. When the solvent evaporates, any solid present should be visible on the glass. Fluorene will be detected as a white solid after evaporation of the solvent. This is a crude visualization technique which allows you to determine when the material is coming off the column and also when most of the material has been eluted. Continue to elute the column with petroleum ether until no fluorene is detected on the watch glass. (Most of the fluorene should have been eluted in 30 to 40 mL petroleum ether.)

When no more fluorene is eluted, change to a second clean 100-mL beaker and collect the intermediate fraction. The first beaker should be placed in a hood so that the solvent may begin to evaporate. While collecting the intermediate fraction it is possible to follow the progress of the yellow fluorenone band as it proceeds down the column. In this case, one of the components is colored and can be visually monitored as the separation progresses. In fact, the term *chromatography* derives from the original use of this technique, which was first used primarily to separate the colored components of flowers and plant materials. Therefore, chromatography means "separation by color."

After several milliliters (about 10 mL) of solvent has been collected, change the solvent to dichloromethane. Dichloromethane is much more polar than petroleum ether. The yellow fluorenone band should now begin to move rapidly down the column. When the yellow color just reaches the bottom of the column, change to a third dry 100-mL beaker. Monitor the progress of the chromatography visually and confirm by using the watch-glass technique. Collect the eluate until it is colorless. You should now have three fractions, the first containing fluorene, the second containing a small amount of fluorene and fluorenone, and the third containing fluorenone.

Check each fraction for purity by tlc. Determine if the separation is complete. Evaporate the first and third fractions by warming on a steam bath (**hood**), which should yield white crystals of fluorene and yellow crystals of fluorenone, respectively. Spectral data for the pure compounds are shown in Figs. 14.3 and 15.3.

2.7 METHODS OF HEATING

There are many methods for heating reaction mixtures and the method which is chosen will be determined by a variety of factors. These factors include the size and shape of the reaction vessel, the reaction temperature, and whether the reaction mixture must be stirred at the same time it is heated. In an undergraduate organic chemistry laboratory the most common heating methods are those which we list below. Note that the information here will not necessarily apply to experiments which are conducted in the research laboratory. In order to determine other and more sophisticated methods of heating, reference should be made to some of the more advanced laboratory manuals listed at the end of the Chemical Literature section.

Free Flame

Free flame can imply a variety of heating devices but generally refers either to a full-sized bunsen burner or to the smaller microburner. The bunsen burner is probably the heating device most commonly used in undergraduate organic laboratories. Bunsen burners are very inexpensive to purchase and operate. In addition, heating may be achieved rapidly and good control may be exercised in heating the reaction mixture. Although a free flame can be a very useful heating device, especially with the proper supervision, there are certain drawbacks to its use which some instructors feel disqualifies it from use in undergraduate laboratories. The principal concern is the danger posed by having open fires burning in the laboratory.

Bunsen burners have been used for decades in both research and undergraduate organic laboratories. In many cases very sophisticated experiments have been conducted with use of this device. The real difficulty arises from the fact that many people are working at the same time in most undergraduate laboratories. People walking about, windows and doors being opened and closed, and a dozen other activities all create significant drafts. A flame may blow first in one direction, then another. It may even blow out and cause a dangerous gas leak. The drafts may also blow heavier-than-air solvent vapors across bench tops right into the flame. Concentrated solvent vapors are dangerous if inhaled, but in the presence of a flame a bad situation can become a disaster.

Nevertheless, the bunsen burner is the only heating device available in many undergraduate laboratories because of the large program and the tremendous expense involved.

When properly used, bunsen burners are completely safe. The key is proper use! If you are going to use a bunsen burner, be certain that (1) there are no flammables near your work area and (2) that you turn the gas off completely when you extinguish the flame. Also be certain that you do not heat too vigorously, as most organic liquids can boil or bump out of the flask and then ignite.

A bunsen burner, a microburner, and a burner with a flame moderator are shown in Fig. 2.33. When possible use the smaller burner; this will increase your control. A protected flame is also advantageous because it is less likely to be blown out by the drafts that are present in every laboratory. Finally, when using a flame be certain that a metal grill is present between the flame and the flask to moderate the intensity of the flame and prevent dangerous local heating of the reaction mixture.

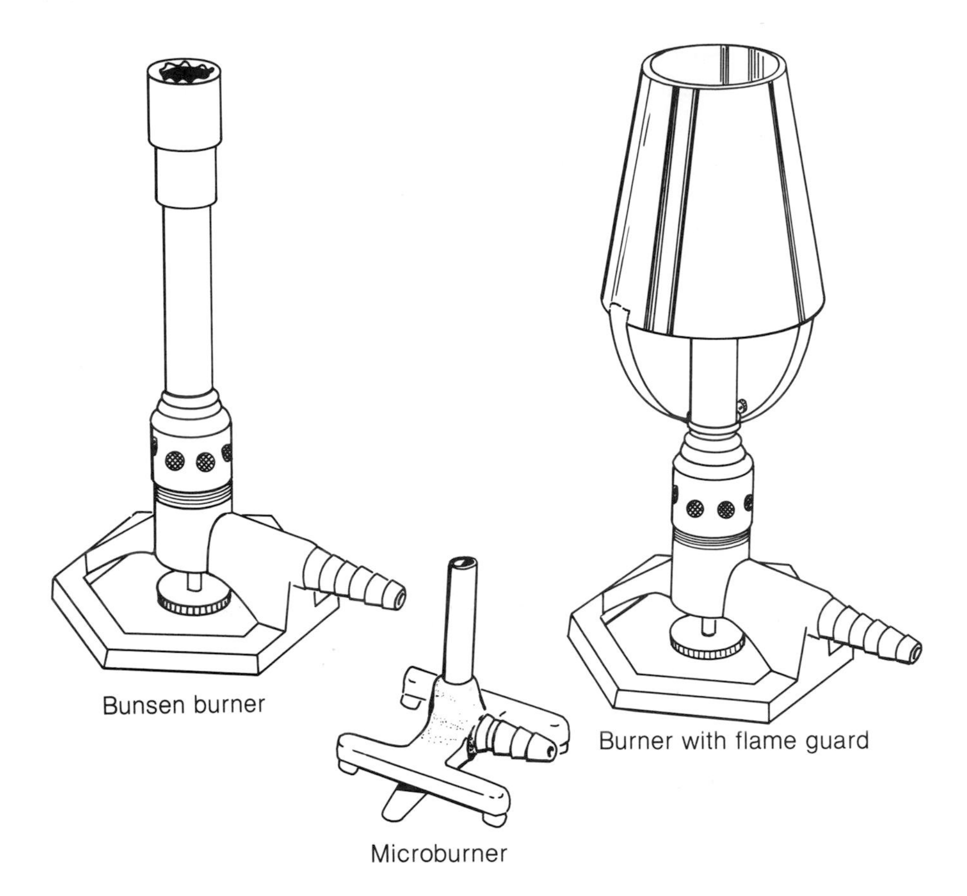

Figure 2.33
Types of bunsen burners.

Steam Bath

For reactions which do not require heating above about 90°C, the steam bath (Fig. 2.34) is the heating source of choice. Steam baths are inexpensive to purchase and operate and are very safe. The important limitation of the steam bath is that the maximum temperature is dictated by the boiling point of water.

The steam bath is much more useful than the free flame for heating low-boiling liquids. Any vapors which may escape from the distillation apparatus will simply mingle harmlessly with water vapor rather than igniting. Since the maximum temperature of steam is 100°C, the boiling is often less vigorous than it might be with a free flame. For refluxing low-boiling organic solvents, a reflux apparatus on top of a steam bath is a convenient way to conduct a reaction. Boiling chips in the reaction mixture usually ensure sufficient agitation that no additional stirring is required.

Because of the distance between the bottom of the flask and the bottom of the steam bath, it is difficult to stir magnetically and use a steam bath at the same time. Overhead stirring is common in research laboratories but uncommon (because of expense) in teaching laboratories. A real advantage to using a

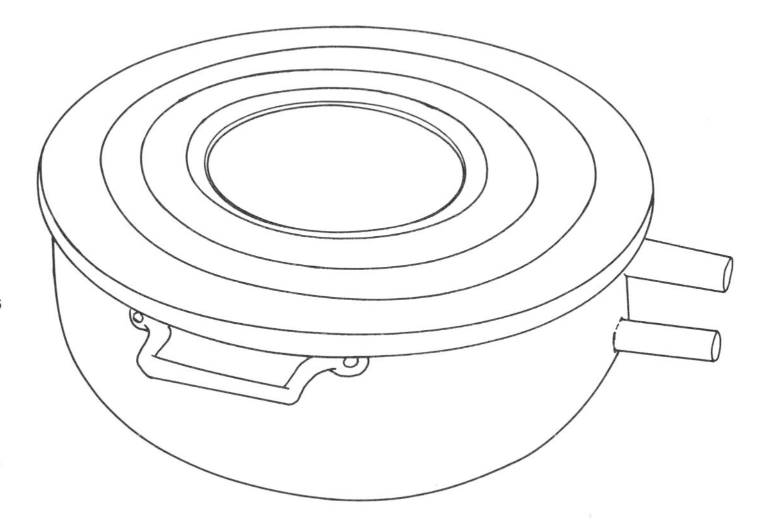

Figure 2.34
Steam bath. To use, remove enough rings so that a round-bottom flask will rest in a ring, or enough so that an Erlenmeyer flask will be exposed to steam without falling through.

steam bath is that if you somehow forget the reaction, the mixture cannot heat above 100°C and usually will not char.

When you use a steam bath, be sure to place the flask on it properly. Steam baths are usually equipped with several concentric rings (see Fig. 2.34). When using a steam bath, remove as many of the rings as necessary for the reaction flask to rest on the largest ring which is slightly smaller than the flask. The flask should not be completely immersed in the steam bath. Adjust the steam pressure so that the flask heats but very little steam escapes. Turning the steam up very high will probably have more effect on you than on the reaction vessel. You can, of course, control the rate of reflux by varying the steam pressure, or by keeping the heat in by use of a towel (Fig. 2.35).

Oil Bath

Oil baths (Fig. 2.36) are particularly useful for heating reaction mixtures. The contact between hot oil and the bottom and sides of a flask is intimate because the vessel is completely surrounded by the hot medium. As a result, heating is even and can be controlled effectively. Oil baths are relatively inexpensive and are usually safe. The safety aspect is twofold. First, no flame is present when an oil bath is used, so ignition of solvent vapors is unlikely. Second, the temperature of a bath will be proportional to the amount of heat put in and usually will not rise above some maximum value. As a result, charring of reaction mixtures is less likely with an oil bath than it is with either an electric mantle or a bunsen burner.

The problems associated with use of an oil bath are: (1) they are often slow to heat; (2) when they are very hot, they fume and can catch fire; and (3) after they are used, they cool slowly. In addition, when a flask is removed from an

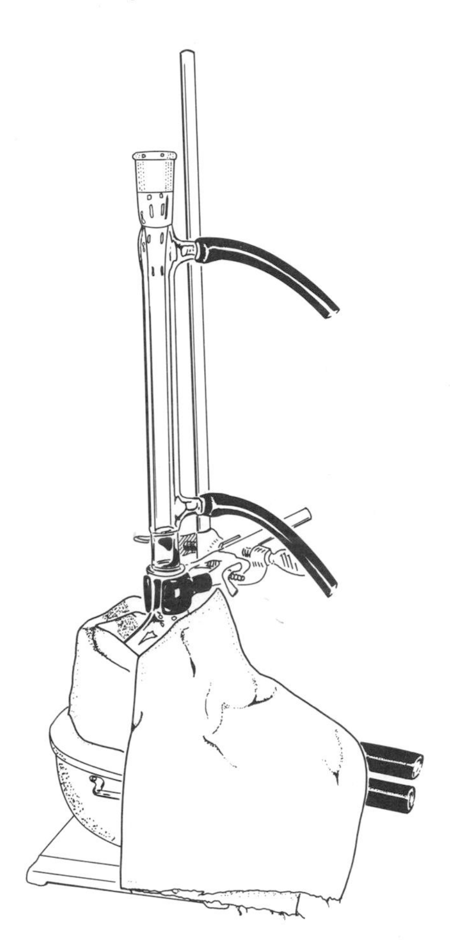

Figure 2.35 Keeping heat in by use of a towel. Reflux on a steam bath is shown here.

oil bath, it will have an oily residue on it. The residue not only is sloppy but if hot can cause burns.

Notwithstanding these difficulties, oil baths are very useful for heating reaction mixtures. The baths are generally glass or metal dishes. Crystallizing dishes of various sizes are particularly useful for oil baths because the glass does not hinder magnetic stirring. Small cooking dishes are also useful, and aluminum or ceramic dishes can be purchased inexpensively at variety stores.

The oils used vary from laboratory to laboratory. Dioctyl phthalate, mineral oil, and silicone electrical fluid are all common. Less common but very useful is 40 to 50 weight motor oil. The disadvantage of using motor oil as compared with dioctyl phthalate or mineral oil is that the former is difficult to see through. This drawback is more than compensated for, in the authors' opinion, by the fact that motor oil is so inexpensive and accessible.

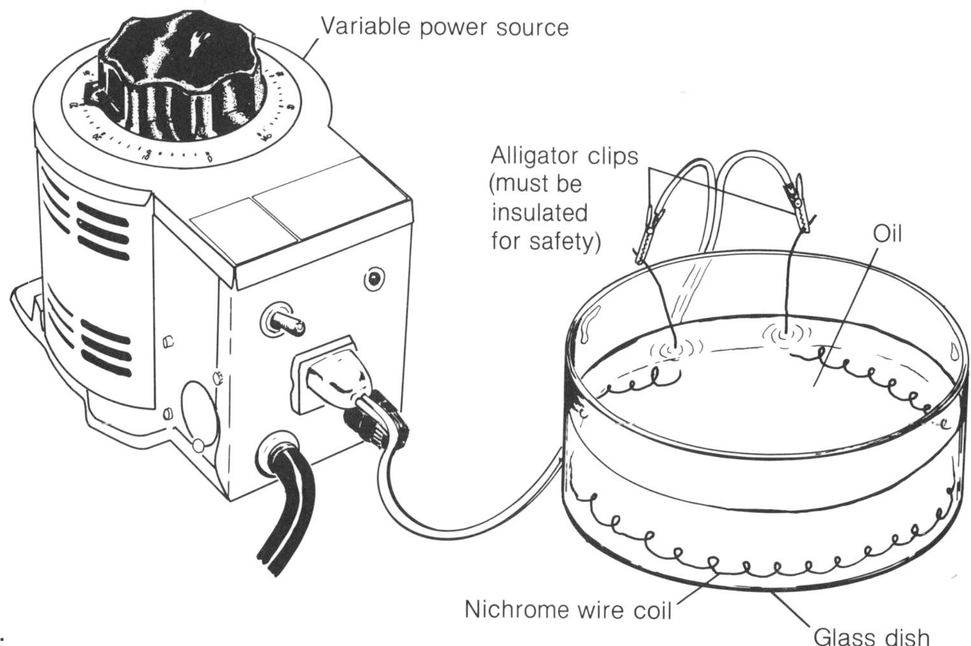

Figure 2.36
Coil-type heater in cooking-pan oil bath.

Tricks for Heating an Oil Bath

Although many methods are available for heating a reaction mixture, use of an oil bath offers many advantages (see Sec. 2.7). A very inexpensive and relatively safe oil bath may be constructed from an immersion heater, designed for use in coffee cups, and a metal pot.

An aluminum saucepan may be purchased at any hardware or variety store. An immersion heater should be suspended over the edge of the pan and taped to the pan holder with electrical tape. The pan should then be filled half to three-quarters full of oil. When the immersion heater is plugged into a rheostat or other temperature controller, an inexpensive and easy-to-control oil bath results. The device is illustrated in Fig. 2.37.

An alternative heating device is a nichrome wire placed in the bottom of an oil bath and heated electrically (as shown in Fig. 2.36). Another alternative is simply to place the reaction vessel in an oil or wax bath on top of a hot plate. The hot plate heats the oil, and the oil heats the reaction vessel. This way there is no direct contact between the electrical heating element and the reaction mixture. If the flask were to break, the contents would deposit in the oil bath rather than directly on the heating element.

Electric Heating Mantles

Heating mantles or electric mantles (Fig. 2.38) are often referred to by the trade name Glas-Col. Heating (or electric) mantles come in a variety of shapes and

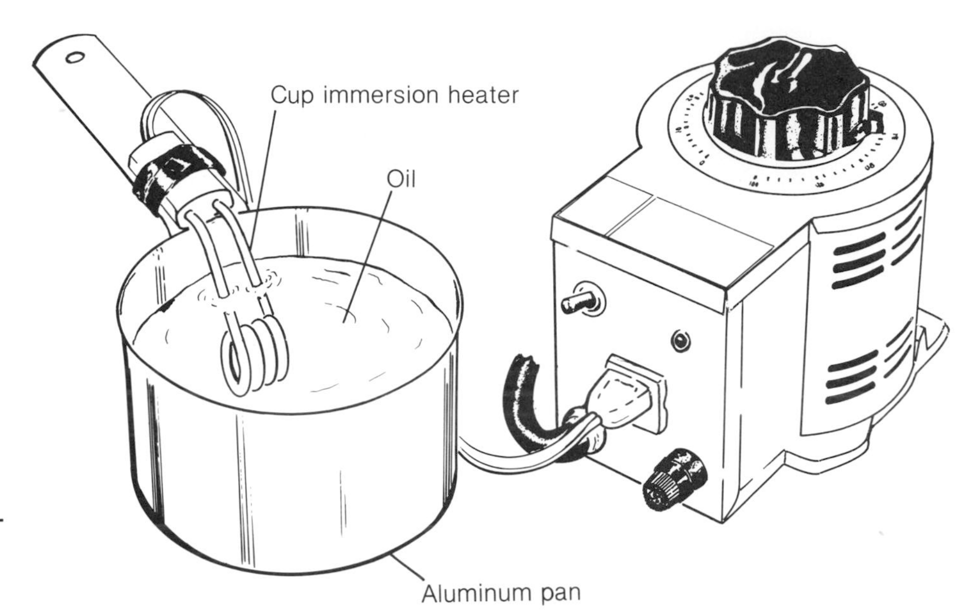

Figure 2.37
A coffee cup–type immersion heater may also be used with an oil bath.

sizes. Commercially available sizes from 10 mL to 12 L are often found in research laboratories, and even larger ones can be purchased. In undergraduate laboratories it is common to find only one or two sizes at each desk, generally those designed to heat 100-, 250-, or 500-mL vessels.

The exterior designs of heating mantles vary to some extent, but they are almost always hemispherical on the inside. The mantles have a heating element

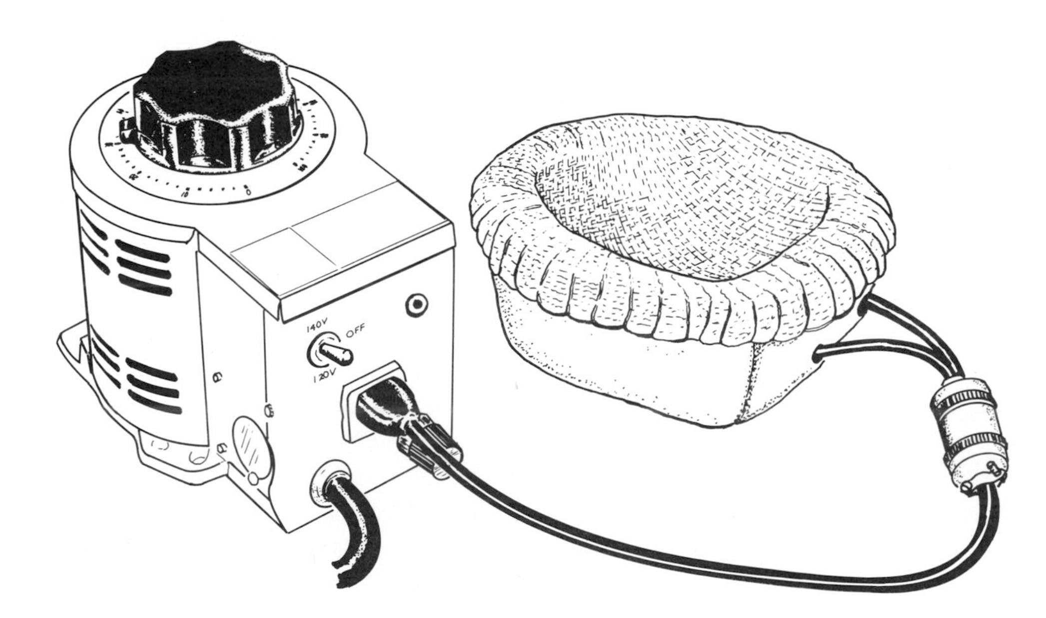

Figure 2.38
Heating mantle.

interwoven with a heat-resistant cloth. The heating element transfers heat to the fabric, and the fabric transfers the heat to the reaction vessel. The two most inexpensive kinds of heating mantles are the soft-bottom type, which has no protective container, and the type which is encased in a hemispherical aluminum cup. These two types of heating mantles permit magnetic stirring of the mixture. A third type, the aluminum-cannister heating mantle, is self-supporting. The aluminum-encased mantles tend to resist breakage better than do the soft-bottom mantles. The disadvantage of using an aluminum-cannister mantle is that magnetic stirring of the mixture is usually difficult, although mechanical stirring may be appropriate with this device. Aluminum mantles resist damage longer but are more expensive.

Heating is usually controlled by use of a voltage regulator. The two most common types are the Variac apparatus and transistorized temperature controllers. The Variac is a large variable resistor (rheostat). Transitorized electronic controllers have on-off switching circuits to control power delivery. Transistorized devices are therefore somewhat more economical to use but have traditionally been more expensive. By varying the voltage across the heating element, the resistance and therefore the thermal output of the mantle changes. It is changing the voltage which changes the temperature.

When using a heating mantle, care should be exercised in setting the voltage. When using a heating mantle, first check to see if there is a label attached which indicates the maximum voltage which can be safely applied. For very small heating mantles this voltage is only about 15 to 20 V. A heating mantle designed to tolerate a maximum of 20 V will burn up quickly if 120 V is applied. Most 100- to 500-mL heating mantles will tolerate a full 120-V input, and some large mantles require two voltage inputs.

When using a heating mantle, be certain that it is the appropriate size for the flask that is going to be heated. The size of the heating mantle is sometimes difficult to judge but is almost always marked on a label. After the mantle has been assembled with the reaction apparatus, plug the mantle into the temperature-controlling device. Then plug the temperature-controlling device into a standard 120-V outlet. With the voltage controller set at 0, the reaction vessel will not heat. If the controller is set at maximum, the mantle will usually heat very rapidly.

There are two approaches useful for initial heating of a mantle. First, you may set the mantle at the value that seems to be appropriate for heating the reaction medium to the desired temperature and wait for it to come to thermal equilibrium. *This is the safest way to use a mantle.* A second, and somewhat riskier, approach is to set the voltage controller for a short time at a value much higher than that required. The mantle and reaction mixture will heat much faster than they would at a lower setting. When the reaction mixture reaches the appropriate temperature, the voltage can be turned down. The danger of using this approach is that if you forget to turn the mantle down, the reaction mixture will overheat and probably be lost. If done carefully, this procedure will allow you to complete the experiment somewhat more rapidly.

A word of caution: **The two most common errors associated with use of a heating mantle are (1) using the wrong size mantle; and (2) plugging the mantle directly into a 120-V outlet. Although in an emergency situation a heating mantle appropriate for one size of flask may be used for a slightly smaller flask if it is done with great care, you will not face this problem in an undergraduate laboratory.** ***Under no circumstances should any heating mantle ever be plugged directly into the wall outlet.*** **If this is done, it will be impossible to control the temperature. Depending on the size of the heating device, the heating element may also burn out. Both these situations should be avoided.**

2.8 REMOVING NOXIOUS GASES

During the conduct of many organic reactions, certain noxious gases are produced as by-products. The reaction of thionyl chloride ($SOCl_2$) with a carboxylic acid, for example, produces a mixture of hydrogen chloride gas (HCl) and sulfur dioxide (SO_2). The reaction of thionyl chloride with a carboxylic acid is a particularly severe example, but it is not an uncommon reaction and is suggested in Sec. 23.6 of this book. Other examples abound in organic chemistry. Sulfur dioxide is released in the Diels-Alder reaction of sulfolene with maleic anhydride (Exp. 9.1). A hydrogen halide gas is released in both experiments in Chap. 10. Hydrogen chloride is released during the hydrolysis of aluminum chloride (Exp. 16.2B). Noxious fumes are also released during many of the reactions described in Chap. 18. It is suggested in the experimental procedures for each of these reactions that the entire procedure be conducted in a good hood (if available). The fact that the hood is utilized makes the procedures safer because noxious fumes exit into a vented fume cupboard, are diluted by large volumes of air, and are expelled away from the laboratory worker, where they are less concentrated and can do less harm. Whether or not a hood is used, there are certain procedures which can be carried out to make reactions somewhat safer. Two alternatives are suggested below.

If a Hood Is Used

If a reaction will be conducted in a hood or fume cupboard, the gases which exit from the reaction mixture will be sucked away into the vent. Nevertheless when the vapors exit from the reaction mixture, they will tend to remain around the reaction apparatus for a short period of time before they can be blown away. Since you may be working around the reaction mixture while these gases are exiting, you may inadvertently come in contact with them. In reactions which give off such gases, it is therefore a good idea to take every precaution to conduct the gases away from the reaction mixture. This has the advantage also of keeping any clamps or ring stands or other metal apparatus from becoming unduly corroded.

If the reaction is conducted in a ground glass apparatus, use the thermometer adapter to fill the joint which would ordinarily be left open. Instead of placing the thermometer in the thermometer adapter, use a piece of 4- or 5-mm glass tubing with a length of approximately 10 cm. Both ends of the tubing should be fire polished, so that when it is inserted into the adapter it cuts neither the rubber nor your hands. Attach to the outside end of the tube a piece of rubber tubing sufficiently long that it will reach to the back of the hood. If you have old rubber tubing in your desk, use a piece of it. If you have only new tubing, select a piece which you will use for noxious gases in the future and reserve it for that application.

Affix the rubber tubing to the glass tubing and run the tube as far toward the back of the hood as possible. The draft exit for many hoods is at the bottom of the hood where it meets the bench. You can determine if this is the draft port by holding a tissue near the bottom. If it is sucked hard toward that opening, this is the exit port. Conduct the tubing from the reaction vessel into that exit port and gases will drift immediately up to the hood without exposing the working area. An alternative but somewhat less satisfactory approach is to run the tubing into the hooded sink drain. Whatever approach you take, the overall idea is to have the gases conducted away from the working area. In positioning the tube, also be careful that it is not exposed to water which can run back up the tube into the reaction mixture.

If a Hood Is Not Used

Sometimes small amounts of noxious fumes are generated in a reaction and it is inconvenient or impossible to conduct the reaction in a hood. When this circumstance arises, a simple device may be used to increase the overall safety of the operation.

If a standard-taper vessel is to be used, place a T-tube in the opening of the thermometer adapter and fix the thermometer adapter into the standard taper opening, which would have been left open for gas to exit. Attach one end of the T-tube to a piece of rubber tubing which is connected to an aspirator trap. Connect the other end of the aspirator trap to a water aspirator. Turn the aspirator on very gently and a slight negative pressure will be created over the reaction mixture. Since one end of the T-tube is completely open, fresh air from the atmosphere will be drawn in and over the reaction mixture. This negative pressure and draft will force noxious fumes away from the reaction mixture into the aspirator trap and eventually into the water stream, where they will be washed away.

If it is necessary to keep the reaction mixture dry, the open end of the T-

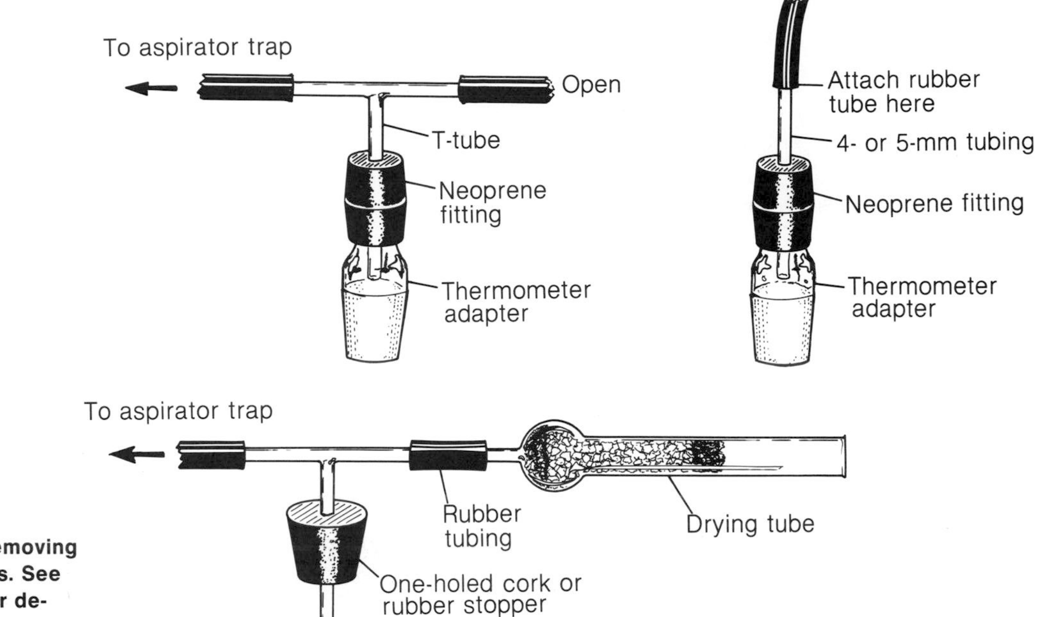

Figure 2.39 Devices for removing noxious fumes. See text for further details.

tube may be connected to a drying tube. The reaction will certainly not stay completely dry because the atmosphere coming through the drying tube will be coming through at too high a rate to be completely dried. Nevertheless, this arrangement will be useful for many purposes.

If a standard-taper apparatus is not used, the T-tube may be fitted into a one-holed cork or a rubber stopper of the appropriate size. In fact, this apparatus may be adapted for use with round-bottom or Erlenmeyer flasks by the appropriate choice of rubber stopper. The two different apparatuses are illustrated in Fig. 2.39.

QUALITATIVE CHARACTERIZATIONS

Any time the preparation of a particular compound is attempted, there is the possibility that not only the product but certain by-products will be isolated. In order to determine whether or not the compound is the appropriate product, certain tests can be conducted. Determination of the melting point, boiling point, refractive index, and other physical properties is often specified in the procedure and carried out by the experimentalist. There are a variety of very rapid tests which can be conducted to determine if the compound has properties which accord with the structure of the material. These properties include color, odor, solubility, and the presence (or absence) of the appropriate elements. In the latter connection, if *n*-butyl alcohol is converted to *n*-butyl bromide (Exp. 10.1), the product must give some indication that it contains bromine or it cannot possibly be the correct compound.

The solubility of a particular compound is characteristic of its functional groups, molecular weight, and the arrangement of its atoms. The prediction of solubility properties is usually easy only if the compound is an acid or a base. For other materials, it is largely experience which allows one to predict whether the compound will be soluble in such a system as acetone, water, toluene, or hexane. Because the phenomenon of solubility is rather more involved than the other topics which will be discussed now, its discussion is deferred to the following chapter.

For the purpose of qualitative characterization of a product, by-product, or even an unknown substance, certain techniques are required. One should be able to readily determine the color, the odor, and the elements present and know how to interpret these observations. It should be possible to perform such tests as the flame and Beilstein tests rapidly, so that a substantial amount of qualitative information may be obtained in a short time.

3.1 COLOR

The color of a compound is determined by, and is therefore characteristic of, its structure. Most organic compounds are colorless (liquids) or white (solids). A liquid such as ethyl alcohol, which has no color associated with it is said to be colorless or "water-white." The liquid may be clear, but clear is not a color. A compound should always be designated as colorless, white, green, blue, orange, or whatever the color may be. "Clear yellow" is descriptive, but use of "clear" alone should be avoided.

Color in pure compounds is almost always the result of conjugation. Saturated hydrocarbons are always colorless or white. As the conjugation increases, however, the colors go through a transition from yellow, to orange, to red, to blue. From the structures below of benzene, naphthalene, anthracene, and tetracene, it should be clear that the extent of double-bond conjugation increases as we go to more and more rings. Benzene is a clear and colorless liquid; naphthalene and anthracene are both white (or colorless) solids. Tetracene, however, is a yellow-orange solid.

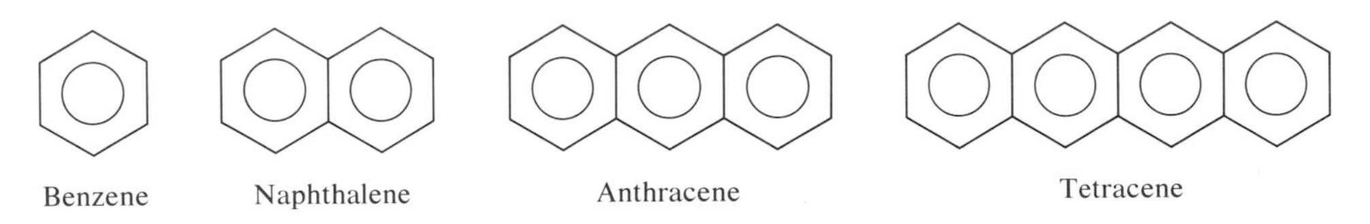

Fluorene, a related ring system, is shown below. Fluorene is transformed to fluorenone by a base-catalyzed air oxidation (Exp. 14.1).

*sp*³ carbon O *sp*² carbon

[ox] →

Fluorene (white) Fluorenone (yellow)

The CH_2 group between the two aromatic rings of fluorene keeps them from being conjugated. Fluorene is therefore not colored; it is a white solid. Fluorenone, which has a carbonyl carbon (sp^2) interposed between the two aromatic rings, is a fully conjugated system. As a result, fluorenone is a yellow solid. If fluorenone is reduced to the corresponding alcohol, fluorenol (Exp. 14.4), the system is no longer fully conjugated. As a result, fluorenol is white. In fact, it is noted in the directions for the synthesis of fluorenol that the reduction may be monitored by the disappearance of the yellow fluorenone color. When the solution no longer shows color, the reaction is presumed to be over. The product isolated must be white, or the qualitative observation does not accord with the presumed structure of the material.

Aryldiazonium ions are conjugated and are very reactive systems, but generally are white solids. These salts simply do not have enough conjugation to afford color. Reaction of benzenediazonium chloride with *N,N*-dimethylaniline, as shown below, yields a substituted azobenzene.

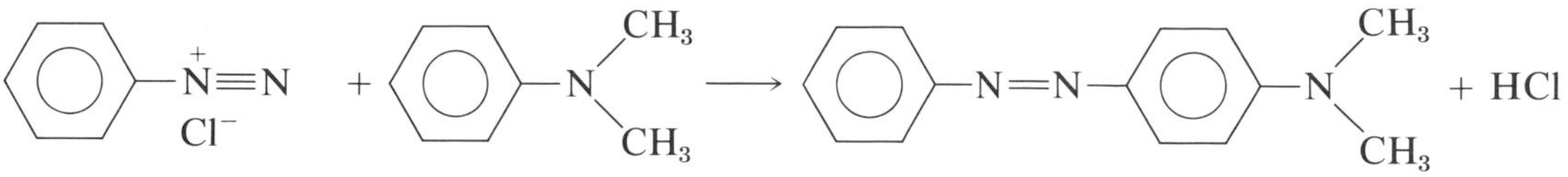

$$C_6H_5-\overset{+}{N}\equiv N\ Cl^- + C_6H_5-N(CH_3)_2 \longrightarrow C_6H_5-N=N-C_6H_4-N(CH_3)_2 + HCl$$

Benzenediazonium chloride (colorless, i.e., white) *N,N*-Dimethylaniline (colorless) 4-(*N,N*-Dimethylamino)azobenzene (deep red)

Some azobenzenes are azo dyes and are highly colored materials. In this case, the two aromatic rings are conjugated through a nitrogen-nitrogen double bond. Once again, extending the conjugation yields a colored product.

A final example of this, and one which we hope will become familiar to you, is the formation of a dinitrophenylhydrazone derivative of a ketone or aldehyde. Formation of acetone dinitrophenylhydrazone (often referred to as acetone DNP) and the formation of cinnamaldehyde dinitrophenylhydrazone are shown in the equations below.

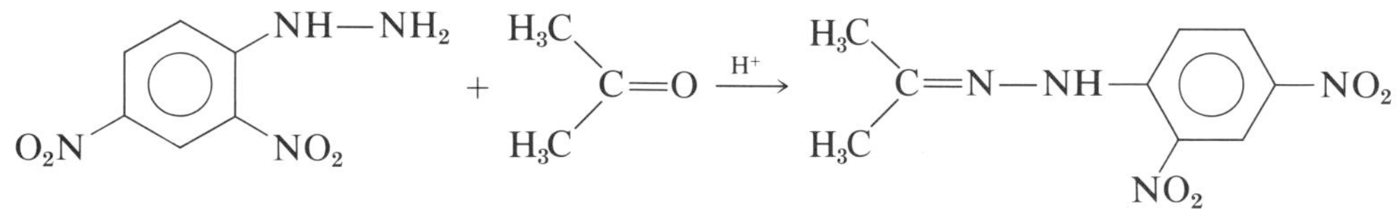

$$(O_2N)_2C_6H_3-NH-NH_2 + (H_3C)_2C=O \xrightarrow{H^+} (H_3C)_2C=N-NH-C_6H_3(NO_2)_2$$

2,4-Dinitrophenylhydrazine Acetone Acetone DNP

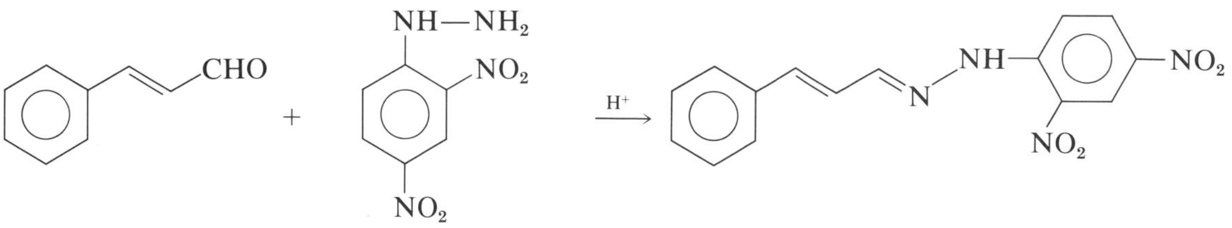

$$C_6H_5-CH=CH-CHO + (O_2N)_2C_6H_3-NH-NH_2 \xrightarrow{H^+} C_6H_5-CH=CH-CH=N-NH-C_6H_3(NO_2)_2$$

Cinnamaldehyde 2,4-Dinitrophenylhydrazine Cinnamaldehyde DNP

Little additional conjugation is added in the formation of acetone DNP, and this derivative is yellow. Formation of cinnamaldehyde DNP essentially joins the conjugation of the two aromatic rings, and therefore this compound is deep red-orange. The color of DNP derivatives is often used in a qualitative way to distinguish aliphatic from aromatic systems.

Color may be present in a sample for reasons other than conjugation. The most common problem is that the hoped-for product is contaminated by some colored impurity. In organic chemistry it is often said that a little color goes a long way. This means that very small amounts of impurity can often add a lot of color to a product.

Aniline (aminobenzene) is a colorless liquid when it is freshly distilled. When placed in a bottle it remains a colorless liquid for several days. Gradually, because of air oxidation, aniline is converted to a variety of products, not all of which have been identified. A transition is gradually observed from yellow to deep yellow to brown and, eventually, to black. If the sample is once again distilled (or otherwise purified, see Exp. 11.5A) almost all the pure, colorless aniline can be recovered, and only a very small amount of colored residue will remain in the distillation flask.

It is sometimes difficult to determine whether a compound is colored because it is impure or because it is highly conjugated. In general, organic compounds (except for certain classes including azo dyes and DNPs) are white to yellow in color. Relatively few organic compounds are highly colored. In the absence of other evidence, a highly colored compound should usually be presumed impure. If the color can be removed, the compound was contaminated. As mentioned above, distillation of aniline usually affords a pure white liquid and a colored residue. Clearly, the color associated with the aniline was due to contamination. Solid 4-nitroaniline is distinctly yellow and on recrystallization remains yellow. The color is due to conjugation. 4-Methoxyaniline (anisidine) is usually obtained as a dark brown, lumpy solid. Purification affords colorless needles. The colored contaminant has been removed.

The color of a compound should be recorded as soon as it is obtained. If purification changes the color, this is an indication of purity and should be so recorded. Generally, an increase in purity of a solid will be accompanied by an increase in melting point. This corroboration is unavailable when purification is effected by distillation.

3.2 ODOR

The odor of a substance depends primarily upon two things: (1) the volatility of the substance, and (2) the shape of the molecule. A substance must be volatile enough for the odor to reach the observer's nose. The volatility of a substance is determined by the shape of the molecule, the molecular weight, and the presence of functional groups (see Sec. 1.2).

The shape of the molecule is also very important in determining the odor of the molecule. It is believed that nasal detection involves certain shape relationships. In other words, many molecules which have so-called camphoraceous odors are those molecules which have an approximately spherical, or ball, shape such as that known for camphor. Camphor has a very distinctive odor despite the fact that it is a high-melting solid. It is quite volatile, as is evidenced by the fact that it sublimes quite rapidly (see Sec. 2.3).

There are many broad classes of odor which have been assigned over the years. A number of excellent reviews on this subject have appeared in the literature; the most appropriate for this level is one which appeared in *Scientific American* (J. E. Amoore et al., "The Stereochemical Theory of Odor," February 1964, p. 210).

In general, one can learn to recognize certain odors simply by experience. It is usually experience, and not the exact shape of the molecule or even more sophisticated relationships, which is of value to most laboratory workers.

We shall discuss odor in somewhat more detail in Sec. 22.2B, particularly with reference to identifying compounds from name, structure, and common sense. We refer you to that section for more detailed discussion. We wish to point out here, however, that if you carry out the esterification reaction described in Sec. 11.2C, the product, isoamyl acetate, should have a strong odor of bananas. If not, you may presume that the reaction has not occurred as desired.

One final note: Be very careful when you smell anything. Refer to Sec. 22.2B for a safe sniffing procedure.

3.3 THE FLAME TEST AND BEILSTEIN TEST

The Flame Test

One of the most important preliminary determinations that one can make on a compound is whether or not it burns. Organic substances, almost without exception, burn. Inorganic substances, almost without exception, do not burn. It is very common to obtain a salt as a by-product in a reaction in which one reactant is sodium or potassium hydroxide (see Secs. 13.3 and 13.4). In many cases, one has the chance of isolating sodium or potassium chloride from the reaction mixture. Sodium chloride is usually easy to identify as it comes out of solution in large cubic crystals. Potassium chloride, however, is more tricky because its crystals are less characteristic. In both cases these solids may be isolated, but since they do not have melting points that can be easily determined and they will not burn, they may usually be discarded.

PROCEDURE

FLAME TEST

A very small amount of a substance is placed on a porcelain or stainless steel spatula and held directly in a bunsen-burner flame. If the material is an organic substance, it will burn. If the material is a salt arising as a by-product of a reaction, it will not burn and can safely be discarded.

If the material burns, the question becomes: Can any information be gleaned from the observation of the flame? The answer to this is a definite yes.

If the material burns cleanly with a smokeless blue flame, the carbon/hydrogen ratio can usually be assumed to be low, and there is probably oxygen present helping to feed the combustion of the substance. The lower the carbon-hydrogen ratio, the hotter and bluer will be the flame. If the carbon-hydrogen ratio is close to 1, as it is for benzene, the compound will burn with a sooty flame indicative of incomplete combustion. Unsaturated aromatic compounds generally burn with a smoky or sooty yellow flame. In addition, ash or residue may remain. The presence of ash is also characteristic of incomplete combustion or of an inorganic element.

$CH_3CH_2CH_2CH_2CH_2CH_2CH_3$

Heptane C/H = 0.44
burns with clean blue flame

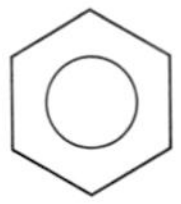

Benzene C/H = 1.00
burns with sooty yellow flame

Burning a small sample by the procedure described above will usually take no more than 30 to 60 s. The observation of the flame will also take no longer than 30 s. In this minute or so, one can often determine whether the compound is an unwanted by-product (such as salt), whether or not the compound has the appropriate burning characteristics for the desired product, or whether the experiment may need to be repeated.

The Beilstein Test

Because of the interaction of copper with organic substances during combustion, one can determine whether or not common halogens (Cl, Br, or I) are present in a compound. This fact is determined by a modification of the flame test above in which a copper wire is substituted for the porcelain or metal spatula above. This modified flame test is named the *Beilstein test* after the German chemist who developed it. In this test a small amount of a compound, either a liquid or solid, is placed on the copper wire loop and the loop is edged into a free flame. If the substance contains chlorine, bromine, or iodine, a bright, billowing green flame will be observed. The presence of a transient, or fugitive, green flame can usually be discounted. Comparison tests, of course, must always be made on known compounds.

PROCEDURE

BEILSTEIN'S FLAME TEST FOR HALOGENS

A 20-cm length of copper wire is bent into a 5-mm diameter loop at one end, and the other end is looped tightly about a cork (see Fig. 3.1). Holding the

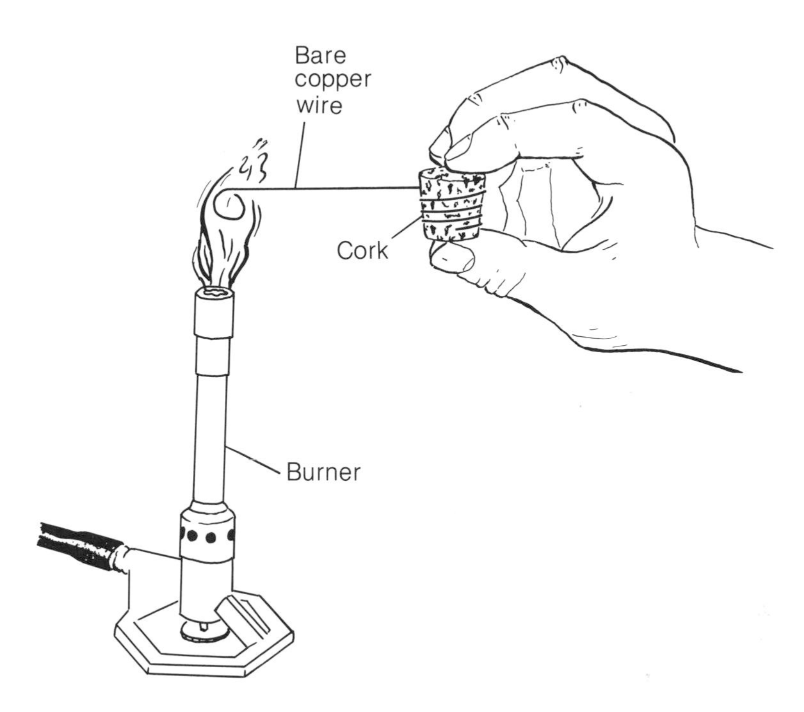

Figure 3.1 Performing a Beilstein test.

cork, place the small loop in a bunsen burner flame and heat to glowing. A faint green coloration may initially appear during this process but should quickly disappear. The copper wire, which has been burned free of contamination, is removed from the flame and allowed to cool to room temperature. The cooled copper wire is then dipped into the sample so that a small amount of the substance is deposited on the open loop.

The sample-bearing end of the copper wire is then edged into the flame. The color of the flame is observed. If a vivid green coloration lasting several seconds is observed, the sample probably contains halogen. If normal combustion is observed and there is either no or only a very transient green flame, it is safe to assume that the compound is halogen-free. Always conduct this test twice and compare the results with those obtained on a known halogen-containing compound.

Note This is a very sensitive test for the presence of halogen. A very small amount of halogen-containing compound will give a positive test. Once the copper wire has been flame-cleaned, it should not be touched, as there is usually enough salt on fingers to give a faintly positive test. The wire should be flame-cleaned immediately before each test to ensure that the wire is free of contamination.

The Beilstein test is a very reliable one. Relatively few compounds present difficulty in interpretation. These are often the dicarboxylic acids, and of these malonic and succinic acids seem to be the worst offenders. There are confirmatory methods, however, which will permit testing for the presence of halogens and, if a neutralization equivalent indicates that the compound is a monoacid, the likelihood that a long-lasting green flame could have resulted from other than the Beilstein-type reaction is very small. Note that the Beilstein test will indicate the presence of halogen, but not *which* halogen is present. It is rare for two compounds to have the same functional group, the same melting or boiling point, and the same derivative but to have different halogen substituents. Other factors will indicate which halogen must be present in the compound, so that, although the precise identification of the halogen will not be known directly from the observation of the green flame, its identity can usually be inferred with confidence.

3.4 ELEMENTAL ANALYSIS

Determining what elements are present in a sample is of crucial importance in analysis. The means by which elements other than hydrogen and carbon are determined is different in each laboratory. Some instructors prefer to inform students if a particular element is present and others prefer to have the student determine this experimentally.

The method which has been most widely used for determining the presence of nitrogen, sulfur, and halogen is the so-called sodium fusion test. This test involves heating a sample with a small piece of sodium metal and then quenching the hot mixture in cold water. There is a certain danger associated with this test, and reliable results are obtained only if the test is carefully and skillfully done. Since this test is rather difficult, we recommend another approach which has recently become available.[1] This test is often referred to as the sodium alloy fusion test.

The sodium alloy fusion test has three advantages over the classical sodium fusion test. The first of these is that sodium metal is not required, so the associated danger is eliminated. The second advantage is that decomposition of the reaction mixture occurs to a moderate extent compared with the classical procedure. Finally, the nitrogen analysis carried out after completion of the alloy modification of the test is usually more reliable than has traditionally been the case.

The sodium fusion or sodium alloy fusion test does not provide direct analysis for the presence of nitrogen, sulfur, or halide. The sodium fusion test is a reductive test; halogen is converted to halide, nitrogen is converted to cyanide, and so on. The presence of these elements is eventually established by detecting their reduced forms.

[1] Vinson, J. A., and W. T. Grabowski, *Journal of Chemical Education*, **54:** 187 (1977).

PROCEDURE

SODIUM ALLOY FUSION TEST

Sodium alloy fusion

Place 0.5 g sodium-lead alloy (Dri-Na, from J. T. Baker Chemical Co.) in a 10 × 75 mm test tube. Clamp the tube in a vertical position and heat with a small flame until the alloy melts and fumes of sodium are seen moving up the walls of the tube. Add two drops of a liquid sample or 15 mg of a solid sample to the test tube, taking care not to get any of the organic material on the side of the tube. Continue heating until the reaction is initiated, remove the flame until the reaction subsides, then heat again to redness for 1 min and allow the test tube to cool. Add 3 mL *distilled* water to the tube after it has cooled and allow the excess sodium to react. Filter the solution if necessary. Dilute the solution with about 2 mL distilled water and proceed with the individual elemental tests.

Test for nitrogen

Place 10 to 15 drops of the aqueous fusion solution obtained above in a small test tube and saturate it with solid sodium bicarbonate (excess solid sodium bicarbonate does not interfere with this reaction). Shake to ensure saturation. Transfer 3 to 5 drops of this solution to another small test tube. Add 10 drops (0.5 mL) of a standard phase-transfer catalyst solution (Sec. 4.7) to the tube, followed by 10 drops (0.5 mL) of a toluene solution containing 1% 4-nitrobenzaldehyde. If cyanide is present, the toluene layer will turn a red-purple color. If cyanide is absent, no color will be observed.

Test for sulfur

Place 10 drops of the aqueous fusion solution in a small test tube and acidify with acetic acid. Add 2 or 3 drops of a 1% aqueous lead acetate solution to the test tube. If sulfur is present, a jet black precipitate will be observed.

Test for halogens

Place 10 drops of aqueous fusion solution in a small test tube and acidify with 5% nitric acid. Boil the solution gently to remove any hydrogen cyanide and hydrogen sulfide. Cool the solution. Add 2 drops of a 2% aqueous silver nitrate solution to the tube. A white-to-yellow precipitate indicates the presence of halogen.

If a precipitate is obtained, the color should be noted. Silver chloride is white, silver bromide is yellowish white, and silver iodide is yellow. Silver chloride will dissolve in concentrated ammonia solution, while the other two precipitates will not.

SOLUBILITY AND REACTIVITY

4.1 INTRODUCTION

There are many requirements for any given chemical reaction to be successful. These may be thermodynamic or kinetic, or they may have their origins in other principles. The most fundamental of these principles, and one which must be abided by in any chemical reaction between two distinct species, is the necessity for collision. In other words, unless two species come into contact, they cannot react with each other. In order for the species to come into contact, they must be in the same phase, and therefore they must be soluble in the same medium. If both substances are vapors, they may obviously react in the gas phase, but most preparative laboratory work is conducted in the liquid phase between substances both of which dissolve in some solvent. The fundamental role of a solvent becomes clear from this requirement. If both substances must be in the same solution in order for a reaction to occur, the solvent is as important as the principle of collision.

4.2 SOLVATION

Let us consider for a moment the question: What exactly is a solvent? This question can be answered in more or less operational terms by saying that a solvent is something which breaks down a lattice, i.e., which can compete with the intermolecular forces which hold molecules together as a solid, or that it is a substance from which some other substance will not precipitate. In both cases we are talking about intermolecular forces between solvent and solute which are stronger than the forces between molecules of the solute itself. If the forces which bind the solute into a crystal lattice exceed those which keep it in solution, it will precipitate; if they are weaker, the crystalline material will go into (or remain in) solution. The same is true of an oil which is insoluble in another medium. In this case, the forces among the various molecules of that oil exceed the forces exerted by the solvent on the oil molecules. In order to understand a solvent in more than operational terms, however, the various kinds of forces which act upon substrates and the various kinds of forces which bind molecules together must be considered.

Most students of chemistry are quite familiar with the distinction between ionic and nonionic substances and ionic and nonionic processes. When dealing with solvents and the forces which exist between solvents and solutes, we deal with two general classes of molecules, those which are *hydrophobic* and those which are *hydrophilic*. The nomenclature used here is analogous to that used in the designations *nucleophilic* and *electrophilic: hydro* means "water," *philic* means "loving," and *phobic* means "fearing." (The Greek word which is the root of *phobic* is the same root from which our English word *phobia* arises.) Any substance having an affinity for water is said to be hydrophilic, and a substance not having such an affinity is called hydrophobic.

There is an alternative approach in describing the solvating ability of various compounds. Instead of being defined in terms of their affinity for or aversion to water, they may be defined in terms of their affinity for fatty or organic-type materials. A general term for neutral fatty substances is "lipids." Therefore, those substances which have an affinity for organic media may be distinguished as *lipophiles* (they are said to be *lipophilic*); those substances which have an aversion to lipophilic media are *lipophobic*. Lipophobic substances obviously may be hydrophilic. This may seem redundant, but there are situations in which one wishes to speak more specifically in terms of hydrocarbon-hydrocarbon interactions rather than in terms of water, which may not be present in the system at all.

Because solvation forces are usually analyzed in terms of either hydrophobic or hydrophilic solvation, the general principle has been expressed: *Similis, simillimus, solutum*. Roughly translated, this means "like dissolves like." In fact, it should be quite clear that a substance which is similar in structure to water should dissolve in water. This is because those forces which hold together molecules of water as a fluid or as ice can also bind some other molecule with similar properties. But what, in general, are these properties?

We will deal here with the following general solvation forces: (1) hydrogen bonding, (2) London forces, and (3) Lewis-base electron-donor forces. Each of these forces is potent indeed and may have an enormous influence on the dissolution of a substance. In concert, these forces serve to define most classes of organic and inorganic solvents.

4.3 HYDROGEN BONDING

As should be clear from the name, the necessary ingredient for a hydrogen bond is a hydrogen atom. To clarify further, a hydrogen bond involves a hydrogen atom shared by two electronegative elements.

Consider the simple case of water, in which a hydrogen atom on an oxygen atom may form a hydrogen bond. Oxygen has an electronegativity of 3.5 and hydrogen an electronegativity of 2.1. This large difference in electronegativity serves to polarize the bond, i.e., the electron density, in the bond will be higher close to the oxygen atom, which is the more electronegative element.

$$\overset{\delta^-}{—\mathrm{O}}—\overset{\delta^+}{\mathrm{H}}$$
$$\longleftarrow\!\!\!+$$

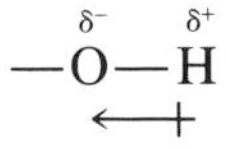

An alternative expression of the same principle is to say that the O—H bond is dipolar. An arrow with a plus sign at one end indicates the direction of the dipole. The hydrogen, as a consequence, possesses a partial positive charge and will coordinate with an atom of a nearby electronegative element. In the case of water, this element is the oxygen atom of another water molecule. Such interactions may be referred to as dipole-dipole interactions. In general, the hydrogen atom will coordinate with the oxygen atom along the axis of the electron pair, thereby regaining some of the electron density lost in the polarization of the single bond toward its parent oxygen atom. Many such hydrogen bonds exist in aqueous solution, and it is this multiplicity of bonds which holds the very low-molecular-weight water molecule (18 amu) in the liquid state; otherwise, water would exist as a vapor at room temperature. Thus it is easy to see why methane (16 amu), for example, is a gas at room temperature, while water is a liquid. Because the electronegativity difference between hydrogen and carbon is relatively small, the intermolecular association by hydrogen bonding is diminished by a very large amount. Water and methane have almost the same molecular weight but their boiling points differ by well over 100°C. This difference in boiling points largely reflects the energy required to break the intermolecular hydrogen bonds or to overcome the dipole-dipole interactions.

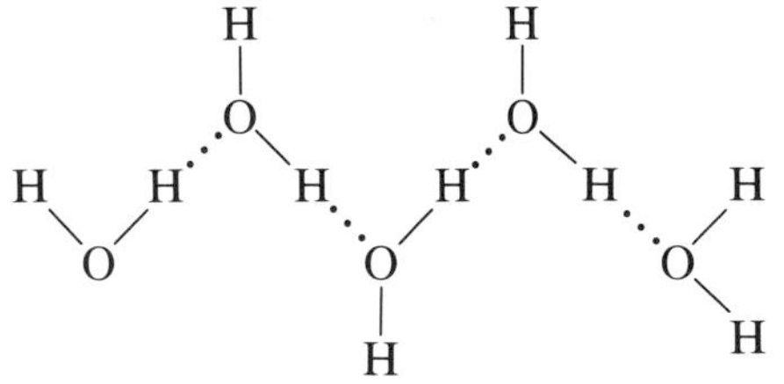

Association of water due to hydrogen bonding

Other organic molecules can form hydrogen bonds as well. Alcohols like methanol can do so but, because only one hydrogen atom is available rather than the two present in water, both the total number of hydrogen bonds and the total force holding the molecules together as a liquid are correspondingly reduced.

A very interesting special case of hydrogen bonding occurs with carboxylic acids. Because a carboxylic acid contains a hydroxyl group, each carboxyl group may function both as a hydrogen bond donor and as an acceptor. When two molecules of a carboxylic acid come together, carboxyl to carboxyl, and are oriented head-to-head with respect to each other, there is the prospect of very good alignment between the hydrogens of the hydroxyl groups and the carboxyls of the two carboxylic groups. This forms essentially an eight-membered ring where the seventh and eighth atoms are actually interposed between two of the oxygens. As it is difficult to discern which of the oxygen atoms actually has the hydrogen atom bonded to it, a symmetrical dimeric association is observed in carboxylic acids. For this reason, the acids are usually higher boiling than the corresponding alcohols, which are capable of forming single hydrogen bonds. In addition, both carboxylic acids and alcohols are higher boiling than the corresponding aldehydes, which do not have a hydrogen atom capable of bonding to an electronegative element.

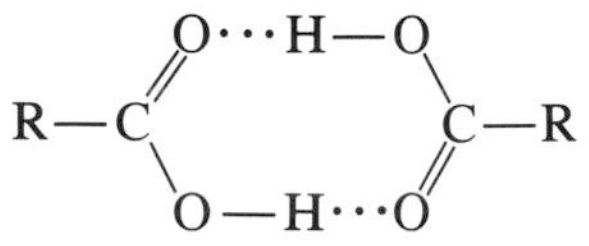

Carboxylic acid dimers

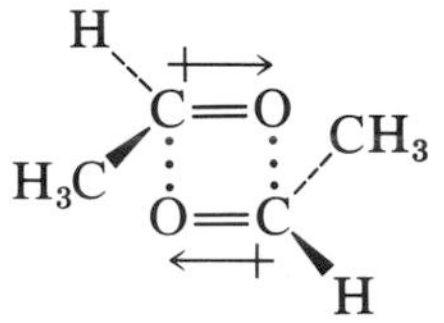

Aldehyde dimers

So far we have considered forces which hold molecules together as liquids. Similar forces keep solutes dissolved in liquid solvent systems. But how does a liquid like water dissolve a solid like sodium chloride? What forces are operating in this case? Chloride ion, Cl^-, is an electron-rich, electronegative element. It should be clear that water can form a hydrogen bond with the chloride ion, just as it can form a hydrogen bond with another molecule of water. Because chloride ion can accept more than one hydrogen bond, a number of water molecules will orient themselves with hydrogen bonds toward the electronegative chloride ion, which they thus solvate. Such interactions are often referred to as ion-dipole interactions. There are a number of water molecules surrounding the chloride ion which serve as a transition into the medium, which in this case is more water.

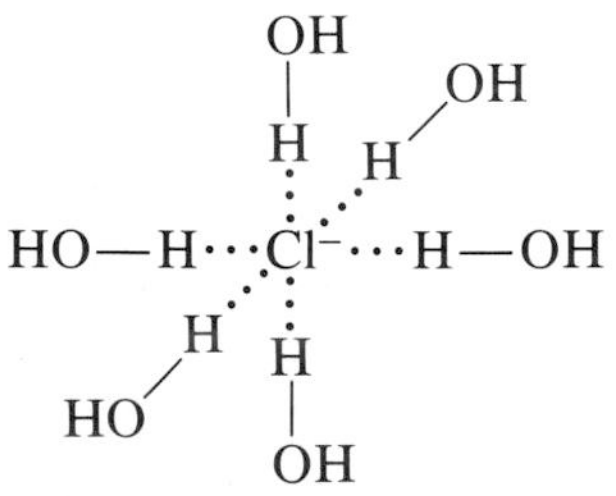

Aqueous solvation of chloride ion

This group of water molecules which directly solvates the chloride ion is called the *solvation shell*. It is fairly easy to see why not only chloride but also iodide, bromide, fluoride, and many other electronegative substances are readily soluble in water because of hydrogen bonding. We have ignored, however, the forces which tend to hold the sodium cation in solution even though it is positively charged and much less electronegative. We shall see later that it is possible to dissolve an entire ion pair by solvating only half of it very well. Water is a special solvent, because it can not only solvate negative ions by hydrogen bonding but also has the ability to solvate cationic species. It is this property which is dealt with in the next section.

4.4 LEWIS BASE PROPERTIES

Recall Gilbert Lewis' definition of a base as an electron-pair donor. From the discussion about sodium chloride it becomes clear that sodium ion is a Lewis acid. This also makes it clear why water can be used to solvate such a variety of substances. Not only does water have two hydrogen atoms bonded to oxygen which are suitable for hydrogen-bond formation but it also has two pairs of electrons on oxygen which are basic in the Lewis sense and can be used as donors to solvate such Lewis acid species as sodium cations. The lone pairs of electrons required to solvate a cation can come from almost any electronegative atom as, for example, from nitrogen in amines or from oxygen in water, alcohols, ethers, and so on. However, if we consider the sodium salts of organic species, we realize that a salt like sodium diethylmalonate is nothing more than a sodium cation which is intimately solvated by a negatively charged organic species. In a sense, then, the anionic portion of the molecule is solvating the sodium cation. This is an ion-ion interaction (an ionic bond). What makes water the remarkable solvent that it is (and it is often called the universal solvent) is the fact that it has both lone pairs of electrons available as Lewis-base donors (negative dipoles) and hydrogen atoms (positive dipoles) available to form hydrogen bonds with Lewis bases. It should be easy to see how water could very readily dissolve a substance like sodium hydroxide. Water can form a hydrogen bond with the oxygen anion of hydroxide (positive dipole-anion interaction) or accept a hydrogen atom from the hydroxyl anion as a Lewis base; water can donate electron pairs to the sodium cation (negative dipole-cation interaction). As a consequence, both species could be well solvated, tending to dissolve readily and remain in solution.

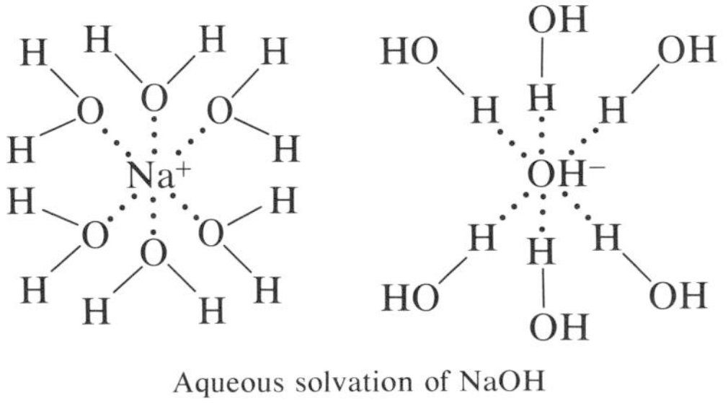

Aqueous solvation of NaOH
by ion-dipole interactions

This example of the dissolution of sodium hydroxide in water is an extreme case in the sense that sodium hydroxide and water complement each other to a remarkable extent. As implied above, effective solvation of half an ion pair can often induce the entire ion pair to dissolve. This is so because electrical neutrality must be maintained. The molecule or ion pair potassium hydroxide (KOH) is readily soluble in acetonitrile (CH_3CN). Acetonitrile has a lone electron pair on its nitrogen atom. Six or more acetonitrile molecules can donate their electrons to a potassium ion to form a solvation shell about the latter. Because the potassium ion is well solvated by acetonitrile, hydroxide anion is readily drawn into solution and, because it is not itself encumbered by a strong solvation shell, the anion is quite reactive. This principle will be dealt with in considerably more detail in the last section of this chapter. For now, consider the third kind of solvation, the interactions of hydrocarbons with each other.

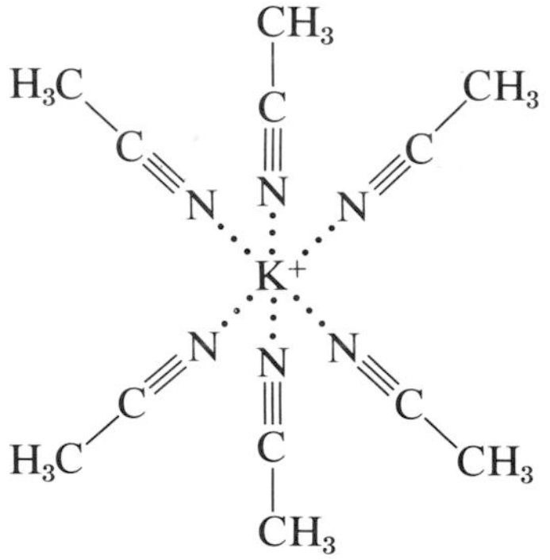

Acetonitrile solvation of potassium ion

4.5 LONDON FORCES

One of the principles discussed in every introductory chemistry class is that there is electronic motion about the nucleus of every atom. Although this motion is centrosymmetric on a time average, it is not momentarily so, i.e., the total electronic circulation does not form a perfect sphere at all times. As a consequence of this, small electric dipoles are set up in molecular systems. Not only are there small electric dipoles associated with each atom in a molecular system, but the electronic motions may couple to give somewhat larger net dipoles. Recall that if all the small dipoles were averaged, the dipole of the entire molecule would be zero, but, because the electronic motions can couple, the net dipole does not at all times average to zero, although in alkanes, for example, the net dipole is small. This coupling of electronic motions was first recognized by London in the 1930s, and as a consequence these dipole-dipole forces are called *London forces*.

Alkane molecules like pentane, hexane, and heptane do not exhibit the properties discussed in Secs. 4.3 and 4.4. Each of the valences of carbon is satisfied by hydrogen or by another carbon atom and, because hydrogen has an electronegativity of 2.1 and carbon an electronegativity of 2.5, there is no appreciable polarization of the σ bonds. Alkanes do not form hydrogen bonds nor do they have any free electron pairs to donate in the Lewis base sense. The solvation forces which alkanes exhibit and those which cause them to remain in solution or to crystallize as solids are largely London forces (the attractions of the small dipoles set up because of noncentrosymmetric electronic motion). Here, perhaps even more than in Secs. 4.3 and 4.4, it should be clear why like dissolves like. The alkanes have relatively limited solvation mechanisms and therefore solvate only organic or lipophilic substrates.

We have referred to small and perhaps transient dipoles which are responsible for the so-called London forces. Consider for a moment what a dipole actually is. From the name it is clear that a dipole must involve two different polarities or charges. Whenever there is an uneven distribution of charge, a dipole is said to exist. If we consider the polarization of the molecule methyl chloride along the axis of the carbon-chlorine bond, it is clear that electron density is polarized towards the more electronegative chlorine. We therefore say that a dipole exists in the molecule, with the positive end on carbon and the negative end on chlorine. This is often symbolized (see below) by drawing an arrow in the direction of the dipole.

$$\underset{\delta^+}{CH_3}-\underset{\delta^-}{Cl} \qquad \underset{+\!\!\longrightarrow}{CH_3-Cl} \qquad \underset{+\!\!\longrightarrow}{CH_3-OH} \qquad \underset{+\!\!\longrightarrow}{CH_3-CN}$$

Polarization in methane derivatives

The arrowhead is always at the negative end of the dipole and the tail end of the arrow, instead of showing feathers, is simply crossed to make a stylized plus charge, which serves as a mnemonic device. Another molecule with a dipole similar to that of methyl chloride is methyl alcohol. The positive end of the dipole is still the methyl group, but this time the negative end of the dipole is oxygen. There is also a dipole in acetonitrile. This case is strictly analogous to the two previous cases, because acetonitrile is methyl cyanide and is often so named. The methyl group is still the positive end of the dipole and the electronegative nitrogen is the negative end of the dipole. Just as the carbon-oxygen bond in methanol and the carbon-chlorine bond in methyl chloride are linear, the carbon-carbon-nitrogen axis in acetonitrile is colinear, i.e., all three atoms lie on a straight line. In acetonitrile it is therefore quite clear that the negative end of the dipole is in the direction of the nitrogen atom.

A more complicated situation arises when there is polarization in more than one direction, i.e., when there is not a single axis. For example, in the molecule dichloromethane it is clear that there is polarization from positive carbon to negative chlorine along two axes. In this particular case the dipole moment (μ) of the molecule will simply be the vector sum of the two dipoles corresponding to each carbon-chlorine bond. If we take this example a step further, we come to trichloromethane, which has the trivial name chloroform. In this case there are three carbon-chlorine bonds, all of which are polarized. The net polarization of the molecule will, therefore, be the vector sum of all three carbon-chlorine bonds. The final extension of this example is carbon tetrachloride, which is tetrahedral and has a very high degree of symmetry. Since all four carbon-chlorine bonds are opposed, i.e., they are directed to the corners of the tetrahedron, the vector sum for the four dipoles cancels and the net dipole moment of carbon tetrachloride is zero. Therefore, carbon tetrachloride, although it has the most carbon-chlorine bonds, is the least polar of the four chloromethanes.

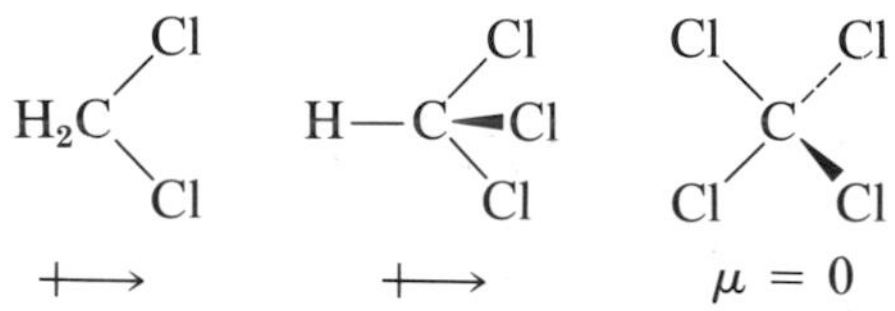

H_2CCl_2 $H{-}CCl_3$ CCl_4

$\longmapsto$ $\longmapsto$ $\mu = 0$

Polarization in chloromethanes

The simple principle of vector addition of dipoles can be extended quite easily for molecules which have different substituents. For example, in chlorobromomethane the dipole of the carbon-chlorine bond will be somewhat greater than will the dipole of the carbon-bromine bond. As a consequence, the net dipole moment of chlorobromomethane will be somewhat smaller than that of dichloromethane. Although there are many minor factors which complicate this simple addition process, it can be used as a rule of thumb for establishing at least the direction of the dipole in far more complicated systems.

Knowing that the dipole moment exists is of more than academic interest. Because the dipole moment of a molecule exists and may be appreciable, we may predict that each end of a bond between different elements will have a partial charge. The positive (less electronegative) end of the dipole will have an affinity for the negative (more electronegative) end of a similar molecule (dipole-dipole interaction) and vice versa. The negative end of the dipole will tend to associate with a positive ion, and the positive end will tend to associate with an anion. Such interactions are referred to as ion-dipole interactions. As a consequence, we can expect methyl chloride to pair itself with another molecule of methyl chloride. The carbon end of the carbon-chlorine bond will associate with the chlorine end of the carbon-chlorine bond of another molecule. The small intermolecular association which results from this accounts for the fact that methyl chloride boils about 20°C higher than propane although its molecular weight is only slightly higher.

Of traditional solvents, water has the highest dipole; typical alkanes like hexane or pentane have the lowest dipole. In this broad range one finds such nonpolar solvents as carbon tetrachloride, chloroform, dichloromethane, carbon disulfide, 1,2-dichloroethane, chlorobenzene, and similar chlorinated substances. In addition, the commonly used hydrocarbons pentane, hexane, heptane, benzene, toluene, xylene, and ethylbenzene all have relatively low dipole moments. Among those substances which have slightly greater dipole moments but which are still primarily nonpolar solvents are ethers, such as diethyl ether, tetrahydrofuran, and dioxane. In stark contrast to these nonpolar solvents are inorganic substances such as water, liquid ammonia, and sulfuric acid, which have high dipole moments and are fine solvents for such inorganic substrates as sodium chloride and sodium bromide. In order for a reaction to occur between an organic substrate and a salt, both substances must be in the same medium. Traditionally, such reactions have been run in boiling water for prolonged periods of time on the presumption that the organic substrate would show some marginal solubility in hot water and would gradually react. Alcoholic media have also been used. Such solvents as methanol and ethanol offer the hydrogen-bonding and electron-pair-donor capabilities of water but also have a lipophilic end which can assist in the solvation of organic substrates. Because of this combination of properties, such solvents as ethyl alcohol can be considered the best of all possible worlds. On the other hand, because ethanol has neither the polarity of water nor the lipophilicity of pentane, it can be considered the worst of all possible worlds. Whether or not a reaction occurs depends on the specific pair of reactants and how effectively they may be solvated by a particular medium.

4.6 DIPOLAR APROTIC SOLVENTS

Solvents like methanol and ethanol which contain dissociable protons are obviously protic solvents. If one wishes to dissolve a substance like sodium hydroxide, there is always the danger, not only of having hydroxide ions present in the solution, but of forming the conjugate base of the alcoholic solvent. Sodium hydroxide dissolved in methanol gives hydroxide ions, which can function both as bases and as nucleophiles, and gives a finite concentration of methoxide ion, which is also basic and nucleophilic. The difference in product that results when either methoxide or hydroxide is used as the base is frequently marginal, but if one is concerned with nucleophilic substitution, the difference is that the product in one case will be an alcohol and in the other case will be a methyl ether.

In recent years this particular difficulty has been overcome by the use of so-called dipolar aprotic solvents. These are solvents which have appreciable polarity and can therefore afford some solvation for polar substances like salts, but because they are aprotic (i.e., do not have a readily dissociable proton) they do not participate in many of the reactions characteristic of methanol, ethanol, or other alcoholic solvents. We have previously called attention to acetonitrile and the means by which acetonitrile may solvate a sodium or potassium cation. It was also mentioned that acetonitrile has an appreciable dipole moment. This substance is one of those known as dipolar aprotic solvents, and its presence in this group is defined by the properties just mentioned. It can rather effectively solvate a cation, which causes an ion pair to be brought into solution. The primary solvation forces exist between acetonitrile and the cation. The anion is relatively free of solvation and, because it is not surrounded by many solvent molecules, is in a reasonably active state. Such molecules as pyridine, dimethyl sulfoxide, acetone, dimethylformamide, and nitromethane all have electron pairs available and are good cation solvators.

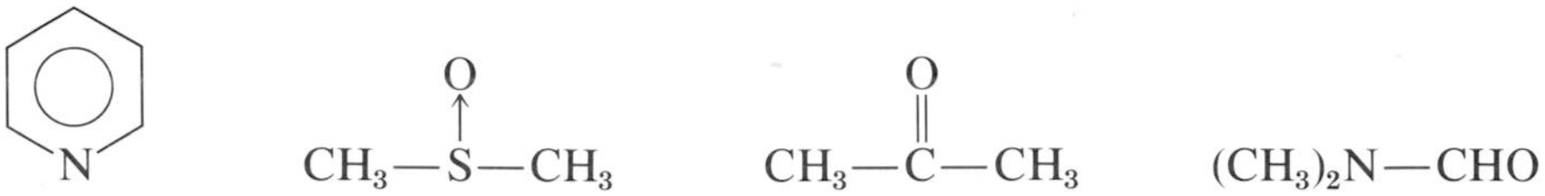

Pyridine (N in six-membered aromatic ring)

$CH_3-\overset{\overset{O}{\uparrow}}{S}-CH_3$
Dimethyl sulfoxide (DMSO)

$CH_3-\overset{\overset{O}{\|}}{C}-CH_3$
Acetone

$(CH_3)_2N-CHO$
Dimethylformamide (DMF)

Traditionally, alcohols have been used in those cases where solubilization of both organic and inorganic substrates is required. Among the several reasons for this are the ready availability of alcoholic solvents and the common knowledge of them, but perhaps the most important reason is that such solvents as dimethyl sulfoxide and dimethylformamide were very difficult to obtain 20 or 30 years ago. Their more recent popularity can be attributed to the advent of chemical supply houses. Until recently, many solvents simply were not readily available to the average practicing chemist. It was much easier to use alcohol, which was readily available, even if it somewhat complicated the reaction, if the alternative was for the chemist to synthesize the solvent. If dimethyl sulfoxide is needed but there is no prospect of obtaining it commercially, the prospect of oxidizing dimethyl sulfide is a gruesome one indeed. Not only is dimethyl sulfide difficult to handle, but it has a very powerful odor of garlic, which many people find unpleasant in high concentrations. The odor is only an advantage if you need to keep vampires away.

The properties of a wide variety of solvents are presented in Table 4.1. Included in it are boiling points, melting points, refractive indexes, and densities, knowledge of which is obviously required for choosing the appropriate solvent for any given reaction. Also included are the dipole moment, which is a good measure of the polarity of the medium, and the dielectric constant, which generally reflects the dipole moment but is much easier to measure. There are many ways of grouping solvents. In this table, they are grouped according to polarity type and listed in ascending order of boiling points. The first series of compounds in the table includes pentane, hexane, heptane, benzene, toluene, ethylbenzene, and xylene, all of which are nonpolar. The polar substances, which are later entries in the table, generally have higher dipole moments, higher boiling points, and higher dielectric constants. These substances make it possible to conduct polar reactions on organic substrates in homogeneous solutions. Because all the reactants are in solution at the same time, the reaction can occur much more readily than when there is only marginal solubility of one component. Furthermore, because the dipolar aprotics generally solvate only half of an ion pair, the other half will exhibit a somewhat greater reactivity than it would were it encumbered by the kind of solvation common to water.

4.7 PHASE-TRANSFER PROCESSES: THE "STANDARD CATALYST" SOLUTION

The obvious method for preparing an octyl cyanide would be to treat *n*-octyl bromide with sodium cyanide. If solid sodium cyanide is added to bromooctane, which is then heated for about 2 weeks, an attempt to isolate octyl cyanide leads to the discovery that virtually no reaction has taken place. It is relatively easy to account for this lack of reactivity: Sodium cyanide and octyl bromide simply do not get into the same phase. Whereas octyl bromide is very hydrocarbon-like or lipophilic, sodium cyanide is lipophobic because it is a salt. Several alternatives are available to resolve this difficulty. If ethyl alcohol is used, both sodium cyanide and octyl bromide enjoy some limited solubility. The reaction will be slow but it should proceed. However, some *n*-octyl ethyl ether may also be obtained because the solvent acts as a nucleophile.

$$C_8H_{17}Br + NaCN \xrightarrow{C_2H_5OH} C_8H_{17}CN + C_8H_{17}OC_2H_5$$

The same potential difficulty would exist if the reaction were attempted in boiling water, but in that case even less reaction would occur because of the extremely limited solubility of octyl bromide in water.

The use of dipolar aprotic solvents has greatly enhanced the organic chemist's ability to conduct reactions which ordinarily would not be easy either in classical organic or classical inorganic media. In this regard organic chemists have a great advantage over nature. Nature, of course, must conduct its reactions in aqueous solution and within a temperature range limited to between 0 and 100°C. In almost all living systems, the temperature range is even more restricted, about 37°C. In the laboratory, a dipolar aprotic solvent like dimethyl sulfoxide can be heated to 150°C. This heat, combined with solvation, will promote a reaction. Use of a pressure device would be necessary to accomplish the reaction in water at this temperature.

The reaction of sodium cyanide with octyl bromide in dimethyl sulfoxide is a reasonably effective one, allowing for the formation of octyl cyanide as desired. The difficulty is that dimethyl sulfoxide is both relatively expensive and

TABLE 4.1
Properties of common solvents

Name	Melting point, °C	Boiling point, °C	Refractive index	Density	Solubility in water*	Dipole moment, μ	Dielectric constant, ϵ	Color	Remarks
				Hydrocarbons					
Pentane	−130	36	1.3580	0.626	0.036	0	1.84	None	**Flammable**
Hexane	−100	69	1.3748	0.659	insol	0	1.89	None	**Flammable** (forms 6% aq azeo)
Cyclohexane	6.5	81	1.4255	0.779	insol	0	2.02	None	**Flammable**
Heptane	−91	98	1.3870	0.684	insol	0	1.98	None	**Flammable** (forms 12% aq azeo)
Methylcyclohexane	−126	101	1.4222	0.77	insol	0	2.02	None	**Flammable**
Benzene **(Caution: toxic)**	5.5	80	1.5007	0.879	0.5	0	2.28	None	**Flammable** (forms 9% aq azeo)
Toluene	−93	111	1.4963	0.865	v.sl.	0.4	2.38	None	**Flammable** (forms 20% aq azeo)
Ethylbenzene	−95	136	1.4952	0.867	insol	0.6	2.41	None	**Flammable** (forms 33% aq azeo)
p-Xylene	12	138	1.4954	0.866	insol	0	2.27	None	**Flammable** (forms 40% aq azeo)
o-Xylene	−24	144	1.5048	0.897	insol	0.6	2.57	None	**Flammable** (forms 40% aq azeo)
				Ethers					
Diethyl ether	−116	35	1.3506	0.715	7	1.2	4.34	None	**Flammable,** peroxides
Tetrahydrofuran (THF)	−108	66	1.4070	0.887	misc	1.6	7.32	None	**Flammable,** peroxides
Dimethoxyethane (glyme, DME)	−69	85	1.3790	0.867	misc	—	—	None	**Flammable,** peroxides
Dioxane	12	101	1.4206	1.034	misc	0	2.21	None	**Flammable,** peroxides
Dibutyl ether	−98	142	1.3988	0.764	insol	—	—	None	**Flammable,** peroxides
Anisole	−31	154	1.5160	0.995	insol	1.4	4.33	None	
Diglyme	−64	162	1.4073	0.937	misc.	—	—	None	
				Chlorohydrocarbons					
Dichloromethane	−97	40	1.4240	1.325	2	1.6	8.9	None	
Chloroform **(caution: toxic)**	−63	61	1.4453	1.492	0.5	1.9	4.7	None	Commercial material contains ethanol (forms 3% aq azeo)
Carbon tetrachloride **(caution: toxic)**	−23	77	1.4595	1.594	0.025	0	2.23	None	(Forms 4% aq azeo)
1,2-Dichloroethane	−35	83	1.4438	1.256	0.9	2.1	10	None	
Chlorobenzene	−46	132	1.5236	1.106	insol	1.7	5.62	None	(Forms 20% aq azeo)
1,2-Dichlorobenzene	−17	178	1.5504	1.305	insol	2.5	9.93	None	

TABLE 4.1 *(Continued)*

Name	Melting point, °C	Boiling point, °C	Refractive index	Density	Solubility in water*	Dipole moment, μ	Dielectric constant, ϵ	Color	Remarks
				Alcohols					
Methanol (wood alcohol)	−98	65	1.3280	0.791	misc	1.7	32.6	None	**Poison: Do not drink**
Ethanol (95% aq azeo)	—	78.2	—	0.816	misc	—	—	None	
Ethanol (anhydrous)	−130	78.5	1.3611	0.798	misc	1.7	24.3	None	Called *alcohol*
2-Propanol (iso)	−90	82	1.3770	0.785	misc	1.7	18.3	None	(Forms 12% aq azeo)
tert-Butanol	25	83	1.3860	0.786	misc	1.7	10.9	None	
n-Propanol	−127	97	1.3840	0.804	misc	1.7	20.1	None	
n-Butanol	−90	118	1.3985	0.810	9.1	1.7	17.1	None	
2-Methoxyethanol	−85	124	1.4020	0.965	misc	2.2	16.0	None	**Poison**
2-Ethoxyethanol	−90	135	1.4068	0.930	misc	2.1	—	None	
Ethylene glycol	−13	198	1.4310	1.113	misc	2.3	37.7	None	**Poison**
				Dipolar aprotics					
Acetone	−94	56	1.3585	0.791	misc	2.9	20.7	None	**Flammable**
Acetonitrile	−48	81	1.3440	0.786	misc	3.94	36.2	None	**Toxic, flammable** (forms 16% aq azeo)
Nitromethane	−29	101	1.3820	1.137	9.1	3.46	38.6	None	Aq solution acidic
Dimethylformamide (DMF)	−61	153	1.4305	0.944	misc	3.7	36.7	None	
Dimethyl sulfoxide (DMSO)	18	189	1.4780	1.101	misc	3.96	47	None	
N,N-Dimethylacetamide	−20	165	1.4375	0.937	misc	3.8	37.8	None	
Formamide	2	210	1.4440	1.134	misc	3.7	110	None	(ϵ = 87 also reported)
Hexamethylphosphoramide (HMPA, HMPT)	7	230	1.4579	1.030	misc	—	—	None	**Toxic**
Tetramethylene sulfone	27	285	1.4840	1.261	misc	4.7	44	None	
				Miscellaneous					
Carbon disulfide	−112	46	1.6270	1.266	0.3	0	2.64	None	**Toxic, flammable** (forms 2% aq azeo)
Ethyl acetate	−84	76	1.3720	0.902	10	1.8	6	None	**Flammable**
Methyl ethyl ketone (MEK)	−86	80	1.3780	0.805	27.5	2.5	18.5	None	IUPAC: 2-butanone
Water	0.00	100	1.3330	1.000	—	1.8	81.5	None	**Caution: Do not drown**
Formic acid	8.5	101	1.3721	1.220	misc	1.41	58	None	**Irritant**
Pyridine	−42	115	1.5090	0.978	misc	2.19	12.3	None	**Depressant?** (forms 43% aq azeo); stench
Acetic acid	16	117	1.3720	1.049	misc	1.7	6.2	None	
Nitrobenzene	5	210	1.5513	1.204	0.2	4.01	35	Light yellow	**Toxic**

* In grams per 100 milliliters unless otherwise specified.
Other abbreviations: insol, insoluble; v.sl., very slightly; misc, miscible in all proportions; aq, aqueous; azeo, azeotrope

difficult to remove when the reaction is completed. In this particular case, the high boiling point of dimethyl sulfoxide, used to advantage in accelerating the rate, makes it difficult to remove the solvent or to separate it from the other substances present.

If the problem is considered in its most basic terms, it is simply that the two reactants are not in the same phase at the same time in the absence of some dipolar aprotic solvent. In order for a substance to go into solution, it must be transferred from one phase to the other; in other words, some process allowing *phase transfer* must be utilized. It is most advantageous to accomplish this catalytically, because in this way only a very small amount of reagent is required to effect reaction. The overall process is therefore called *phase-transfer catalysis*.

The idea of using a catalyst to assist the dissolution of an otherwise insoluble substance dates back many years. It has been only in the last decade, however, that the process has been recognized, formulated in some detail, dealt with theoretically, and exemplified. Early work in this field was conducted by a broad, international group of scientists in Australia, Poland, Sweden, and the United States. The process involved can be explained quite readily by a consideration of the sodium cyanide–octyl bromide example. If an aqueous reservoir of sodium cyanide is put in contact with a hexane solution of octyl bromide, each of the individual components will be soluble in its respective dissolving medium, but the two phases, although in contact, will not be mutually soluble. Because there is no mutual solubility, no reaction will occur unless some catalyst is present to help get the bromooctane into water or, more conveniently, to get the cyanide ion into hexane. A cationic detergent, i.e., a substance which has a positively charged head and a lipophilic tail, is ideal for this purpose. The most common cationic detergents are quaternary ammonium compounds, commonly called *quats* and abbreviated Q^+.

Addition of a small amount of a quaternary ammonium chloride to the two-phase mixture leads to the following sequence of events: The Q^+Cl^- seeks the phase boundary. Some of it, of course, is dissolved in water because it has a charge, and some is dissolved in the organic medium because it has a lipophilic end. The anion is exchanged for cyanide because there is far more cyanide present than there is chloride. The new ion pair Q^+CN^- is also soluble in both organic and inorganic media. In the organic solution, cyanide displaces bromide to form octyl cyanide. The process of crossing the phase boundary is the phase-transfer process. The cyanide ion is quite reactive in hexane because it is not well solvated by hexane. The products of this transformation are octyl cyanide and Q^+Br^-. The new salt is soluble in both the aqueous and the organic phases. In the aqueous medium, bromide exchanges for cyanide because CN^- is present in excess. This process is repeated until all octyl bromide is converted to octyl cyanide.

$$
\begin{array}{lcl}
\text{QCN} + \text{R—Br} \longrightarrow \text{R—CN} + \text{QBr} & & \text{Organic phase} \\
\downarrow\uparrow \qquad\qquad\qquad\qquad \downarrow\uparrow & & \text{Phase boundary (interface)} \\
\text{QCN} + \text{NaBr} \rightleftharpoons \text{NaCN} + \text{QBr} & & \text{Aqueous phase}
\end{array}
$$

The phase-transfer process

This reaction is enormously faster in the presence of the quaternary ammonium catalyst than in its absence. This rate acceleration is a combination of two factors: (1) the nucleophile is in solution in a reasonable concentration and can react; and (2) because it is poorly solvated, it is relatively more reactive than it would be in aqueous solution.

The quaternary ammonium compound which has a methyl group and three octyl groups bonded to nitrogen is a common catalyst and is the one employed in this textbook. Its solution in toluene (100 mg catalyst per milliliter of solution) is the standard catalyst solution used throughout this book. Addition of a small amount of this solution to a two-phase medium in the aldehyde test and in various other reactions facilitates a reaction where no reaction was observed in its absence.

Another method used for anion activation requires quite a different kind of cation solvator. In the case of quaternary ammonium compounds, solubility is achieved simply by exchanging the cation; in other words, no sodium ions are transferred into the organic phase and the only cations present are quaternary ammonium cations. An alternative is to complex the sodium ion, i.e., to surround it by a lipophilic skin so that the new complex is soluble in hydrocarbon-like media. The anion, which is dragged into solution to maintain electrical neutrality, can now function as a nucleophile, as it did in the previous phase-transfer process.

Macrocyclic polyethers are a group of compounds which have been eminently successful in phase-transfer applications of this sort. The most commonly used examples of the macrocyclic polyethers are those illustrated below, which are called 18-crown-6 and dicyclohexano-18-crown-6, respectively. Their equilibria and complexation are also illustrated in the equations.

$$\text{18-Crown-6} + \text{KBr} \rightleftharpoons [\text{18-Crown-6}\cdot\text{K}]^{+}\ \text{Br}^{-}$$

18-Crown-6

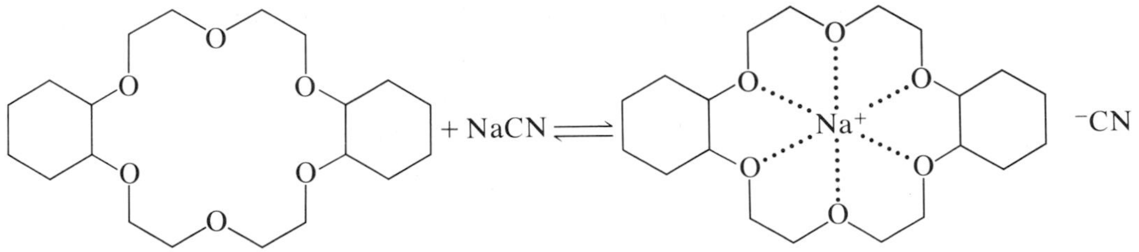

$$\text{Dicyclohexano-18-crown-6} + NaCN \rightleftharpoons [\text{crown}\cdot Na^+]\ ^-CN$$

Dicyclohexano-18-crown-6

The advantage of crown ether catalysis over quaternary ammonium catalysis is that no aqueous reservoir of a salt is required in order to carry out many transformations. One of the best examples can be found in the formation of the phenacyl ester of a carboxylic acid. In order to make a derivative of a carboxylic acid, the sodium or potassium salt of this substance can be generated and treated with the very lipophilic phenacyl bromide.

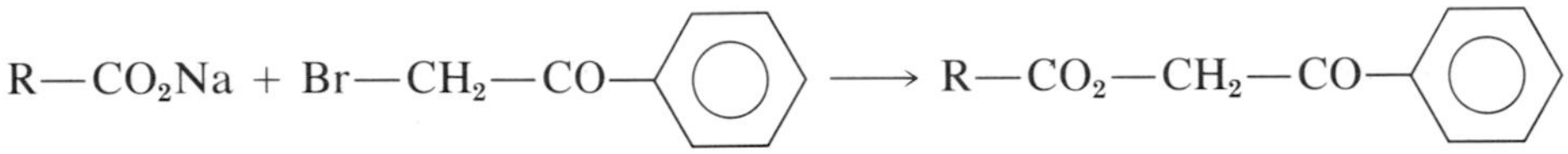

$$R—CO_2Na + Br—CH_2—CO—C_6H_5 \longrightarrow R—CO_2—CH_2—CO—C_6H_5$$

Phenacyl ester formation

This reaction frequently fails, however, because the salt and the bromoketone are not mutually soluble. A further difficulty is encountered if an alcoholic solution is used because ether formation would tend to occur simultaneously. As the carboxylate salt is not a potent nucleophile, the alcohol would very effectively compete with it. In an aqueous reservoir, the small amount of water which would be present in the organic phase would also tend to hydrolyze the phenacyl bromide and, if an aqueous reservoir of our unknown acid were required, the process would be inefficient in terms of the amount of material needed. The alternative is to prepare the acid salt by the procedure described in Sec. 23.6C and then transfer the solid directly into an organic phase by complexing the cation and making the ion pair soluble. There is then no competition by water or other medium.

There are many examples, indeed, of phase-transfer processes utilizing both macrocyclic polyethers (known as crowns), quaternary ammonium salts, and phosphonium salts. No attempt has been made to discuss these processes in more detail than necessary to understand why certain tests and derivative formations are conducted in a specific way. For more detailed information on this subject you are directed to the reference section at the end of this chapter. It is hoped that the discussion of solvents has afforded you a somewhat better understanding of what is occurring at a molecular level and why certain reactions always seem to fail. Occasionally, the reasons for failure are obvious to an experienced chemist but not always to the organic chemistry student. We hope that, in addition to obtaining a better understanding of solubilization phenomena, you also better understand why certain catalysts are added here in reactions that have not always proved reliable.

REFERENCES

Starks, C. M.: *J. Amer. Chem. Soc.*, **93:** 195 (1971).

Gokel, G. W., and H. D. Durst: *Synthesis*, 168 (1976).

Weber, W. P., and G. W. Gokel: *Phase Transfer Catalysis in Organic Synthesis*, Springer-Verlag, New York, 1977.

Gokel, G. W., and W. P. Weber: *J. Chem. Educ.*, **55:** 350, 429 (1978).

V SPECTROSCOPIC IDENTIFICATION OF ORGANIC COMPOUNDS

For eons the only electromagnetic radiation available for the identification of any substance was visible light. This meant that only the gross physical properties of a material could be determined. As science became more sophisticated, increasing advantage was taken of the various ranges of electromagnetic radiation for determining details about structures and reactivity. Electromagnetic radiation is often discussed in terms of the energy involved, which is inversely proportional to the wavelength of the radiation in question.

$$E = h\nu = \frac{hc}{\lambda} \tag{5.1}$$

where E = energy
h = Planck's constant
ν = frequency
c = speed of light
λ = wavelength

Organic chemists make use of various wavelengths of electromagnetic radiation, as well as of mass spectroscopy (a relatively recent advance, discussion of which is deferred to Sec. 5.4, as it is unrelated to the present discussion). The common spectroscopic techniques, listed in order of decreasing energy, are ultraviolet, infrared, microwave, and nuclear magnetic resonance spectroscopy. As the energy decreases, more and more subtle structural information can be obtained.

Of these methods the one utilizing the highest energies is ultraviolet (uv) spectroscopy, a technique which involves electronic transitions. The energy values involved are approximately 100 to 200 kcal/mol for the vacuum uv; 80 to

150 kcal/mol for the ultraviolet region (quartz uv); and 30 to 80 kcal/mol for the visible region.

Infrared (ir) spectroscopy records the bending and stretching of interatomic bonds, a relatively low-energy phenomenon. The energies involved in these bends and stretches are generally on the order of 2 to 15 kcal/mol and absorption is observed in the range of 2.5 to about 16 μm. This range corresponds to approximately 4000 to 600 reciprocal centimeters (cm^{-1}). The ir spectrum results from relatively low-energy interactions, and as a consequence the spectrum is much more sensitive to changes in molecular arrangement or geometry than is either the uv or visible spectrum. This principle also applies to the microwave region, where changes in rotations or rotational force constants are observed at a wavelength of about 1 cm. The rotational energies involved here are really quite small (about 10^{-4} kcal/mol).

Finally, the nuclear magnetic resonance (nmr) technique involves energies with wavelengths in the radio frequency or radar region. The wavelengths involved are on the order of meters, with energies of approximately 10^{-6} kcal/mol. This low-energy spectral technique requires sensitive instrumentation but yields very detailed structural information. Overall, then, the energies of these systems decrease from uv through ir and microwave to nmr. In many ways, the utility of a method as a structural identification technique increases as the energy decreases.

5.1 ULTRAVIOLET (UV) SPECTROSCOPY

Background

Ultraviolet (uv) spectroscopy is based on the detection of electronic transitions, i.e., the promotion of electrons from one energy level to another. Those transitions most often observed involve movement of an electron from a σ to a σ^* orbital, from a nonbonding, or *n*, orbital to a π^* orbital, or from a π to a π^* orbital. Less common transitions are n to σ^* and π to σ^*. These electronic transitions involve energies in the range of 80 to 150 kcal/mol, which is also the range of single-bond energies.

As a consequence of the very large energies involved in uv spectroscopy, absorption is usually observed as a rather broad band. This band is commonly referred to as an envelope or peak. Fine structure in the band results from the influence of molecular vibration and rotation.

Figure 5.1 shows the energy levels for both the ground state and the excited state of a simple alkene. When electromagnetic radiation of an energy equal to the energy level separation is absorbed by the molecule, an electron is promoted from an energy level in the ground state to an energy level in the excited state. These electronic transitions take approximately 10^{-16} s. The relationship between the speed of an electronic transition and a vibration (10^{-13} s) is a statement of the Frank-Condon principle. In other words, the energy absorption occurs so rapidly it does not affect the structure of the molecule. As a result,

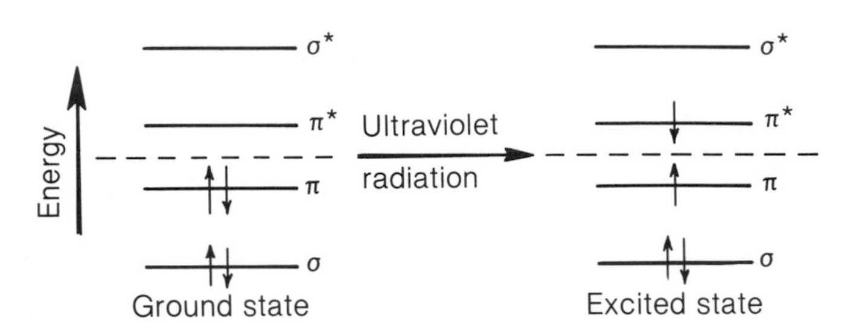

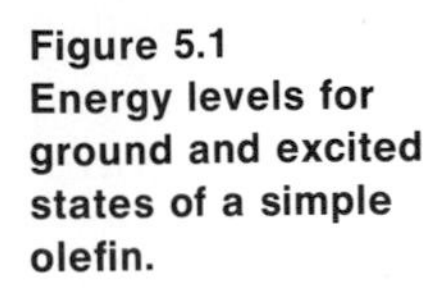

Figure 5.1 Energy levels for ground and excited states of a simple olefin.

one can often learn about the structure from the wavelength (or the energy) of the absorption.

The lines across the bottom of the potential-energy well in Fig. 5.2 represent vibrational energy levels, i.e., the various modes in which the molecule can vibrate. At room temperature, the thermal energy available is relatively small, and most of the molecules will be in the so-called vibrational ground state or lowest-energy state. As a consequence, the statistically most favorable electronic transition is the so-called ν_0 to ν'_0 transition. This transition is actually the promotion of an electron in one orbital to a higher energy orbital.

Beer-Lambert Law

It is known that the quantity of light absorbed is proportional to the amount of the absorbing medium which is present. The statement of this was offered many years ago by Beer and is still called Beer's law. Further information was added by Lambert who stated that light absorption at a given wavelength is in-

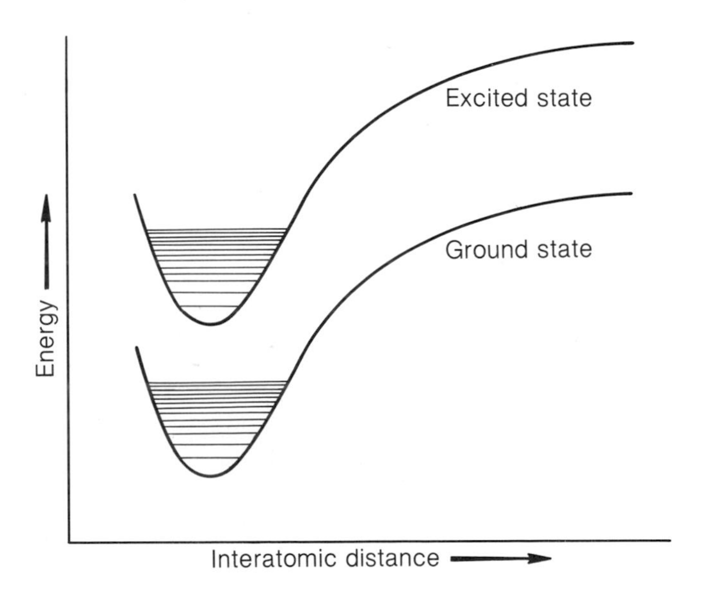

Figure 5.2 Vibrational energy levels.

dependent of the intensity of the source. A combination of Beer's law and Lambert's law leads to the following statement:

$$\log \frac{I_0}{I} = \epsilon c l = A \tag{5.2}$$

where I_0 is the intensity of incident monochromatic light; I is the intensity of emergent monochromatic light; ϵ is the molar absorptivity (sometimes called the extinction coefficient); c is the concentration: l is the path length (in centimeters); and A is the absorbance or optical density. It should be apparent that

$$\epsilon = \frac{A}{cl} \tag{5.3}$$

which is in the units of square centimeters per mole. Therefore, the extinction coefficient is a kind of photon-capture cross section. The larger the cross section (i.e., the larger the extinction coefficient), the greater will be the intensity of the absorption.

The Instrumentation of uv Spectroscopy

In order to obtain a uv spectrum, it is necessary to determine how much energy is absorbed at each wavelength scanned. This information is usually easy to obtain for a liquid sample, but the uv spectrum of a solid sample must be determined in solution.

Many solvents which are used to dissolve solid organic compounds also absorb small amounts of uv radiation. The uv spectrum which is obtained in solution actually compares the spectrum of a material in solution with that of the solvent itself. A convenient instrument must therefore be capable of simultaneously recording and comparing the energy absorption of two samples at once. To do this a sample cell containing the material in some solvent and a reference cell containing the solvent alone are placed in the optical path of an instrument lamp. The uv lamp's radiation first passes through a dispersion grating to obtain the monochromatized radiation, which in turn passes through a beam splitter (generally a mirror-and-prism system) so that the beam may pass through both samples simultaneously. The emergent beams from the two samples are focused alternately on a rotating-sector mirror (sometimes called a chopper), and are then alternately focused on a photoelectric detector. Any difference in absorptions between the samples will be observed wherever there is an intensity difference between the sample and the reference cell. This procedure allows the spectrum of the sample only to be recorded on chart paper. In so-called single-beam instruments, subtraction of solvent absorption must be done manually, but this is effected automatically in modern double-beam instruments. A block diagram of such an instrument is shown in Fig. 5.3.

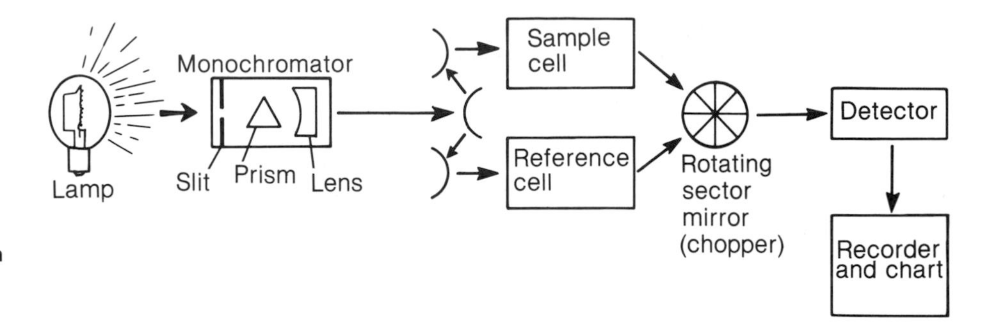

Figure 5.3
Block diagram of an ultraviolet spectrophotometer.

Electronic Absorption and Structure

Many electronic transitions are possible within a molecule. In general, however, the utility of these for structure identification will accord with the extent of conjugation. The energy transition which is commonly of highest energy is a π to σ^* transition. This is a transition from an electron in a double- or triple-bond orbital to a σ antibonding orbital, generally observable only in the so-called vacuum uv, that is, below about 180 nm. The process involved is electron promotion from a double bond to a continuum. As such a process involves very high energy, it is of relatively little value for structural analysis in organic chemistry.

A more useful electronic transition is that from π to π^*, which involves promotion of an electron from a double- or triple-bond orbital to a π antibonding orbital, as shown in Fig. 5.4.

Such alkenes as ethylene or tetramethylethylene undergo electronic absorption to give discernible uv bands. The absorption maximum (λ_{max}, longest wavelength at which a maximum peak is observed) for ethylene is observed at about 165 nm. The absorption maximum for tetramethylethylene, which is a more highly substituted system, is centered at 185 nm. The extinction coefficients are similar in these two systems (about 1000 cm^2/mol).

Even more common than the π to π^* transition is the n to π^* transition.

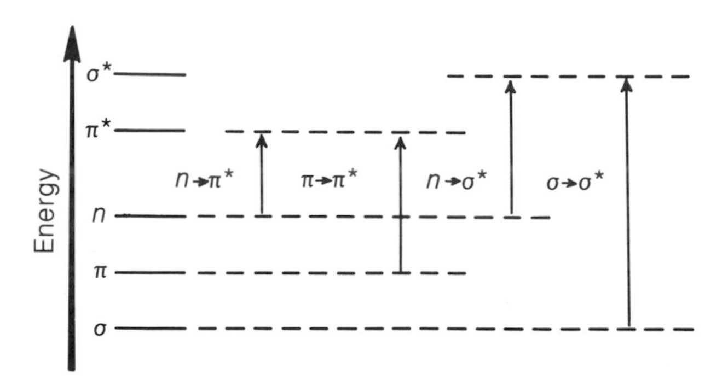

Figure 5.4
Some electronic transitions.

This transition involves excitation of an electron which is in a nonbonding orbital (such as is present on oxygen or nitrogen) to a π antibonding orbital. The chromophore involved most often in this is the carbonyl group. For a molecule such as acetone, both π to π^* and n to π^* transitions are possible. The π to π^* transition occurs at about 188 nm and its extinction coefficient is about 900. The n to π^* transition, on the other hand, is observed at much longer wavelengths (much lower energy) and occurs at about 279 nm, with an extinction coefficient of approximately 15. The n to π^* absorption is distinguishable, although it is much weaker than that arising from the π bond.

Since the fundamental energy relationship is $E = h\nu$, the position of the absorption must be directly related to the energy of the absorption. Since ν is related to the wavelength by the speed of light [see Eq. (5.1)], the closer two energy levels are to each other, the smaller the energy of the absorption will be. Clearly, as the energy decreases, the wavelength increases, and vice versa. For an isolated double bond (e.g., in ethylene), the transition involves a relatively high energy, because the π and π^* orbitals are widely separated. In a conjugated diene (a delocalized π system), the excited state is stabilized compared with that of an isolated double bond. The separation of the energy levels therefore decreases, and the wavelength at which the uv absorption occurs increases. In general, as the conjugation increases, the wavelength of the absorption also increases, whether carbon-carbon bonds or carbon-heteroatom bonds are involved.

Structural Analysis

We have stressed that the energies involved in uv spectroscopy are relatively high. As a result of this, subtle structural information cannot be obtained by this technique. What should be obvious from the discussion above is that the position and intensity of absorption will vary as the extent of conjugation changes. In general, the more conjugated a system is, the longer will be the wavelength at which absorption is observed. Compounds which are structurally similar but which have different arrangements of chromophores may often be distinguished by uv spectroscopy.

It is probably easiest to illustrate this point by use of an example. In the structures shown below are two unsaturated ketones. Each of these compounds has the molecular formula $C_9H_{14}O$. Combustion analysis would reveal that they are isomers but would not indicate anything about the relative positions of the double bond and carbonyl group. If the uv spectrum were recorded for each compound, the results would be as recorded in Fig. 5.5*a* and 5.5*b*. Note that compound 5.5*a* gives a relatively weak absorption in the 220-nm region. The other compound has a band which not only is much more intense but appears at longer wavelength (about 225 nm). One would have to conclude that

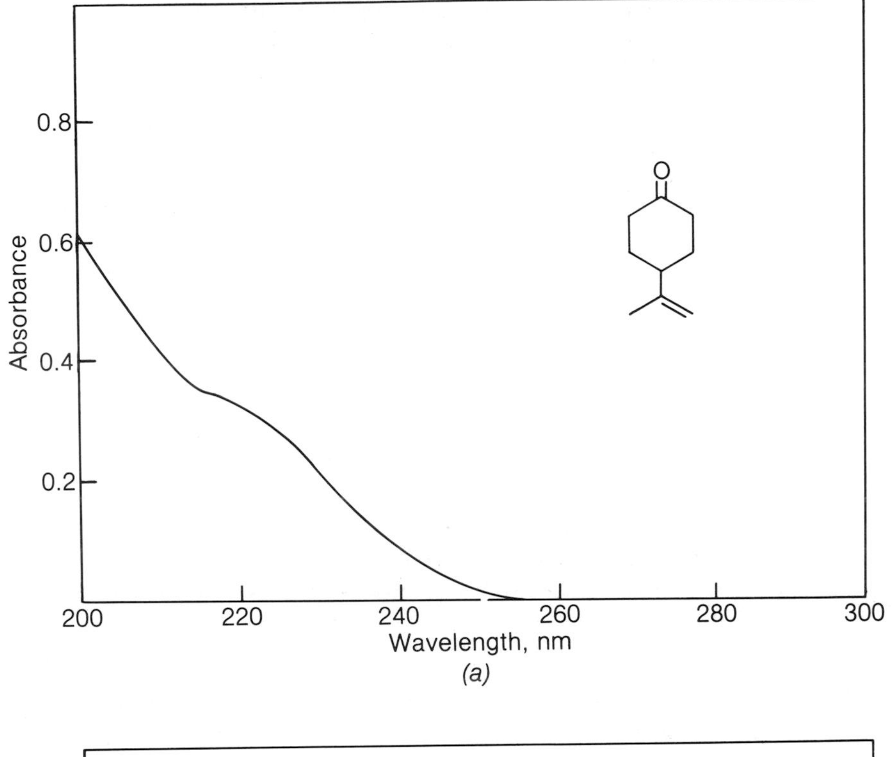

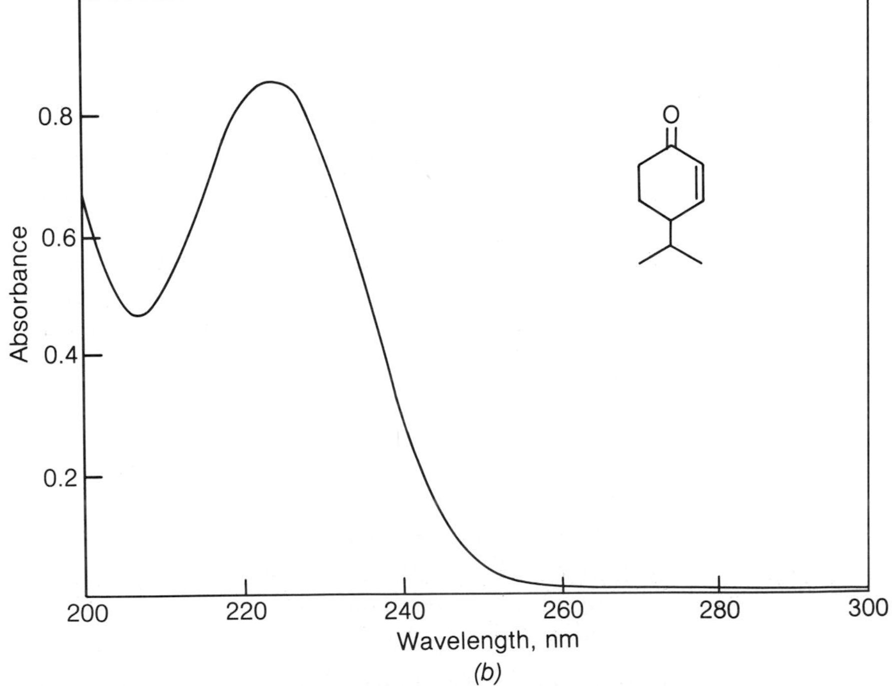

Figure 5.5
The uv spectra of (*a*) 4-isopropenylcyclohexanone and (*b*) 4-isopropylcyclohex-2-enone.

the conjugated isomer 5.5*b* is the one whose uv spectrum is observed at longer wavelength. This, indeed turns out to be the case.

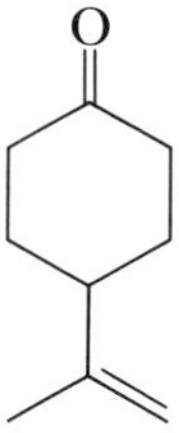

5.5*a* 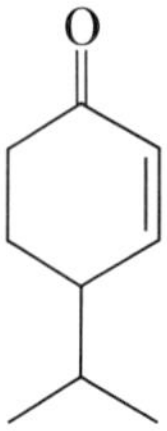 5.5*b*

A number of empirical correlations have been worked out over the years which relate the position of a uv band to the structure of the absorbing compound. These rules are approximate, but for certain types of compounds may give quite useful and detailed information. Prominent among these rules are the so-called Woodward-Fieser rules. If the use of this technique and application of such rules seem desirable, refer to one of the more detailed references cited at the end of this chapter.

To emphasize the fact that electronic spectra are relatively insensitive to minor structural perturbations, especially if they do not affect the nature of the chromophore, we have included in this section the uv spectra of benzene and ethylbenzene. Note how similar their spectra are because the absorbing groups (a benzene ring in each case) are the same. A table of typical uv absorptions is presented in Table 5.1.

An advantage of the uv absorption technique which relates to structure determination is that a spectrum can be obtained on very little material. A uv

TABLE 5.1
Comparison of uv chromophores

Functional group	Typical absorption position, nm	Extinction coefficient, ϵ
C=C	165–185	1,000
C=C—C=C	215–235	20,000
C=O	170–190	1,000
C=C—C=O	220–250	20,000
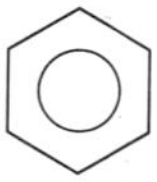	255	200
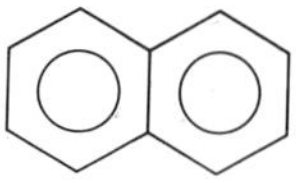	275 314	6,000 250

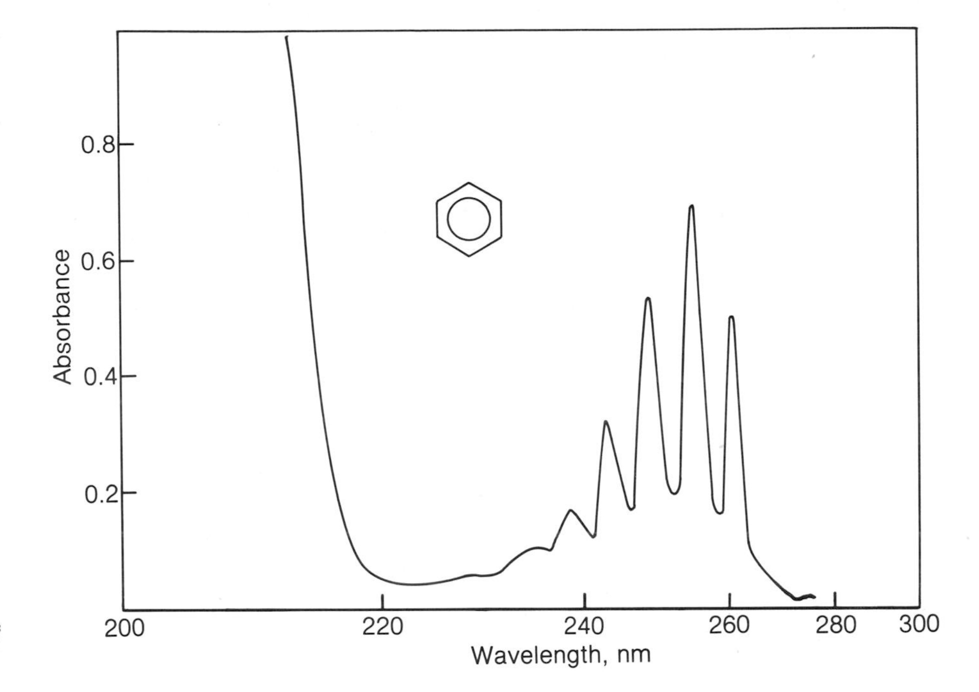

Figure 5.6
The uv spectrum of benzene.

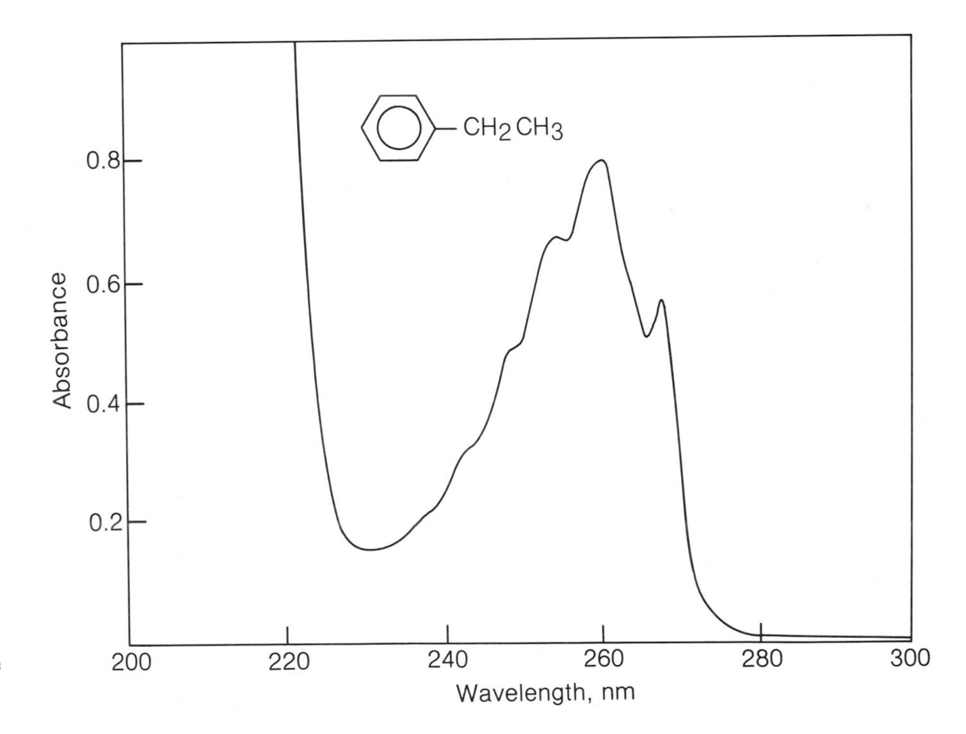

Figure 5.7
The uv spectrum of ethylbenzene.

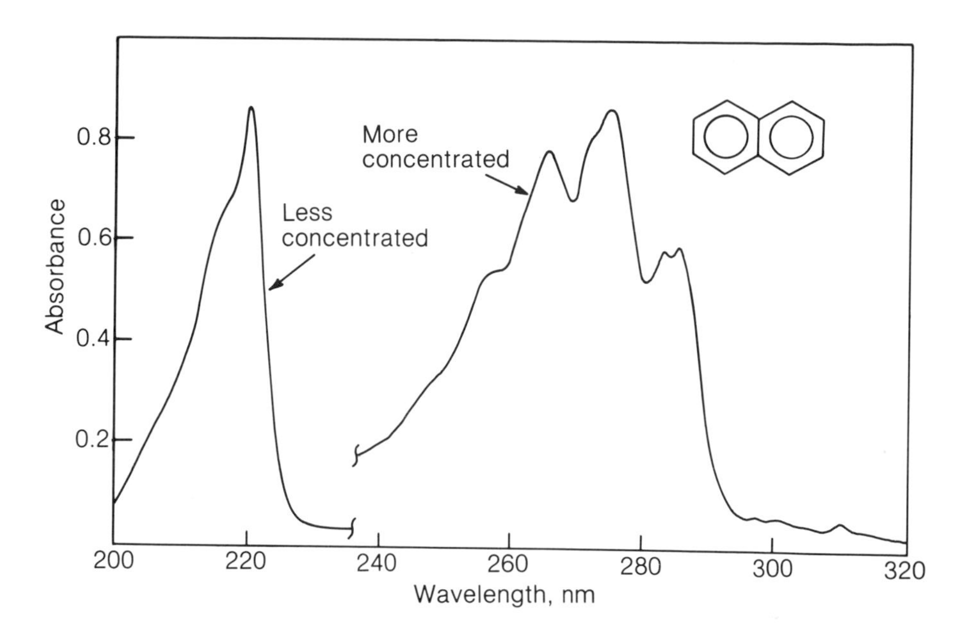

Figure 5.8
The uv spectrum of naphthalene.

spectrum can often be obtained on as little as 1 mg of material. This is because a compound whose extinction coefficient is about 10,000 requires only a 0.0004 molar (M) solution for the spectrum to be obtained. That this is so may be confirmed by checking the Beer-Lambert equation [Eq. (5.2)]. Note also that a molecule of molecular weight 200 would require only 0.02 mg of material dissolved in 5 mL of solvent to afford a usable spectrum.

The spectra of benzene, ethylbenzene, and naphthalene are given for comparative purposes in Figs. 5.6, 5.7, and 5.8, respectively.

5.2 INFRARED (IR) SPECTROSCOPY

Theory

The observation of an infrared (ir) spectrum is the result of electromagnetic radiation of the appropriate wavelength interacting with the vibrations (stretching and bending) of atoms within a molecule. Because there are many possible motions within a molecule, the ir spectrum is correspondingly complex and gives rise to many, many peaks.

An understanding of the ir spectroscopy process can be gleaned by considering the mechanics of vibration. If two atoms, A and B, are attached to each other through a spring, then the ends of the spring may vibrate back and forth with a certain periodicity. The vibrational period will be determined by the distance between the two atoms and the force of the spring. If the atoms are pressed too closely together, the spring will try to push them apart; if they are stretched apart, it will try to pull them together. Repeating this process many times leads to a vibration of A and B about some equilibrium point. This is essentially a statement of Hooke's law:

$$F = k\Delta r \tag{5.4}$$

in which F is force, k is the restoring force, and Δr is the separation of the vibrating bodies. This can be represented easily by the potential energy diagram shown in Fig. 5.9*a*.

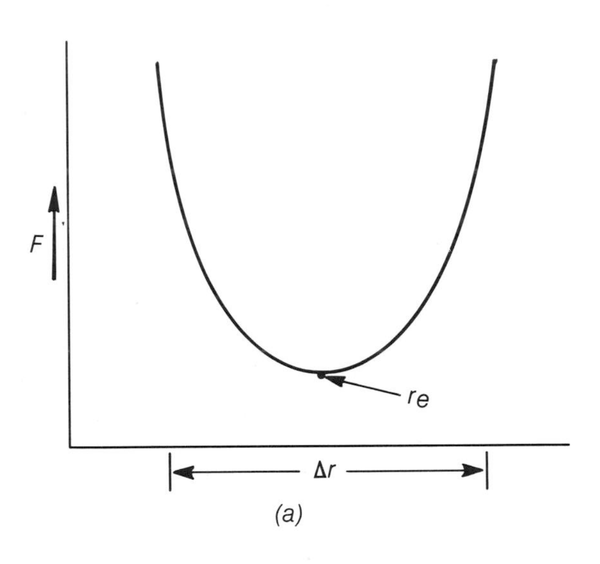

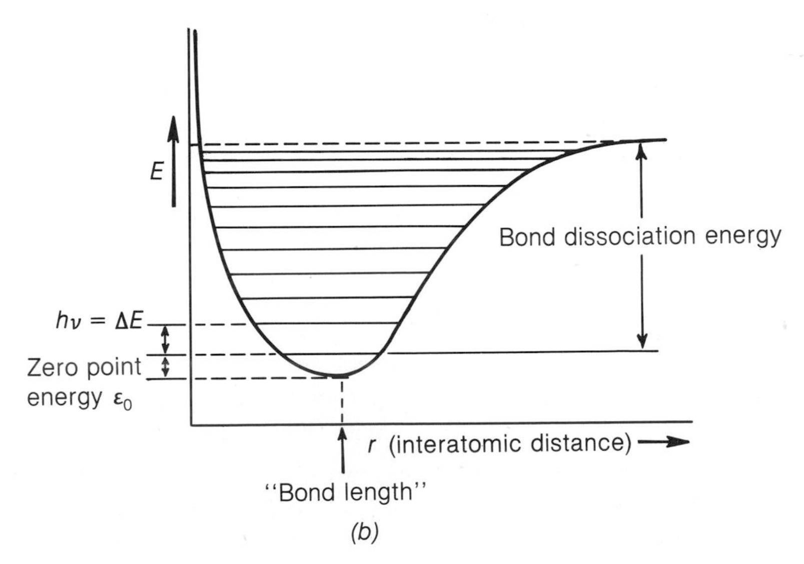

Figure 5.9 Potential energy diagrams. See text for details.

Classical mechanics would lead us to believe that a molecule may have any of numerous vibrations (that there will be a continuum of vibrational levels). Quantum mechanics, however, places restrictions on microphysical systems, the results of which are discrete energy levels. Vibrational energy can be described by saying that E_v is the energy

$$E_v = h\nu(v + \tfrac{1}{2}) \tag{5.5}$$

of the vth vibrational state, h is Planck's constant, ν is the fundamental vibrational frequency, and v is the vibrational quantum number for the various states. These are integral values 0, 1, 2, . . . , y. Remember that the zero point energy of a molecule is always

$$E_0 = h\nu \times \tfrac{1}{2} \qquad \text{or} \qquad E_0 = \tfrac{1}{2}h\nu \tag{5.6}$$

There are several significant stipulations concerning the absorption of energy by a vibrating system (Fig. 5.9*b*). Most important among these is that, in order for molecules to absorb ir radiation as vibrational excitation energy, there must be a change in the dipole moment of the molecule as it vibrates. Recall that a dipole moment is simply a charge times a distance, so that any nonsymmetrical bond undergoes a change in dipole moment as the distance between the two vibrating atoms changes. The molecule HCN is an excellent example of this. As the molecule vibrates, the bond distances and the charge separation change, with a resulting change in dipole moment. Unless there is a change in the dipole moment, no coupling to and absorption of electromagnetic (ir) radiation will occur. The bending of the carbon-hydrogen bond also changes the dipole moment in HCN; the vibrations of both are observed in the ir spectrum. The dipole moment change is essential because an interaction between the dipole and the electromagnetic radiation can occur only when there is an oscillating electric field. When the interaction of these two oscillating electric systems does occur, energy is transferred and absorption occurs.

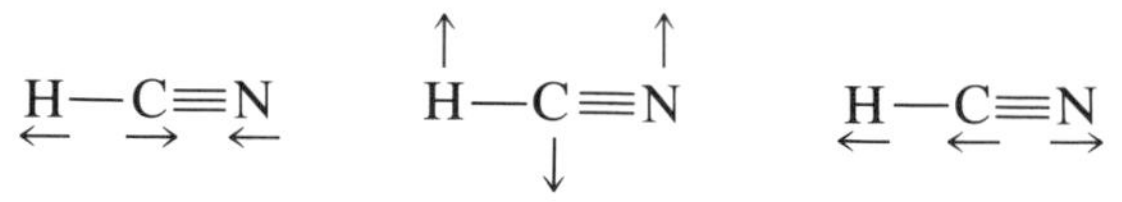

H—C≡N (← → ←)	H—C≡N (↑ on H and N, ↓ on C)	H—C≡N (← ← →)
C—H stretch: 3312 cm^{-1}	C—H bend: 712 cm^{-1}	C≡N stretch: 2089 cm^{-1}

(Arrows indicate the changes in dipole)

The second requirement for the observation of an ir spectrum is that only transitions of $\Delta\nu$ are allowed, which means that the most commonly observed vibrational transition will be from the zero vibrational ground state (v_0) to the first excited vibrational state (v_1). This is because at room temperature only the lowest vibrational energy level v_0 is significantly populated.

Since vibrations are not perfectly harmonic, overtones are observed. These overtones are usually transitions from v_0 to v_2, although those from v_0 to v_3 may also occur. The v_0 to v_2 transition is called the first overtone, and the absorption will be observed at about twice the frequency of the fundamental (v_0 to v_1) vibration. The spacing in energy between v_0 and v_1 corresponds to the equation $E = h\nu$, where ν represents the frequency of the vibration.

The difference in energy (ΔE) between two adjacent levels, E_v and E_{v+1}, is shown in Eq. (5.7):

$$\Delta E = \frac{h}{2\pi}\sqrt{\frac{k}{\mu}} \tag{5.7}$$

in which k is the force constant for the bond and μ is the reduced mass. The reduced mass (μ) is given for a molecule AB by

$$\mu = \frac{M_A \times M_B}{M_A + M_B} \tag{5.8}$$

where M_A and M_B are the atomic masses of atoms A and B, respectively. For the carbon monoxide molecule, the value of the reduced mass would be $(12 \times 16)/(12 + 16)$ or approximately 6.86. For carbon monoxide the vibrational frequency of the carbon-oxygen double bond is given by $\nu = 2143\ \text{cm}^{-1}$. Thus, the force constant (k) can be calculated from the following relationship:

$$\nu = \frac{1}{2\pi}\sqrt{\frac{k}{\mu}} \tag{5.9}$$

For carbon monoxide, $k = 18.7 \times 10^5$ dyn/cm. In general, the force constants for bonds will increase as the order of the bond (single bond, double bond, triple bond) increases. This trend will generally parallel the bond strength. As a rule, single bonds will have force constants in the range of 4×10^5 to 6×10^5 dyn/cm; double bonds, 9×10^5 to 12×10^5 dyn/cm; and triple bonds, 15×10^5 to 18×10^5 dyn/cm. The same trend is observed for bond strength: Typical single bonds are weaker (approximately 100 kcal) than double bonds (approximately 160 kcal) or triple bonds (approximately 220 kcal).

So far, we have dealt with molecules so simple that one might imagine an ir spectrum consisting of a single absorption. This is certainly not the case. Molecules which have many atoms have many possible arrangements of those atoms within the molecule and many orientations of the molecule in space.

Molecules which contain n atoms will need $3n$ coordinates to define their positions and all the interatomic relationships. Some of the possible orientations (degrees of freedom) correspond to motion of the molecule as a whole.

These motions are translation and rotation. The remaining degrees of freedom correspond to vibrations of the molecule.

In order to define the position of a carbon dioxide molecule, three translational degrees of freedom, corresponding to the three cartesian coordinates, must be specified. The rotation of a molecule generally requires three additional degrees of freedom, although for linear molecules it requires only two. Since carbon dioxide is a linear molecule, rotation along the oxygen-carbon-oxygen axis is obviously undetectable. The number of vibrational degrees of freedom (or vibrational modes) for any molecule can now be specified. For a molecule containing n atoms, the number of possible vibrations will be $3n - 6$, or, if the molecule is linear, $3n - 5$. The linear molecule carbon dioxide has three atoms so it must have $3 \times 3 - 5 = 4$ degrees of freedom. The four vibrations characteristic of CO_2 are shown below

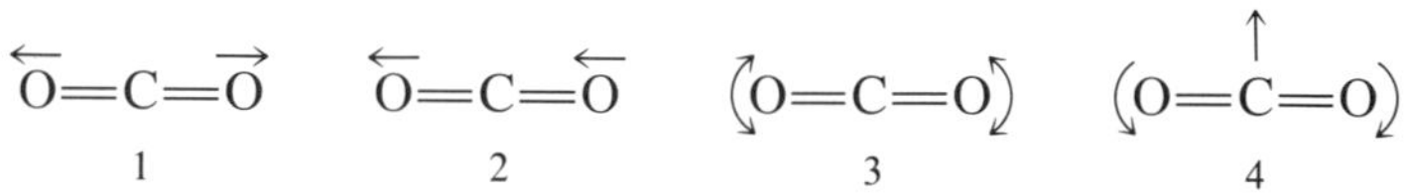

O=C=O O=C=O O=C=O O=C=O

1 2 3 4

Vibrational modes of CO_2

Note that the dipole moment of the CO_2 molecule at rest is zero. This is because the carbon-oxygen dipole in one direction exactly cancels the carbon-oxygen dipole in the opposite direction. As the symmetrical vibration represented by structure 1 occurs, each individual carbon-oxygen dipole changes, but the overall dipole moment for the molecule remains zero. Since no net change in dipole moment occurs, no energy absorption occurs, and this molecular vibration is said to be ir-inactive. Three of the vibrations do undergo changes in dipole moment and are ir-active. It is coincidental that the active and inactive bonds correspond to symmetrical and unsymmetrical vibrations, respectively.

Three generalizations are particularly important for the interpretation of ir spectra. First, a heavy atom vibrates with lower frequencies than does a light atom bonded to the same element. For example, a carbon-carbon bond vibrates at about 1200 cm^{-1}, but a carbon-hydrogen bond vibrates at about 3000 cm^{-1}. Second, stronger bonds have higher force constants and higher vibrational frequencies. Typical carbon-carbon single bonds vibrate at 1200 cm^{-1}, carbon-carbon double bonds at 1600 cm^{-1}, and carbon-carbon triple bonds at 2200 cm^{-1}. Third, the more polar the bond, i.e., the greater the difference in electronegativity between the two bonded atoms, the stronger (or more intense) will be the absorption. (This is very qualitative, but it is a good general rule.) Since it is generally more difficult to quantify the strength of a peak corresponding to a vibration than it is to quantify an electronic (uv) transition, ir peaks are usually categorized simply as strong, medium, or weak.

The Instrumentation of Infrared Spectroscopy

A standard double-beam ir spectrometer differs from a uv instrument only in that the radiation is generated by a different source, since a different frequency range is required. The ir radiation source is usually a zirconium oxide or silicon carbide globar, which is heated electrically. The resulting ir radiation is emitted into a monochromator, which separates the light into its component parts. The monochromatized radiation goes into a beam splitter (mirror-and-prism system) so that equivalent beams of relatively narrow wavelength range can be passed through a sample and a reference cell simultaneously. The emergent light is again focused through a prism-and-mirror system onto a rotating-sector mirror. The energy transmitted by the sample and the reference cell are detected in a bridge system as a voltage difference. Note that the block diagram shown in Fig. 5.10 is essentially the same as that for a uv spectrometer. The important difference is the wavelength of light required for a vibrational transition compared to that required for an electronic (uv) transition.

Sampling Techniques

Infrared spectra of liquid (or oily) samples are most conveniently determined when the samples are placed between two salt plates. Solids must generally be dissolved in such solvents as chloroform, dichloromethane, carbon tetrachloride, or carbon disulfide. An alternative for solids is to mix the sample with potassium bromide or a similar salt and press the mixture into a pellet. Solids may also be finely ground with mineral oil (often called Nujol mulls, for the trade name of mineral oil). The use of mineral oil (Nujol mull) is indicated when used for spectra presented in this book because bands due to C—H and C—C vibrations of the hydrocarbon solvent also appear and might confuse the student.

Functional Group Absorptions

A number of possible generalizations about the absorption of ir radiation are shown in Table 5.2. Note that although bonds may be formed between carbon

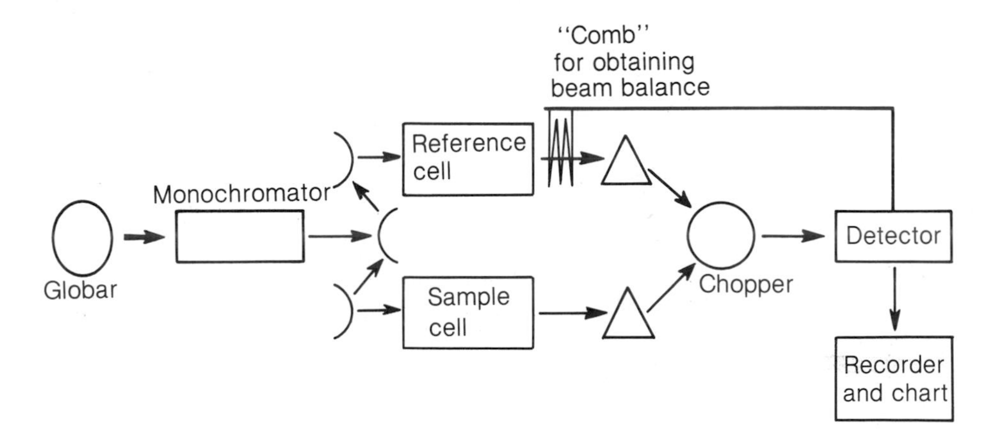

Figure 5.10
Block diagram of an infrared spectrometer.

TABLE 5.2
Typical infrared absorptions

Functional group	Band position, cm^{-1}
C—H stretching	2800 to 3100
N—H, O—H stretching	3400 to 3600
C≡C, C≡N	2000 to 2400
C=C, C=N, C=O (see text for details)	1500 to 1900
Single-bond stretching	1000 to 1400
Bending vibrations	600 to 1600

and hydrogen, carbon and carbon, carbon and oxygen, or carbon and sulfur, the vibrations generally fall into relatively narrow ranges, i.e., most of the stretching vibrations of hydrogen-bearing atoms will be observed in the 3000 to 4000 cm^{-1} range (those of carbon-hydrogen bonds in the 2800 to 3100 cm^{-1} region; those of OH and NH bonds in the 3400 to 3600 cm^{-1} region). Triple bonds between such elements as carbon and carbon, carbon and oxygen, or carbon and nitrogen are observed in the 2000 to 2400 cm^{-1} region. Double-bonded systems (carbon-nitrogen, carbon-oxygen, oxygen-nitrogen, etc.) are generally found in the 1500 to 1900 cm^{-1} region, and various single-bond vibrations occur in the range of 1000 to 1400 cm^{-1}. A wide range of bending and coupled vibrations found in the 600 to 1600 cm^{-1} region are due to combination and overtone bands, which are often difficult to interpret. The presence of bands in the indicated regions is evidence for the presence of a certain functional group. The absence of absorption is often good evidence for the absence of the functional group.

Certain vibrations, such as the carbonyl (C=O) stretching vibration, are dramatically affected by substituents. The carbonyl vibration of acetone (or of the strain-free cyclohexanone system) is commonly referred to as the carbonyl standard. Its absorption in a nonpolar solvent is observed at 1715 cm^{-1}. If strain is introduced by, for example, forcing the carbonyl group into a small ring, the vibrational frequency increases. The carbonyl vibration of cyclopentanone is observed at 1745 cm^{-1}. Likewise, the carbonyl group of cyclopropanone (the three-membered ring ketone) absorbs at 1820 cm^{-1}.

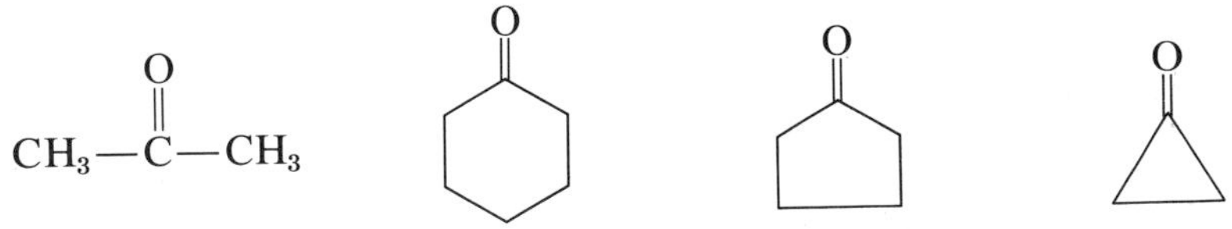

$CH_3—C(=O)—CH_3$

Acetone
$\nu_{C=O} = 1715\ cm^{-1}$

Cyclohexanone
$\nu_{C=O} = 1715\ cm^{-1}$

Cyclopentanone
$\nu_{C=O} = 1740\ cm^{-1}$

Cyclopropanone
$\nu_{C=O} = 1820\ cm^{-1}$

As might be imagined, conjugation with a carbonyl group will cause a decrease in electron density in the double bond, with a resultant decrease in the bond strength or force constant. The vibrational frequency decreases with diminishing force constant because the energy required to cause the vibration also falls. If the carbonyl frequencies of acetophenone and benzophenone are compared with that of acetone, their values decrease in the order acetone 1715 cm^{-1}, acetophenone 1690 cm^{-1}, and benzophenone 1665 cm^{-1}.

$CH_3-C(=O)-CH_3$

Acetone	Acetophenone	Benzophenone
$\nu_{C=O}$ = 1715 cm^{-1}	$\nu_{C=O}$ = 1690 cm^{-1}	$\nu_{C=C}$ = 1665 cm^{-1}

When a simple ketone is changed into an ester by placing an oxygen adjacent to the carbonyl, inductive electron release into the carbonyl group occurs, which strengthens the bond and increases the vibrational frequency. The carbonyl group of typical aliphatic esters is thus observed at about 1740 cm^{-1}, compared with 1715 cm^{-1} for acetone. The carbonyl group of α,β-unsaturated (i.e., conjugated) esters is observed at a lower frequency for the reasons mentioned above (generally, approximately 1720 cm^{-1}). A heteroatom conjugated with an adjacent carbon-oxygen double bond tends to lower the carbonyl vibrational frequency. Because of this conjugation, amides are usually observed at much lower frequencies, generally in the 1690 to 1670 cm^{-1} region. A number of typical ir absorption spectra are shown in Figs. 5.11 through 5.23.

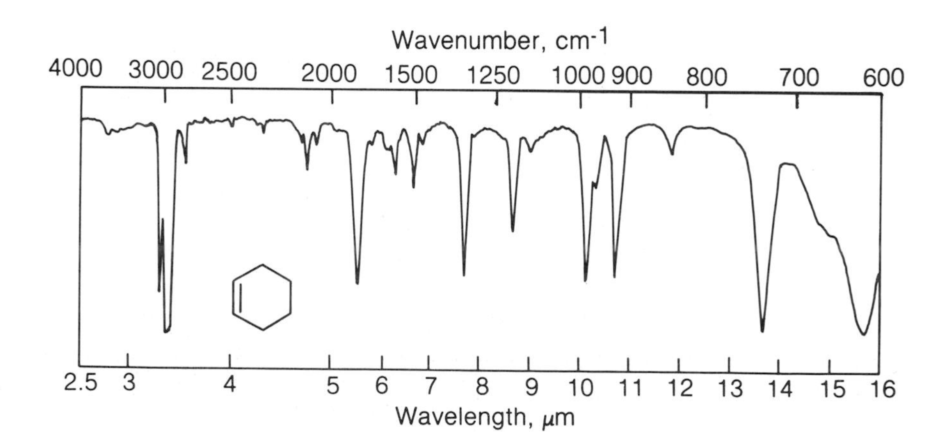

Figure 5.11
The ir spectrum of cyclohexene, a typical alkene.

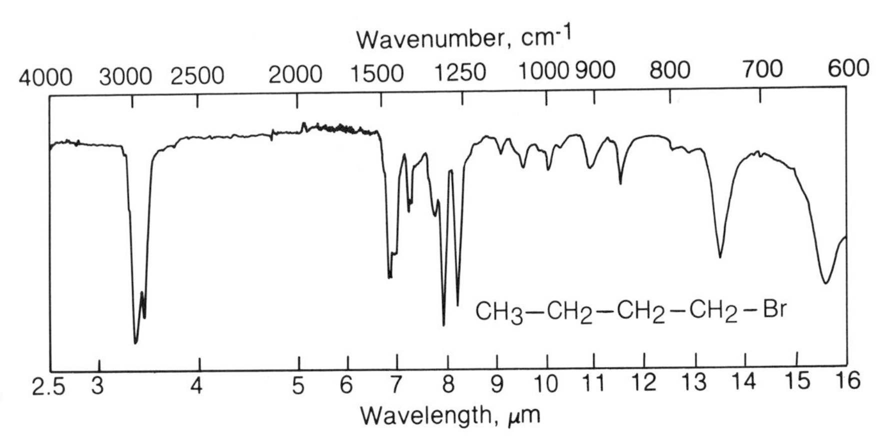

Figure 5.12
The ir spectrum of 1-bromobutane, a typical alkyl bromide.

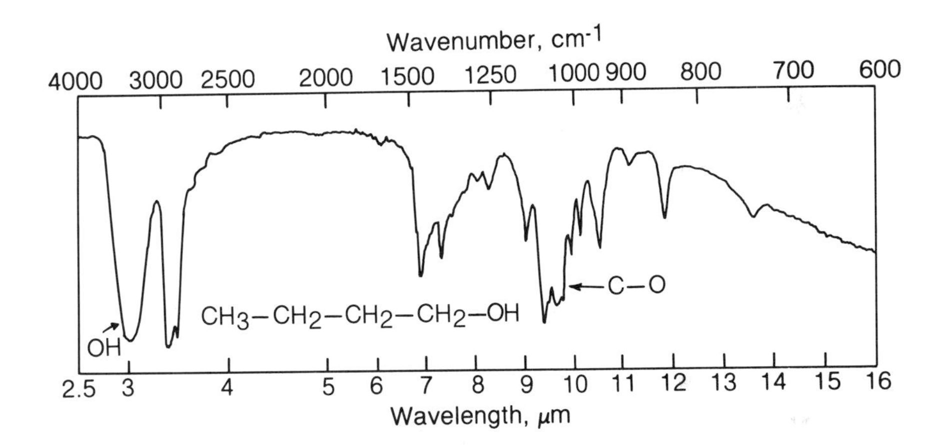

Figure 5.13
The ir spectrum of *n*-butanol, a typical alcohol.

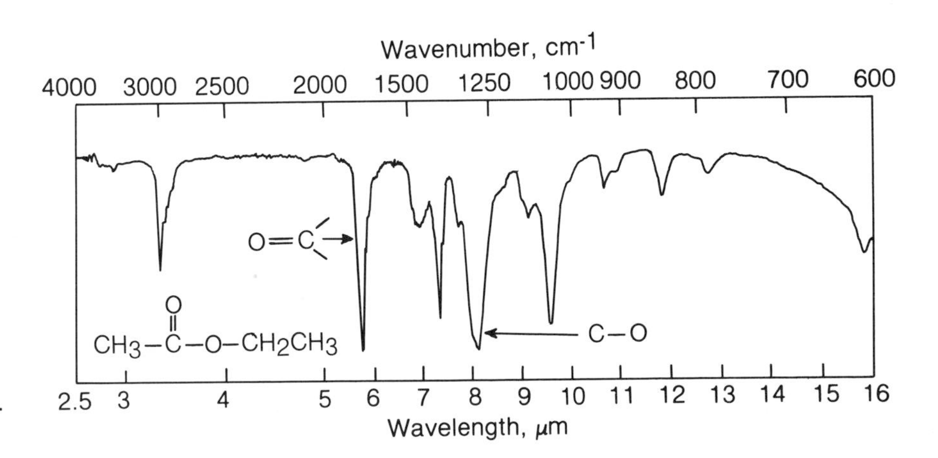

Figure 5.14
The ir spectrum of ethyl acetate, a typical ester.

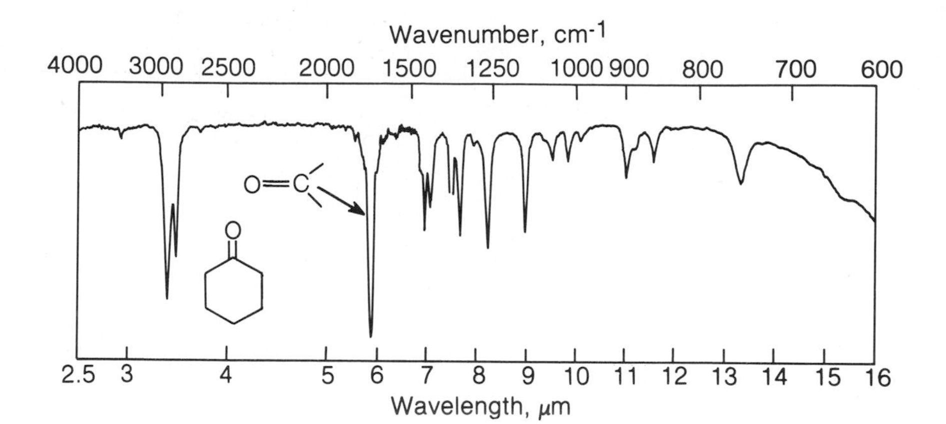

Figure 5.15
The ir spectrum of cyclohexanone, a typical dialkyl ketone.

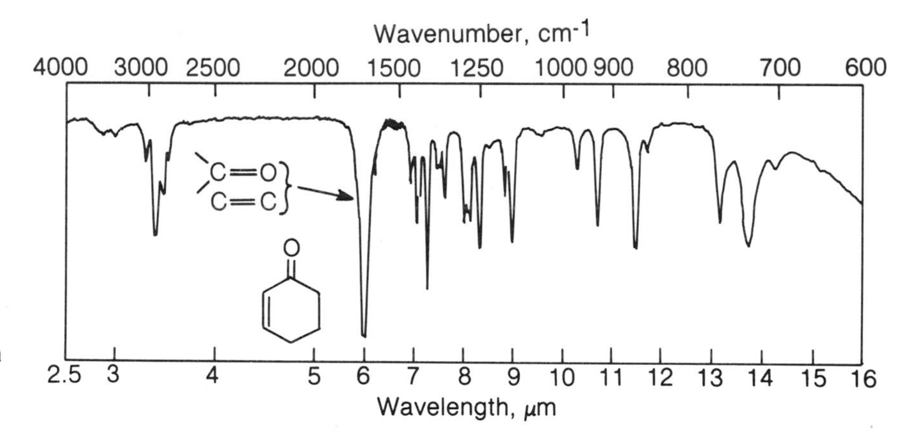

Figure 5.16
The ir spectrum of 2-cyclohexen-1-one, a typical *α*, *β*-unsaturated ketone.

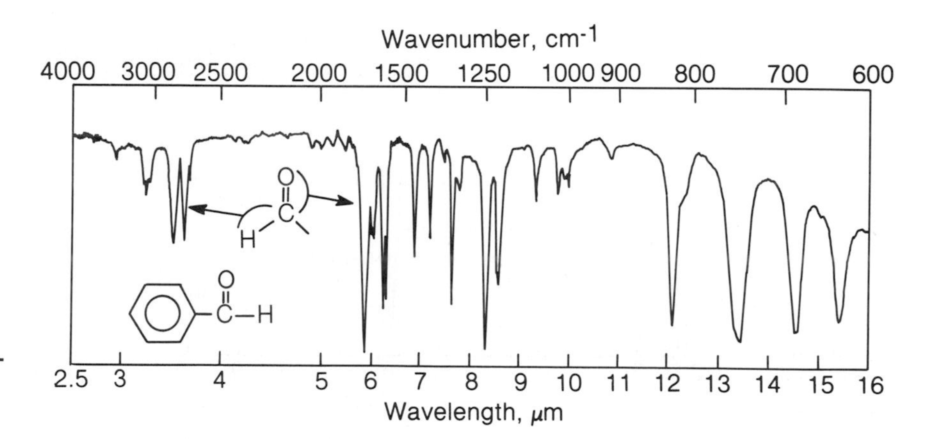

Figure 5.17
The ir spectrum of benzaldehyde, a typical aromatic aldehyde.

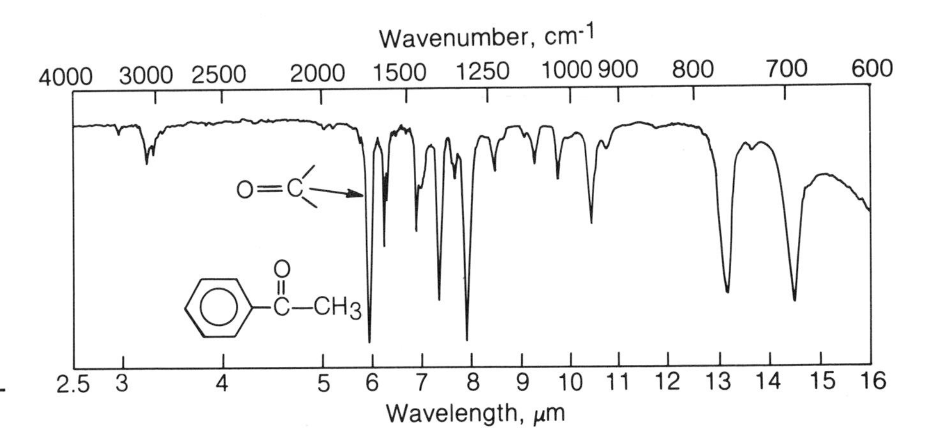

Figure 5.18
The ir spectrum of acetophenone, a typical aryl ketone.

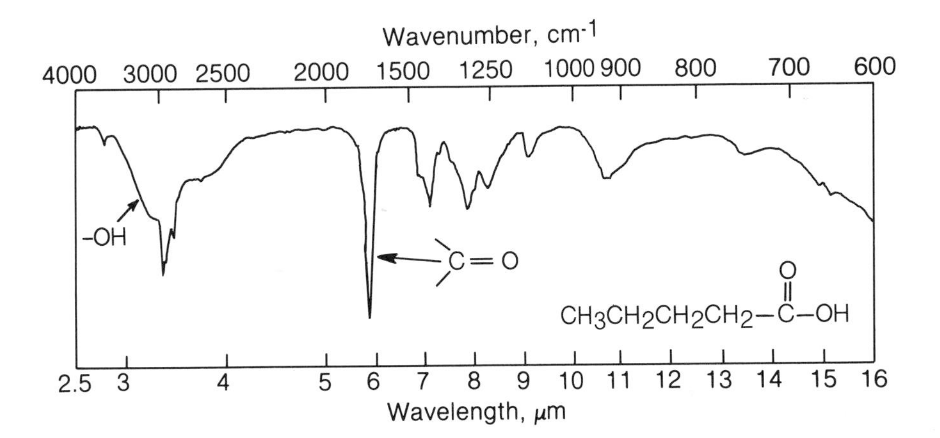

Figure 5.19
The ir spectrum of pentanoic (valeric) acid, a typical aliphatic acid.

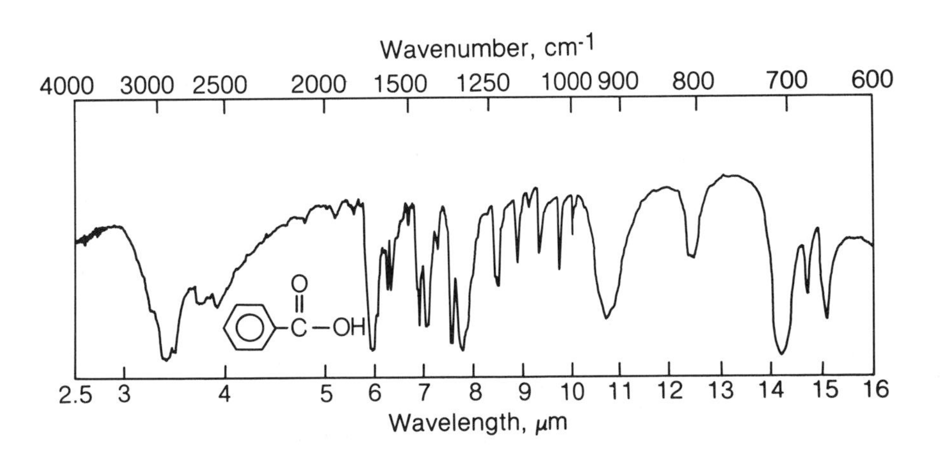

Figure 5.20
The ir spectrum of benzoic acid, a typical aromatic acid.

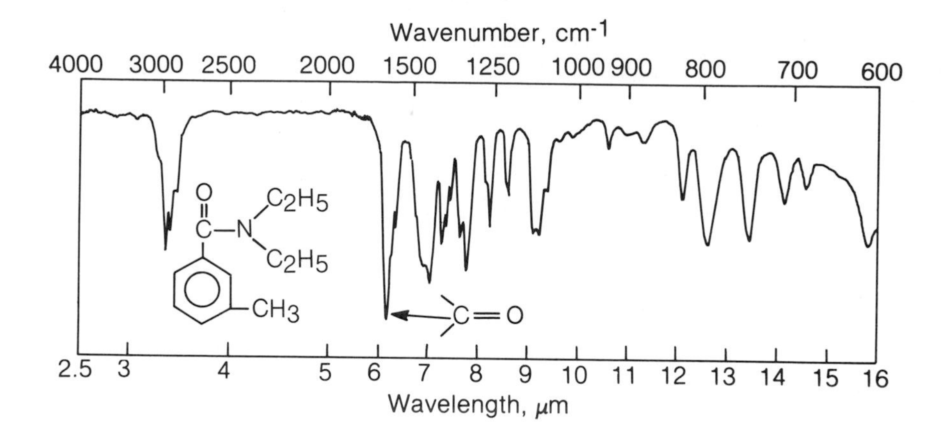

Figure 5.21
The ir spectrum of *N,N*-diethyl-*m*-toluamide, a typical amide.

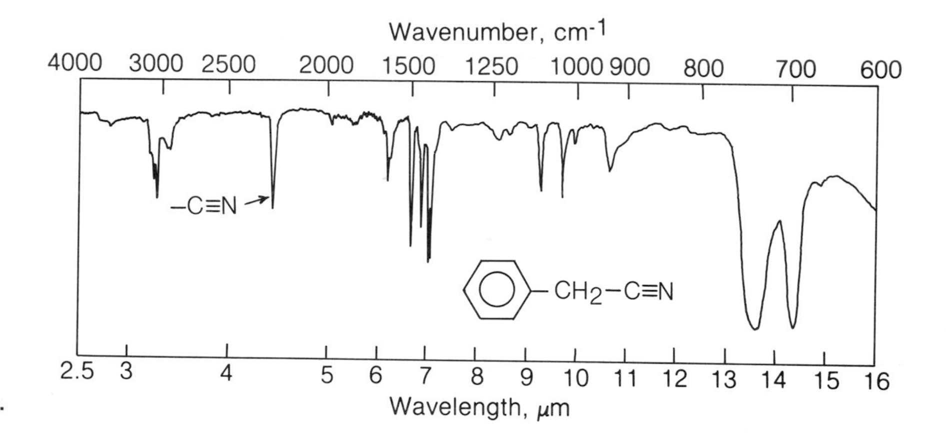

Figure 5.22
The ir spectrum of benzyl cyanide (phenylacetonitrile), a typical alkyl nitrile.

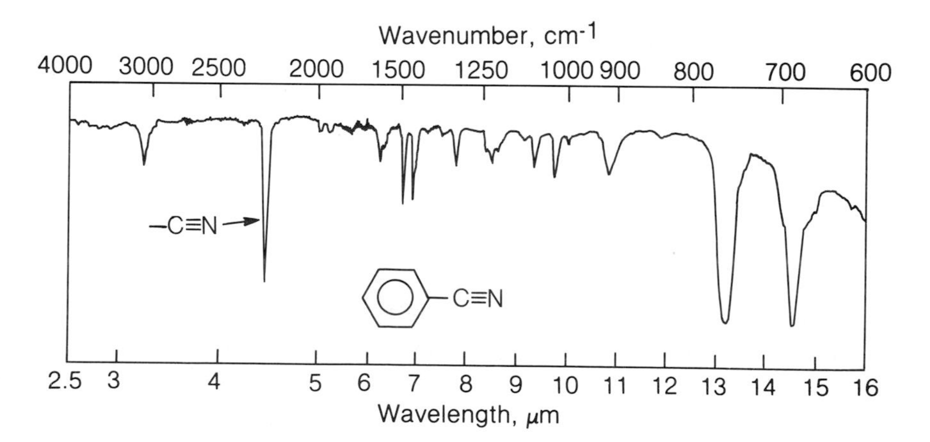

Figure 5.23
The ir spectrum of benzonitrile, a typical aryl nitrile.

5.3 NUCLEAR MAGNETIC RESONANCE (NMR)

Introduction

Nuclear magnetic resonance (nmr) spectroscopy is a recent development, even by chemical standards. The first nmr signals were observed in 1945 by Felix Bloch at Stanford (octane) and Edward Purcell at Harvard (water). The three-line spectrum of ethanol was reported in 1951, and in 1953 Bloch and Purcell shared the Nobel prize. By that year, Varian Associates had delivered three nmr machines to Exxon, DuPont, and Shell.

Theory

It is known that a moving electric charge creates a magnetic field. Atomic nuclei, which are known to have a charge, should also create a magnetic field if they spin. Many isotopes have what appears to be a mechanical spin, to which a spin angular momentum is assigned. All microphysical systems are quantized, and it is the spin number which concerns us here. The spin number is the maximum observable angular momentum for the nucleus. For purposes of this discussion, it will suffice to say that certain nuclei exhibit this property. For example, ^{1}H, ^{13}C, ^{15}N, ^{19}F, and ^{31}P all have spins of $\frac{1}{2}$. Among the commonly encountered elements which have a spin of 1 are ^{2}H (deuterium, a hydrogen isotope) and ^{14}N. Other nmr-active nuclei include lithium, boron, chlorine, and one of the isotopes of oxygen. Frequently encountered nuclei which have no spin are ^{12}C, ^{16}O, and ^{32}S.

Every isotope with a spin not equal to zero will be characterized by a nuclear magnetic moment, which is represented by a symbol μ. This can be thought of as a bar magnet with a strength μ. If the nucleus (bar magnet) is placed in a magnetic field, there will obviously be an interaction. Like a bar magnet, the nucleus must be either attracted to or repelled by the magnetic field. Since only two possibilities exist for a system with a spin of $\frac{1}{2}$, there are only two possible orientations in the magnetic field, referred to as plus and minus. Thus it is clear that the nmr method requires a magnetic field as well as an external energy source. This is different from the ir and uv techniques, which require only a sample and incident radiation.

The result of some simple mathematics (not discussed here) reveals that an energy transition from a minus to a plus state (i.e., from $-\frac{1}{2}$ to $+\frac{1}{2}$) is equal to $\gamma H_0 h/2\pi$. This is further equal to $h\nu_0$ on the general principle that $E = h\nu$.

$$\Delta E = \frac{\gamma H_0 h}{2\pi} = h\nu_0 \tag{5.10}$$

where γ = the magnetogyric ratio, a nuclear characteristic
H_0 = the applied magnetic field
h = Planck's constant

By rearrangement, it can be seen that $2\pi\nu_0 = \gamma H_0$. This is the so-called resonance condition. While it is not important to memorize this equation, remember that if there is no magnetic field, i.e., if $H_0 = 0$, there will be no difference in energy levels, the whole equation will reduce to zero, and no nmr phenomenon will be observed. The relationship between energy separation and magnetic field may be illustrated by a Zeeman diagram, shown in Fig. 5.24.

If this analogy is taken further, we note that the larger H_0 is, the larger will be the frequency difference ν_0. Obviously, it will be easier to observe a spectrum when a larger magnetic field is applied. The current maximum possible is 60,000 to 70,000 gauss (G), but it is very difficult to maintain magnetic field uniformity at these strengths. The most common compromise between sensitivity and economy results in the use of a 10,000-G field and a radio frequency of 60 million cycles/sec (60 MHz).

It is relatively easy to conceptualize what happens in the nmr experiment. In the absence of a magnetic field, the nuclear spins are randomized in all possible directions. When a magnetic field is applied, the spins tend to be oriented either in the same direction as the applied magnetic field (low energy state) or opposite to it (high energy state). As the molecule encounters incident radiation, energy absorption occurs and one of the spins flips direction, i.e., a nucleus in a low energy state changes orientation and goes to a high energy state. This energy absorption is what the nmr system detects.

As energy cannot be absorbed indefinitely, there must be some mechanism by which those spins in a high energy state can lose energy. This energy loss, or *relaxation,* cannot occur by fluorescence or phosphorescence, as is possible in uv and ir spectroscopy. Energy is lost by excited nuclei through a mechanism called *spin-lattice relaxation.* As the concepts involved here are beyond the scope of this book, the reader is referred to any of the standard texts on magnetic resonance for a comprehensive discussion of this phenomenon (see References).

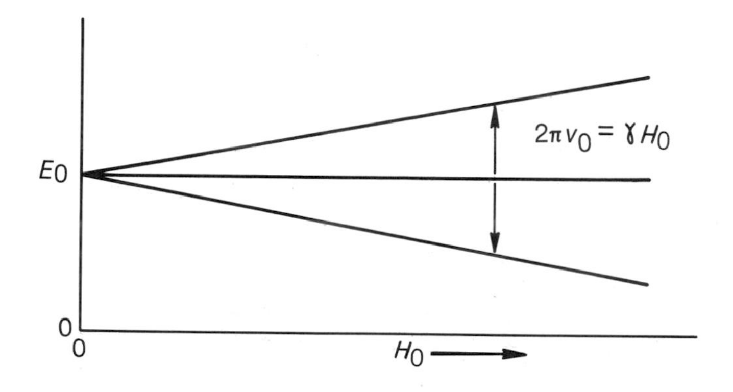

Figure 5.24 Zeeman diagram. Relationship of energy separation and magnetic field.

Instrumentation

The apparatus required for an nmr experiment is shown in Fig. 5.25. In the center of the diagram is a sample which is in the presence of a uniform magnetic field. Energy is put in (radio-frequency transmitter) and the resulting energy loss is detected (radio-frequency receiver) and recorded on a chart as a peak. In principle, the spectrum can be observed by varying either the magnetic field strength or the radio frequency of the spectrometer. In practice, it is generally easier to vary the magnetic field than to alter the frequency.

The Chemical Shift

According to the discussion above, it might be presumed that all protons absorb at the same position or have the same resonance condition. Fortunately, the actual absorption positions vary somewhat depending on the local electronic environment. Circulating electrons generate a local magnetic field opposite to that of the applied magnetic field. This local field tends to shield the nucleus from the applied magnetic field. The electron density about a hydrogen nucleus (or any other nucleus, for that matter) will depend to a first approximation on the inductive effect of the other groups attached to the same atom. This means that local environmental factors will change the resonance position; therefore, not all protons are observed at the same combination of field and frequency. These chemical factors account for the so-called chemical shift. (The chemical shift is really of a relatively small magnitude, approximately 600 Hz in a total of more than 60 MHz for ^{1}H.) The range of chemical shifts is approxi-

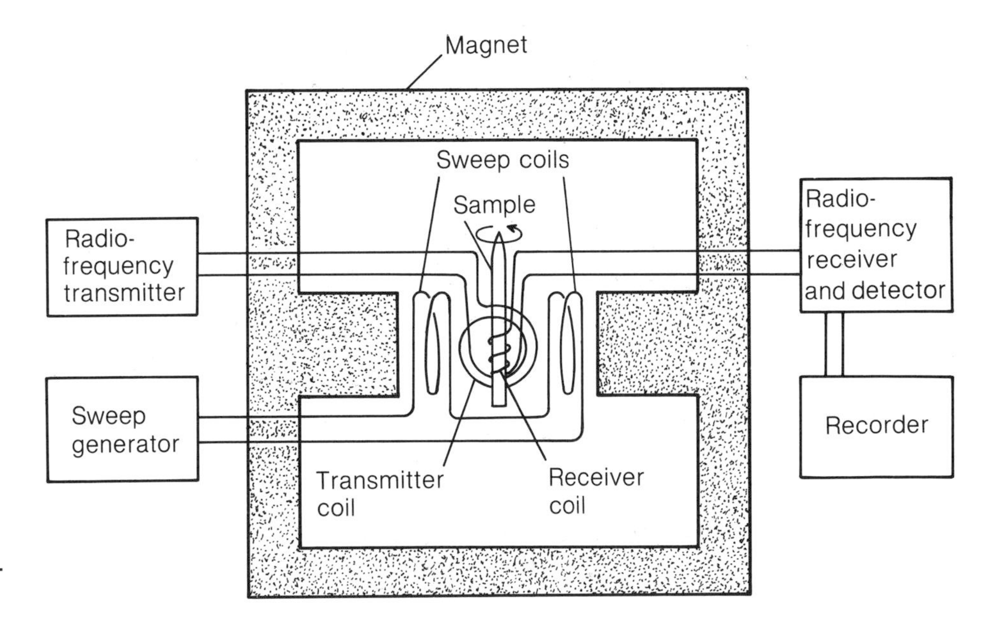

Figure 5.25 Schematic diagram of an nmr spectrometer.

mately 600 Hz/(60 × 10^6) Hz = 10 parts in a million = 10 ppm. Because it is difficult to accurately measure this small difference, an internal standard is required for comparison. Tetramethylsilane (TMS), which is illustrated below,

$$\begin{array}{c} CH_3 \\ | \\ CH_3{-}Si{-}CH_3 \\ | \\ CH_3 \end{array}$$

is the most commonly used standard, and proton positions are assigned relative to it. The chemical shift of acetone, for example, is 2.05 ppm downfield from TMS. Chemical shift values are given in parts per million downfield from TMS which is sometimes called the delta (δ) scale. Another common scale is the τ scale (τ = 10 − δ).

The general trends in nmr can be illustrated quite easily. Electronegative elements tend to deshield (move downfield) the resonance of a proximate proton. The chemical shifts for a series of related compounds are: methyl fluoride 4.26; methyl chloride 3.05; methyl bromide 2.68; and methyl iodide 2.16 ppm.

The electronegativity effect is also approximately additive. For example, the chemical shifts for a series of related compounds are: methane approximately 1.0 ppm; chloromethane 3.05; dichloromethane 5.28; and trichloromethane (chloroform) 7.28 ppm. In addition to electronegativity effects, proton positions are determined by adjacent electron clouds. Double bonds and triple bonds will typically cause shielding and deshielding effects. The shielding cones for acetylene, benzene, the carbonyl group, and carbon-carbon double bonds are shown in Fig. 5.26. Everywhere a plus (+) is drawn, shielding (i.e., an upfield shift) is observed, and everywhere a minus (−) is shown, deshielding (i.e., a downfield shift) occurs.

The Coupling Constant

Another fortunate complication in nmr spectroscopy is the phenomenon known as *spin-spin coupling*. In essence, this is the coupling of proton spins due to the intervention of bonding electrons. For a proton where there are two possible

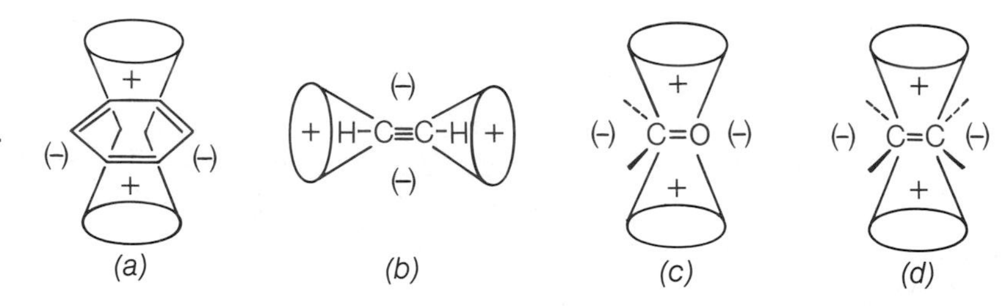

Figure 5.26 Shielding cones for (*a*) benzene, (*b*) acetylene, (*c*) the carbonyl group, and (*d*) carbon-carbon double bonds.

spin states ($\pm\frac{1}{2}$), the nuclear spin will tend to align itself with that of the bonding electron adjacent to it. The pair of bonding electrons which is influenced by the nuclear spin will tend to orient the spin of the adjacent nucleus. As a consequence, the orientation of the spin in the second nucleus will respond to the orientation of the first nuclear spin through this series of orientation effects. This, as mentioned above, is known as *spin-spin interaction* or spin-spin coupling. The mechanism suggested by the discussion above and the drawing below is the *Fermi contact mechanism*.

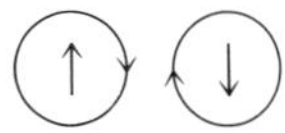

Any change in the orientation of one of the spins will influence the orientation of the other spin. For two nonidentical nuclei, the spin alignment will be affected slightly by the adjacent spins. For an absorption of energy E, the different spin orientations will cause a slight splitting of the resonance. The energy separation caused by this orientation effect is called the coupling constant, and is represented by the symbol J. Generally (to a first approximation), a nucleus coupled to a second nucleus will show a spin multiplicity in accordance with the formula $2nI + 1$, where n is the number of adjacent nuclei and I is the nuclear spin. This is illustrated quite simply for the molecule 1,1,2-trichloroethane (Fig. 5.27). The proton on the carbon bearing two chlorines is adjacent to two

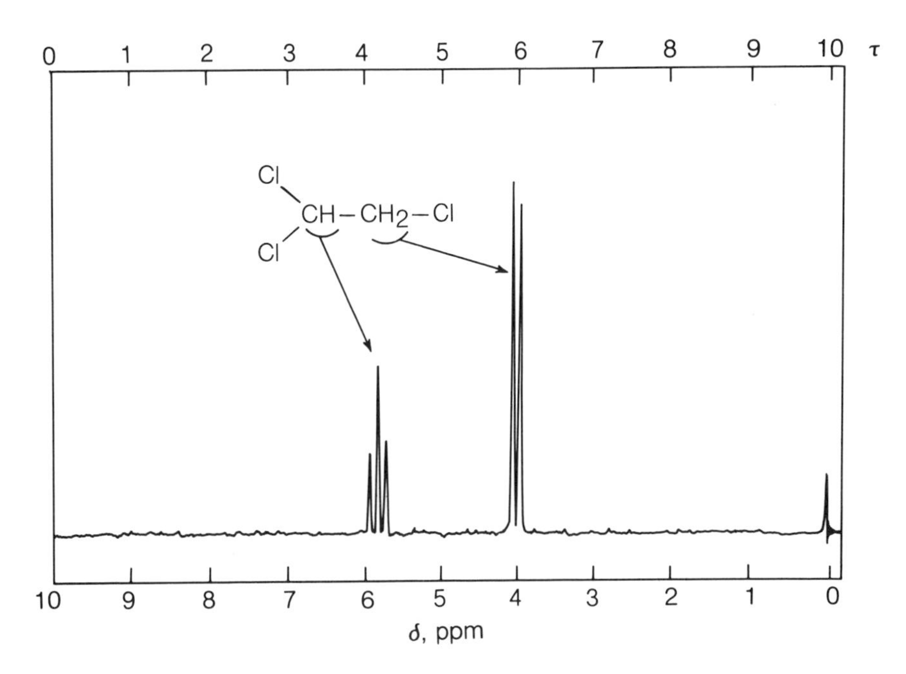

Figure 5.27
The nmr spectrum of 1,1,2-trichloroethane.

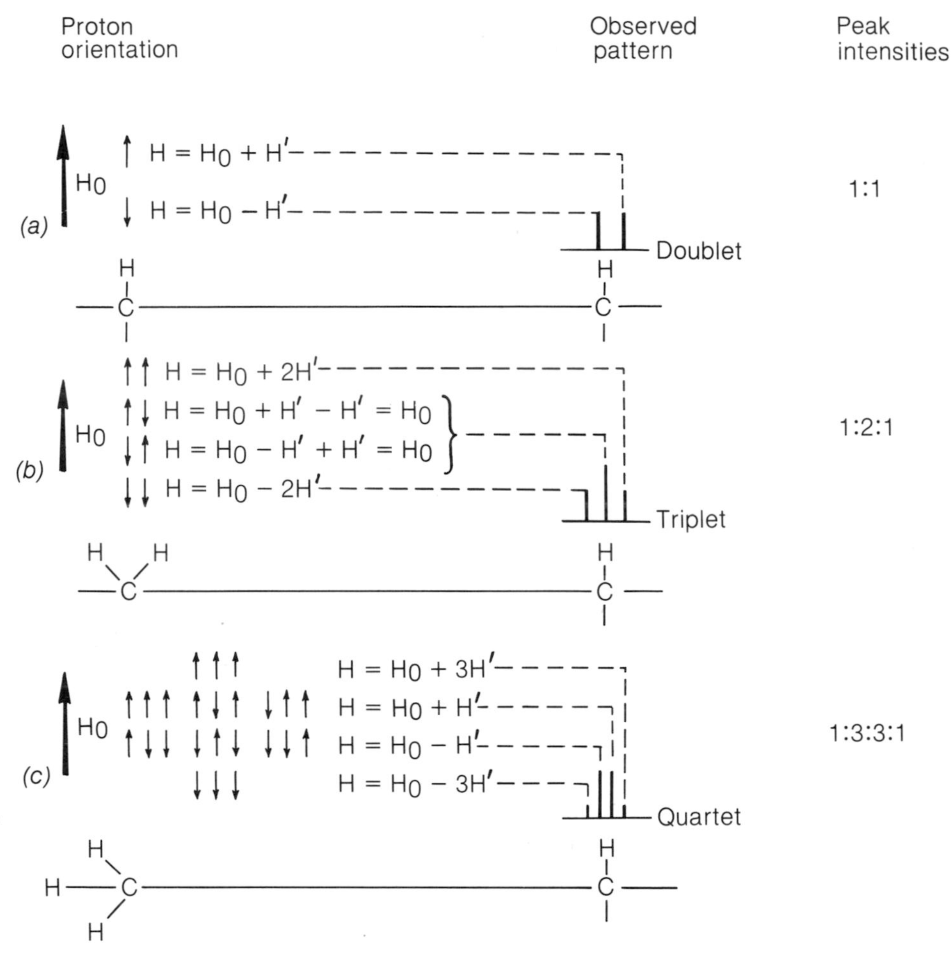

Figure 5.28 Different orientation of nuclear spins (left) causes splitting pattern observed on the right.

equivalent protons. Remember the spin number for the proton is $\frac{1}{2}$ ($I = \frac{1}{2}$); therefore the spin multiplicity should be $2 \times 2 \times \frac{1}{2} + 1 = 3$. This corresponds to the possible orientations of the two proton spin states. As shown in Fig. 5.28*b*, they can be down-down (opposed to the magnetic field), up-down and down-up, or up-up (aligned with the magnetic field). These three possible orientations are all separated in energy by the coupling constant J. The chemical shift pattern this produces will be three lines in the intensity ratio 1:2:1 or, corresponding to the spin orientations, 2:4:2 (which reduces to the same thing). These intensities reflect the fact that the paired spins are twice as probable as either of the other orientations. For simple systems, the spin multiplicity given by the formula $2nI + 1$ will show intensity ratios that conform to the coefficients of a binominal expansion. These are given by Pascal's triangle (Fig. 5.29) as expansions of the expression $(A + B)^n$.

```
                1
              1   1
            1   2   1
          1   3   3   1
        1   4   6   4   1
      1   5  10  10   5   1
```

Figure 5.29
Pascal's triangle.

The spectra shown later in this chapter (Figs. 5.31 through 5.40) are for ethyl iodide, *n*-propyl iodide, isopropyl iodide, diethyl ether, di-*n*-propyl ether, di-isopropyl ether, triethyl orthoformate, 1,3-dibromopropane, propiophenone, and *n*-butyrophenone. Note that in all these systems the spin multiplicities appear as would be anticipated from the discussion above. Much more complicated systems are certainly known, but they are beyond the scope of this discussion.

Decoupling a Resonance

Sometimes it is very difficult to interpret an nmr spectrum because the coupling is complex. In such cases it is sometimes possible to electronically *decouple* nuclei. In this technique, one of a pair of coupled nuclei is irradiated and the other is observed. The energy input causes the possible orientations of one nucleus to randomize. In this situation, the adjacent nucleus will not be influenced by different orientations and will be observed as a singlet (unless coupled to a third nucleus). This method of simplifying a spectrum is valuable but requires special equipment and skill.

Another means of simplifying a spectrum involves exchange with D_2O. This process is a chemical decoupling technique which is very useful in detecting acidic protons, particularly those on alcohols (ROH) and carboxylic acids (RCOOH). This technique is carried out as follows.

After a spectrum has been recorded, the sample tube is removed from the spectrometer and a drop of D_2O is added. After brief shaking, the spectrum is again recorded. Acidic protons will exchange with D_2O, and the resonance due to their presence will no longer be observed at the previous position. A new peak due to partially deuterated water (HOD) will be observed. This technique is often used for identifying an acidic proton. *Note:* In the nmr spectra presented in this chapter and elsewhere in this book, the solvent is not specified. To avoid confusion, certain solvent-related impurity peaks have been deleted.

Determining a Structure from a Spectrum

Typical chemical shifts are shown in Fig. 5.30. By using this chart and the information presented above, it should be possible to determine the structure of most simple compounds. The best approach to use is to correlate the line posi-

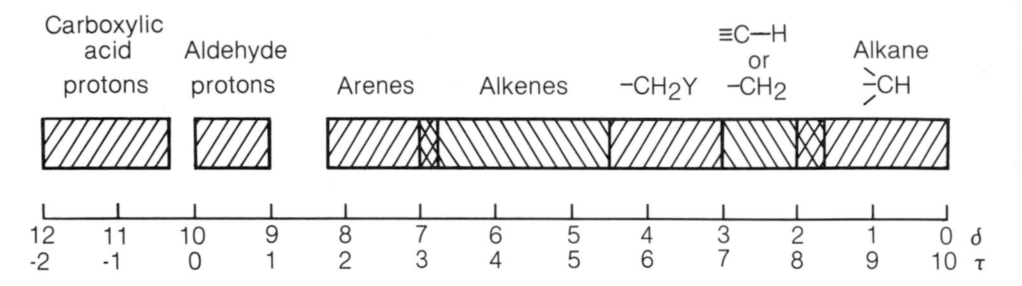

Figure 5.30 Proton chemical shift ranges. Y is an electronegative element or group such as oxygen, the nitro group, or a halogen.

tions with various functional groups by using the chart and comparing with the spectra shown in this chapter (Figs. 5.27 and 5.31 to 5.40). Once the resonances have been tentatively identified, examine the coupling to see if the number of protons and their couplings correlate. By trial and error, most simple structures can be identified.

More detailed information is available by consulting one or more of the references listed at the end of this chapter. In addition, some information concerning the effect of functional groups on spectra is available in Chaps. 23 to 27. Finally, most texts used for basic organic chemistry include a full chapter detailing the use of spectral methods in structural analysis.

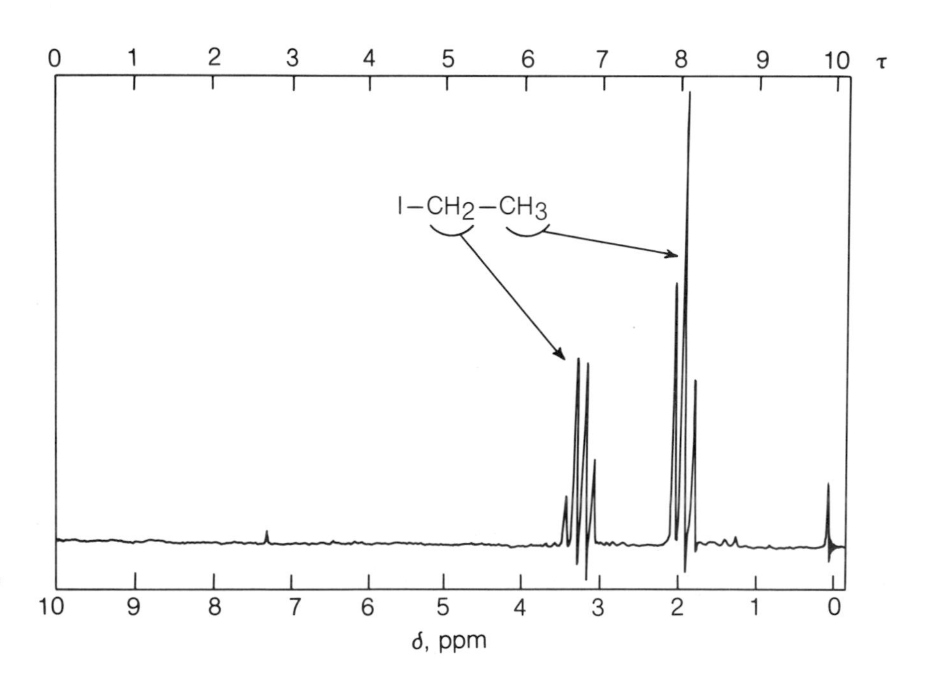

Figure 5.31 The nmr spectrum of iodoethane (ethyl iodide).

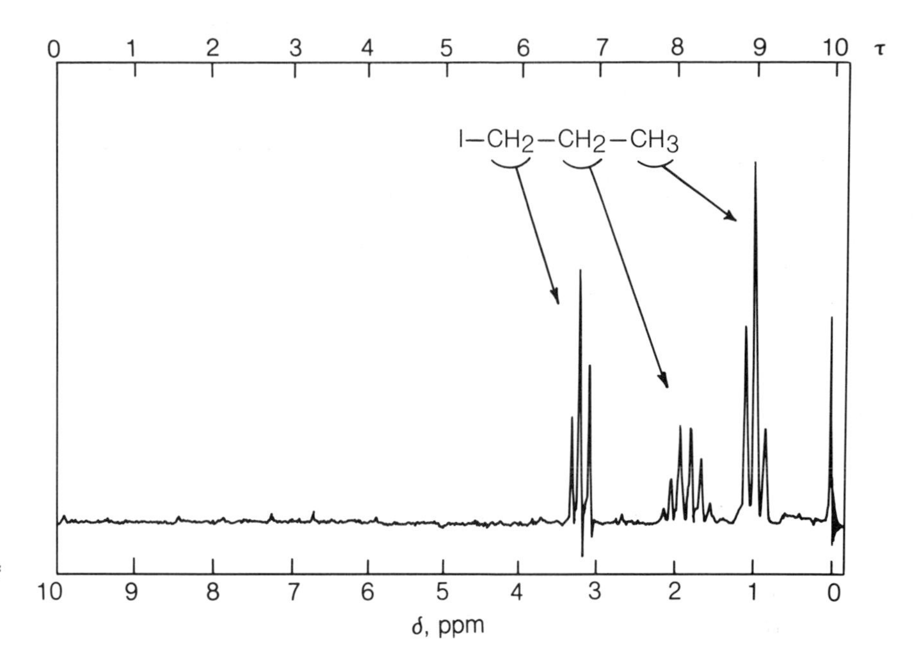

Figure 5.32
The nmr spectrum of 1-iodopropane (*n*-propyl iodide).

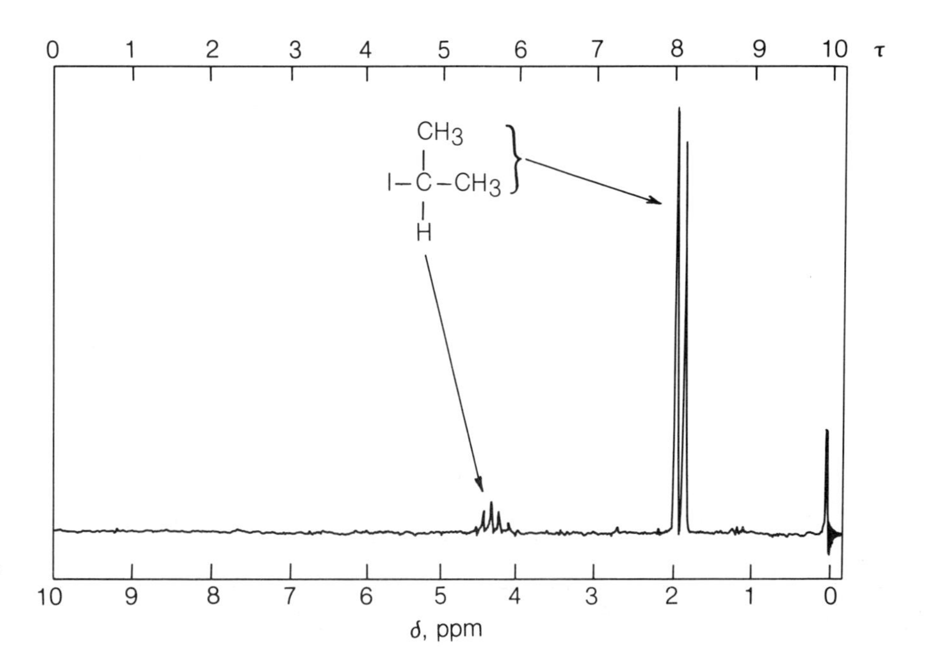

Figure 5.33
The nmr spectrum of 2-iodopropane (isopropyl iodide).

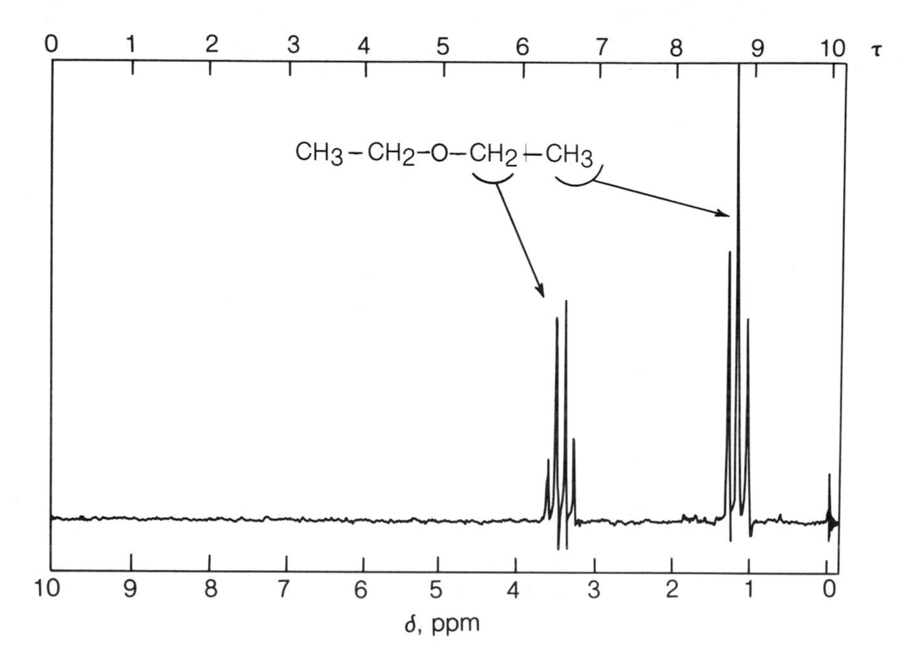

Figure 5.34
The nmr spectrum of ethyl ether.

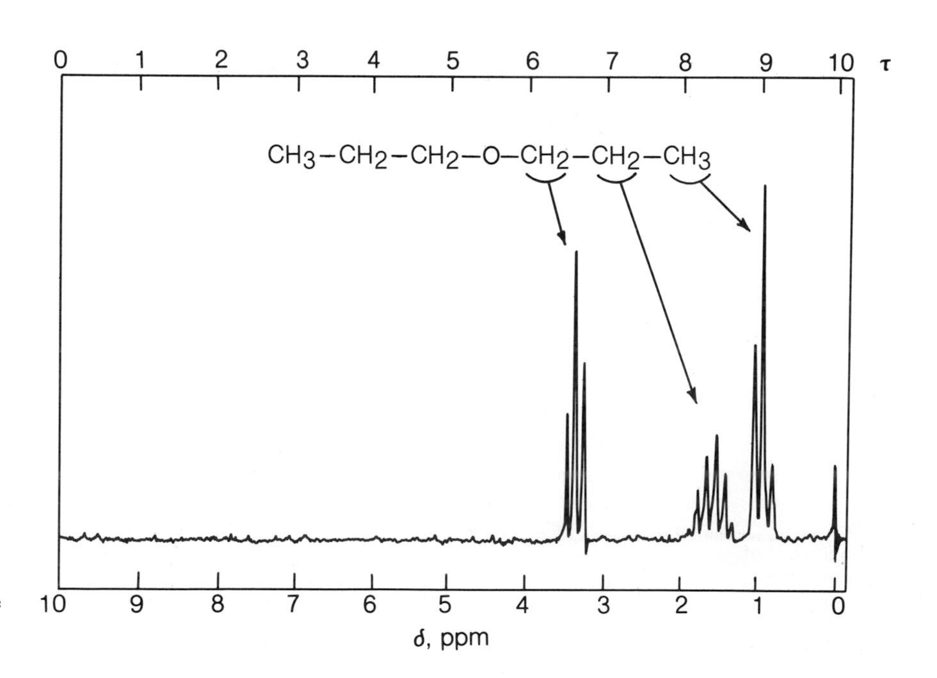

Figure 5.35
The nmr spectrum of *n*-propyl ether.

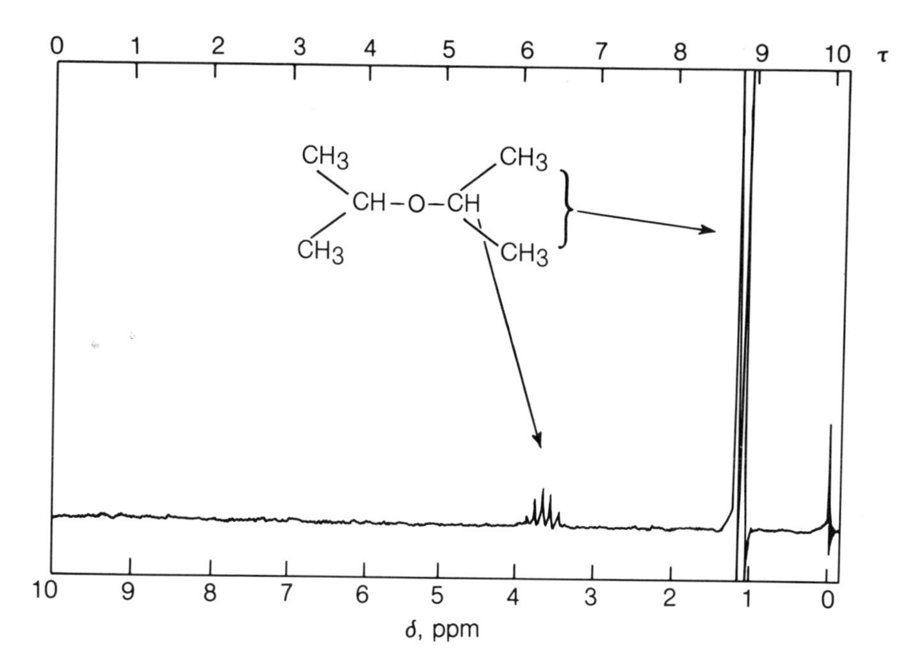

Figure 5.36
The nmr spectrum of isopropyl ether.

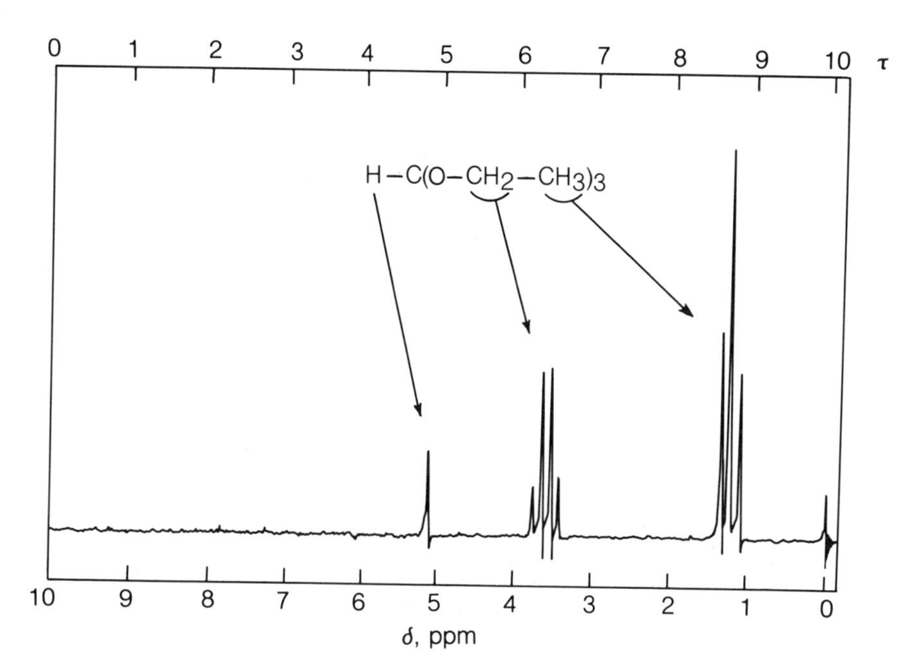

Figure 5.37
The nmr spectrum of triethyl orthoformate.

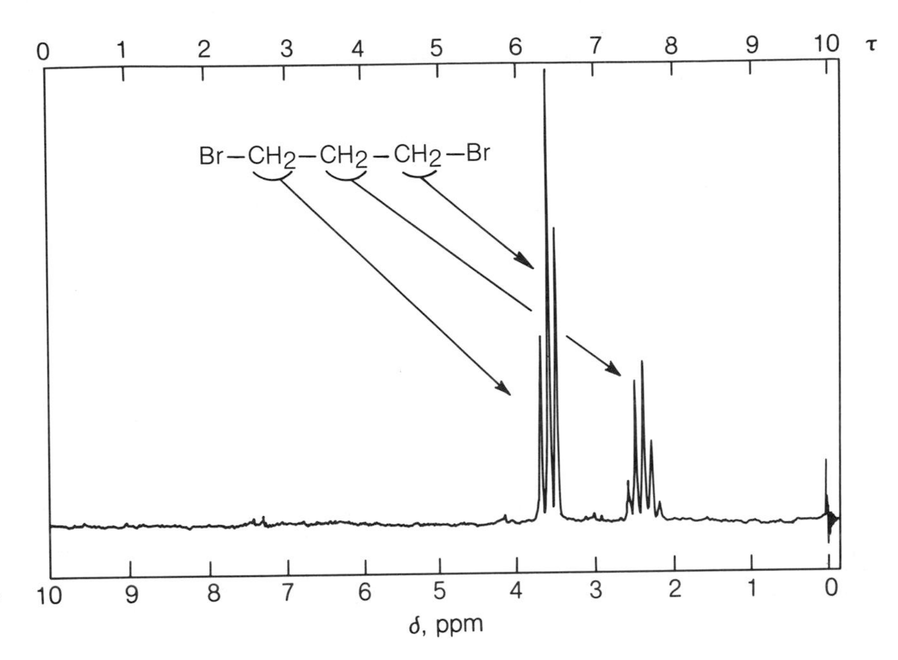

Figure 5.38
The nmr spectrum of 1,3-dibromopropane.

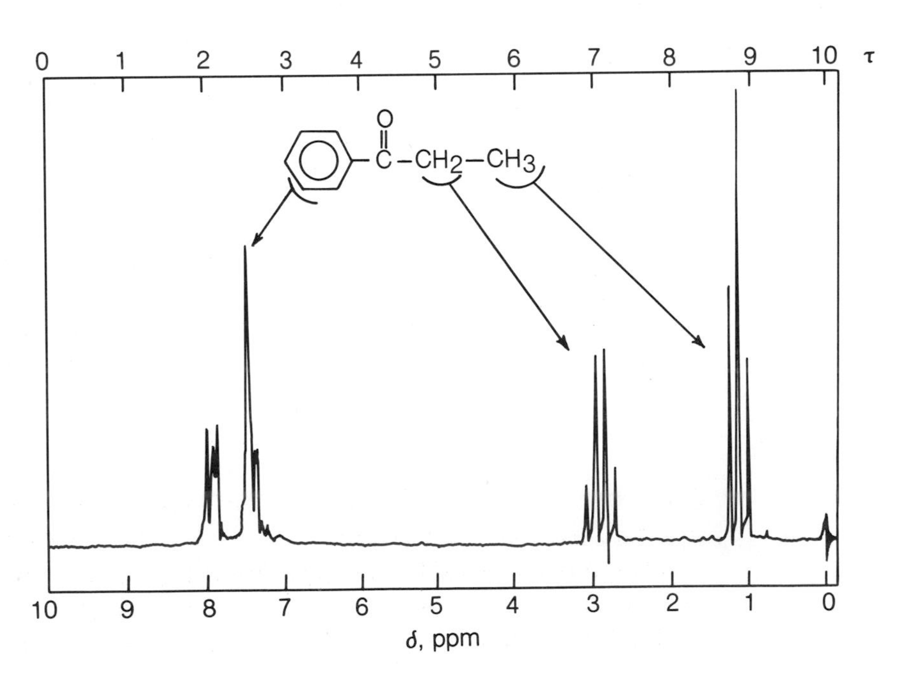

Figure 5.39
The nmr spectrum of propiophenone.

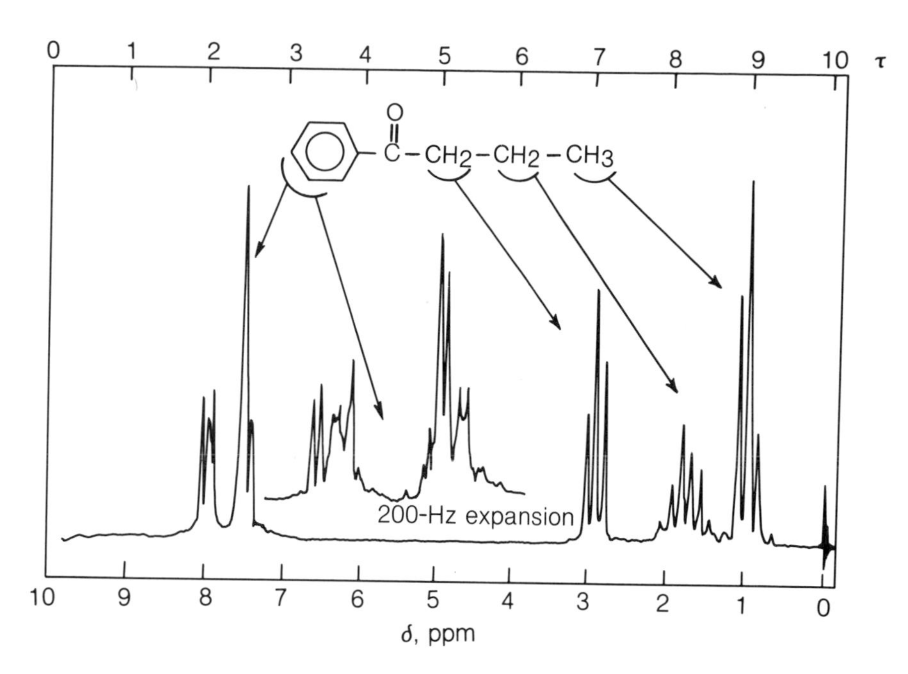

Figure 5.40
The nmr spectrum of *n*-butyrophenone.

5.4 MASS SPECTROMETRY

Introduction

Mass spectrometry differs from infrared, ultraviolet, and nuclear magnetic resonance spectroscopy in a fundamental way. The ir, uv, and nmr techniques all involve the absorption of electromagnetic radiation. Mass spectrometry, on the contrary, owes its utility to the impact of high-energy electrons, which cause ionization of molecules when they impinge on the sample.

Theory

A molecular sample, represented by M in the equation below, is vaporized. The gaseous sample of M is bombarded by electrons, represented in the equation by *e*. An electron is "knocked off" the sample to produce an ion, represented by M^+. The molecule which has lost an electron but no organic fragment is referred to either as the molecular ion (M) or the parent ion (P).

$$e + M \rightarrow M^+ + 2e$$

The molecular ion is a highly energetic and unstable species. This species dissipates its high energy by undergoing a series of fragmentation reactions to form smaller ions and neutral molecules. The mass/charge ratio (m/e) of any ion can be detected and recorded by the mass spectrometer. From these values, the molecular weight and structure of a compound may often be determined.

Instrumentation

A number of different devices exist for ionizing molecules and then recording their mass fragmentation patterns. These devices all go by the name *mass spectrometer*. Any of these devices must have, at a minimum, an ionizing chamber, an ion accelerator, an ion-focusing device, and a detector coupled to a recorder. Since the instrumentation of mass spectrometry is less standard than that of other spectrometric techniques, no illustration is included here. The reader is referred to one of the more detailed books on this subject which are cited at the end of this chapter. Suffice it to say that the ions which are produced in a mass spectrometer can be detected and recorded. These ions can then be identified, and structural information can be derived therefrom.

Molecular Weight Determination

The ions which are detected are based on whole atomic weights. Most elements consist of more than one isotope, so atomic weights are usually not whole numbers. For example, atomic chlorine consists of a mixture of chlorine isomers of weights 35 and 37 amu in a 3:1 ratio. The atomic weight of the isotope mixture is 35.5. Since the mass spectrometer detects individual ions, only exact masses are detected.

Consider, for example, the molecule methyl chloride. The molecular weight of methyl chloride is 50.5 amu. In fact, there are two kinds of methyl chloride in this mixture, one arising from chlorine 35 and the other arising from chlorine 37. The mass spectrometer will detect a nearly 3:1 mixture of the two compounds as molecular ions (M^+) at m/e 50 and m/e 52.

The molecular ion usually appears at the highest mass (m/e) in the spectrum. Using exact atomic weights, the molecular weight can be calculated. The exact atomic weights used in mass spectrometry are: hydrogen, 1; carbon, 12; nitrogen, 14; oxygen, 16; and fluorine, 19. Chlorine is a mixture of isotopes of atomic weights 35 and 37, and bromine is a mixture of isotopes of atomic weights 79 and 81. If the latter two elements are present, two peaks should be observed for any molecular ion (or other ion) containing them.

When working problems involving mass spectrometry, other spectral evidence should be used to suggest structures and these either confirmed or discarded on the basis of molecular weight information.

Other Fragmentation Processes

The highly energetic molecular ion loses energy by undergoing cleavage and rearrangement reactions. The details of this process are beyond the scope of this book, but a typical fragmentation reaction series is shown on p. 187 for 3-methyl-4-phenyl-2-butanone. Note that a series of ions is formed and these may be detected. The fragmentation reactions all occur according to chemical principles, and a structure may often be reconstructed from an analysis of the ions detected. For further information about this important and sophisticated technique refer to one of texts cited at the end of this section.

Cation	Observed mass peak
Parent ion	162
A: $^{+}O{:}\equiv C-CH(CH_3)-CH_2C_6H_5$	147 (162 − 15)
B: $CH_3-C\equiv\overset{+}{O}{:}$	43
C: $C_6H_5-CH_2^{+}$	91

$CH_3 \,|\, C(=O) \,|\, CH(CH_3) \,|\, CH_2-C_6H_5$ (cleavage sites A, B, C)

3-Methyl-4-phenyl-2-butanone
(MW 162)

QUESTIONS AND EXERCISES

5.1 Although a student had carefully washed all the glassware with acetone before distilling the product, a forerun was obtained in the distillation. The principal constituent of this forerun was found to be a colorless liquid, bp 56°C. The refractive index was determined to be 1.359. A single peak was observed in the nmr spectrum at 2.05δ, and a strong band at 1715 cm^{-1} was observed in the ir spectrum. What might have been the contaminant?

5.2 A compound was subjected to a variety of tests and determined to be either phenylacetone, 4-ethylbenzaldehyde, 3-phenylpropionaldehyde, or propiophenone. Infrared analysis showed a strong absorption at 1715 cm^{-1}, and the nmr spectrum showed a broad singlet at 7.21δ (5 H), a singlet at 5.4δ (2 H), and a singlet at 2.1δ (3 H). Which of the four possibilities is consistent with the data?

5.3 A compound was obtained from a sample of gasoline and presumably was a substance added by the manufacturer to increase the octane rating of the fuel. It had the following properties: bp 83°C; density 0.785 g/mL; and refractive index 1.386. Two peaks were observed in the nmr spectrum, a large singlet at 1.1δ and a small singlet at 3.2δ. Addition of D_2O to the sample tube followed by a second scan of the spectrum indicated that only the small peak disappeared. A strong band between 3400 and 3600 cm^{-1} was observed in the ir spectrum. What might this octane-improving additive be?

5.4 Cyclopropane reacted with bromine at room temperature to form a mixture of products. Among these was a compound with the following properties: bp 166 to 169°C, density 1.989, and refractive index of 1.5214. Other than C—H absorption, the ir spectrum showed no strong peak between 3600 and 1500 cm^{-1}. The nmr spectrum of this oil showed only a triplet centered at 3.6δ and a pentet centered at 2.4δ. Reaction of this compound with zinc gave cyclopropane as the major product. What might the compound in question be?

5.5 The nmr spectrum of diethyl benzylphosphonate prepared for use in the Wittig reaction is shown in Sec. 20.1. Examine the spectrum and assign all the peaks in the spectrum. Pay special attention to the splitting pattern you observe.

5.6 An oil which had the odor of peaches was isolated by steam distillation from a plant source. The compound boiled at 206°C and had a density about the same as that of water. The refractive index was measured and found to be 1.500. Three singlets were observed in the nmr spectrum at 7.2, 5.1, and 2.1 ppm downfield from TMS. From the integral it was found that these peaks had an intensity ratio of 5:2:3, respectively. A strong peak was observed at 1745 cm^{-1} in the ir spectrum. A parent ion was detected in the mass spectrum at m/e 150. When this compound was heated with aqueous KOH solution, the product mixture smelled of vinegar. What might the peachy-smelling compound be?

5.7 When benzyl chloride ($C_6H_5CH_2Cl$) was heated with 50% aqueous sodium hydroxide solution, a very slow reaction occurred. The product obtained from this reaction was insoluble in water (although its density was the same as that of water) but dissolved readily in toluene, and had bp 295 to 300°C and refractive index 1.561. When the Beilstein test (Sec. 3.3) was conducted on this compound, a smoky yellow flame was observed. Infrared analysis failed to reveal any absorption in either the 3200 to 3600 cm^{-1} or the 1650 to 1750 cm^{-1} region. Two peaks were observed in the nmr spectrum, a broad singlet at 7.3δ (5 H) and a sharp singlet at 4.5δ (2 H). Suggest a structure for this compound which is consistent with both the spectral and chemical data.

5.8 When diphenylacetylene (tolan, Exp. 8.1) is reduced by one method, a crystalline compound, mp 125°C, is formed. If diphenylacetylene is reduced by another method, the product obtained is an oil. Both reduction products react with bromine and potassium permanganate, and tests indicate that a nonaromatic double bond is present in both (see Secs. 7.2 and 26.4B). The nmr spectrum of each compound showed only complex absorptions between 7 and 8δ. Ultraviolet spectral analysis gave a λ_{max} at 301 nm for the solid and a λ_{max} at 280 nm for the oil. Elemental analysis of both the solid and the oil indicates the empirical formula $C_{14}H_{12}$. Give structures for the oil and solid consistent with the above data.

5.9 A pleasant-smelling oil (A) was isolated from a plant source by steam distillation (Sec. 2.2). The oil was a water-white, mobile liquid which had the following physical properties: bp 197 to 200°C, density 1.1 g/mL, and refractive index 1.516. No absorption was apparent in the ir spectrum in the 3200 to 3600 cm^{-1} region, but there was a strong, sharp peak at 1726 cm^{-1}. The substance burned with a yellow, sooty flame and the Beilstein test (Sec. 3.3) was negative. The nmr spectrum exhibited three bands: a sharp singlet at 3.0δ (3 H), a complex absorption at 7.5δ (3 H), and another complex absorption centered at 8.0δ (2 H). When this compound was heated for 1 h with dilute, aqueous sodium hydroxide solution and then treated with HCl, a white solid (B), mp 120 to 122°C, was obtained. The solid was almost insoluble in cold water, more soluble in hot water, and completely soluble in 5% aqueous sodium bicarbonate solution. Suggest structures for compounds A and B.

5.10 In an exploratory reaction conducted in an industrial laboratory, propylene oxide, $H_2C\overset{O}{\frown}CH—CH_3$, was treated with a catalyst in the presence of water. The material actually isolated had the following properties: bp 46 to 50°C, density 0.805 g/mL, and refractive index 1.3650. Infrared analysis revealed a strong absorption at 1725 cm^{-1} but no absorption between 3200 and 3600 cm^{-1}. The material was water soluble and had a sharp, unpleasant odor. Are the physical characteristics of the above compound consistent with the expected product? If not, suggest a material whose chemical characteristics are consistent (use derivative tables, Chap. 28).

5.11 Treatment of acetone with barium hydroxide slowly converted the acetone into another compound, bp 166°C. The nmr spectrum was complex and contained several absorptions between 1 and 3.5δ. Infrared spectral analysis showed a strong peak at 3400 to 3600 cm^{-1} and a sharp, strong peak at 1702 cm^{-1}. Treatment of this compound with acid converted it to another compound, bp 129°C. Infrared spectral analysis of this material revealed no absorption at 3400 to 3600 cm^{-1}, but a strong peak at 1680 cm^{-1} was observed. Suggest structures for both compounds whose properties are discussed above.

5.12 An unknown solid, mp 50 to 52°C, was given to a student as an unknown. The material burned with a sooty flame, and when it was burned in the presence of copper (Beilstein test, Sec. 3.3) a very bright green flame was observed. Mass spectral analysis showed two equally intense peaks (parent ions) at m/e 198 and 200. Infrared analysis showed a strong absorption at 1690 cm^{-1}. The nmr spectrum showed a singlet at 2.6δ (3 H), a doublet at 7.6δ (2 H), and another doublet at 7.85δ (2 H). When the unknown reacted with 2,4-dinitrophenylhydrazine reagent (see Secs.

24.4A and 25.5A), a red precipitate (mp 230°C) was obtained. Suggest a structure for the unknown compound.

REFERENCES

General

Creswell, C. J., O. Runquist, and M. M. Campbell: *Spectral Analysis of Organic Compounds,* 2d ed., Burgess, Minneapolis, 1972.

Drago, R. S.: *Physical Methods in Inorganic Chemistry,* Reinhold, New York, 1965.

Ewing, G. W.: *Instrumental Methods of Chemical Analysis,* 4th ed., McGraw-Hill, New York, 1975.

Gouw, T. H.: *Guide to Modern Methods of Instrumental Analysis,* Wiley-Interscience, New York, 1972.

Kemp, W.: *Organic Spectroscopy,* Halsted-Wiley, New York, 1975.

Lambert, J. B., H. F. Shurvell, L. Verbit, R. G. Cooks, and G. H. Stout: *Organic Structural Analysis,* Macmillan, New York, 1976.

Laszlo, P. L., and P. J. Stang: *Organic Spectroscopy,* Harper and Row, New York, 1975.

Parikh, V. M.: *Absorption Spectroscopy of Organic Molecules,* Addison-Wesley, Reading, Mass., 1974.

Pasto, D. J., and C. R. Johnson: *Organic Structure Determination,* Prentice-Hall, Englewood Cliffs, N.J., 1969.

Pecsok, R. L., and L. D. Shields: *Modern Methods of Chemical Analysis,* Wiley, New York, 1976.

Silverstein, R. M., G. C. Bassler, and T. C. Morrill: *Spectrometric Identification of Organic Compounds,* 3d ed., Wiley, New York, 1974.

Williams, D. H., and I. Fleming: *Spectroscopic Methods in Organic Chemistry,* 2d ed., McGraw-Hill, New York, 1973.

Ultraviolet Spectroscopy

Gillam, A. E., and E. S. Stern: *An Introduction to Electronic Absorption Spectroscopy in Organic Chemistry,* Edward Arnold, London, 1967.

Jaffe, H. H., and M. Orchin: *Theory and Application of Ultraviolet Spectroscopy,* Wiley, New York, 1962.

Rao, C. N. R.: *Ultraviolet and Visible Spectroscopy,* 2d ed., Butterworths, London, 1967.

Infrared Spectroscopy

Avram, M., and G. H. D. Mateescu: *Infrared Spectroscopy,* Wiley, New York, 1972.

Bellamy, L. J.: *The Infrared Spectra of Complex Molecules,* 3d ed., Wiley, New York, 1975.

——: *Advances in Infrared Group Frequencies,* Methuen, London, 1968.

Conley, R. T.: *Infrared Spectroscopy,* 2d ed., Allyn and Bacon, Boston, 1972.

Nakanishi, K., and P. H. Solomon: *Infrared Absorption Spectroscopy,* Holden-Day, San Francisco, 1977.

Pouchert, C. J.: *The Aldrich Library of Infrared Spectra,* 2d ed., Aldrich Chemical Co., Milwaukee, 1975.

Szymanski, H. A., and R. E. Erickson: *Infrared Band Handbook,* 2nd ed., Plenum, New York, 1970.

Nuclear Magnetic Resonance Spectroscopy

Abraham, R. S., and P. Loftus: *Proton and Carbon-13 NMR Spectroscopy,* Heyden and Son, Philadelphia, 1978.

Becker, E. D.: *High Resolution NMR,* Academic Press, New York, 1969.

Bersohn, M., and J. C. Baird: *An Introduction to Electron Paramagnetic Resonance,* Benjamin-Cummings, Reading, Mass., 1966.

Bible, R. H.: *Interpretation of NMR Spectra: An Empirical Approach,* Plenum, New York, 1965.

Bovey, F.: *Nuclear Magnetic Resonance Spectroscopy,* Academic Press, New York, 1969.

Carrington, A., and A. D. McLachlan: *Introduction to Magnetic Resonance.* Harper and Row, New York, 1967.

Emsley, J. W., J. Feeney, and L. H. Sutcliffe: *High Resolution Nuclear Magnetic Resonance Spectroscopy,* Pergamon Press, Elmsford, N. Y., 1965.

Jackman, L. M., and S. Sternhell: *Applications of Nuclear Magnetic Resonance Spectroscopy in Organic Chemistry,* 2nd ed., Pergamon Press, Elmsford, N. Y., 1969.

Johnson, L. F., and W. C. Jankowski: *Carbon-13 NMR Spectra,* Wiley-Interscience, New York, 1972.

Levy, G. C., and G. L. Nelson: *Carbon-13 Nuclear Magnetic Resonance for Organic Chemists,* Wiley-Interscience, New York, 1972.

Mathieson, D. W.: *Nuclear Magnetic Resonance for Organic Chemists,* Academic Press, New York, 1967.

Pople, J. A., W. G. Schneider, and H. J. Bernstein: *High-Resolution Nuclear Magnetic Resonance,* McGraw-Hill, New York, 1959.

Pouchert, C. J., and J. R. Campbell: *The Aldrich Library of NMR Spectra,* vols. I–X, Aldrich Chemical Co., Milwaukee, 1975.

Roberts, J. D.: *An Introduction to the Analysis of Spin-Spin Splitting in Nuclear Magnetic Resonance,* Benjamin-Cummings, Reading, Mass., 1962.

——: *Nuclear Magnetic Resonance,* McGraw-Hill, New York, 1959.

Stothers, J. B.: *Carbon-13 NMR Spectroscopy,* Academic Press, New York, 1972.

Mass Spectrometry

McDowell, C. A.: *Mass Spectrometry,* McGraw-Hill, New York, 1963.

McLafferty, F. W.: *Interpretation of Mass Spectra,* Benjamin-Cummings, Reading, Mass., 1973.

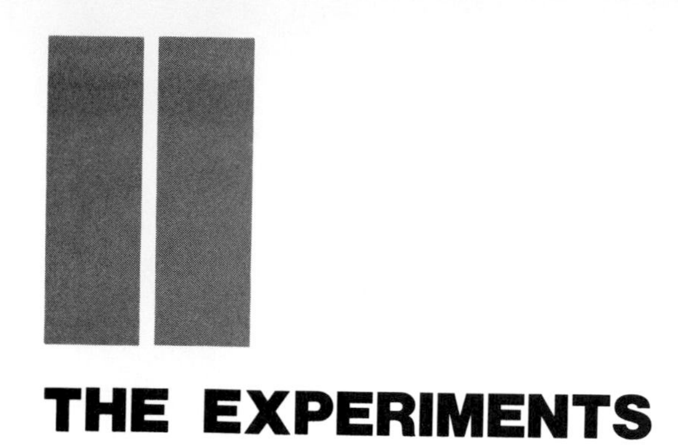

THE EXPERIMENTS

ALKANES

6.1 INTRODUCTION

Alkanes are the fully saturated organic compounds comprised solely of carbon and hydrogen. As such, they are the simplest class of organic compounds, being chains of carbon saturated with hydrogen. The general formula for alkanes is C_nH_{2n+2}.

Most alkanes are obtained by the distillation of various crude petroleum fractions. The distillation process may take place at a refinery, or preliminary separation of the gaseous fractions may be done at the wellhead. More than half of all natural gas is methane and in fact, natural gas may be almost pure methane. The gaseous alkanes are methane, ethane, propane, and butane; all these have boiling points below room temperature. The alkanes containing five or more carbon atoms are either liquids or solids. In Table 6.1, you will see listed the first eight normal (straight-chain) alkanes with their boiling points, empirical formulas, the number of possible isomers corresponding to each formula, and their densities.

It can be seen from Table 6.1 that the boiling points of the alkanes increase in a regular way with increasing molecular weight, although the increase is very much larger from methane to ethane than it is from heptane to octane. In general, among the liquid and solid alkanes the boiling points differ by 25 to 30°C for each additional CH_2 unit. Going down the table from gaseous to liquid alkanes, we see not only that the heavier compounds are liquid but that the number of possible structures associated with each formula begins to increase. Butane has two possible isomers: *n*-butane, the so-called normal or straight-chain alkane, and isobutane, which has three methyl groups bonded to a single carbon. Pentane has three isomers, hexane five, heptane nine, and so on. The number of possible isomers approximately doubles with each additional $—CH_2—$ unit. Pentadecane ($C_{15}H_{32}$) has more than 4000 possible isomers.

TABLE 6.1
Properties of normal alkanes

Name	bp, °C (normal isomer)	Formula	Number of isomers	Density, g/mL (normal isomer)
Methane	−161	CH_4	1	—
Ethane	−88	C_2H_6	1	—
Propane	−42	C_3H_8	1	—
Butane	−0.5	C_4H_{10}	2	—
Pentane	36	C_5H_{12}	3	0.645
Hexane	69	C_6H_{14}	5	0.660
Heptane	98	C_7H_{16}	9	0.684
Octane	121	C_8H_{18}	18	0.703

The lower-molecular-weight, gaseous alkanes are generally utilized as fuels in stationary power plants. By stationary power plants we mean furnaces, stoves, and similar appliances. Fuel lines containing these gases must be run from tanks or from various storage facilities directly to the place where the fuel is burned. It is much less convenient to fuel a mobile power plant such as an automobile with a gaseous alkane. Even propane and butane are often liquefied under pressure to facilitate storage and transfer.

The low-boiling alkanes in the next group are of considerable use as fuels also. Generally, alkanes boiling in the 100 to 150°C range can be utilized for gasoline. The boiling points must be sufficiently high that the gasoline does not boil out of an automobile engine when standing on a hot day. Likewise it must be sufficiently volatile that it can be injected into a carburetor and undergo complete combustion. Isooctane (2,2,4-trimethylpentane) is considered the standard for gasoline performance. In a test engine a pure sample of isooctane has a so-called octane number of 100 (by definition). Actually, pure isooctane is not utilized for high-performance gasoline, but mixtures of aliphatic and aromatic hydrocarbons are used to achieve the same high-performance properties associated with pure isooctane. Heptane is given the value 0 on the octane scale. Therefore a gasoline which has an octane rating of 90 has the same burning characteristics as a mixture of 90% isooctane and 10% heptane. As a general rule, the more branching in an alkane chain, the higher the octane rating. One of the most important tasks of the petroleum refining industry is to convert straight-chain alkanes into more highly branched materials, thus raising the octane number. The higher petroleum fractions are utilized as diesel fuel, jet fuel, solvents, heating oil, waxes, and so on. The higher alkanes may also be broken down into smaller units and used as chemical feedstocks.

Of particular interest to the organic chemist are the five-, six-, and seven-carbon alkanes. Mixtures of isomers of these materials are often used as organic solvents. As alkanes, they obviously dissolve many organic substances, particularly those which do not have polar functional groups such as carboxyl, nitro, or hydroxyl. Even in the presence of polar functional groups, if there are many carbons in the molecule relative to the number of functional groups, the alkanes will dissolve them.

Pentane is often used as a solvent because it dissolves many organic materials and it also can be quite easily removed by evaporation. Its boiling point is 36°C. Somewhat more common is hexane, which is a little easier to store and handle, yet still quite easy to remove.

We noted that the number of isomers begins to increase quite substantially from pentane to hexane to heptane and on up. Mixtures of alkane isomers are generally obtained in the distillation and cracking of petroleum. These light fractions of petroleum are usually mixtures of pentane, hexane, heptane, and their isomers and are often referred to as *light petroleum*. Equivalent names for these fractions are *petroleum ether* and *ligroin*.

Many organic preparations call for the use of hexane or pentane as a solvent. Almost as many call for the use of petroleum ether or ligroin as solvent. It is a common misconception that petroleum ether is an oxygen-containing compound because of the designation *ether*. In this case *ether* is used to imply volatility rather than the presence of oxygen. It is a name which is of historical significance and is less than systematic. It should also be noted that, if a preparation calls for pentane or hexane, either petroleum ether or ligroin may be substituted in most cases if it has the appropriate boiling range. For example, if a preparation calls for the use of hexane as a solvent, a petroleum ether fraction designated by its 60 to 80°C boiling range could also be a suitable solvent. There is a danger, however, in the substitution of petroleum ether or ligroin for hexane or heptane, and vice versa. Petroleum ether is always a mixture of isomers and sometimes also contains alkenes. As a result, its solvent properties may be slightly different from those of the pure alkane, and this may affect a sensitive crystallization or other delicate process. In general, however, the light petroleum fractions, petroleum ether (sometimes called *pet ether*), or ligroin may usually be substituted for the appropriate monoisomeric hydrocarbons.

6.2 PREPARATION

Although most simple alkanes are obtained from distillation of crude oil, they may nevertheless be prepared from a variety of starting materials. Chemical transformations which occur in organic compounds involve certain functional groups, or arrangements of atoms substituted in alkanes, which impart a certain level of reactivity to these systems. Alkanes are often called *paraffins* from the Latin *parum affinis,* meaning "too little affinity." The designation comes from the obvious chemical properties of alkanes: They are relatively inert. The synthesis of an inert material is generally an undesired reaction because the product cannot be further transformed. As a consequence, the synthesis of alkanes is of less interest in an organic course than one might otherwise think.

Numerous methods do exist, particularly in the industrial sector, for the synthesis of alkanes because these materials are of such great value as fuels. Several reactions for the synthesis of alkanes are shown below. Notice that in all four of these reactions gaseous alkanes are produced by methods which are very difficult to carry out in a normal undergraduate organic laboratory. Gases are difficult to handle without special equipment, and the synthesis of a pure alkane will, after all, lead to a material which undergoes further reaction with difficulty. For this reason we have not included any preparations of alkanes among the laboratory exercises of this book.

$$(CH_3—CO—O—)_2Ba + NaOH \xrightarrow{\Delta} CH_3—CH_3 + CO_2 + \text{salts}$$

$$CH_3—CH_2—CH_2—CH_2—MgBr + H_2O \longrightarrow CH_3—CH_2—CH_2—CH_3 + Mg(OH)(Br)$$

$$H—C{\equiv}C—H + 2\,H_2 \longrightarrow CH_3—CH_3$$

$$2CH_3—CH_2—Br + 2Na \longrightarrow CH_3—CH_2—CH_2—CH_3 + 2NaBr$$

Although alkanes yield useful reaction products only at high temperatures, under free-radical or other special conditions, their physical and chemical properties are of considerable interest. In the procedures below, we attempt to assess the solubility and reactivity of some alkanes and contrast these properties with those of some more polar compounds.

6.3 SOLUBILITY

Alkanes, or paraffins, are not only inert to most reagents but they are also insoluble in some common solvents. Alkanes will dissolve similar substances but often will not dissolve substances containing polar functional groups. Obtain from your instructor a sample of an alkane (hexane, heptane, cyclohexane, or octane will be fine) and determine its solubility in the following materials.

PROCEDURE

SOLUBILITY OF ALKANES

In each of nine small (10 × 75 mm) test tubes, place 1 mL of the following solvents: water, methanol, *n*-butyl alcohol, ethylene glycol, acetone, kerosene, ethyl acetate, toluene, and dichloromethane. Add to each of these test tubes 0.5 mL (amount approximate) of the alkane to be tested; stir, swirl, or shake as seems appropriate; and determine whether or not the material dissolves. If the material is insoluble, heat gently on the steam bath. In your notebook, construct a table with each solvent listed on the far left, and indicate to the right whether each substance was soluble, partially soluble, or insoluble in the cold or hot medium. From these tests you should be able to determine what kinds of solvents will be useful for dissolving alkanes.

From the observations obtained in the part above, you should be able to predict whether sodium benzoate is soluble in a hydrocarbon solvent and you ought to be able to predict whether or not the salt is soluble in water (see Chap. 4). Obtain a sample of sodium benzoate from your instructor and determine its solubility in water and in hexane. Indicate whether or not your observations accord with your predictions.

PROCEDURE

SOLUBILITY OF ALKANES AND ALKENES IN SULFURIC ACID

Obtain a small amount of hexane and a small amount of hexene (cyclohexane and cyclohexene will be appropriate substitutes). Place 0.5 mL concentrated sulfuric acid (**Caution: Corrosive**) in a small test tube. Add 0.5 mL of alkane to one test tube and 0.5 mL of alkene to the other tube. Swirl both tubes. Indicate which of the two substances you believe to be less reactive in the presence of sulfuric acid and tell what qualitative information this gives you.

QUESTIONS AND EXERCISES

6.1 Write structures for all the possible isomers of hexane.

6.2 Would you expect 2-methyldecane or 2,2,7-trimethyloctane to be a better automobile fuel? Would either be better than 2,2,4-trimethylpentane? Why?

6.3 Inexperienced laboratory workers often "discover" that pentane is insoluble in methanol. Experienced workers know that pentane is soluble in anhydrous methanol. What is the role of water in these differing observations?

6.4 A compound, C_5H_{12}, was found to be a low-boiling liquid, bp 9.5°C. The nmr spectrum of this material showed only a single resonance line at 1.05 ppm. Suggest a structure for this compound.

6.5 Gaseous alkanes are used as fuels in stationary power plants. What difficulties would attend the use of ethane as an automobile fuel?

ALKENES

INTRODUCTION

Alkenes are actually quite closely related to alkanes. The principal difference between alkenes and alkanes is that the former contain sites of unsaturation (double bonds). The designation *ene* should indicate the presence of unsaturation to you. An alkene and an alkane which differ only by the presence or absence of a double bond have very similar formulas. The only difference will be the absence of two hydrogen atoms in the alkene, giving the general formula C_nH_{2n}. Recall that the formula for an alkane is C_nH_{2n+2}. The reactivity of alkanes and alkenes differs, even though most of their physical properties are about the same. For example, ethylene and ethane are both gases but only ethylene will react with chlorine to produce an oily product. Since alkenes form oils in the presence of electrophiles, they are often referred to as olefins (oil formers).

Properties

The odor of a simple alkene will generally be similar to that of the corresponding alkane. The principal difference is that simple straight-chain alkanes have pleasant, almost sweet odors, whereas alkenes are usually more pungent. Many structurally diverse alkenes have very distinctive odors. Prominent among these is the lemon odor of limonene (isolated in Sec. 2.2 from caraway seeds). Likewise, the so-called terpenoid substances cedrene, pinene, caryophyllene, and safrole all have very distinctive odors.

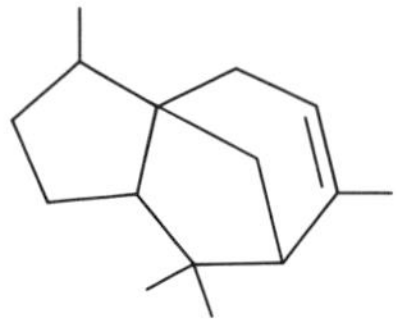

Cedrene 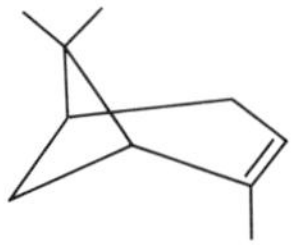 Pinene 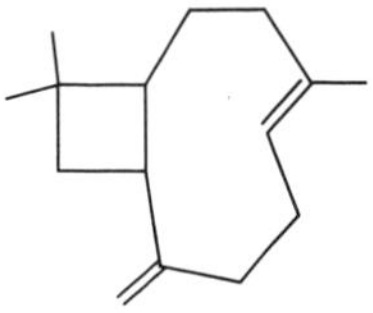 Caryophyllene 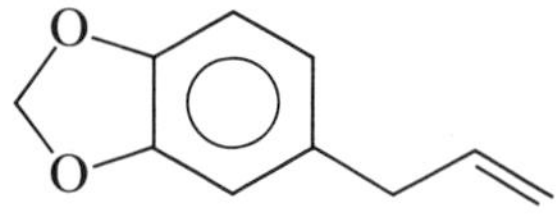 Safrole

The odors of cedrene (from cedar) and pinene (from pine) are suggested by their names.

The boiling point of a simple alkene is generally similar to that of the corresponding alkane. The melting points of alkenes may be similar as well, but the presence of cis and trans isomerism can dramatically affect these physical properties. If we think about a long-chain hydrocarbon such as octadecane ($C_{18}H_{38}$), whose melting point is 29 to 30°C, we can see that it should form a regular array of molecules when it packs in a crystal.

Alkanes

If a trans double bond is formed in this alkane anywhere along the chain, the general orientation of the molecule in a crystal would not be changed, and we could expect the melting point to be quite similar to that of the alkane.

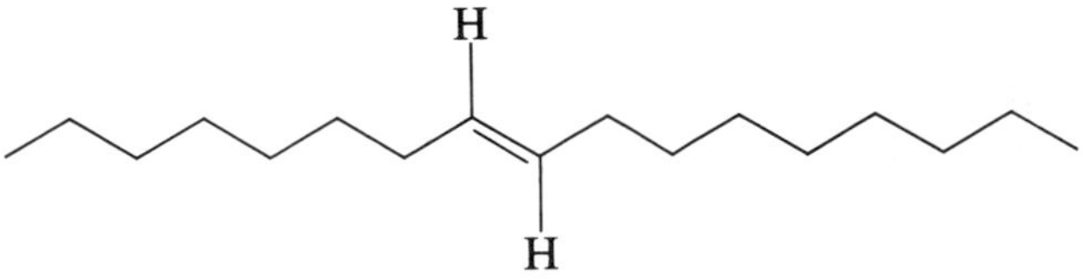

H

H

trans alkene

On the other hand, if a cis double bond is present, there will be a distinct bend in the molecule, and it can no longer form the same sort of crystals.

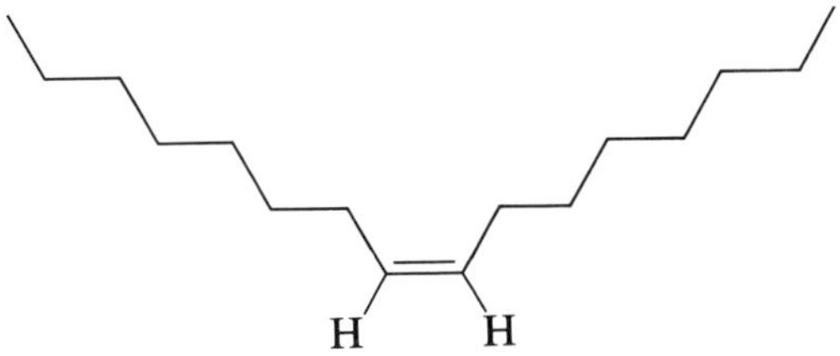

H H

cis alkene

Thus one would expect that, in general, cis isomers would have lower melting points than the corresponding trans isomers (see Secs. 1.1 and 2.4). This is found to be the case in many compounds.

This geometric characteristic is of great commercial importance in the margarine industry. As you may know, margarine is a mixture of glycerides which have long hydrocarbon chains, as shown below.

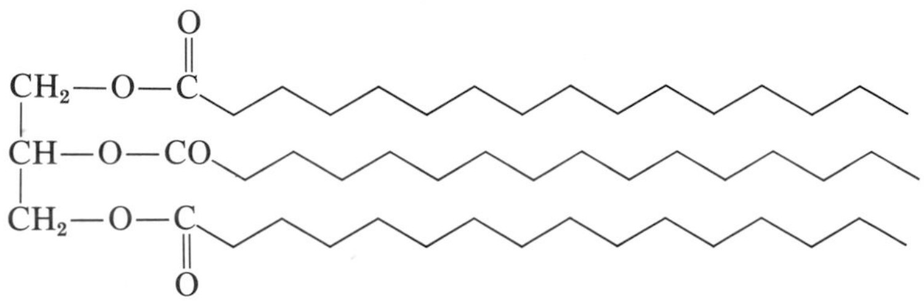

$$
\begin{array}{l}
\quad\quad\quad\quad\;\; O \\
\quad\quad\quad\quad\;\; \| \\
CH_2-O-C\!\sim\!\sim\!\sim \\
| \\
CH-O-CO\!\sim\!\sim\!\sim \\
| \\
CH_2-O-C\!\sim\!\sim\!\sim \\
\quad\quad\quad\quad\;\; \| \\
\quad\quad\quad\quad\;\; O
\end{array}
$$

The same kinds of structures are found in cooking oils. In order to convert the oily material which contains double bonds of both cis and trans types into a solid, an orderly lattice must be formed. By hydrogenating the double bonds, the relative order increases because the molecules can align more favorably. Where there was once an oil, there is now a solid. The hydrogenation of the double bonds causes this solidification of oil. Commercially this is called *hardening* of oil.

The geometries of an alkane, a cis alkene, and a trans alkene are shown by space-filling molecular models in Fig. 7.1. You should prepare models of these three substances as an exercise and compare them for yourself. Convince yourself that the trans alkene and the alkane have similar shapes but the cis alkene cannot align as the alkanes do.

Reactivity

The reactivity of simple alkenes containing an isolated double bond differs appreciably from that of the corresponding alkanes. A double bond has a very much higher electron density than a normal single bond, and excess electron density implies nucleophilic behavior. As a consequence of a double bond's nucleophilicity, it reacts readily with electrophiles. Double bonds not only react with acid but react readily with the halogens chlorine, bromine, and iodine. In fact, the addition of iodine to olefins is used as a standard method for determining how many double bonds are present in a compound (the *iodine number*).

Stable double bonds are known to exist between two carbon atoms, carbon and oxygen, carbon and nitrogen, and carbon and sulfur, and a number of unstable double bonds are known with other elements, e.g., silicon. We are concerned in this chapter only with the carbon-carbon double bonds, while the C=O and C=N bonds are discussed later.

Alkenes are generally formed by the elimination of a small molecule from a saturated system. An example is the loss of water from cyclohexanol (see Exp.

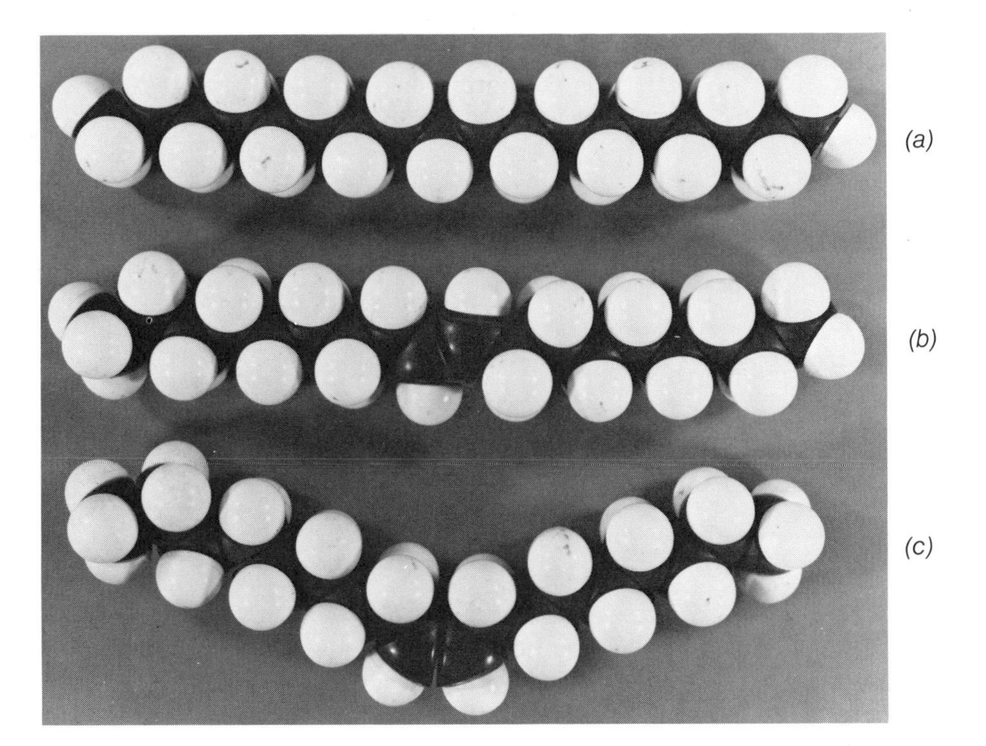

Figure 7.1
Space-filling models of (*a*) an alkane, (*b*) a trans alkene, and (*c*) a cis alkene.

7.1). The process in which double bonds are formed can often be reversed by treatment with different reagents. For example, cyclohexanol can be prepared by adding water to cyclohexene in the presence of an acid catalyst. Chlorides can be prepared by the addition of hydrochloric acid to the corresponding alkenes. Dichlorides result from the addition of chlorine to the alkene. The alkene may be converted into an alkane (i.e., reduced) by reaction with hydrogen gas (in the presence of a catalyst). The first three of these reactions can be carried out conveniently in most laboratories. Hydrogenation requires slightly more sophisticated equipment as well as an expensive catalyst. Several addition and elimination reactions are shown below.

Addition Reactions to Alkenes and
Elimination Reactions Which Form Alkenes

+ H—OH ⇌ OH, H

+ H—Cl ⇌ Cl, H

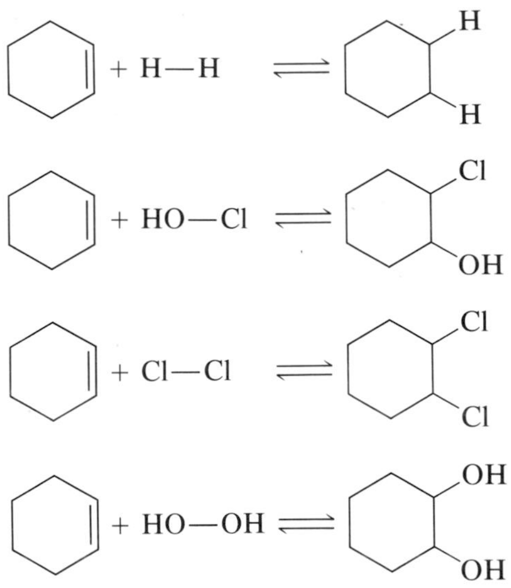

$$\text{cyclohexene} + H{-}H \rightleftharpoons \text{1,2-}H,H\text{-cyclohexane}$$

$$\text{cyclohexene} + HO{-}Cl \rightleftharpoons \text{2-chlorocyclohexanol (Cl, OH)}$$

$$\text{cyclohexene} + Cl{-}Cl \rightleftharpoons \text{1,2-dichlorocyclohexane (Cl, Cl)}$$

$$\text{cyclohexene} + HO{-}OH \rightleftharpoons \text{1,2-cyclohexanediol (OH, OH)}$$

The addition of bromine to a double bond is similar to the addition of any of the other electrophilic reagents in that it converts the alkene to a 1,2-disubstituted alkane. The approach is shown in the equation below:

$$R{-}HC{=}CH{-}R' + Br_2 \longrightarrow R{-}CHBr{-}CHBr{-}R'$$

The addition of bromine is a particularly valuable qualitative test because, in contrast to the reactions described above, a color change is observed during the reaction. A solution of a typical alkene is colorless whereas 2% bromine solution is deep red. Addition of red bromine solution to a colorless alkene will result in formation of a colorless 1,2 dibromide. The disappearance of color will therefore indicate the presence of a reactive double bond.

Not every double bond reacts in the same way with bromine solution. In aromatic compounds, each bond has only partial double-bond character and participates in a conjugated system. The reactivity of each double bond is consequently very much lower that it is in isolated systems. The logical extension of this is that, if a double bond is conjugated to other double-bond systems or to an aromatic system, its electron density will likewise be reduced and its reactivity should be lower than that found, for example, in cyclohexene.

In the first part of this procedure the reactivity of cyclohexene, toluene, cinnamic acid, and limonene are all compared. Cyclohexene and limonene both react very readily, as they are isolated double-bond systems. In order for toluene to react with bromine, an electrophilic catalyst, such as iron tribromide, is required. Cinnamic acid will react with bromine, but the reaction is slow.

PROCEDURE

PREPARATION OF BROMINE SOLUTION

Bromine solution for qualitative analysis is prepared by diluting 1 mL bromine (**Caution: corrosive, gloves**) with 99 mL dichloromethane. Bromine is very dense and the resulting solution will be about 2% bromine (by weight) in dichloromethane. Bromine solution has traditionally been utilized in carbon tetrachloride solution. Carbon tetrachloride is believed not to be safe in high concentrations, and so dichloromethane is recommended here as a substitute. Dichloromethane is considerably more volatile than carbon tetrachloride, so solutions containing it should be tightly stoppered.

PROCEDURE

BROMINE ADDITION TO ALKENES

In each of four small (10 × 75 mm) test tubes place 1 mL dichloromethane and add about 5 drops (or a 5-mm mound) of cyclohexene, toluene, limonene, or cinnamic acid (one to each tube). Swirl or stir with a spatula blade until each material is fully dissolved. Add bromine solution (**hood**) dropwise and observe each solution. Record all observations in your notebook. If no reaction occurs, stopper the test tubes, leave them in your desk, and examine them during the next laboratory period.

In the second part of this experiment, prepare three more test tubes, each containing 1 mL dichloromethane, and add 5 drops acetone, benzonitrile, and phenol, respectively. Again add bromine solution (5 drops) to each test tube. Observe the reactivity of bromine with these three compounds, record your observations, and indicate what conclusions you would draw from them.

7.1 DEHYDRATION OF CYCLOHEXANOL TO CYCLOHEXENE

One of the most common methods for the preparation of a carbon-carbon double bond is to dehydrate the corresponding alcohol. The term *dehydrate* means to remove water. The removal of water from an alcohol is probably used more often than any other method for the preparation of a double bond, both in the chemical laboratory and on an industrial scale.

The dehydration reaction actually involves two distinct steps, although these occur in rapid succession in the reaction mixture. The first is protonation of the hydroxyl group to provide the molecule with a suitable leaving group. After protonation has occurred, the water molecule is lost and a carbonium ion is formed, at least transiently. Loss of water is an equilibrium process and water may return to the cation, eventually regenerating the alcohol. If, instead of adding water back, the carbonium ion loses a proton, the alkene is formed. This sequence of events is illustrated in the equation below:

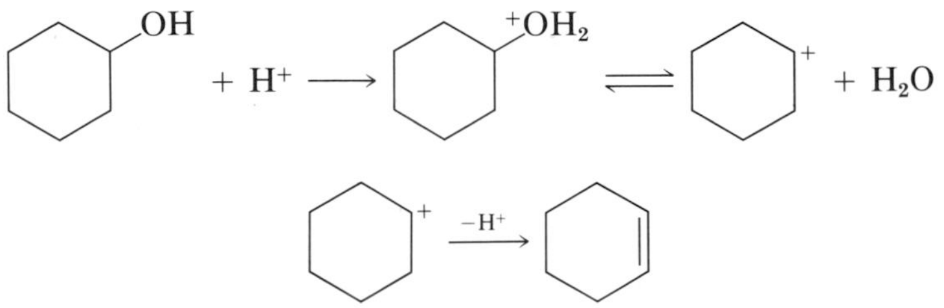

$$C_6H_{11}OH + H^+ \longrightarrow C_6H_{11}OH_2^+ \rightleftharpoons C_6H_{11}^+ + H_2O$$

$$C_6H_{11}^+ \xrightarrow{-H^+} C_6H_{10}$$

In the procedure described here, water is driven off and the equilibrium is driven to the right.

A variety of reagents are used to effect the dehydration of alcohols in different situations. The most common acids used for this purpose are phosphoric acid (this experiment), sulfuric acid, and the anhydride of phosphoric acid, phosphorus pentoxide (P_2O_5). Other reagents used for this purpose include thionyl chloride ($SOCl_2$), phosgene ($COCl_2$), and either alumina or silica gel at high temperature. The latter four reagents dehydrate alcohols by a somewhat different mechanism than that shown above.

EXPERIMENT 7.1 DEHYDRATION OF CYCLOHEXANOL TO CYCLOHEXENE

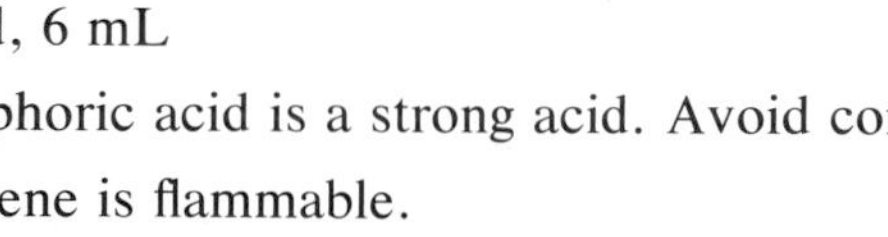

$$C_6H_{11}OH \xrightarrow{H_3PO_4} C_6H_{10} + H_2O$$

Time 3 h

Materials Cyclohexanol, 25 mL (MW 100, d 0.96 g/mL, bp 160 to 161°C)
85% Phosphoric acid, 6 mL

Precautions Phosphoric acid is a strong acid. Avoid contact with skin.

Hazards Cyclohexene is flammable.

Experimental Procedure

Place 25 mL (0.250 mol) cyclohexanol in a 100-mL round-bottom single-neck flask. Add 6 ml 85% phosphoric acid and several boiling chips. Swirl the flask to mix the layers. Affix a Claisen head and distillation apparatus to the flask as shown in Fig. 7.2. Use a 50-mL round-bottom flask as the receiver and immerse it in an ice-water bath to keep the receiver as cool as possible.

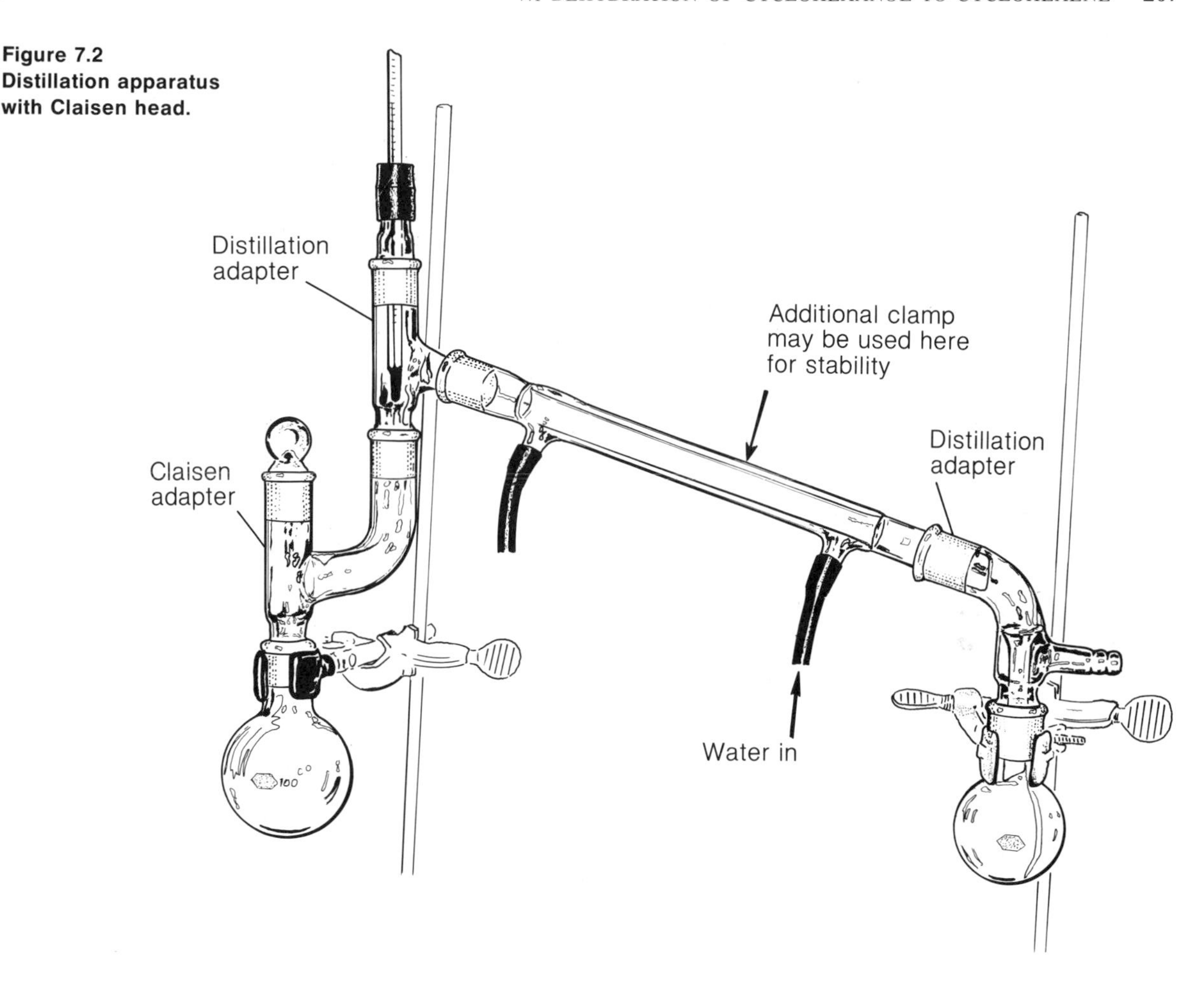

Figure 7.2
Distillation apparatus with Claisen head.

Heat the mixture gently with a small flame (bunsen burner or microburner) so that cyclohexene slowly distills. The head temperature should not be allowed to exceed 100 to 105°C. Continue distilling the mixture until only approximately 5 to 10 mL of liquid remains in the distillation flask.

Extinguish all flames. Transfer the material in the receiver flask to a small separatory funnel and wash with one 5-mL portion of 10% sodium carbonate solution, then with two 10-mL portions of saturated salt solution. Transfer the organic liquid to a 25-mL Erlenmeyer flask and dry over anhydrous $CaCl_2$ or Na_2SO_4. Decant the clear liquid into a 50-mL round-bottom flask and distill through a *dry* simple distillation apparatus (see Fig. 2.2). The receiver should be cooled in an ice-water bath, as before. Collect all the material which distills between 80 and 83°C. Cyclohexene is usually obtained as a clear, water-white liquid in 70 to 80% yield.

Test a few drops of the product with potassium permanganate test solution (Sec. 25.5F) and with 2% bromine solution.

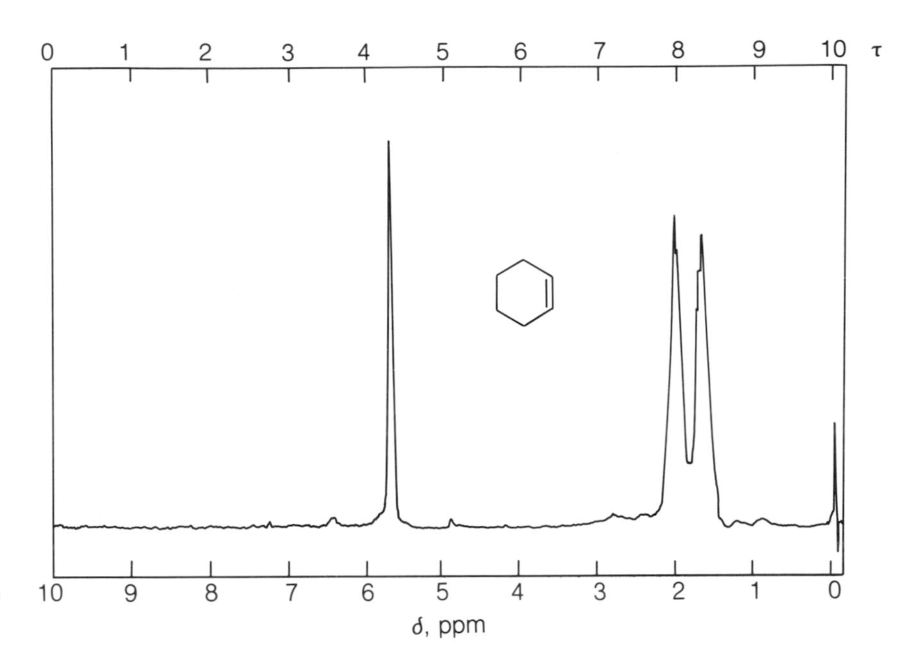

Figure 7.3
The nmr spectrum of cyclohexene.

The nmr spectrum of cyclohexene is shown in Fig. 7.3. Note that the vinyl protons are observed at 5.7 ppm and all the aliphatic protons are grouped near 2 ppm. None of the resonance lines disappear when D_2O is added, which indicates that no hydroxyl is present.

7.2 BROMINATION OF *trans*-STILBENE

The addition of bromine to a double bond is initiated by attack of the nucleophilic double bond on the electrophilic reagent. A complex forms between the two, and a cyclic bromonium ion subsequently forms from the complex. The geometry of the double bond is maintained because addition of bromine begins at one side of the double bond. The bromide ion present in solution then attacks the bromonium ion to give a trans dibromide. This overall sequence of reactions is illustrated in the equation below.

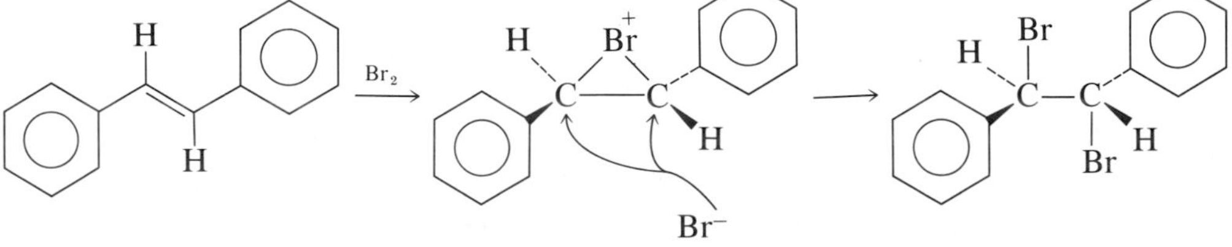

The product of this reaction, 1,2-dibromo-1,2-diphenylethane, has two chiral centers. As a result, there are four possible isomers, a *dl* pair and two meso forms. Trans addition of bromine to a trans double bond affords only the meso compounds. This has been ascertained independently but is clear from the melting points of the products. The *dl* pair melts at 114°C.

EXPERIMENT 7.2

BROMINATION OF trans-STILBENE

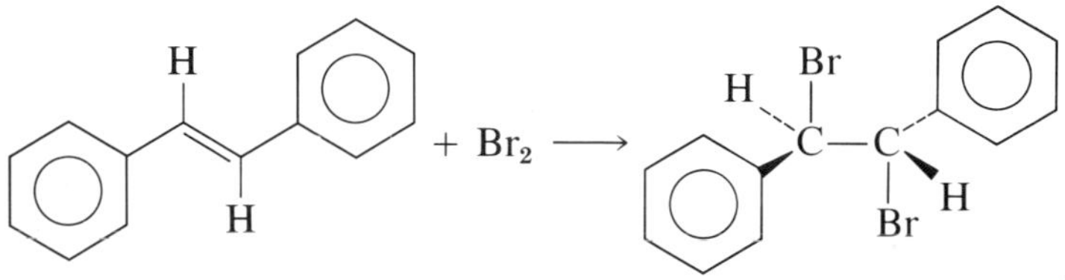

$$+ \ Br_2 \longrightarrow$$

Time 1.5 h

Materials *trans*-Stilbene, 5 g (MW 180, mp 122 to 124°C)
10% Bromine-dichloromethane solution (20% by weight), 15 mL[1]
Dichloromethane, 100 mL
Cyclohexene, several drops

Precautions Wear gloves. Do all parts of this experiment in a good hood.

Hazards Bromine is a toxic corrosive compound. Great care should be taken to avoid breathing its vapors or coming into skin contact with it.

Experimental Procedure

Weigh out 5 g (0.028 mol) *trans*-stilbene and add it to a 250-mL Erlenmeyer flask. Add 40 mL dichloromethane and swirl the flask to dissolve all the stilbene.

After the stilbene has dissolved, *carefully* (**wear gloves**) measure out 15 mL of a 10% bromine-dichloromethane solution[1] into a 25-mL graduated cylinder. Add 5 mL of this bromine solution to the flask and swirl. The bromine color will rapidly discharge, and after a few minutes of swirling a white finely crystalline solid will precipitate. Continue to swirl the flask and add another 5 mL of the bromine solution. After the color has discharged a second time, add the remainder of the bromine solution. After several minutes of vigorous swirling, the bromine color should remain. If this is not the case, add bromine solution 1 mL at a time until the bromine color persists. A thick crystalline mass should now be observed in the flask.

Several drops of cyclohexene should be added to the flask to *just* discharge the bromine color. When the color is gone, cool the flask briefly in an

[1] The bromine solution is made by carefully pouring 10 mL bromine (**gloves**) into a 100-mL graduated cylinder and then diluting to 100 mL with dichloromethane.

ice bath and then filter the suspension with the aid of a Buchner funnel and flask. Wash the crystalline mass with 15 mL cold dichloromethane and air dry. The dibromostilbene (1,2-dibromo-1,2-diphenylethane) obtained in this way is usually isolated as an off-white solid (mp 241 to 243°C) in 70 to 80% yield.

7.3 ISOMERIZATION OF MALEIC ACID TO FUMARIC ACID

Many compounds which contain carbon-carbon double bonds exist as both cis and trans isomers. These compounds have distinctly different physical properties and often have spectacularly different biological properties. From both the biological point of view and the point of view of synthesis, it is often desirable to convert the cis to the trans isomer. It is simpler to go in this direction than the other because the trans product is usually the thermodynamically more stable one.

In the experiment described in this section, maleic acid (a cis acid) is converted into the corresponding trans isomer, fumaric acid. Reactions of this sort which are carried out with an acid catalyst usually proceed by one of two related mechanisms. In the first step, acid adds to the double bond to form a carbonium ion. The carbonium ion may deprotonate and form not the cis, but the trans isomer.

Instead of simple deprotonation, the carbonium ion may add the acid anion (Cl^- if HCl is used) and form a chloride. Elimination of hydrogen chloride from this substance will usually form the thermodynamically more favored product, i.e., the trans isomer.

$$(R)(H)C{=}C(R)(H) + H^+ \rightleftharpoons (R)(H)\overset{+}{C}{-}C(R)(H)(H) \rightleftharpoons H^+ + (R)(H)C{=}C(H)(R)$$

EXPERIMENT 7.3 ISOMERIZATION OF MALEIC ACID TO FUMARIC ACID

$$\text{HOOC–CH=CH–COOH (cis, maleic acid)} \xrightleftharpoons{HCl} \text{HOOC–CH=CH–COOH (trans, fumaric acid)}$$

Time 2 h

Materials Maleic acid, 10 g (MW 116, mp 134 to 136°C)
[Maleic anhydride, 10 g (MW 98, mp 54 to 56°C)]
24% Hydrochloric acid solution, 30 mL[2]

Precautions Avoid breathing of or contact with maleic acid or maleic anhydride dust. Wear gloves when weighing and handling either the acid or the anhydride.

Hazards HCl fumes and maleic anhydride dust are irritants.

Experimental Procedure

Place 10 g (0.087 mol) maleic acid in a 125-mL Erlenmeyer flask. [Alternatively, 10 g maleic anhydride (0.1 mol) may be used in place of the maleic acid.] Add 30 mL 24% hydrochloric acid to the flask, swirl, and heat gently on the steam bath to effect solution of the solid. When a clear solution is obtained, place the flask on a steam bath and heat vigorously for 30 min. After 5 to 10 min of heating, a white precipitate should begin to appear in the reaction mixture. After 30 min of heating on the steam bath, remove the flask from the heat source and allow it to cool to room temperature.

Use a Buchner funnel and flask to filter the solid crystalline mass and wash the residue with two 25-mL portions of cold distilled water. Air dry the white, crystalline mass and determine its melting point. The melting point of the fumaric acid (295 to 300°C) should be obtained in a closed capillary tube, as the fumaric acid will sublime before its melting point is reached unless it is so contained. A metal-block melting apparatus is more advantageous than an oil-containing one, as oil heated to 300°C can be dangerous. The product should be close to analytical purity. If a low melting point is observed (below 285°C), the crystalline mass may be crystallized from 1 *N* HCl solution. Usually only one crystallization is needed to obtain pure product.

It should be noted that the fumaric acid may also be obtained from 10 g (0.1 mol) maleic anhydride, which in this reaction sequence rapidly hydrolyzes to maleic acid and then isomerizes to fumaric acid. The purity and yield of product from this starting material correspond directly to the results obtained with maleic acid. A problem is that maleic anhydride is usually obtained in chunks; therefore after it is weighed and added to the Erlenmeyer flask, the large chunks should be very gently crushed before heating on the steam bath is commenced. A safer but less convenient method is to crush the chunks with a mortar and pestle. If the latter method is used, the maleic anhydride should be reweighed before commencing the reaction.

The nmr spectra of both maleic acid and fumaric acid are shown in Fig. 7.4. Note that the only difference between the two spectra is in the position of the protons on the double bond. If you had only one spectrum, it would be very hard to tell which compound you had. This is an example of the importance of comparing the spectra of starting materials and products.

[2] Prepared by adding 2 vol of concentrated HCl to 1 vol of distilled water.

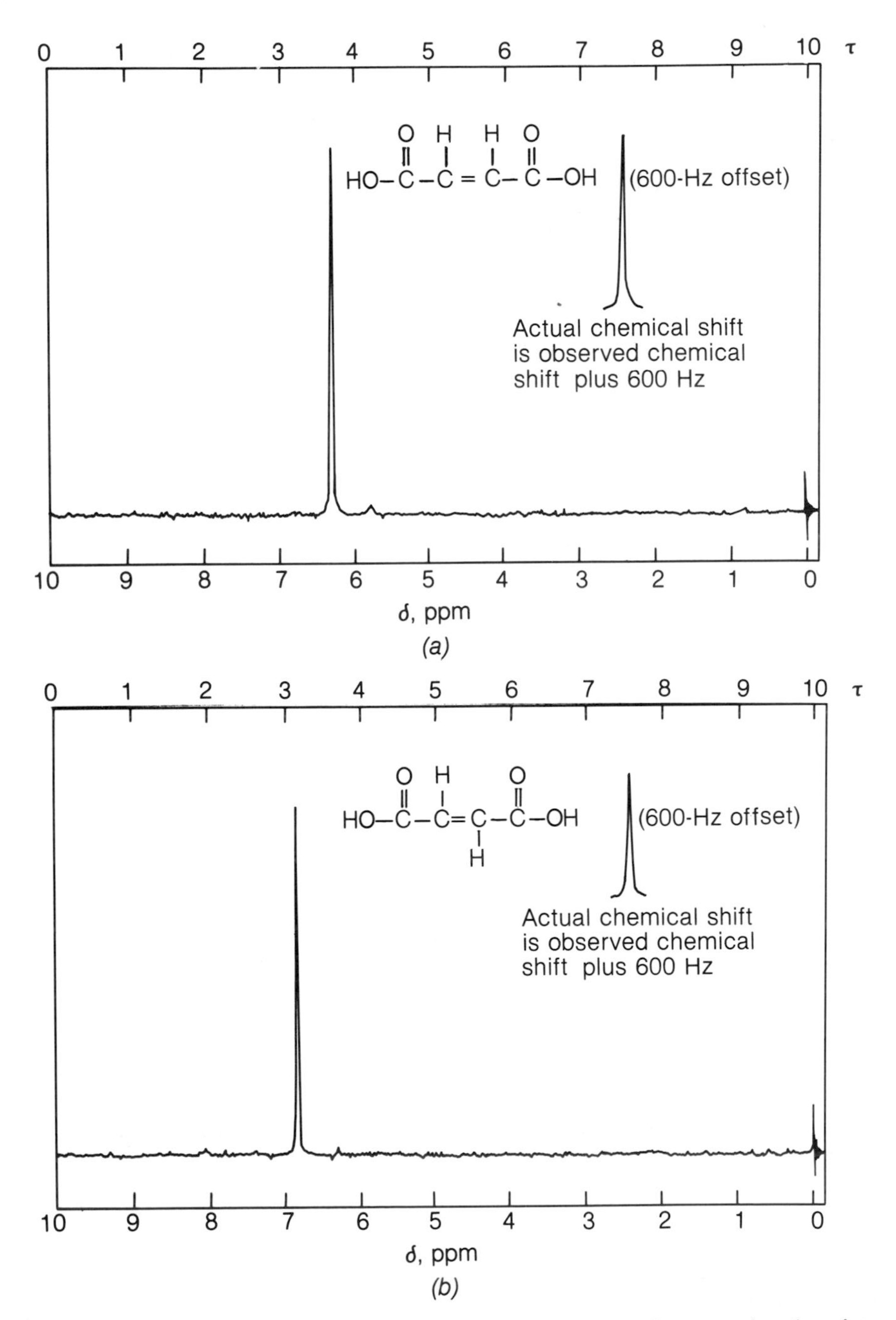

Figure 7.4
The nmr spectra of (*a*) maleic acid and (*b*) fumaric acid.

7.4 DICHLORO-CARBENE ADDITION TO CYCLOHEXENE

The three reactive species one usually encounters in organic chemistry are cations, anions, and radicals. A fourth and very important species is the *carbenes*. Carbenes are compounds of formally divalent carbon. These molecules consist of a carbon atom with an electron pair and two substituents. Although neutral overall, they are electron-deficient and therefore quite electrophilic.

The extreme reactivity of the dichlorocarbene species is illustrated by the rapidity of its addition to cyclohexene. Dichlorocarbene is so reactive that it is hydrolyzed by water to form carbon monoxide. The only reason that it is possible to prepare dichlorocarbene as we do in this experiment in the presence of water is that we use the trick of phase-transfer catalysis. This important technique is described in detail in Sec. 4.7. Phase-transfer catalysis allows hydroxide to be used as a base without interference in the carbene reaction by water.

In the first step in the phase-transfer carbene reaction, chloroform is deprotonated to form trichloromethide anion. This anion rapidly loses chloride to form the carbene, which reacts with the only available substrate, cyclohexene. The product of the reaction is 7,7-dichloronorcarane.

EXPERIMENT 7.4 DICHLOROCARBENE ADDITION TO CYCLOHEXENE (ptc)

$$Cl_3CH + OH^- \rightleftharpoons H_2O + CCl_3^-$$

$$CCl_3^- \rightleftharpoons :CCl_2 + Cl^-$$

$$\text{cyclohexene} + :CCl_2 \longrightarrow \text{7,7-dichloronorcarane}$$

Time 3 h

Materials Cyclohexene, 12.5 mL (MW 82, bp 83°C, d 0.8 g/mL)
Chloroform, 10 mL (MW 119, bp 61°C, d 1.5 g/mL)
50% Sodium hydroxide solution, 25 mL (from 17 g NaOH)
Tetrabutylammonium hydrogensulfate, 0.4 g (MW 339) or benzyltriethylammonium chloride, 0.4 g, as prepared in Exp. 13.6
Dichloromethane, 50 mL

Precautions Do all operations in a good hood. Keep an ice bath at hand to control the temperature.

Hazards Carbon monoxide, a minor by-product in the carbene reaction, is toxic.

Experimental Procedure

Place 17 g sodium hydroxide in a 250-mL Erlenmeyer flask and add 17 mL distilled water. Swirl the flask to dissolve the base and then cool the flask to room temperature with the aid of an ice bath or a cold-water tap.

Using either a small graduated cylinder or graduated pipet and bulb (**no lips**), transfer 12.5 mL (0.127 mol) cyclohexene and 10 mL (0.127 mol) chloroform to a 50-mL Erlenmeyer flask. Swirl the flask to mix the solutions.

After the sodium hydroxide solution has returned to room temperature, add 0.4 g tetrabutylammonium hydrogensulfate (or other quaternary ammonium salt catalyst, such as benzyltriethylammonium chloride, which may be suggested by your instructor) to the aqueous solution, followed immediately by the mixture of cyclohexene and chloroform. Now grasp the flask in one hand, cradling the neck between your thumb and forefinger (Fig. 7.5). Swirl the flask vigorously, using the motion of your fingers to move the bottom of the flask in a circle. Swirl vigorously enough to form a thick emulsion in the flask, but be careful not to let any liquid swirl up the sides of the flask and out the neck. If a magnetic stirrer is available, it will make this operation easier. As the swirling continues, the temperature will gradually rise and increase more rapidly as the reaction proceeds. Monitor the temperature of the reaction mixture with a thermometer, and maintain an internal temperature of 50 to 60°C by intermittent use of an ice bath. After 10 to 15 min the reaction will subside and the solution will gradually cool.

After the reaction mixture has reached room temperature, dilute with 100 mL distilled water and transfer the entire mixture to a separatory funnel. Extract with two 25-mL portions of dichloromethane. Combine the organic layers and wash with two 25-mL portions of water. Dry the organic layer with granular anhydrous sodium sulfate. After drying, decant the organic layer to separate it from the drying agent and remove the dichloromethane by evaporation on the steam bath.

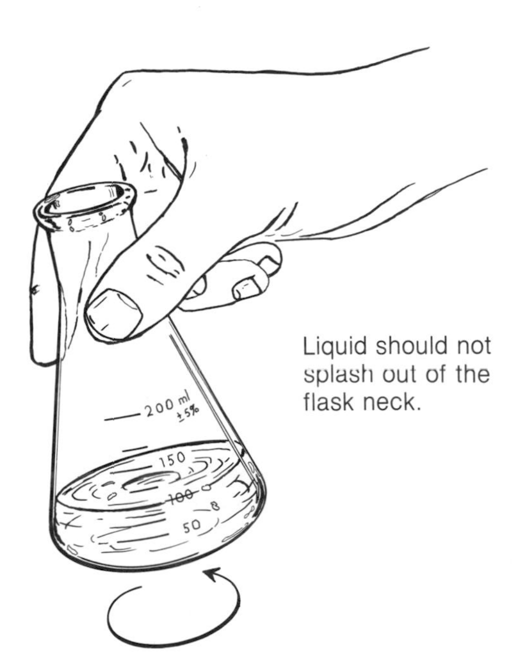

Figure 7.5
Swirling a flask.

After the solvent has evaporated, transfer the residual oil to a 50-mL round-bottom distillation flask and obtain the product by simple distillation over a flame (Fig. 2.2). The first fraction, boiling below 120°C, consists of unreacted cyclohexene and chloroform and should be discarded after collection. The second fraction, bp 190 to 200°C, should be 7,7-dichlorobicycloheptane. The product is a water-white liquid which is usually obtained in 60 to 70% yield.

QUESTIONS AND EXERCISES

7.1 How would you expect the presence of a terminal double bond to affect the overall physical properties of an alkene?

7.2 Most naturally occurring oils contain alkenes which exist in the cis configuration. Assuming that nature had the option of making all trans, all cis, or a mixture of the two possibilities, explain why you think that nature has chosen a predominance of the cis configuration over the trans configuration in many natural products.

7.3 An exception to the general rule that most naturally occurring oils contain cis alkenes is found in palm oil, in which the trans configuration predominates. From this information, can you think of an explanation as to why palm oil is important in the production of margarines?

7.4 Although cooking oils and fats are indistinguishable in their net effect on cooked foods, fats have been used much more often in cooking than have oils. Can you think of any explanation, either historical or chemical, to account for this predominant use of fats in cooking?

7.5 If one equivalent of bromine (Br_2) is added to the hydrocarbon butadiene, a mixture of dibrominated alkenes is obtained. Predict the structure of these alkenes and design a mechanism for their production.

7.6 Cyclohexane and cyclohexene differ only by the absence of two hydrogens in cyclohexene. Cyclohexane melts at about 6°C whereas cyclohexene melts at −104°C. How can you account for this extraordinary difference in melting point?

7.7 When *cis*-2-butene is treated with hydrogen peroxide in the presence of acid, 2,3-dihydroxybutane is obtained as a *dl* pair. On the other hand, when *cis*-2-butene is treated with permanganate, the product is 2,3-dihydroxybutane, a meso compound. How do you account for this difference in product formation?

7.8 Predict the stereochemistry (meso or *dl*) of the addition of bromine to *cis*-stilbene.

7.9 Maleic acid readily forms an anhydride (loses water) when heated. Fumaric acid, on the other hand, does not form an anhydride even at high temperature. Explain this observation.

7.10 Make a model of cyclohexene. Would it be possible for cyclohexene to exist in the trans configuration?

7.11 Make a model of the dichlorocarbene addition product of cyclohexene. What is the stereochemistry of the ring junction, i.e., is it cis or trans?

VIII

SYNTHESIS OF DIPHENYLACETYLENE (TOLAN) FROM STILBENE DIBROMIDE

INTRODUCTION

Hydrocarbons which differ from alkanes and alkenes by the presence of a triple bond as the only functional group are known as *alkynes*. In the Geneva nomenclature system the triple bond is designated as *yne* and the double bond is designated as *ene*. The general formula for simple alkynes is C_nH_{2n-2}. Generally, the alkynes have physical properties similar to those of alkenes and alkanes. For example, alkynes exhibit volatility similar to that of alkenes. Recall that the five-carbon alkane *n*-pentane has a boiling point of 36°C. The terminal alkyne 1-pentyne has a boiling point of about 40°C, and the internal four-carbon alkyne 2-butyne has a boiling point of about 27°C; 2-pentyne boils at 55°C.

The only commercially important alkyne is acetylene (C_2H_2). Acetylene is important for two reasons: (1) it trimerizes to benzene (often explosively), and (2) when burned, its energy content is enormous. As a consequence, a mixture of acetylene and oxygen produces an extremely hot flame. Oxyacetylene torches, so-called because the acetylene flame is fed by oxygen, produce a flame with a temperature over 3000°C.

In connection with the importance of acetylene, it should also be noted that there is a considerable danger associated with its handling. Because the energy content of acetylene is so high, its mixtures with air will react explosively over a wide range of acetylene/oxygen or acetylene/air ratios. The gas acetylene should therefore be handled only by experienced workers, and by them with extreme caution.

Reactivity

The triple bond of an alkyne reacts in very much the same way as a double bond does. The difference in reactivity arises because a triple bond is really a double double bond. In other words, the two carbons contain two double bonds between them. One or both of the double bonds may react in the nucleophilic sense, so that either 1 or 2 mol hydrogen or bromine may be added under the appropriate conditions. These processes are shown schematically below.

$$R{-}C{\equiv}C{-}R' + H_2 \longrightarrow R{-}CH{=}CH{-}R' + H_2 \longrightarrow R{-}CH_2{-}CH_2{-}R'$$

$$R{-}C{\equiv}C{-}R' + Br_2 \longrightarrow R{-}CBr{=}CBr{-}R' + Br_2 \longrightarrow R{-}CBr_2{-}CBr_2{-}R'$$

Terminal alkynes also react as nucleophiles, but because they have a hydrogen atom bonded to the carbon terminus of the triple bond, their reactivity is slightly different. The *sp* carbon at the terminus of the acetylenic linkage has a relatively high electronegativity and will support a negative charge. Terminal acetylenes, under the appropriate conditions, can therefore split off a proton (act as an acid) to yield an alkyne (or acetylide) salt. The reaction of an acetylene with sodium metal produces sodium acetylide, as indicated in the equation below.

$$2\ R{-}C{\equiv}C{-}H + 2\ Na \longrightarrow 2\ R{-}C{\equiv}C{-}Na + H_2 \uparrow$$

From the chemical point of view, the principal utility of acetylenic compounds is the use of acetylenic salts as reactive intermediates in organic synthesis. The products which result are unsaturated and can be further transformed. Internal acetylenes, once formed, may be used to generate specific (i.e., cis or trans) double bonds. Hydrogenation of an acetylene using an appropriate catalyst will yield a cis alkene. Hydrogenation via a dissolving-metal reduction (usually with sodium in liquid ammonia) will give the opposite stereochemistry (a trans alkene). The alkyne may thus be used in stereospecific alkene syntheses. An example of acetylene's versatility is shown in the synthesis of muscalure, the sex attractant of the common housefly (*Musca domestica,* see below).

$$H{-}C{\equiv}C{-}H + Na \longrightarrow H{-}C{\equiv}C{-}Na$$

$$H{-}C{\equiv}C{-}Na + C_{13}H_{27}{-}Br \longrightarrow C_{13}H_{27}{-}C{\equiv}C{-}H$$

$$C_{13}H_{27}{-}C{\equiv}C{-}H + Na \longrightarrow C_{13}H_{27}C{\equiv}C{-}Na$$

$$C_{13}H_{27}{-}C{\equiv}C{-}Na + C_8H_{17}{-}Br \longrightarrow C_{13}H_{27}{-}C{\equiv}C{-}C_8H_{17}$$

$$C_{13}H_{27}{-}C{\equiv}C{-}C_8H_{17} + H_2 \xrightarrow{\text{Cat.}} \underset{\text{Muscalure}}{C_{13}H_{27}{-}\overset{H}{\overset{|}{C}}{=}\overset{H}{\overset{|}{C}}{-}C_8H_{17}}$$

Preparation

Acetylene itself may be prepared simply by the addition of water to calcium carbide (CaC_2). Acetylene evolves as a gas from the surface of the salt. Acetylene could therefore be generated very easily in the laboratory, but because its mixtures with air are so dangerous and so prone to explode, no directions are given for such a preparation here.

Most higher acetylenes are prepared by a double elimination reaction. In each case, the elimination of HX or XX occurs as it would in forming an alkene, but *two* moles of reagent are lost in this case, generating the acetylene. In the experiment described below diphenylacetylene, commonly called tolan, is generated by the base-catalyzed elimination of two moles of hydrogen bromide from 1,2-dibromo-1,2-diphenylethane (stilbene dibromide). Stilbene dibromide is prepared by bromination of stilbene (Exp. 7.2), which in turn may be prepared by a Wittig reaction from benzaldehyde (Exp. 21.2).

The double elimination of hydrogen bromide from a substrate is usually more difficult that the simple elimination of hydrogen bromide to give an alkene. As a consequence, high temperatures and strong bases are usually required for the completion of this reaction.

EXPERIMENT 8.1 SYNTHESIS OF DIPHENYLACETYLENE (TOLAN)

$$C_6H_5\text{—CHBr—CHBr—}C_6H_5 + 2KOH \longrightarrow C_6H_5\text{—C}\equiv\text{C—}C_6H_5$$

Time 2.5 h

Materials Stilbene dibromide, 4 g (MW 340, mp 241 to 243°C)
Potassium hydroxide, 2 g (MW 56)
Ethylene glycol, 20 mL

Hazards Ethylene glycol boils at 200°C; the reaction mixture will be hot.

Experimental Procedure

Weigh out 2 g (0.036 mol) of potassium hydroxide and place it in a 250-mL round-bottom flask. Measure 20 mL ethylene glycol with a graduated cylinder and pour this over the potassium hydroxide. Swirl the flask and warm on a steam bath until all the potassium hydroxide dissolves.

Weigh out 4 g (0.012 mol) stilbene dibromide (made in Exp. 7.2) and add it to the flask in one portion. Place a reflux condenser with lightly greased joints in place, but do *not* run water through the condenser (see Fig. 8.1). Add two or three boiling chips to the flask and slowly heat to reflux with a small flame (bunsen burner) or heating mantle. Maintain a gentle reflux for 20 min.

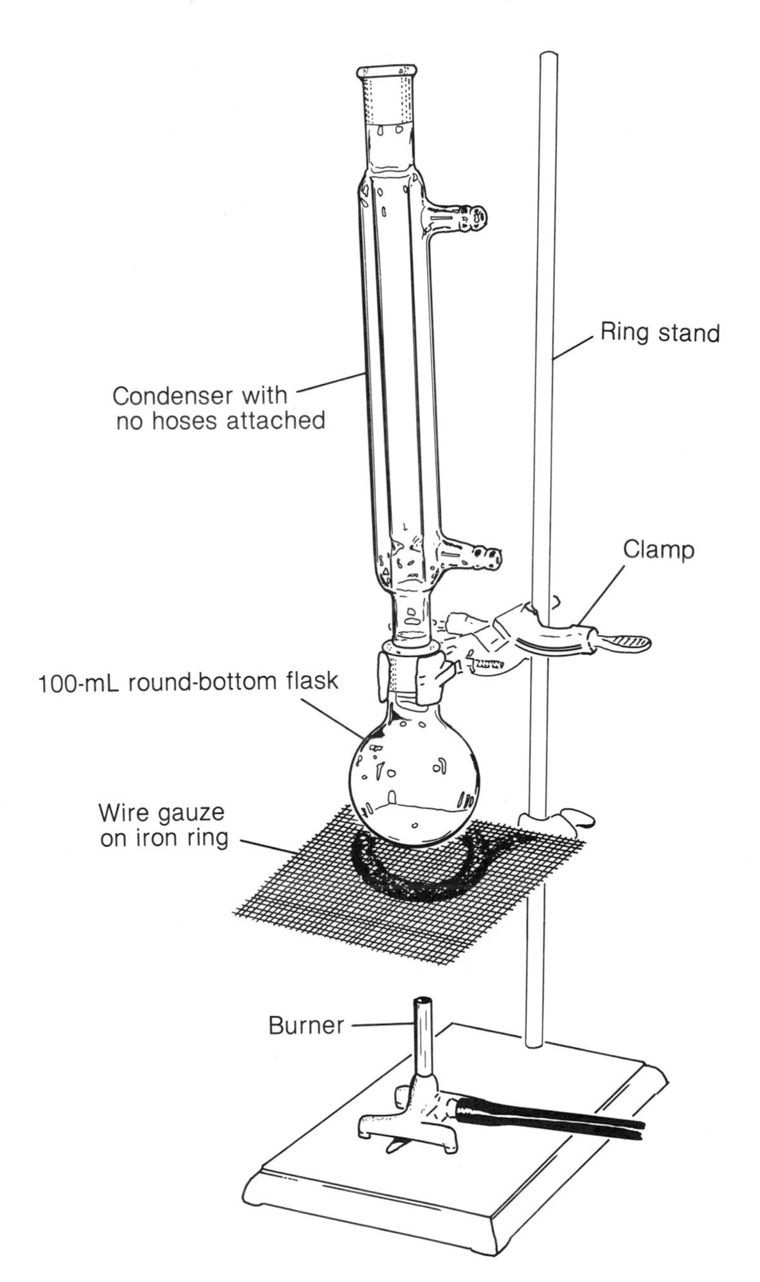

Figure 8.1
Reflux using an air-cooled condenser.

After the reflux period the hot solution is poured (**hold the flask with the clamp**) into a 250-mL Erlenmeyer flask and allowed to cool to room temperature. Now 150 mL cold water is slowly added while swirling the flask. The product separates as a gummy, yellow-orange, semicrystalline mass. The aqueous mixture is allowed to stand undisturbed for 10 to 15 min to allow all the product to coagulate. The larger chunks are broken up with a glass rod and the suspension filtered with the aid of a Buchner funnel. The flask is rinsed with one 15-mL portion of cold water, and this liquid is poured over the solid just collected.

The solid material is taken up in hot ethanol, treated with decolorizing carbon if necessary, and crystallized by the addition of water (the final solution should be about 60 to 65% ethanol). The solution is allowed to cool to room temperature. The product first forms an oil and then slowly crystallizes. Filtra-

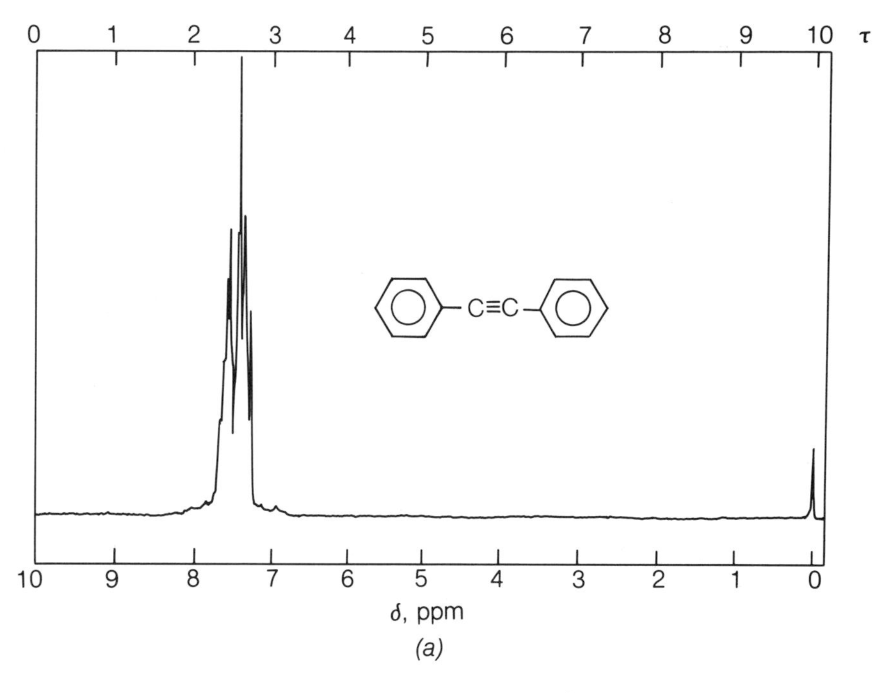

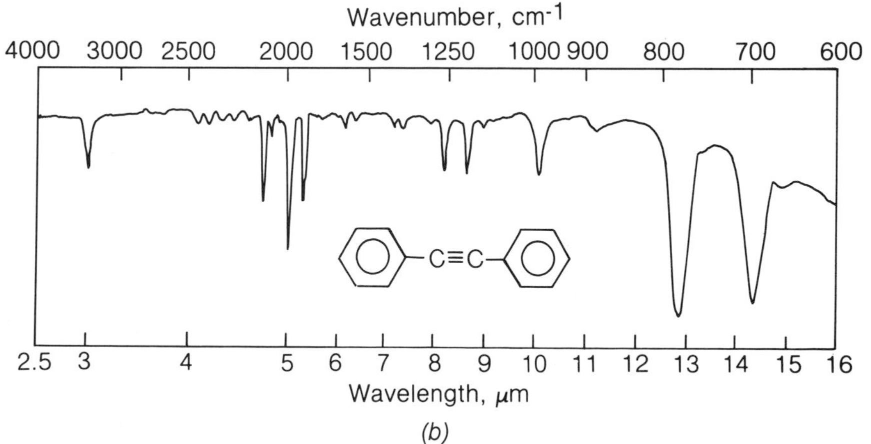

Figure 8.2
The (*a*) nmr and (*b*) ir spectra of diphenylacetylene (tolan).

tion with a Buchner funnel and flask gives chunky pale yellow crystals whose melting point is 58 to 60°C. The yield of diphenylacetylene (tolan) is 65 to 75%. The purity of this product can be assessed by thin-layer chromatography (silica gel, 15% dichloromethane-hexane).

The product may be recrystallized from 95% ethanol (5 mL) to give off-white needles, mp 59 to 61°C. Extreme care should be exercised in this process as the diphenylacetylene crystallizes very slowly from pure ethanol and the recovery is poor. [*Note:* If sublimation equipment is available, the tolan may be purified by that method (see Sec. 2.3). Check with your instructor.]

The nmr and ir spectra of tolan are shown in Fig. 8.2. Note that, since the C≡C bond in tolan is symmetrical, no absorption is observed for it in the ir spectrum.

QUESTIONS AND EXERCISES

8.1 The triple bond of an alkyne can add one molecule of hydrogen (H_2) by several mechanisms. In one experiment an alkyne was hydrogenated over a catalyst and a liquid product was obtained. When the same species was reduced by using a dissolving-metal medium (sodium in liquid ammonia, another method for reducing alkynes), a solid alkene was obtained. What would you presume the stereochemistry of each of these hydrogenation processes must be to account for the observed results?

8.2 Based on your answer to question 8.1, indicate whether you think the same stereochemistry of product would result if 2-butyne were first catalytically hydrogenated and then brominated and if it were first brominated and then catalytically hydrogenated.

8.3 If stilbene dibromide is treated with potassium hydroxide, the product is diphenylacetylene. If 1,2-dibromocyclohexane is treated with potassium hydroxide, the product is 1,3-cyclohexadiene. Explain the difference in these two observations.

IX THE DIELS-ALDER REACTION

Much of our understanding of organic chemistry derives from research conducted by the early natural product chemists working in Europe. Many natural products contain six-membered rings, and the construction of six-membered rings has therefore been a long-standing synthetic goal. The two major methods which have historically been utilized for their construction are the Robinson annelation reaction and the Diels-Alder reaction. Otto Diels and Kurt Alder, working in Germany in the early 1900s, discovered that appropriately substituted olefins (alkenes) and dienes would react to form two new carbon-carbon bonds. The reaction illustrated below is for the addition of ethylene to butadiene. The electronic shifts are symbolized by the three arrows, which indicate that the three double bonds form two new carbon-carbon sigma bonds and a new double bond. This reaction can be carried out as indicated to give cyclohexene, but the yield is not very satisfactory.

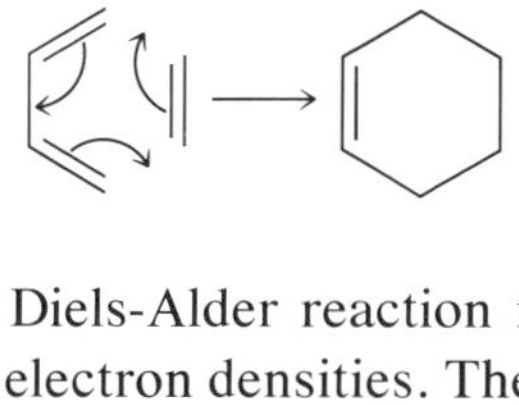

A characteristic of the Diels-Alder reaction is that the reacting partners must be of different relative electron densities. The alkene (often called the *ene* component, or *dienophile*) is ordinarily so substituted that it is electron-deficient. The diene component is often so substituted that it is electron-rich. The greater the difference in electronegative substitution between the ene and the diene, the more efficient the reaction. The most important electronic requirement for the reaction is that the electron densities of the two components should be different, and not specifically that the ene component should be electron-deficient. Examples of the other situation are also known but are less common.

A second requirement for the reaction to occur is that the diene must be in the (*S*)-cis conformation, i.e., the two double bonds must be on the same side of the single bond. Because there is delocalization in the diene component, the single bond between the two double bonds has some π character, causing rotation about it to be somewhat restricted. Two isomers are possible, one corresponding to the trans arrangement, the other to a cis arrangement about the single bond. Because, formally, the single bond is not capable of exhibiting cis-trans isomerism, this special case of partial double-bond character is often referred to as (*S*)- (for single bond) cis or trans isomerism. The structure for butadiene illustrated is the (*S*)-cis isomer. A molecule such as 1,3-cyclohexadiene, which contains a double-bond system constrained to the cis geometry, is generally much more reactive than one which has the option of being either cis or trans.

(*S*)-*cis*-Butadiene 1,3-Cyclohexadiene

Of the three reactions for which experimental procedures are given in this chapter, only one of the diene systems is constrained to a five-membered ring. It is interesting to note that cyclopentadiene is so reactive that it undergoes a Diels-Alder reaction with itself to give a dimer.

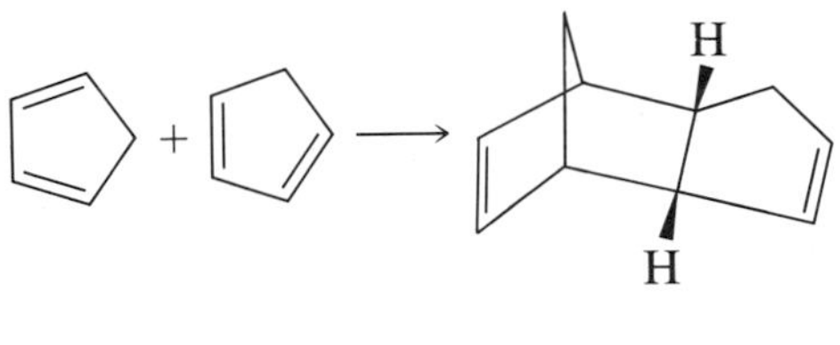

H

H

Cyclopentadiene, on standing, forms dicyclopentadiene, which is the stable form of this material. Dicyclopentadiene must be heated to effect a so-called reverse, or retro, Diels-Alder reaction in order to obtain the monomeric diene, which can then participate in the Diels-Alder reaction with an appropriate dienophile.

A similar situation is encountered with the molecule called sulfolene. Note the equilibrium reaction drawn below. If SO_2 and butadiene are allowed to stand together, they undergo a Diels-Alder type reaction to yield a five-membered ring. On heating, the reverse reaction is effected, SO_2 is lost, and the diene component is regenerated. If the reaction is performed in a mixture containing a dienophile other than SO_2, the Diels-Alder reaction ensues.

:SO_2 ⇌ SO_2

X → X

Note that maleic anhydride is used as the ene component in all three of these reactions. The double bond is electron-deficient because of conjugation to both the carbonyl functions. In addition, the carbonyl groups, since they are contained in the five-membered ring, are relatively remote from the reaction site and therefore they do not sterically hinder the Diels-Alder reaction. Finally, hydrolysis and elimination of CO_2 from the acid can yield useful ring systems without any remnant of the carboxyl function.

9.1 REACTION OF SULFOLENE AND MALEIC ANHYDRIDE

As indicated above, the reaction between sulfolene and maleic anhydride is more complex than it might appear to be from the stoichiometry. There are really two closely related but separate stages involved. First, sulfolene must undergo the ''reverse-'' or ''retro-Diels-Alder'' reaction to form sulfur dioxide (as a by-product) and butadiene, which then reacts with maleic anhydride. Since sulfolene (2,5-dihydrothiophene-1,1-dioxide) yields butadiene during the course of the reaction it is often referred to as a ''masked'' butadiene.

It may seem unnecessarily complicated to use sulfolene rather than butadiene in this reaction. There is a good reason for it, however. Butadiene itself is a gas at room temperature (bp −4°C) and is therefore much harder to handle then sulfolene which is a stable white solid. When sulfolene is heated at the boiling point of xylene (the reaction solvent) it falls apart to butadiene and SO_2. Since this reaction is the direct reverse of the Diels-Alder reaction, the butadiene which is produced must be in the (*S*)-cis conformation. Since this is the appropriate conformation of butadiene for a cycloaddition, the addition of maleic anhydride occurs readily. The reaction is a relatively clean one because the only by-product is SO_2, which is lost as a gas.

When the cycloaddition reaction occurs, the two new bonds form at the same time so the diene and dienophile must be aligned for reaction as shown on the next page.

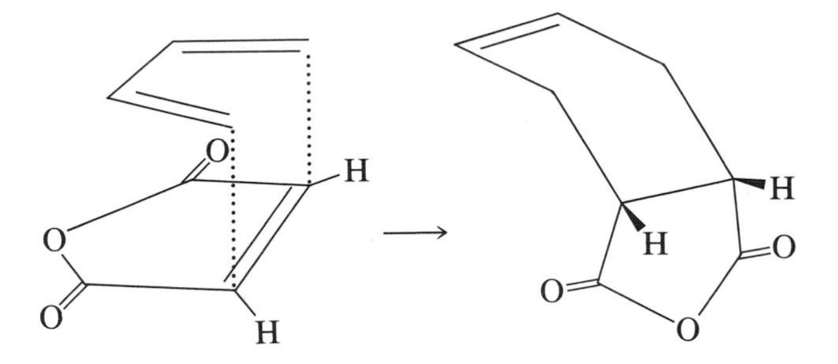

This alignment leads exclusively to the cis product (i.e., the product in which the two hydrogens where the rings meet are on the same side).

EXPERIMENT 9.1

SYNTHESIS OF 4-CYCLOHEXENE-1,2-DICARBOXYLIC ANHYDRIDE

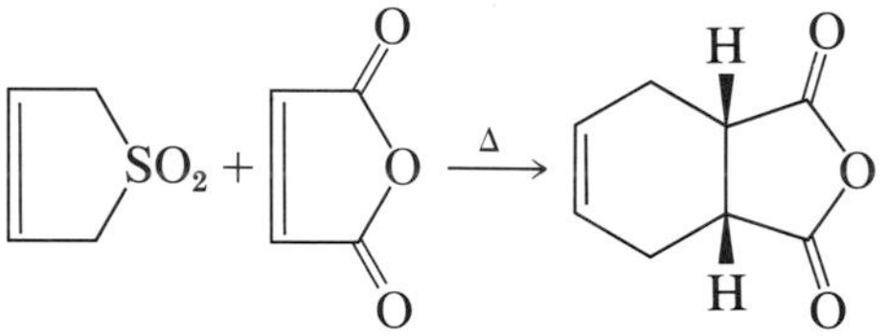

Time 2.5 h

Materials Sulfolene (butadiene sulfone), 10 g (MW 118, mp 65 to 66°C)
Maleic anhydride, 5 g (MW 98, mp 52 to 54°C)
Xylene, 10 mL
Toluene, 50 mL

Precautions Do not contact maleic anhydride dust. Keep all solvents and apparatus dry.

Hazards Xylene is toxic in high concentrations. Maleic anhydride is an irritant.

Experimental Procedure

Work in hood or use gas trap (Sec. 2.8).

Butadiene sulfone (10 g) and maleic anhydride (5 g, powdered) are placed in a 100-mL round-bottom flask. The solid is covered with 10 mL xylene, and a lightly greased reflux condenser is placed on top of the round-bottom flask (Fig. 9.1).

The mixture is heated with a free flame (very gently at first) and gradually refluxed. The heating process should be especially gentle at first because the

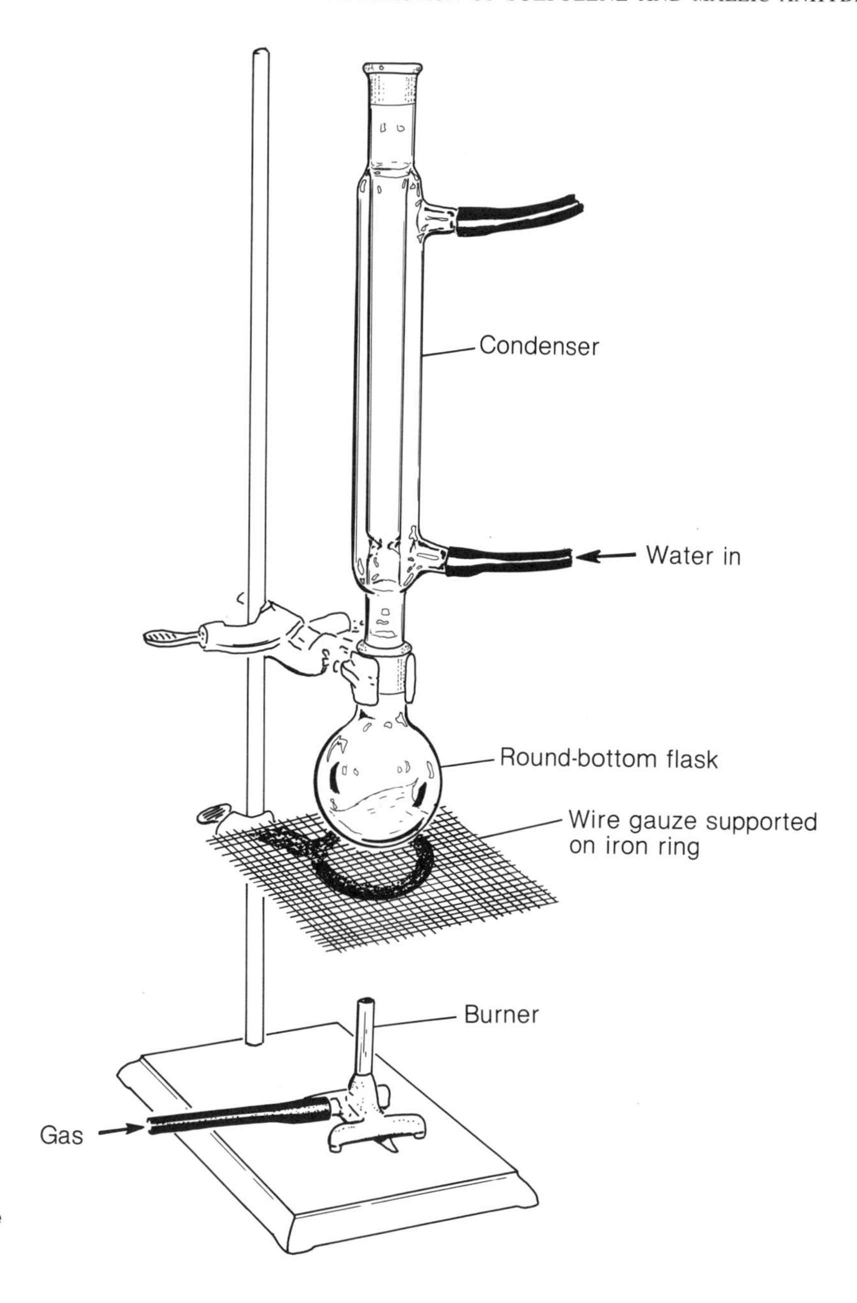

Figure 9.1
Apparatus for simple reflux.

reverse Diels-Alder reaction of butadiene sulfone gives SO_2 as a by-product. The SO_2 fumes will be coming out of the condenser. The reaction mixture should be refluxed with the free flame for 45 min after reflux commences.

After the reflux period is complete, cool the reaction mixture to room temperature and add 50 mL anhydrous toluene. Add a squirt of decolorizing carbon to the darkly colored mixture, warm, and filter through a pad of Celite (filter aid). Transfer the filtrate to a 125-mL Erlenmeyer flask and place it on the steam bath. Add 25 to 30 mL petroleum ether or hexane and warm the mixture to about 80°C. Remove the hot petroleum ether–toluene mixture from the steam bath and allow it to stand. As the mixture cools, colorless crystals are deposited. (More crystals are deposited if the solution is cooled in an ice-water bath.) Collect the solid by suction filtration. The yield of 4-cyclohexene-1,2-dicarboxylic anhydride is 7 to 8 g, mp 102 to 104°C. Its nmr and ir spectra are shown in Fig. 9.2.

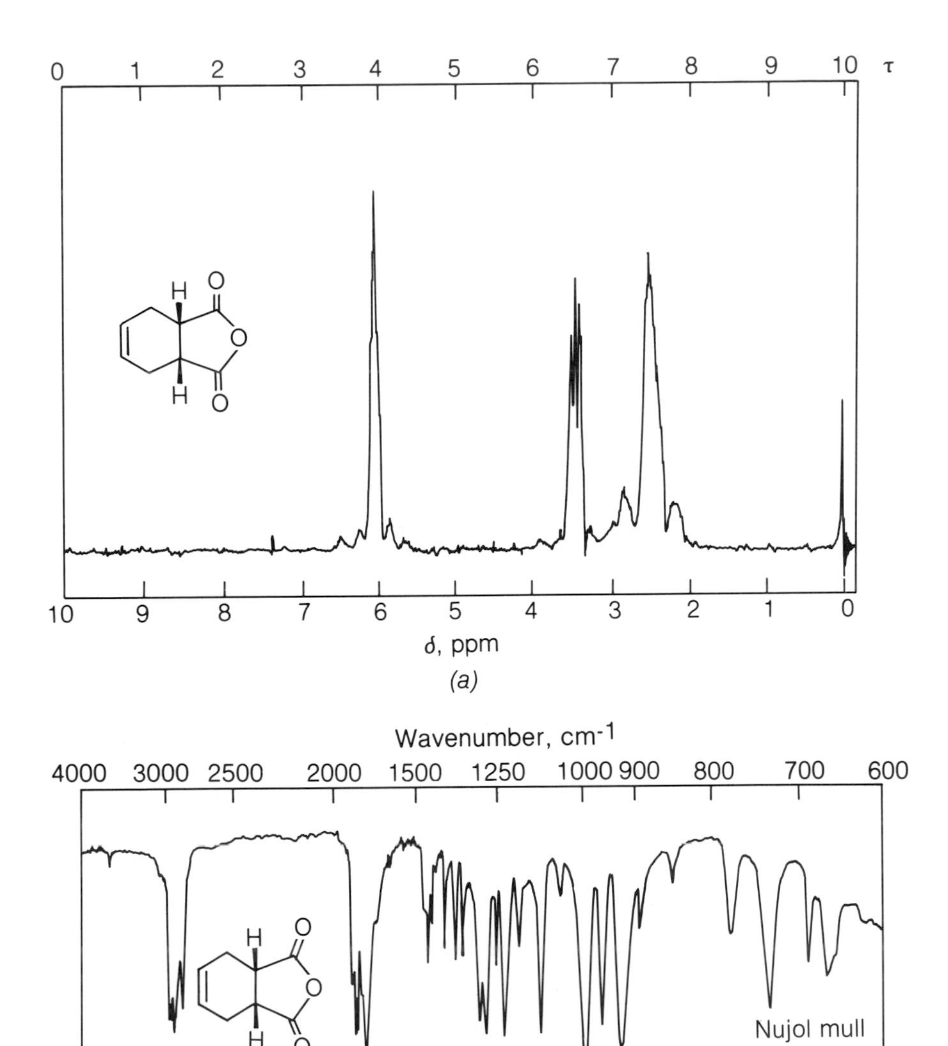

Figure 9.2
The (*a*) nmr and (*b*) ir spectra of 4-cyclohexene-1,2-dicarboxylic anhydride.

9.2 REACTION OF CYCLOPENTADIENE WITH MALEIC ANHYDRIDE

As described in the introduction to this chapter, cyclopentadiene is such a reactive system that it forms a dimer by a direct Diels-Alder reaction. Cyclopentadiene is sold commercially as the dimer which is called dicyclopentadiene. This dimer must be heated to effect the "retro-" or "reverse-Diels-Alder" reaction. As a result, the synthesis described in this section really involves two steps: the "cracking" of dicyclopentadiene to give the reactive diene, followed by the Diels-Alder reaction to give the product.

Cyclopentadiene is an electron-rich diene and is especially reactive because it is held in the required (*S*)-cis conformation by the ring structure. Because of the great reactivity of cyclopentadiene, it must be kept cold until used and used as quickly as possible.

Because cyclopentadiene is a cyclic diene, the Diels-Alder reaction with the cyclic dienophile affords a tricyclic system. The alignment of diene with dienophile (see Sec. 9.1) makes only a cis ring junction possible. There is a new complication, however. The product can have the anhydride group either on the same or the opposite side as the $—CH_2—$ bridge. The two possibilities are referred to as *exo*- or *endo*-, respectively. The alignment of reacting partners is such that the endo rather than the exo product is obtained.

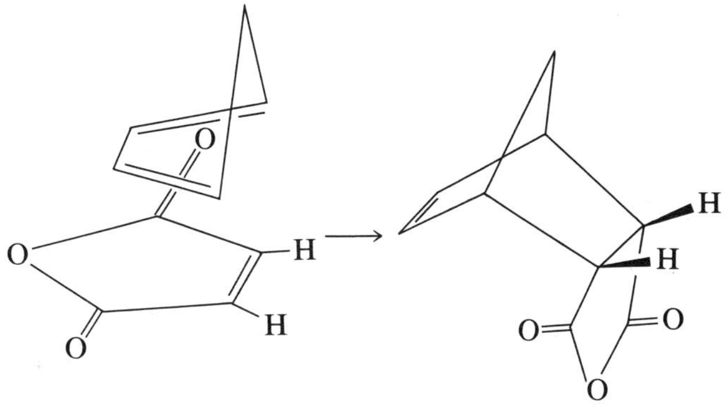

O O O H H → H H O O O

endo

O H H O O → O O O H H O

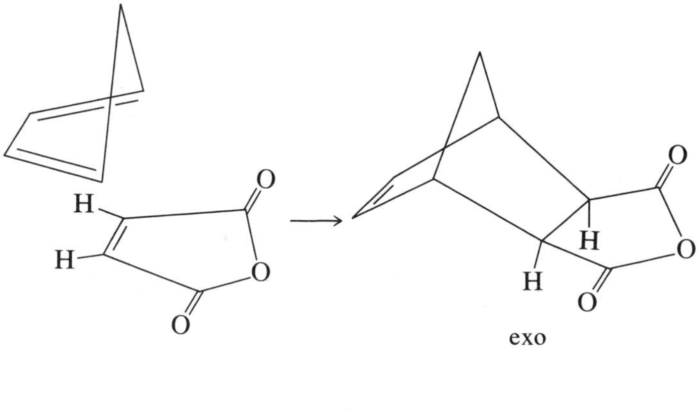

exo

EXPERIMENT 9.2

SYNTHESIS OF cis-NORBORNENE-5,6-endo-DICARBOXYLIC ANHYDRIDE

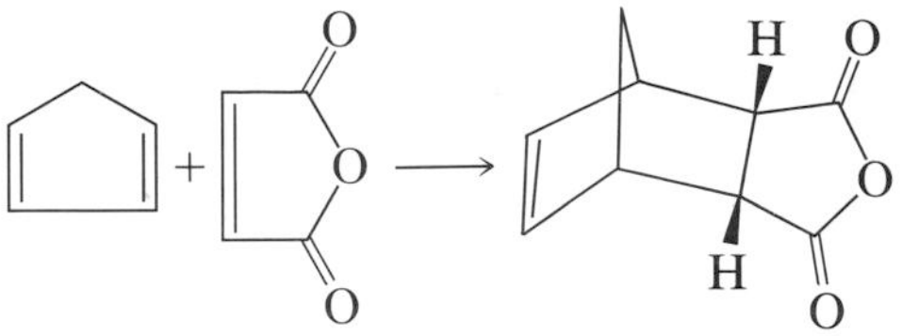

Time 2.5 h

Materials Dicyclopentadiene, 30 mL (MW 132)
Maleic anhydride, 6 g (MW 98, mp 52 to 54°C)
Toluene, 30 mL
Petroleum ether, 35 mL

Precautions Keep all glassware and solvents dry. Avoid contact with maleic anhydride dust. **Conduct this procedure in a good hood.**

Hazards Maleic anhydride is a skin irritant. Cyclopentadiene and toluene fumes are toxic in high concentrations.

Experimental Procedure

Assemble a fractional distillation apparatus consisting of a 250-mL round-bottom flask, a packed distillation column, and a condenser and distillation head (see Fig. 2.5). Charge the round-bottom flask with 30 mL dicyclopentadiene and reflux over a small flame, so that the cyclopentadiene is gradually distilled off. The head temperature during this process should be approximately 45°C. Collect the cyclopentadiene which distills over in an ice-cooled flask. Stopper the flask, set it in an ice bath, and keep it there until the remaining materials are prepared.

Charge a 125-mL Erlenmeyer flask containing 6.0 g powdered maleic anhydride with 25 mL toluene. Swirl until the maleic anhydride dissolves in the toluene. Cool the toluene mixture in an ice bath so that the internal temperature becomes approximately 10°C (check with a thermometer). Dissolve 6 mL cyclopentadiene in 5 mL toluene. Add this solution in small portions to the Erlenmeyer flask which is immersed in the ice bath. The flask should be vigorously swirled during the addition to dissipate the heat of reaction. A paste will gradually begin to form. After the flask has been in the ice bath approximately 20 min, warm it on a steam bath and add 35 mL petroleum ether while swirling. Cool the hot petroleum ether solution in an ice bath. The product will crystallize out and should be collected by suction filtration.

Dissolve the product in the smallest amount of hot dichloromethane necessary and dilute the resulting solution with twice its volume of warm petroleum ether or hexane. Set the flask aside. As the solution cools, colorless crystals

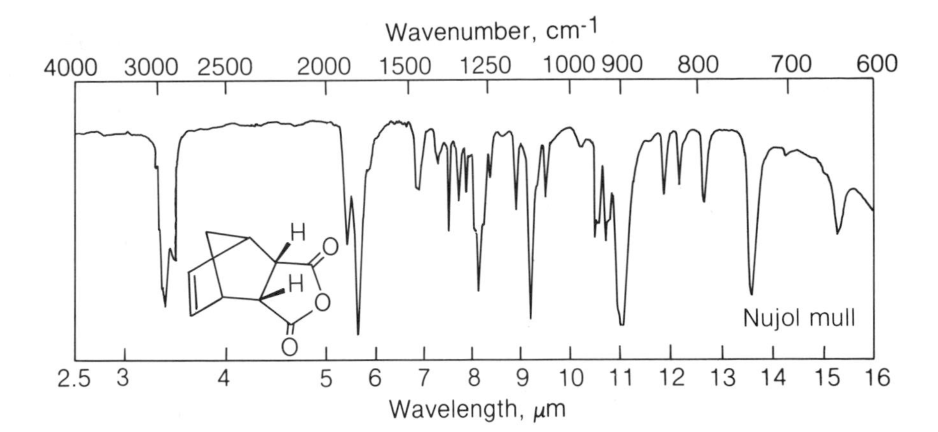

Figure 9.3
The ir spectrum of *cis*-norbornene-5,6-*endo*-dicarboxylic anhydride.

(mp 166 to 168°C) may be collected. A second crystallization may be required from ethyl acetate and petroleum ether. The total yield of pure product is approximately 5 g.

Note the appearance of two bands (a doublet) in the carbonyl region of the ir spectrum (Fig. 9.3). Would this have been observed in the starting material or is it characteristic of the product?

QUESTIONS AND EXERCISES

9.1 Only one isomer of cyclopentadiene is possible. Cyclohexadiene, however, may have the double bonds in either the 1,3- or 1,4-positions. Only one of these isomers undergoes the Diels-Alder reaction. Predict which one and tell why.

9.2 While doing Exp. 9.2, two students were working together as laboratory partners. One student was given the job of distilling enough cyclopentadiene for use by both students in this reaction. One partner did the distillation and the other partner went out for coffee. The partner doing the distillation finished the distillation of cyclopentadiene and immediately began the cycloaddition reaction. When the second partner returned from the coffee break and proceeded to carry out the reaction, a strange result was observed. One of the partners obtained the cycloaddition product in high yield, while the other partner obtained very little product. Which partner obtained the product in high yield and which partner did not, and why?

9.3 Maleic anhydride reacts readily with cyclopentadiene to give a norbornene anhydride. 2,3-Dimethylmaleic anhydride does not afford such a product under identical conditions. Can you account for this difference in reactivity?

9.4 In Exp. 9.1 the solvent used is xylene. Suggest two reasons why benzene would not be a good solvent for this reaction.

X

ALKYL HALIDES

A primary alkyl bromide can be prepared by heating a primary alcohol with an aqueous solution of sodium bromide and excess sulfuric acid. These reagents form an equilibrium mixture containing hydrobromic acid.

$$NaBr + H_2SO_4 \longrightarrow NaHSO_4 + HBr$$

$$HBr \rightleftharpoons H^+ + Br^-$$

$$CH_3—CH_2—CH_2—CH_2—OH + H_2SO_4 \longrightarrow CH_3—CH_2—CH_2—CH_2—\overset{+}{O}H_2 + {}^-OSO_3H$$

$$CH_3—CH_2—CH_2—CH_2—\overset{+}{O}H_2 + Br^- \xrightarrow[S_N2]{\text{rate-determining step}} CH_3—CH_2—CH_2—CH_2—Br + H_2O$$

The procedure given below calls for certain amounts of each reagent and specifies other conditions such as temperature and reaction time. Although one mole of *n*-butyl alcohol theoretically requires one mole each of sodium bromide and sulfuric acid for complete reaction, the procedure below calls for use of a slight excess of the bromide and a 100% excess of the acid. The excess acid serves to shift the equilibrium to the right, i.e., to form more protonated alcohol.

n-Butyl bromide is formed from *n*-butyl alcohol by a nucleophilic substitution reaction. No carbonium ion can be involved in this reaction because a primary carbonium ion is too unstable to be formed readily. A direct substitution of hydroxyl by bromide ion is not effective because hydroxyl ion is a very poor leaving group.

In order to facilitate the loss of hydroxyl from the alcohol, sulfuric acid is required. Sulfuric acid and the hydrobromic acid generated in the reaction mixture serve as the proton source. The hydrogen ion protonates the hydroxyl group, forming water as the leaving group. Bromide ion can now attack the primary carbon from the back side in the S_N2 fashion and water, not hydroxyl, is lost as a neutral leaving group. It is often the case that loss of poor, negatively charged leaving groups can be facilitated by the presence of a proton. Water is present in very small concentrations in the reaction medium and is generally protonated in this solution anyway, so it is too poor a nucleophile to reverse the reaction to any appreciable extent. At the end of the reaction virtually all the alcohol has been converted to the bromide, which can be obtained by distillation.

The by-products in this reaction are 1-butene, dibutyl ether, and the starting material. The alkene and ether may be removed by distillation, but some fraction of each may remain. All three possible by-products can be eliminated by extraction of the reaction mixture with concentrated sulfuric acid. The mechanism by which sulfuric acid removes these by-products probably involves protonation of the various compounds to form various charged species. These species are soluble in the concentrated sulfuric acid itself. 1-Bromobutane is resistant to this proton-donating ability of concentrated sulfuric acid because bromine is too weakly basic to accept a proton. This procedure is used extensively in qualitative organic analysis to determine unsaturation (see Chap. 22). The reaction with each product is shown below:

$$R{-}CH{=}CH_2 + H_2SO_4 \rightleftharpoons R{-}\overset{+}{C}H{-}CH_3 + {}^{-}OSO_3H$$

$$R{-}O{-}R + H_2SO_4 \rightleftharpoons R{-}\overset{\overset{H}{|}}{\underset{+}{O}}{-}R + {}^{-}OSO_3H$$

$$R{-}O{-}H + H_2SO_4 \rightleftharpoons R{-}\overset{\overset{H}{|}}{\underset{+}{O}}{-}H + {}^{-}OSO_3H$$

$$R{-}Br + H_2SO_4 \not\longrightarrow \text{no reaction}$$

An S_N1 reaction differs from an S_N2 reaction in the order or molecularity of the reaction which occurs. S_N2 reactions are direct nucleophilic displacements which involve both nucleophile and leaving group in the rate-controlling step (two species generally correspond to second order). In an S_N1 reaction, the rate-controlling step is formation of the intermediate carbonium ion. This carbonium ion is usually formed by the loss of some leaving group from the substrate. Since no other reactant is involved, it is presumably a unimolecular, or first-order, reaction.

As was the case with *n*-butyl alcohol, the hydroxide ion of *tert*-butyl alcohol is too poor a leaving group to effectively be lost under mild reaction conditions. In the presence of strong acids such as HCl, the hydroxyl group is protonated and then can readily be lost as water. The intermediate carbonium ion which is formed from *tert*-butyl alcohol is tertiary and therefore has a substantial intrinsic stability. The most common nucleophile available in solution is chloride ion, and the carbonium ion readily combines with it. In some cases water will add back to the carbonium ion but this will simply regenerate *tert*-butyl alcohol, which can then be recycled to the chloride. Gradually all the *tert*-butyl alcohol is converted to *tert*-butyl chloride, which is then recovered by distillation. The mechanism of this reaction is shown below:

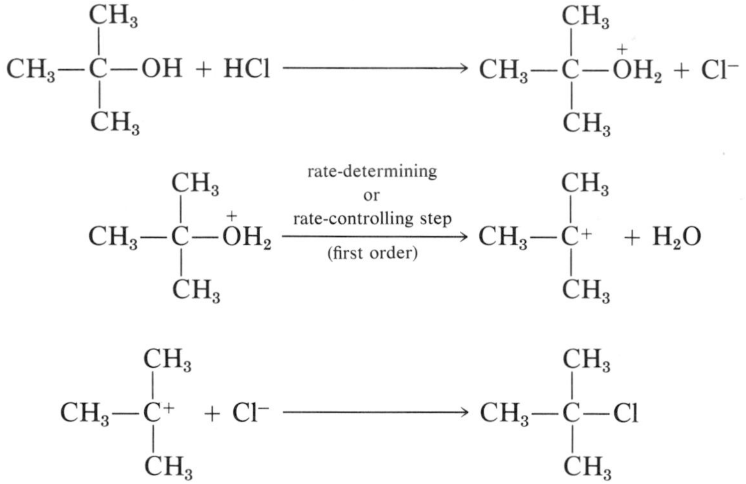

$$(CH_3)_3C{-}OH + HCl \longrightarrow (CH_3)_3C{-}\overset{+}{O}H_2 + Cl^-$$

$$(CH_3)_3C{-}\overset{+}{O}H_2 \xrightarrow[\text{(first order)}]{\text{rate-determining or rate-controlling step}} (CH_3)_3C^+ + H_2O$$

$$(CH_3)_3C^+ + Cl^- \longrightarrow (CH_3)_3C{-}Cl$$

EXPERIMENT 10.1 SYNTHESIS OF *n*-BUTYL BROMIDE BY AN S_N2 REACTION

$$CH_3{-}(CH_2)_2{-}CH_2{-}OH + NaBr + H_2SO_4 \longrightarrow CH_3{-}(CH_2)_2{-}CH_2{-}Br + H_2O + NaHSO_4$$

Time 2 h

Materials *n*-Butyl alcohol, 20 mL (MW 74, bp 118°C, d 0.81 g/mL)
Sodium bromide, 27 g (MW 103)
Concentrated H_2SO_4, 23 mL (MW 98, d 1.84 g/mL)

Precautions Perform experiment in hood. Wear gloves when handling sulfuric acid.

Hazards Sulfuric acid is a strong dehydrating acid. Alkyl halide fumes are toxic in high concentrations.

Experimental Procedure

Place 27.0 g sodium bromide, 30 mL water, and 20 mL *n*-butyl alcohol in a 250-mL round-bottom flask. Cool the mixture in an ice-water bath and slowly add 23 mL concentrated sulfuric acid while swirling and cooling the flask. Mount the flask over a burner (or place it in a heating mantle) and fit it with a water-cooled condenser (Fig. 9.1). Heat until the reaction mixture is boiling, note the time, and adjust the flame or mantle to maintain a brisk, steady reflux. The upper layer that soon separates is the alkyl bromide, since the aqueous solution of inorganic salts has a higher specific gravity. Reflux for 30 min, remove the source of heat, and let the condenser drain for a few minutes. Remove the condenser and set up a simple distillation apparatus as shown in Fig. 2.2, using a 125-mL Erlenmeyer flask as a receiver. Distill the mixture. Make frequent readings of the temperature and distill until no more water-insoluble droplets come over. By this time the temperature should have reached 115°C (collect a few drops of distillate in a test tube and add distilled water to see if it is soluble). The boiling point gradually rises because of azeotropic distillation of butyl bromide with water containing increasing amounts of sulfuric acid.

Pour the distillate into a separatory funnel, add about 20 mL water, stopper, and shake. Note that butyl bromide now forms the *lower* layer. A pink coloration in this layer due to a trace of bromine can be discharged by adding a pinch of sodium bisulfite ($NaHSO_3$) and shaking again. Drain the lower layer of butyl bromide into a fresh Erlenmeyer flask. Clean and dry the separatory funnel and return the butyl bromide to it. Thoroughly cool 20 mL concentrated sulfuric acid in an ice bath and add it to the funnel, shake well, and allow 5 min for settling of the layers. Water and *n*-butyl bromide have densities of 1.0 and 1.3 g/mL, respectively. An empirical method for telling the layers apart is to draw off a few drops of the lower layer into a test tube and see whether the material is soluble (H_2SO_4) or insoluble (butyl bromide) in water. Separate the layers, allow 5 min for further drainage, and separate again. Then wash the butyl bromide with 20 mL 10% sodium hydroxide solution to remove traces of acid, separate, and be careful to save the proper layer.

Dry the cloudy butyl bromide by adding 5 g calcium chloride and warming the mixture gently on the steam bath, with swirling, until the liquid clears. Decant the dried liquid into a 50-mL flask through a funnel fitted with a small loose plug of cotton, add a boiling stone, distill, and collect the material boiling in the range 99 to 103°C. The yield is 21 to 25 g. Note and record in your notebook the approximate volumes of forerun and residue.

The nmr and ir spectra of *n*-butyl bromide are shown in Fig. 10.1. The high-field triplet observed in the nmr spectrum is due to the methyl group protons. The low-field triplet is due to the protons closest to electronegative bromine.

Note that if alcohol were a contaminant in *n*-butyl bromide, an ir spectral band in the 3400 to 3600 cm^{-1} region would be observed. Is it?

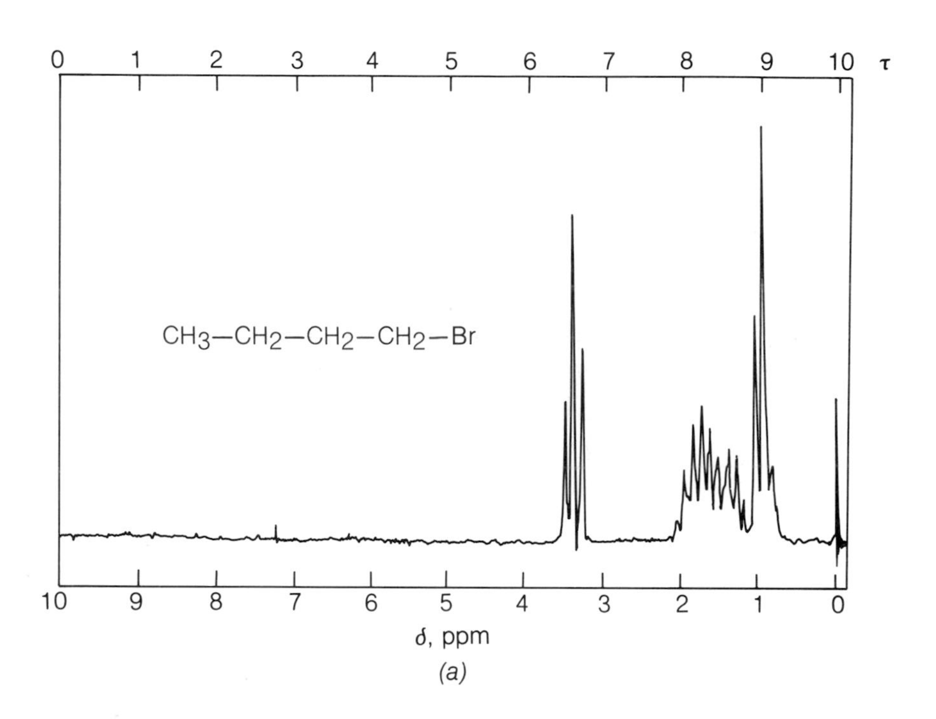

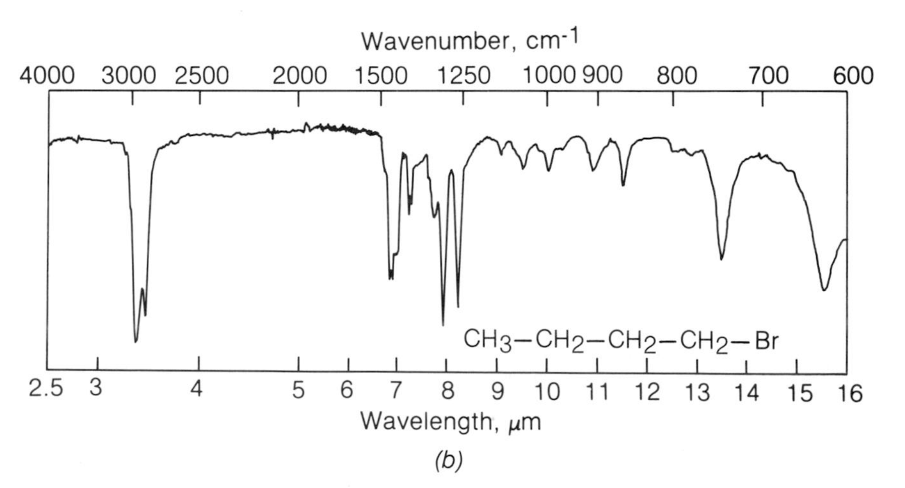

Figure 10.1
The (*a*) nmr and (*b*) ir spectra of *n*-butyl bromide.

EXPERIMENT 10.2

SYNTHESIS OF tert-BUTYL CHLORIDE BY AN S_N1 REACTION

$$CH_3-\overset{\overset{\large CH_3}{|}}{\underset{\underset{\large CH_3}{|}}{C}}-OH + HCl \longrightarrow CH_3-\overset{\overset{\large CH_3}{|}}{\underset{\underset{\large CH_3}{|}}{C}}-Cl + H_2O$$

Time 2 h

Materials *tert*-Butyl alcohol, 25 mL (MW 74, bp 83°C, d 0.786 g/mL)
12 *N* Hydrochloric acid, 100 mL (36% w/w, d 1.18 g/mL)

Precautions Perform experiment in hood.

Hazard Alkyl halide fumes are toxic in high concentrations.

Experimental Procedure

Measure 25 mL *tert*-butyl alcohol (graduated cylinder) and add it to a 250-mL separatory funnel held in a ring stand in the hood. To this *slowly* add 100 mL 12 *N* HCl and swirl to mix the layers.

Place a lightly greased stopper on the separatory funnel and vigorously shake the funnel for 2 to 3 min. The pressure is released during this process through the stopcock in the usual way. Continue to shake the separatory funnel intermittently for 10 min. After shaking, allow the mixture to stand for 7 to 10 min (or until the layers have separated cleanly), and remove (and discard) the *lower* layer.

Wash the liquid remaining in the separatory funnel with two 10-mL portions of 5% K_2CO_3 solution, removing the *lower* layer in each case. Be sure to vent the separatory funnel after each shaking (Sec. 2.5). Transfer the liquid to a 125-mL Erlenmeyer flask and dry over $CaCl_2$ for 10 min. Decant the liquid away from the drying agent through a plug of glass wool into a *dry* 50-mL round-bottom flask. Set up an apparatus for distillation as in Fig. 2.2.

Using a steam bath, heating mantle, or very small flame (check with your instructor), distill the above liquid material and collect two fractions; the first should contain all material which distills up to 48°C, and the second should contain all material which boils between 48 and 54°C. After the second fraction is collected, stop the distillation and discard the first fraction and the material left in the boiling flask. Weigh the second fraction and calculate the yield of the reaction. Store the clear, colorless liquid obtained in a tightly stoppered flask until it is needed in another procedure or submit it to your instructor, as directed.

The nmr and ir spectra of *tert*-butyl chloride are shown in Fig. 10.2. What

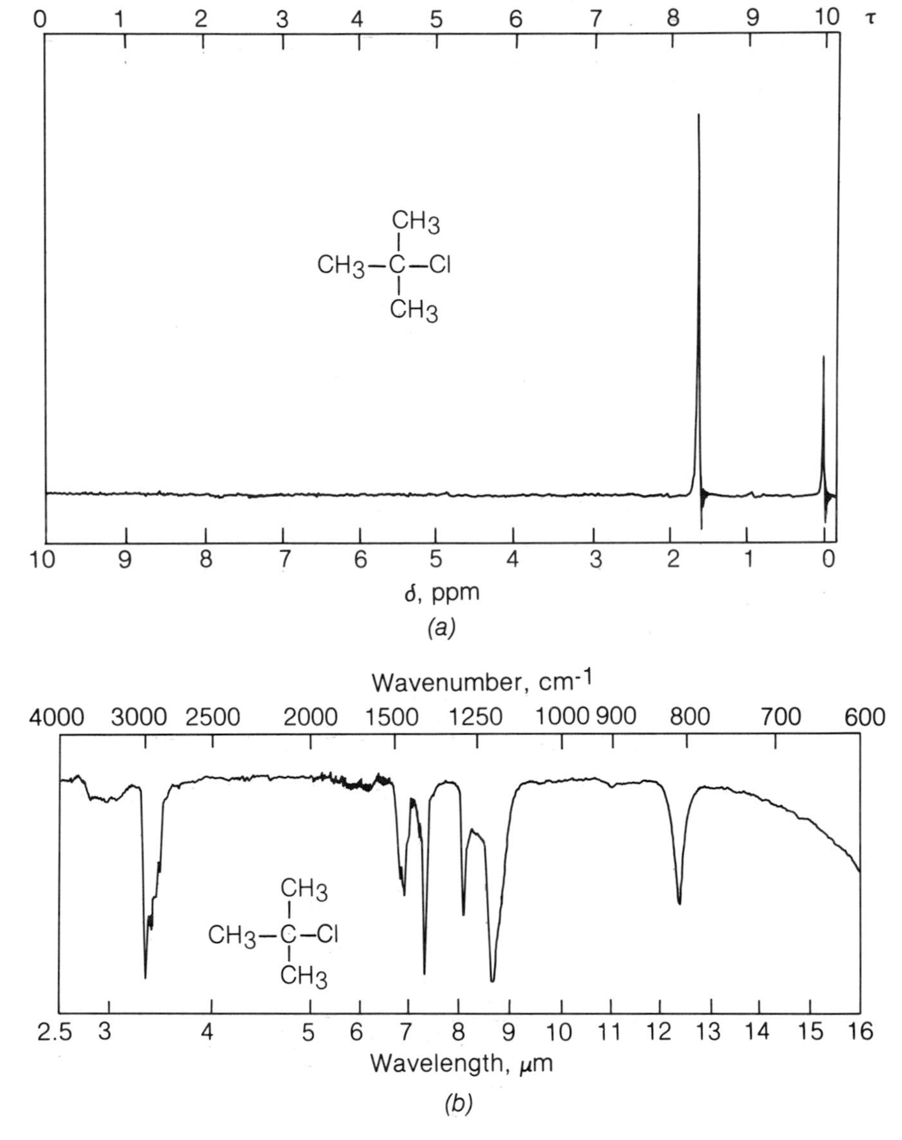

Figure 10.2
The (*a*) nmr and (*b*) ir spectra of *tert*-butyl chloride.

extra absorptions would you expect to see in the spectra if starting material were present? (See spectra for previous experiment.)

QUESTIONS AND EXERCISES

10.1 A student who was making *n*-butyl bromide was momentarily distracted at the beginning of the experiment and neglected to add sodium bromide to the reaction mixture. The material isolated after the required period of time turned out to be an alcohol rather than a bromide. The student expected that the isolated alcohol would be starting material but found on analysis that this was not so. Although the material was an alcohol, it was an alcohol which was different from the starting material. What do you suppose the identity of this unknown alcohol was, and can you suggest a mechanism for its formation?

10.2 In the discussion which introduces the experiments contained above, the principal by-products mentioned as accompanying the formation of *n*-butyl bromide are (1) 1-butene, (2) dibutyl ether, and (3) starting material. Can you think of an explanation for the formation of 1-butene and dibutyl ether in the reaction sequence?

10.3 The procedures for preparation of butyl bromide from a primary alcohol and for preparation of *tert*-butyl chloride differ drastically in their acid concentrations. If you had an unknown alcohol, how could you determine which of the two procedures listed above might be used to convert the alcohol into its corresponding halide?

10.4 Freshly prepared *tert*-butyl chloride was found to be neutral. After standing for several weeks on the shelf, however, the same sample was found to be acidic. What mechanism can you suggest to account for this apparent change?

10.5 In the nmr spectrum of *n*-butyl bromide the most downfield resonance is at about 3.5 ppm. The most downfield resonance in *tert*-butyl chloride, however, is at only about 1.5 ppm. Account for this difference.

10.6 If in an experiment you required anhydrous hydrogen bromide gas and none were available from the stockroom, based on the experimental procedure above how might you generate hydrogen bromide for the purpose needed?

ESTERS AND AMIDES

INTRODUCTION
Esters

Esters are often formed from the corresponding carboxylic acid and alcohol in the equilibrium reaction illustrated below:

$$RCOOH + R'OH \overset{H^+}{\rightleftharpoons} RCOOR' + H_2O$$

In the ester formation reaction, water is eliminated in the reaction between a molecule of a carboxylic acid and a molecule of alcohol. The proton over the arrow indicates that the process is acid-catalyzed. Also note that one molecule of water is produced, which will tend to reverse the equilibrium, i.e., to hydrolyze the ester back to starting material. The Fischer esterification process, which uses a very large excess of an alcohol, is most commonly used for the formation of methyl and ethyl esters of carboxylic acids. The large excess of alcohol favors the equilibrium formation of ester (Le Chatelier's principle). The excess of alcohol is economically acceptable in the case of methanol and ethanol because these are very inexpensive alcohols. In the formation of methyl benzoate, for example, benzoic acid is dissolved in a large molar excess of methyl alcohol, and a drop of a strong mineral acid such as sulfuric acid is added as a catalyst. The mixture is refluxed for 1 to 2 h and, after removal of the residual methanol, methyl benzoate can be obtained by distillation. The Fischer esterification is particularly advantageous when the alcohol being used for the solvent is very inexpensive and volatile.

In principle, the formation of an ester from one equivalent each of a carboxylic acid and an alcohol can be performed by the formal elimination of water in a previous step. For example, if the carboxylic acid is neutralized with silver oxide, the silver salt of the carboxylic acid is produced and water eliminated. Further reaction with an alkyl chloride gives an alkylation reaction which produces insoluble silver chloride and the ester as the products. The silver salt reaction is important but expensive.

$$2RCOOH + Ag_2O \longrightarrow 2RCOOAg + H_2O$$

$$RCOOAg + R'Cl \longrightarrow RCOOR' + AgCl$$

A procedure which would allow an alkyl chloride to be used with a sodium or potassium salt as a nucleophile would be a very expeditious way of generating esters. The traditional drawback to such a procedure has been the lack of nucleophilicity in the sodium and potassium salt, particularly when the reaction is run in aqueous solution. This difficulty can be circumvented by the use of phase-transfer catalysis. The use of a nonpolar medium minimizes the solvation forces, which generally reduce the activity of carboxylates so that simple nucleophilic substitution becomes effective. The procedures for two experiments are given below. The first is the esterification of benzoic acid using methanol, which is the traditional Fischer esterification procedure. The second is a procedure for the preparation of *n*-butyl benzoate from butyl bromide (prepared in Exp. 10.1) and sodium benzoate.

Organic esters are characterized by the functional group RCO_2R'. The structure is shown below:

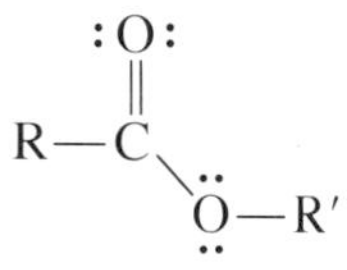

```
      :O:
       ‖
  R—C
       \ ..
        O—R'
        ..
```

The groups represented by R and R′ in the drawing may be alkyl, aryl, or even heterocyclic. The key feature of the ester functionality (see Chap. 27) is really that it contains a carbonyl group with an adjacent oxygen. The presence of the second oxygen in the ester affords this functional group reactivity quite different from that of a ketone for two reasons. First, the electron-releasing effect of oxygen on the carbonyl carbon reduces the electrophilicity of the carbonyl group, thereby making it somewhat less susceptible to nucleophilic addition reactions than are ketones (see Chaps. 14, 15, and 25 for a discussion of ketone reactivity). Second, when a nucleophile does add to the carbon, the tetrahedral intermediate can re-form a carbon-oxygen double bond in two ways: (1) by expelling the added nucleophile to re-form the starting material; and (2) by retaining the nucleophile in the product and expelling alkoxide from the ester, leading to the formation of a product. Such reactions are particularly advantageous if R′ is an aromatic group, as the leaving group will be phenoxide. The very important consequence of carbonyl adjacent to an alkoxy group will be discussed in Chap. 12.

Amides

The amides and esters of carboxylic acids are closely related in the sense that each of them contains a carbonyl group with an adjacent heteroatom. Esters are described by the general formula R—CO—OR′ and amides by the formula R—CO—NHR′. The R′ group on either the oxygen or the nitrogen may be hydrogen (in NH_2), alkyl, or aryl. In the case of amides the second hydrogen atom may be replaced with an alkyl or aryl group (NR′R″), giving another class of substitution. Notice however, that the carbonyl function remains intact even though the possibilities for substitution on the nitrogen atom are enormous.

The presence of nitrogen causes the reactivity of an amide to be very different in some ways from that of an ester. If we think about the resonance forms which can exist for esters and amides (see below), it is clear that in the amide resonance forms the positive charge is more localized on nitrogen. This is a more favorable situation than having the positive charge localized on the more electronegative oxygen of an ester. As a result, the charged resonance form is a more significant contributor to the overall structure of an amide than to an ester.

$$\mathrm{R{-}C({=}O){-}N(H){-}R} \longleftrightarrow \mathrm{R{-}C(O^-){=}\overset{+}{N}(H){-}R} \qquad \text{versus} \qquad \mathrm{R{-}C({=}O){-}O{-}R} \longleftrightarrow \mathrm{R{-}C(O^-){=}\overset{+}{O}{-}R}$$

If we think about what the consequences of this resonance interaction are, it should be clear that the reactivity of the carbonyl group will be lower in an amide, because in esters the double-bond character of the carbonyl group is lower. It should also be clear that rotation about the nitrogen-carbon bond should be very much more restricted than rotation about the oxygen-carbon bond in esters for the same reason. Both these predictions are verified by experimental observation. In particular, esters react much more readily with hydroxide ions than do amides. There is another factor which compounds this reactivity problem: the fact that an alkoxide ion (RO^-) is a better leaving group than an amide anion (R_2N^-). As a consequence of all these factors, amides are very much more difficult to hydrolyze than are esters.

Esters and amides can in principle be formed by the same kinds of reactions. From previous discussions, it is clear that simple esters can be formed from a carboxylic acid and an excess of the alcohol in the presence of an acidic catalyst. This approach is not as effective for the formation of simple amides. There are several reasons for this. First of all, methylamine, which is the amine corresponding to methyl alcohol, is a gas. Ethylamine is also a gas. It would be very difficult, therefore, to use these as solvents for amide formation. Second, the esterification reaction is acid-catalyzed. In the presence of an amine, however, the proton will be in residence exclusively on the nitrogen. If the electron pair of the nitrogen is involved in the formation of a salt, the amine will no longer be a nucleophile and the reaction will not proceed.

Because of these difficulties, some sort of activating group is necessary for the formation of an amide. In general, amides are formed from amines and either acid halides (see Secs. 23.6A and 24.7B) or acid anhydrides. In the experiments described below, both acid chloride and acid anhydride have been utilized. In these reactions, the amine remains nucleophilic and attacks the carbonyl group, which is activated by a potent leaving group; when the tetrahedral intermediate (see below) re-forms the double bond, it is not the amide anion which is expelled but rather HCl from the acid halide or acetic acid from acetic anhydride. This process is shown below.

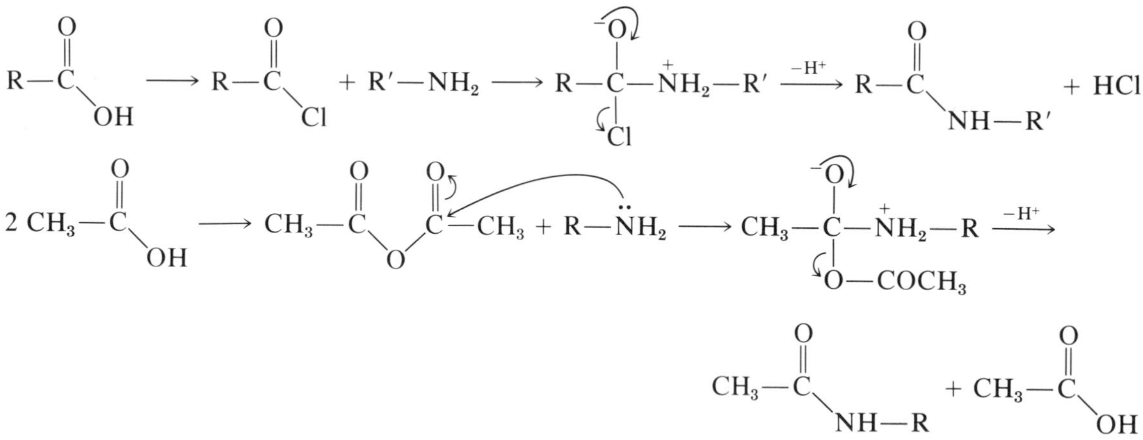

$$R\text{—}C(=O)\text{—}OH \longrightarrow R\text{—}C(=O)\text{—}Cl + R'\text{—}NH_2 \longrightarrow R\text{—}C(O^-)(Cl)\text{—}\overset{+}{N}H_2\text{—}R' \xrightarrow{-H^+} R\text{—}C(=O)\text{—}NH\text{—}R' + HCl$$

$$2\,CH_3\text{—}C(=O)\text{—}OH \longrightarrow CH_3\text{—}C(=O)\text{—}O\text{—}C(=O)\text{—}CH_3 + R\text{—}\ddot{N}H_2 \longrightarrow CH_3\text{—}C(O^-)(O\text{—}COCH_3)\text{—}\overset{+}{N}H_2\text{—}R \xrightarrow{-H^+}$$

$$CH_3\text{—}C(=O)\text{—}NH\text{—}R + CH_3\text{—}C(=O)\text{—}OH$$

Several examples of amide formation are presented in Secs. 11.4 and 11.5. In the first of these, *N,N*-diethyl-*m*-toluamide, the insect repellent DEET, is prepared. In the subsequent preparations, details are presented for the formation of acetanilide and of the analgesic phenacetin.

11.1 SYNTHESIS OF *n*-BUTYL BENZOATE

The formation of esters from acid salts and alkyl halides has been fraught with a number of problems (see above). A solution to this problem involves the modern method of phase-transfer catalysis (see Sec. 4.7). By using this technique, carboxylic acid salts may efficiently be used as nucleophiles. In the reaction sequence described here, sodium benzoate, in the presence of a tetraalkylammonium halide catalyst ($R_4N^+Cl^-$) exchanges cations to afford the tetraalkylammonium benzoate. The latter salt is quite nucleophilic and facilitates the S_N2 reaction by which the ester is formed.

The experimental conditions suggested for this reaction may seem a little unusual. The presence of water can in some cases reverse ester formation. In this reaction, the displacement occurs in the organic phase (*n*-butyl bromide), where no water is present. At the end of the reaction, excess benzoate salt, the catalyst, and water are all washed away.

EXPERIMENT 11.1 SYNTHESIS OF n-BUTYL BENZOATE BY CARBOXYLATE ION ALKYLATION

$$C_6H_5C(=O)ONa + CH_3CH_2CH_2CH_2Br \xrightarrow{R_4N^+Cl^-} C_6H_5C(=O)OCH_2CH_2CH_2CH_3$$

Time 2.25 h

Materials Sodium benzoate, 15 g (MW 144)
n-Butyl bromide, 10 mL (MW 137, bp 100°C, d 1.276 g/mL)
Water, 50 mL
Aliquat 336 or equivalent phase-transfer catalyst, 1 g

Precautions Carry out procedure in hood.

Hazards *n*-Butyl bromide is an irritant and is toxic in high concentrations.

Experimental Procedure

A 100-mL round-bottom distilling flask is charged with tricaprylmethylammonium chloride (Starks' catalyst, Aliquat 336, 1.0 g), *n*-butyl bromide (10 mL, 93.2 mmol), sodium benzoate (15 g, 110 mmol), and water (25 mL). A reflux condenser with lightly greased joints is placed on top of the flask, and reflux with a free flame is commenced (Fig. 8.1). The reaction mixture is boiled for 75 min, the flame removed, and the reaction mixture allowed to cool. After cooling, the residue is poured into a separatory funnel containing 50 mL water. Any residual material remaining in the round-bottom flask is rinsed into the separatory funnel with a total of 30 mL dichloromethane. The layers are separated, and the organic layer is first washed with two 15-mL portions of half-saturated sodium chloride and then dried over sodium sulfate. The dichloromethane solution is filtered and poured into a 250-mL Erlenmeyer flask, and the solvent is evaporated on the steam bath.

The crude product, an oil, is poured into a 50-mL round-bottom distilling flask set up for simple distillation with an air condenser [this is the same as a normal distillation (Fig. 2.5) except that no water is circulated through the condenser]. The oil is distilled using a free flame, and all the material which distills below 190°C is collected. The receiver flask is then changed and all the material distilling above 190°C is collected. **Do not distill to dryness in this or any other distillation.** In this experiment about 2 mL oil should be left in the flask.

The product obtained in the fraction boiling above 190°C is a clear, water-white liquid weighing about 12 g. The *n*-butyl benzoate is obtained in about 75% yield.

Note in the ir spectrum (Fig. 11.1*a*) that a strong carbonyl peak is observed

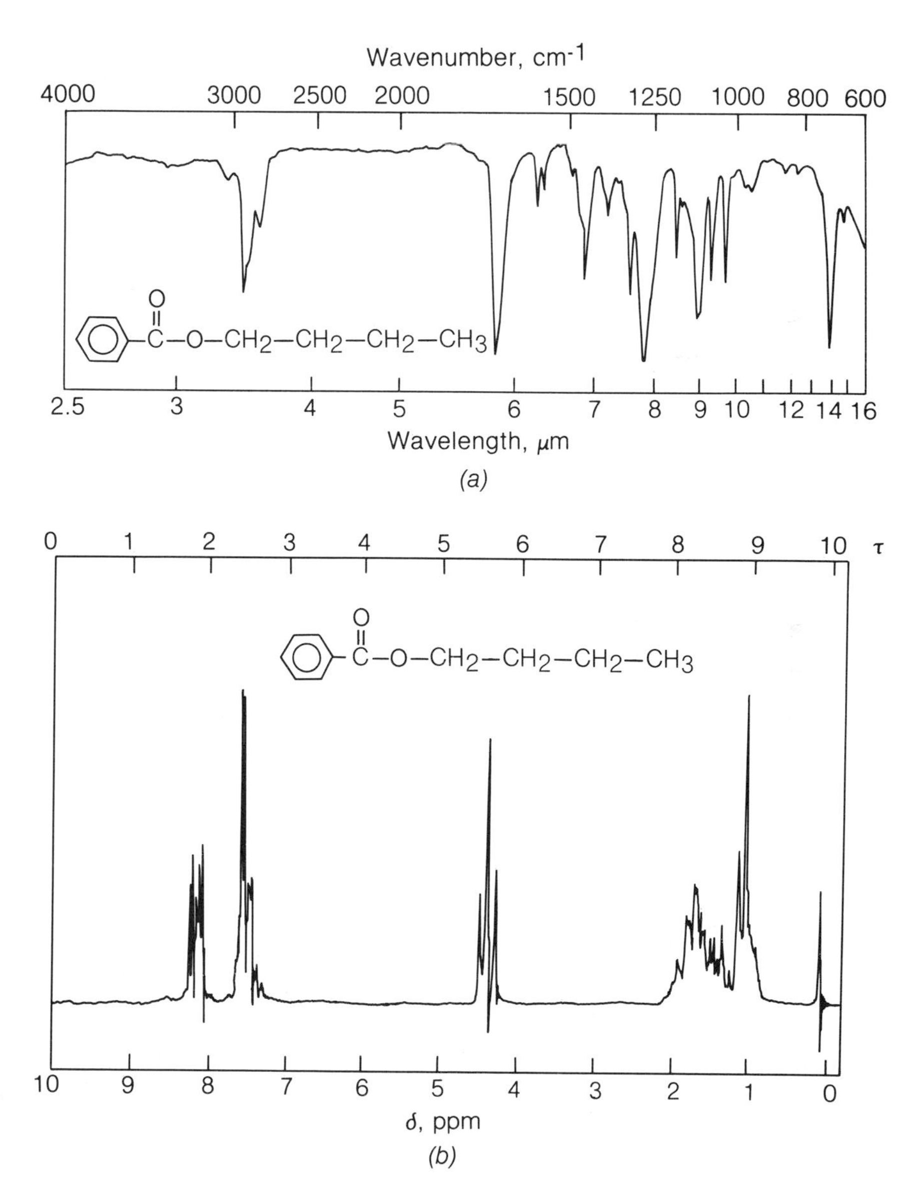

Figure 11.1
The (*a*) ir and (*b*) nmr spectra of *n*-butyl benzoate.

at 1720 cm^{-1} and that there is no hydroxyl peak observed in the 3400 cm^{-1} region. The position of the carbonyl peak is interesting because the effect of the heteroatom adjacent to the carbonyl is to increase the force constant of the carbon-oxygen double bond and thereby its vibrational frequency. However, this increase in vibrational frequency is compensated by the conjugative effect of the aromatic ring, which makes the net vibrational frequency for this molecule very close to that of a normal carbonyl compound such as acetone.

The nmr spectrum of *n*-butyl benzoate is shown in Fig. 11.1*b*.

11.2 SYNTHESIS OF ESTERS BY THE FISCHER ESTERIFICATION

The classical Fischer esterification reaction involves the conversion of a carboxylic acid to an ester by heating with an excess of alcohol and a mineral acid catalyst (usually sulfuric acid). As discussed in the introduction, this reaction is an equilibrium process and depends for its success on a large excess of alcohol.

The preparation of methyl benzoate and methyl 4-chlorobenzoate are parallel preparations. Methyl benzoate is an oil used in perfumery under the name peau d'Espagne and is isolated by distillation. Methyl 4-chlorobenzoate is a solid and is isolated by crystallization. As a solid, the latter compound is not as odoriferous as methyl benzoate and has not found a similar application in perfumery.

The second preparation presents an advantage over the first in that the preparation requires less time. Either ester may be used in the preparation of a triarylcarbinol (Exp. 12.2B). Only methyl benzoate will yield triphenylcarbinol. Methyl 4-chlorobenzoate will yield the closely related chlorophenyldiphenylcarbinol.

The third preparation in this series involves chemistry similar to that above and yields the strongly fragrant ester isoamyl acetate. This substance is often referred to as banana oil or pear oil. The crude material is often dissolved in water to produce pear-flavored syrups. An especially interesting application of this compound has been to mask the unpleasant odor of shoe polish.

EXPERIMENT 11.2A SYNTHESIS OF METHYL BENZOATE

$$C_6H_5COOH + CH_3OH \overset{H^+}{\rightleftharpoons} C_6H_5COOCH_3 + H_2O$$

Time 4.5 h (two laboratory periods)

Materials Benzoic acid, 12.2 g (MW 122, mp 121°C)
Sulfuric acid, 4 mL
Methanol (anhydrous), 50 mL

Hazards Sulfuric acid is a strong dehydrating acid; contact with skin or clothes should be avoided. Methanol vapors are toxic in high concentrations.

Experimental Procedure

A 100-mL round-bottom distilling flask is charged with benzoic acid (12.2 g, 100 mmol) and *anhydrous* methanol (50 mL). To this mixture is added *slowly* (**considerable heat is evolved**) sulfuric acid (4 mL). (The methanol may come to a boil as the last of the sulfuric acid is added.) A reflux condenser with lightly greased joints and drying tube is placed on top of the flask (Fig. 9.1) and the mixture is gently heated with a free flame for 1 h. After this reflux period the flame is removed and the reaction mixture is allowed to come to room temperature. After cooling, the mixture is poured into a separatory funnel containing water (50 mL). Any residual material in the round-bottom flask is washed into the separatory funnel with dichloromethane (50 mL). The layers are shaken, separated, and the lower, organic layer removed. The organic layer is washed with water (25 mL), followed by saturated $NaHCO_3$ (two 25-mL portions) and then by half-saturated brine (two 25-mL portions). The organic layer is then dried over sodium sulfate for several minutes and filtered into a 250-mL Erlenmeyer flask, and the solvent is evaporated on the steam bath.

After all the solvent has been removed, the residual oil is taken up in toluene (20 mL) and transferred to a dry 50-mL Erlenmeyer flask. Anhydrous $CaCl_2$ chips (approximately 2 g) are added and the mixture is tightly stoppered with a cork. This flask is set aside until the next laboratory period.

Flame dry a simple distillation apparatus (Fig. 2.2). Filter the toluene solution into the cool 50-mL round-bottom flask and carefully distill (free flame) the methyl benzoate–toluene mixture. Collect at least two fractions: toluene (bp 111°C) and pure methyl benzoate. (*Note:* Make certain that the thermometer bulb is *below* the level of the side projection of the distillation head to ensure that the boiling point you are reading is accurate.) Turn off the condenser water (but do not drain it) as the temperature increases above 150°C or you run the risk of cracking the condenser. Collect the pure water-white methyl benzoate (everything above 190°C) in a *dry,* tared Erlenmeyer flask, weigh the flask containing the product (determine the yield by subtracting the tare weight), and use this material in the Grignard reaction.

Notice in the ir spectrum shown in Fig. 11.2*a* that the carbonyl frequency is the same as that observed in the butyl benzoate case. Also note that only two types of protons, methyl and aromatic, are observed in the nmr spectrum (Fig. 11.2*b*) obtained from this material.

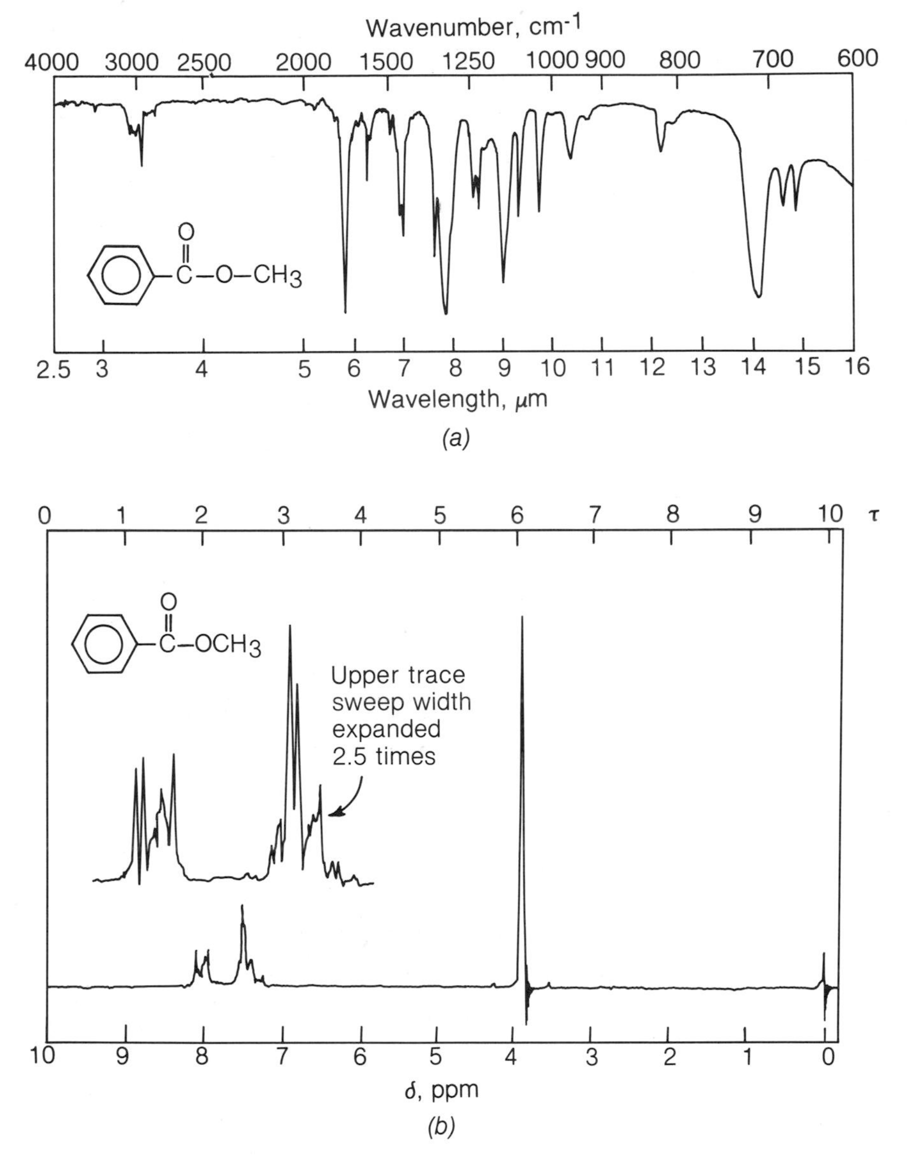

Figure 11.2
The (*a*) ir and (*b*) nmr spectra of methyl benzoate.

EXPERIMENT 11.2B

SYNTHESIS OF METHYL 4-CHLOROBENZOATE

$$4\text{-}Cl\text{-}C_6H_4\text{-}C(=O)OH + CH_3OH \xrightarrow{H^+} 4\text{-}Cl\text{-}C_6H_4\text{-}C(=O)OCH_3 + H_2O$$

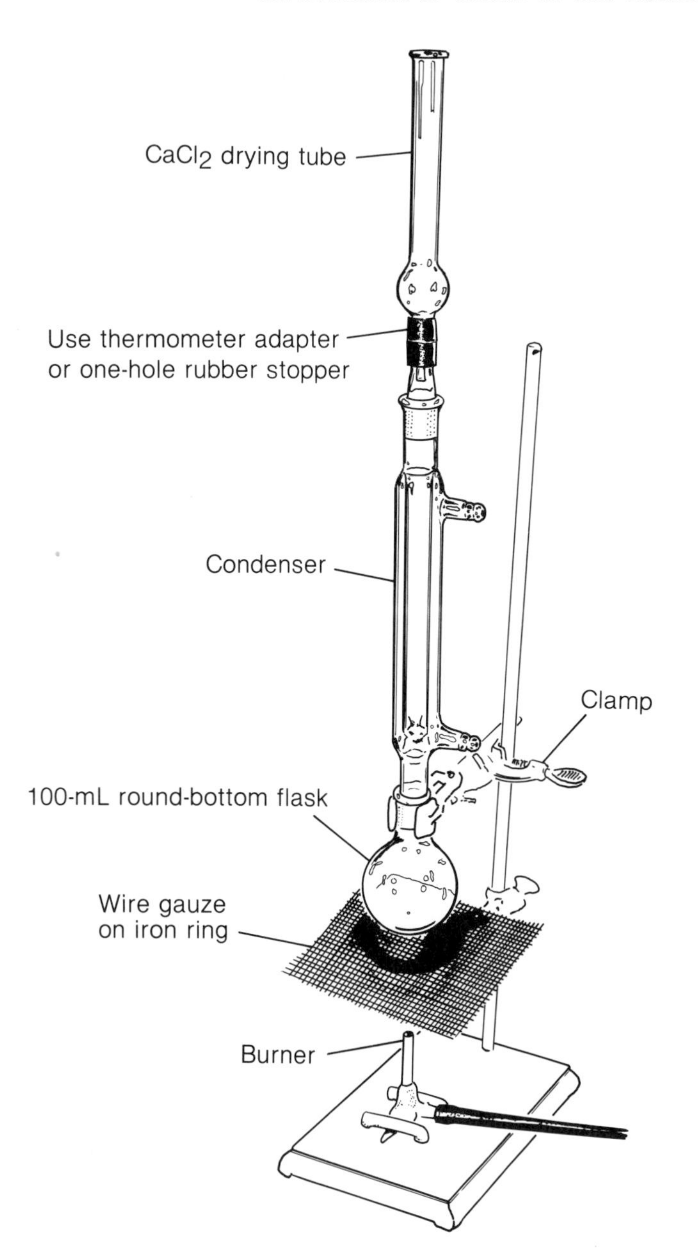

Figure 11.3
Reflux with exclusion of moisture.

Time 2 h

Materials 4-Chlorobenzoic acid, 5 g (MW 156.57, mp 239 to 241°C)
Sulfuric acid, 4 mL (MW 98)
Methanol (anhydrous), 50 mL

Hazards Sulfuric acid is a strong dehydrating acid; contact with skin or clothes should be avoided. Methanol vapors are toxic in high concentrations.

Experimental Procedure

A 100-mL round-bottom distilling flask is charged with 4-chlorobenzoic acid (5 g, 32 mmol) followed by *anhydrous* methanol (50 mL). To this mixture is added *slowly* (**considerable heat is evolved**) sulfuric acid (4 mL). (The methanol may come to a boil as the last of the sulfuric acid is added.) A reflux condenser with lightly greased joints and drying tube is placed on top of the flask and the mixture is heated with a free flame for 1 h (see Fig. 11.3). As the reaction proceeds the 4-chlorobenzoic acid, which is not initially soluble in the methanol, goes into solution. After about 25 to 30 min the methanol solution should be clear.

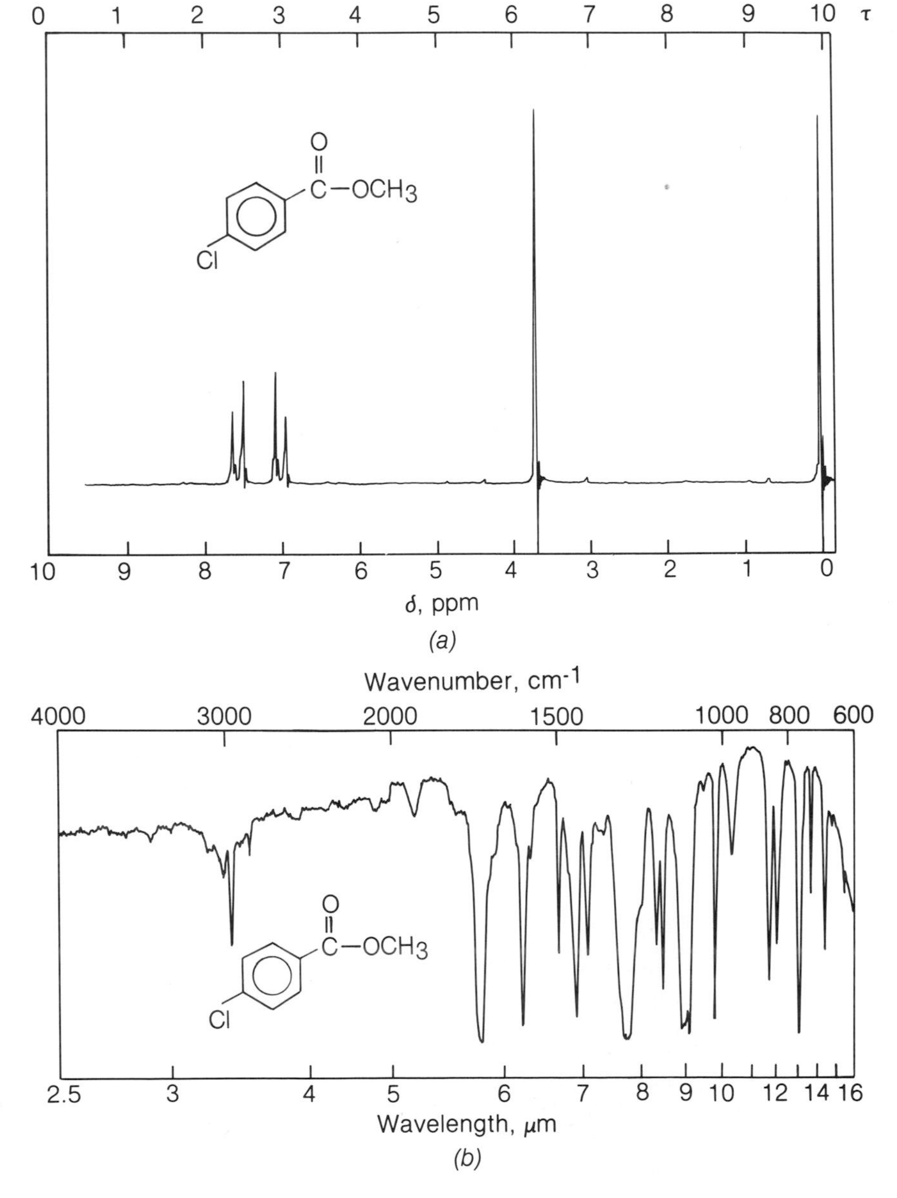

Figure 11.4
The (*a*) ir and (*b*) nmr spectra of methyl 4-chlorobenzoate.

After this reflux period the flame is extinguished and the reaction mixture is allowed to cool. After cooling, the mixture is poured into a separatory funnel containing water (50 mL). Any residual material in the round-bottom flask is washed into the separatory funnel with dichloromethane (25 mL). The layers are shaken and separated and the organic layer is drawn off. The aqueous solution is then extracted with another portion of dichloromethane (25 mL) and the organic layers are combined. The organic layer is washed with water (25 mL) followed by saturated $NaHCO_3$ (two 25-mL portions) and then by half-saturated brine (two 25-mL portions). The organic layer is then dried over sodium sulfate for several minutes and filtered into a 250-mL Erlenmeyer flask, and the solvent evaporated on the steam bath.

After all the solvent has been removed, the remaining yellow oil (approximately 5 g) is dissolved in methanol (10 mL) and warmed briefly on the steam bath. Water (2 mL) is added to the warm methanol solution, which is allowed to cool, first to room temperature and then in an ice bath. The pure ester crystallizes as small needles from the cold methanol solution. The solid is filtered rapidly under vacuum using a Hirsch funnel (see Fig. 2.15) and the solid material on the filter is rinsed with 5 mL ice-cold aqueous methanol [80% v/v (1 mL H_2O, 4 mL CH_3OH)]. The crystalline material is air dried for several minutes. The yield of methyl 4-chlorobenzoate, mp 41 to 43°C, is approximately 4 g (about 70%).

The ester may also be recrystallized from hexane, which affords material of equal purity but in somewhat lower yield.

The ir and nmr spectra of methyl 4-chlorobenzoate are shown in Fig. 11.4.

EXPERIMENT 11.2C

SYNTHESIS OF ISOAMYL ACETATE (PEAR OIL)

$$CH_3-C(=O)-OH + (H_3C)_2CH-CH_2-CH_2-OH \xrightarrow{H^+} CH_3-C(=O)-O-CH_2-CH_2-CH(CH_3)_2 + H_2O$$

Time 3 h

Materials Acetic acid, 25 mL (MW 60, bp 116 to 118°C)
Isoamyl alcohol, 20 mL (MW 88, bp 130°C)
Concentrated sulfuric acid, 5 mL (MW 98)

Precautions Wear gloves when pouring concentrated sulfuric acid. Pour acetic acid in the hood.

Hazards Sulfuric acid is a strong dehydrating agent. Acetic acid can burn the skin. Avoid contact of either with skin or clothes.

Experimental Procedure

To a 100-mL round-bottom flask, add 25 mL (0.420 mol) glacial acetic acid (**caution: hood**) followed by 20 mL (0.185 mol) isoamyl alcohol (3-methyl-1-butanol). Swirl the flask to mix the layers. To the solution add (**carefully, gloves**) 5 mL concentrated sulfuric acid. Swirl the flask as the sulfuric acid is added (**heat generated**).

Add several boiling chips to the flask, then place a reflux condenser with lightly greased joints on the flask as shown in Fig. 11.3. Bring the solution to boiling with a flame (bunsen burner) or electric mantle and reflux the solution for 1 h.

After the reflux period is completed, allow the solution to cool to room temperature. Transfer the entire solution to a separatory funnel and add 50 mL distilled water. Swirl the solutions, allow the layers to separate, and remove the lower aqueous layer. Add another 25-mL portion of distilled water, shake the flask, and separate and remove the lower aqueous layer.

The organic layer is then extracted with three 25-mL portions of 5% aqueous sodium bicarbonate solution (to remove excess acetic acid). (*Note:* Be careful, as carbon dioxide is given off during the extraction.) Test the last extract and if the aqueous phase is not basic (pH paper), extract the organic layer with two more 25-mL portions of sodium bicarbonate solution. After removal of the acetic acid, wash the organic layer with two 5-mL portions of saturated salt solution. Transfer the organic layer to a 50-mL Erlenmeyer flask and dry over granular anhydrous sodium sulfate (or magnesium sulfate).

After drying (liquid should be clear), the organic layer is decanted into a 50-mL round-bottom distillation flask. Assemble a simple distillation apparatus, as shown in Fig. 2.2. Add several boiling chips and distill (with a free flame or mantle). Cool the receiver in an ice bath. Collect the fraction which distills between 135 and 143°C. The clear, colorless product has an overpowering odor of bananas, and is obtained in 80 to 90% yield.

The nmr and ir spectra of the pure ester are shown in Fig. 11.5. The acetate methyl group is observed in the nmr spectrum as a characteristic singlet at 2.05 ppm. Note the position of the carbonyl group in the ir spectrum and the absence of any hydroxyl absorption in the 3400 to 3600 cm^{-1} range.

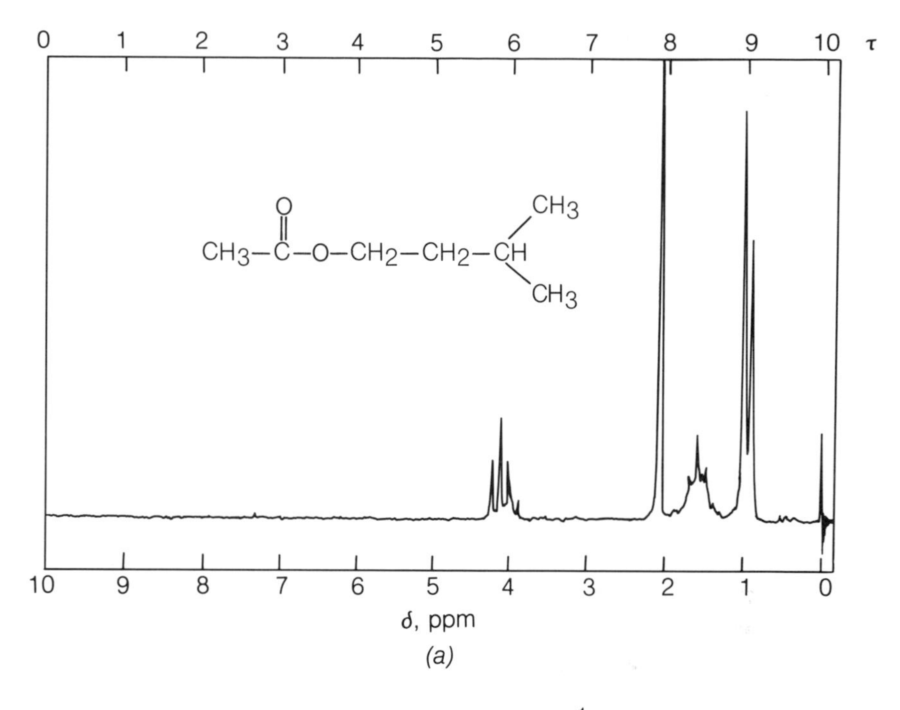

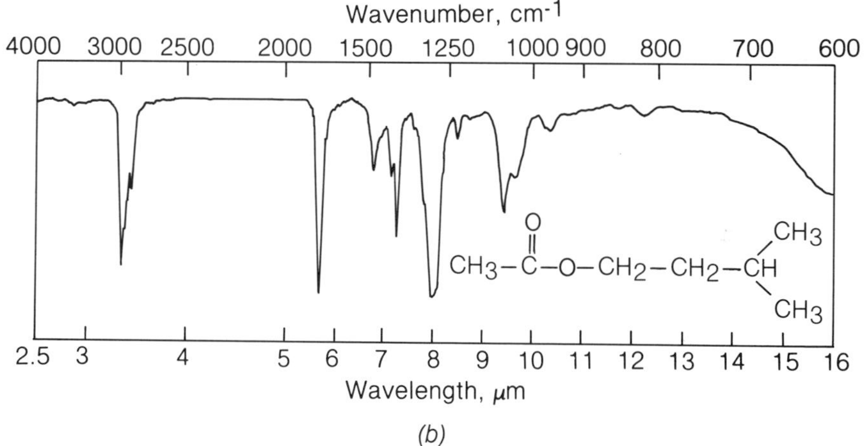

Figure 11.5
The (*a*) nmr and (*b*) ir spectra of isoamyl acetate.

11.3 ASPIRIN

Aspirin is the most widely used drug in modern society. Billions of tablets of aspirin are ingested annually in the United States alone. It is probably the most widely used intentional drug, as opposed to such nominal drugs as caffeine (found in coffee and tea) and ethanol (found in liquor, wine, and beer). Its name derives from its structure, acetylsalicylic acid. In earlier times salicylic acid was known as spiraeic acid (from the meadowsweet family) and so aspirin was actually acetylspiraeic acid, whence the name.

Aspirin is the most frequently used of the so-called analgesics (pain killers). Aspirin is also a powerful antipyretic (fever-reducing) and anti-inflammatory (swelling-reducing) substance. It is most important as an effective drug for the symptomatic relief of various painful afflictions, including arthritis. While many compounds have analgesic properties, only aspirin has antipyretic and anti-inflammatory properties in conjunction with its ability to relieve pain. Aspirin is therefore used in many preparations and in conjunction with many other compounds.

Salicylic acid is itself an analgesic. In fact, it is in this form that the drug is extracted from plant sources associated with pain relief. The early medical administration of the drug in its pure form was as the sodium salt. When used and administered in this way, however, the material had unpleasant side effects, and early chemists sought a modification which would retain the analgesic and anti-inflammatory properties while decreasing the adverse side effects. The modification of salicylic acid with acetic anhydride satisfied this requirement, and acetylsalicylic acid is as effective as sodium salicylate but does not have as many adverse side effects. The same strategy was used later to modify morphine, another powerful analgesic. In this case it was the addictive properties of morphine which presented a problem. The solution was to acetylate morphine, converting it into diacetylmorphine, commonly known as heroin. Needless to say, the strategy was not as successful with morphine as it was with salicylic acid.

It is interesting that aspirin is acetylsalicylic acid and oil of wintergreen is methyl salicylate. The structures of these two compounds are very similar, but oil of wintergreen is largely a flavoring agent, whereas aspirin is a pain killer. Oil of wintergreen is used in many liniments, however, as it is absorbed through the skin. Once absorbed it may be hydrolyzed to salicylic acid; thus it is a source of pain relief, albeit localized pain relief. The structures of the two esters, aspirin and methyl salicylate, are shown below.

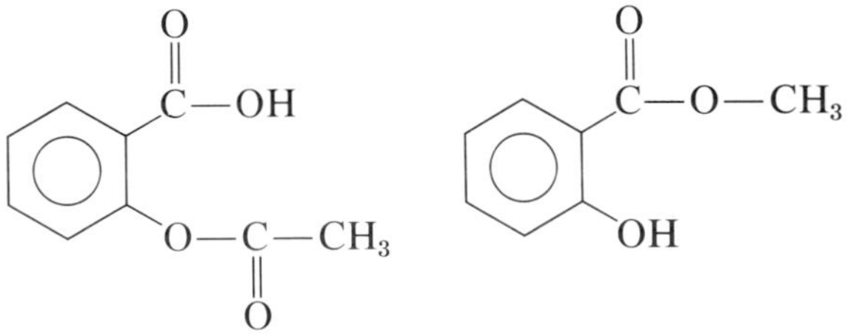

Aspirin (acetylsalicylic acid)

Oil of wintergreen (methyl salicylate)

The industrial synthesis of aspirin can actually be started at several points. A consideration of the structure of aspirin will indicate that several features must be incorporated into the molecule. In the preparation recorded here, salicylic acid will be acetylated with acetic anhydride (aspirin is a phenolic ester of acetic acid). Before the acetylation, however, the carboxyl function, the hydroxyl function, and the aromatic ring must be available. In practice salicylic acid, even though it may be extracted from plants of the birch or meadowsweet family, is usually synthesized by Kolbe carboxylation of phenol. In the Kolbe reaction, a phenol is transformed into a phenolic acid by the action of base and carbon dioxide. This reaction is very effective when run on an industrial scale. Phenol can likewise be (and is) prepared from several aromatic compounds (including benzene, chlorobenzene, and isopropylbenzene) produced from crude oil. Thus the price of oil may ultimately determine the cost and availability of drugs like aspirin and flavoring agents like oil of wintergreen.

In contrast to the previous examples of ester formation, an excess of acetic anhydride is used to acylate the phenolic hydroxyl group. The reaction occurs as shown below:

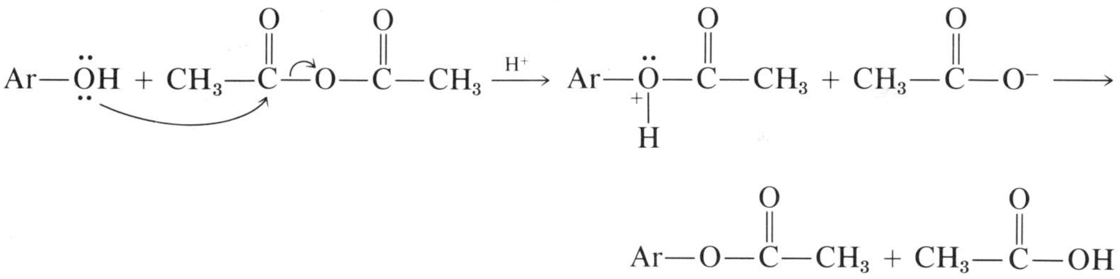

$$\text{Ar—}\ddot{\text{O}}\text{H} + \text{CH}_3\text{—}\overset{\text{O}}{\overset{\|}{\text{C}}}\text{—O—}\overset{\text{O}}{\overset{\|}{\text{C}}}\text{—CH}_3 \xrightarrow{\text{H}^+} \text{Ar—}\underset{\text{H}}{\underset{+|}{\ddot{\text{O}}}}\text{—}\overset{\text{O}}{\overset{\|}{\text{C}}}\text{—CH}_3 + \text{CH}_3\text{—}\overset{\text{O}}{\overset{\|}{\text{C}}}\text{—O}^- \longrightarrow$$

$$\text{Ar—O—}\overset{\text{O}}{\overset{\|}{\text{C}}}\text{—CH}_3 + \text{CH}_3\text{—}\overset{\text{O}}{\overset{\|}{\text{C}}}\text{—OH}$$

Notice that salicylic acid contains two hydroxyl groups. If the carboxyl group attacked acetic anhydride, a new anhydride would be formed, which would eventually acylate a phenol. The product of this reaction is a half-acid ester.

EXPERIMENT 11.3 SYNTHESIS OF ASPIRIN

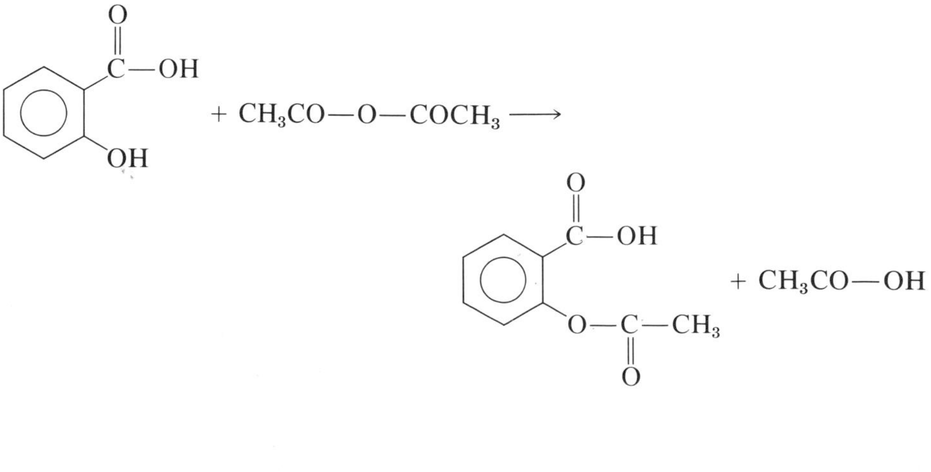

$$\text{C}_6\text{H}_4(\text{COOH})(\text{OH}) + \text{CH}_3\text{CO—O—COCH}_3 \longrightarrow \text{C}_6\text{H}_4(\text{COOH})(\text{O—}\overset{\|}{\underset{\text{O}}{\text{C}}}\text{—CH}_3) + \text{CH}_3\text{CO—OH}$$

Time 2.5 h

Materials Salicylic acid, 3 g (MW 138, mp 158 to 160°C)
Acetic anhydride, 6 mL (MW 102, bp 138 to 140°C)
85% phosphoric acid, 6 to 8 drops
Ethanol, 10 mL

Precautions The reaction may be initially exothermic. Carry out all steps in a good hood.

Hazards Acetic anhydride is a corrosive liquid and quite flammable. Avoid breathing vapors and carry out all transfers in a hood.

Experimental Procedure

Place 3 g (0.022 mol) salicylic acid in a 125-mL Erlenmeyer flask. Add 6 mL acetic anhydride (**caution: hood**) and then add 6 to 8 drops 85% phosphoric acid. Swirl the flask gently to mix all the reagents and then place the flask in a beaker of warm water (at 70 to 80°C) or on the top of a steam bath for about 15 min. Remove the flask from the hot-water or steam bath and while it is still warm, carefully add about 1 mL cold water a drop at a time, swirling the flask after each addition. (*Note:* Acetic anhydride reacts vigorously with water and the reaction mixture may spatter.)

After the first milliliter of water has been added, 20 mL distilled water may be added rapidly and the flask cooled in an ice bath. The product should gradually begin to crystallize. If the material does not appear as a solid or if an oil appears, grasp the flask in one hand while still holding it in the ice bath and gently scratch the inside surface of the flask with a glass rod. After the material has crystallized, it should be collected by suction filtration with a Buchner funnel. The flask and the product should be washed with a small quantity of cold distilled water.

Acetylsalicylic acid can be purified by a mixed-solvent recrystallization. Place the crude aspirin in a 125-mL Erlenmeyer flask and add 8 to 10 mL ethanol. Heat the flask gently on a steam bath until all the crystals dissolve. Slowly add 25 mL distilled water and continue heating on the steam bath until the solution is almost boiling. Remove the flask from the steam bath and set it aside. As the solution cools, crystals should gradually begin to appear. Once again, if crystals do not appear, gently scratch the inner surface of the flask with a glass rod or use a few seed crystals to initiate crystallization. Cool the mixture in an ice bath to be certain that all the product has crystallized. Collect the product by suction filtration as above and wash it with a small amount of cold distilled water. Place the product on a piece of filter paper, place another piece of filter paper gently over the crystals, and press with a cork, so that the water is transferred from the crystals to the filter paper. Remove the top piece of filter paper and allow the crystals to dry in air. Weigh the air-dried product and determine the yield. The melting point of the air-dried product should be 138 to 140°C.

The nmr and ir spectra of acetylsalicylic acid are shown in Fig. 11.6. The acetyl methyl group is observed as a distinct singlet in the nmr spectrum but slightly downfield of its position in Fig. 11.5. The ir spectrum is particularly informative: Two carbonyl absorptions are apparent and the hydroxyl group absorption is observed near 2600 cm^{-1} (see Secs. 5.2 and 23.8).

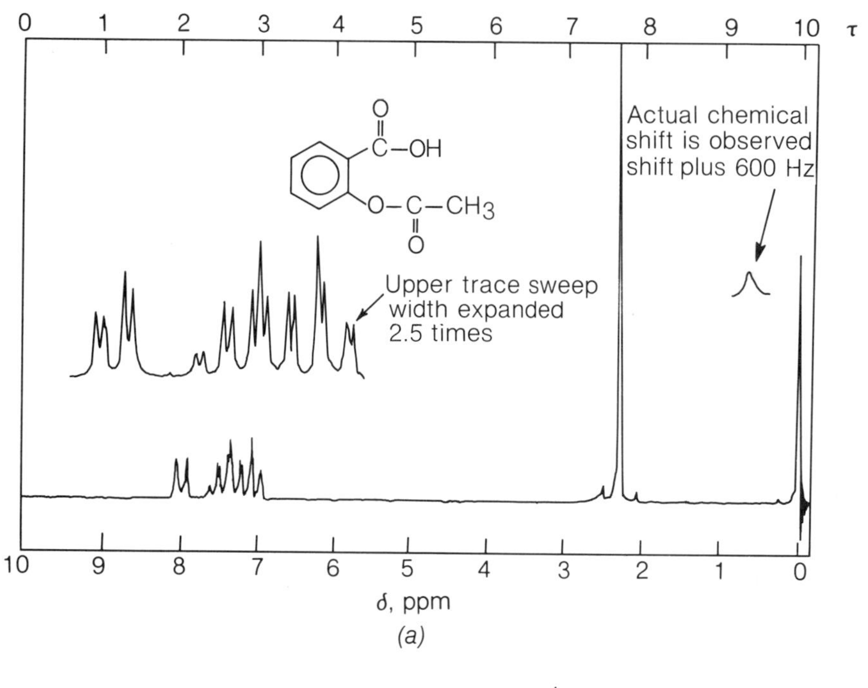

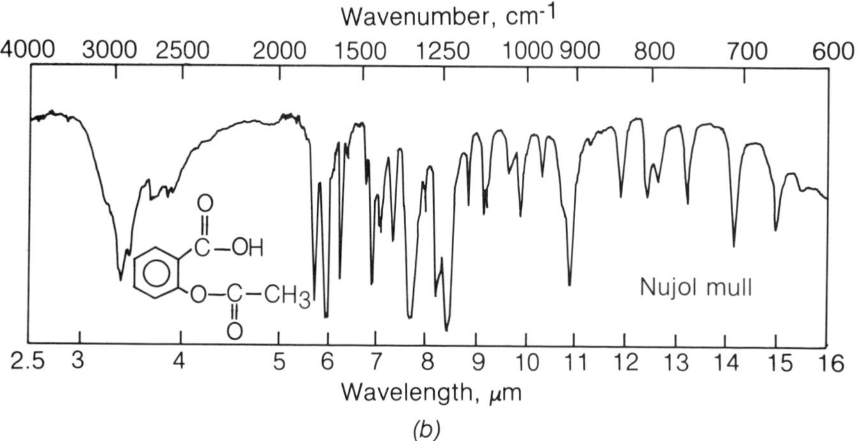

Figure 11.6
The (*a*) nmr and (*b*) ir spectra of acetylsalicylic acid.

11.4 SYNTHESIS OF *N,N*-DIETHYL-*m*-TOLUAMIDE: FORMATION OF AN AMIDE FROM AN ACID CHLORIDE

The compound prepared in this experiment is *N,N*-diethyl-*m*-toluamide, generally referred to by the initials DEET. It is a typical amide but it is unusual in the sense that it is biologically active. Biological activity may take many forms. Aspirin is an analgesic, i.e., its biological activity is as a pain killer. DEET has a rather unusual biological effect: it is a potent insect repellent and is the principal constituent of most commercial insect repellents. This particular compound is interesting in that it does not have a toxic effect on insects; it just smells bad to them.

The experiment described in this section is actually done in two stages. In the first of these, *m*-toluic acid is converted to the corresponding acid chloride by reaction with thionyl chloride. The details of acid chloride formation and the special role of dimethylformamide as a catalyst may be found in Sec. 23.6B and should be read carefully before attempting this experiment. In particular, the cautionary notes concerning the use of DMF should be observed.

Once the acid chloride is formed, the amine is added, and rapid addition to the carbonyl group takes place. The by-product of this reaction is HCl, which rapidly protonates the amine, as shown in the equation.

$$m\text{-}CH_3C_6H_4C(=O)Cl + \ddot{N}H(CH_2CH_3)_2 \longrightarrow m\text{-}CH_3C_6H_4C(=O)N(CH_2CH_3)_2 + H_2\overset{+}{N}(CH_2CH_3)_2Cl^-$$

The amine salt is unable to attack the carbonyl group because the lone pair of electrons is unavailable. In order for the reaction to go to completion, sodium hydroxide must be added to the reaction medium to neutralize the acid formed. Since amines are much more nucleophilic and more soluble than hydroxide, very little of the acid chloride is lost by simple hydrolysis.

EXPERIMENT 11.4 SYNTHESIS OF N,N-DIETHYL-m-TOLUAMIDE (DEET)

$$m\text{-}CH_3C_6H_4C(=O)OH \longrightarrow m\text{-}CH_3C_6H_4C(=O)Cl \longrightarrow m\text{-}CH_3C_6H_4C(=O)N(CH_2CH_3)-CH_2CH_3$$

Time 3 h

Materials *m*-Toluic acid, 9 g (MW 136, mp 108 to 110°C)
Thionyl chloride, 7 mL (MW 119, d 1.655 g/mL)
Dimethylformamide (DMF), 1 drop
Diethylamine, 7 mL (MW 78, d 0.71 g/mL)
20% sodium hydroxide solution, 35 mL
Solvent ether

Precautions All procedures should be carried out in a hood. Do not add DMF to warm thionyl chloride under any circumstances. Use gloves when measuring out thionyl chloride.

Hazards Thionyl chloride is highly corrosive and is a lachrymator. Avoid breathing of $SOCl_2$, SO_2, or HCl vapors and skin contact with any of these.

Experimental Procedure

Add 9 g (0.066 mol) *m*-toluic acid to a 100-mL round-bottom single-neck flask. Now add *1 drop* dimethylformamide [DMF, the catalyst (see Sec. 23.6B)] together with several boiling chips.

Warning: The DMF is a catalyst which enhances conversion of *m*-toluic acid to the acid chloride. If you forget to add the drop of DMF and you assemble and start the reaction described above, under no circumstances are you to add the DMF once the reaction has commenced. If you do, the product will have to be isolated from the roof of the hood.

A reflux condenser with lightly greased joints and fitted with a calcium chloride drying tube is added to the top of the round-bottom flask as shown in Fig. 11.3.

Through the top of the reflux condenser add 7 mL thionyl chloride. The calcium chloride drying tube is replaced in the top of the reflux condenser and the mixture is *slowly* heated with a very small flame from a bunsen burner, in an oil bath, or with an electric mantle.

(*Note:* **Thionyl chloride is a highly corrosive material and all transfers should be made in a hood. You should carefully avoid breathing the vapors or exposing your skin to this material. If an accidental spill occurs, notify your laboratory instructor *immediately*.**)

Thionyl chloride is a very volatile material. Since heating which is too strongly applied will drive the material out of the reflux condenser before the reaction is complete, the heat source should be adjusted to obtain a very gentle reflux. Although a steam bath may be used, heat from that source is more difficult to control and the thionyl chloride is more likely to be lost. After the reaction mixture has started to reflux, continue the heating for 45 min.

During the reflux period described above, measure out 35 mL 20% NaOH solution and place this material in a 250-mL Erlenmeyer flask loosely fitted with a cork or rubber stopper. Cool this aqueous basic solution in an ice bath. When the internal temperature reaches approximately 15°C, add 7 mL (0.066 mol) diethylamine (**hood**) to the aqueous layer. Stopper the Erlenmeyer flask and continue cooling the solution in the ice bath.

At the end of the 45-min reflux the acid should have been converted to a liquid acid chloride. Extinguish and remove all heating sources. Cool the acid chloride solution in an ice bath until its internal temperature is approximately

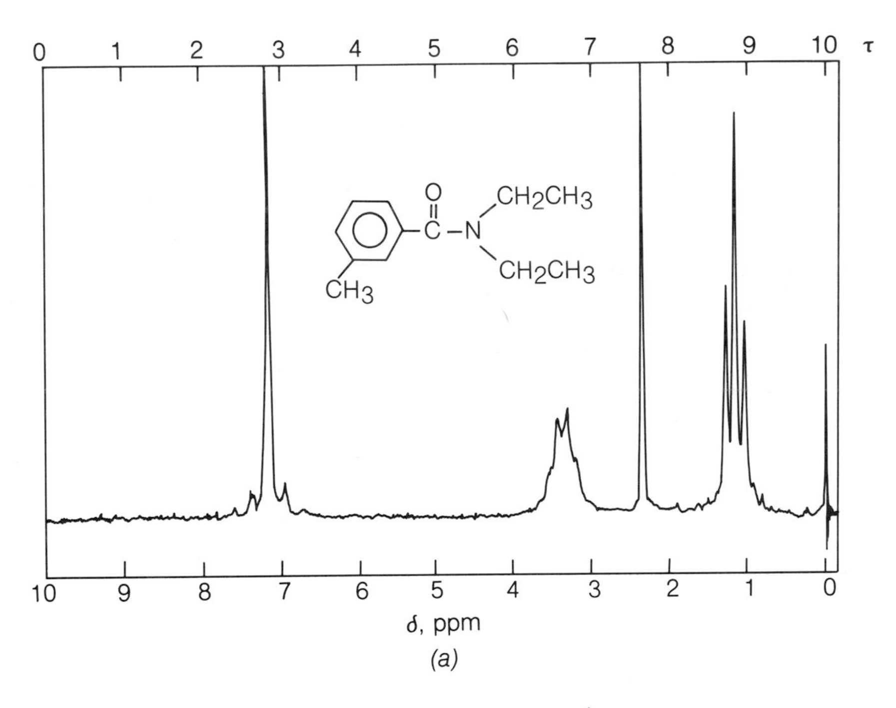

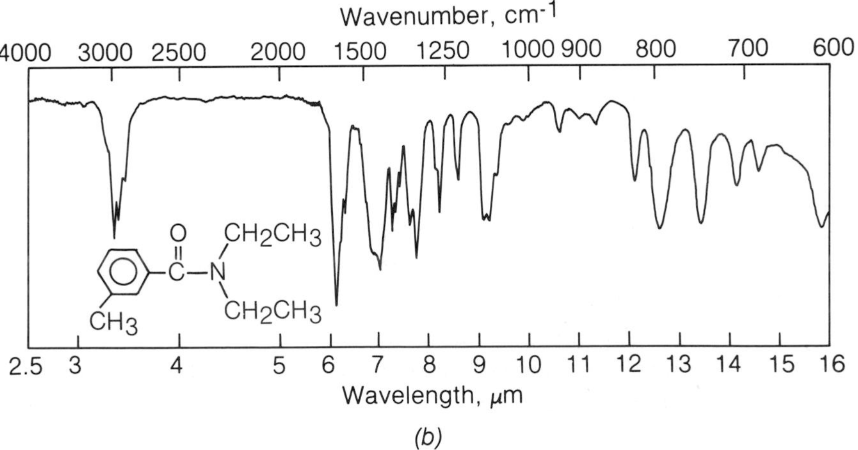

Figure 11.7
The (*a*) nmr and (*b*) ir spectra of *N,N*-diethyl-*m*-toluamide (DEET).

15°C. The cooled acid chloride is added dropwise to the mixture of bases by using a disposable or Pasteur pipet.

Warning: Rapid addition of the acid chloride to the mixture of the amine and sodium hydroxide solution can result in violent evolution of gases and must be avoided. Controlled gas evolution will be noted during the dropwise addition, and you should avoid breathing these vapors.

After each 2-mL addition, carefully swirl the mixture; then stopper the flask and swirl vigorously. Keep the flask in the ice bath except during the vigorous swirling. When addition of the acid chloride is complete, again swirl the mixture in the Erlenmeyer flask for 5 min. A yellow-brown, two-phase mixture should be observed at this point.

Transfer this two-phase mixture to a separatory funnel and extract the aqueous solution with three 50-mL portions of solvent-grade ether. Combine the ether layers and wash with 50 mL saturated salt solution (brine). Transfer the yellow ether layer to a 250-mL Erlenmeyer flask and dry with solid anhydrous magnesium sulfate (or sodium sulfate). After a 10-min drying period, gravity filter the mixture into a 250-mL Erlenmeyer flask, using small portions of fresh ether to wash out the flask containing the drying agent. Remove the colored impurity by adding a few boiling chips and a small amount of decolorizing charcoal (Norite, 0.2 g) to the cold ether solution, and heat this mixture lightly on a steam bath with constant swirling. After about 5 min of warming, gravity filter the hot mixture into a dry Erlenmeyer flask. Add boiling chips and evaporate the ether on the steam bath. You should obtain about an 85% yield of a very light, clear yellow oil which has *no* odor of diethylamine.

The nmr and ir spectra of DEET are shown in Fig. 11.7. The ir spectrum is particularly informative. No OH absorption is detected in the 3400 to 3600 cm^{-1} region, which indicates that no hydroxyl group remains. The carbonyl vibration is observed near 1640 cm^{-1}, which is characteristic of an amide rather than a carboxylic acid.

11.5 SYNTHESIS OF ACETANILIDE AND PHENACETIN: ANHYDRIDE ACYLATION OF AMINES

In the preparations which follow (Exps. 11.5A and B), acetanilide is formed from aniline and acetic anhydride, and phenacetin is prepared by the same approach from 4-ethoxyaniline (phenetidine). Phenacetin is an analgesic (like aspirin) and is often formulated with aspirin in common drugstore pain remedies. Those who have been in the armed forces will know this drug as the APC tablet (A is aspirin, P phenacetin, and C caffeine). It is sometimes alleged that APC stands not for the chemical constituents, but for "all-purpose capsule." Acetanilide itself is an analgesic, but is not an effective drug because it has the side effect of causing liver damage in people who take it over an extended period of time. As a matter of fact, another product which can be made from acetanilide, 4-bromoacetanilide, is a much better analgesic in this respect. The bromination of acetanilide to form 4-bromoacetanilide is described in Sec. 18.2.

In these two reactions, an amide is prepared from an amine and an anhydride. In the absence of vigorous heating, some activation of the carboxyl group is required for reaction to occur between the amine and the acid. In the preceding experiment, the acid is activated by conversion to an acid chloride. In Exps. 11.5A and 11.5B commercially available acetic anhydride is used as the acylating agent. This reagent is inexpensive and especially convenient to use because acetic acid is the only by-product.

EXPERIMENT 11.5A SYNTHESIS OF ACETANILIDE

$$C_6H_5NH_2 + CH_3\overset{O}{\overset{\|}{C}}-O-\overset{O}{\overset{\|}{C}}CH_3 \longrightarrow C_6H_5NH-\overset{O}{\overset{\|}{C}}-CH_3 + CH_3\overset{O}{\overset{\|}{C}}-OH$$

Time 2 h

Materials Aniline, 10 mL (MW 93, bp 184°C, d 1 g/mL)
Acetic anhydride, 12 mL (MW 102, bp 138 to 140°C)
Sodium acetate trihydrate, 15 g (MW 136)
Concentrated hydrochloric acid, 10 mL

Precautions Carry out all steps in a good hood.

Hazards Aniline is an irritant. Acetic anhydride and hydrochloric acid are corrosive liquids. Avoid breathing the vapors from all these compounds, use gloves, and make all transfers in the hood. Keep flames away from acetic anhydride.

Experimental Procedure

Measure out 10 mL (0.108 mol) aniline and transfer it to a 250-mL Erlenmeyer flask. Add 200 mL distilled water to the flask, followed by 10 mL concentrated HCl (**hood**). Swirl the water layer to dissolve the aniline. Treat the solution with decolorizing carbon and filter. *Before going on* you should have a *clear, water-white* aqueous acidic solution. If any color persists, treat the solution again with decolorizing carbon. (*Note:* If you proceed and colored impurities remain, you will transfer these impurities to the product, where they will be much harder to remove.) Weigh out 15 g (0.110 mol) sodium acetate trihydrate and place it in a 125-mL Erlenmeyer flask. Dissolve the salt in 50 mL distilled water. Set the solution aside.

Transfer the acidic solution of aniline (aniline hydrochloride) to a 500-mL Erlenmeyer flask. Add 12 mL acetic anhydride to the aqueous solution, swirl once, and then add the sodium acetate solution all at once, following this addition with vigorous swirling. The reaction is very rapid and product starts coming out of solution almost immediately.

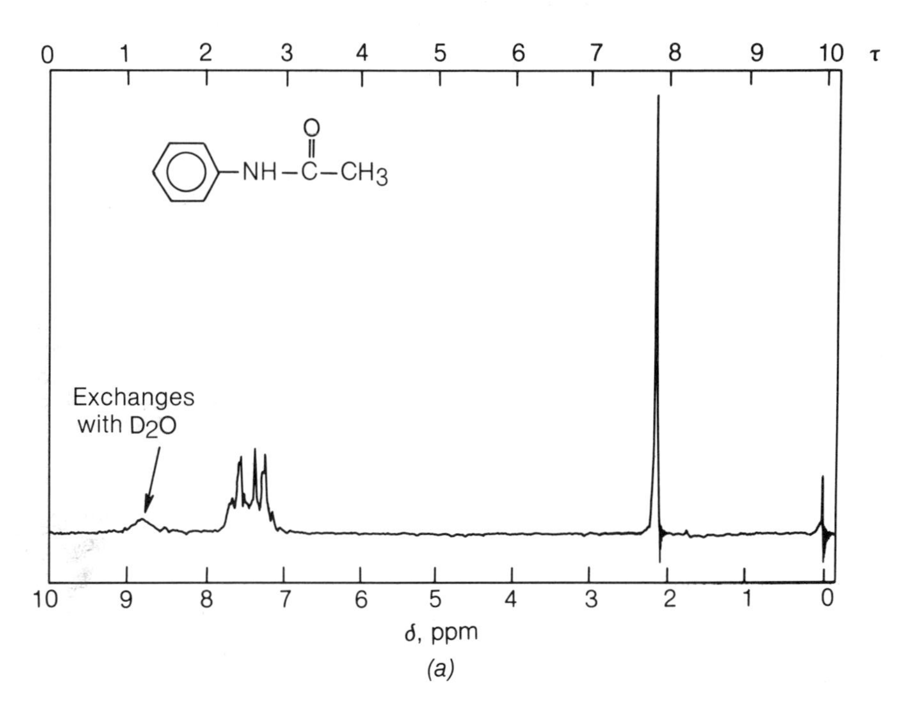

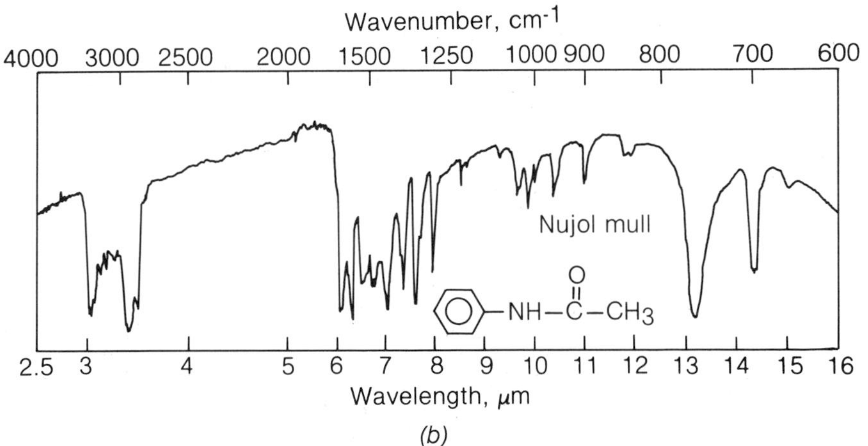

Figure 11.8 The (*a*) nmr and (*b*) ir spectra of acetanilide.

The flask is swirled for 10 min and then cooled in an ice bath for 10 to 15 min. The solid material is collected by suction filtration with a Buchner funnel and flask. The solid product is washed with cold water and air dried for a few minutes.

The crude material may be recrystallized from hot water or from an ethanol-water mixture. The crude solid should be white. After recrystallization, water may be removed from the crystals by the filter paper technique described in the aspirin synthesis. The air-dried pure material has a melting point of 113 to 115°C.

The nmr and ir spectra of acetanilide are shown in Fig. 11.8. In the nmr spectrum, the N—H resonance is observed as a broad band. This is often the case for amides. The acetyl methyl group is observed as a single sharp line at 2.1 ppm. Notice that the N—H stretching band is clearly discernible near 3310 cm^{-1} and that the C=O and C=C (benzene ring) region of the ir is partially obscured by absorptions due to Nujol.

EXPERIMENT 11.5B

SYNTHESIS OF PHENACETIN (4-ETHOXYACETANILIDE)

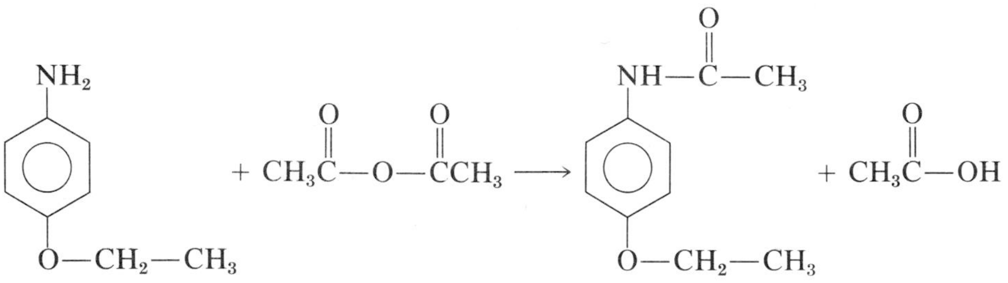

$$4\text{-}(CH_3CH_2O)C_6H_4NH_2 + CH_3C(=O)\text{—}O\text{—}C(=O)CH_3 \longrightarrow 4\text{-}(CH_3CH_2O)C_6H_4NH\text{—}C(=O)\text{—}CH_3 + CH_3C(=O)\text{—}OH$$

Time 2 h

Materials 4-Ethoxyaniline, 10 mL (MW 137, bp 250°C, d 1 g/mL)
Acetic anhydride, 8.5 mL (MW 102, bp 138 to 140°C)
Sodium acetate trihydrate, 11 g (MW 136)
Concentrated hydrochloric acid, 6.5 mL

Precautions Carry out all steps in a good hood. Wear gloves when transferring liquids.

Hazards 4-Ethoxyaniline is an irritant. Acetic anhydride and hydrochloric acid are corrosive liquids. Avoid breathing the vapors from all these compounds, use gloves, and make all transfers in the hood. Acetic anhydride is flammable.

Experimental Procedure

Measure out 10 mL (0.073 mol) 4-ethoxyaniline and transfer it to a 250-mL Erlenmeyer flask. Add 200 mL distilled water, followed by 6.5 mL concentrated HCl (**hood**). Then swirl the mixture to dissolve the 4-ethoxyaniline. Treat the solution with decolorizing carbon and filter. *Before going on* you should have a *clear, water-white* aqueous acidic solution. If any color persists, treat the solution again with decolorizing carbon. (*Note:* If you proceed and colored impurities remain, you will transfer these impurities to the product where they will be *much* harder to remove.)

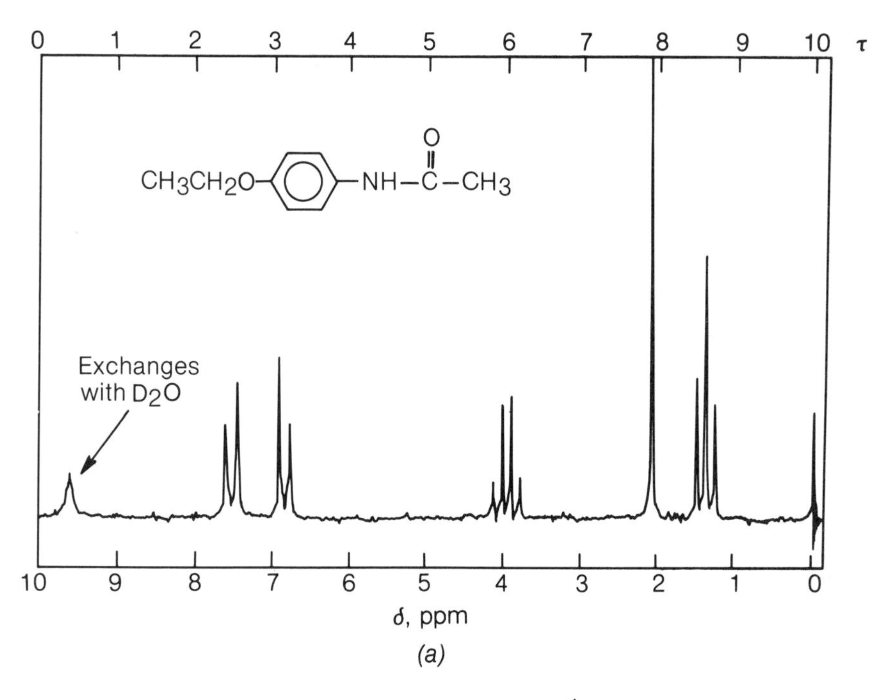

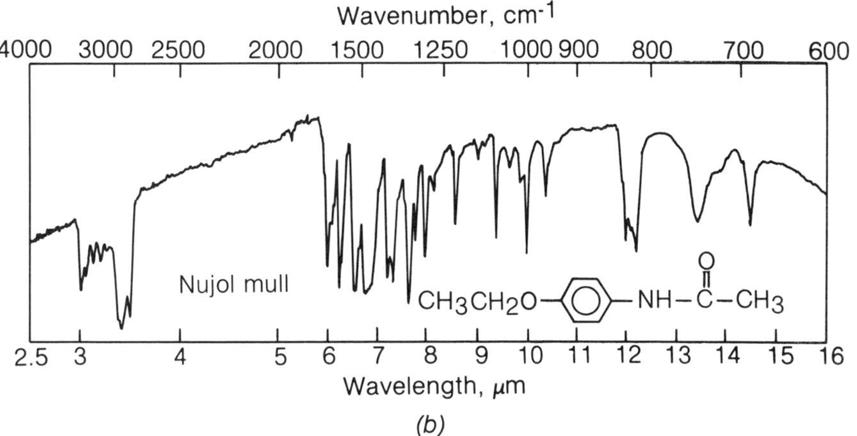

Figure 11.9
The (*a*) nmr and (*b*) ir spectra of phenacetin.

Weigh out 11 g (0.081 mol) sodium acetate trihydrate and place it in a 125-mL Erlenmeyer flask. Dissolve the salt in 50 mL distilled water. Set the solution aside.

Transfer the acidic solution of 4-ethoxyaniline (4-ethoxyaniline hydrochloride) to a 500-mL Erlenmeyer flask. Add 8.5 mL acetic anhydride to the aqueous solution, swirl once, and then add the sodium acetate solution all at once and swirl vigorously. The reaction is very rapid and product starts coming out of solution almost immediately.

Swirl the flask for 10 min and then cool it in an ice bath for 10 to 15 min. Collect the solid material by suction filtration on a Buchner funnel. Wash the solid product with cold water and air dry it for a few minutes.

The crude material may be recrystallized from ethanol or from ethanol-water. The crude solid should be white. After recrystallization, the product should be isolated as white crystals, mp 137 to 138°C. It may be air dried until the next lab period if it is too damp to give a satisfactory melting point.

The nmr and ir spectra of phenacetin are shown in Fig. 11.9. The nmr spectrum is not simple but is easy to interpret. The ethyl group is observed as a triplet and quartet, the NH proton is observed near 9.5 ppm, and the acetyl group is observed as a single sharp line near 2.1 ppm. See the discussion in the previous experiment for notes on the ir spectrum.

QUESTIONS AND EXERCISES

11.1 The preparation of certain esters is described in Exps. 11.1 and 11.2. The esters are prepared by entirely different methods in these two sections. Which of these two methods would you use to prepare an ester from an expensive alcohol component and an inexpensive acid component? Why?

11.2 Occasionally a freshly opened bottle of aspirin will have a distinct odor of vinegar about it. What does the odor of vinegar suggest to you about the purity of the aspirin contained in your sample? What would you expect the effect of the aspirin would be when you ingested it?

11.3 If one had a mixture containing one part phenyl acetate, one part *p*-nitrophenyl acetate, and one part *p*-methoxyphenyl acetate, and this mixture were hydrolyzed with one-third of an equivalent of base, what would the product of this hydrolysis be? Would only one of the esters hydrolyze, or would all the esters hydrolyze? If only one ester hydrolyzed, why would this be the only reaction? If all these esters hydrolyzed, why is there no selectivity?

11.4 Would you expect the methyl ester of pivalic acid (trimethylacetic acid) to hydrolyze more slowly or more rapidly than the corresponding ester of the isomeric acid methyl valerate (methyl pentanoate)? Why?

11.5 In the Fischer esterification process, sulfuric acid is almost always used as the acid catalyst. Concentrated hydrochloric acid is as strong as sulfuric acid but is almost never used. Account for the choice of sulfuric acid in this process.

11.6 One of the most efficient methods for preparing methyl esters of acids is to heat the acid with methyl alcohol (as solvent) using an acidic catalyst. Why is the same procedure not used in the synthesis of methylamides from the corresponding methylamine and an acid?

11.7 If two students were conducting hydrolysis experiments side-by-side, one hydrolyzing methyl benzoate and the other *N,N*-dimethylbenzamide, who would achieve the highest yield of benzoic acid first?

11.8 During World War II, the unavailability of silk made the new nylon (a synthetic polyamide) stockings very desirable commodities. A problem which arose, however, was that women in metropolitan areas who wore nylon stockings discovered that they were prone to unravel, whereas their country cousins did not observe this problem. How do you account for this difference in stability in nylon stockings in a city versus a rural environment?

11.9 When the kitchen sink stops up, it is usually a combination of dirt, some fat (esters), and some hair (amides) which causes the clog. The principal ingredient in most drain-cleaning preparations is sodium hydroxide. What reaction(s) are involved in clearing the drain by using these commercial preparations?

XII REACTIONS OF THE GRIGNARD REAGENT

The carbonyl group has often been described as the key functional group in organic synthesis because the many reactions it undergoes are often key steps in the syntheses of a wide variety of compounds. The carbonyl group is particularly versatile because it may have carbon substituents on both sides (ketones) or a carbon substituent on one side and hydrogen on the other (aldehydes). In addition, the carbonyl group may be adjacent to heteroatoms (as in esters and amides) or it may be attached to a halogen, such as chlorine, in acyl halides. The importance of base-catalyzed condensation chemistry to the carbonyl group is discussed in Chap. 15, and you will have the opportunity there to perform a number of functional-group transformations involving these substances.

The condensation chemistry of aldehydes and ketones has been known and widely studied since the late 1800s. At about the time German and English chemists were working on the development of condensation chemistry, Barbier and his student Grignard were studying the chemistry of organomagnesium compounds. Barbier discovered that a new reagent could be formed from alkyl chlorides and magnesium. Grignard, who utilized these reagents and explored their chemistry very extensively, was awarded the Nobel prize in 1912 for his accomplishments. He discovered that the new reagent behaved as if there were a negative charge on carbon and a positive charge on magnesium, i.e., almost as if the compound were a carbanion salt (R^-M^+). He found these materials to be quite nucleophilic and to add easily to aldehydes and ketones. The addition of the Grignard reagent to an aldehyde or ketone is, in a sense, limited by the fact that carbon and hydrogen are poor leaving groups. As a consequence, only one equivalent of Grignard reagent adds to each carbonyl. When an aldehyde reacts with a Grignard reagent, a secondary alcohol results; ketones yield tertiary alcohols (see below):

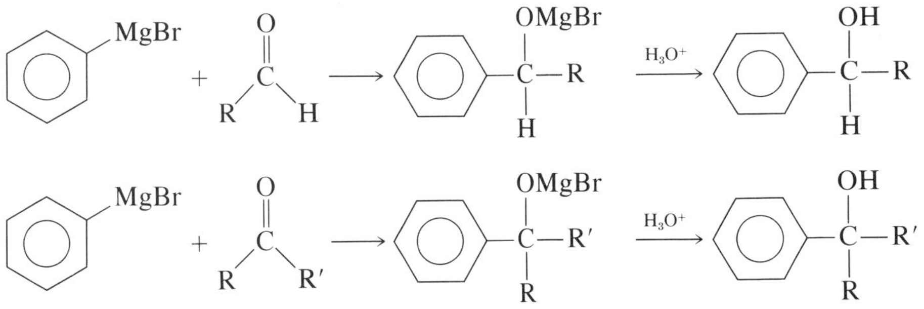

$$C_6H_5MgBr + RC(=O)H \longrightarrow C_6H_5\text{–}C(OMgBr)(R)(H) \xrightarrow{H_3O^+} C_6H_5\text{–}C(OH)(R)(H)$$

$$C_6H_5MgBr + RC(=O)R' \longrightarrow C_6H_5\text{–}C(OMgBr)(R')(R) \xrightarrow{H_3O^+} C_6H_5\text{–}C(OH)(R')(R)$$

Addition of a Grignard reagent to either an ester or ketone carbonyl begins in the same way. The intermediate species in the ester reaction loses alkoxide, forming a ketone during the reaction. This ketone rapidly adds a second mole of Grignard reagent, producing a tertiary alcohol. Two of the substituents in the product are identical, because both are derived from the Grignard reagent. This process is illustrated below for the reaction of a methyl Grignard reagent with ethyl benzoate.

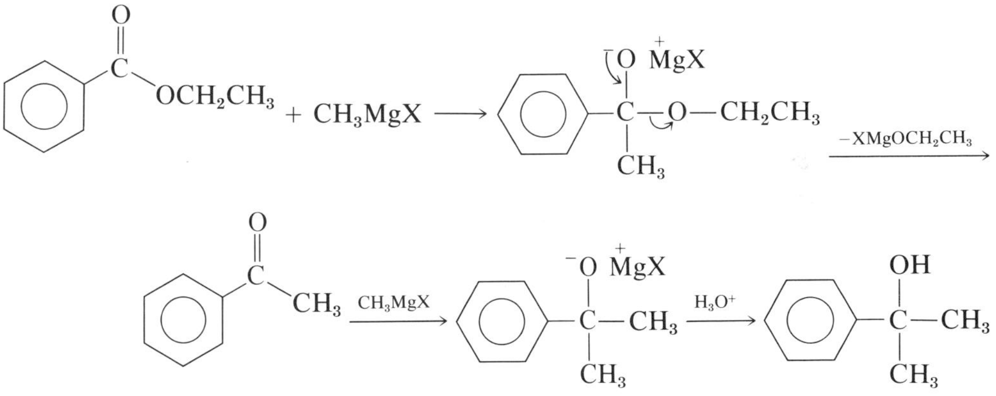

$$C_6H_5C(=O)OCH_2CH_3 + CH_3MgX \longrightarrow C_6H_5\text{–}C(^-O\ \overset{+}{M}gX)(CH_3)\text{–}O\text{–}CH_2CH_3 \xrightarrow{-XMgOCH_2CH_3}$$

$$C_6H_5C(=O)CH_3 \xrightarrow{CH_3MgX} C_6H_5\text{–}C(^-O\ \overset{+}{M}gX)(CH_3)\text{–}CH_3 \xrightarrow{H_3O^+} C_6H_5\text{–}C(OH)(CH_3)\text{–}CH_3$$

Several important features of Grignard reagents are worth noting. Not only are these reagents nucleophilic, but they are also basic species and react readily with any water or other acid present in the reaction medium. The presence of trace water in the solvent or reagent presents a particular problem, since water has a very low molecular weight and even a drop is a substantial quantity on a molar basis. Water contains an acidic proton which reacts with a Grignard reagent to give a hydrocarbon (corresponding to the starting halide) and hydroxides of magnesium.

$$R{-}Cl + Mg \longrightarrow R{-}Mg{-}Cl \xrightarrow{H_2O} R{-}H + Cl{-}Mg{-}OH$$

This destruction of a Grignard by an acid is called the Zerewittenoff reaction and has been used historically to analyze for the presence of acidic hydrogens.

Ethers are usually the favored solvents for Grignard reactions. Ethers like diethyl ether or tetrahydrofuran (THF) are good solvents for Grignard reagents and fairly easy to dry. They are also nonacidic.

If a Grignard reaction is to be successful, moisture must be rigorously excluded from both the solvent and starting materials. If the reaction involves an ester, care must also be taken that it is not contaminated by any of the acidic catalysts used in its preparation. The presence of either water or acid may significantly reduce the yield in the Grignard reaction.

The Grignard reaction apparently begins on the surface of the magnesium. Oxidation on the surface of the magnesium can retard the reaction or even prevent it from starting at all. When this difficulty arises, a very reactive alkyl halide (such as iodomethane or 1,2-bromoethane) or molecular iodine may be added to promote reaction. Alternatively, a glass rod may be used to gently crush the magnesium pieces, thus exposing enough new surface to initiate the Grignard reaction (see caution in Exp. 12.1).

The reactivity of ketones and esters can be compared in the experiments described below. Triphenylcarbinol (triphenylmethyl alcohol, Exp. 12.2A) is prepared by Grignard addition of phenylmagnesium bromide (Exp. 12.1) to benzophenone. Note in the equation below that only one equivalent of phenylmagnesium bromide reacts with the benzophenone to form the carbinol.

$$C_6H_5MgBr + C_6H_5{-}\overset{O}{\overset{\|}{C}}{-}C_6H_5 \xrightarrow{H_3O^+} (C_6H_5)_3C{-}OH$$

If methyl benzoate and two equivalents of phenyl Grignard reagent are used, a double addition occurs and the product is exactly the same, triphenylcarbinol. Either methyl benzoate (Exp. 11.2A) or butyl benzoate (Exp. 11.1) may be used to prepare triphenylcarbinol, because either methoxide or butoxide will be lost in the reaction. The alcohol component of the ester starting material is inconsequential in this case.

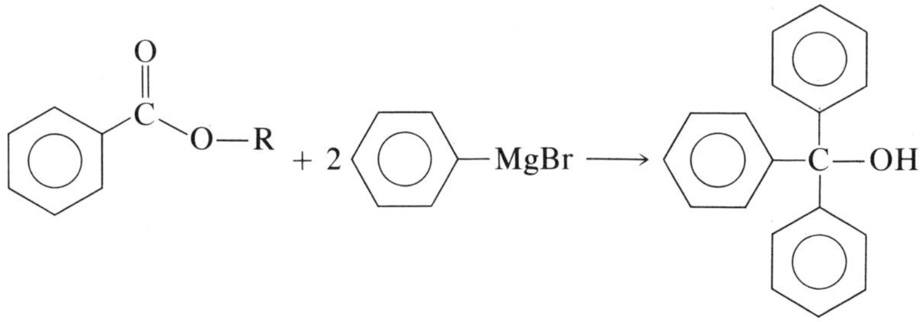

$$C_6H_5-C(=O)-O-R + 2\,C_6H_5-MgBr \longrightarrow (C_6H_5)_3C-OH$$

Finally, there are procedures for the addition of a Grignard reagent to two other electrophiles. One of these, carbon dioxide (Exp. 12.5), is available in the form of dry ice. The condensation of Grignard reagent with CO_2 yields an acid salt, which is hydrolyzed to the free carboxylic acid.

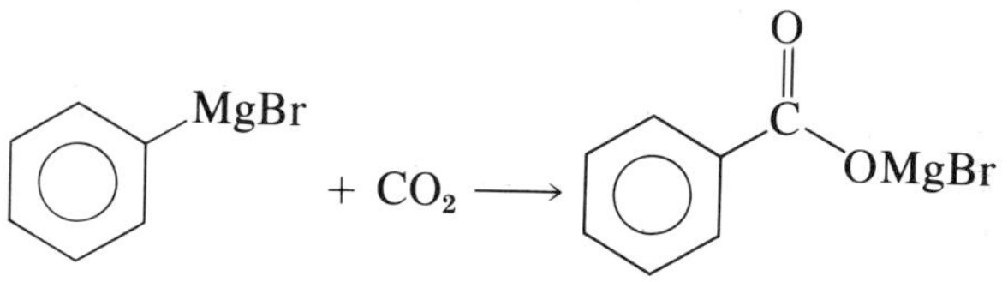

$$C_6H_5-MgBr + CO_2 \longrightarrow C_6H_5-C(=O)-OMgBr$$

Addition of a Grignard reagent to a carbon-nitrogen triple bond (as in a nitrile) generally yields a monoaddition product. The imine salt, shown in the equation below, is readily hydrolyzed in aqueous acid to give a ketone. Two equivalents of Grignard reagent will not add because the intermediate is stable. This reaction is illustrated below:

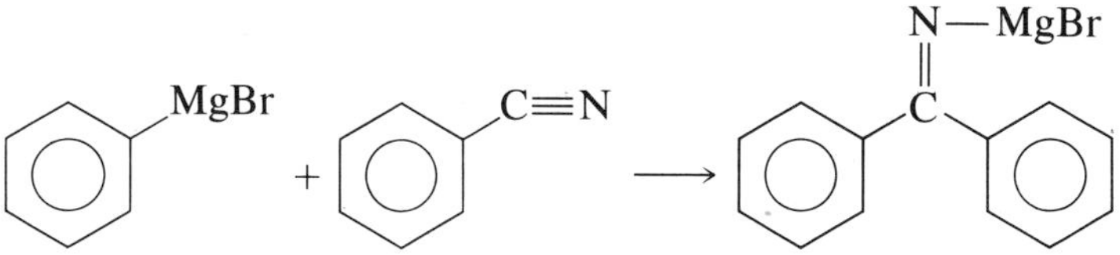

$$C_6H_5-MgBr + C_6H_5-C\equiv N \longrightarrow C_6H_5-C(=N-MgBr)-C_6H_5$$

12.1 SYNTHESIS OF PHENYLMAGNESIUM BROMIDE

The formation of a Grignard reagent is illustrated by the preparation of phenylmagnesium bromide. In this reaction, magnesium metal is formally inserted into the carbon-halogen bond. This insertion is much more complex than it appears to be on paper, and the details of the process are still under investigation. It is clear, however, that ether plays a key role in the reaction by strongly solvating magnesium in the complex. In fact, it appears that for most Grignard reagents, two molecules of the reagent are bound together with two molecules of ether solvent.

When organic chemists talk about "ether" they are generally referring to diethyl ether. This ether is the solvent used most often for formation of Grignard reagents because it is relatively inexpensive, nonacidic, and fairly easy to dry and to keep anhydrous. A drawback of diethyl ether is its low boiling point (33°C). If a higher temperature is required to initiate a Grignard reaction, the somewhat higher boiling ether tetrahydrofuran (THF, bp 65°C) is usually used.

Once the Grignard reagent is formed, some electrophilic reagent will be added to it in the hope that reaction will occur. The more concentrated a solution is, the faster a reaction between two substances in it usually occurs. The needed concentration of a Grignard reagent is usually determined by consideration of different factors. Since the organomagnesium compound reacts very rapidly with most electrophiles, the reactions can be run under more dilute reaction conditions than can many other bimolecular reactions. The important consideration in Grignard reactions is whether the reactive organomagnesium reagent will couple with itself, forming in this case biphenyl. The more dilute the Grignard reaction mixture, the less serious will be the coupling problem.

The preparation of phenylmagnesium bromide described below should be carried out in glassware which is kept as dry as possible. Once formed, the reagent should be used as quickly as possible.

If difficulty is encountered in starting the reaction, there are several tricks which may be used. These are mentioned in the instructions for the preparation. Be sure to read them carefully, and if the operation should be performed by your laboratory instructor, do not attempt it yourself.

Finally, ether is very flammable. Keep the laboratory completely free of flames during the entire operation.

EXPERIMENT 12.1 SYNTHESIS OF PHENYLMAGNESIUM BROMIDE

$$C_6H_5-Br + Mg \xrightarrow{\text{ether}} C_6H_5-Mg-Br$$

Time 2.0 h

Materials Bromobenzene, 6 mL (MW 157; bp 156°C; d 1.49 g/mL)
Magnesium, 1.2 g (MW 24)
Anhydrous ether, 50 mL

Precautions Perform experiment in hood. Avoid flames when using ether.

Hazards Ether is flammable and a powerful narcotic. All flame drying should be complete before *any* solvent vessel is opened. Bromobenzene fumes are toxic in high concentrations.

Experimental Procedure

The apparatus for this experiment consists of a 250-mL round-bottom single-neck flask, a Claisen head, a reflux condenser, a 125-mL addition funnel, and drying tubes filled with calcium chloride atop both the condenser and the addition funnel (as shown in Fig. 12.1).

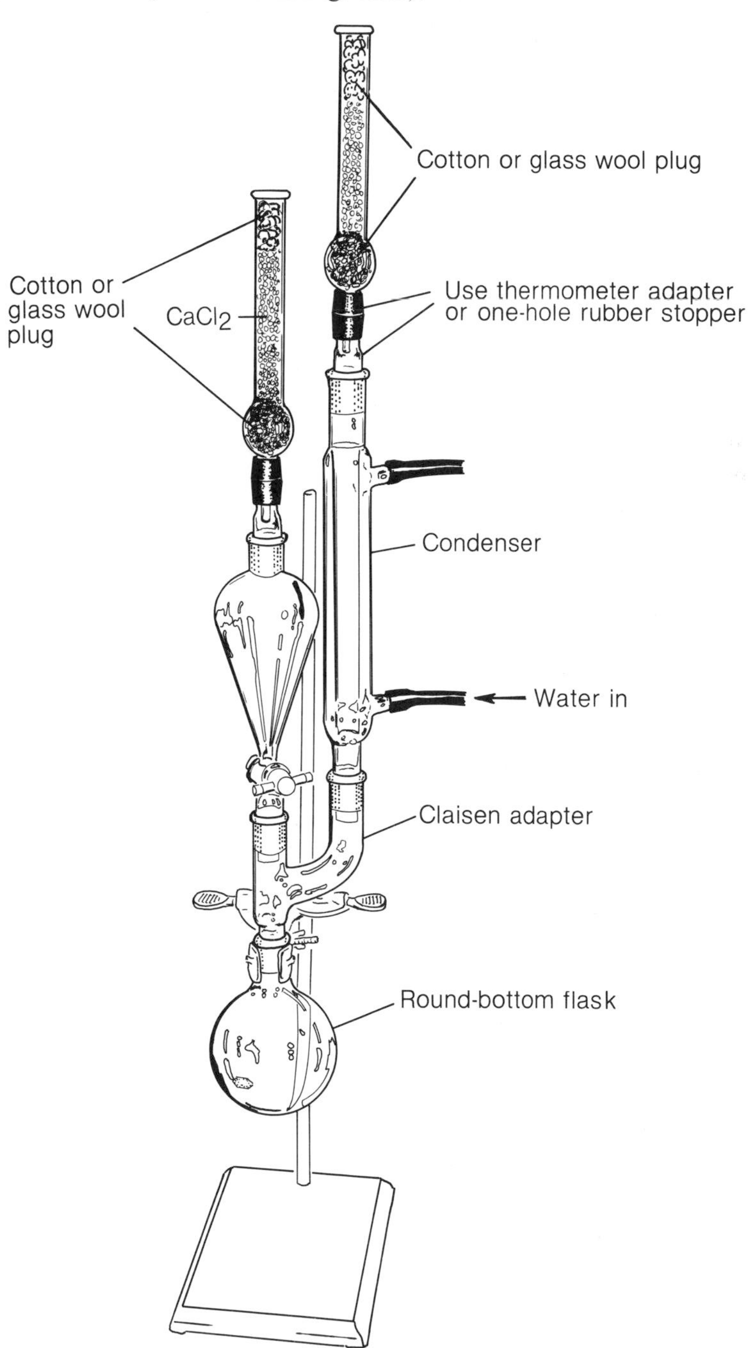

Figure 12.1 Apparatus for preparing the Grignard reagent.

Weigh out 1.2 g (0.05 mol) magnesium turnings and place them in a 250-mL round-bottom flask. Remove the drying tube from the top of the condenser (*do not* run water through the condenser) and flame dry[1] the entire apparatus by brushing the outside of the glass with a free flame. The water vapor will be visible as it is driven from the apparatus. (*Note:* Be very careful not to heat the Teflon stopcock directly with the flame.) As soon as the flame-drying procedure is complete, restopper the addition funnel with a lightly greased glass stopper and return the drying tube to its original position atop the condenser to prevent any water vapor from being sucked in as the glass and air inside the apparatus cool.

When the apparatus has returned to room temperature and all flames in the laboratory have been extinguished, begin water flow through the condenser and add sufficient (approximately 50 mL) anhydrous ether to the round-bottom flask to cover the magnesium turnings. Use a clean, *dry* 10-mL graduated cylinder and pour 6 mL (8.9 g, 0.057 mol) bromobenzene into it. Note that this is a slight excess of bromobenzene; magnesium is therefore the limiting reagent. Pour the bromobenzene into the addition funnel and then add approximately 1 mL of it to the reaction mixture. If the ether, the other reagents, and the apparatus are all dry, the reaction will start immediately. The magnesium will darken, and bubbles will be produced which eventually lead to reflux of the ether. If the reaction does not start after the addition of 1 mL bromobenzene either a crystal of iodine or a small amount of 1,2-dibromoethane should be added. If neither of these expedients is successful in initiating the reaction, consult your instructor, who may attempt to very carefully grind some of the magnesium surface with a glass rod. *Do not attempt this yourself* without supervision, as it is very easy to push the glass rod through the bottom of the unprotected flask.

After the reaction has begun, the remainder of the bromobenzene is added in approximately 1-mL portions at such a rate that a gentle reflux is maintained throughout the addition. As the reaction proceeds, note that a light precipitate begins to form, a coating appears on the magnesium, and the solution turns dark and grayish.

After the addition of bromobenzene is complete, gently swirl the flask, then allow it to stand until reflux stops. Note that some magnesium appears not to have reacted despite the fact that bromobenzene is present in excess. This is not harmful to the reaction, and the Grignard reagent is now ready for further reaction.

[1] *Special Note:* If glassware is cleaned during the previous lab period and allowed to dry overnight, flame drying can be avoided. In most cases, flame drying will not be necessary unless the laboratory atmosphere is extremely humid. Ask your instructor for directions.

12.2 GRIGNARD SYNTHESIS OF ALCOHOLS

Primary, secondary, and tertiary alcohols may all be synthesized by the addition of a Grignard reagent to the appropriate aldehyde or ketone. The three possibilities are illustrated below.

$$C_6H_5\text{—}MgBr + CH_2O \longrightarrow C_6H_5\text{—}CH_2OH \qquad (1°)$$

$$C_6H_5\text{—}MgBr + R\text{—}CHO \longrightarrow C_6H_5\text{—}CH(OH)\text{—}R \qquad (2°)$$

$$C_6H_5\text{—}MgBr + R\text{—}CO\text{—}R \longrightarrow C_6H_5\text{—}C(OH)(R)\text{—}R \qquad (3°)$$

Note that a primary alcohol can be formed only by the reaction of a Grignard with formaldehyde. This reaction presents experimental difficulties, since formaldehyde is a gas and is usually found in organic laboratories as an aqueous solution (called formalin). Since gases are hard to handle and water destroys Grignard reagents, primary alcohols are usually prepared instead by reduction of aldehydes (see Exp. 14.4A).

Tertiary alcohols may be formed either by addition of one equivalent of a Grignard reagent to a ketone or by addition of two equivalents of reagent to an ester. In the latter case, the first addition of organomagnesium compound is followed by loss of the alcohol portion of the ester as the alkoxide, as shown below.

$$RMgX + R'COOR'' \longrightarrow R\text{—}\overset{\overset{\displaystyle O}{\|}}{C}\text{—}R' + R''OMgX$$

The intermediate ketone then rapidly adds another equivalent of the Grignard reagent to form a tertiary alcohol in which two of the three substituents are the same.

The ester-based synthesis of tertiary alcohols is usually the best choice for the preparation of a tertiary alcohol when two substituents are identical and when the ester component is readily available. There is some flexibility in the choice of ester components. Because the alkoxide residue will be lost in the reaction, almost any ester can be used in the reaction. In Exp. 12.2B, either methyl or *n*-butyl benzoate, prepared in Exps. 11.2A and 11.1 respectively, may be used to form the same product.

EXPERIMENT 12.2A

SYNTHESIS OF TRIPHENYLCARBINOL FROM BENZOPHENONE

$$(C_6H_5)_2C{=}O + C_6H_5MgBr \xrightarrow[\text{2. } H_3O^+]{\text{1. ether}} (C_6H_5)_3C{-}OH$$

Time 3.0 h

Materials Benzophenone, 8.0 g (MW 182 g, mp 46 to 48°C)
Anhydrous ether, 25 mL
10% H_2SO_4, 50 mL

Precautions Perform experiment in hood. Extinguish all flames before opening any solvent vessel. Avoid flames when using ether.

Hazards Ether is flammable and is a powerful narcotic.

Experimental Procedure

Prepare phenylmagnesium bromide on a 0.05-mol scale as described in Exp. 12.1.

Weigh out 8 g (0.044 mol) pure, dry benzophenone and place it in a 125-mL Erlenmeyer flask. Pour in 25 mL anhydrous ether. Swirl until all the benzophenone has gone into solution. If a small residue of benzophenone remains, an additional 5 mL anhydrous ether may be added so that solution is complete. The homogeneous solution is then poured into the addition funnel which held bromobenzene during preparation of the Grignard reagent (see Fig. 12.1).

Adjust the stopcock on the addition funnel so that a drop at a time of benzophenone solution falls into the Grignard reagent. During the addition, a vigorous reaction will occur and the solution will boil. Dropwise addition fosters a controlled reaction. (*Note:* Reflux should always be kept at a gentle level.) During the addition, a precipitate forms in the reaction solution. After all the benzophenone has been added, place the 250-mL round-bottom flask on a steam bath and reflux the entire mixture for 30 min.

Just before the reflux period is complete, prepare an acid solution by pouring 50 mL of 10% aqueous sulfuric acid into a 250-mL Erlenmeyer flask containing approximately 50 g cracked ice. As the Grignard reaction mixture is cooling, the sulfuric acid solution will also be cooling. Pour approximately 10 mL of this cold aqueous acid solution into a graduated cylinder for later use. When the reaction mixture has reached room temperature, pour it into the Erlenmeyer flask containing the very cold acid. When the addition is complete, swirl vigorously so that the ether and aqueous acid solutions intimately mix and all salts hydrolyze. Now add the reserved 10 mL of acid solution to thc 250-mL round-bottom flask to hydrolyze any salts which remained in it after the original transfer (use a small amount of ether if necessary to dissolve most of the material).

When the hydrolysis of all the salts in both containers is complete, transfer the combined solutions to the separatory funnel. Place a glass stopper in the separatory funnel and gently shake the mixture. Draw off the lower, more dense, acid layer and wash the remaining ether layer once with 25 mL water, twice with 25 mL saturated sodium bicarbonate solution, and finally with two 25-mL portions of saturated aqueous sodium chloride solution (brine). After all these washings, a relatively dry ether solution remains, which should be run into a 125-mL Erlenmeyer flask. Add anhydrous sodium sulfate (10 to 15 g) to complete the drying process. After gravity filtering the solution into a 250-mL Erlenmeyer flask, evaporate the ether on a steam bath (**hood**). When most of the ether has evaporated, add 75 mL hexane and heat the solution until the precipitation of considerable solid material is observed. Cool the solution to room temperature and filter using a Buchner funnel to obtain the crude product.

The crude triphenylcarbinol (or triphenylmethanol as it is sometimes called, mp 155 to 160°C) can be purified by recrystallization from hexane. The recrystallization procedure works well because triphenylcarbinol is very soluble in ether and dichloromethane but very insoluble in hexane. Biphenyl, the major by-product of the reaction, is formed when the Grignard reagent couples with itself. Biphenyl contaminates the crude triphenylcarbinol but is quite soluble in hexane. Recrystallization can therefore be carried out by dissolving the crude mixture in the smallest possible amount of boiling dichloromethane (heated on a steam bath) and diluting it with four times its volume of hexane. Since it is difficult to determine in advance exactly how much dichloromethane will be required, it is best to fill a 25-mL graduated cylinder with dichloromethane and add it a little at a time to the crude solid while heating. As soon as all the material has dissolved, check the graduated cylinder to determine by difference how much dichloromethane has been used. From this, determine the volume of hexane which is required.

As the solution cools, colorless crystals of triphenylmethanol deposit and are collected on a Buchner funnel by suction filtration. Pure triphenylcarbinol has a melting point of 162°C. The nmr and ir spectra for this product are shown in Fig. 12.2.

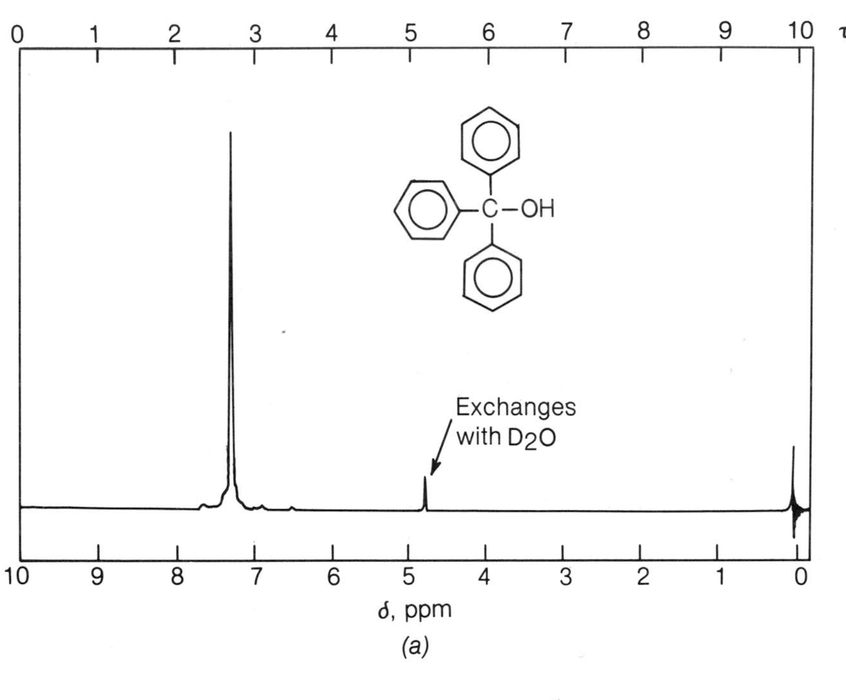

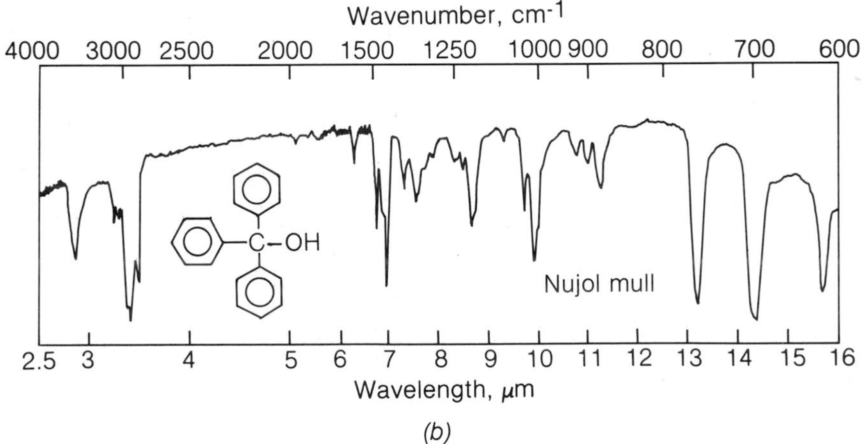

Figure 12.2
The (*a*) nmr and (*b*) ir spectra of triphenylcarbinol.

EXPERIMENT 12.2B

SYNTHESIS OF TRIPHENYLCARBINOL FROM METHYL OR n-BUTYL BENZOATE

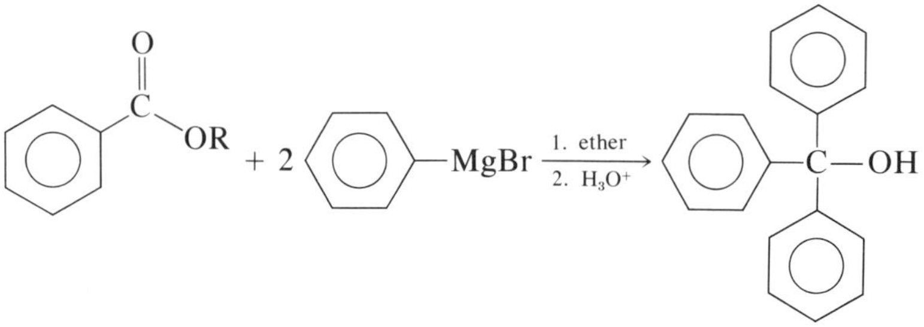

$$C_6H_5C(=O)OR + 2\ C_6H_5MgBr \xrightarrow[2.\ H_3O^+]{1.\ \text{ether}} (C_6H_5)_3C{-}OH$$

Time 3.0 h

Materials *n*-Butyl benzoate (Sec. 11.1), 8 mL (MW 178; bp 245°C; d 1 g/mL)
[Methyl benzoate (Sec. 11.2), 6 mL (MW 136; bp 200°C; d 1 g/mL)]
10% H_2SO_4, 50 mL
Anhydrous ether, 50 mL

Precautions Perform experiment in hood. Extinguish all flames before opening any solvent vessel. Avoid flames when using ether.

Hazards Ether is flammable and is a powerful narcotic.

Experimental Procedure

Prepare phenylmagnesium bromide on a 0.10-mol scale, *double* that described in Exp. 12.1.

Pour 8 mL (0.045 mol) *n*-butyl benzoate [or 6 mL (0.044 mol) methyl benzoate] into a *dry* 10-mL graduated cylinder. Pour the ester into a 50-mL Erlenmeyer flask and add 25 mL anhydrous ether. Swirl until all the *n*-butyl benzoate has dissolved. Pour the homogeneous solution into the addition funnel which held bromobenzene in Exp. 12.1B (see Fig. 12.1).

Adjust the stopcock on the addition funnel so that a drop at a time of *n*-butyl benzoate (or methyl benzoate) solution falls into the Grignard reagent. During the addition, a vigorous reaction will occur and the solution will reflux. Dropwise addition fosters a controlled reaction. (*Note:* Reflux should always be kept at a gentle level.) During the addition, a precipitate may form in the reaction solution. This precipitate (more often observed when using methyl benzoate), usually appears during the second half of the addition. After all the ester solution has been added, place the 250-mL round-bottom flask on a steam bath and reflux the entire mixture for 30 min.

Just before the reflux period is complete, prepare an acid solution by pouring 50 mL of 10% aqueous sulfuric acid into a 250-mL Erlenmeyer flask containing approximately 50 g cracked ice. As the reaction mixture is cooling, the sulfuric acid solution will also be cooling. Pour about 10 mL of this cold, aqueous acid solution into a graduated cylinder for later use. When the reaction mixture has reached room temperature, pour it into the Erlenmeyer flask containing the very cold acid. When the addition is complete, swirl vigorously so that the ether and aqueous acid solutions intimately mix and all salts hydrolyze. Now add the reserved 10 mL of acid solution to the 250-mL round-bottom flask to hydrolyze any salts which remained in it after the original transfer (use a small amount of ether if necessary to dissolve most of the material).

When the hydrolysis of all the salts in both containers is complete, transfer the combined solutions to the separatory funnel. Place a glass stopper in the separatory funnel and gently shake the mixture. Draw off the lower, more dense, acid layer and wash the remaining ether layer once with 25 mL water, twice with 25 mL saturated sodium bicarbonate solution, and finally with two 25-mL portions of saturated aqueous sodium chloride solution (brine). After all these washings, a relatively dry ether solution remains, which is run into a 125-mL Erlenmeyer flask. Add anhydrous sodium sulfate (10 to 15 g) to complete the drying process. After gravity filtering the solution into a 250-mL Erlenmeyer flask, evaporate the ether on a steam bath (**hood**). When most of the ether has evaporated, add 75 mL hexane and heat the solution until precipitation of considerable solid material is observed. Cool the solution to room temperature and filter on a Buchner funnel to obtain the crude product.

The crude triphenylcarbinol (or triphenylmethanol as it is sometimes called, mp 155 to 160°C) can be purified by recrystallization from hexane. The recrystallization procedure works well because triphenylcarbinol is very soluble in ether and dichloromethane but very insoluble in hexane. Biphenyl, the major by-product of the reaction, contaminates the crude triphenylcarbinol and is quite soluble in hexane. Recrystallization can therefore be carried out by dissolving the crude mixture in the smallest possible amount of boiling dichloromethane (heated on a steam bath) and diluting it with four times its volume of hexane. Since it is difficult to determine in advance exactly how much dichloromethane will be required, it is best to fill a 25-mL graduated cylinder with dichloromethane and add it a little at a time to the crude solid while heating. As soon as all the material has dissolved, check the graduated cylinder to determine by difference how much dichloromethane has been used. From this, determine what volume of hexane is required.

As the solution cools, colorless crystals of triphenylcarbinol deposit and are collected on a Buchner funnel by suction filtration. Pure triphenylcarbinol is found to have a melting point of 162°C.

The nmr and ir spectra of triphenylcarbinol are shown in Fig. 12.2. Note that all the aromatic hydrogen atoms appear at 7.3 ppm, even though they are not chemically identical. This is a case of accidental chemical shift equivalence. Note that the hydroxyl proton is barely discernible in the nmr spectrum (4.8 ppm) but very prominent in the ir spectrum (3485 cm^{-1}). Residual ester could be detected by an ir absorption at about 1720 cm^{-1}. Note that this is absent in the spectrum of the pure product.

EXPERIMENT 12.2C

SYNTHESIS OF BENZHYDROL FROM BENZALDEHYDE

$$C_6H_5CHO + C_6H_5MgBr \xrightarrow[2.\ H_3O^+]{1.\ ether} (C_6H_5)_2CHOH$$

Time 3.0 h

Materials Benzaldehyde, 4.5 mL (MW 106; bp 183°C; d 1 g/mL)
10% H_2SO_4, 50 mL
Anhydrous ether, 25 mL

Precautions Perform experiment in hood. Extinguish all flames before opening any solvent vessel.

Hazards Ether is flammable and is a powerful narcotic.

Experimental Procedure

Prepare the phenylmagnesium bromide Grignard reagent on a 0.05-mol scale as described in Exp. 12.1.

In a *dry* 10-mL graduated cylinder pour out 4.5 mL (0.045 mol) benzaldehyde. Pour this into a 50-mL Erlenmeyer flask, followed by 25 mL anhydrous ether. Swirl until all the benzaldehyde has gone into solution. The homogeneous solution is then poured into the addition funnel which was used for the bromobenzene in preparing the Grignard reagent (see Fig. 12.1).

Adjust the stopcock on the addition funnel so that a drop at a time of the benzaldehyde solution falls into the Grignard reagent. During the addition, a vigorous reaction will occur and the solution will reflux. Dropwise addition maintains a controlled reaction. (*Note:* Reflux should always be kept at a gentle level.) During the addition a precipitate may form in the reaction solution. This precipitate, if observed, usually appears during the second half of the addition. After all the aldehyde solution has been added, place the 250-mL round-bottom flask on a steam bath and reflux the entire mixture for 30 min.

Just before the reflux period is complete, prepare an acid solution by pouring 50 mL of 10% aqueous sulfuric acid into a 250-mL Erlenmeyer flask containing approximately 50 g cracked ice. As the reaction mixture is cooling, the sulfuric acid solution will also be cooling. Pour about 10 mL of this cold aqueous acid solution into a graduated cylinder for later use. When the reaction mixture has reached room temperature, pour it into the Erlenmeyer flask containing the very cold acid. When the addition is complete, swirl vigorously so that the ether and aqueous acid solutions intimately mix and all salts hydrolyze. Now add the reserved 10 mL of acid solution to the 250-mL round-bottom flask to hydrolyze any salts which remained in it after the original transfer (use small amounts of solvent ether if necessary to dissolve most of the material).

When the hydrolysis of all the salts in both containers is complete, transfer the combined solutions to the separatory funnel. Place a glass stopper in the separatory funnel and gently shake the mixture. Draw off the lower, more dense, acid layer and wash the remaining ether layer once with 25 mL water, twice with 25 mL saturated sodium bicarbonate solution, and finally with two 25-mL portions of saturated aqueous sodium chloride solution (brine). After all these washings, a relatively dry ether solution remains, which is run into a 125-mL Erlenmeyer flask. Add anhydrous sodium sulfate (10 to 15 g) to complete the drying process. After gravity filtering the solution into a 250-mL Erlenmeyer flask, evaporate the ether on a steam bath. After all the ether has been removed, add 75 mL hexane (ligroin) and heat the solution to reflux on the steam bath. Cool the solution to room temperature, then continue to cool the solution in an ice bath until the internal temperature is approximately 10°C. Filter the crystalline material on a Buchner funnel. (*Note:* Benzhydrol tends to form an oil as the solution cools in the ice bath. If a source of seed crystals is available, several should be added while cooling. It also helps to swirl the flask vigorously immediately after observing the first crystals.)

The crude benzhydrol is purified by recrystallization from cold hexane as described above. The product is collected on a Buchner funnel and washed with *cold* hexane. The material is then air dried. Pure benzhydrol has a melting point of 67 to 69°C.

Thin-layer chromatographic analysis of the crude material (silica gel, 30% dichloromethane-hexane) shows benzhydrol and a small amount of benzaldehyde. Thin-layer chromatography of the purified material (same conditions) shows *only* benzhydrol. The yield of pure material should be 5 to 6 g.

The nmr and ir spectra of benzhydrol are shown in Fig. 12.3. Hydroxyl protons usually exchange too rapidly for coupling to be observed (see Sec.

5.3). Note that in the nmr spectrum of benzhydrol, the R_2CH—OH is coupled, giving rise to two doublets. Examine the nmr spectrum of triphenylcarbinol (preceding section) and decide which peak corresponds to the hydroxyl proton. Note that no benzaldehyde (C=O peak in the ir at about 1700 cm^{-1}) is detected in the ir spectrum of pure benzyhydrol.

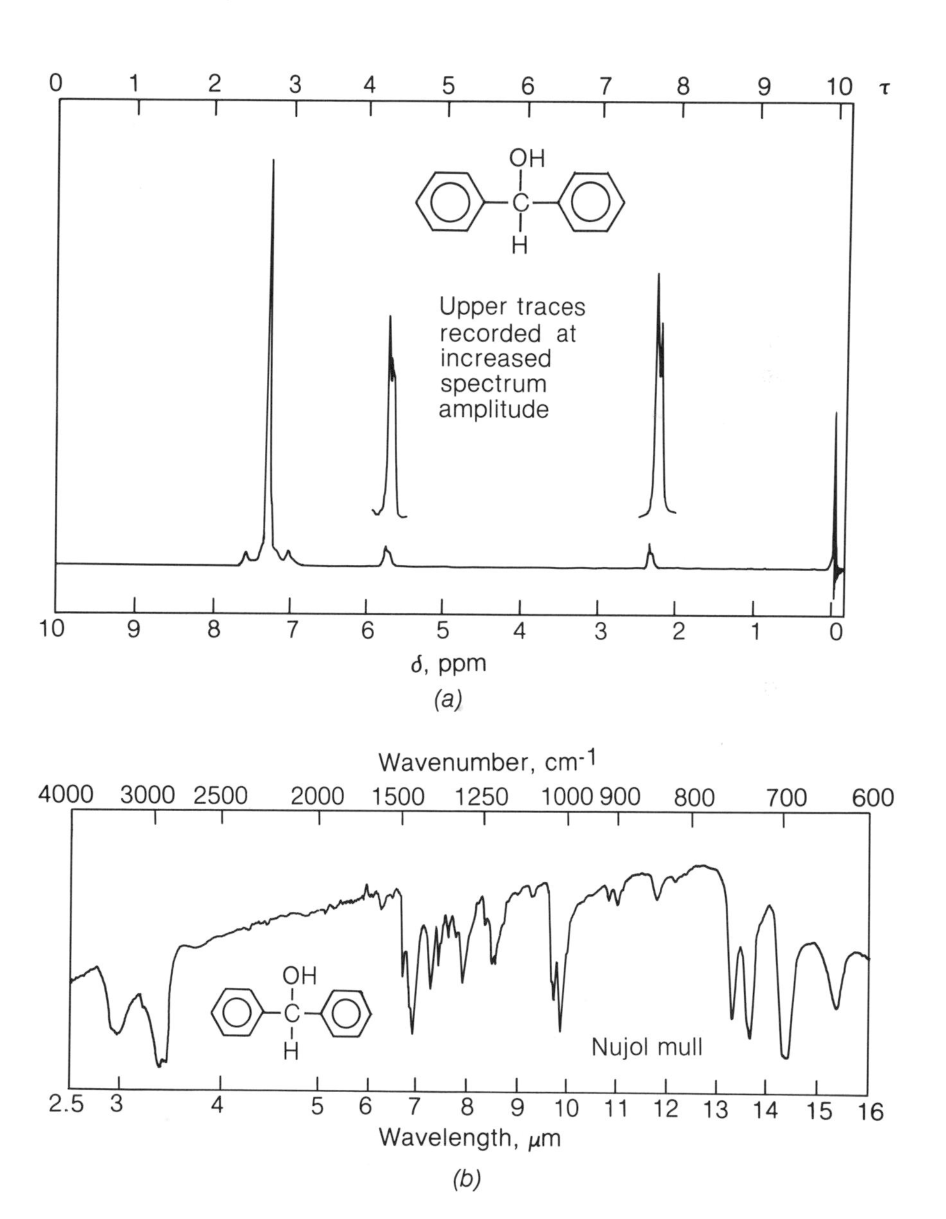

Figure 12.3 The (*a*) nmr and (*b*) ir spectra of benzhydrol.

EXPERIMENT 12.2D

SYNTHESIS OF 4-CHLOROBENZHYDROL FROM 4-CHLOROBENZALDEHYDE (to be followed by oxidation, Exp. 14.3B)

$$4\text{-}ClC_6H_4CHO + C_6H_5MgBr \xrightarrow[\text{2. } H_3O^+]{\text{1. ether}} 4\text{-}ClC_6H_4CH(OH)C_6H_5$$

Time 3.0 h

Materials 4-Chlorobenzaldehyde, 6 g (MW 141; mp 47°C)
10% H_2SO_4, 50 mL
Anhydrous ether, 25 mL

Precautions Perform experiment in hood. Extinguish all flames before opening any solvent vessel. Avoid flames when using ether.

Hazards Ether is flammable and is a powerful narcotic.

Experimental Procedure

Prepare the phenylmagnesium bromide Grignard reagent on a 0.05-mol scale as described in Exp. 12.1.

Weigh out 6 g (0.043 mol) 4-chlorobenzaldehyde and place it in a *dry* 125-mL Erlenmeyer flask. Pour in 25 mL anhydrous ether. Swirl until all the 4-chlorobenzaldehyde has gone into solution. (Slight warming on a steam bath will help this process. Be careful, however, not to introduce any moisture into the flask at this point.) The homogeneous solution is then poured into the addition funnel which was used for the bromobenzene in preparing the Grignard reagent (see Fig. 12.1).

Adjust the stopcock on the addition funnel so that a drop at a time of the 4-chlorobenzaldehyde solution falls into the Grignard reagent. During the addition, a vigorous reaction will occur and the solution will reflux. Dropwise addition maintains a controlled reaction. (*Note:* Reflux should always be kept at a gentle level.) During the addition a precipitate may form in the reaction solution. This precipitate, if observed, usually appears during the second half of the addition. After all the 4-chlorobenzaldehyde solution has been added, place the 250-mL round-bottom flask on a steam bath and reflux the entire mixture for 30 min.

Just before the reflux period is complete, prepare an acid solution by pouring 50 mL of 10% aqueous sulfuric acid into a 250-mL Erlenmeyer flask containing approximately 50 g cracked ice. As the reaction mixture is cooling, the sulfuric acid solution will also be cooling. Pour about 10 mL of this cold aqueous acid solution into a graduated cylinder for later use. When the reaction mixture has reached room temperature, pour it into the Erlenmeyer flask containing the very cold acid. When the addition is complete, swirl vigorously so that the ether and aqueous acid solutions intimately mix and all salts hydrolyze. Now add the reserved 10 mL of acid solution to the 250-mL round-bottom flask to hydrolyze any salts which remained in it after the original transfer (use small amounts of solvent ether if necessary to dissolve most of the material).

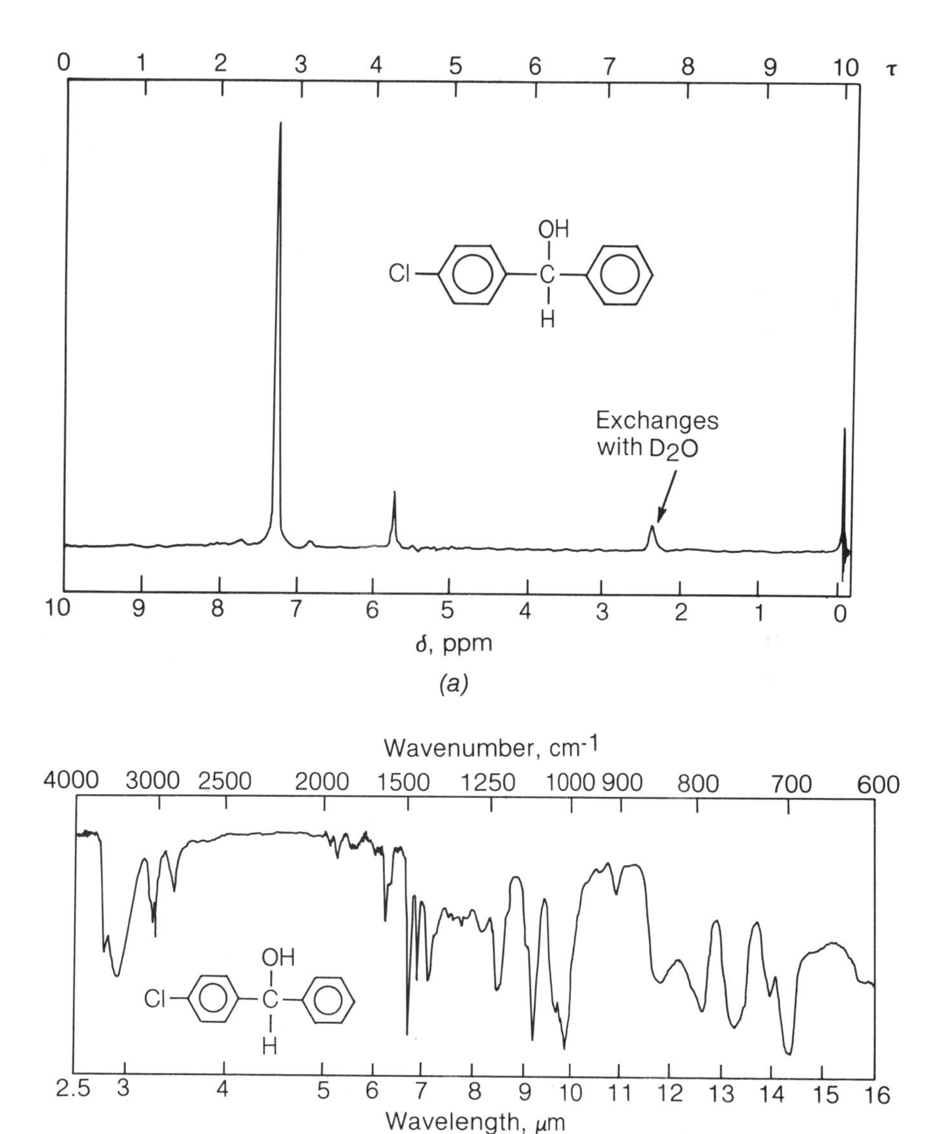

Figure 12.4
The (*a*) nmr and (*b*) ir spectra of 4-chlorobenzhydrol.

When the hydrolysis of all the salts in both containers is complete, transfer the combined solutions to the separatory funnel. Place a glass stopper in the separatory funnel and gently shake the mixture. Draw off the lower, more dense, acid layer and wash the remaining ether layer once with 25 mL water, twice with 25 mL saturated sodium bicarbonate solution, and finally with two 25-mL portions of saturated aqueous sodium chloride solution (brine). After all these washings, a relatively dry ether solution remains, which is run into a 125-mL Erlenmeyer flask. Add anhydrous sodium sulfate (10 to 15 g) to complete the drying process. After gravity filtering the solution into a 250-mL Erlenmeyer flask, evaporate the ether on a steam bath (**hood**). After all the ether has been removed, a viscous, glassy material will be obtained which resists solidification. (Pure 4-chlorobenzhydrol has a recorded melting point of 58 to 60°C.) This is not important, however, as this material may be oxidized directly to 4-chlorobenzophenone, as described in Exp. 14.3B.

Thin-layer chromatographic analysis of the crude material (silica gel, 30% dichloromethane-hexane) shows 4-chlorobenzhydrol and a small amount of 4-chlorobenzaldehyde. If the material is used in the preparation of 4-chlorobenzophenone, the residual aldehyde will be oxidized and the acid removed during purification.

The nmr and ir spectra of 4-chlorobenzhydrol are shown in Fig. 12.4. Notice that chlorine and hydrogen must have similar effects on adjacent hydrogens because all the aromatic protons are observed in about the same place (7.2 to 7.3 ppm, some splitting). The $R_2CH—OH$ system does not exhibit the coupling observed in the nmr spectrum of benzhydrol (Exp. 12.2C). This is probably due to the presence of a small amount of water or acidic impurity in the sample. The pattern of peaks observed in the ir spectrum from 1660 to 2000 cm^{-1} is characteristic of a para-substituted benzene ring. Notice that this pattern is absent in the ir spectrum of benzhydrol (Exp. 12.2C).

12.3 CARBONATION OF GRIGNARD REAGENTS

Grignard reagents react as sources of carbanionic carbon. As a result, they react with a large number of electrophiles. In earlier sections, reaction with aldehydes, ketones, and esters has been discussed. In addition, the presence of water as an electrophile has been mentioned. In most research laboratories, a Grignard reagent would be prepared and utilized under a blanket of nitrogen gas. This inert atmosphere would prevent unwanted side reactions of the Grignard reagent with atmospheric moisture (water) and with oxygen (whose reactions with this reagent are somewhat more complex).

One of the atmospheric gases which is usually not worried about too much is carbon dioxide. The concentration of CO_2 in the atmosphere is very low, and if a reaction did occur between a Grignard reagent and CO_2, an acid would result. Since most products from the Grignard reaction are neutral (like alcohols), the acidic by-products can be removed rather easily. The formation of a carboxylic acid is shown in the equation below.

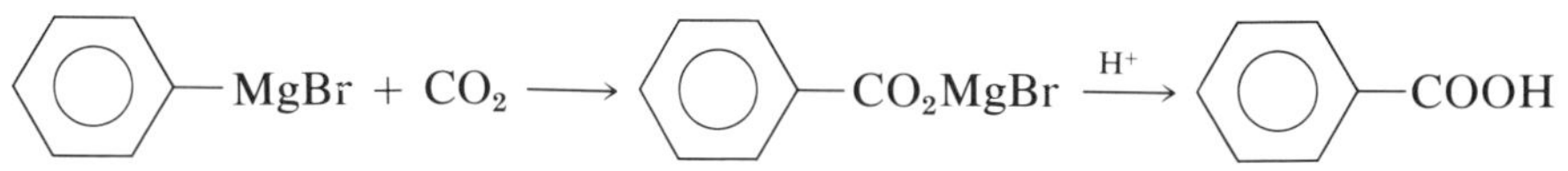

$$C_6H_5\text{—MgBr} + CO_2 \longrightarrow C_6H_5\text{—}CO_2MgBr \xrightarrow{H^+} C_6H_5\text{—COOH}$$

A Grignard reagent can be forced to react with carbon dioxide if this is the only electrophile available. In the two reactions which follow, dry ice (solid carbon dioxide) is used for this purpose. Although the reaction is really quite simple, some practical details need to be kept in mind.

The reaction of a Grignard reagent in ether solution with solid carbon dioxide will be a heterogeneous (surface) reaction. The dry ice should be ground (powdered) as finely as possible, so that when the reagent solution is poured over it, the desired reaction can occur easily. A problem arises, however, because dry ice is very cold. Atmospheric moisture (water) will condense readily on the surface of the cold CO_2 and form a layer of "wet" ice. Since the Grignard reagent reacts with water, some hydrocarbon by-product can be expected from the reaction. In order to minimize this side reaction, the dry ice should be powdered and prepared for reaction immediately before the prescribed reflux period is over.

The starting material for Exp. 12.3B is 2-bromo-1,4-dimethylbenzene (2-bromo-*p*-xylene). This bromide is commercially available at modest cost but may be prepared readily by the bromination of *p*-xylene, as described in Exp. 18.2A.

EXPERIMENT 12.3A

SYNTHESIS OF BENZOIC ACID BY CARBONATION OF A GRIGNARD REAGENT

$$C_6H_5\text{MgBr} + CO_2 \xrightarrow[2.\ H_3O^+]{1.\ \text{ether}} C_6H_5\text{—C(=O)—OH}$$

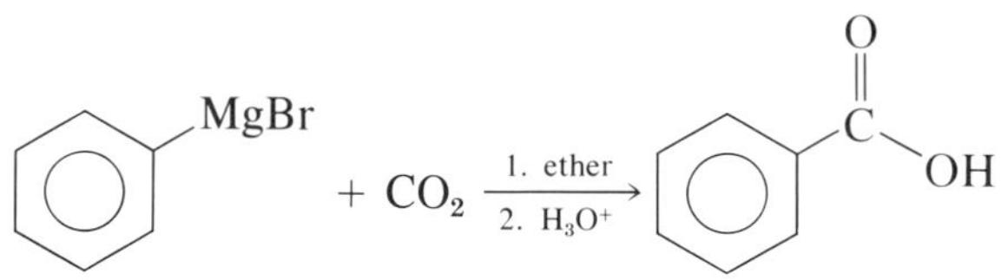

Time 3.0 h

Materials Bromobenzene, 12 mL (MW 157; bp 200°C; d 1.49 g/mL)
Magnesium, 2.4 g (MW 24)
Dry ice, 100 g
Anhydrous ether, 100 mL

Precautions Perform experiment in hood. Extinguish all flames before opening any solvent vessel.

Hazards Ether is flammable and is a powerful narcotic. Bromobenzene is toxic in high concentration.

Experimental Procedure

Prepare phenylmagnesium bromide on *double* the scale described in Exp. 12.1.

Disconnect the reflux condenser, the Claisen head, and the addition funnel from the 250-mL round-bottom flask. Quickly pour the contents of the flask into a 600-mL beaker containing approximately 100 g freshly crushed dry ice. While adding the Grignard reagent, stir the slush with a glass rod in order to maximize contact between the Grignard reagent and the solid carbon dioxide. Do all of this as quickly as possible because at the temperature (−78°C) of dry ice water condenses rapidly. Water reacts with the Grignard reagent to produce benzene instead of the desired product. Note that the mixture becomes viscous after all the Grignard reagent has been added. Be careful to keep the 600-mL beaker well into the hood, and allow all the dry ice to sublime. Some of the ether will evaporate at the same time, so **it is very important to keep flames away and not to breathe the atmosphere immediately surrounding the beaker.**

After the dry ice has sublimed, carefully add 100 mL cold 10% aqueous sulfuric acid and swirl as vigorously as possible, taking care not to splash the acid solution onto your hand. Transfer the resulting mixture of ether and water to a separatory funnel. Place an ungreased stopper in the top of the separatory funnel and shake gently, so that the hydrolysis process continues. Remove the stopper and carefully draw off the denser aqueous acid layer. Wash the remaining ether layer once with 25 mL distilled water and extract three times with 25-mL portions of 10% aqueous sodium hydroxide solution.

Be very careful when shaking the ether with sodium hydroxide because considerable heat may be evolved. Be sure to vent the separatory funnel, as shown in Fig. 2.21.

After each washing, run the basic aqueous phase into a 125-mL Erlenmeyer flask. After the third wash, the material should be strongly basic and the flask nearly full.

Remove the thoroughly extracted ether solution from the separatory funnel and transfer the basic water layers to the funnel. Wash them with a 25-mL portion of ether. Transfer to a 250-mL Erlenmeyer flask and, using an eye dropper or disposable pipet, add 12 *N* HCl dropwise to the basic solution while swirling continuously. Continue adding HCl until the reaction mixture is strongly acidic. The pH should be less than 1 (as indicated by hydrion paper, or by blue litmus instantly turning bright red). Considerable heat will be evolved during the acidification process.

The benzoic acid which separates from the aqueous solution as the reaction mixture cools to room temperature may be collected on a Buchner funnel. The crude benzoic acid remaining on the filter should be washed with several small portions of cold water (benzoic acid has a small but finite solubility in warm water). Recrystallize the crude benzoic acid from water. Collect the crystals on a Buchner funnel and allow them to air dry. After recrystallization, approximately 6 g (about a 50% yield) of material with mp 120 to 122°C should be obtained.

The nmr and ir spectra of benzoic acid are shown in Fig. 12.5. All the prod-

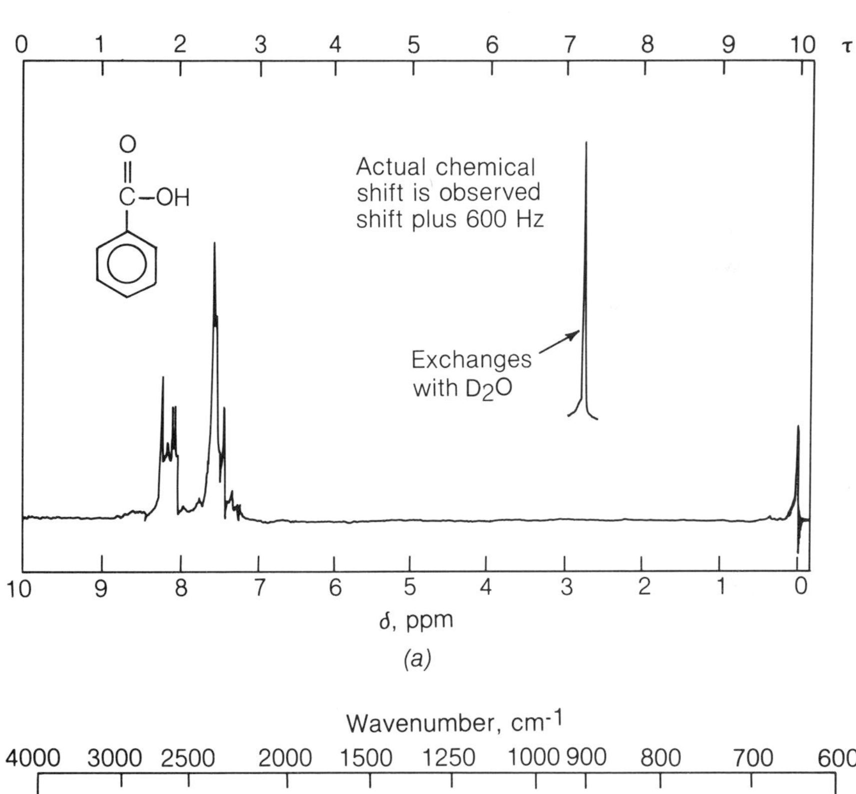

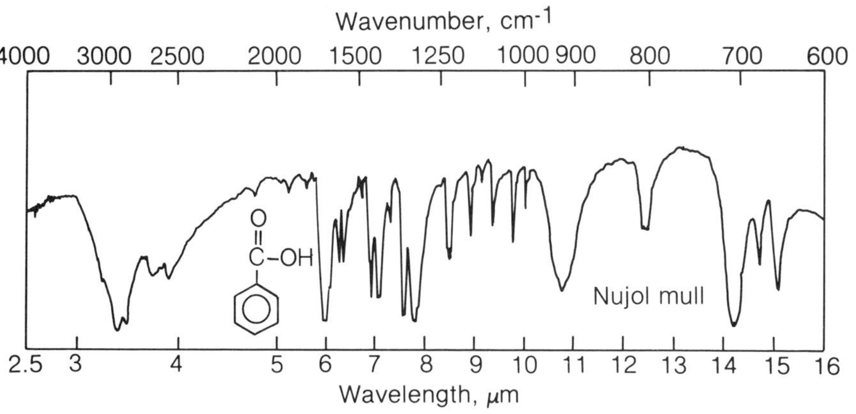

Figure 12.5
The (*a*) nmr and (*b*) ir spectra of benzoic acid.

ucts prepared in earlier parts of this chapter had aromatic rings and hydroxyl groups. The aromatic ring of benzoic acid is adjacent to a double bond and gives the characteristic pattern shown here. In previous compounds, the aromatic protons appeared almost as a singlet. Notice also that the acidic hydroxyl group is very far downfield [2.7 ppm + 600 Hz (= 10 ppm) = 12.7 ppm].

The hydroxyl vibration is visible in the ir spectrum, but its characteristic shape is quite different from that of the hydroxyl group in benzhydrol (Exp. 12.2C).

EXPERIMENT 12.3B SYNTHESIS OF 2,5-DIMETHYLBENZOIC ACID BY CARBONATION OF A GRIGNARD REAGENT

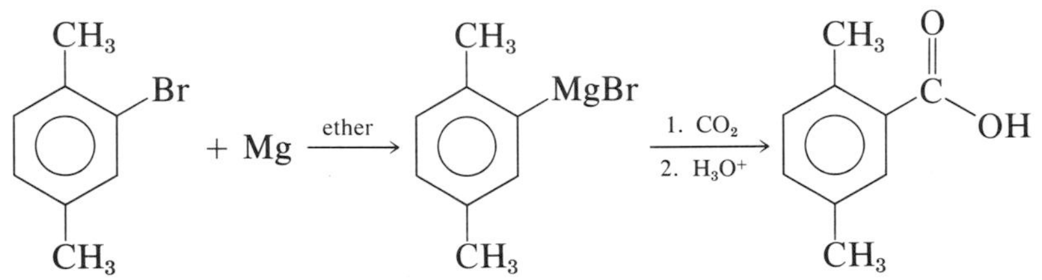

$$\text{2-bromo-}p\text{-xylene} + Mg \xrightarrow{\text{ether}} \text{2,5-}(CH_3)_2C_6H_3MgBr \xrightarrow[2.\ H_3O^+]{1.\ CO_2} \text{2,5-}(CH_3)_2C_6H_3COOH$$

Time 3 h

Materials 2-Bromo-1,4-dimethylbenzene (2-bromo-*p*-xylene), 7 mL (MW 185, bp 200°C, d 1.34 g/mL)
Magnesium, 1.2 g (MW 24)
Dry ice, 50 to 60 g
Anhydrous ether, 50 mL

Precautions Perform experiment in hood. Extinguish all flames before opening any solvent vessel.

Hazards Ether is highly flammable and is a powerful narcotic. High concentrations of 2-bromo-*p*-xylene are toxic.

Experimental Procedure

Set up an apparatus consisting of a 250-mL round-bottom flask equipped with a Claisen head, reflux condenser, drying tube, and addition funnel, as described for the preparation of phenylmagnesium bromide (Exp. 12.1, see Fig. 12.1).

Flame dry the apparatus and allow it to cool (as described in Exp. 12.1). Place 1.2 g (0.050 mol) magnesium turnings in the flame-dried and cooled apparatus and add 50 mL ether. Add 7 mL (9.25 g, 0.050 mol) 2-bromo-*p*-xylene dropwise from the addition funnel until the reaction begins. Continue the addition in approximately 1-mL portions at a rate sufficient to maintain a gentle but continuous reflux. After the final portion of bromide has been added, continue reflux for an additional 15 min by heating on a steam bath.

Remove the steam bath and allow the apparatus to cool. Dilute the Grignard solution with 50 mL dry ether and pour the resulting mixture into a 600-mL beaker containing 50 to 60 g crushed dry ice. As the Grignard solution is added, stir the dry ice vigorously with a glass rod or spatula in order to maximize contact between the solid and solution. A viscous paste will form as the solution is stirred. Add anhydrous ether as needed to maintain the material as a liquid. When the addition is complete, allow the reaction mixture to stand in the hood until all the dry ice has sublimed and then add 100 mL cold 10% aqueous sulfuric acid. Swirl the beaker with sufficient vigor to ensure hydrolysis of the magnesium salts, but **be very careful that the acidic solution does not slosh onto your hand. It is often useful to wear gloves during this part of the procedure.**

When the solution has cooled, transfer the mixture of water and ether to a separatory funnel and use a little additional ether to rinse any product remaining in the beaker into the funnel. Stopper the separatory funnel and shake gently. When the layers separate, remove the stopper from the funnel and draw off the bottom aqueous phase. Wash the remaining ether layer once with 25 mL distilled water and then three times with 25 mL portions 10% aqueous sodium hydroxide. Be very careful when shaking the ether with sodium hydroxide because considerable heat may be evolved, and this may cause pressure to build up in the separatory funnel. Be careful to hold the separatory funnel with both hands, one cupped over and retaining the stopper, the other manipulating the stopcock (see Fig. 2.20). After shaking, turn the bottom of the separatory funnel upward and away from you and open the stopcock to safely release any pressure which has built up.

Combine all three basic layers and wash once with 25 mL ether. Acidify the basic aqueous layer by the cautious addition of 12 *N* HCl. The material is sufficiently acidified when its pH is below 1 (as measured by hydrion paper) or when it instantly turns blue litmus paper bright red. Use a Buchner funnel to filter the solid which precipitates during this acidification, and wash the solid with distilled water. Air dry the solid acid and recrystallize it from petroleum ether (bp 65 to 70°C). The pure acid is obtained in 50 to 60% yield (approximately 3 to 4 g) and should have a mp of 130 to 132°C.

The nmr and ir spectra of the acid are shown in Fig. 12.6. The ir spectrum looks quite similar to that for benzoic acid (Exp. 12.3A), but the "monosubstitution bands" between 1660 and 2000 cm^{-1} are missing.

It is clear from the nmr spectrum that the acid has two distinctly different kinds of aromatic hydrogen atoms (in a ratio of 2:1) and two different kinds of methyl groups.

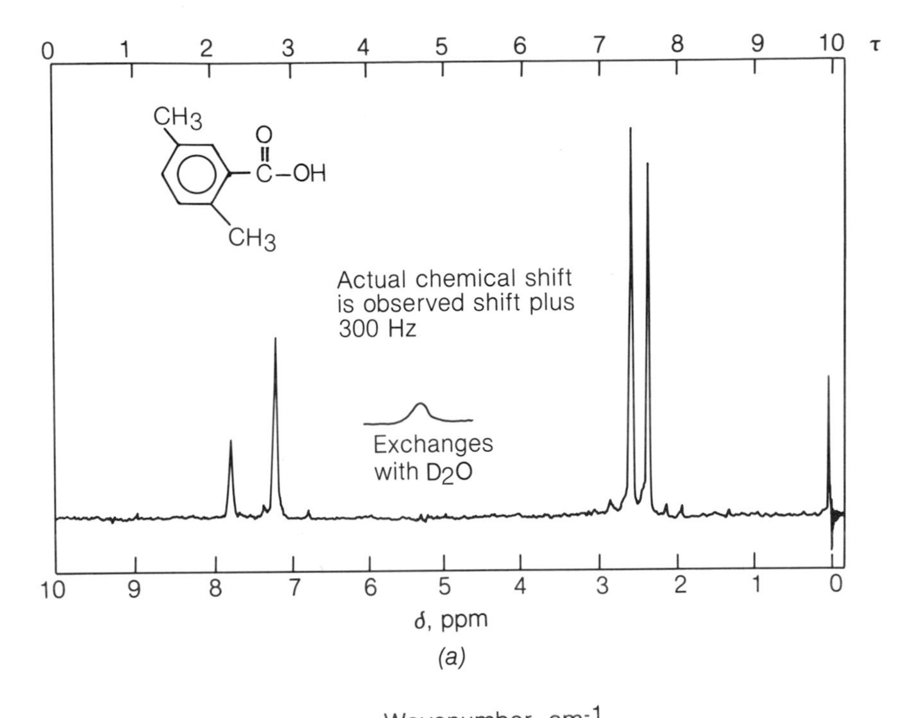

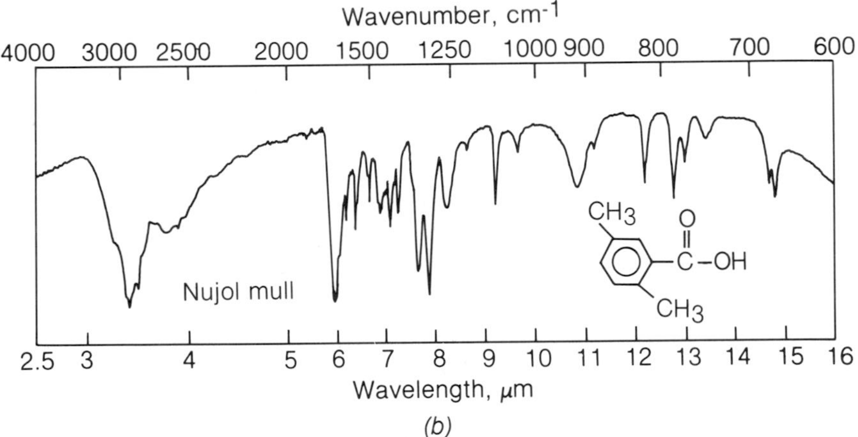

Figure 12.6
The (*a*) nmr and (*b*) ir spectra of 2,5-dimethylbenzoic acid.

12.4 SYNTHESIS OF INSECT PHEROMONES BY GRIGNARD REACTIONS

Most human communication involves either the written or spoken word. Although current psychological theory suggests that a considerable amount of nonverbal communication takes place ("body language"), most people rely on what they see and hear for information. This is not the case in the insect kingdom. On the contrary, insects rely primarily on chemical signals for the transmittal of information. These chemical messengers are referred to as *pheromones*.

Insects have the remarkable ability to synthesize, emit, receive, and decipher chemical pheromones. It is these signals which warn (alarm pheromones), attract (sex pheromones), or cause insects to group (aggregation pheromones). The pheromones may be simple or complex chemicals, and related substances may be entirely different in their effects on the same insects. For example, the sex attractant of the sugar beet wireworm (Exp. 12.4A) is pentanoic (valeric) acid. The wireworm is unaffected by the sex attractant of the Pennsylvania mushroom-infesting fly [*Lycoriella mali* (Fitch)], whose sex pheromone is heptadecane. The alarm pheromone of the honey bee is isoamyl acetate, which we have already seen (Exp. 11.2C) and which is known commercially as banana oil or pear oil.

It should be obvious that, since these chemical messenger systems are used by insects to communicate, only small quantities of compounds must be involved. Humans think of a "drop" (about 0.05 or 0.1 mL) as a very small quantity. It appears that the amount of materials involved in insect communication is smaller than that by a factor of 10^{10} to 10^{15}.

Extensive research has gone on over the past 30 or 40 years aimed at discovering what compounds attract and repel insects. The development of *N*,*N*-diethyl-*m*-toluamide (DEET, Sec. 11.4) for use as an insect repellent resulted from this general program. Note that DEET is a repellent, not an insecticide. The latter kills rather than repels the insects. It has been the long-standing hope of such research that insects could be drawn away from a site of infestation by the appropriate attractant. Another hope has been that all the males could be attracted to a location by use of the female sex attractant and thereby prevented from mating with the female of the species. The result of this ploy would be a reduction in the number of offspring and thereby in the insect population without the application of any insecticide. Such hopes have not generally been realized, but promising research in this area continues.

The compounds whose preparations are discussed in this section are aliphatic products of Grignard reactions presented before. They are particularly interesting compounds because they are "natural products" (see Chap. 20). These are compounds which have been identified in and isolated from living organisms and are known to serve a role as chemical messengers. In Exp. 12.4A valeric acid, the sex attractant of the sugar beet wireworm (*Limonium californicum*) is synthesized; in Exp. 12.4B 4-methyl-3-heptanol, the aggregation pheromone of the European elm beetle (*Scolytus multistriatus*) is prepared.

There is one special note necessary before the synthesis of valeric acid can be started. Valeric acid sends a signal to the wireworm, but it also sends a strong signal to humans in the form of its odor. Its odor is described in the literature as "unpleasant." If you carry out this experiment, be sure to shower with a basic soap when you return home. The basic soap will help neutralize the acid and thereby the odor.

EXPERIMENT 12.4A

SYNTHESIS OF VALERIC ACID BY CARBONATION OF A GRIGNARD REAGENT

$$CH_3CH_2CH_2CH_2Br + Mg \xrightarrow{\text{ether}} CH_3CH_2CH_2CH_2MgBr \xrightarrow[\text{2. } H_3O^+]{\text{1. } CO_2} CH_3CH_2CH_2CH_2\overset{\overset{\displaystyle O}{\|}}{C}OH$$

Time 3.0 h

Materials *n*-Butyl bromide (Sec. 10.1), 11 mL (MW 137; bp 100 to 104°C; d 1.276 g/mL)
Magnesium, 2.4 g (MW 24)
Anhydrous ether, 100 mL
Carbon dioxide (dry ice), 100 g

Precautions Carry out all reactions in a good hood. Extinguish all flames before opening any solvent vessel. Avoid breathing ether.

Hazards *n*-Butyl bromide is toxic in high concentrations. Ether is flammable and is a powerful narcotic.

Experimental Procedure

Before beginning this procedure, refer to Sec. 12.3.

Assemble an apparatus consisting of a 250-mL round-bottom flask, Claisen head, addition funnel with stopper, and reflux condenser with drying tube on top (see Fig. 12.1).

Weigh out 2.4 g (0.100 mol) magnesium turnings and place it in the 250-mL round-bottom flask. Flame out the apparatus as described in Exp. 12.1. Put a drying tube on top of the condenser as the apparatus cools and connect the water to the reflux condenser. Add 100 mL ether to cover the solid and place the *n*-butyl bromide (11 mL, 0.1 mol) in the addition funnel. Run in approximately 1 mL of the halide. The reaction will usually commence immediately. Addition of *n*-butyl bromide should be carried out at such a rate that an even and gentle reflux is maintained. After all the halide has been added, reflux the reaction mixture for an additional 10 min (steam bath) so that all the magnesium is converted into *n*-butylmagnesium bromide.

Cool the reaction mixture to room temperature. Disassemble the apparatus and pour the solution into a 600-mL beaker which contains 100 g crushed dry ice. Vigorously stir the slurry with a spatula or heavy glass rod so that there is good contact between the dry ice and the ether solution. When the addition is complete, allow the dry ice to sublime. Hydrolyze the remaining paste by slowly adding 100 mL cold 10% aqueous sulfuric acid. Again, stir vigorously with a spatula or heavy glass rod in order to ensure good mixing and complete hydrolysis. Transfer the resulting mixture to a separatory funnel and wash in any material remaining in the beaker, using a small amount of ether if necessary. Stopper the separatory funnel, shake, and then remove the aqueous layer. Draw off the ether layer and return the aqueous layer to the separatory funnel. Extract the aqueous phase three times with 50-mL portions of ether. Combine all the ether layers and extract (wash) them with 50 mL cold distilled water.

Carefully extract the ether layer with three 25-mL portions 10% aqueous

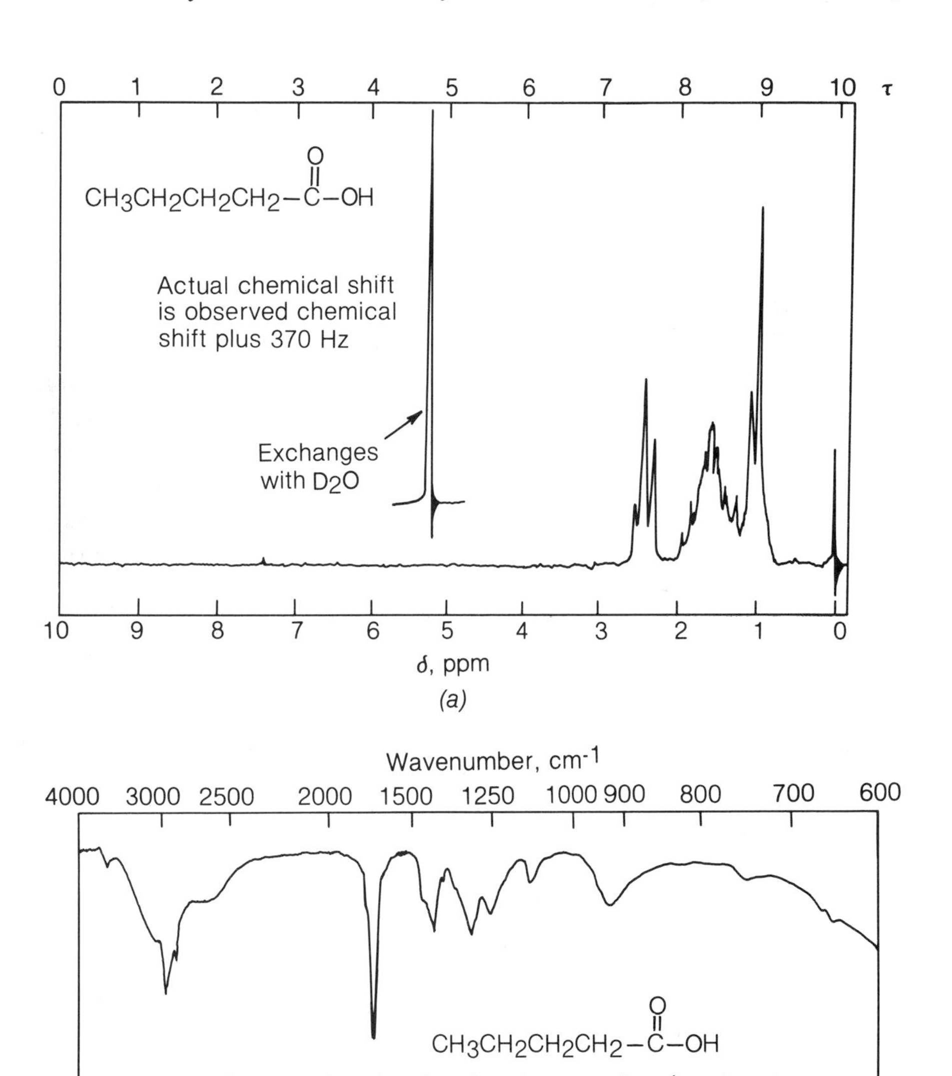

Figure 12.7
The (*a*) nmr and (*b*) ir spectra of valeric acid.

sodium hydroxide. Add the first portion of sodium hydroxide to the funnel and shake gently. Hold the funnel with one hand cupped over the stopper and the other hand controlling the stopcock (as shown in Figs. 2.20 and 2.21). Heat will be evolved, which may cause some ether to vaporize. Turn the funnel so that its bottom is pointed upward and away from you and open the stopcock to safely release the pressure. Repeat this procedure twice more; then combine all the aqueous base solutions and check the pH. (It should be strongly basic.)

Wash the solution with two 25-mL portions of ether and then carefully acidify with HCl until blue litmus paper is instantly turned bright red or hydrion paper shows a pH less than 1. Allow the solution again to cool to room temperature and extract with four 25-mL portions dichloromethane. Combine the dichloromethane layers and wash with 25 mL half-saturated sodium chloride solution made by diluting 25 mL saturated sodium chloride solution with an equal volume of water. Dry the dichloromethane solutions over sodium sulfate. Gravity filter to remove the drying agent and then evaporate the dichloromethane on a steam bath. Transfer the residual oil to a simple distillation apparatus. Distill valeric acid with a free flame (**caution: no ether in the area**) and collect the product distilling between 170 and 180°C. Valeric acid is a water-white, odoriferous liquid. The yield of pure valeric acid is 5 to 6 g (about 50%). The odor of this compound is strong, so all operations should be conducted in the hood. Some people find the odor offensive and so it would be well to shower after the experiment with a relatively basic soap.

The nmr and ir spectra of valeric acid are shown in Fig. 12.7. Note that the hydroxyl peak (nmr) is observed at 165 Hz + 550 Hz = 715 Hz. This corresponds to a chemical shift (715/60) of 11.9 ppm. Both the carbonyl band (1700 cm^{-1}) and the hydroxyl band (centered near 3000 cm^{-1}) are visible in the ir spectrum.

EXPERIMENT 12.4B SYNTHESIS OF 4-METHYL-3-HEPTANOL BY A GRIGNARD REACTION

$$CH_3CH_2CH_2CH(Br)CH_3 + Mg \xrightarrow{\text{ether}} CH_3CH_2CH_2CH(MgBr)CH_3 \xrightarrow[\text{2. } H_3O^+]{\text{1. } CH_3-CH_2-CHO} CH_3CH_2CH_2CH(CH_3)-CH(OH)CH_2CH_3$$

Time 3.0 h

Materials 2-Bromopentane, 23 mL (MW 151, bp 116 to 118°C, d 1.22 g/mL)
Magnesium, 7.3 g (MW 24)
Propionaldehyde (propanal), 11.6 g (MW 58; bp 46 to 50°C; d 0.81 g/mL)
Anhydrous ether, 100 mL

Precautions Carry out all steps in a good hood. Extinguish all flames before opening any solvent vessel.

Hazards High concentrations of 2-bromopentane are toxic. Ether is extremely flammable and is a powerful narcotic.

Special Instructions

Recall that Grignard reactions are very sensitive to the presence of water and acids. Both propionaldehyde and 2-bromopentane should be freshly distilled and kept dry. Each sample should be colorless. Propionaldehyde, which is especially prone to undergo autoxidation and self-condensation, should be distilled immediately prior to use. 2-Bromopentane will not require distillation unless water is visible in it or unless it is yellow in color.

If it is necessary to distill the 2-bromopentane, be extremely careful, because this distillation will require a flame at a time many other students may be setting up their Grignard reactions. The Grignard reaction uses ether, which is very flammable. If a free flame is near a reaction in progress, there is a very serious fire and explosion hazard.

Experimental Procedure

Assemble the apparatus for the Grignard reaction by placing a Claisen head in a 250-mL round-bottom flask. Fit the Claisen head with a reflux condenser and an addition funnel (see Fig. 12.1). Charge the flask with 7.3 g (0.30 mol) magnesium and flame dry the apparatus as described in Exp. 12.1. Stopper the addition funnel with a lightly greased glass stopper and place a drying tube in a lightly greased thermometer adapter in the top of the reflux condenser (see Fig. 12.1). Begin the flow of water in the condenser only after flame drying is complete. When the apparatus has cooled completely, cover the magnesium in the round-bottom flask with 100 mL anhydrous ether by removing the addition funnel and carefully pouring the ether into the Claisen head. Place 28 g 2-bromopentane in the addition funnel and then add 2 to 3 mL of it to the ether-magnesium mixture. The reaction should begin rapidly. If it does not, refer to the expedients and cautions described in Exp. 12.1. Continue the addition of 2-bromopentane at such a rate that a gentle but constant reflux is maintained. When all the 2-bromopentane has been added, continue reflux with a steam bath for an additional 15 min.

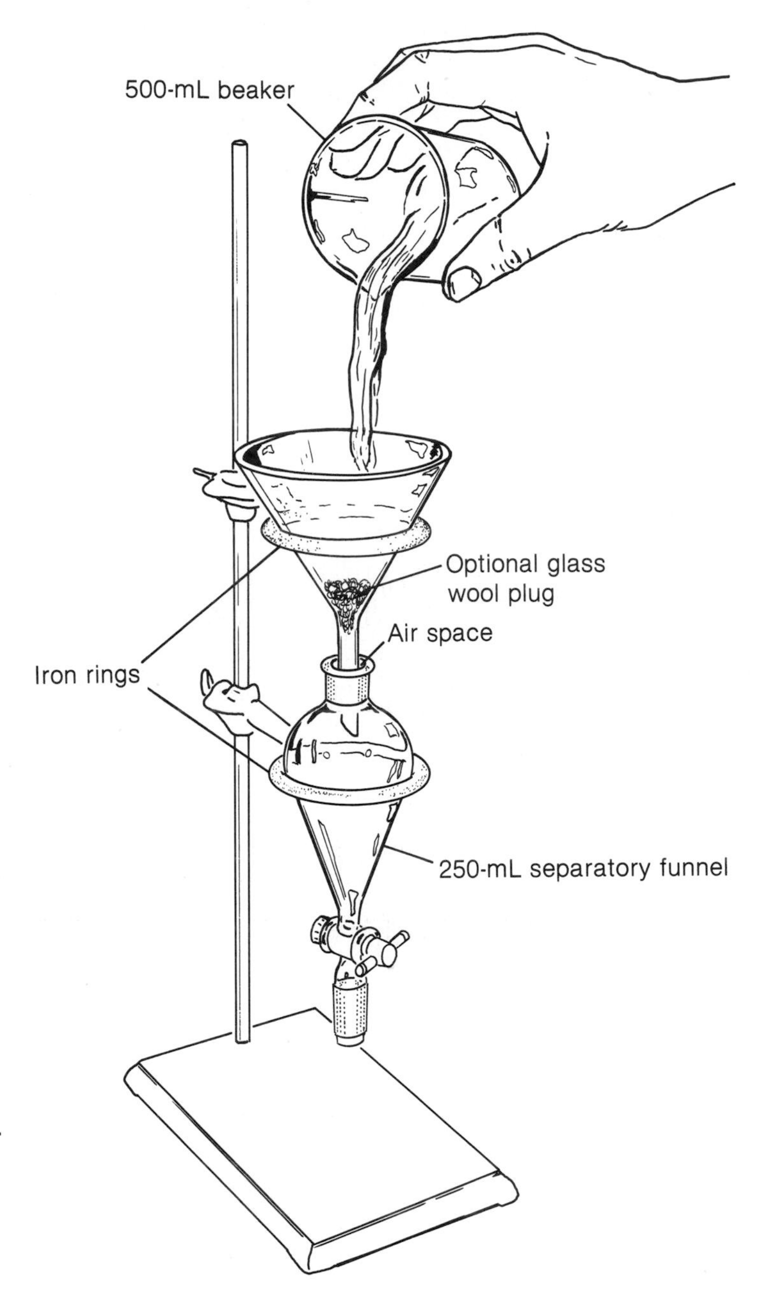

Figure 12.8 Apparatus for Experiment 12.4B. Use of a glass wool plug to remove excess magnesium from a mixture being decanted.

Once Grignard formation is complete, transfer 11.6 g (0.2 mol, 14.5 mL) of freshly distilled propionaldehyde to the addition funnel and rinse the graduated cylinder with 20 mL ether. Add this ether rinse to the graduated cylinder. Add the propionaldehyde solution from the addition funnel at such a rate that a gentle but constant reflux is maintained. Reflux the mixture for 30 min, allow it to cool briefly, and then pour it into a 500-mL beaker containing a mixture of 100 g cracked ice and 100 mL 10% aqueous sulfuric acid. All the salts should dissolve. The mixture should be stirred continuously with a heavy glass rod or a spatula during the addition. After dissolution of most of the salt, carefully decant the mixture into the separatory funnel (a glass wool plug may be used in the funnel to remove excess magnesium) and separate the layers (see Fig. 12.8). Remove and discard the lower layer. Wash the ether layer (which contains the product) with two 25-mL portions of 5% aqueous sodium hydroxide solution and then with two 25-mL portions of saturated aqueous sodium chloride solu-

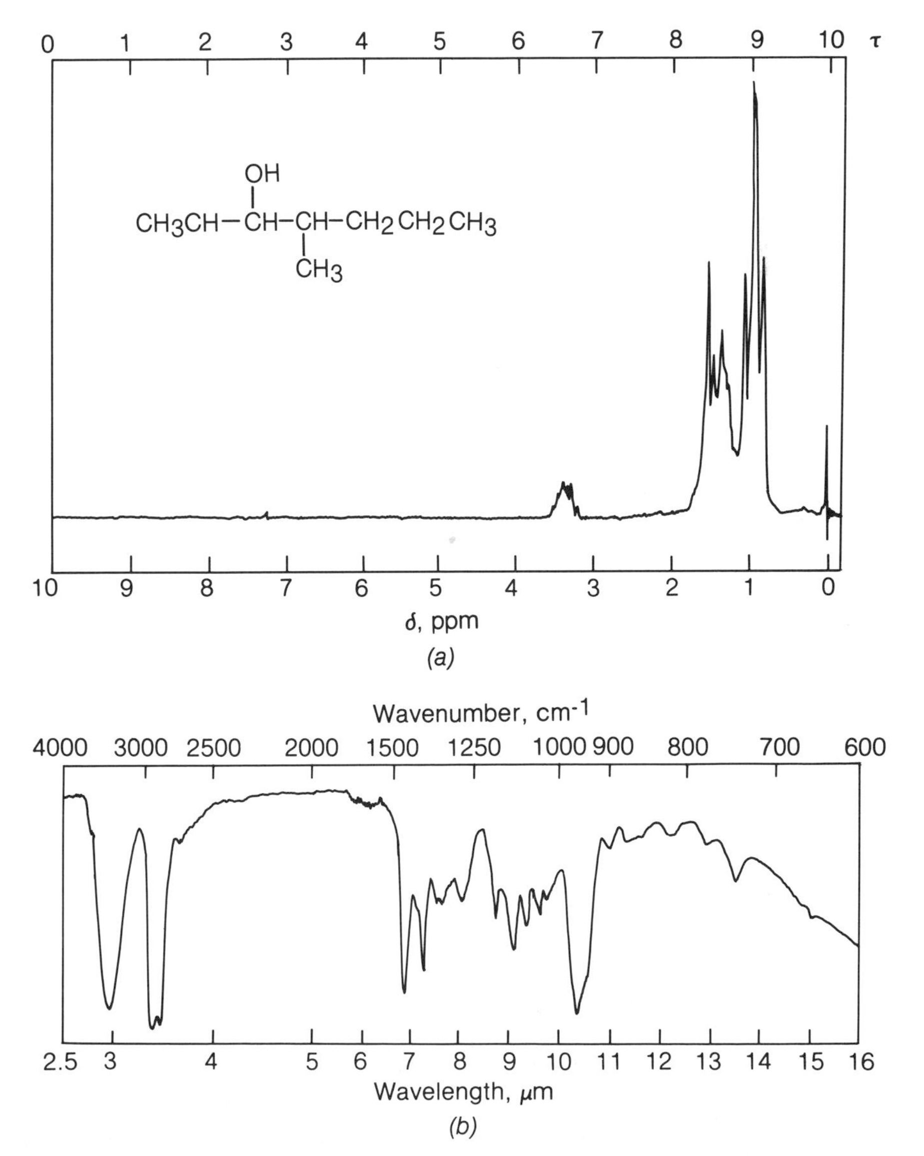

Figure 12.9
The (*a*) nmr and (*b*) ir spectra of 4-methyl-3-heptanol.

tion. Dry the resulting ether solution over sodium sulfate and then filter into a 250-mL Erlenmeyer flask to remove the sodium sulfate. Place the Erlenmeyer flask on a steam bath to evaporate the ether. Transfer the resulting oil with the smallest amount of ether possible into a simple distillation apparatus and distill with a free flame.

Check to be sure that no one near you is handling ether or other flammable solvent when you begin your distillation.

Collect the material which distills between 150 and 165°C. The product (approximately 12 g) should be a water-white liquid. Its nmr and ir spectra are shown in Fig. 12.9.

QUESTIONS AND EXERCISES

12.1 Triphenylcarbinol (triphenylmethanol) can, in theory, be prepared by adding 1 equiv of phenyl Grignard to benzophenone or 2 equiv of phenyl Grignard to methyl benzoate. If you had to make a decision about which of these methods to choose, which method would you choose and what factors would you consider in making that choice?

12.2 If 1 equiv of 4-bromo-1-chlorobenzene were treated with 1 equiv of magnesium in ether solution and the resulting reagent allowed to react with benzaldehyde, what would you expect the product to be?

12.3 A nitrile is like an ester in the sense that it can be hydrolyzed to a carboxylic acid. A tertiary alcohol is formed if the ester is treated with excess Grignard reagent, whereas if a nitrile is treated with excess Grignard reagent, a ketone is ultimately isolated. Can you suggest a reaction mechanism which accounts for this reactivity?

12.4 In the preparation of triphenylcarbinol one of the principal by-products of the reaction is biphenyl. Can you think of any analogy in alkyl halide chemistry which would be similar to the production of biphenyl from bromobenzene and magnesium?

12.5 If either *n*-butyl benzoate or methyl benzoate reacts with phenylmagnesium bromide, the product isolated is triphenylcarbinol. During the reaction, however, a precipitate appears when methyl benzoate is used as the substrate, whereas no precipitate appears when butyl benzoate is used. How might you account for these observations?

12.6 In the synthesis of 4-chlorobenzhydrol, 4-chlorobenzaldehyde is added to phenylmagnesium bromide. Why does the presence of chlorine not interfere with the reaction?

12.7 If you were an industrial chemist and needed benzoic acid for a process, you would obviously consider many preparations for this compound. Among other methods, benzoic acid can be made by the hydrolysis of benzoate esters, by the hydrolysis of benzonitrile, or by the addition of a phenyl Grignard reagent to carbon dioxide. If these were the only three methods which were available to you, which of them would you choose and what factors would you take into consideration in making your choice?

12.8 If phenylmagnesium bromide remains in the reaction mixture at the end of a Grignard reaction and is hydrolyzed, benzene, not phenol, is the product observed. Explain this observation.

12.9 If air is bubbled through a solution of phenylmagnesium bromide for several hours and the mixture then hydrolyzed, phenol, not benzene, is the product obtained. Can you explain this observation?

12.10 If you wished to make triphenylmethyl chloride from triphenylcarbinol, would you use the procedure described earlier for the synthesis of *n*-butyl bromide (Exp. 10.1) or *tert*-butyl chloride (Exp. 10.2)? Why?

XIII NUCLEOPHILIC SUBSTITUTION AT SATURATED CARBON

Unsubstituted alkanes are characterized by their general lack of reactivity. Even the trivial name (paraffin) suggests this lack of reactivity (see Chap. 6). In order for an organic molecule to be reactive, it must ordinarily have excess electron density, a deficiency of electron density, or a bond which is highly polarized. A molecule which contains excess electron density will generally function either as a base or as a nucleophile. A molecule which is electron-deficient will usually function either as an electrophile or as a Lewis acid. If the molecule contains a highly polarized bond, it may be attacked either by nucleophilic or electrophilic species, depending on the direction of polarization. Examples of nucleophiles include alcohols, amines, and carbanions. Examples of electrophiles include protons and carbonium ions. Species which have highly polarized bonds are typified by ethers and alkyl halides.

Reactions which occur at saturated carbon atoms usually do so either by an S_N1 or S_N2 mechanism. Both of these reactions occur by displacement or nucleophilic substitution mechanisms. The numbers following the letter designations refer to the kinetics of the process: 1 indicates first-order kinetics and 2 indicates a second-order reaction. The meaning of such designations and examples of each may be found in Chap. 10.

The reactions presented in this section all occur by the S_N2 reaction mode. Such reactions generally require a highly polarized bond. That is to say that carbon (electronegativity 2.5) must be bonded to a more electronegative element like chlorine (3.5). The polarization of a carbon-halogen bond will favor attack of the nucleophile at the carbon atom to which the halogen atom (Cl, Br, etc.) is attached. If the carbon atom is primary, direct displacement will also be favored. First, a primary carbon atom is usually less sterically hindered than a secondary or tertiary carbon atom. The lower degree of substitution also encourages the second factor: A carbonium ion is unlikely to form at a primary carbon atom, excluding an S_N1 mechanism.

Direct displacement will also be favored if the atom which polarizes carbon is also capable of supporting a negative charge. If the atom, like bromine, can be lost as a stable negative ion, the incoming nucleophile will be able to displace it all the more readily. Halide ions are probably the most common leaving groups, but anions of alcohols (alkoxides), phenols (phenoxides), and carboxylic acids (carboxylates) are not at all unusual.

All the reactions presented in this chapter involve substitution at a primary carbon atom. Several of the examples are substitution reactions which occur at a benzylic ($ArCH_2$) carbon atom. Although when formed the benzyl cation is more stable than an ordinary primary carbonium ion, the reactions described here undoubtedly exhibit second-order kinetics. The formation of a dibenzyl ether (Sec. 13.4) is actually a two-step reaction, but each step occurs by the S_N2 mechanism.

13.1 SYNTHESIS AND HYDROLYSIS OF PHENYLACETONITRILE

A number of methods exist for the preparation of carboxylic acids. Among these are the carbonation of a Grignard reagent (Secs. 12.3 and 12.4), oxidation of an aldehyde (Sec. 25.7E), hydrolysis of an ester (Secs. 13.2 and 27.7B), and the hydrolysis of a nitrile, mentioned here and in Sec. 27.7H. The hydrolysis of nitriles is a particularly attractive synthesis of an acid because the nitriles are readily available from alkyl halides by a simple nucleophilic substitution.

The example described in Exp. 13.1 is particularly informative. Phenylacetic acid could be prepared by the oxidation of 2-phenylethanol, but care would have to be taken to avoid oxidation at the carbon bonded to the benzene ring. An alternative synthesis might be the carbonation of the Grignard reagent formed from benzyl chloride (see Chap. 12). This approach should afford the desired acid directly, but benzylmagnesium chloride is so prone to undergo coupling (to form 1,2-diphenylethane) that the Grignard approach is virtually useless for the formation of phenylacetic acid.

The cyanide ion is a good nucleophile but, as is characteristic of salts, functions best in aqueous solution. In the reaction described here, the phase-transfer catalytic method (Sec. 4.7) is used to facilitate this reaction. In this reaction, first benzyl chloride and then the product phenylacetonitrile serve as solvent for the reaction. The phase-transfer catalyst is tetrabutylammonium hydrogensulfate (although any other suitable catalyst may be used, including benzyltriethylammonium chloride, Sec. 13.6). This catalyst assists the dissolution of cyanide ion in the organic phase and enhances the reaction rate.

Once the nitrile is formed, it may be separated from the by-products by a combination of extraction and distillation. The pure nitrile may be hydrolyzed directly to phenylacetic acid. Hydrolysis of this nitrile may be effected by heating with either acid or base. In the hydrolysis described here, the addition of water across the nitrile group is catalyzed by acid. The addition of one molecule of water yields an amide, which is then further hydrolyzed to the carboxylic acid. The overall reaction is shown below.

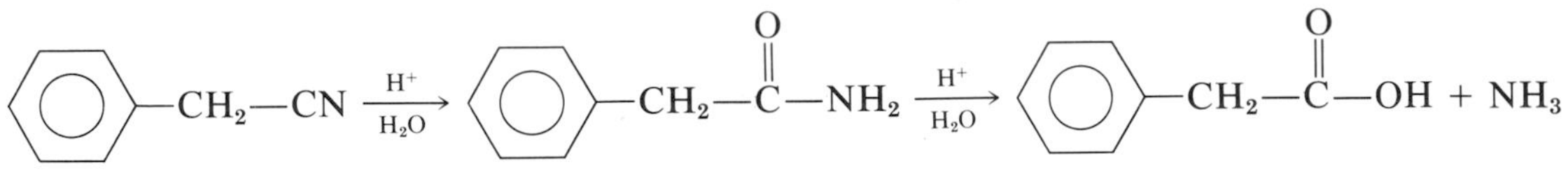

$$C_6H_5-CH_2-CN \xrightarrow[H_2O]{H^+} C_6H_5-CH_2-\overset{\overset{\displaystyle O}{\|}}{C}-NH_2 \xrightarrow[H_2O]{H^+} C_6H_5-CH_2-\overset{\overset{\displaystyle O}{\|}}{C}-OH + NH_3$$

Since the hydrolysis is conducted with acid, it is important to purify the nitrile before beginning this part of the sequence. Any cyanide ion which might remain will react with acid to form HCN gas. Hydrogen cyanide gas is the gas which has been used historically in prison gas chambers, so the danger of forming it should be obvious. All parts of this experiment should be conducted with care because of the hazard posed by the use of cyanide ion. The rules listed below should be carefully observed:

1 **Do not allow your skin or eyes to contact cyanide in any form.**
2 **Do not *under any circumstances* allow cyanide to contact acid.**
3 **Pour any residual cyanide-containing solution directly into the drain and flush it down with water.**
4 **Follow your laboratory instructor's directions explicitly during this experiment.**

There is always reason for concern about the potential danger of using cyanide ion in a large number of preparations carried out simultaneously. Some instructors may feel that because of their circumstances or facilities the preparation should not be conducted. In such situations, the hydrolysis step may be carried out independently on commercial phenylacetonitrile. If this is unavailable, the same preparation may be carried out on benzonitrile by adjusting the quantity of starting material used.

EXPERIMENT 13.1 SYNTHESIS OF PHENYLACETONITRILE FROM BENZYL CHLORIDE AND ITS HYDROLYSIS TO PHENYLACETIC ACID

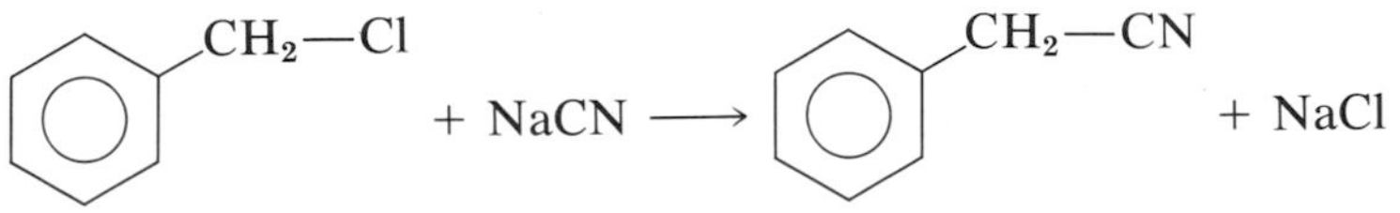

$$C_6H_5-CH_2-Cl + NaCN \longrightarrow C_6H_5-CH_2-CN + NaCl$$

$$C_6H_5CH_2{-}CN + H_2O \xrightarrow{H^+} C_6H_5CH_2{-}C(=O){-}OH + NH_3$$

Time 6 h (two laboratory periods)

Materials Benzyl chloride, 11.5 mL (MW 126.5, bp 177 to 181°C, d 1.1 g/mL)
Sodium cyanide, 7.5 g (MW 49)
Tetrabutylammonium hydrogensulfate, 1.5 g (MW 339)
Glacial acetic acid, 10 mL
Concentrated sulfuric acid, 10 mL
Ether, 50 mL

Precautions Carry out all operations in a good hood. Wear gloves during the entire first part of the reaction sequence (be especially careful when weighing sodium cyanide). Do not breath the cyanide dust and **clean up spills immediately.** Small amounts of cyanide waste should be flushed down the drain in a hood with lots of water. Do not allow any NaCN to contact acid (**HCN gas produced**).

Hazards Sodium cyanide is an extremely toxic and hazardous material. Sodium cyanide or its solutions may be absorbed through the skin. Exposure of cyanide to acid produces HCN, a toxic gas. Any dust on the skin should be washed off **immediately** with copious amounts of water. Benzyl chloride and phenylacetonitrile are lachrymators and skin irritants. Avoid breathing vapors or skin contact.

Experimental Procedure

Synthesis of phenylacetonitrile

Pour 11.5 mL benzyl chloride (0.1 mol) into a 250-mL Erlenmeyer flask (**hood, gloves**). To the liquid in the Erlenmeyer flask add 1.5 g (0.0045 mol) tetrabutylammonium hydrogensulfate.

Now, very carefully, weigh out 7.5 g (0.15 mol) sodium cyanide (**hood, gloves**) and add this material to the Erlenmeyer flask above (**in the hood**). Immediately after the addition of the sodium cyanide to the Erlenmeyer flask, add 15 mL distilled water to the flask and swirl it vigorously for several minutes to ensure mixing of all the phases.

Place the Erlenmeyer flask on a steam bath and heat, with occasional swirling, for 1 h. During this reaction period, the color of the organic layer will change from colorless to yellow to orange-red. During this same period a precipitate of sodium chloride will appear in the water layer.

After the 1-h heating period, remove the Erlenmeyer flask from the steam bath and allow it to cool **in the hood** until it reaches room temperature. Add to the Erlenmeyer flask 50 mL distilled water and swirl the flask. Transfer the entire mixture to a separatory funnel (**hood, gloves**) and rinse the Erlenmeyer flask with 30 mL ether. Transfer the ether to the separatory funnel and extract the aqueous layer. Draw off the lower aqueous cyanide layer from the separatory funnel and **immediately** flush it down the hood drain with copious amount of water. Extract the colored ether layer with three 25-mL portions of distilled water, separate the layers, and immediately flush each aqueous wash down the drain as above. Wash the ether layer with two 25-mL portions of saturated salt solution, again flushing each aqueous wash down the drain immediately. Transfer the ether layer to a 125-mL Erlenmeyer flask and dry for several minutes over anhydrous sodium sulfate.

Filter the ether layer away from the drying agent into a clean, dry 125-mL Erlenmeyer flask. Remove all the ether by evaporation on a steam bath (**hood**). Transfer the liquid remaining in the flask after removal of the ether to a 50-mL round-bottom boiling flask. Assemble an apparatus for simple vacuum distillation (see Fig. 2.7) and distill the dark liquid under aspirator vacuum. About 8 to 10 mL pure phenylacetonitrile (benzyl cyanide) is obtained after distillation as a clear, water-white liquid, bp 110 to 115°C at 15 torr. Gas chromatographic analysis on a 20% SE-30 column at 230°C shows only one product and none of the starting benzyl halide.

Hydrolysis

Prepare a solution for hydrolysis as follows. For each milliliter of phenylacetonitrile to be hydrolyzed, charge a 125-mL Erlenmeyer flask with 1 mL distilled water, followed by 1 mL concentrated sulfuric acid (98%) and then by 1 mL glacial acetic acid. If 10 mL phenylacetonitrile is to be hydrolyzed, the flask should contain 10 mL water, 10 mL concentrated sulfuric acid (98%), and 10 mL glacial acetic acid. Swirl the mixture (**caution: exotherm**) and transfer the hydrolysis solution to a 100-mL round-bottom flask. Add the phenylacetonitrile. Affix a reflux condenser (with lightly greased joints) to the round-bottom flask and reflux (flame or oil bath) for 1 h. Allow the reaction mixture to cool briefly and then pour it in a thin stream, with stirring, into a 600-mL beaker containing 100 mL distilled water. Stir the mixture vigorously (with a glass rod or spatula) while it is cooling, then filter the crude phenylacetic acid. Wash the solid with several small (10-mL) portions of water. The crude phenylacetic acid may be crystallized from hot water or petroleum ether. The yield of pure acid, mp 77°C, is approximately 50%.

If the crude acid appears to be contaminated with phenylacetamide, it may be purified as follows. Dissolve the crude acid in excess sodium carbonate solution (to convert the acid to the sodium salt), filter to remove any solid material, and reprecipitate the phenylacetic acid by cautiously acidifying with dilute sulfuric acid. If the acid is beige or brown, the aqueous sodium salt solution should be treated with decolorizing carbon to remove the colored impurities. Reprecipitation of the acid by cautious treatment with dilute acid should afford pure product.

The nmr and ir spectra of phenylacetonitrile and phenylacetic acid are shown in Figs. 13.1 and 13.2, respectively. Note how similar the nmr spectra are to each other. The ir spectra are much more informative. The cyanide band at 2260 cm^{-1} (Fig. 13.1) is not observed in Fig. 13.2, but carbonyl and hydroxyl vibrations are obvious.

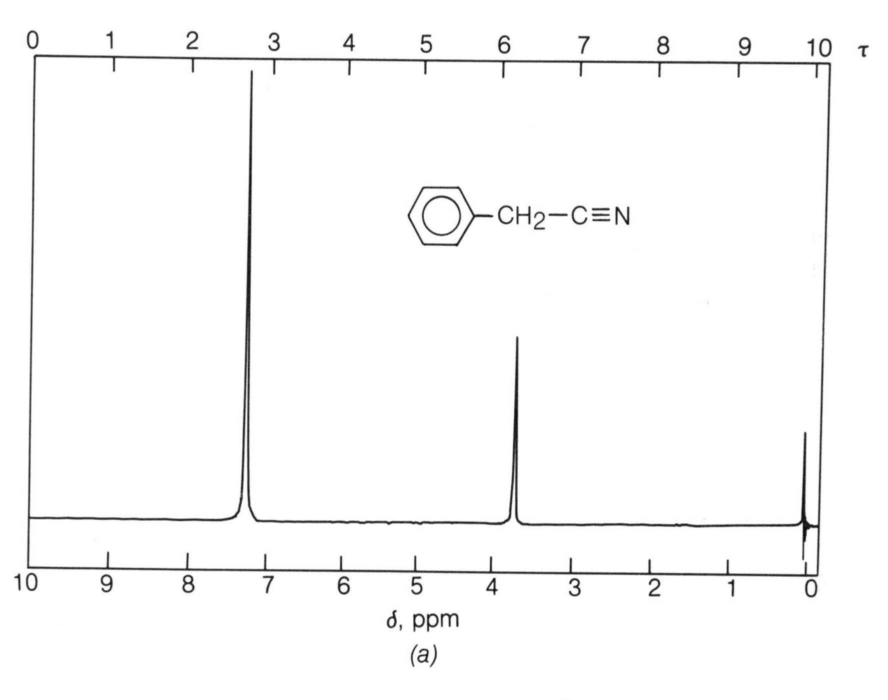

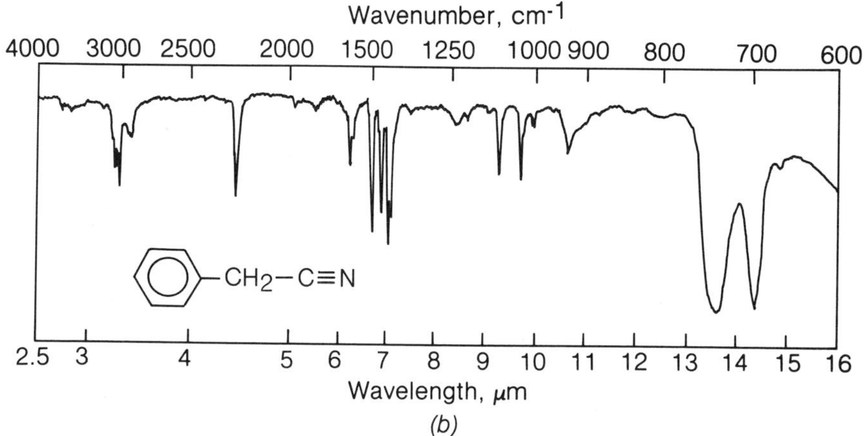

Figure 13.1
The (*a*) nmr and (*b*) ir spectra of phenylacetonitrile.

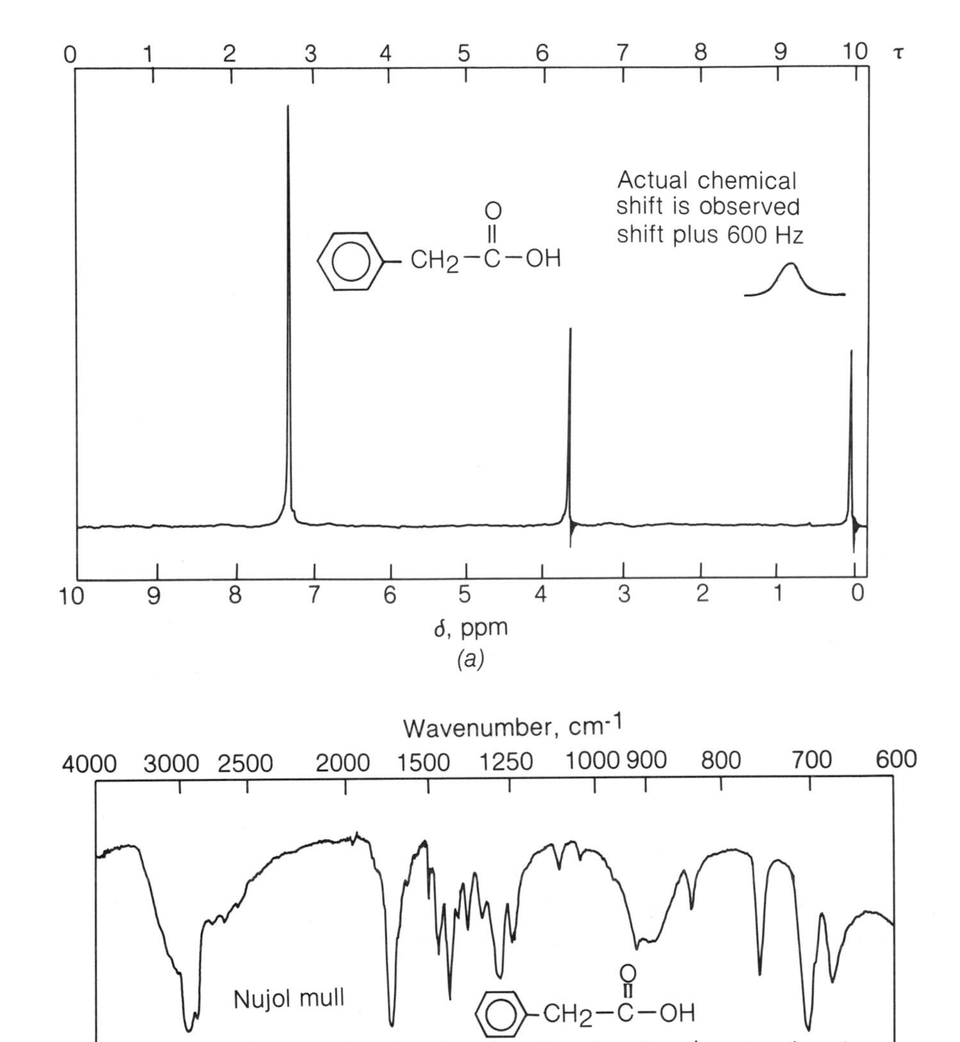

Figure 13.2
The (*a*) nmr and (*b*) ir spectra of phenylacetic acid.

13.2 SYNTHESIS AND HYDROLYSIS OF 4-CHLOROBENZYL ACETATE

Esters may be formed by a variety of methods. A number of these methods are discussed in other parts of this book, including Secs. 11.1 through 11.3, 23.6, 23.7, 26.6B, and 27.7C. The most common method for the synthesis of esters is the Fischer esterification, in which an excess of alcohol is heated with the acid in the presence of an acid catalyst. An alternative method is illustrated for the synthesis of *n*-butyl benzoate in Sec. 11.1. The latter example is similar to the present one, but the product of that reaction is intended to be used in subsequent Grignard reactions rather than in the formation of an alcohol. Nevertheless, Sec. 11.1 should be consulted for additional information.

The reaction sequence described here is for the synthesis of an ester, which is then hydrolyzed to a carboxylic acid and the corresponding alcohol. Sometimes this apparently involved route is the most efficient method for the synthesis of an alcohol from a halide, because direct hydrolysis may yield an ether. In this case, the acetate which is formed cannot react further until sodium hydroxide is added. If direct hydrolysis of the chloride were attempted, both starting chloride and alcohol (or alkoxide) would be present in the reaction mixture. It would not be surprising if the two reacted to yield an ether. This ether is exactly the one which is prepared in Sec. 13.4, and the alcohol prepared here is intended to be used in that experiment.

The advantage of this experimental approach is that by-products are minimized in the formation of both the ester and the alcohol. In a sense, the acetate group serves to protect the benzyl carbon from other reactions, and as such may be considered a "protecting group."

In this experiment, the phase-transfer catalytic technique (Sec. 4.7) is used, so that acetate ion and not hydroxide ion or water will be the nucleophile. Some of the advantages of ester formation by this technique are discussed in Sec. 11.1.

If both parts of this experiment will not be run, but the hydrolysis of an ester is to be carried out, commercially available benzyl acetate may be substituted for 4-chlorobenzyl acetate, and the hydrolysis procedure described in Exp. 13.2 may be used as written.

EXPERIMENT 13.2 SYNTHESIS OF 4-CHLOROBENZYL ACETATE FROM 4-CHLOROBENZYL CHLORIDE AND ITS HYDROLYSIS TO 4-CHLOROBENZYL ALCOHOL

$$4\text{-}Cl\text{-}C_6H_4CH_2Cl + NaOAc \longrightarrow 4\text{-}Cl\text{-}C_6H_4CH_2OAc$$

$$4\text{-}Cl\text{-}C_6H_4CH_2OAc + NaOH \longrightarrow 4\text{-}Cl\text{-}C_6H_4CH_2OH$$

Time 3.0 h

Materials 4-Chlorobenzyl chloride, 10 mL (MW 161, mp 28 to 30°C)
Aliquat 336 or other phase-transfer catalyst, 1 g
Sodium acetate trihydrate, 25 g
Sodium hydroxide, 10 g (MW 40)

Precautions Carry out this procedure in a good hood. Wear gloves when transferring benzyl chloride.

Hazards 4-Chlorobenzyl chloride is a lachrymator and skin irritant. Avoid breathing vapors or skin contact.

Experimental Procedure

Charge a 100-mL round-bottom flask with 1 g tricaprylylmethylammonium chloride (Aliquat 336, Starks' catalyst). Add 10 mL (12.5 g, 0.078 mol) 4-chlorobenzyl chloride to the flask, followed by 25 g (0.184 mol) solid sodium acetate trihydrate and several boiling chips. Attach a reflux condenser (with lightly greased joints) to the flask (Fig. 9.1) and heat the mixture to boiling over a free flame or oil bath for approximately 45 min.

As heat is applied, the solid sodium acetate gradually liquefies to form a separate lower layer in the reaction vessel. As the reaction proceeds, sodium chloride gradually precipitates from solution.

After the reflux period is over, cool the flask and its contents to room temperature; then add 30 mL distilled water to the mixture. Transfer the mixture to a separatory funnel, draw off and discard the aqueous layer (upper or lower?). Add 30 mL 95% ethanol to the oil that remains in the separatory funnel. Swirl the separatory funnel to be certain all the oil is dissolved, and then run the solution back into the original 100-mL round-bottom flask. Add to this ethanol solution of crude ester a solution of aqueous sodium hydroxide (made by dissolving 10 g sodium hydroxide in 25 mL water) and reflux this new mixture, as before, for 35 to 40 min. As the hydrolysis begins, the reaction mixture will appear light yellow but will gradually turn rather dark.

After the reflux period is over, allow the reaction mixture to cool as before, add 50 mL water, and transfer the entire mixture to a separatory funnel. Extract the mixture with 35 mL dichloromethane. Separate the organic layer (note the density of dichloromethane relative to water); then wash the organic layer with distilled water (15 mL), followed by 5% aqueous hydrochloric acid (15 mL), and dry over granular sodium sulfate. After drying for about 5 min, filter the solution away from the drying agent and remove the solvent by evaporation on a steam bath. (*Note:* Ethyl alcohol is extracted along with the benzyl alcohol. Most of the ethanol should be removed during the evaporation of the solvent on the steam bath.)

The residue which remains after evaporation of the solvent should be crystallized from a mixture of 5% acetone-hexane (about 20 mL).[1] After cooling, collect the crystallized material on a Buchner funnel by suction filtration. The product obtained from the recrystallization should appear as long white needles with mp 71 to 73°C. After air drying, the yield should be approximately 9 to 10 g.

[1] 1 mL acetone + 19 mL hexane.

The nmr and ir spectra of 4-chlorobenzyl acetate and 4-chlorobenzyl alcohol are shown in Figs. 13.3 and 13.4, respectively. The nmr spectrum of the alcohol (Fig. 13.4*a*) shows only benzylic protons, the hydroxyl group, and the aromatic ring protons. Because chlorine and hydroxymethyl have a very similar influence on the chemical shift of the aromatic protons, only a broad singlet is observed in the aromatic region of the spectrum.

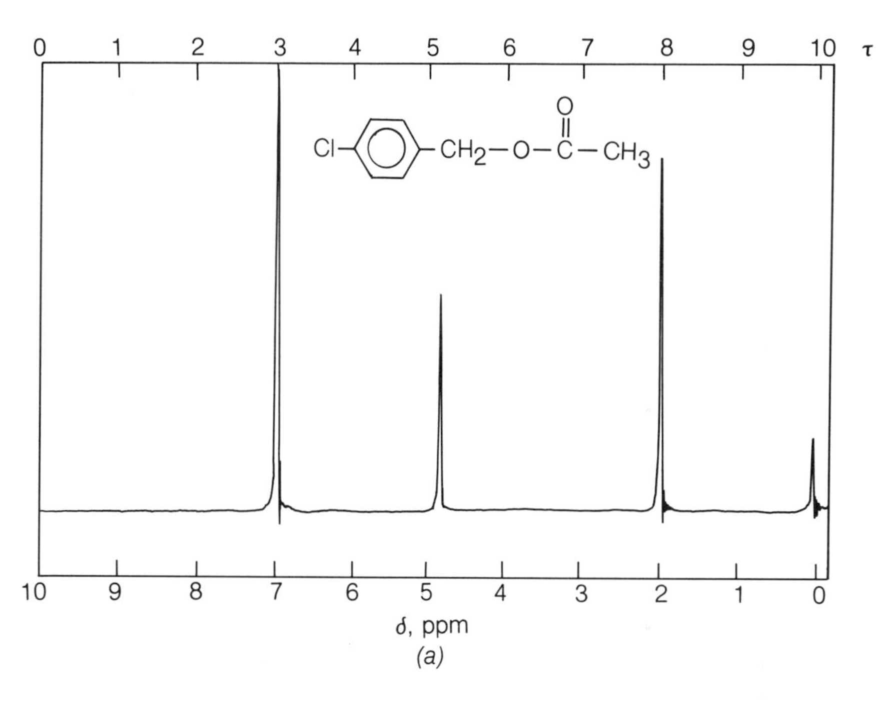

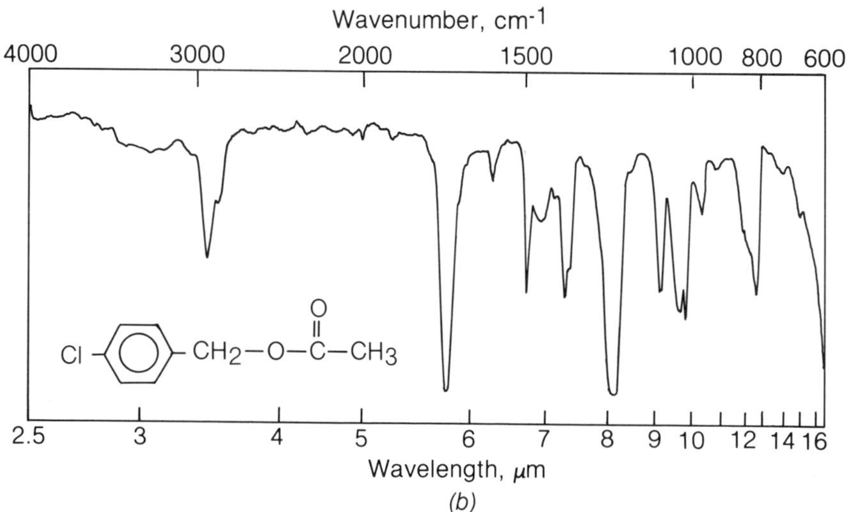

Figure 13.3
The (*a*) nmr and (*b*) ir spectra of 4-chlorobenzyl acetate.

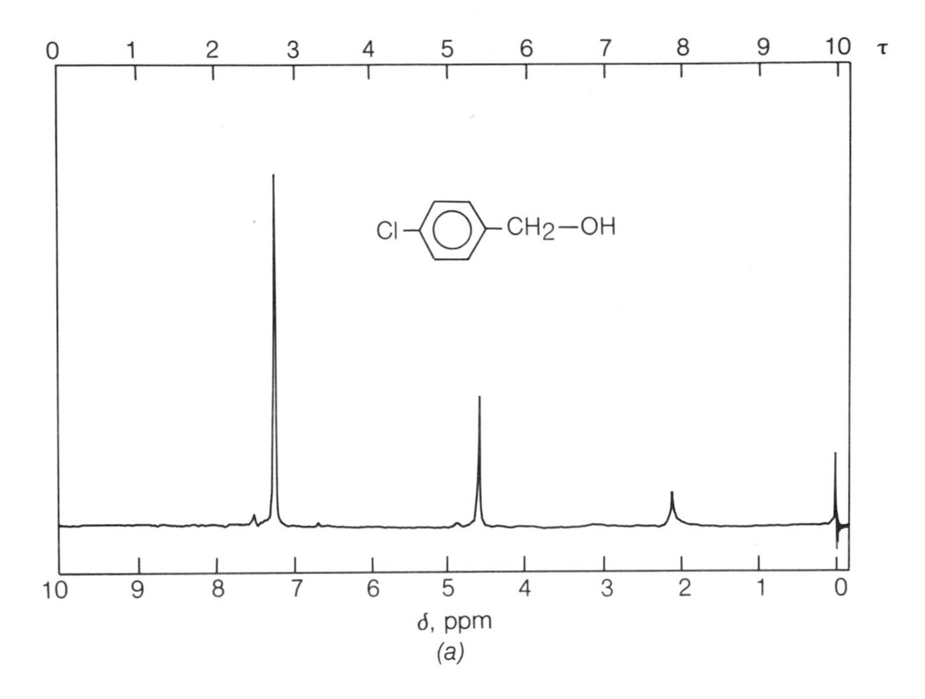

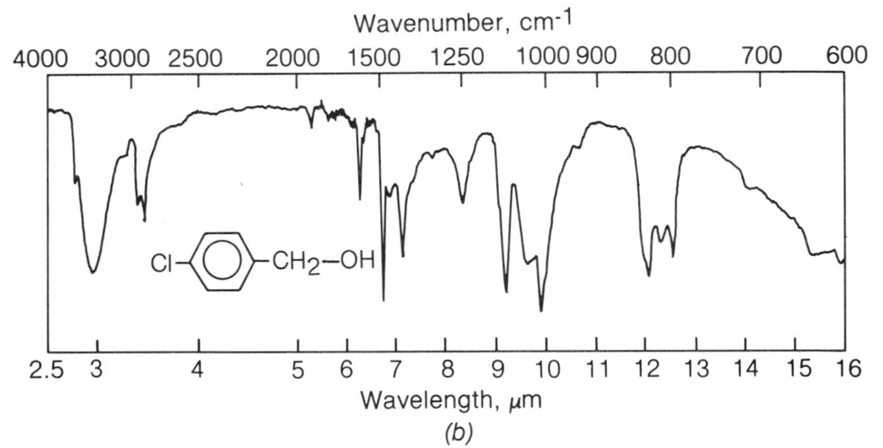

Figure 13.4
The (*a*) nmr and (*b*) ir spectra of 4-chlorobenzyl alcohol.

13.3 SYNTHESIS OF 4-METHYLPHENOXYACETIC ACID AND 2,4-DICHLOROPHENOXYACETIC ACID

Phenols are more acidic than alcohols by several orders of magnitude. This is because phenoxide anions are more stable than alkoxide ions (see Secs. 23.3 and 23.4). Because of the greater acidity of phenols, their anions may be generated by treatment with hydroxide ion without having to resort to a much stronger base, such as sodium methoxide. Once formed, the phenoxide anion is quite nucleophilic and participates in a variety of nucleophilic substitution reactions. An example which demonstrates how easily phenol can be made to react is found in the formation of phenyl methyl ether (anisole). In this reaction, phenol is dissolved in aqueous sodium hydroxide solution and the strong methylating agent dimethyl sulfate [$(CH_3)_2SO_4$] is added. After only a few minutes, the desired ether may be isolated by distillation.

The two reactions described in this section are similar to the simple example mentioned above. There are some differences, however. If a phenoxide anion added to chloroacetic acid, the net reaction would be nothing more than protonation of phenoxide by the acetic acid derivative, as shown below.

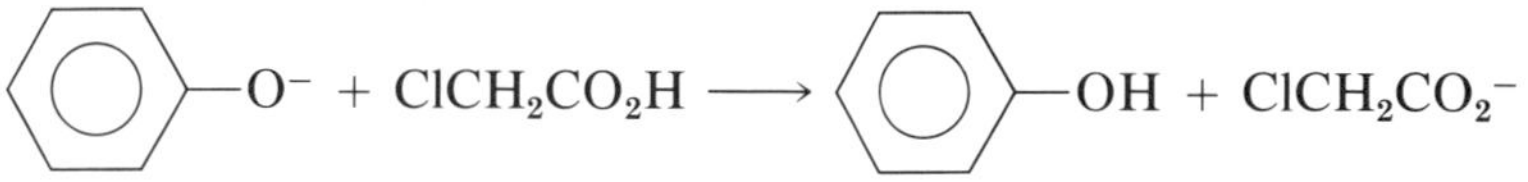

$$C_6H_5\text{—}O^- + ClCH_2CO_2H \longrightarrow C_6H_5\text{—}OH + ClCH_2CO_2^-$$

In doing this reaction, enough base must be added so that both the phenol and chloroacetic acid are present as their anions. When the two are brought together, a rapid nucleophilic substitution takes place to yield a class of compounds known as *aryloxyacetic acids*.

Many phenols are low-melting solids which are difficult to obtain as pure crystals for characterization. Over the years, the formation of an aryloxyacetic acid derivative has facilitated characterization of phenols because the former compounds are almost always crystalline solids. More information concerning this application can be obtained by referring to Sec. 23.7B.

The experiments which are described in this section are relatively simple reactions. In each case, a substituted phenoxyacetic acid derivative is formed by a simple S_N2 reaction. The aryloxyacetate salt which is obtained from the reaction mixture is protonated and then recrystallized to purity. Such reactions are also easy to run on a large scale. The compound prepared in Exp. 13.3B, 2,4-dichlorophenoxyacetic acid (2,4-D), is prepared commercially on a rather large scale. It was found some years ago that 3-indoleacetic acid (IAA) is a plant-growth-regulating hormone. The structurally similar 2,4-dichlorophenoxyacetic acid was found to have an herbicidal effect on broad-leaved plants but not on most grasses. As a result of this biological activity, 2,4-D has been used as a defoliating agent.

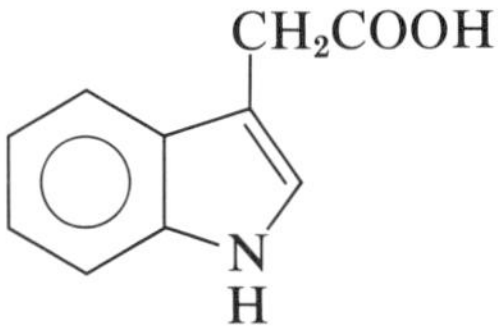

CH_2COOH (indole ring, N–H)

IAA

Cl (2-position), Cl (4-position), OCH_2COOH

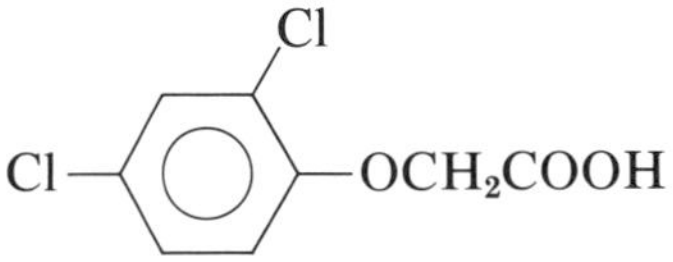

2,4-D

EXPERIMENT 13.3A PREPARATION OF 4-METHYLPHENOXYACETIC ACID FROM 4-METHYLPHENOL

$$4\text{-}H_3C\text{-}C_6H_4\text{-}OH + ClCH_2\text{-}C(=O)\text{-}OH \xrightarrow{KOH} 4\text{-}H_3C\text{-}C_6H_4\text{-}O\text{-}CH_2\text{-}C(=O)\text{-}OH$$

Time 3 h

Materials 4-Methylphenol (*p*-cresol), 5 g (MW 108, mp 32 to 34°C, bp 202°C)
50% 2-Chloroacetic acid, 14 mL (MW 94.5)
Potassium hydroxide, 10 g (MW 56)
Ether, 100 mL
5% Potassium carbonate, 75 to 100 mL

Precautions Wear gloves when weighing phenol. Carry out the entire procedure in a hood.

Hazards Both the phenol and the acid are skin irritants. Avoid contact whenever possible. Ether is flammable.

Experimental Procedure

Add 10 g potassium hydroxide (0.180 mol) to a 125-mL Erlenmeyer flask. Add 20 mL distilled water to the potassium hydroxide and swirl the solution vigorously until the hydroxide pellets dissolve.

When the aqueous solution of hydroxide has cooled somewhat, add all the base solution to a 125-mL Erlenmeyer flask containing 5 g 4-methylphenol (*p*-cresol). Swirl the solution vigorously until all the phenol has dissolved in the aqueous base. A clear aqueous solution should result.

Add the aqueous solution of the phenol through a plastic funnel to a 250-mL round-bottom flask. (*Note:* Be careful to avoid contact between the ground glass joint and the basic solution.) Place a reflux condenser (with lightly greased joints) in the top of the round-bottom flask and assemble the apparatus for reflux (see Fig. 9.1). Before heating is commenced, pour 7 mL of a 50% aqueous 2-chloroacetic acid solution slowly down the top of the reflux condenser into the aqueous base solution. After this has been done, add several boiling chips to the flask and commence heating so that the aqueous solution refluxes at a moderate rate. Reflux the mixture for 15 min and then add an additional 7 mL 2-chloroacetic acid *slowly* through the top of the reflux condenser. After this addition is complete, reflux the mixture for another 15 min.

After the reflux period, pour the warm aqueous mixture into a 250-mL Erlenmeyer flask containing 50 g cracked ice. Use small portions of distilled water to rinse out the round-bottom flask and add the wash to the flask above. Acidify the entire solution by slowly adding concentrated hydrochloric acid to the Erlenmeyer flask (use approximately 10 mL), and cool the mixture in an ice bath (if needed) to return it to room temperature. Once the acidification is completed (pH less than 2), transfer the mixture to a 250-mL separatory funnel. Extract the solution with 50 mL ether, separate the phases, and extract the aqueous layer with another 25 mL ether. Combine the ether layers in the separatory funnel and wash with 25 mL distilled water.

At this point the ether solution contains phenol and the phenoxyacetic acid. To separate the two compounds, extract the ether solution with 50 mL 5% potassium carbonate solution. (*Note:* A solid material may form in the separatory funnel at this point. If this happens, add more distilled water and potassium carbonate solution.) Extract the ether solution again with 25 mL fresh potassium carbonate solution and combine the aqueous extracts. Wash this aqueous solution with a 25-mL portion of fresh ether. Transfer the aqueous solution to a 250-mL Erlenmeyer flask.

Acidify the aqueous solution *slowly* (carbon dioxide is given off with much effervescence) with concentrated hydrochloric acid (about 5 to 7 mL). After the acidification, a solid residue should remain in the aqueous solution. Cool this solution in an ice bath and filter using a Buchner funnel. Wash the solid residue with small portions of cold water and then air dry. After air drying, a solid of mp 135 to 136°C should be obtained in approximately 70% yield.

This crude material (4-methylphenoxyacetic acid) is easily recrystallized

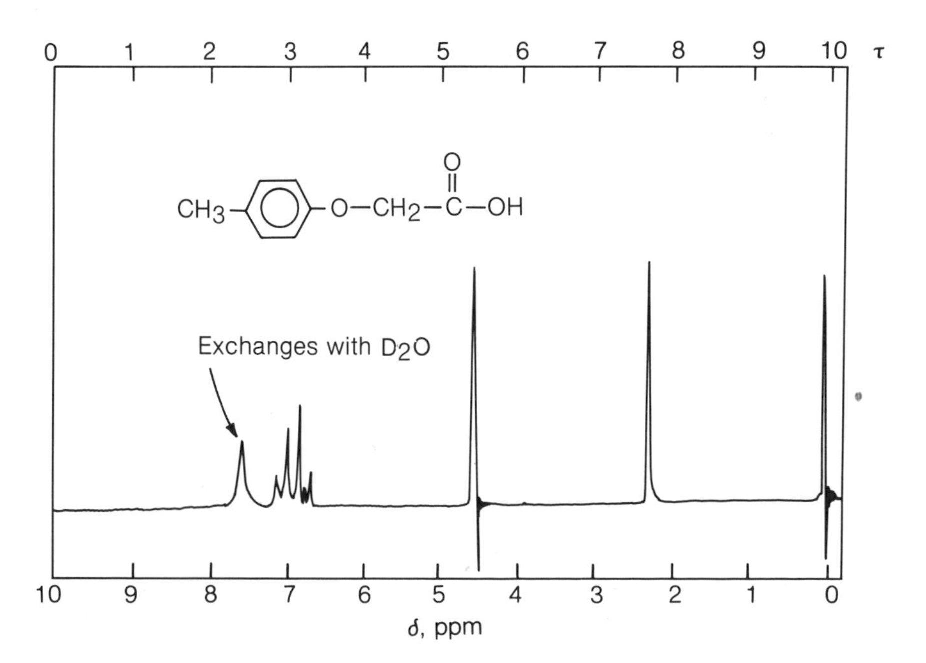

Figure 13.5 The nmr spectrum of 4-methylphenoxyacetic acid.

from methanol-water or ethanol-water with the aid of a few seed crystals. The melting point after one recrystallization is 136 to 137°C.

The nmr spectrum of *p*-methylphenoxyacetic acid is shown in Fig. 13.5. Compare this spectrum with that shown in Fig. 13.3 to determine which of the upfield singlets is due to the methyl group. Check your assignment by comparing with Fig. 13.6. The most downfield resonance is due to the hydroxyl group and some water which was not removed during purification.

EXPERIMENT 13.3B

SYNTHESIS OF 2,4-DICHLOROPHENOXYACETIC ACID FROM 2,4-DICHLOROPHENOL

$$2,4\text{-}Cl_2C_6H_3OH + ClCH_2COOH \xrightarrow{KOH} 2,4\text{-}Cl_2C_6H_3O{-}CH_2{-}COOH$$

Time 3 h

Materials 2,4-Dichlorophenol, 5 g (MW 163, mp 42 to 44°C)
50% 2-Chloroacetic acid, 14 mL (MW 94.5)
Potassium hydroxide, 10 g (MW 56)
Ether, 100 mL
5% Potassium carbonate, 75 to 100 mL

Precautions Wear gloves when weighing phenol. Conduct the entire procedure in the hood.

Hazards Both the phenol and the acid are skin irritants. Avoid contact whenever possible. Ether is flammable; extinguish flames before extraction.

Experimental Procedure

Add 10 g potassium hydroxide (0.180 mol) to a 125-mL Erlenmeyer flask. Add 20 mL distilled water to the potassium hydroxide pellets and swirl the solution vigorously as the hydroxide pellets dissolve.

When the aqueous solution has cooled somewhat, add 5 g 2,4-dichlorophenol all at once to the Erlenmeyer flask. Swirl the solution vigorously until all the phenol has dissolved in the aqueous base solution. A clear aqueous solution should result.

Add the aqueous solution of the phenol through a plastic funnel to a 250-mL round-bottom flask. (*Note:* Be careful to avoid contact between the ground glass joint and the basic solution.) Place a reflux condenser (with lightly greased joints) in the top of the round-bottom flask and assemble the apparatus for reflux (see Fig. 9.1). Before heating is commenced, pour 7 mL of a 50% aqueous 2-chloroacetic acid solution slowly down the top of the reflux condenser into the aqueous base solution. After this has been done, add several boiling chips to the flask and commence heating (free flame) so that the aqueous solution refluxes at a moderate rate. Reflux the mixture for 20 min and then add an additional 7 mL 2-chloroacetic acid *slowly* through the top of the reflux condenser. After this addition is complete, reflux the mixture for another 20 min.

After the reflux period, pour the warm aqueous mixture into a 250-mL Erlenmeyer flask over 50 g cracked ice. Use small portions of distilled water to rinse out the round-bottom flask and add the wash water to the flask above. Acidify the entire solution by slowly adding concentrated hydrochloric acid to the Erlenmeyer flask (use approximately 10 mL) and cool the mixture in an ice bath (if needed) to return it to room temperature. Once the acidification is completed (pH less than 2), transfer the mixture to a 250-mL separatory funnel. Extract the solution with 50 mL ether, separate the phases, and extract the aqueous layer with another 25 mL ether. Combine the ether layers in the separatory funnel and wash with 25 mL distilled water.

At this point the ether solution contains phenol and the phenoxyacetic acid. To separate the two compounds, the ether solution is extracted with 50 mL 5% potassium carbonate solution. (*Note:* A solid material may form in the separatory funnel at this point. If this happens, add more distilled water and potassium carbonate solution.) Extract the ether solution again with 25 mL

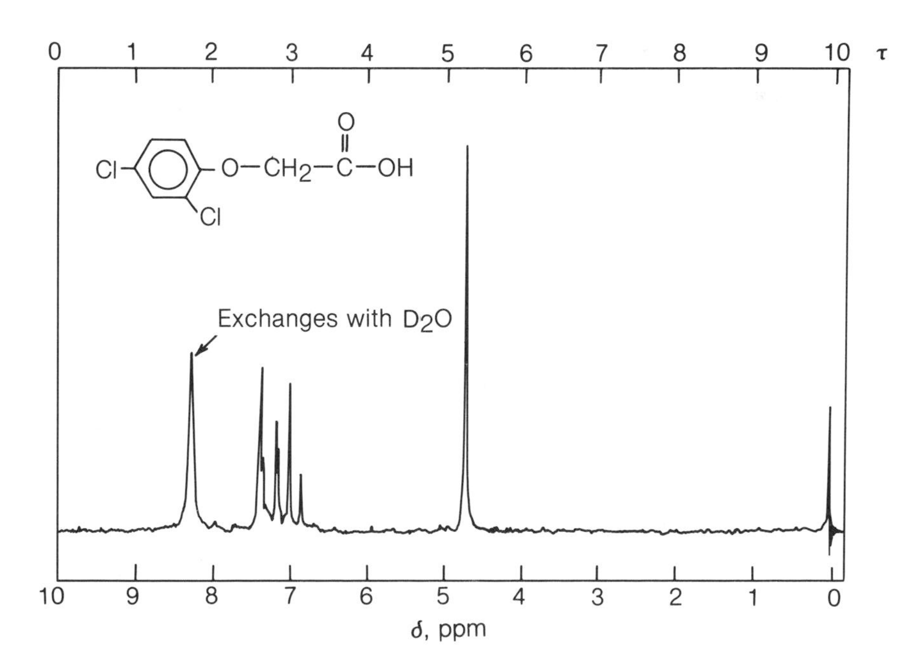

Figure 13.6
The nmr spectrum of 2,4-dichlorophenoxyacetic acid.

fresh potassium carbonate solution and combine the aqueous extracts. Wash this aqueous solution with a 25-mL portion of ether. Transfer the aqueous solution to a 250-mL Erlenmeyer flask.

Acidify the aqueous solution *slowly* (carbon dioxide is given off with much effervescence) with concentrated hydrochloric acid (about 5 to 7 mL). After the acidification, a solid residue should remain in the aqueous solution. Cool this solution in an ice bath and filter, using a Buchner funnel. Wash the solid residue with small portions of cold water and then air dry. After air drying, a solid of mp 135 to 138°C should be obtained in approximately 65% yield.

This crude material is easily recrystallized from methanol-water or ethanol-water with the aid of a few seed crystals. The melting point after one recrystallization is 136 to 138°C. The nmr spectrum of 2,4-dichlorophenoxyacetic acid is shown in Fig. 13.6. Compare the aromatic region in this spectrum with that shown in Fig. 13.5. Why is this spectrum more complex?

13.4 PREPARATION OF BIS-4-CHLOROBENZYL ETHER BY THE WILLIAMSON ETHER SYNTHESIS

There are two principal methods for the formation of ethers. The first involves the partial dehydration of two moles of alcohol. In this reaction, both halves of the ether arise from the same precursor, and the ether which is formed must necessarily be symmetrical. This reaction is illustrated below:

$$R—OH + R—OH \xrightarrow{H^+} R—O—R + H_2O$$

The second general approach to ether synthesis involves the so-called Williamson reaction. This reaction occurs by nucleophilic substitution rather than by the dehydration scheme shown above. Because separate nucleophiles and electrophiles are used in this reaction, it is possible to form either symmetrical ethers as above or unsymmetrical ethers which cannot be prepared by the dehydration approach. The Williamson ether reaction is characterized by the general equation shown below.

$$R—OH + R'—X \xrightarrow{\text{base}} R—O—R' + X^-$$

In the equation, X is intended to represent any leaving group, such as chloride or bromide. The bases used most often to deprotonate the alcohol are very strong ones, such as potassium *tert*-butoxide.

In the procedure given here, use of a very strong base is avoided by application of the phase-transfer catalytic technique described in Sec. 4.7. Sodium hydroxide is a sufficiently strong base for this reaction to occur because the small amount of alkoxide formed by reaction with sodium hydroxide dissolves quickly in the organic phase, where reaction occurs with great speed. The principal advantage of the phase-transfer technique in this reaction is that the reaction can be run without resorting to strong bases or long reaction times.

The formation of bis-4-chlorobenzyl ether is included in this book for several reasons. First, the starting material, 4-chlorobenzyl chloride, is inexpensive and readily available. Second, the alcohol component is also inexpensive and it may, if desired, be prepared as described in Exp. 13.2. Finally, the product is a very stable ether and is easy to isolate. Most dialkyl ethers are liquids and undergo air oxidation to form dangerously explosive peroxides. The solid ethers are much less prone to do so, since less surface area at which reaction can occur is exposed in a solid.

EXPERIMENT 13.4 PREPARATION OF BIS-4-CHLOROBENZYL ETHER BY THE WILLIAMSON ETHER SYNTHESIS

$$4\text{-}ClC_6H_4CH_2Cl + 4\text{-}ClC_6H_4CH_2OH + NaOH \longrightarrow 4\text{-}ClC_6H_4CH_2\text{-}O\text{-}CH_2C_6H_4\text{-}4\text{-}Cl$$

Time 2.5 h

Materials 4-Chlorobenzyl chloride, 3.3 g (MW 161, mp 28 to 30°C)
4-Chlorobenzyl alcohol, 2.8 g (MW 142, mp 70 to 72°C)
Phase-transfer catalyst, 0.5 g
30% Sodium hydroxide, 10 mL

Precautions Do the entire experiment in a good hood.

Hazards 4-Chlorobenzyl chloride is a lachrymator and skin irritant. Avoid breathing its vapors or skin contact.

Experimental Procedure

Place approximately 0.5 g tricaprylylmethylammonium chloride (0.5 mL, Aliquat 336, Starks' catalyst), in a 100-mL round-bottom boiling flask. Add to the flask 3.3 g 4-chlorobenzyl chloride (0.021 mol), followed by 2.8 g 4-chlorobenzyl alcohol (0.020 mol), 10 mL aqueous sodium hydroxide solution (30%), and two boiling chips. Attach a lightly greased reflux condenser and heat the mixture for 1 h using a free flame (see Fig. 9.1). After the reflux period is complete, cool the solution nearly to room temperature. Transfer the mixture to a separatory funnel and add 20 to 30 mL water. Add 30 mL dichloromethane, swirl gently, and draw off the aqueous layer. Wash the organic layer with water (10 mL), followed by 10% aqueous hydrochloric acid (10 mL) and two 10-mL portions of half-saturated sodium chloride solution, and

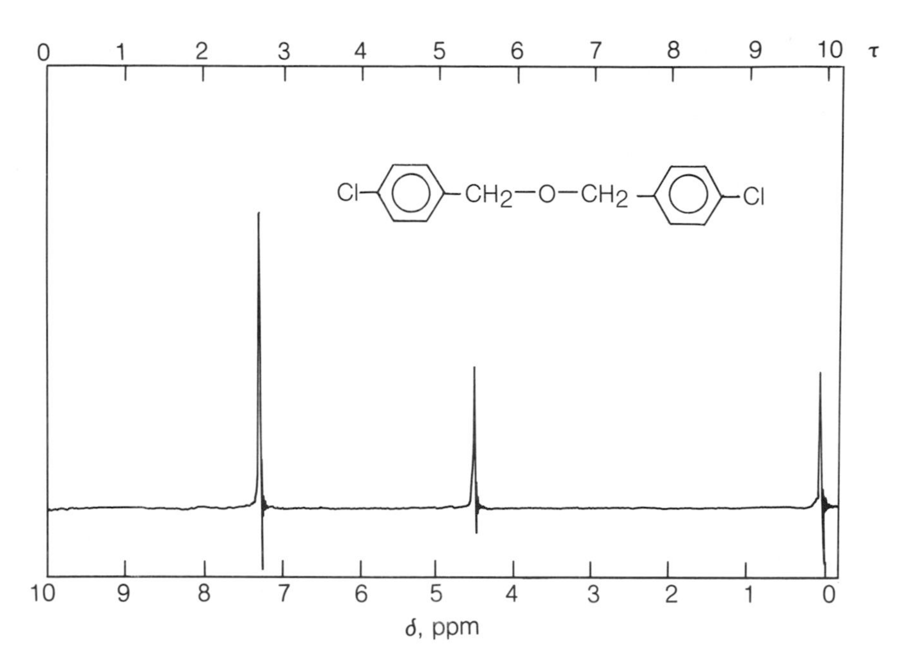

Figure 13.7
The nmr spectrum of bis-4-chlorobenzyl ether.

finally dry it over sodium sulfate. After the organic layer has stood for 5 to 10 min in contact with sodium sulfate, filter it to remove the drying agent and evaporate the dichloromethane by heating on a steam bath.

Dissolve the oil which remains in 20 mL hexane, warm the solution on a steam bath, and then allow it to cool to room temperature. Place the solution in an ice bath for a few minutes, at which time white needles should begin to separate. Filter this crystalline material using a Buchner funnel. Air dry the product for several minutes. 4,4′-Dichlorodibenzyl ether is obtained in a yield of approximately 3.5 g (about 60%) and has a 52 to 54°C mp. Its very simple nmr spectrum is shown in Fig. 13.7.

Careful concentration of the mother liquors to approximately half the original volume will yield an additional 0.5 g of material.

An alternative purification procedure is to crystallize the crude oil from methanol. The yield of the first crop of crystals in this case is very similar to that obtained in hexane.

13.5 THE MALONIC ESTER AND ACETOACETIC ESTER CONDENSATIONS

Many of the reactions presented in this manual are functional group transformations of the type alcohol to ether, alcohol to halide, or alcohol to acid. These reactions are important and are widely used in synthetic organic chemistry. Nevertheless, it is usually the formation of carbon-carbon bonds which is of paramount importance to synthetic organic chemists. For this purpose, the key functional group in the synthetic chemist's arsenal of methods is the carbonyl group.

A proton on a carbon atom adjacent to a carbonyl group is acidified by the adjacent electronegative group. The anion which results is stabilized by the formation of an enolate ion. There are two resonance forms of this enolate ion as shown below.

$$-CH_2-\overset{\overset{O}{\|}}{C}- \underset{H^+}{\overset{\text{base}}{\rightleftharpoons}} -\overset{-}{C}H-\overset{\overset{O}{\|}}{C}- \longleftrightarrow -CH=\overset{\overset{O^-}{|}}{C}- \underset{\text{base}}{\overset{H^+}{\rightleftharpoons}} -CH=\overset{\overset{OH}{|}}{C}-$$

When there are two carbonyl groups next to a $-CH_2-$ (methylene) group, the proton is lost even more easily, because the carbanion will be stabilized by enolate ion formation involving not one, but two carbonyl groups. The possible structures for malonic ester enolate are shown below.

$$EtO-\overset{\overset{O}{\|}}{C}-CH_2-\overset{\overset{O}{\|}}{C}-OEt \xrightarrow{\text{base}} EtO-\overset{\overset{O}{\|}}{C}-\overset{-}{C}H-\overset{\overset{O}{\|}}{C}-OEt \longleftrightarrow$$

$$EtO-\overset{\overset{O^-}{|}}{C}=CH-\overset{\overset{O}{\|}}{C}-OEt \longleftrightarrow EtO-\overset{\overset{O}{\|}}{C}-CH=\overset{\overset{O^-}{|}}{C}-OEt$$

It is a direct consequence of this additional stabilization that the pK_a of malonic and acetoacetic esters is well below that of simple ketones.

When either acetoacetic or malonic ester is treated with base, the enolate ion is formed. In the presence of an electrophile such as *n*-butyl bromide (Sec. 10.1), the enolate ion reacts as a nucleophile to displace the leaving group and form a carbon-carbon bond. In either case, the *n*-butyl group ends up substituted on the methylene group and flanked by carbonyl groups.

$$CH_3-\overset{\overset{O}{\|}}{C}-CH_2-\overset{\overset{O}{\|}}{C}-OEt \xrightarrow{\text{base}} CH_3-\overset{\overset{O}{\|}}{C}-\overset{-}{C}H-\overset{\overset{O}{\|}}{C}-OEt \xrightarrow{n\text{-}C_4H_9Br} CH_3-\overset{\overset{O}{\|}}{C}-\underset{\underset{C_4H_9}{|}}{C}H-\overset{\overset{O}{\|}}{C}-OEt$$

$$EtO-\overset{\overset{O}{\|}}{C}-CH_2-\overset{\overset{O}{\|}}{C}-OEt \xrightarrow{\text{base}} EtO-\overset{\overset{O}{\|}}{C}-\overset{-}{C}H-\overset{\overset{O}{\|}}{C}-OEt \xrightarrow{n\text{-}C_4H_9Br} EtO-\overset{\overset{O}{\|}}{C}-\underset{\underset{C_4H_9}{|}}{C}H-\overset{\overset{O}{\|}}{C}-OEt$$

The importance of the malonic ester condensation lies in the fact that an acetic acid derivative may be formed from it. The direct alkylation of acetic acid requires very strong bases, and technology appropriate to ensure the success of this reaction has only recently been developed. Historically, advantage has been taken of the effect of the two carbonyl groups in acidifying the methylene group and the subsequent removal of one of them by a hydrolysis-decarboxylation scheme, as shown in the equations below. Decarboxylation of acetoacetic ester yields an acetone derivative.

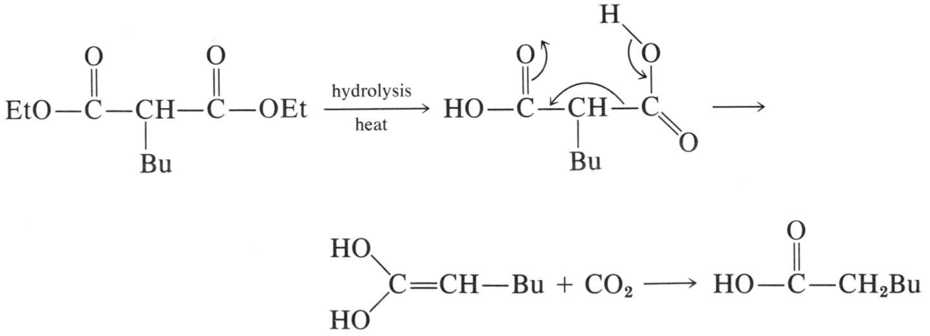

$$\mathrm{EtO{-}C({=}O){-}CH(Bu){-}C({=}O){-}OEt \xrightarrow[\text{heat}]{\text{hydrolysis}} HO{-}C({=}O){-}CH(Bu){-}C({=}O){-}O{-}H} \longrightarrow$$

$$\mathrm{(HO)_2C{=}CH{-}Bu + CO_2 \longrightarrow HO{-}C({=}O){-}CH_2Bu}$$

An important advantage of both malonic ester and acetoacetic ester is that these starting materials are readily available. Acetoacetic ester is formed from ethyl acetate by the so-called Claisen condensation and malonic acid is prepared from chloroacetic acid, as shown below.

$$\mathrm{2CH_3{-}CO{-}OEt \xrightarrow{Na} \underset{\text{Acetoacetic ester}}{CH_3{-}C({=}O){-}CH_2{-}C({=}O){-}OEt} + NaOEt}$$

$$\mathrm{NC^- + Cl{-}CH_2{-}CO_2^- \longrightarrow NC{-}CH_2{-}CO_2^- + Cl^- \xrightarrow{H_3O^+} \underset{\text{Malonic acid}}{HOOC{-}CH_2{-}COOH}}$$

EXPERIMENT 13.5A SYNTHESIS OF ETHYL n-BUTYLACETOACETATE BY THE ACETOACETIC ESTER CONDENSATION

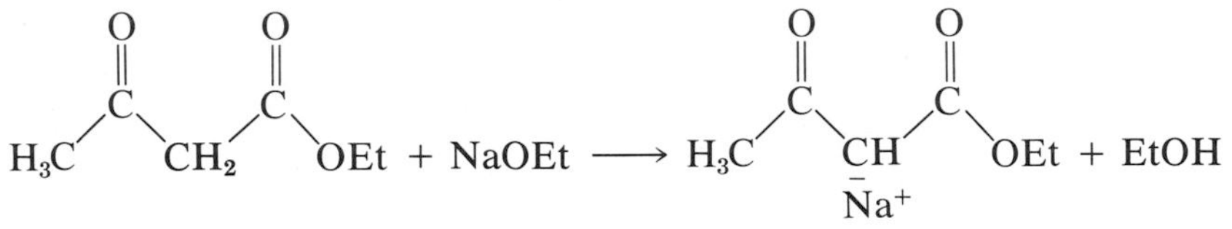

$$\mathrm{H_3C{-}C({=}O){-}CH_2{-}C({=}O){-}OEt + NaOEt \longrightarrow H_3C{-}C({=}O){-}\overset{-}{C}H(Na^+){-}C({=}O){-}OEt + EtOH}$$

$$H_3C{-}C({=}O){-}\overset{-}{C}H(Na^+){-}C({=}O){-}OEt + CH_3{-}CH_2{-}CH_2{-}CH_2{-}Br \longrightarrow$$

$$H_3C{-}C({=}O){-}CH(CH_2{-}CH_2{-}CH_2{-}CH_3){-}C({=}O){-}OEt + NaBr$$

Time 6.0 h (two laboratory periods)

Materials Ethyl acetoacetate, 13 mL (MW 130, bp 181°C, d 1.021 g/mL)
Sodium metal, 2.4 g (MW 23, Na as spheres, available from Matheson, Coleman & Bell Co.)
Potassium iodide, 1 g
n-Butyl bromide, 10 mL (MW 137, bp 100°C, d 1.276 g/mL)
Absolute ethanol, 50 mL
Ether, 30 mL

Precautions Carry out all reactions in the hood. Be careful not to allow sodium to come into contact with water.

Hazards Sodium metal and water react very vigorously to produce hydrogen gas, heat, and a caustic solution (NaOH). Avoid skin contact with the metal. *n*-Butyl bromide is toxic in high concentrations. Avoid breathing vapors.

Experimental Procedure

Preparing sodium ethoxide and sodium ethyl acetoacetate solution
Preweigh a *dry* 50 mL beaker with a watch glass on top. This beaker and watch glass will be used to weigh out metallic sodium.

Using a pair of long tweezers, pick up a small sphere of sodium from the appropriate bottle (sodium spheres usually have a diameter of 2 to 6 mm). The sodium spheres should be stored under a hydrocarbon solvent (usually xylene) in the bottle. After picking up a small sodium sphere, rapidly blot dry (with a paper towel) the sodium sphere from the hydrocarbon solvent and add it to the beaker, replacing the watch glass on top. Each sodium sphere usually weighs between 200 and 250 mg. Continue to pick up, blot dry, and add the spheres of sodium to the beaker until you have weighed out approximately 2.4 g of sodium metal (any weight between 2.4 and 2.45 g is acceptable). Remove the beaker from the balance and add 5 mL anhydrous toluene to protect the sodium metal during storage.

Place in the hood a 250-mL round-bottom flask fitted with a reflux condenser and drying tube. Using a *dry* 50-mL graduated cylinder, measure 50 mL anhydrous (200 proof) ethanol and add it to the round-bottom flask. Immediately replace the reflux condenser (with lightly greased joints) and drying tube (Fig. 11.3). Using tweezers, take the sodium you have just weighed and, one piece at a time, place the sodium in the round-bottom flask by lifting off the condenser and dropping the sodium into the absolute alcohol. Be careful to blot dry the sodium briefly before adding it to the ethanol. As the sodium spheres contact the ethanol, a reaction which produces hydrogen bubbles will begin. As the sodium reacts with the alcohol (to form sodium ethoxide), the ethanol solution will heat up substantially. Allow the reaction to continue, with occasional swirling, until all the sodium has reacted. You should obtain a clear solution. Allow the reaction mixture to cool to room temperature before proceeding with the next step. (The addition of sodium and its conversion to sodium ethoxide usually takes about 20 to 30 min.) If a small amount of sodium remains in the flask at the end of 30 min, warm the solution *gently* on a steam bath to complete the conversion to sodium ethoxide.

Once the reaction solution has returned to room temperature, add 1 g *anhydrous powdered* potassium iodide (KI), needed to catalyze the alkylation reaction. The easiest way to do this is to remove the reflux condenser, place a powder funnel in the neck of the round-bottom flask, add the potassium iodide through the powder funnel, and then replace the reflux condenser. Care should be taken not to allow any of the potassium iodide to contact the greased joint, as this will prevent a good fit between condenser and flask.

Swirl the solution several times and then place the round-bottom flask on a steam bath. Commence heating with occasional swirling until the solution almost reaches reflux. Add, through the top of the reflux condenser, 13 mL ethyl acetoacetate (0.1 mol) in four equal portions, allowing the vigorous reaction which occurs after each addition to subside before adding the next portion. After the addition of the entire 13 mL ethyl acetoacetate, refit the drying tube to the top of the reflux condenser and heat the mixture to reflux with occasional swirling for several minutes.

Note: Many chemical companies (such as the Aldrich Chemical Company) supply the sodium salt of ethyl acetoacetate as a reagent material (MW 152, mp 168 to 171°C). If this material is available, the syntheses of sodium ethoxide and sodium ethyl acetoacetate may be eliminated and the following procedure used.

Quickly weigh 15.5 g (0.102 mol) sodium ethyl acetoacetate into a 250-mL round-bottom flask. (This salt is very hygroscopic. It is best to use a freshly opened bottle.) Using a *dry* 50-mL graduated cylinder, measure

50 mL anhydrous ethanol and add it to the flask. Immediately fit the flask with a reflux condenser (with lightly greased joints) and a drying tube. Swirl the apparatus for several minutes to ensure good mixing. Add 1 g *anhydrous powdered* potassium iodide (KI) to the flask, using the directions for this procedure above. Swirl the solution several times, place the flask on a steam bath, and proceed as directed below.

Reaction of sodium ethyl acetoacetate with n-butyl bromide

To the flask containing the sodium ethyl acetoacetate (commerical or prepared in situ) add, through the top of the reflux condenser, 10 mL *dry n-*butyl bromide in three equal portions. Allow the reaction to subside after the addition of each portion and replace the drying tube. Gently reflux the entire mixture on a steam bath for 1 h. During the reaction period, a yellow color and a white crystalline material will appear in the solution.

After the reflux period is over, remove the reflux condenser and equip the flask for simple distillation (Fig. 2.1). Heat the flask vigorously on the steam bath to distill as much ethanol as possible. Collect at least 40 mL ethanol distillate. The flask will contain fine, white solid salts along with the product (a yellow oily material). (*Note:* Bumping will occur many times during this distillation but this can generally be ignored during the removal of ethanol.)

After removal of the ethanol, disassemble the apparatus and allow the residue in the round-bottom flask to cool to room temperature. Add to the residue 50 mL distilled water which contains 1 mL concentrated hydrochloric acid. Swirl the flask vigorously to decompose all the salts. Test the aqueous solution with pH paper to ensure that the aqueous solution is acidic. If the water layer is not acidic, add another 0.5 mL concentrated hydrochloric acid, swirl, and test. After the hydrolysis of the salts, transfer the entire mixture to a separatory funnel. Add 30 mL ether to the round-bottom flask, swirl the flask, and add the ether to the aqueous solution in the separatory funnel. Shake the separatory funnel and draw off the lower, aqueous phase. The ether layer should be colorless to slightly yellow.

Wash the ether layer with two 10-mL portions of distilled water, followed by one 10-mL portion of 5% sodium bicarbonate and three 10-mL portions of saturated salt solution. Transfer the ether layer to a 125-mL Erlenmeyer flask and dry with several grams of anhydrous sodium sulfate. Filter the ether layer into a 125-mL Erlenmeyer flask and evaporate the ether by heating on a steam bath. The yellow, oily residue should be ethyl *n*-butylacetoacetate.

Note: The procedure may be stopped at this stage. Be certain to stopper the Erlenmeyer flask tightly until the distillation can be performed.

Assemble an apparatus for simple distillation under reduced pressure. Place the crude product in a 50-mL round-bottom flask and vacuum distill the material using either a mechanical vacuum pump or a water aspirator. The product should be a water-white liquid, which distills between 115 and 130°C at 30 torr, 100 and 120°C at 20 torr, and 80 and 95°C at 1 torr. The yield of this material is usually 8 to 10 mL.

Gas chromatographic analysis of this distilled oil on 20% SE-30 at 180°C

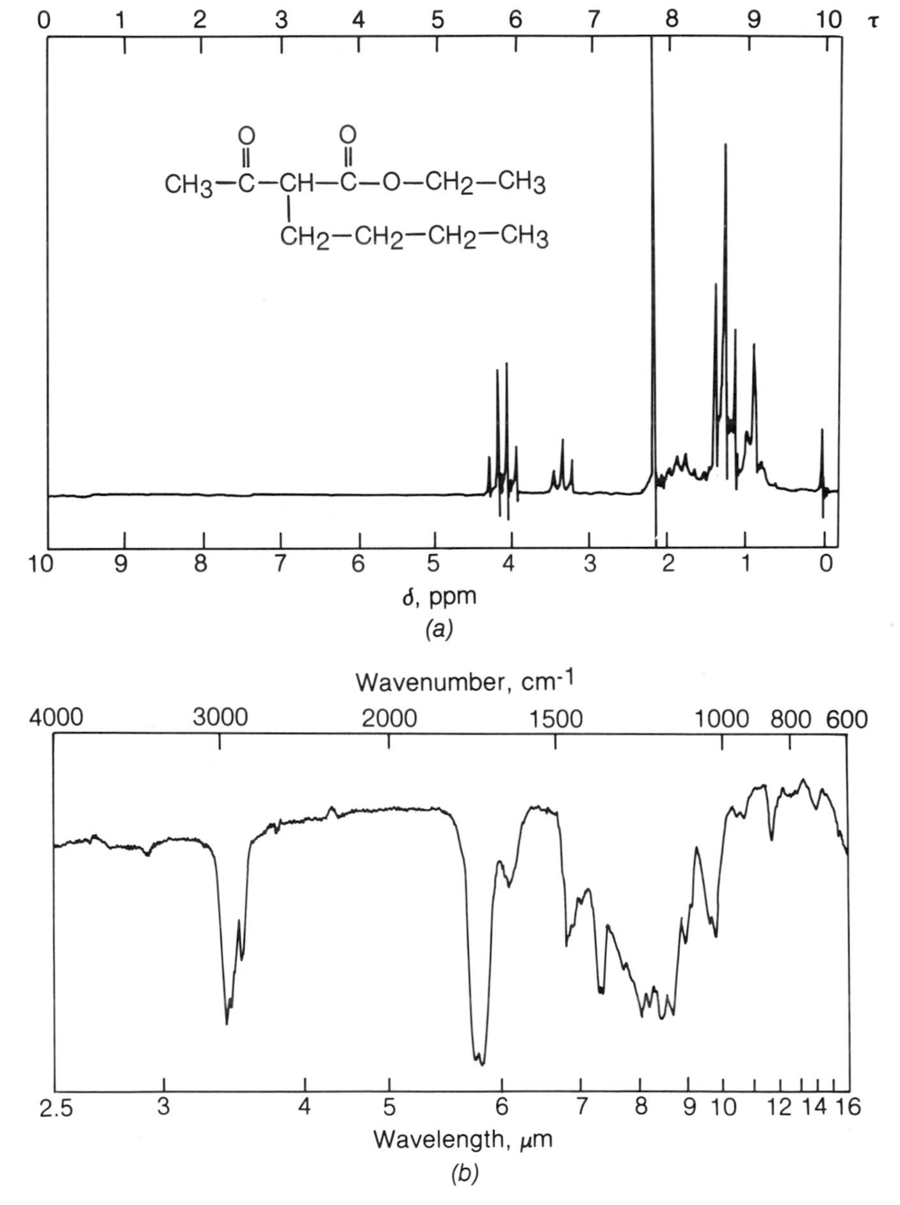

Figure 13.8
The (*a*) nmr and (*b*) ir spectra of ethyl *n*-butylacetoacetate.

shows a material which is more than 94% pure. It is contaminated by starting material and dialkylation product.

The nmr and ir spectra of ethyl *n*-butylacetoacetate are given in Fig. 13.8.

EXPERIMENT 13.5B SYNTHESIS OF DIETHYL n-BUTYLMALONATE BY THE MALONIC ESTER CONDENSATION

$$EtO-C(=O)-CH_2-C(=O)-OEt + NaOEt \longrightarrow EtO-C(=O)-\overset{-}{C}H(Na^+)-C(=O)-OEt + EtOH$$

$$EtO-C(=O)-\overset{-}{C}H(Na^+)-C(=O)-OEt + CH_3-CH_2-CH_2-CH_2-Br \longrightarrow$$

$$EtO-C(=O)-CH(CH_2-CH_2-CH_2-CH_3)-C(=O)-OEt + NaBr$$

Time 6 h (two laboratory periods)

Materials Diethyl malonate, 16 mL (MW 160, bp 199°C, d 1.055 g/mL)
Sodium metal, 2.4 g (MW 23, Na as spheres available from Matheson, Coleman & Bell Co.)
Potassium iodide, 1 g
n-Butyl bromide, 10 mL (MW 137, bp 100°C, d 1.276 g/mL)
Absolute ethanol, 50 mL
Ether, 30 mL

Precautions Carry out all reactions in the hood. Be careful not to allow sodium to come into contact with water.

Hazards Sodium metal and water react very vigorously to produce hydrogen gas, heat, and a caustic solution (NaOH). Avoid skin contact with the metal. *n*-Butyl bromide is toxic in high concentrations. Avoid breathing vapors.

Experimental Procedure

Preparing sodium ethoxide solution

Preweigh a *dry* 50 mL beaker with a watch glass on top. This beaker and watch glass will be used to weigh out metallic sodium.

Using a pair of long tweezers, pick up a small sphere of sodium from the appropriate bottle (sodium spheres usually have diameters between 2 and 6 mm). The sodium spheres should be stored under a hydrocarbon solvent (usually xylene) in the bottle. After picking up a small sodium sphere, rapidly blot dry (with a paper towel) the sodium sphere from the hydrocarbon solvent and add it to the beaker, replacing the watch glass on top. Each sodium sphere usually weighs between 200 and 250 mg. Continue to pick up, blot dry, and add the spheres of sodium to the beaker until you have weighed out approximately 2.4 g sodium metal (any weight between 2.4 and 2.45 g is acceptable). Remove the beaker from the balance and add 5 mL anhydrous toluene to protect the sodium metal during storage.

Place in the hood a 250-mL round-bottom flask fitted with a reflux condenser and drying tube (Fig. 11.3). Using a *dry* 50-mL graduated cylinder, measure 50 mL anhydrous (200 proof) ethanol and add it to the round-bottom flask. Immediately replace the reflux condenser (with lightly greased joints) and drying tube. Using tweezers, take the sodium you have just weighed and, one piece at a time, place the sodium in the round-bottom flask by lifting off the condenser and dropping the sodium into the absolute ethanol. Be careful to blot dry the sodium briefly before adding it to the ethanol. As the sodium spheres contact the ethanol, a reaction which produces hydrogen bubbles will begin. As the sodium reacts with the alcohol (to form sodium ethoxide), the ethanol solution will heat up substantially. Allow the reaction to continue, with occasional swirling, until all the sodium has reacted. You should obtain a clear solution. Allow the reaction mixture to cool to room temperature before proceeding with the next step. (The addition of sodium and its conversion to sodium ethoxide usually takes about 20 to 30 min.) If a small amount of sodium remains in the flask at the end of 30 min, warm the solution *gently* on a steam bath to complete the conversion to sodium ethoxide.

Alkylation of diethyl malonate

Once the reaction solution has returned to room temperature, add 1 g *anhydrous powdered* potassium iodide (KI) to catalyze the alkylation reaction. The easiest way to do this is to remove the reflux condenser, place a powder funnel in the neck of the round-bottom flask, add the potassium iodide through the powder funnel, and then replace the reflux condenser. Care should be taken not to allow any of the potassium iodide to contact the greased joint, as this will prevent a good fit between condenser and flask.

Swirl the solution several times and then place the round-bottom flask on a steam bath. Commence heating with occasional swirling until the solution almost reaches reflux. Add, through the top of the reflux condenser, 16 mL diethyl malonate (0.100 mol) in four equal portions, allowing the vigorous reaction which occurs after each addition to subside before adding the next portion. After the addition of the entire 16 mL diethyl malonate, refit the drying tube to the top of the reflux condenser and heat the mixture to reflux with occasional swirling for several minutes.

After addition of the ester, add, through the top of the reflux condenser, 10 mL *dry n*-butyl bromide in three equal portions. Allow the reaction to subside after the addition of each portion and replace the drying tube. Gently reflux the entire mixture on a steam bath for 1 h. During the reaction period, a yellow color, along with a white crystalline material, will appear in the solution.

After the reflux period is over, remove the reflux condenser and equip the flask for simple distillation. Heat the flask vigorously on the steam bath to distill as much ethanol as possible. Collect at least 40 mL ethanol distillate. The flask will contain fine, white solid salts along with the product (a yellow oily material). (*Note:* Bumping will occur many times during this distillation but this can generally be ignored during the removal of ethanol.)

After removal of the ethanol, disassemble the apparatus and allow the residue in the round-bottom flask to cool to room temperature. Add to the residue 50 mL distilled water which contains 1 mL concentrated hydrochloric acid. Swirl the flask vigorously to decompose all the salts. Test the aqueous solution with pH paper to ensure that the aqueous solution is acidic. If the water layer is not acidic, add another 0.5 mL concentrated hydrochloric acid, swirl, and test. After the hydrolysis of the salts, transfer the entire mixture to a separatory funnel. Add 30 mL ether to the round-bottom flask, swirl, and add the ether to the aqueous solution in the separatory funnel. Shake the separatory funnel and draw off the lower, aqueous phase. The ether layer should be colorless to slightly yellow.

Wash the ether layer with two 10-mL portions of distilled water, followed by one 10-mL portion of 5% sodium bicarbonate and three 10-mL portions of saturated salt solution. Transfer the ether layer to a 125-mL Erlenmeyer flask and dry with several grams of anhydrous sodium sulfate. Filter the ether layer into a 125-mL Erlenmeyer flask and evaporate the ether by heating on a steam bath. The yellow, oily residue should be diethyl *n*-butylmalonate.

Note: The procedure may be stopped at this stage. Be certain to stopper the Erlenmeyer flask tightly until the distillation can be performed.

Assemble an apparatus for simple distillation under reduced pressure. Place the crude product in a 50-mL round-bottom flask and vacuum distill the material using either a mechanical vacuum pump or a water aspirator. The product should be a water-white liquid, which distills between 115 and 130°C at

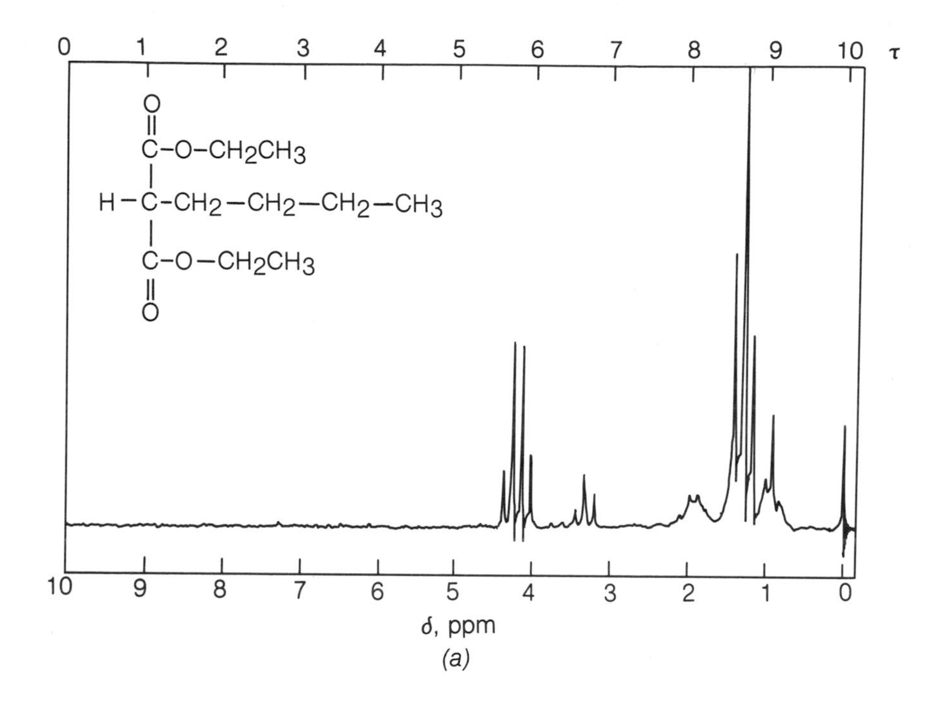

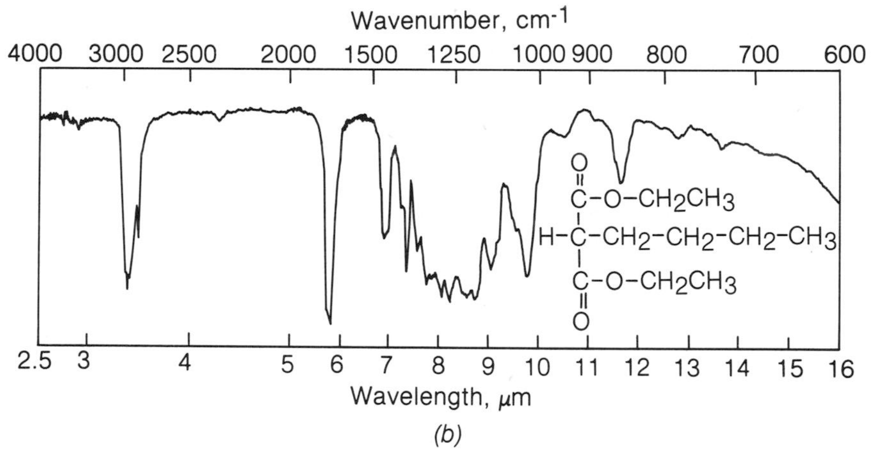

Figure 13.9
The (*a*) nmr and (*b*) ir spectra of diethyl *n*-butylmalonate.

30 torr, 100 and 120°C at 20 torr, or 80 and 95°C at 1 torr. The yield of this material is usually 12 to 15 mL.

Gas-chromatographic analysis of this distilled oil on 20% SE-30 at 180°C shows a material which is more than 97% pure. It is contaminated by starting material and dialkylation product.

The nmr and ir spectra of diethyl *n*-butylmalonate are shown in Fig. 13.9.

13.6 SYNTHESIS OF BENZYLTRIETHYLAMMONIUM CHLORIDE: A PHASE-TRANSFER CATALYST

A number of the reactions described in this book require a phase-transfer catalyst. The compounds Aliquat and tetrabutylammonium hydrogensulfate are often recommended to fulfill the role described in Sec. 4.7. The common feature of each of these phase-transfer catalysts is that they are quaternary ammonium salts. The salt prepared in this experiment, benzyltriethylammonium chloride, was one of the earliest phase-transfer catalysts to enjoy wide use and may be substituted for any of those suggested in this book.

Quaternary ammonium salts are usually formed by the Menschutkin reaction. A tertiary amine, such as triethylamine, functions as the nucleophile, and an alkyl halide such as benzyl chloride undergoes nucleophilic substitution. The lone pair electrons on nitrogen attack the polarized carbon atom and form a nitrogen-carbon bond. Since a positive charge is relatively stable on nitrogen, and since there is no good leaving group present, the reaction stops at the quaternary ammonium salt stage.

In the reaction described here, salt formation is relatively slow. Since a charged product is being formed from two neutral reactants, the reaction is favored by polar solvents. Acetone is only moderately polar and is not the best choice of solvent if a fast reaction is desired. Acetone affords an important advantage, however, because the product is insoluble in it. As the reaction proceeds, the quaternary salt gradually separates from the reaction mixture, and at the end of the week-long reaction period, the product may be isolated by filtration. The reaction would be faster in water, but the product would be much harder to isolate.

EXPERIMENT 13.6 SYNTHESIS OF BENZYLTRIETHYLAMMONIUM CHLORIDE

$$C_6H_5{-}CH_2{-}Cl + \ddot{N}(CH_2{-}CH_3)_3 \longrightarrow C_6H_5{-}CH_2{-}\overset{+}{N}(CH_2{-}CH_3)_3 \quad Cl^-$$

Time 1 h, over a 2-week period

Materials Benzyl chloride, 5 mL (MW 126.59, bp 177 to 181°C, d 1.1 g/mL)
Triethylamine, 6.5 mL (MW 101, bp 89°C, d 0.726 g/mL)
Acetone, 100 mL

Precautions Use gloves when measuring and carry out all transfers in a good hood.

Hazards Benzyl chloride is a skin irritant. Benzyl chloride and triethylamine are lachrymators. Avoid contact with skin and eyes and avoid breathing vapors.

Experimental Procedure

Measure 5 mL (**hood, gloves**) benzyl chloride (0.043 mol) in a *dry* 10-mL graduated cylinder and pour into a *dry* 125-mL Erlenmeyer flask. Measure 10 mL acetone using a graduated cylinder and add it to the Erlenmeyer flask. Measure 6.5 mL (**hood, gloves**) triethylamine (0.047 mol) in a second *dry* 10-mL graduated cylinder and add it to the 125-mL Erlenmeyer flask. Measure 10 mL acetone in this graduated cylinder and add it to the flask. Swirl the flask to ensure good mixing. After swirling, add an additional 30 mL acetone to the flask (total volume of acetone, 50 mL). Swirl again to mix the solution.

Tightly stopper the Erlenmeyer flask with a cork (care should be taken to ensure a good seal). Place the flask in your desk drawer and allow to stand at room temperature for 2 weeks.

At the end of this time period the flask will be filled with needlelike crystals. Filter the crystalline mass with the aid of a Buchner funnel and immediately wash the crystals with 50 mL acetone. Air dry the crystalline mass briefly. (*Note:* Benzyltriethylammonium chloride is somewhat hygroscopic. Prolonged air drying of the crystalline mass will lower the melting point substantially as a result of water pickup, especially in humid areas.) The yield product, mp 182 to 185°C, is 7 to 9 g (70 to 90%).

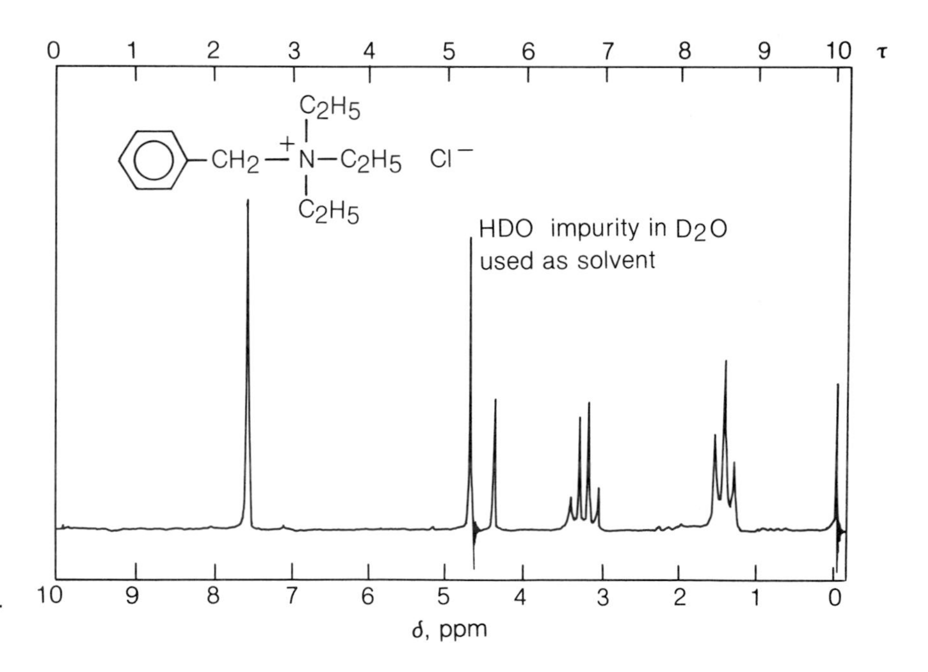

Figure 13.10 The nmr spectrum of benzyltriethylammonium chloride.

Note: This experiment may be terminated at the end of 1 week instead of 2 weeks. The only variation observed is a lower yield (usually around 50 to 60%). Extending the reaction time to more than 2 weeks is of no practical value.

The hygroscopic product should be stored in a tightly sealed bottle until needed. An nmr spectrum of the salt is shown in Fig. 13.10. Notice the tall peak at about 4.6 ppm. This peak is a contaminant and is due to the presence of partially deuterated water, HDO.

QUESTIONS AND EXERCISES

13.1 In carrying out the basic hydrolysis of phenylacetonitrile, an experimenter observed that at first a colorless oil floated on the top of the aqueous base. As the reaction proceeded, however, the oil gradually disappeared and the reaction mixture became homogeneous. What was the oil and what was its fate?

13.2 One thoughtful student reasoned that instead of treating benzyl chloride with cyanide and then hydrolyzing the product, the entire reaction sequence could be circumvented by treating benzyl chloride directly with sodium formate. What was wrong with this student's reasoning and what would the result of this reaction probably be?

13.3 A student who was supposed to prepare the ether from 4-chlorobenzyl alcohol and 4-chlorobenzyl chloride used the correct alcohol but added it to benzyl alcohol. What was the product of the reaction, and how might the properties of this material differ from those of the expected product?

13.4 A student attempting Exp. 13.3A used benzyl chloride instead of 4-methylphenol. Do you think that any product might be obtained from this reaction, and if so, what might its structure be?

13.5 Draw structures for all the possible by-products of the alkylation of ethyl acetoacetate (Exp. 13.5A) with a large excess of *n*-butyl bromide.

13.6 During the attempted synthesis of benzyltriethylammonium chloride (Exp. 13.6) very wet acetone was used. After the reaction should have been over, only a small amount of solid product had been obtained. The student evaporated the mother liquors to try to obtain a higher yield and found a colorless liquid, bp 204°C, to be present. What was the structure of this compound and what are two possible mechanisms by which it might have formed?

XIV OXIDATION AND REDUCTION

Transformations which are formally oxidations or reductions involve the gain or loss of electrons. If electrons are gained in a reaction, the process is termed a reduction; if electrons are lost, the process is referred to as an oxidation. In the redox system involving ferrous ion and ferric ions, the reaction is an oxidation from left to right in the equation shown below and a reduction if carried out as indicated by the left-pointing arrow in the equilibrium.

$$Fe^{2+} \rightleftharpoons Fe^{3+} + e$$

In organic chemistry most oxidation and reduction reactions are characterized by either the loss or gain of hydrogen. This may occur by the addition of electrons and then protons, by hydrogen atoms, or by hydride ions, but the results are about the same. Direct loss of an electron from a species is rare but known in organic chemistry, and is difficult to bring about under the best of circumstances. The direct addition of electrons by electrochemical methods is more convenient and is often carried out in the research laboratory. Examples of the simple oxidation-reduction transformations encountered in organic chemistry are shown below.

Reduction	**Oxidation**
$—CH{=}CH— + H_2 \longrightarrow —CH_2—CH_2—$	$—CH_2— \longrightarrow —CH(OH)—$
$—CH{=}O + H_2 \longrightarrow —CH_2—OH$	$—CH{=}O \longrightarrow —COOH$

The reagents which effect oxidation and reduction are many and varied. Occasionally the reagents are simply oxygen or inorganic electron donors, but most oxidation and reduction reactions involve more sophisticated reagents. For example, in Sec. 14.5 electrons transferred directly from tin reduce nitrobenzene, and in Sec. 14.1 the oxidizing agent is oxygen. In the other three sections, the oxidizing and reducing agents take on a very different appearance.

The use of these rather more sophisticated reagents, each of which operates by a different mechanism, is required so that selectivity can be achieved. Not every oxidizing or reducing agent is amenable to the same experimental conditions. For example, reduction of nitrobenzene is effected with metal and acid (electrons followed by protons). If reduction with sodium borohydride were attempted in acidic solution, the results would be disastrous. The many different reagents and sets of conditions have been developed so that oxidation or reduction can be carried out while differences in structure and reactivity are accommodated.

The different methods of oxidation and reduction are discussed individually in the text preceding each experiment.

14.1 AIR OXIDATION OF FLUORENE TO FLUORENONE

Hydrogen atoms in the benzyl positions of many aromatic hydrocarbons are acidic enough to undergo proton transfer to a suitable strong base. The resulting carbanion can then react as a nucleophile. One of the characteristic reactions of carbanions is their oxidation. Although this is usually an unwanted side reaction, when the reaction can be controlled it can be an effective and economical method for oxidation. One aromatic hydrocarbon which can be readily oxidized under appropriate conditions is fluorene. Loss of a proton from the methylene (CH_2) group in fluorene results in the formation of a carbanion which is stabilized by resonance with both aromatic rings. Oxidation of the carbanion to a ketone transforms white fluorene into the yellow ketone fluorenone, according to the overall equation below.

$$\text{fluorene } (C_{13}H_{10}) + O_2 \xrightarrow{\text{NaOH}} \text{fluorenone } (C_{13}H_8O) + H_2O$$

Oxidation-reduction reactions are always electron-transfer reactions. In an oxidation, one or more electrons are given up by the substrate undergoing oxidation. In a reduction, one or more electrons are gained by the substrate undergoing reduction. In the case of fluorene, the carbanion which is generated adds to oxygen in the air, cleaves, and undergoes net loss of electrons. It is therefore being oxidized. Oxygen, on the other hand, is accepting electrons and is reduced. In a redox reaction, by definition, something must always be oxidized and something must always be reduced. As the old song says, ''You can't have one without the other.'' The designation of a particular reaction as an oxidation or reduction in reality depends on which side of the process interests the chemist. If attention is focused on the conversion of fluorene to fluorenone, the process is an oxidation. If the primary interest is in the conversion of oxygen into water, the reaction is a reduction (of oxygen).

There is one further confusing factor in this nomenclature system. That is the designation of a compound as an oxidizing agent or a reducing agent. In the example above, oxygen is the oxidizing agent (on the left-hand side of the equation). Fluorene is the reducing agent (again on the left-hand side of the equation). As a general rule the following definitions apply (and should be committed to memory):

The *oxidizing agent* is always reduced during the reaction.

The *reducing agent* is always oxidized during the reaction.

The mechanism for the oxidation of fluorene is given in detail below. In the first step, base deprotonates the hydrocarbon, yielding a resonance-stabilized carbanion. This carbanion attacks atmospheric oxygen and forms a hydroperoxide anion. This hydroperoxide anion is a strong enough base to remove the second proton on the fluorene ring. Now a β-elimination process can occur, with ejection of hydroxide and formation of fluorenone.

H H $\xrightarrow{OH^-}$ H $\longleftrightarrow$ H $\longleftrightarrow$ etc. $\xrightarrow{O_2}$

H O—O$^-$ $\xrightarrow{H_2O}$ H—O—O H $^-$OH $\longrightarrow$ O + $^-$OH

Note that in this reaction hydroxide is catalytic. Also note that, in general, aqueous hydroxide ion is not a strong enough base to deprotonate a hydrocarbon such as fluorene. In nonpolar media in the presence of a phase-transfer catalyst, hydroxide is an effective base for this reaction.

In the following experiment two alternative procedures are presented. One is the complete oxidation of fluorene to fluorenone and the other is the partial oxidation of fluorene to fluorenone. If your instructor directs you to use the partial oxidation sequence, you may gain experience in separating mixtures by using chromatographic techniques (see Sec. 2.6). The nonpolar hydrocarbon is easily separated from the polar, colored ketone by simple column chromatography.

EXPERIMENT 14.1A

AIR OXIDATION OF FLUORENE TO FLUORENONE

$$\text{Fluorene} + O_2 \xrightarrow{NaOH} \text{Fluorenone} + H_2O$$

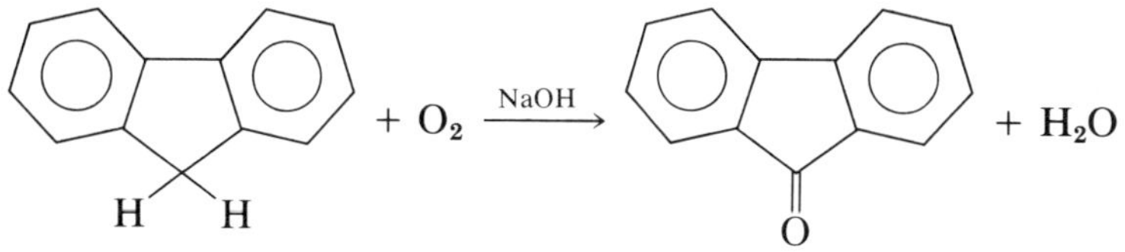

Time 2.5 h

Materials Fluorene, 4 g (MW 166, mp 112 to 115°C)
Phase-transfer catalyst, 750 mg (3 mL catalyst solution)
Petroleum ether, (bp 140 to 160°C), 30 mL
Cyclohexane, 10 mL
Sodium hydroxide, 7.5 g (MW 40)

Precautions Use no flames and use a hood if available.

Hazards The solvents used in this experiment are flammable. Sodium hydroxide solution is caustic; avoid contact with skin and eyes.

Experimental Procedure

Place 7.5 g solid sodium hydroxide in a 250-mL Erlenmeyer flask followed by 15 mL distilled water. Swirl the mixture until all the sodium hydroxide dissolves (3 to 4 min). Allow the solution to cool to room temperature before the next step. (Swirling the flask in cold tap water or an ice bath will speed this cooling.)

Add 5 g fluorene (practical grade will do) to the sodium hydroxide solution, followed by 30 mL high-boiling petroleum ether (trade name of the Ashland Chemical Company product is High-Flash, which has a boiling range of 140 to 160°C). Add to this solution 3 mL (250 mg/mL) of Aliquat 336 or equivalent phase-transfer catalyst in cyclohexane with a 5-mL pipet. Add a 1.5-in magnetic stirring bar (smaller stirring bars are less efficient) and adjust the stirring rate to as rapid a rate as is possible without splashing the solution from the Erlenmeyer flask (see Fig. 14.9). This rate is usually indicated by a froth on the surface of the swirling liquid. If magnetic stirring apparatus is unavailable, the partial oxidation of fluorene may be carried out as described in Exp. 14.1B.

Note that the fluorene does not completely dissolve in the petroleum ether. There is, however, enough solubility to initiate the reaction. As the reaction proceeds, the fluorene will slowly dissolve and then react in solution to produce fluorenone. Toward the end of the reaction, fluorenone will start to crystallize from the reaction mixture. During the course of the reaction, the liquid will turn from colorless to yellow to green. At the end of the reaction the mixture is usually a deep green color with a yellow tinge. The mixture will usually heat up during the reaction and some solvent may be lost because of evaporation. Replace the solvent as needed to maintain the total volume in the organic layer near 30 mL. As the reaction nears completion, the heating will abate and fluorenone will crystallize from solution.

After the reaction period (usually about 1 h) turn off the stirrer and cool the solution in an ice bath until the internal temperature is approximately 15°C. Remove as much of the lower aqueous NaOH layer as possible by using a pipet and bulb. Filter the remaining organic phase (with solid fluorenone suspended in it) and collect the product on a Hirsch funnel. Briefly (2 to 4 min) air dry the resulting solid material (greenish yellow crystals) and then transfer it to a 125-mL Erlenmeyer flask. Add 40 mL dichloromethane to the product and swirl the solution to dissolve the solid. Transfer this material to a separatory funnel and wash the organic solution with two 25-mL portions of 5% HCl. The *upper* aqueous wash should be discarded after each extraction. Next, wash the organic solution with two 25-mL portions of half-saturated sodium chloride solution, draw off the organic layer, and dry over granular anhydrous sodium sulfate. Filter the organic solution to remove the sodium sulfate and rinse the sodium sulfate with several small (5- to 7-mL) portions of dichloromethane. Combine the organic layers and remove the solvent by evaporation on the steam bath. Add 10 to 12 mL cyclohexane to the golden yellow oil and bring the solution to a boil by heating on the steam bath. Cool the solution slowly to room temperature and collect the crystals on a Hirsch funnel. Yield: 3.2 to 4.4 g, mp 79 to 81°C.

The above reaction may be monitored by thin-layer chromatography (tlc), using silica gel as an adsorbent with 20% methylene chloride in hexane as the eluting solvent. Fluorene has R_f 0.9, while fluorenone has R_f 0.45 in this solvent mixture. Recrystallization from 10 mL cyclohexane gives beautiful yellow crystals, which are pure by tlc analysis.

EXPERIMENT 14.1B

PARTIAL OXIDATION OF FLUORENE

$$\text{Fluorene} + O_2 \xrightarrow{NaOH} \text{Fluorenone} + H_2O$$

Time 3 h

Materials Fluorene, 4 g (MW 166, mp 112 to 115°C)
Phase-transfer catalyst, 750 mg in 2 mL toluene
Toluene, 20 to 45 mL
Cyclohexane, 10 mL
Sodium hydroxide, 5 g (MW 40)

Precautions Use no flames and use a hood if available.

Hazards The solvents used in this experiment are flammable. Sodium hydroxide solution is caustic; avoid contact with skin and eyes.

Experimental Procedure

Place 5 g sodium hydroxide pellets in a 50-mL round-bottom flask and add 10 mL distilled water. Swirl the flask until the sodium hydroxide dissolves (3 to 4 min); then cool the aqueous solution nearly to room temperature. Add 4 g (0.024 mol) fluorene (practical grade) to the flask, followed by 20 mL toluene. Swirl the flask so that the fluorene dissolves in the toluene layer. Add 2 mL of phase-transfer catalyst solution after the fluorene has dissolved.

Assemble the reaction apparatus as illustrated in either Fig. 14.1 or 14.2 or as your instructor indicates. Turn on the aspirator and adjust the air flow so that a vigorous rolling agitation is effected in the flask. Continue the agitation for approximately 1 h 20 min to obtain about a 50:50 mixture of starting material and product. During the agitation, some toluene will be lost by evaporation; additional toluene should be added, at 5- to 10-min intervals, so that a volume approximately equal to that at the start is maintained. Commonly, an additional 25 mL toluene will be needed.

At the end of 1.5 h (or 2.25 h if total conversion is desired), transfer the solution to a separatory funnel and remove the aqueous layer. Wash the organic solution with 5 mL 10% HCl and then with two 5-mL portions of saturated salt solution. Draw off the aqueous layer and discard it after each washing. Transfer the toluene solution to a 100-mL beaker and evaporate the toluene by heating on a small hot plate in the hood. The product will be obtained as a

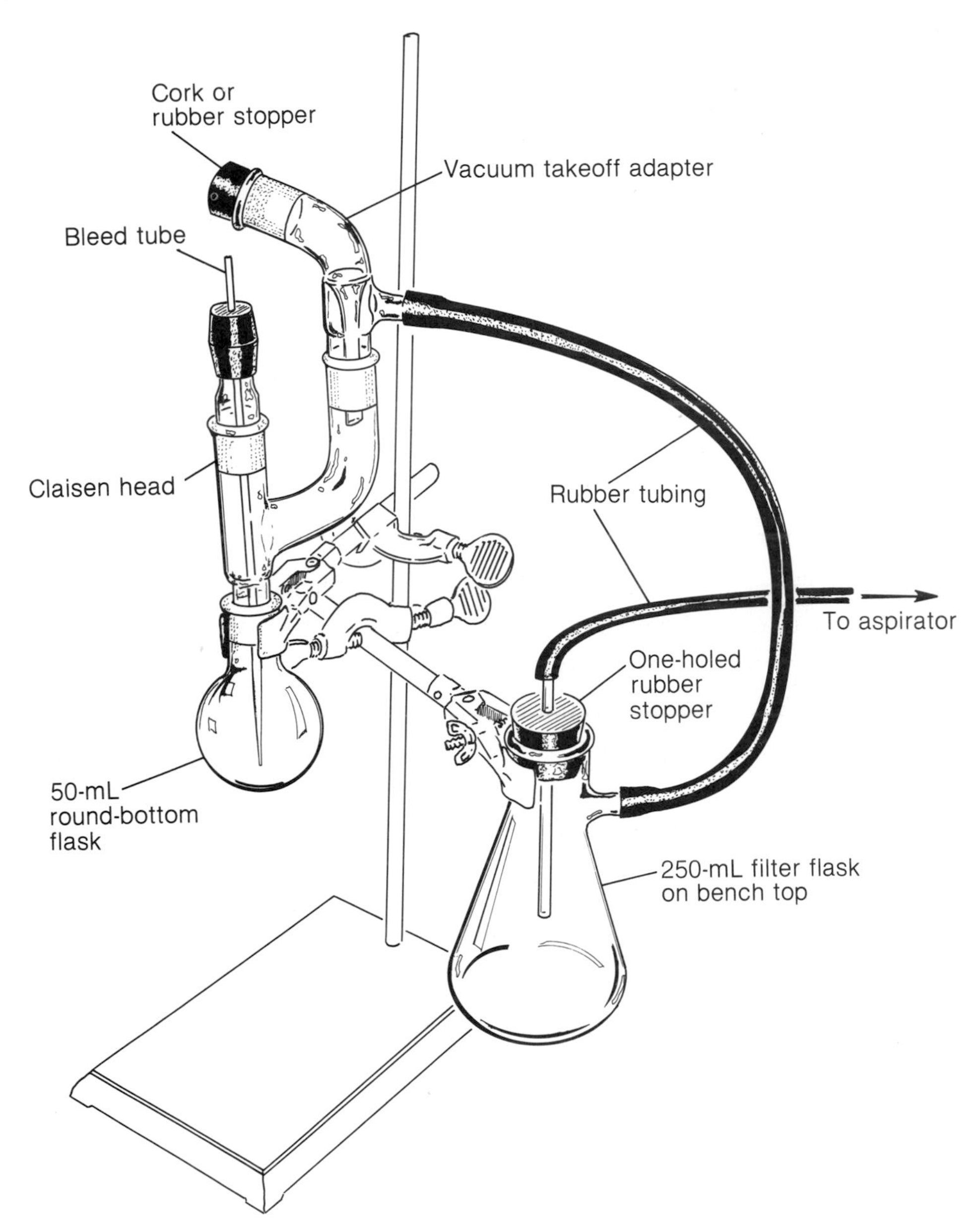

Figure 14.1
Apparatus for the oxidation of fluorene.

heavy yellow liquid, which may start to crystallize when it cools to room temperature. Add 10 mL cyclohexane to the oil and heat briefly on the steam bath. Cool the solution in an ice-water bath. Crystals should deposit which are a mixture of fluorene and fluorenone. This mixture is suitable for the chromatographic separation procedure described in detail at the end of Sec. 2.6.

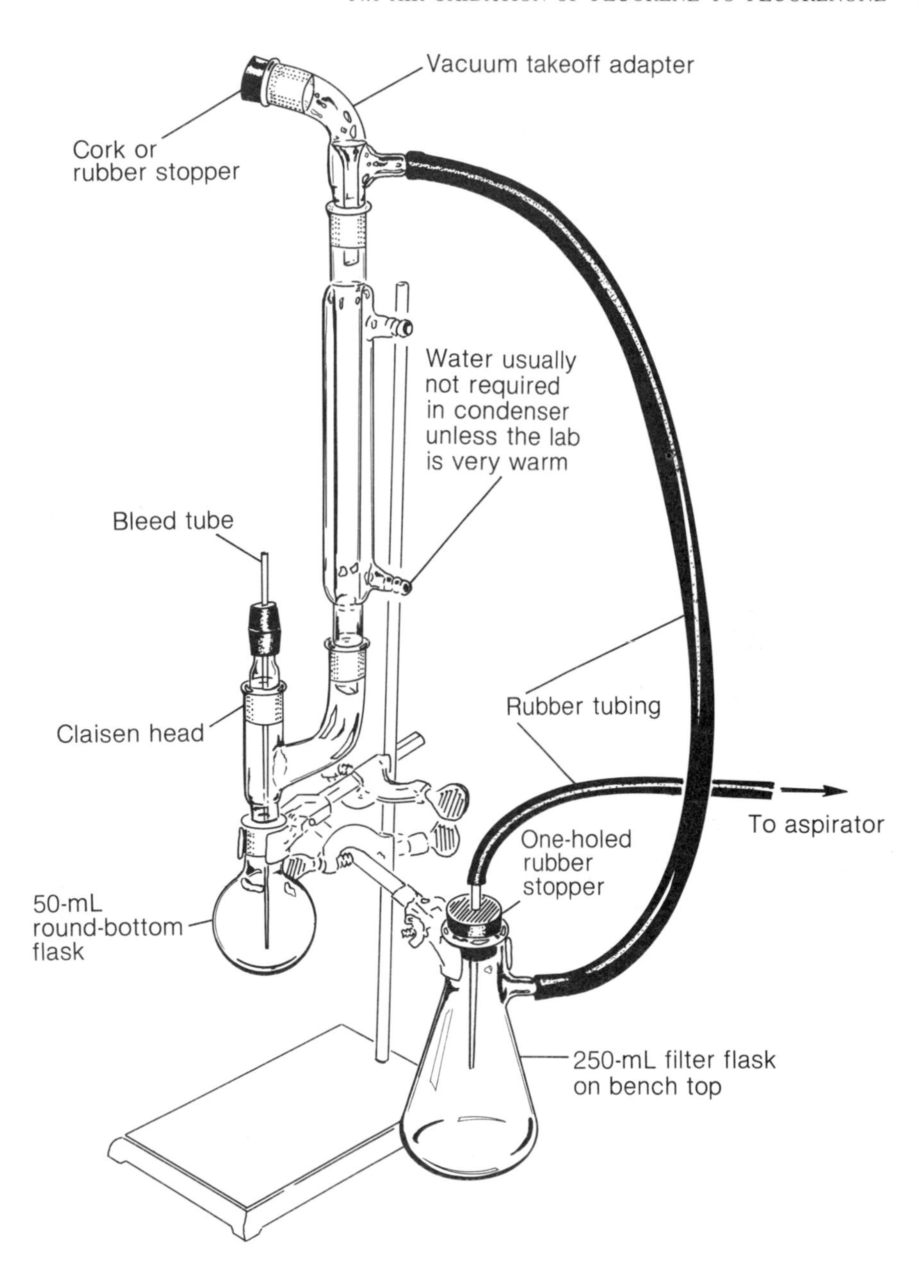

Figure 14.2 Alternate set-up for oxidation of fluorene.

The nmr spectrum of pure fluorenone is shown in Fig. 14.3. Compare it with the nmr spectrum of fluorene, which is shown in Fig. 15.3. By comparing the integral of the upfield methylene signals with that of the aromatics, the percentage conversion can be calculated.

Compare this spectrum with that of benzophenone in Fig. 14.4.

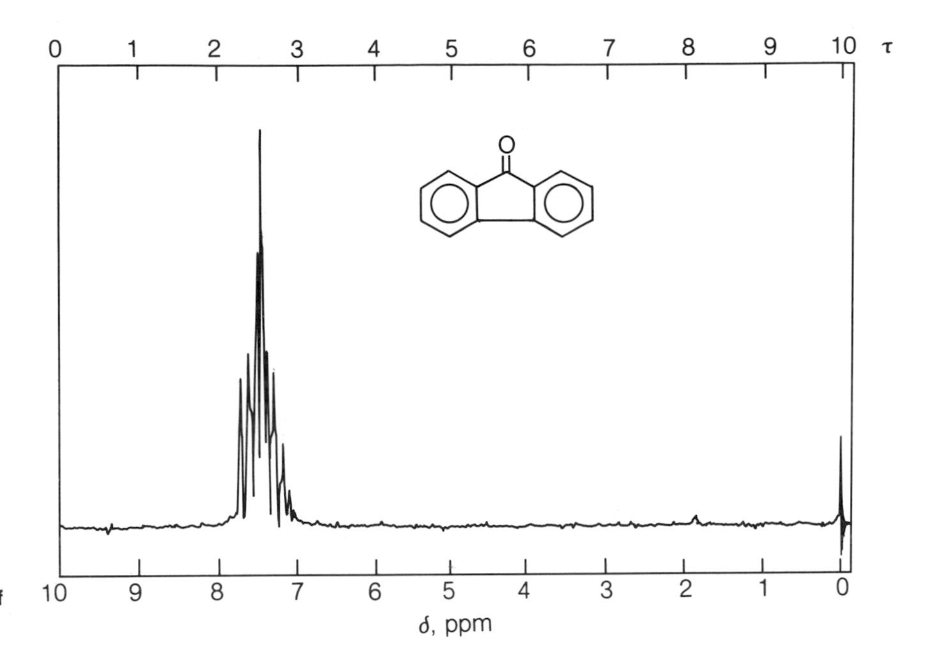

Figure 14.3
The nmr spectrum of fluorenone.

14.2 CHROMIUM TRIOXIDE OXIDATION OF BENZHYDROL AND ISOBORNEOL

In each of the two experiments described below, alcohols are oxidized to ketones by the action of chromium trioxide in acidic acetone solution. The mechanism of each of these reactions is the same. Each alcohol is oxidized while the metal is reduced. In each case the hydrogen atom bonded to the carbon atom bearing the hydroxyl group is lost, with formation of a ketone. Further oxidation does not occur because chromium trioxide in aqueous acetone is not a powerful enough oxidizing agent. It is for this reason that acetone itself can be used. If oxidation of ketones were observed, acetone would be a foolish choice for the solvent.

The mechanism of chromium trioxide oxidation is really esterification followed by β elimination. In the first step the alcohol adds to chromium trioxide in the presence of protons to give a protonated chromate ester. Base (which may be water or other solvent) attacks the β proton, inducing an elimination reaction. The net effect of this process is that the alcohol is oxidized to a ketone and the chromium is reduced (i.e., goes to a lower oxidation state). Note also that a stoichiometric amount of chromium is required for this oxidation, while the proton acts as a catalyst.

$$(C_6H_5)_2C(H)\text{—}OH + HCrO_4^- + H^+ \rightleftharpoons (C_6H_5)_2C(H)\text{—}O\text{—}CrO_3H + H_2O$$

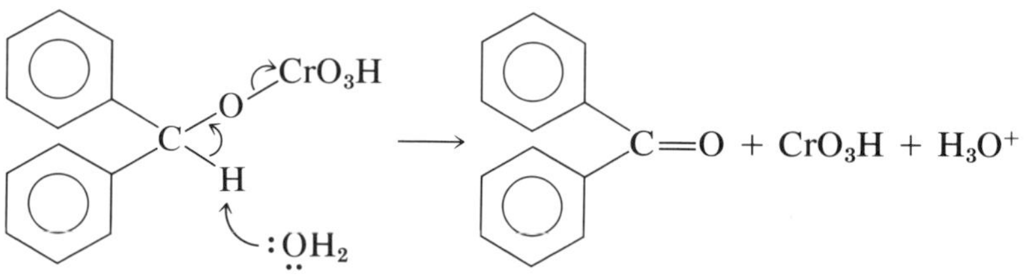

$$(C_6H_5)_2C(H)-O-CrO_3H \;\; (:\ddot{O}H_2) \longrightarrow (C_6H_5)_2C{=}O + CrO_3H + H_3O^+$$

The chromic acid oxidation is not as simple as we have pictured it. The change in oxidation state for chromium is from +6 to +3, a net overall change of three electrons. The alcohol, however, changes by only the equivalent of two electrons. Thus, for every mole of chromium which is reduced, 1.5 mol of alcohol has been converted to the ketone. Although the detailed mechanism of this transformation is complicated (and in fact is not well understood), the simple mechanism above is sufficient to explain many of the observations associated with oxidations by chromate ion.

The structures of both isoborneol and benzhydrol are shown below. Note again that each of them can oxidize under mild conditions only to a ketone, and that much more vigorous conditions would be required to cleave the carbon-carbon bond and effect a further oxidation to the acid. Note also that the mechanism illustrated for the oxidation of an alcohol to a ketone is not readily adaptable to the oxidation of an acid.

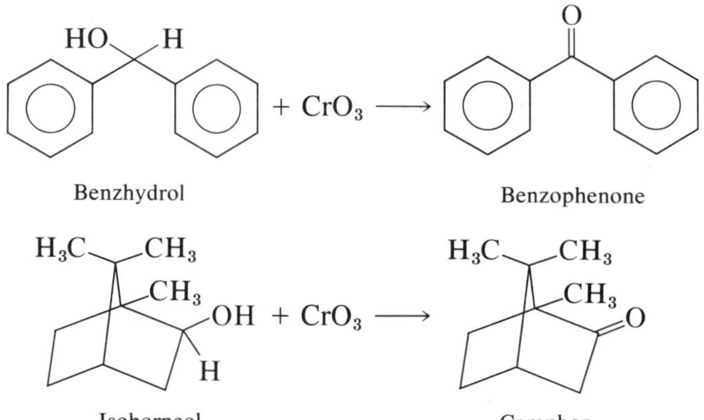

$$\text{Benzhydrol} + CrO_3 \longrightarrow \text{Benzophenone}$$

Benzhydrol — Benzophenone

$$\text{Isoborneol} + CrO_3 \longrightarrow \text{Camphor}$$

Isoborneol — Camphor

EXPERIMENT 14.2A CHROMIUM TRIOXIDE OXIDATION OF BENZHYDROL TO BENZOPHENONE

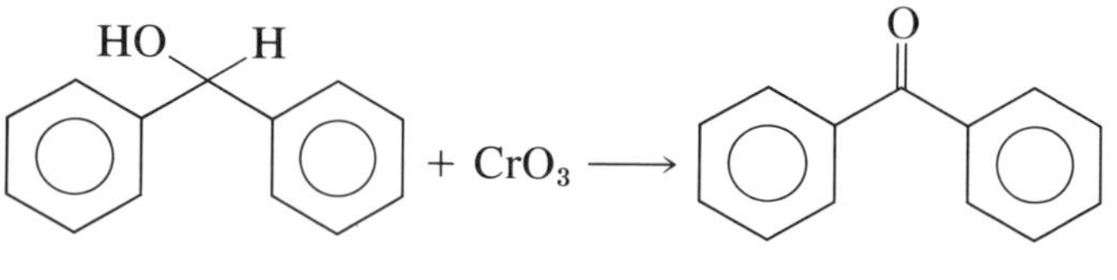

$$\text{Benzhydrol} + CrO_3 \longrightarrow \text{Benzophenone}$$

Time 2.5 h

Materials Benzhydrol, 5 g (MW 184, mp 69°C)
2.7 *M* Chromic acid solution,[1] 9 mL
Acetone, 75 mL
Ether, 75 mL
2-Propanol, 5 mL

Precautions Wear gloves when handling chromic acid.

Hazards Chromic acid is a strong oxidizing agent. Avoid contact with skin or clothes.

Experimental Procedure

Weigh 5 g (0.028 mol) benzhydrol and place this material in a 250-mL Erlenmeyer flask. Add 50 mL acetone and cool the flask in an ice-water bath until the internal temperature is approximately 10°C.

Pour 9 mL 2.7 *M* chromic acid solution into a 10-mL graduated cylinder (**gloves**). When the acetone mixture reaches approximately 10°C, slowly add the chromic acid solution to it in 1-mL portions. Swirl the acetone mixture vigorously after each addition. The reddish yellow chromium solution will turn dark green very soon after addition to the cold acetone layer. As more oxidizing solution is added, a granular, dark green precipitate will deposit from the acetone solution. Toward the end of the addition (the last 2 mL), the dark green acetone solution will take on a yellowish tint. After all the chromic acid solution has been added, swirl the Erlenmeyer flask vigorously for 10 min while maintaining the internal temperature at around 10°C by judicious use of an ice-water bath.

Add one 5-mL portion of 2-propanol to the flask and again swirl (10 min) the reaction mixture vigorously while cooling. The alcohol will destroy any excess oxidizing agent which remains. The reaction mixture should no longer appear yellow but should have a dark green tint, and green precipitated solids should be visible.

Filter the reaction mixture through a thin layer of Celite (filter aid) using a Buchner funnel. Wash the gummy dark green precipitate with 25 mL cold acetone and suck as dry as possible. The filtered acetone layer should now be clear and colorless or have only a slight greenish tint.

Transfer the acetone filtrate from the filter flask to a 250-mL Erlenmeyer flask, add several boiling chips, and remove the acetone by heating on a

[1] The chromic acid solution is made by placing 27 g chromium trioxide in a 125-mL Erlenmeyer flask, followed by 50 mL distilled water. Swirl the aqueous solution to dissolve the solid, then carefully add (**caution: exotherm**) 23 mL concentrated (98%) sulfuric acid. Swirl the flask and allow it to cool to room temperature. Transfer the solution to a 100-mL graduated cylinder and dilute to 100 mL with distilled water. This solution is now 2.7 *M*. The solution should be stored in a brown bottle.

steam bath. After the acetone is removed, add 50 mL distilled water to the residue and transfer the material to a separatory funnel. Rinse the Erlenmeyer flask with 50 mL ether, and add the ether solution to the separatory funnel. The ether layer will cause the aqueous layer to separate so that it can be extracted with another 25-mL portion of ether. Combine the ether extracts, return them to the separatory funnel, and wash with two 25-mL portions of distilled water. The second 25-mL portion of distilled water should be tested with pH paper and should be neutral. If not, repeat the process. Finally, wash the

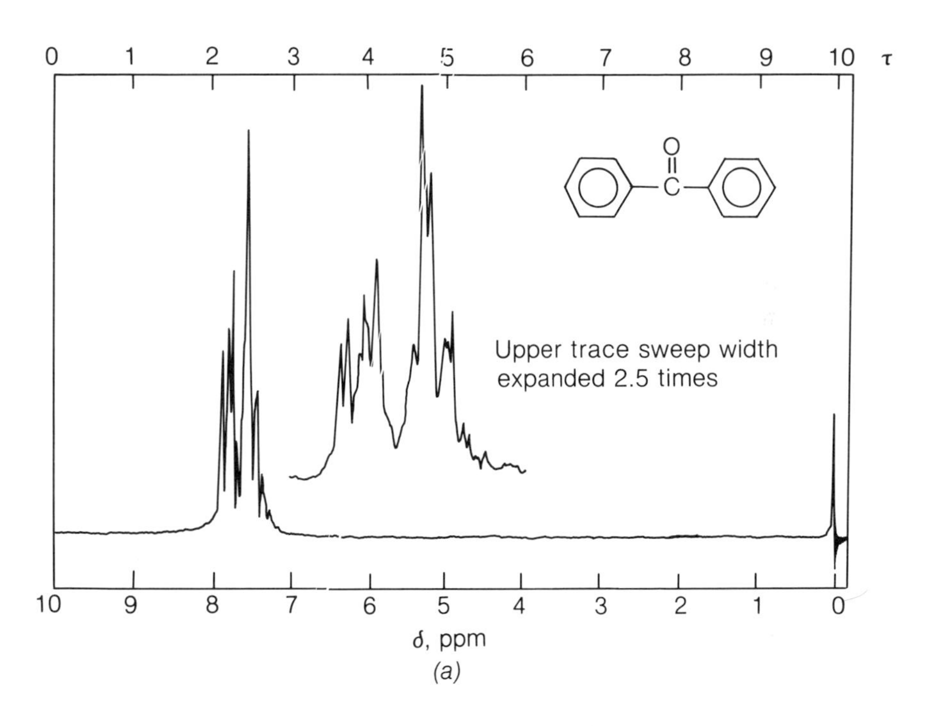

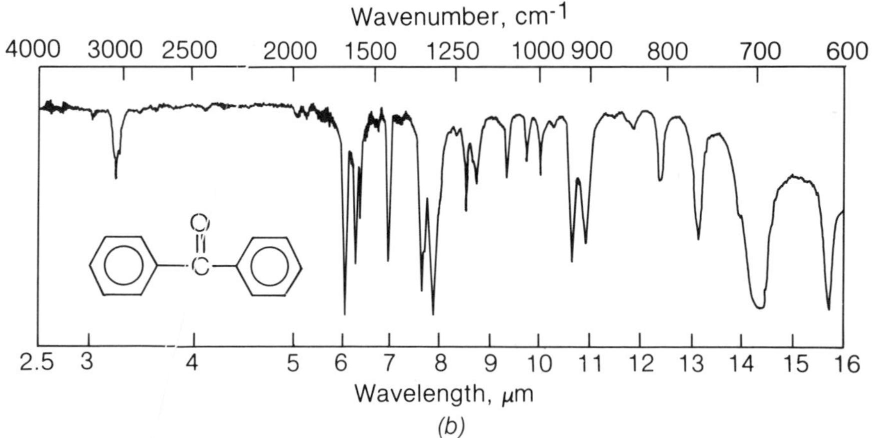

Figure 14.4 The (*a*) nmr and (*b*) ir spectra of benzophenone.

ether layer with 25 mL saturated salt solution (brine) and dry over granular anhydrous sodium sulfate.

Decant the ether from the drying agent and transfer to a 125-mL Erlenmeyer flask. Remove the ether by heating on a steam bath. After evaporating all the ether, remove the Erlenmeyer flask from the heat source. The remaining liquid should crystallize on cooling. The yield of crude benzophenone is usually 80 to 90%.

Dissolve the crude product in 25 mL methanol and warm on a steam

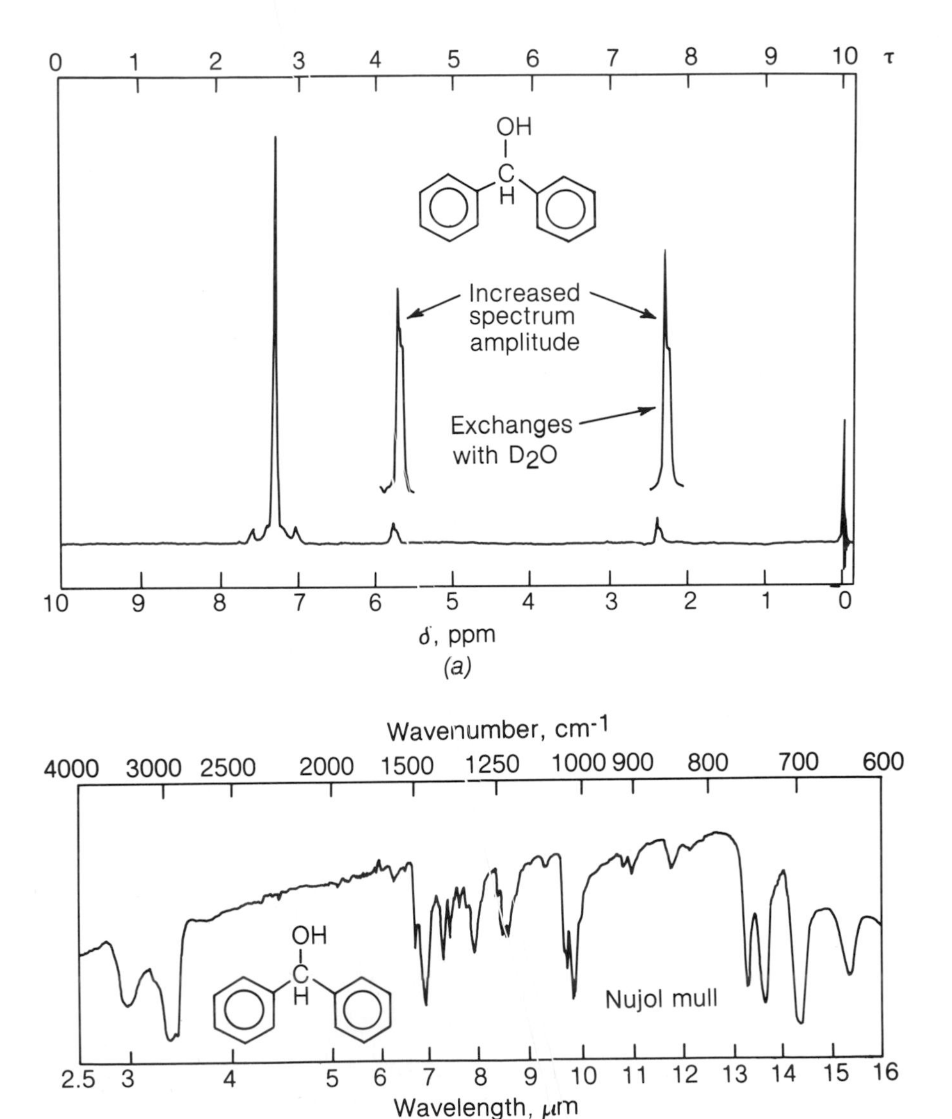

Figure 14.5 The (*a*) nmr and (*b*) ir spectra of benzhydrol.

bath to the boiling point (65°C). Add approximately 10 mL distilled water and continue heating until the solution turns cloudy. Allow the solution to cool. Benzophenone usually deposits as an oil and then crystallizes. Vigorous swirling and the addition of a few seed crystals at the cloud point will often be of great help. Cool the aqueous methyl alcohol solution in an ice bath and filter (Buchner funnel). Air dry the crystals on the funnel. The crystalline product should have a melting point of 46 to 48°C and weigh approximately 3.5 to 4.0 g.

Thin-layer chromatography of this crystalline material should show a single pure material. If the melting point is slightly below that given above, the product may be recrystallized from methanol-water or from petroleum ether to obtain a purer product.

The nmr and ir spectra of benzophenone and benzhydrol are shown in Fig. 14.4 and 14.5, respectively. Compare the spectra of benzophenone with those of fluorenone (Fig. 14.3).

EXPERIMENT 14.2B CHROMIUM TRIOXIDE OXIDATION OF ISOBORNEOL TO CAMPHOR

H_3C CH_3 CH_3 OH H + CrO_3 ⟶ H_3C CH_3 CH_3 O

Time 2.5 h

Materials DL-Isoborneol, 5 g (MW 154, mp 212 to 214°C)
2.7 *M* chromic acid solution,[2] 10 mL
Acetone, 75 mL
Ether, 75 mL
2-Propanol, 5 mL

Precautions Wear gloves when handling chromic acid.

Hazards Chromic acid is a strong oxidizing agent. Avoid contact with skin or clothes.

[2] The chromic acid solution is made by placing 27 g chromium trioxide in a 125-mL Erlenmeyer flask, followed by 50 mL distilled water. Swirl the aqueous solution to dissolve the solid, then carefully add (**caution: exotherm**) 23 mL concentrated (98%) sulfuric acid. Swirl the flask and allow it to cool back to room temperature. Transfer the solution to a 100-mL graduated cylinder and dilute to 100 mL with distilled water. This solution is now 2.7 *M*. The solution should be stored in a brown bottle.

Experimental Procedure

Weigh 5 g (0.033 mol) isoborneol and transfer to a 250-mL Erlenmeyer flask. Add 50 mL acetone and cool the flask in an ice bath until the internal temperature is approximately 10°C.

Pour 10 mL 2.7 *M* chromic acid solution into a 10-mL graduated cylinder (**gloves**). When the acetone mixture reaches approximately 10°C, slowly add the chromic acid solution to it in 1-mL portions. Swirl the acetone mixture vigorously after each addition. The reddish yellow chromium solution will turn dark green very soon after addition to the cold acetone layer. As more oxidizing solution is added, a granular, dark green, gummy precipitate will appear in the acetone solution. Toward the end of the addition (the last 2 mL), the dark green acetone solution will take on a yellowish tint. After all the chromic acid solution has been added, swirl the Erlenmeyer flask vigorously for 10 min and break up the larger chunks of solid (chromium salts) while maintaining the internal temperature at about 10°C (ice-water bath).

Now add one 5-mL portion of 2-propanol to the flask and again (10 min) swirl the reaction mixture vigorously while cooling. The alcohol will destroy any excess oxidizing agent which remains. The reaction mixture should no longer appear yellow but should have a dark green tint, and green precipitated solids should be visible.

Filter the reaction mixture through a thin layer of Celite (filter aid), using a Buchner funnel and a filter flask. Wash the gummy, dark green precipitate with 25 mL cold acetone and suck as dry as possible. The filtered acetone layer should now be clear and colorless or have only a slight greenish tint.

Transfer the acetone filtrate from the filter flask to a 250-mL Erlenmeyer flask, add several boiling chips, and remove the acetone by heating on a steam bath. After the acetone is removed, add 50 mL distilled water to the residue and transfer the material to a separatory funnel. Add 50 mL ether to the Erlenmeyer flask, swirl, and transfer the ether solvent mixture to the separatory funnel. The ether layer will cause the aqueous layer to separate so that it can be extracted with another 25-mL portion of ether. Combine the ether extracts, return to the separatory funnel, and wash with two 25-mL portions of distilled water. The second 25-mL portion of distilled water should be tested with pH paper and should be neutral. If not, repeat the process. Finally, wash the ether layer with 25 mL saturated salt solution (brine) and dry over granular anhydrous sodium sulfate.

Decant the ether from the drying agent and transfer to a 125-mL Erlenmeyer flask. Remove the ether by heating on a steam bath. After evaporating all the ether, remove the Erlenmeyer flask from the heat source. Do not heat too long after the ether is gone, or camphor will sublime and be lost. The remaining liquid should crystallize on cooling. The yield of crude camphor is usually about 80%.

Camphor can be purified most efficiently by sublimation (Sec. 2.3). Place

2 g crude camphor in a 250-mL filter flask fitted with a cold finger, as shown in Fig. 14.6. Place ice and a little water in the test tube and attach the flask to a water aspirator. Conduct the sublimation at the lowest pressure possible under these conditions (usually about 30 torr). If the bottom of the filter flask is heated with a small hot plate or steam bath, the camphor will sublime quite rapidly onto the cold finger (also to the upper reaches of the filter flask). This sublimation is best performed with the temperature of the flask bottom between 100 and 125°C. The temperature will obviously be a little lower if a steam bath is used.

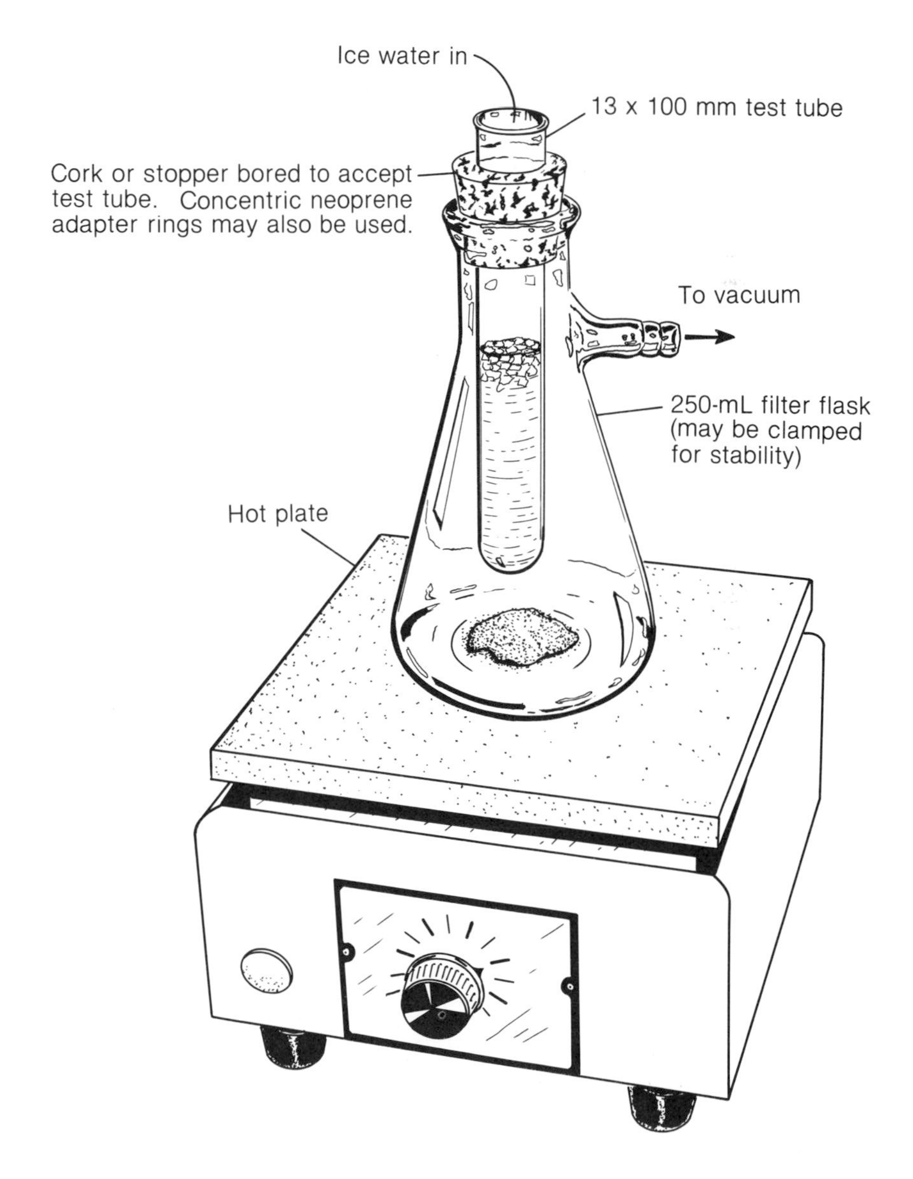

Figure 14.6 Apparatus for sublimation of camphor.

After most of the material has collected on the cold finger, remove the flask from the heat source, allow it to cool, and break the vacuum at the flask. Remove the cold finger and scrape off the camphor using a flat-blade spatula. The melting point of the camphor is usually 172 to 174°C (the melting point must be determined in a sealed capillary tube as the camphor will sublime out of an open tube before reaching the melting point).

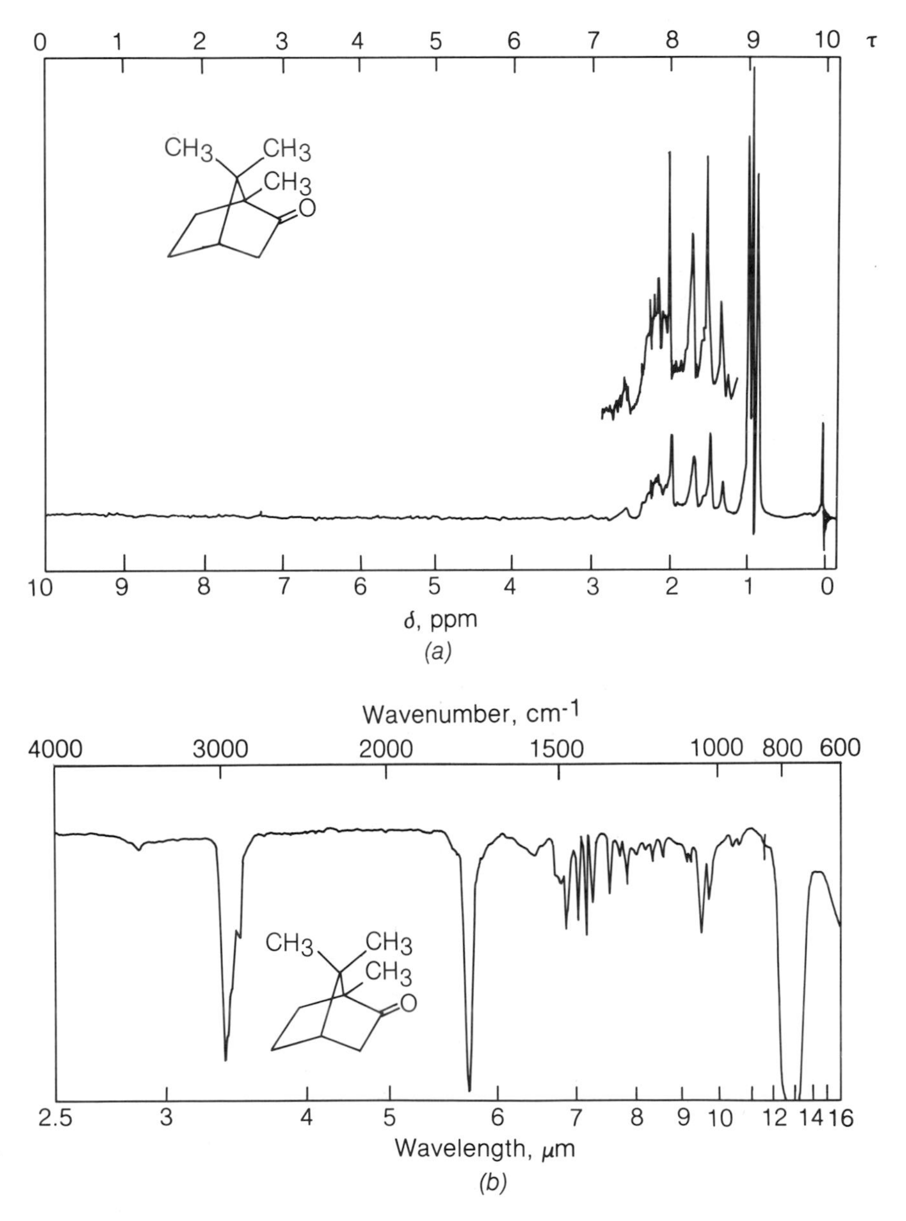

Figure 14.7
The (*a*) nmr and (*b*) ir spectra of camphor.

The melting point of camphor is extremely sensitive to small amounts of contaminants (see Sec. 2.3). If the oxidation is incomplete, isoborneol will co-sublime with camphor and the melting point of the "purified" material will be 165 to 170°C. If an ir instrument is available, confirm the purity of your product by the absence of an OH group absorption in the product. The nmr and ir spectra of both compounds are shown in Figs. 14.7 and 14.8 for reference.

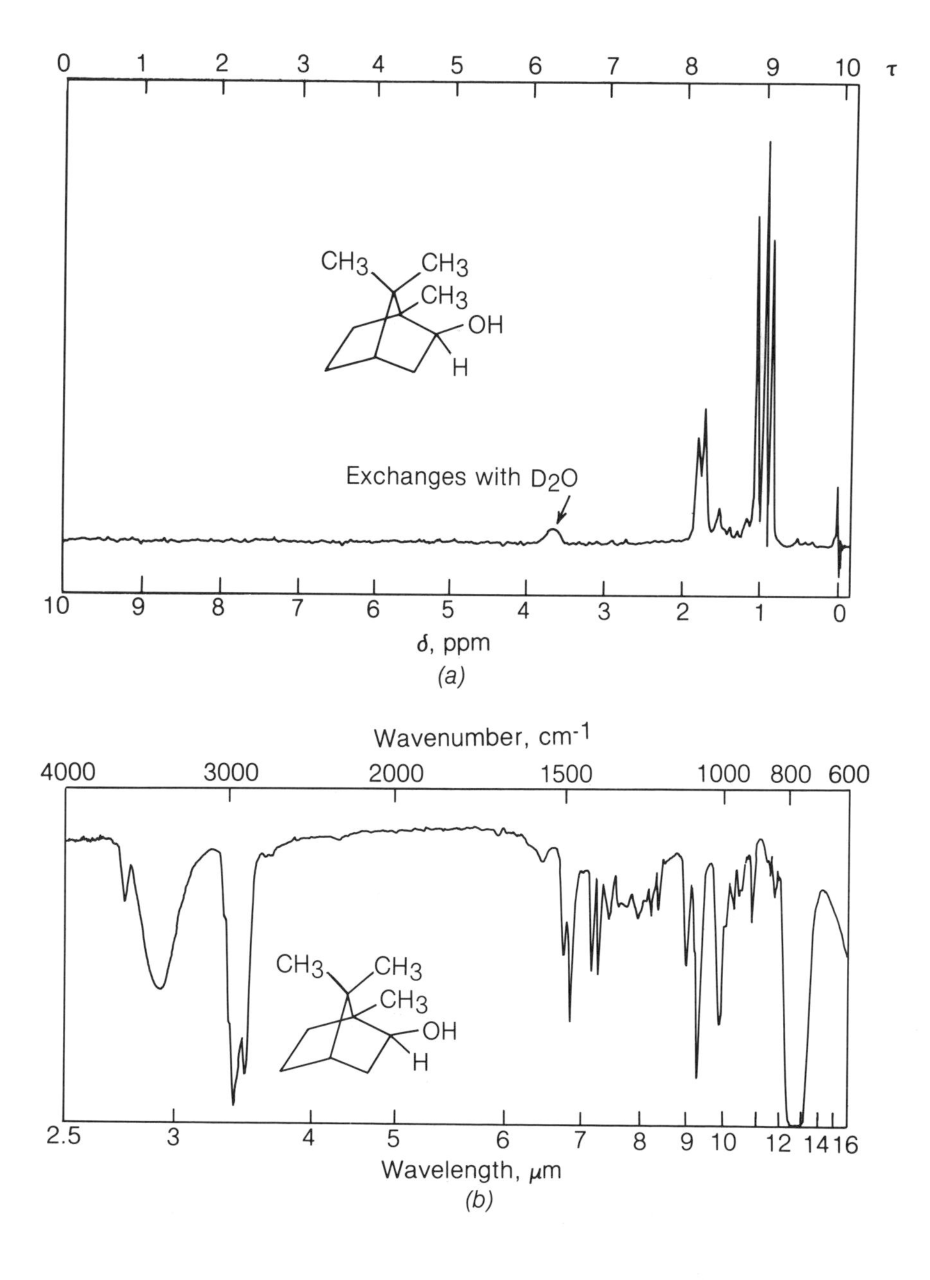

Figure 14.8
The (*a*) nmr and (*b*) ir spectra of isoborneol.

14.3 HYPOCHLORITE OXIDATION OF BENZHYDROL AND 4-CHLORO-BENZHYDROL

Sodium hypochlorite (NaOCl) is a strong oxidizing agent which has been utilized both industrially and in the home for many years. Aqueous 4% sodium hypochlorite solution is laundry bleach, often referred to interchangeably by its trade name Clorox. An aqueous 10% sodium hypochlorite solution is the bleach used for cleaning swimming pools. The strength of this oxidant and the fact that it is soluble in water solutions have kept it from being useful for mild oxidations in organic laboratories. It has, however, been utilized for some time in the haloform reaction of methyl ketones (see Sec. 25.5I). In the latter reaction the oxidant may also be bromine or iodine. The formation of sodium hypochlorite occurs when chlorine is dissolved in aqueous base. The overall reaction is shown below:

$$Cl_2 + 2NaOH \longrightarrow NaOCl + NaCl + H_2O$$

In hypochlorite oxidations, the actual oxidizing agent is effectively Cl^+ (chloronium ion). Cl^+ is reduced in the reaction; electrons are transferred to it to form chloride Cl^-. Some other agent such as an alcohol (or other oxidizable substrate) must supply these electrons and thus undergo oxidation. The overall transformation is illustrated in the equation below.

$$R_2CH{-}OH + ClO^- \longrightarrow R_2C{=}O + H_2O + Cl^-$$

A more detailed mechanism for this reaction could be written, but in fact the complete details of the mechanism are not well understood.

Under phase-transfer conditions one might expect that further oxidation could easily take place. This is apparently not the case, so further speculation will not be undertaken at this point. The purpose of the phase-transfer agent is simply to obtain a solution of the inorganic hypochlorite anion in the organic phase. Further discussion of the phase-transfer process may be found in Sec. 4.7.

EXPERIMENT 14.3A HYPOCHLORITE OXIDATION OF BENZHYDROL TO BENZOPHENONE

$$(C_6H_5)_2CH{-}OH + NaOCl \xrightarrow{\text{catalyst}} (C_6H_5)_2C{=}O + NaCl + H_2O$$

Time 2.5 h

Materials Benzhydrol, 5 g (diphenylmethanol, MW 184, mp 69°C)
10% Sodium hypochlorite, 50 mL (swimming-pool bleach)
Ethyl acetate, 50 mL
Tetrabutylammonium hydrogensulfate, 500 mg

Precautions Sodium hypochlorite is concentrated bleach and skin contact should be avoided.

Hazards Sodium hypochlorite is a strong oxidizing agent. Ethyl acetate is a narcotic in high concentrations.

Special Instructions

This experiment requires a magnetic stirring apparatus (Fig. 14.9). In the absence of a magnetic stirring apparatus only vigorous agitation for protracted periods will allow the reaction to succeed and even then yields will be poor.

Figure 14.9 Magnetic stirring apparatus.

Experimental Procedure

Place a magnetic stirring bar, 5 g benzhydrol (0.027 mol), and 50 mL ethyl acetate in a 250-mL Erlenmeyer flask. Add 50 mL concentrated sodium hypochlorite solution (swimming pool bleach, 10% aqueous NaOCl), followed by 500 mg tetrabutylammonium hydrogensulfate.

Stir the mixture vigorously for 1 h. If the flask begins to heat up, place it briefly in an ice-water bath until its temperature returns to ambient. The mixture should be maintained at room or ambient temperature throughout the reaction.

After the stirring period is over, transfer the mixture to a separatory funnel. Separate the layers and wash the ethyl acetate solution once with 25 mL distilled water, then twice with 25 mL 5% sodium bicarbonate, and finally once with 25 ml saturated aqueous sodium chloride solution. Place the solution in a 125-mL Erlenmeyer flask on a steam bath and allow the ethyl acetate to evaporate.

Dissolve the residue which remains after this treatment in 25 mL methyl alcohol and warm almost to the boiling point (65°C) of the alcohol. Add distilled water a little at a time with continued heating until the solution turns cloudy (usually 9 to 10 mL will suffice). Allow the solution to cool so that benzophenone deposits as an oil and then crystallizes. Vigorous swirling and the addition of a seed crystal assist the crystallization. Cool the aqueous methanol solution in an ice bath and filter. Approximately 4.5 g crude benzophenone should be obtained.

The crude material may be recrystallized from aqueous methyl alcohol or from petroleum ether (bp 65 to 75°C). The product which is obtained from the recrystallization will have a mp of 46 to 48°C and weigh approximately 3.5 to 4 g (which corresponds to a 70 to 80% yield). Thin-layer chromatography of this material shows a single pure material. (The tlc analysis can be run on a silica gel plate with 30% dichloromethane in petroleum ether as solvent.)

EXPERIMENT 14.3B HYPOCHLORITE OXIDATION OF 4-CHLOROBENZHYDROL TO 4-CHLOROBENZOPHENONE

4-Chlorobenzhydrol (Cl–C₆H₄–CH(OH)–C₆H₅) + NaOCl $\xrightarrow{\text{catalyst}}$ 4-Chlorobenzophenone (Cl–C₆H₄–CO–C₆H₅) + NaCl + H_2O

Time 2.5 h

Materials 4-Chlorobenzhydrol, 6 g (MW 218.68, mp 58 to 60°C)
10% Sodium hypochlorite, 50 mL (swimming-pool bleach)
Ethyl acetate, 50 mL
Tetrabutylammonium hydrogensulfate, 500 mg

Precautions Sodium hypochlorite is concentrated bleach and skin contact should be avoided.

Hazards Sodium hypochlorite is a strong oxidizing agent. Ethyl acetate is a narcotic in high concentrations.

Special Instructions

This experiment requires a magnetic stirring apparatus (Fig. 14.9). In the absence of a magnetic stirring apparatus only vigorous agitation for protracted periods will allow the reaction to succeed and even then yields will be poor.

Experimental Procedure

Place a magnetic stirring bar, 6 g 4-chlorobenzhydrol (0.027 mol, see Sec. 12.4) and 50 mL ethyl acetate in a 250-mL Erlenmeyer flask. Add 50 mL concentrated sodium hypochlorite solution (swimming-pool bleach, 10% aqueous NaOCl), followed by 500 mg tetrabutylammonium hydrogensulfate.

Stir the mixture vigorously for 1 h. If the flask begins to heat up, place it briefly in an ice-water bath until its temperature returns to ambient. The mixture should be maintained at room or ambient temperature throughout the reaction.

After the stirring period is over, transfer the mixture to a separatory funnel. Separate the layers and wash the ethyl acetate solution once with 25 mL distilled water, then twice with 25 mL of 5% sodium bicarbonate, and finally once with 25 mL saturated aqueous sodium chloride solution. Place the solution in a 125-mL Erlenmeyer flask on a steam bath and allow the ethyl acetate to evaporate.

Dissolve the residue which remains after this treatment in 50 mL methyl alcohol and warm almost to the boiling point (65°C) of the alcohol. Allow the solution to cool to room temperature and then place the flask in an ice-water bath until 4-chlorobenzophenone crystallizes. Vigorous swirling and the addition of a seed crystal assists the crystallization. Filter the solution (Buchner funnel) to obtain 4 to 4.5 g crude 4-chlorobenzophenone.

The crude material may be recrystallized from methyl alcohol (as above) or from petroleum ether (bp 65 to 75°C). The product which is obtained from the recrystallization will have a mp of 75 to 77°C and weigh approximately 3.5 to 4 g (which corresponds to 70 to 80% yield). Thin-layer chromatography of this material shows a single pure compound. (The tlc analysis can be run on a silica gel plate with 50% dichloromethane in petroleum ether as solvent.)

The nmr spectra of 4-chlorobenzhydrol and 4-chlorobenzophenone are shown in Figs. 14.10*a* and 14.11*a*, respectively. Compare the 4-chlorobenzophenone spectrum with that of benzophenone (Fig. 14.4*a*) obtained by the oxidation of benzhydrol (Fig. 14.5*a*). Although the aromatic regions of these two

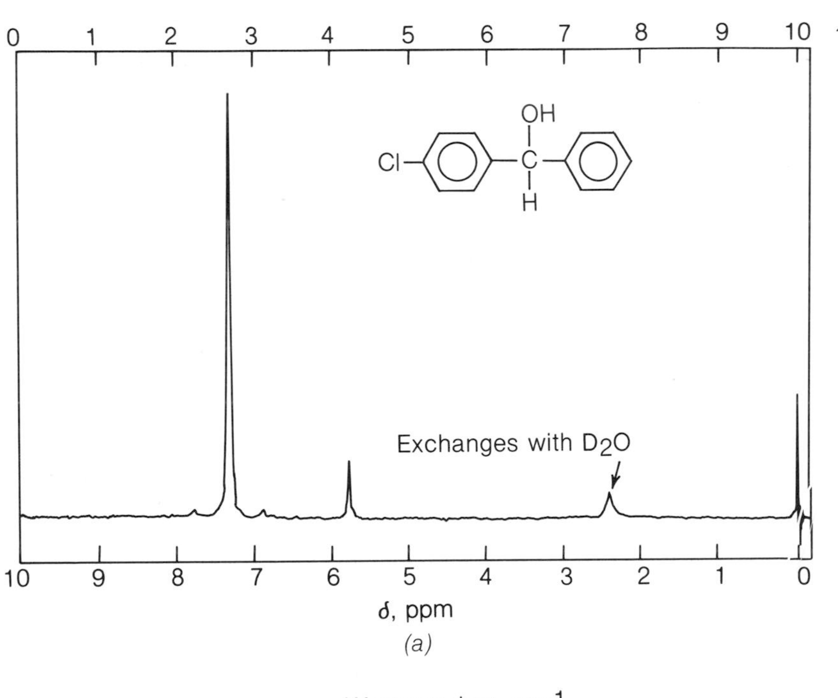

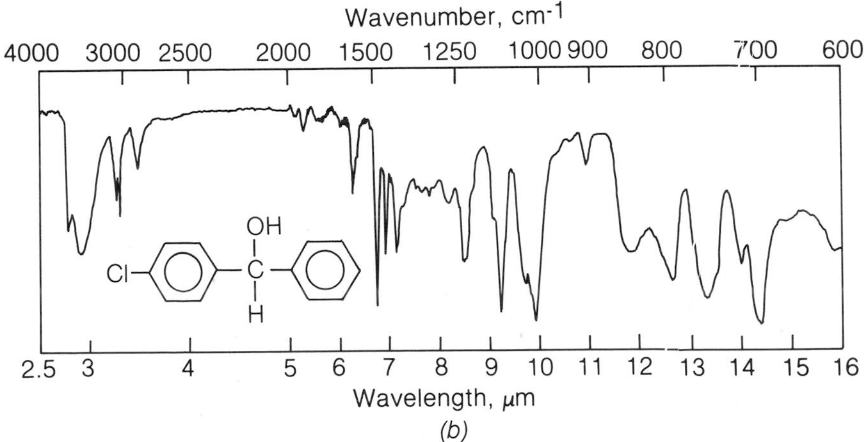

Figure 14.10
The (*a*) nmr and (*b*) ir spectra of 4-chlorobenzhydrol.

compounds look similar, a four-line pattern resulting from the substituted phenyl can be discerned.

An examination of the ir spectra of starting material (Fig. 14.10*b*) and product (Fig. 14.11*b*) would reveal the disappearance of the hydroxyl vibration (a broad band near 3400 cm^{-1}) of benzhydrol and the appearance of the ketone vibration band (at 1656 cm^{-1}).

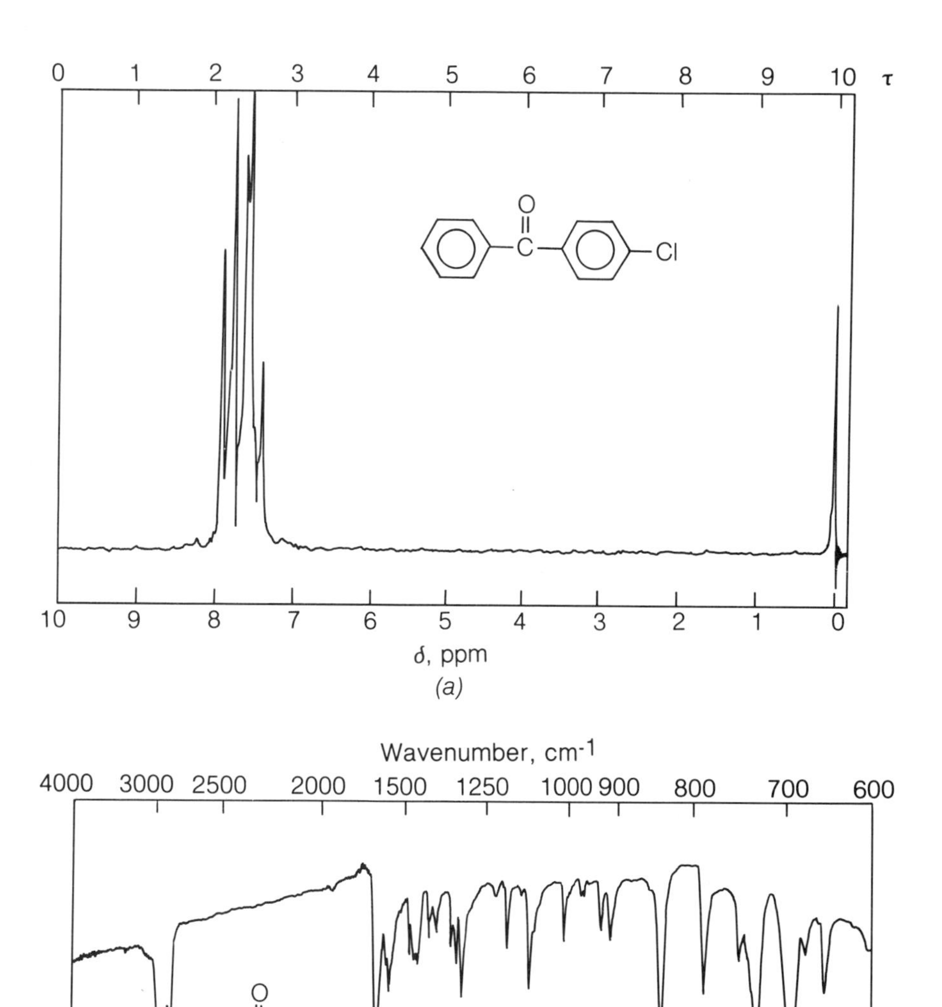

Figure 14.11
The (*a*) nmr and (*b*) ir spectra of 4-chlorobenzophenone.

14.4 SODIUM BOROHYDRIDE REDUCTION OF 4-CHLOROBENZALDEHYDE AND FLUORENONE

Two of the most important reducing agents now used in the organic chemistry laboratory were virtually unknown 30 years ago. Sodium borohydride ($NaBH_4$) and lithium aluminum hydride ($LiAlH_4$) were discovered in the 1940s, but exploitation of these reagents did not begin until after World War II.

Hydride ion itself, H^-, is a very poor nucleophile and a very potent base. It can deprotonate even very weak carbon acids but does not effectively add as a nucleophile to a carbonyl group. Sodium hydride (NaH) is therefore a poor reducing agent. The hydrogen-boron bond is strongly polarized toward hydrogen ($^-H—B^+$) and can effectively add to carbonyl groups ($R_2C=O$), resulting in the formation of a carbon-oxygen single bond ($R_2CH—OB$).

The overall reduction of an aldehyde with sodium borohydride is illustrated in the equation below.

$$4R—CHO + NaBH_4 \longrightarrow 4R—CH_2—O^- \xrightarrow{4H^+} 4R—CH_2—OH + H_3BO_3$$

It should be clear from this equation that 1 mol of sodium borohydride can reduce 4 mol of a carbonyl compound. The intermediate which actually forms is one in which hydride is added to carbon and an oxygen-boron bond forms. Hydrolysis of the oxygen-boron bond ultimately results in the formation of alcohol. The overall reduction of carbonyl requires two hydrogens, but only one of these comes from the reagent (that on carbon); the other hydrogen comes from acid present (e.g., as solvent) during workup.

Sodium borohydride has a molecular weight of 38. It therefore has massive molar reducing power relative to its molecular weight. Coincidentally, lithium aluminum hydride also has a molecular weight of 38 and therefore has similar reducing power. The principal difference between these two reagents is that an aluminum-hydrogen bond is far more reactive than a boron-hydrogen bond. For this reason lithium aluminum hydride can be utilized only in aprotic solvents (such as ether). It is a more powerful reducing agent; it can reduce esters to alcohols, whereas sodium borohydride cannot. Sodium borohydride, because it is a mild reducing agent, can be used in aqueous or alcoholic (usually methanol and ethanol) solvents. Its principal utility as an organic synthetic tool is in the reduction of aldehydes and ketones to primary and secondary alcohols.

In the overall reaction, borohydride is oxidized to boric acid. The reduction step is the conversion of the carbonyl group to the alcohol function. In the experiments described below, 4-chlorobenzaldehyde and fluorenone are reduced to 4-chlorobenzyl alcohol and fluorenol, respectively. Both compounds are soluble in methanol, and sodium borohydride can conveniently be used for either of these reduction reactions. Lithium aluminum hydride could be used to achieve the same end, but it is too reactive to be used in alcoholic solution.

Warning: Lithium aluminum hydride is so reactive that it can ignite on contact with water, even if the water is only atmospheric moisture. Great care must be exercised when using this reagent.

EXPERIMENT 14.4A REDUCTION OF 4-CHLOROBENZALDEHYDE TO 4-CHLOROBENZYL ALCOHOL

$$4\text{-}ClC_6H_4\text{-}CHO + NaBH_4 \longrightarrow 4\text{-}ClC_6H_4\text{-}CH_2\text{-}O\text{-}H$$

Time 2.0 h

Materials 4-Chlorobenzaldehyde, 7 g (MW 140.57, mp 44 to 47°C)
Sodium borohydride, 0.750 g (MW 38)
Methanol, 40 mL

Precautions Conduct reaction in a good hood. Weigh sodium borohydride rapidly and reclose the bottle quickly as the reagent is a very hygroscopic solid.

Hazards Sodium borohydride produces a caustic material on hydrolysis. Wash any spilled material immediately with water. Hydrolysis also produces hydrogen gas. Avoid breathing methanol vapors.

Experimental Procedure

Place 7 g (0.050 mol) 4-chlorobenzaldehyde in a 125-mL Erlenmeyer flask. Add 40 mL methanol and swirl (slight warming may be necessary) to dissolve the aldehyde; then allow the solution to come to room temperature.

Weigh 0.750 g (0.021 mol) sodium borohydride quickly into a small (3-dram) vial.[3] Stopper the vial to protect the hygroscopic reagent from moisture. Add the reducing agent to the methanol solution in one portion (at room temperature) and swirl vigorously to dissolve it. Allow the solution to stand, swirling every few minutes, for 20 min at room temperature.

After the reaction period, add 15 mL distilled water and place the methanol solution on the steam bath. A white precipitate should appear. Heat the aqueous methanol solution just to boiling (65°C) and then remove it from the steam bath. Swirl the solution vigorously. Return the solution to the steam bath intermittently during 5 min so the solution stays near 65°C.

After the 5-min hydrolysis period, allow the mixture to cool to room temperature and then pour it into a 500-mL separatory funnel containing 200 mL cold distilled water. Wash any material remaining in the Erlenmeyer flask into the separatory funnel with a 25-mL portion of dichloromethane.

[3] If a small vial is unavailable, a 10 × 75 mm test tube may be used instead. Stand the test tube up in either a 50-mL beaker or 50-mL Erlenmeyer flask during weighing, and then stopper the test tube with either a cork or a rubber stopper.

Shake the mixture, separate the layers, and then draw off the organic portion. Extract the aqueous solution twice more with 25-mL portions of dichloromethane.

Wash the combined organic fractions twice with saturated $NaHCO_3$ and then dry over anhydrous sodium sulfate for several minutes. Gently filter the organic layer into a 250-mL Erlenmeyer flask and evaporate the solvent by heating on a steam bath. Dissolve the residual oil in 4 vol % acetone in hexane (25 mL), heat just to boiling, and then allow the solution to cool to room

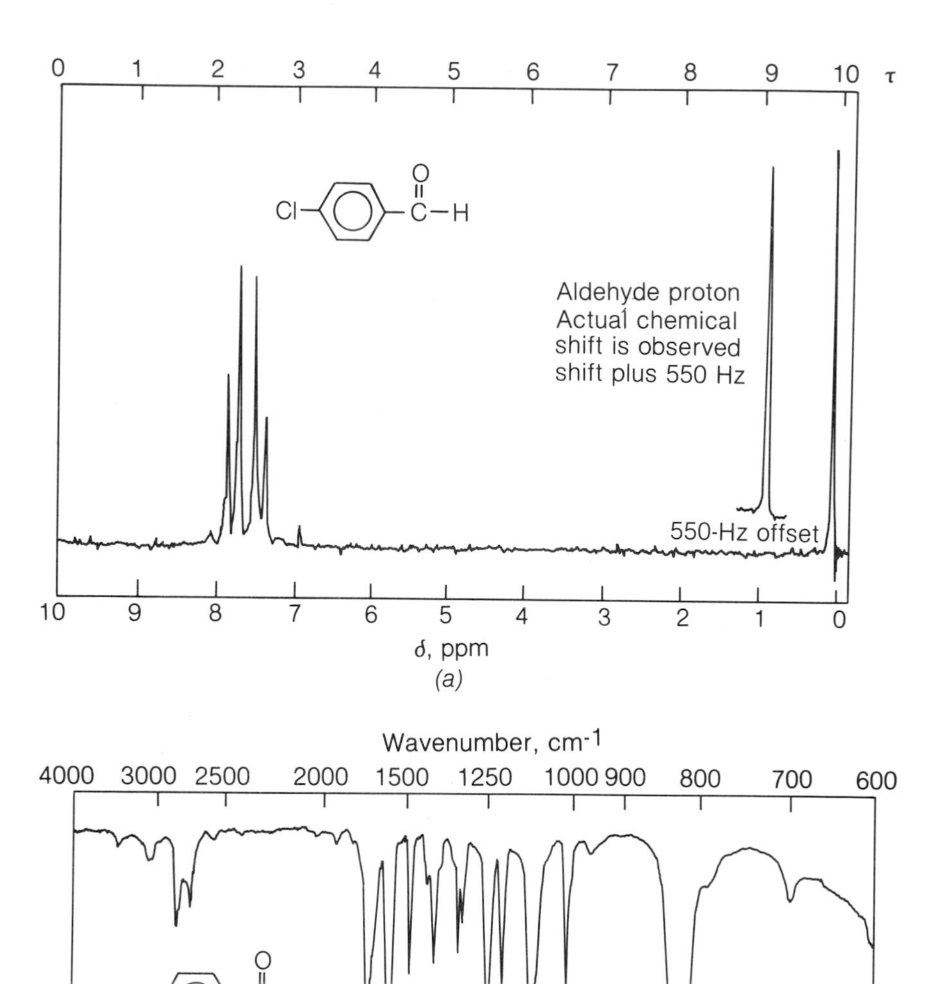

Figure 14.12
The (*a*) nmr and (*b*) ir spectra of 4-chlorobenzaldehyde.

temperature. Collect the crystals on a Buchner funnel and wash them with ice-cold hexane (10 to 15 mL). After air drying for several minutes, the crystalline 4-chlorobenzyl alcohol (approximately 5.5 g, 78%) should have mp 70 to 72°C.

The nmr and ir spectra of 4-chlorobenzaldehyde are shown in Fig. 14.12 and those of 4-chlorobenzyl alcohol in Fig. 14.13. Note that the A_2B_2 pattern characteristic of para substitution is obscured in the nmr spectrum because the chemical shift difference of the aromatic protons is too small. If the alco-

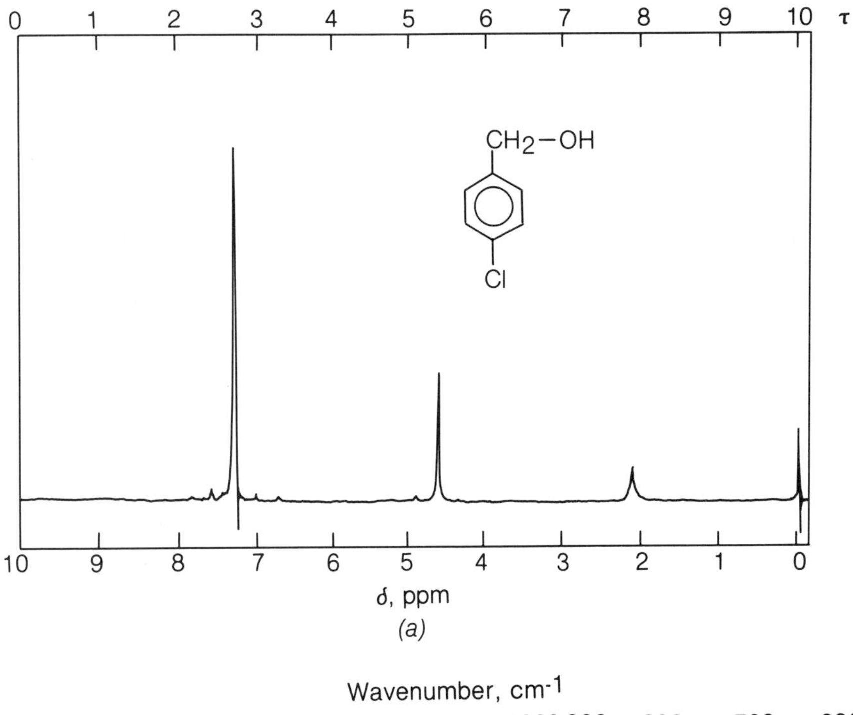

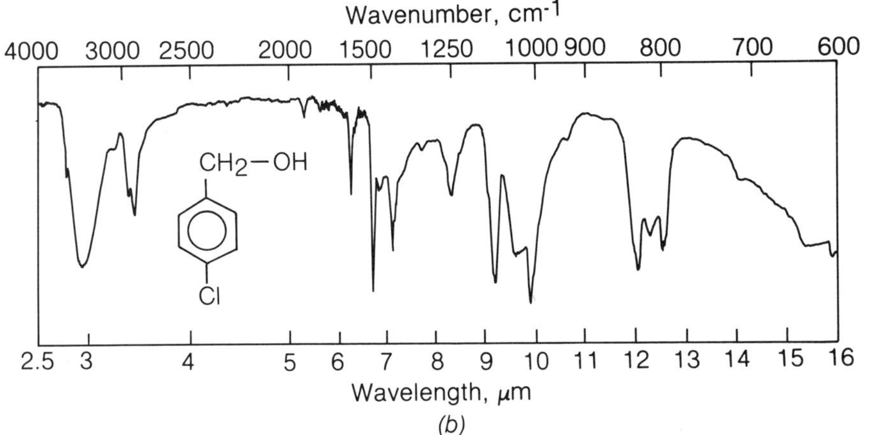

Figure 14.13
The (*a*) nmr and (*b*) ir spectra of 4-chlorobenzyl alcohol.

hol is very pure, the benzylic protons will appear as a doublet due to nonexchange of the alcohol proton. If D_2O is added to the tube, the doublet changes into a singlet. Note that this observation is made in the next preparation, that of fluorenol. In many cases highly pure alcohols will show this splitting.

Note in the ir spectra that the carbonyl band characteristic of the starting material disappears and a strong hydroxyl vibration becomes clearly visible in the spectrum of the product.

EXPERIMENT 14.4B *REDUCTION OF FLUORENONE TO FLUORENOL*

$$\text{Fluorenone (C=O)} + NaBH_4 \longrightarrow \text{Fluorenol (HO, H)}$$

Time 2.0 h

Materials Fluorenone, 3 g (MW 180, mp 82°C)
Sodium borohydride, 0.250 g (MW 38)
Methanol, 25 mL

Precautions Conduct reaction in a good hood. Weigh sodium borohydride rapidly and reclose the bottle quickly as the reagent is a very hygroscopic solid.

Hazards Sodium borohydride produces a caustic material upon hydrolysis. Wash any spilled material immediately with water. Hydrolysis also produces hydrogen gas. Avoid breathing methanol vapors.

Experimental Procedure

Place 3 g (0.017 mol) fluorenone in a 125-mL Erlenmeyer flask. Add 25 mL methanol and swirl (slight warming may be necessary) to dissolve the ketone; then allow the solution to come to room temperature.

Weigh 0.250 g (0.0066 mol) sodium borohydride quickly into a small (3-dram) vial.[4] Stopper the vial to protect the hygroscopic reagent from moisture. Add the reducing agent to the methanol solution in one portion (at room temperature) and swirl vigorously to dissolve it. Allow the solution to

[4] If a small vial is unavailable, a 10 × 75 mm test tube may be used instead. Stand the test tube up in either a 50-mL beaker or 50 mL Erlenmeyer flask during weighing, and then stopper the test tube with either a cork or rubber stopper.

stand, swirling every few minutes, for 20 min at room temperature. During the reaction the yellow color of fluorenone should gradually fade.

After the reaction period add 10 mL cold distilled water and place the methanol solution on the steam bath. A white precipitate should appear. Heat the aqueous methanol solution just to boiling (65°C) and then remove it from the steam bath. Swirl the solution vigorously. Return the solution to the steam bath intermittently during 5 min so the solution stays near 65°C. The solid which precipitated should redissolve during the heating process.

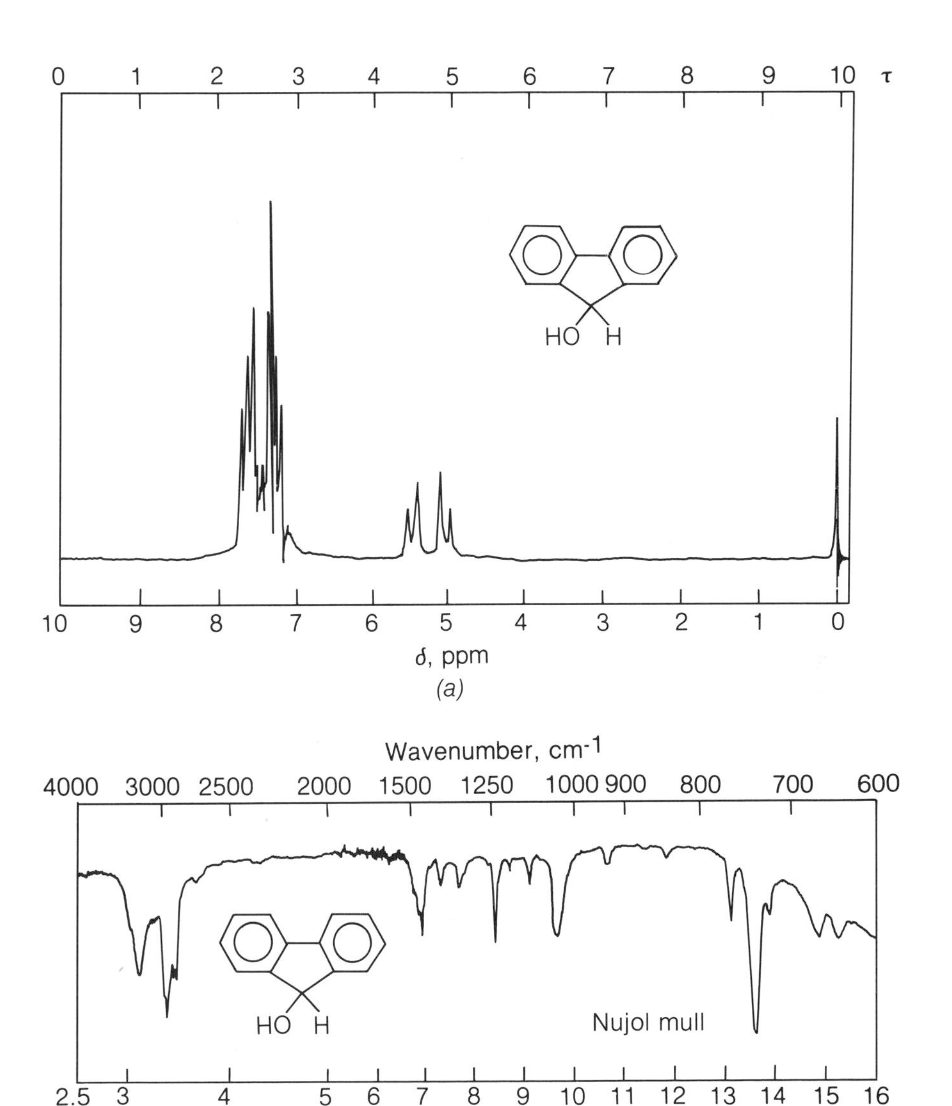

Figure 14.14
The (*a*) nmr and (*b*) ir spectra of fluorenol.

After the 5-min hydrolysis period remove the flask from the steam bath and allow the solution to cool slowly to room temperature. Collect the crystalline product on a Buchner funnel. Wash the product with 5 mL ice-cold 50% aqueous methanol. Briefly air dry the crystals to obtain 1.7 to 2.0 g (60 to 65% yield) fluorenol of mp 150 to 154°C.

The product may be recrystallized, if necessary, from methanol-water as above. The mother liquors may be reduced to half the original volume by heating on a steam bath. The second crop of crystals gives 0.5 to 0.75 g of additional fluorenol which is almost as pure as the first crop (total yield 80 to 85%).

The nmr and ir spectra of fluorenol are shown in Fig. 14.14. Notice in the nmr spectrum of fluorenol that the hydroxyl proton is split into a doublet (slow exchange). Addition of several drops of deuterium oxide (D_2O) will collapse this doublet into a singlet (fast exchange). The ir spectrum shows a typical aromatic alcohol, with no evidence of starting ketone. This fact may be confirmed by tlc analysis on silica gel using dichloromethane as eluant. Fluorenone has an R_f of 0.95 in this system, whereas fluorenol has an R_f of 0.5.

14.5 REDUCTION OF NITROBENZENE TO ANILINE

We have described above an oxidation process as one involving a loss of electrons. Conversely, a reduction is the addition of electrons to some reducible substance. Formally, one can think of a hydride reduction (with $NaBH_4$) as the addition of a proton and an electron pair. The electron pair obviously functions as the reducing agent.

In metal-ion reductions, it is an electron in the coordination sphere of the metal which is transferred to the reducible compound. In the example chosen here, a nitrogen-oxygen bond is reduced by electron transfer from a metal (in this case tin). The reducing agent, i.e., the source of electrons, is metallic tin. Tin may be used directly in these reductions without amalgamation (alloying with mercury). Other metals may be used in this procedure, although tin and zinc are used most often because they have large reduction potentials and are relatively inexpensive. A further consideration is the fact that the metal should be stable in aqueous acid solution. Other metals, such as sodium, magnesium, aluminum, and iron, may be used as reducing agents if the appropriate conditions are chosen.

One final but interesting point is the fact that the reduction conducted in this experiment is a heterogeneous reaction. In other words, the reduction does not occur in the liquid phase but rather at the metal surface. Another example of a heterogeneous reduction reaction is the metal-catalyzed transfer of hydrogen to a double bond by a platinum or palladium catalyst. The tin reduction described below can be easily formalized as the transfer of electrons followed by transfer of a proton. In hydrogenation with a palladium catalyst, the proton and the electrons happen to be transferred together.

EXPERIMENT 14.5

REDUCTION OF NITROBENZENE TO ANILINE

$$C_6H_5\overset{+}{N}(=O)O^- + Sn + H^+ \longrightarrow C_6H_5NH_2$$

Time 4 h (may be done in two lab periods)

Materials Nitrobenzene, 12.5 mL (MW 123, bp 210°C, d 1.2 g/mL)
Granulated tin, 30 g (MW 119)
Concentrated (36%) hydrochloric acid, 70 mL
Sodium hydroxide, 80 g (MW 40)
Sodium chloride, 40 to 50 g

Precautions Wear gloves and carry out all transfers in a good hood. Have an ice bath handy at all times.

Hazards Nitrobenzene and aniline are toxic materials which have high vapor pressures and can be absorbed through the skin. Avoid breathing vapors of either compound and avoid contact with skin or eyes.

Special Instructions: Apparatus

A 500-mL flask is a little too bulky for many people to handle easily. Since swirling is required in this experiment, it may be useful to securely clamp the central neck of the flask (see Fig. 14.15). When the clamp is attached to the ring stand, it is secured for the addition of HCl in 5-mL portions. After each addition, the clamp may be disconnected from the ring stand and the flask may be held while being swirled. The flask may also be outfitted with a stopper in one side neck and a thermometer and thermometer adapter in the other side neck. When the flask is swirled, the liquid will tend to rise in the two side necks but not in the central neck, so it is not crucial that this neck be stoppered. Nevertheless, there is another danger. The flask will contain small, granular pieces of tin and a relatively limited amount of liquid. The reaction mixture must not be swirled too vigorously, or the granular tin pieces may strike the thermometer bulb and break it. The mercury which would escape into the reaction mixture if this should happen would not hinder the reaction, but such an accident would be expensive for the operator.

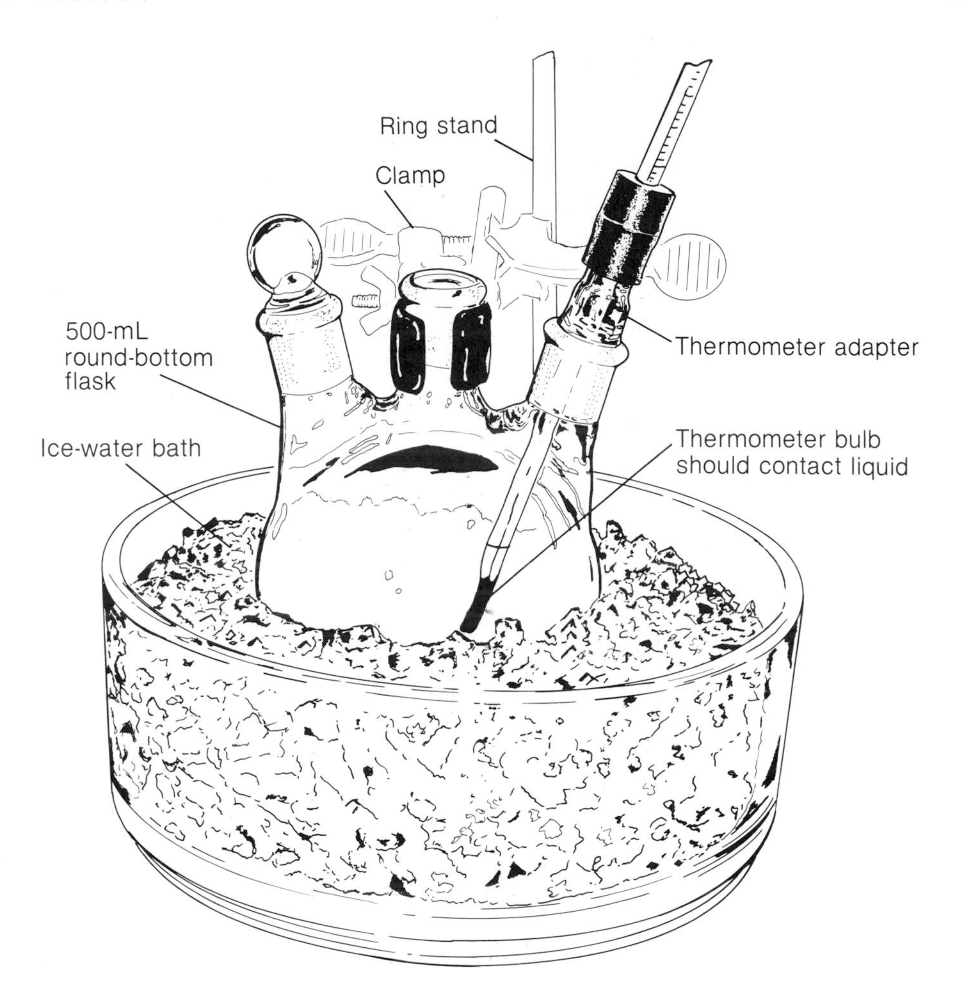

Figure 14.15 Apparatus for Exp. 14.5, the synthesis of aniline.

One way around this potential danger is to suspend the thermometer above the liquid, swirl so that the liquid does not contact the thermometer, and then slide the thermometer down so that it contacts the liquid after the swirling has ceased. This way, the temperature may be recorded but the thermometer remains out of danger. It is important to know that the vapors above the liquid will not be anywhere near the temperature of the liquid itself. As a result the liquid must contact the thermometer bulb to obtain an accurate reading.

This entire experiment can be greatly facilitated if a magnetic stirring apparatus is available. In setting up the apparatus, slide the stir bar through the side neck and gently down the side of the empty flask at the beginning of the setup operation (to prevent punching a hole in the bottom of the round-bottom flask). Add the tin and nitrobenzene as described below. Begin stirring and notice to what level the liquid rises. The thermometer may now be suspended in such a way that it touches the liquid but does not contact the stirring bar. If an ice bath is used to cool the flask, the ice bath must be made of glass, plastic, or aluminum. Small aluminum cooking pans available in variety stores for about 50 to 99¢ are often useful as ice baths.

After the HCl has been added, fix the reflux condenser to the central neck and continue as described in the experimental procedure below.

Experimental Procedure

Place 30 g (0.254 mol) of granulated tin in a 500-mL, round-bottom, three-neck flask. Add 12.5 mL nitrobenzene (**hood, gloves**). Prepare an ice bath large enough to fit around the flask and keep it next to the reaction apparatus. Add 70 mL concentrated hydrochloric acid in 5-mL portions. After each addition of acid, swirl the flask and keep the internal temperature of the reaction at about 60°C by using the ice bath (see Fig. 14.15). An obvious reaction occurs after each 5-mL portion of acid is added to the reaction mixture. After all the hydrochloric acid has been added, swirl the mixture for several minutes and then attach a condenser (with lightly greased joints) to the flask and heat the reaction mixture on a steam bath for 30 min. Remove the round-bottom flask from the steam bath and cool to room temperature. If the thermometer is still attached, remove it as well. Add slowly, with vigorous swirling and ice-bath cooling, a concentrated sodium hydroxide solution (made from 80 g sodium hydroxide and 125 mL water). The solution in the round-bottom flask should now be strongly basic. Fit the flask with a Claisen head (with lightly greased joints) and a separatory funnel filled with water. Using a flame, steam distill the mixture while adding water by means of the separatory funnel to keep the volume in the round-bottom flask at about 300 mL (refer to Sec. 2.2 for additional information). Continue to steam distill until the distillate becomes clear; then collect 50 mL more distillate. Saturate the aqueous layer with solid sodium chloride (use approximately 20 g sodium chloride per 100 mL distillate). Transfer the solution to a separatory funnel and extract the aqueous layer with three 25-mL portions of dichloromethane. Combine the dichloromethane layers and dry over granular sodium sulfate. Filter the organic layer to remove the drying agent and evaporate the dichloromethane by heating on a steam bath. (When extracting a concentrated sodium chloride solution with dichloromethane, shake more gently than usual. This will help avoid the formation of an emulsion). If this procedure is to be done in two laboratory periods, place the crude aniline in a 50-mL flask, stopper, and conduct the second distillation during the following period.

Transfer the remaining oil to a 50-mL round-bottom flask and assemble a simple distillation apparatus. Distill at atmospheric pressure using a flame or oil bath until the temperature of the distillate reaches 150°C. At this point turn off the flame, stop the water flow in the condenser, drain the water from the condenser into a sink, and continue the distillation. Collect (in a tared 25-mL Erlenmeyer flask) all the product which distills between 180 and 185°C. Pure aniline is a water-white liquid which boils at 184°C. Weigh the flask and calculate the yield of aniline.

The nmr spectra of nitrobenzene (Fig. 14.16*a*) and aniline (Fig. 14.17*a*) are

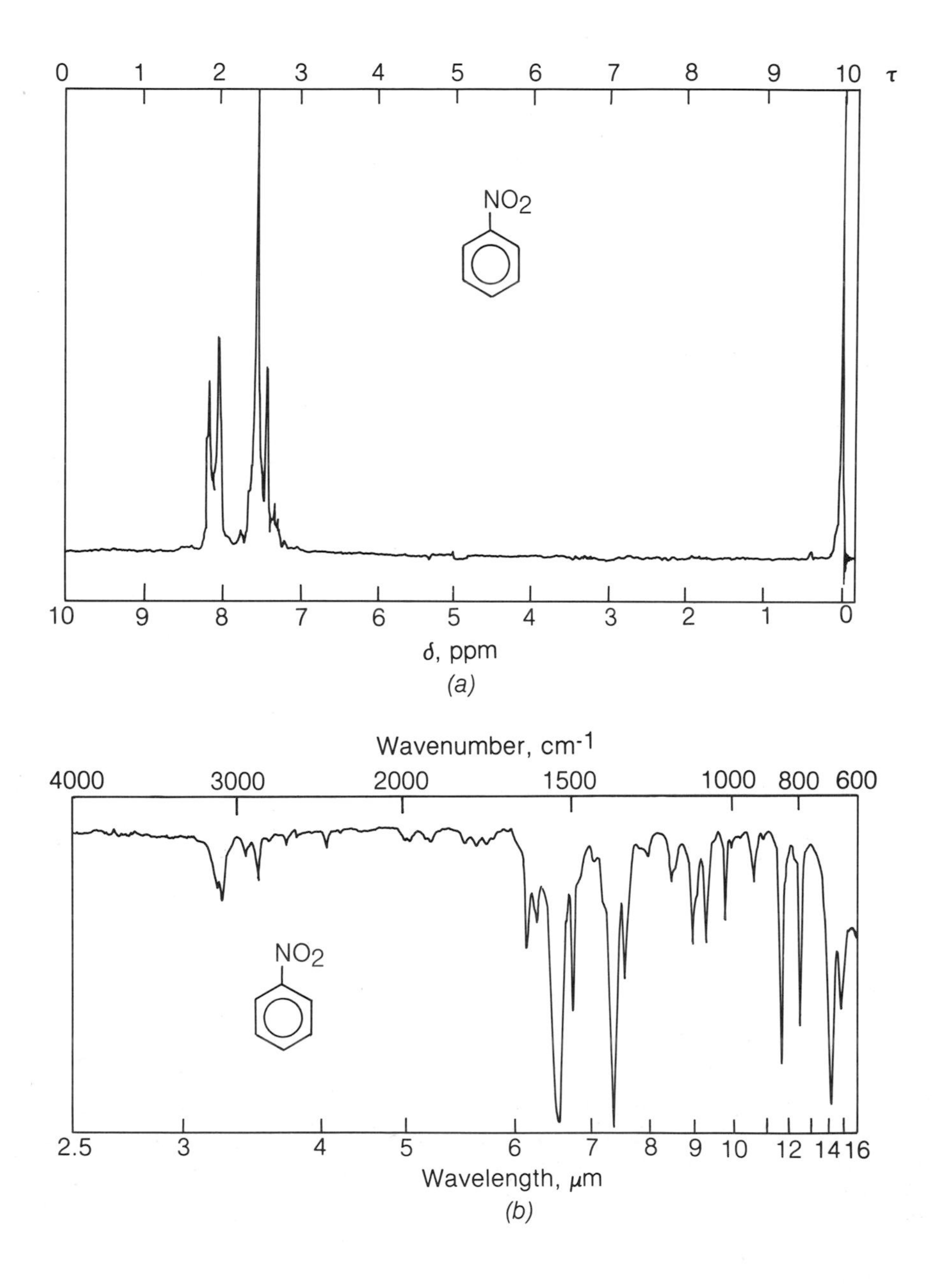

Figure 14.16 The (*a*) nmr and (*b*) ir spectra of nitrobenzene.

particularly informative. The protons ortho to the functional group in each case appear at very different positions, whereas the remaining protons are not affected as much. Can you correlate the nmr spectra with the relative electron-releasing and electron-withdrawing abilities of the NO_2 and NH_2 groups?

The ir spectra of nitrobenzene and aniline are shown in Figs. 14.16*b* and 14.17*b*, respectively.

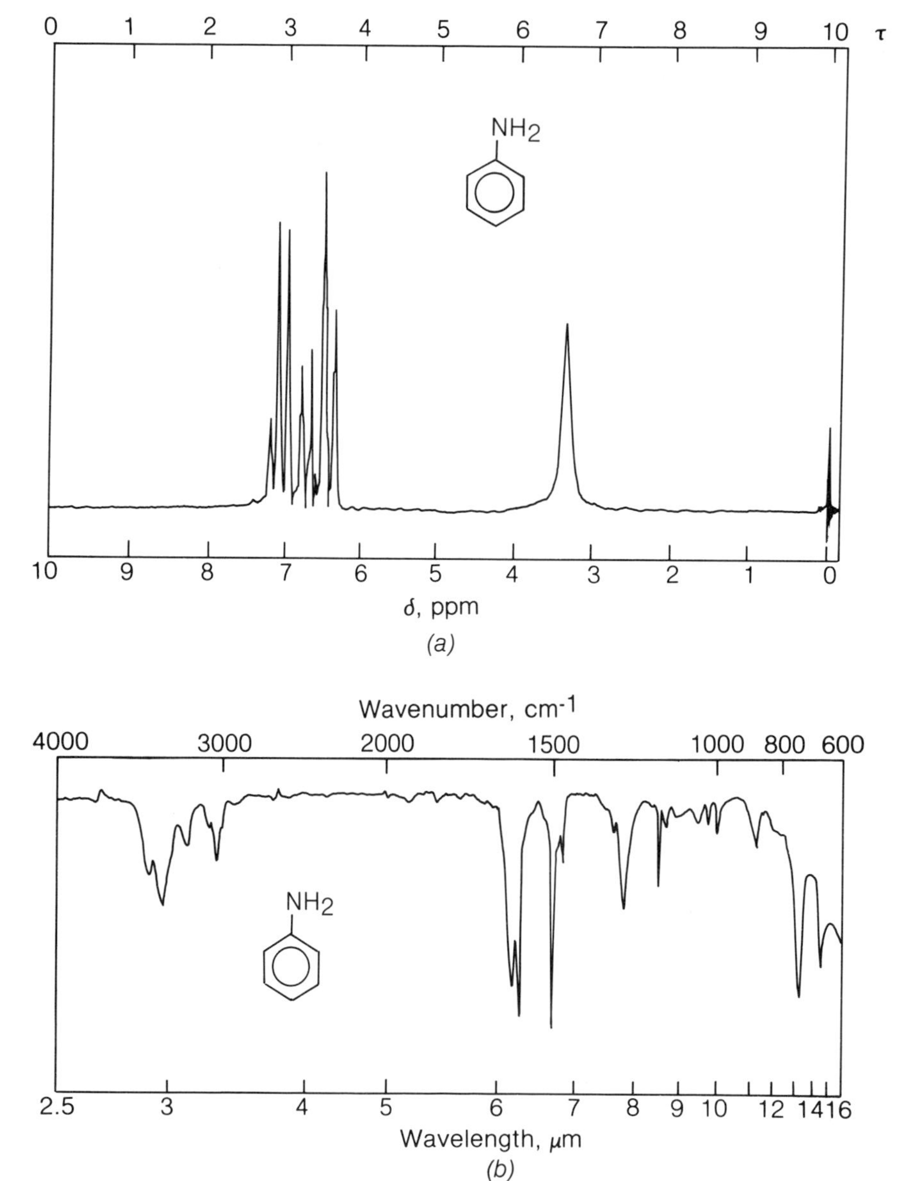

Figure 14.17
The (*a*) nmr and (*b*) ir spectra of aniline.

QUESTIONS AND EXERCISES

14.1 Compare the nmr spectrum of fluorenone (Fig. 14.3) with that of fluorenol (Fig. 14.14*a*). Would you be able to detect any fluorenol as an intermediate in the hydrocarbon oxidation? Is the presence of fluorenol predicted by the mechanism discussed in Sec. 14.1?

14.2 Thin-layer chromatographic analysis of fluorenone-fluorenol mixtures is made easy by the large difference in R_f exhibited by these compounds (see Exp. 14.4B). Why do you suppose the R_f values are so different? What would you predict for the R_f values of 4-chlorobenzaldehyde and 4-chlorobenzyl alcohol (Exp. 14.4A)? What would you predict for the R_f values of fluorene and fluorenol?

14.3 The reduction of nitrobenzene to aniline may be carried out by using either silver or tin. Why is tin usually chosen for this reduction?

14.4 The air oxidation of fluorene to fluorenone requires an hour or so to go to completion if a magnetic stirrer is used. How would you expect the rate of reaction to change if oxygen were used instead of air? Would air or oxygen oxidation be better for the commercial preparation of fluorenone by this reaction?

14.5 Refer to the oxidation mechanism illustrated in Sec. 14.2. In one step, water appears to initiate an elimination reaction. It appears that the CrO_3^- fragment is lost in this step. Is chromium undergoing an oxidation or reduction in this step?

14.6 If chromium trioxide oxidation of both benzhydrol and isoborneol were carried out in the same flask but with only half the required amount of oxidant present, what would the product mixture be like?

14.7 Two preparations of benzophenone are presented in this chapter. One involves oxidation of the alcohol with hypochlorite and the other involves oxidation with chromium trioxide. If you had to carry out this oxidation reaction on the 100,000-ton scale, which method would you choose and why?

14.8 The reagent lithium aluminum hydride is mentioned in Sec. 14.4. It is stated there that reaction of this reagent with water can lead to a fire. What chemical reaction must take place for this to occur? Would you expect reaction of lithium aluminum hydride to be more vigorous with water or with *tert*-butyl alcohol?

14.9 The reaction of sodium borohydride with 4-chlorobenzaldehyde is described in Exp. 14.4A. What reaction would you expect to occur with 4-chlorobenzoic acid?

XV
CONDENSATIONS OF ALDEHYDES AND KETONES

The carbonyl group has been characterized as the central or key functional group in organic chemistry. If so, base-catalyzed condensations involving carbonyl compounds are probably the most important set of reactions for forming carbon-carbon bonds. Among the many important base-catalyzed reactions are the Perkin, Claisen, and Knoevenagel condensations, illustrated below.

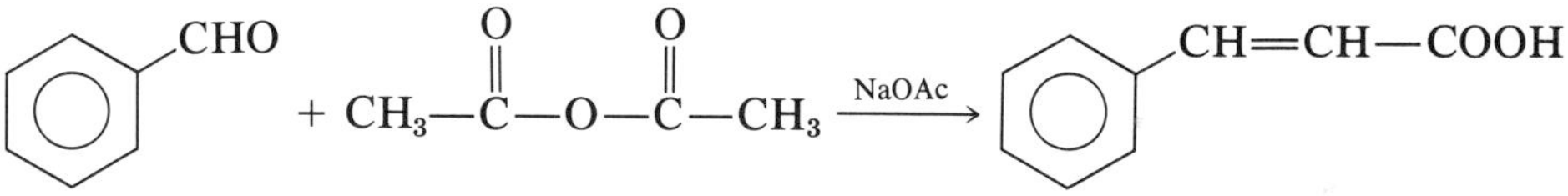

$$C_6H_5-CHO + CH_3-\overset{O}{\overset{\|}{C}}-O-\overset{O}{\overset{\|}{C}}-CH_3 \xrightarrow{NaOAc} C_6H_5-CH{=}CH-COOH$$

Perkin condensation

$$CH_3-\overset{O}{\overset{\|}{C}}-OC_2H_5 + CH_3-\overset{O}{\overset{\|}{C}}-OC_2H_5 \xrightarrow{NaOC_2H_5} CH_3-\overset{O}{\overset{\|}{C}}-CH_2-\overset{O}{\overset{\|}{C}}-OC_2H_5$$

Claisen condensation

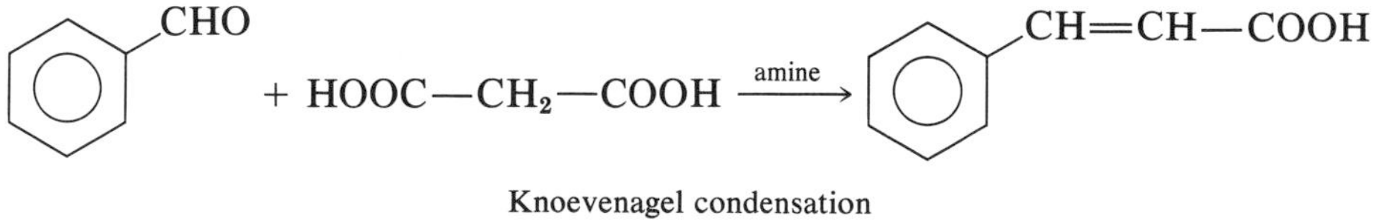

$$C_6H_5-CHO + HOOC-CH_2-COOH \xrightarrow{amine} C_6H_5-CH{=}CH-COOH$$

Knoevenagel condensation

All these related condensation reactions are characterized by two things: a carbonyl compound which serves as the electrophile, and a carbon acid from which a nucleophilic carbanion is formed. If a carbonyl compound is used as the carbon acid, the acidity is usually attributable to its ability to form a stable enolate ion. The delocalization of charge in the enolate ion affords it considerable stability. (Discussion of the effect of two adjacent carbonyl groups may be found in Sec. 13.5.)

$$\underset{\text{Keto form}}{-CH_2-\overset{O}{\overset{\|}{C}}-} \rightleftharpoons \underset{\text{Enol form}}{-CH=\overset{OH}{\overset{|}{C}}-} \underset{+H^+}{\overset{-H^+}{\rightleftharpoons}} -CH=\overset{O^-}{\overset{|}{C}}- \longleftrightarrow -\bar{C}H-\overset{O}{\overset{\|}{C}}-$$

Enolate ion

Once the enolate ion has been formed by deprotonation of the carbonyl compound, it reacts with some electrophile present in solution. If we consider what reaction might occur between acetaldehyde (ethanal) and base, we see that both the electrophile and the carbon acid are present in the same molecule. The reaction sequence can be formulated as follows:

$$H-\overset{O}{\overset{\|}{C}}-CH_3 + NaOH \rightleftharpoons H-\overset{O}{\overset{\|}{C}}-CH_2^-Na^+ + H_2O$$

$$H-\overset{O}{\overset{\|}{C}}-CH_2^-Na^+ + H-\overset{O}{\overset{\|}{C}}-CH_3 \longrightarrow H-\overset{O}{\overset{\|}{C}}-CH_2-\underset{H}{\underset{|}{\overset{O^-Na^+}{\overset{|}{C}}}}-CH_3$$

$$H-\overset{O}{\overset{\|}{C}}-CH_2-\underset{H}{\underset{|}{\overset{O^-Na^+}{\overset{|}{C}}}}-CH_3 \xrightarrow{H_3O^+} H-\overset{O}{\overset{\|}{C}}-CH_2-\overset{OH}{\overset{|}{C}H}-CH_3$$

3-Hydroxybutanal (aldol)

The base-catalyzed reaction of acetaldehyde with itself is the simplest base-catalyzed condensation. The product, 3-hydroxybutanal, is given the trivial name *aldol,* and this name has been generalized to the reaction type. Many reactions between a carbon acid and a carbonyl compound are referred to as aldol condensations, although the term may not be strictly appropriate. The general mechanism for the reactions which are discussed in this chapter may be represented by the equation below, in which Z is an electronegative group such as C=O, NO_2, or CN.

$$Z-CH_3 + :\text{base} \rightleftharpoons Z-CH_2^- + H-\text{base}$$

$$Z-CH_2^- + R-\overset{O}{\overset{\|}{C}}-R' \rightleftharpoons Z-CH_2-\underset{R}{\underset{|}{\overset{O^-}{\overset{|}{C}}}}-R' \xrightarrow[-H_2O]{+H^+} Z-CH=C\begin{matrix}R'\\R\end{matrix}$$

The last step, in which water is eliminated, often occurs spontaneously, especially if Z, R, or R′ is some group capable of conjugating with the double bond.

The richness of base-catalyzed carbonyl condensation chemistry is apparent from a consideration of how many different groups and structures are represented by Z, R, and R′. Only a few examples are included in this chapter.

15.1 THE ALDOL CONDENSATION

In this section two traditional aldol condensations are described. In the first of the two experiments, the compound known as dibenzalacetone is prepared from benzaldehyde (electrophile) and acetone (nucleophile, carbon acid) by using sodium hydroxide as the base. In the course of the reaction a β-hydroxyketone is produced as the initial product, just as in the aldol reaction above. In this case the reaction does not stop at the β-hydroxycarbonyl stage but continues to the dehydrated product. The only major difference between this reaction and the condensation of acetaldehyde itself is this dehydration step to form an α,β-unsaturated carbonyl compound.

$$C_6H_5CHO + CH_3-\overset{O}{\overset{\|}{C}}-CH_3 \rightleftharpoons C_6H_5CH(OH)CH_2COCH_3 \rightleftharpoons$$

$$C_6H_5CHO + C_6H_5CH=CHCOCH_3 \rightleftharpoons$$

$$C_6H_5CH(OH)CH_2COCH=CHC_6H_5 \rightleftharpoons C_6H_5CH=CHCOCH=CHC_6H_5$$

Dibenzalacetone

Notice that the system produced is conjugated through the carbonyl system to the aromatic ring, making loss of water energetically favorable.

Observe that there are two sets of acidic hydrogen atoms in the acetone molecule, i.e., the methyl groups on both sides of the ketone carbonyl. After one condensation has been performed, the same reaction may occur on the other side of the molecule. This happens, and the final product, dibenzalacetone, is isolated. This reaction is a double aldol condensation between two molecules of benzaldehyde (electrophile) and one molecule of acetone (nucleophile).

Exactly the same mechanism operates in the second example, the synthesis of chalcone.

$$C_6H_5\text{–}CHO + CH_3\text{–}CO\text{–}C_6H_5 \underset{}{\overset{\text{base}}{\rightleftharpoons}} C_6H_5\text{–}CH(OH)\text{–}CH_2\text{–}CO\text{–}C_6H_5 \underset{+H_2O}{\overset{-H_2O}{\rightleftharpoons}} C_6H_5\text{–}CH{=}CH\text{–}CO\text{–}C_6H_5$$

Chalcone

The only variation in the mechanism is that benzaldehyde may react only once with acetophenone as there is only one site of carbon acidity. The production of the α,β double bond is also enhanced by the conjugation to the aromatic ring.

It should be emphasized that all steps in these reactions are reversible. Therefore one may treat an α,β-unsaturated carbonyl compound such as dibenzalacetone with sodium hydroxide in order to obtain benzaldehyde and acetone, although this reaction is almost never carried out in the laboratory. The relative concentrations of the starting materials and products are very dependent on the stabilities of the various compounds and the conditions under which the reaction is carried out. For example, a good rule of thumb observed in these types of reactions is that low temperatures tend to favor the aldol condensation (equilibrium shifted to the right side of the equation above), while high temperatures tend to favor the retro (or reverse) aldol reaction (equilibrium shifted to the left side of the equation above).

The reversibility of the aldol reaction may seem at first to be a disadvantage compared with other types of chemical reactions. This is not so. Notice that in either direction carbon-carbon bonds are formed and broken under mild conditions. Notice also that by controlling the conditions the reaction can be driven either way, thus offering tremendous flexibility in process design. The best example of this flexibility is found in biochemistry. The chemistry of sugar and fat metabolism abounds with examples of aldol and retroaldol-type reactions. The product distribution in these reactions is controlled by enzyme catalysis. The examples shown on the next page occur in biochemical processes and are arranged to show the relationship with the aldol reaction.

$$\begin{array}{l} CH_2{-}OP \\ | \\ C{=}O \\ | \\ CHOH \\ | \\ CHOH \\ | \\ CHOH \\ | \\ CH_2{-}OP \end{array} \rightleftharpoons \begin{array}{l} CH_2{-}OP \\ | \\ C{=}O \\ | \\ CH_2{-}OH \end{array} + \begin{array}{l} H \\ \quad \backslash \\ \quad C{=}O \\ \quad | \\ \quad CHOH \\ \quad | \\ \quad CH_2{-}OP \end{array}$$

Fructose 1,6-diphosphate — Dihydroxyacetone phosphate — Glyceraldehyde phosphate

$$\begin{array}{l} CO_2^- \\ | \\ C{=}O \\ | \\ CH_2 \\ | \\ CO_2^- \end{array} + CH_3{-}\overset{\overset{\displaystyle O}{\|}}{C}{-}SCoA \rightleftharpoons HO{-}\begin{array}{c} CH_2{-}CO_2^- \\ | \\ C{-}CO_2^- \\ | \\ CH_2{-}CO_2^- \end{array}$$

Acetylcoenzyme A — Citric acid

$$CH_3{-}\overset{\overset{\displaystyle O}{\|}}{C}{-}SCoA + CH_3{-}\overset{\overset{\displaystyle O}{\|}}{C}{-}SCoA \longrightarrow CH_3{-}\overset{\overset{\displaystyle O}{\|}}{C}{-}CH_2{-}\overset{\overset{\displaystyle O}{\|}}{C}{-}SCoA$$

Acetylcoenzyme A (thioesters) — Acetoacetic acid thioester

EXPERIMENT 15.1A SYNTHESIS OF DIBENZALACETONE BY THE ALDOL CONDENSATION

$$2\,C_6H_5{-}CHO + CH_3{-}\overset{\overset{\displaystyle O}{\|}}{C}{-}CH_3 \xrightarrow{NaOH} C_6H_5{-}CH{=}CH{-}\overset{\overset{\displaystyle O}{\|}}{C}{-}CH{=}CH{-}C_6H_5 + 2H_2O$$

Time 2.0 h

Materials Benzaldehyde, 10.5 mL (MW 106, bp 182°C, d 1 g/mL)
Acetone, 2.9 g (3.63 mL, MW 58, d 0.79 g/mL)
Sodium hydroxide, 5 g (MW 40)
95% Ethanol, 25 mL

Precautions Use no flame.

Hazards Ethyl acetate is flammable and an irritant to the eyes and respiratory system. Avoid breathing vapors and contact with eyes.

Experimental Procedure

Add 5 g (0.0125 mol) NaOH to 25 mL water in a 250-mL Erlenmeyer flask. Swirl to effect solution. Add 25 mL 95% ethanol, swirl, and allow the solution to come nearly to room temperature. Add 2.9 g (3.63 mL, 0.05 mol) acetone and then 10.5 mL (10.6 g, 0.1 mol) benzaldehyde. The solution quickly turns yellow to orange (depending on the purity of the benzaldehyde used) and also warms up some. A yellow precipitate begins to appear almost immediately.

After about 15 min of occasional swirling, the reaction mixture is filtered on a Buchner funnel. The product is washed with cold alcohol and allowed to suck dry briefly. The yellow mass is recrystallized from a minimum amount of ethyl acetate.

After recrystallization, a yellow crystalline product of mp 112°C is obtained. The nmr spectrum of dibenzalacetone is shown in Fig. 15.1. The two most upfield lines are due to protons on the carbon-carbon double bonds. From the value of the coupling constant (17 Hz), the trans(*E*)-geometry can be assigned.

Put a small amount of your pure dibenzalacetone in a small test tube and dissolve it in a small amount of dichloromethane. In a separate test tube, dissolve a small amount of cyclohexene in a small amount of dichloromethane, and then add a few drops of bromine-dichloromethane solution to each tube. What does the difference in reactivity tell you about the reactivity of these two olefins? How might you account for this difference?

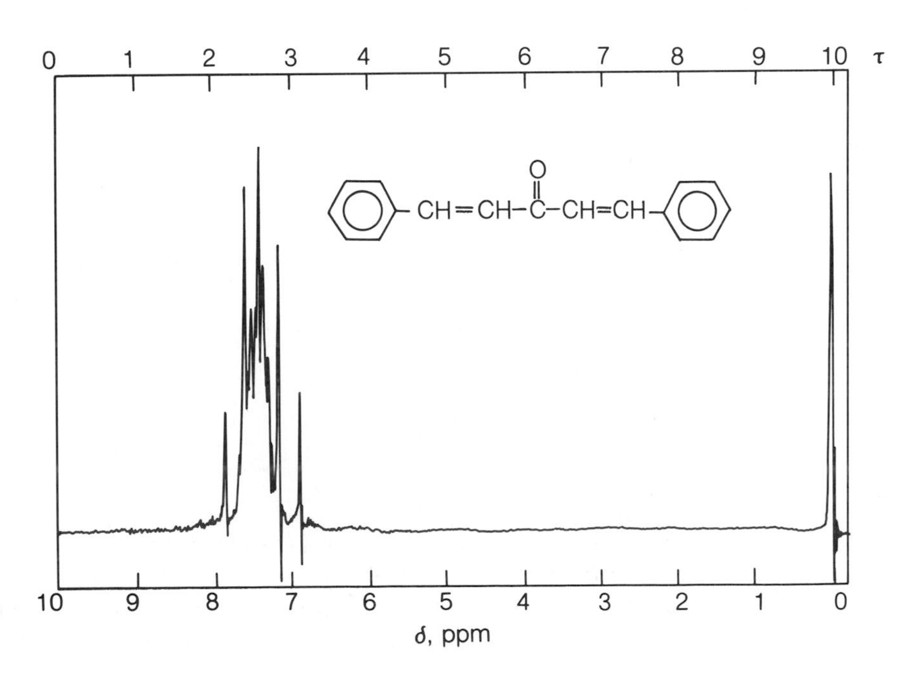

Figure 15.1
The nmr spectrum of dibenzalacetone.

EXPERIMENT 15.1B

SYNTHESIS OF BENZALACETOPHENONE (CHALCONE) BY THE ALDOL CONDENSATION

$$C_6H_5COCH_3 + C_6H_5CHO \xrightarrow{NaOH} C_6H_5CO{-}CH{=}CH{-}C_6H_5$$

Time 2.5 h

Materials Benzaldehyde, 10.5 mL (MW 106, bp 182°C, d 1 g/mL)
Acetophenone, 12 mL (MW 120, bp 202°C, d 1 g/mL)
Sodium hydroxide, 5 g (MW 40)
95% Ethanol, 50 to 75 mL

Precautions Wear gloves during this entire experiment. Be very careful to carry out all operations in a good hood and immediately wash off any spilled product from skin and clothes.

Hazards The product, chalcone, appears to be a strong skin irritant to some people. It should not be allowed to come in contact with skin or eyes. On repeated exposure, a sensitization reaction sometimes occurs; therefore all manipulations, including the recrystallization, should be done in *one* laboratory period. Hand in all samples to your laboratory instructor at the end of the period.

Experimental Procedure

Add 5.0 g (0.125 mol) sodium hydroxide to 40 mL distilled water in a 250-mL Erlenmeyer flask. After the base has dissolved (swirl), add 30 mL 95% ethanol and swirl the flask to effect mixing of the liquids. Check the temperature of the liquid by inserting a thermometer. The liquid should be between 15 and 30°C. If it is warmer than that, place the flask in an ice-water bath and swirl it until the temperature falls to about 20°C. Add 12 mL (0.1 mol) acetophenone. After briefly swirling the flask, add 10.5 mL (0.1 mol) benzaldehyde and swirl the flask often during the next 0.5 to 1 h while maintaining the temperature between 15 and 30°C by judicious use of an ice-water bath.

Eventually the reaction mixture will appear as a light yellow paste at the bottom of the flask, covered with an almost clear solution. Using a glass rod, touch the yellow mass to see if it is indeed solid. If so, filter it using a Buchner funnel. Allow the mass to suck dry, then carefully break up the mass and transfer all but a few very small pieces of the solid to another 250-mL flask. Recrystallize the solid from 95% ethanol. The small bits of crude product which were put aside can now be used as seed crystals. The pure yellow product, mp 57°C, is obtained by suction filtration and should amount to greater than 50% yield.

Collect the crystals, record the melting point in your notebook, and **turn in all materials to your instructor *before* you leave the laboratory.**

The nmr and ir spectra of chalcone are presented in Fig. 15.2. Note at what wavelength the carbonyl vibration is observed in the ir spectrum. The carbonyl

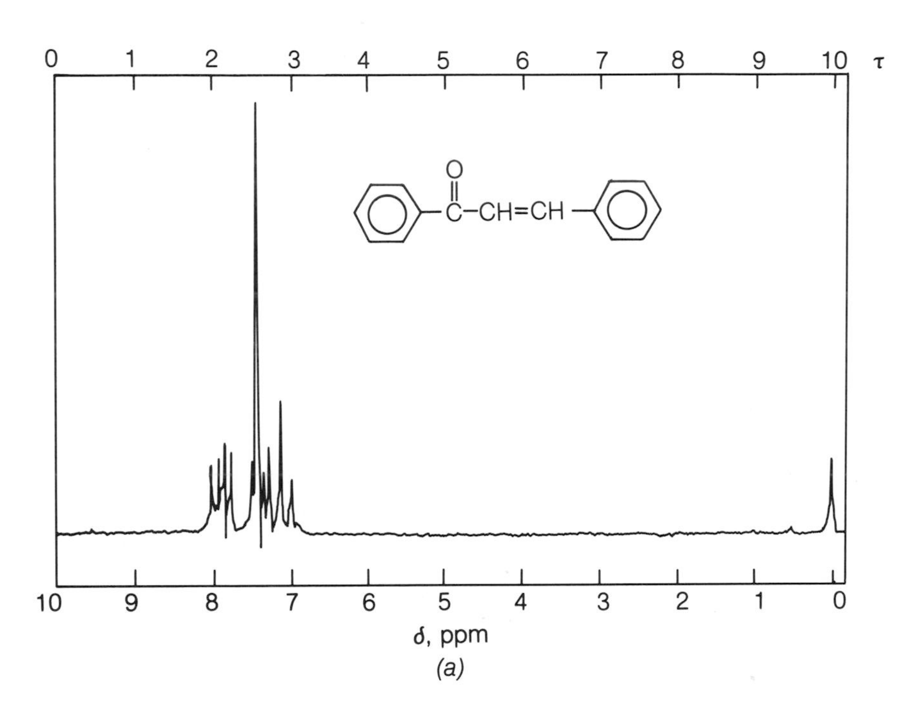

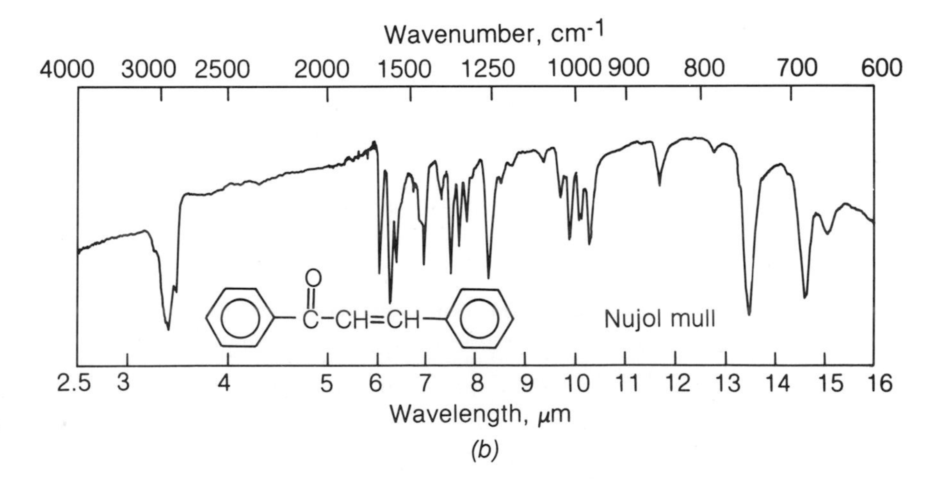

Figure 15.2
The (*a*) nmr and (*b*) ir spectra of benzalacetophenone (chalcone).

bands in benzaldehyde and acetophenone are observed at 1704 and 1686 cm^{-1}, respectively.

15.2 REACTION OF ACTIVATED HYDROCARBONS

In our discussion of the aldol condensation, we characterized the reaction in general terms by saying that an electrophile reacts with the anion derived from a carbon acid to form a β-hydroxy compound, which may then be dehydrated to a double-bonded species. In the first experiment (15.1A) the electrophile is benzaldehyde and the carbon acid is acetone, the product being dibenzalacetone. In the second experiment (15.1B) the electrophile is again benzaldehyde, while the carbon acid is acetophenone and the product is chalcone. Even though ketones are used as nucleophiles in these two experiments, there is no reason why this reaction should not occur with other carbon acids. The only requirement imposed by the aldol reaction is that a carbon acid, the anion of which is usually stabilized by resonance of some type, be present.

Nitromethane is a fairly reactive carbon acid. The acidity is due to resonance stabilization of the anion by the nitro group (a nitrogen-oxygen double bond rather than a carbon-oxygen double bond). One might expect nitromethane to participate in the aldol reaction, and this is observed experimentally.

$$C_6H_5CHO + CH_3-\overset{+}{N}(=O)O^- \xrightarrow{\text{base}} C_6H_5CH=CH-\overset{+}{N}(=O)O^-$$

This reaction is important in the synthesis of drugs. For example, a β-nitrostyrene molecule may be reduced to a biologically active phenethylamine. The molecule may then be used as a starting material for other medicinally important compounds:

$$C_6H_5CH=CHNO_2 + H_2 \text{ (reduction)} \longrightarrow C_6H_5CH_2CH_2NH_2 \xrightarrow[\text{steps}]{\text{several}} \text{substituted tetrahydroisoquinoline (N-R, C-H, R')}$$

Phenethylamine

Substituted tetrahydroisoquinoline

Another carbon acid which has been discussed before is the aromatic compound fluorene. As noted in Sec. 14.1, fluorene is acidic because its anion is resonance-stabilized by the aromatic rings. One might expect that fluorene, under the right conditions, would act as a carbon acid in an aldol sense. This is indeed the case. An example is the condensation of fluorene with benzaldehyde to give 9-benzalfluorene (see below).

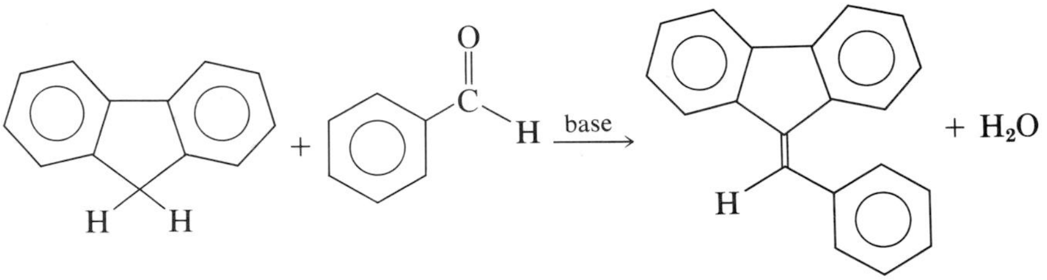

9-Benzalfluorene is a typical aldol-type product in which the β-hydroxy intermediate has been dehydrated to form the resonance-stabilized double bond.

There is another interesting reaction observed in this system. Under appropriate conditions, the double bond in 9-benzalfluorene will accept a hydride, giving the reduced product 9-benzylfluorene. This hydride transfer reaction is similar to the one which occurs in the Cannizzaro reaction (see Sec. 15.5). The hydride source here is the solvent, benzyl alcohol. During the course of the reaction, benzyl alcohol is oxidized to benzaldehyde while 9-benzalfluorene is reduced to 9-benzylfluorene. The procedure below describes the synthesis of 9-benzylfluorene from fluorene in benzyl alcohol.

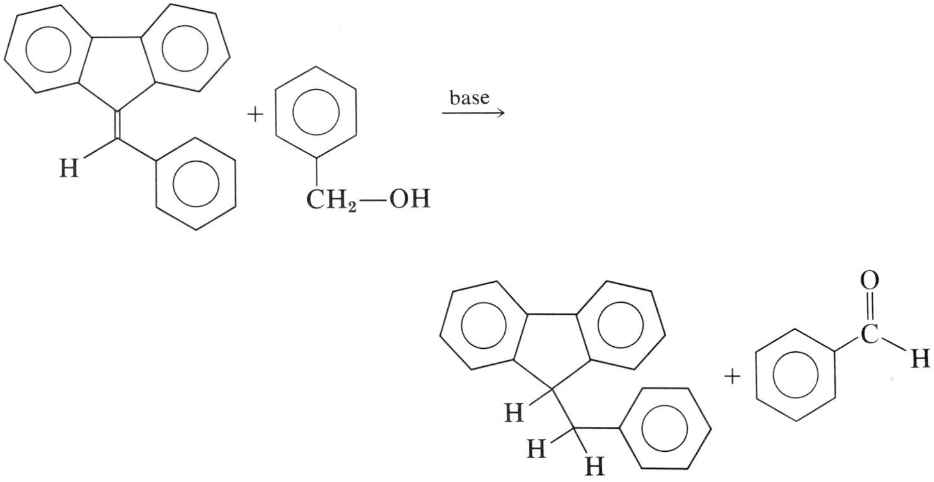

The intermediate in this reaction is 9-benzalfluorene, which is reduced to the saturated material. Only a few drops of benzaldehyde are needed at the start of the reaction because the formation of 9-benzylfluorene is accompanied by formation of more benzaldehyde, which immediately reacts with more fluorene to carry on the reaction.

EXPERIMENT 15.2A

SYNTHESIS OF 9-BENZALFLUORENE FROM FLUORENE AND BENZALDEHYDE

$$\text{fluorene} + \text{PhCHO} \xrightarrow{\text{base}} \text{9-benzalfluorene} + H_2O$$

Time 3 h

Materials Fluorene, 5 g (MW 166, mp 112 to 115°C)
Benzaldehyde, 5 mL (MW 106, bp 182°C, d 1 g/mL)
Benzyl alcohol, 25 mL (MW 108, bp 205°C, d 1 g/mL)
Potassium hydroxide, 2.5 g (MW 56)

Hazards Potassium hydroxide is a caustic material. Avoid contact with skin and eyes.

Experimental Procedure

Place 5 g (0.030 mol) fluorene in a 500-mL Erlenmeyer flask. Add 5 mL (0.047 mol) benzaldehyde, 25 mL benzyl alcohol, and 2.5 g (0.045 mol) potas-

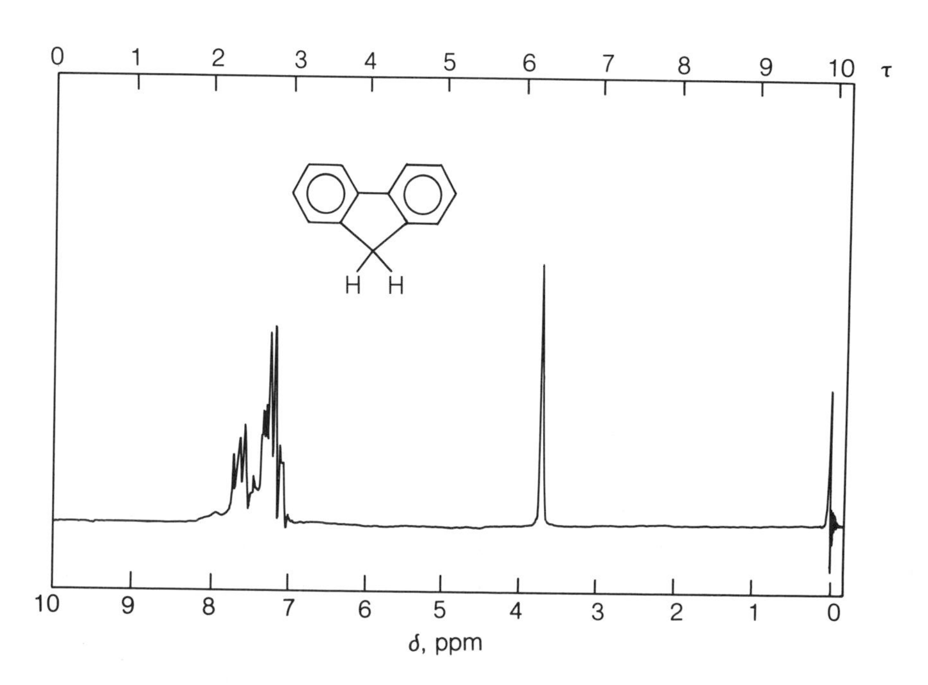

Figure 15.3
The nmr spectrum of fluorene.

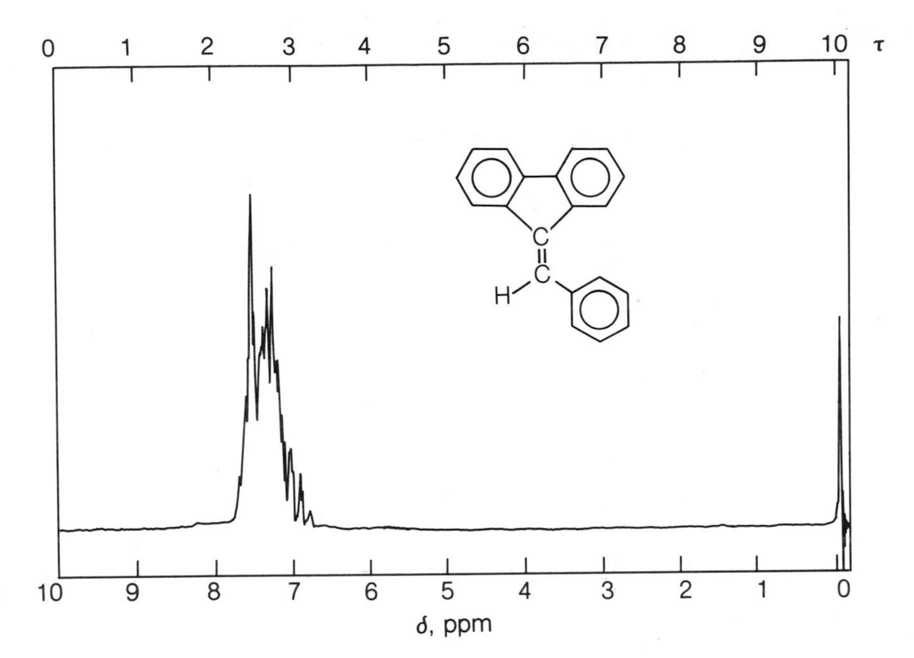

Figure 15.4
The nmr spectrum of 9-benzalfluorene.

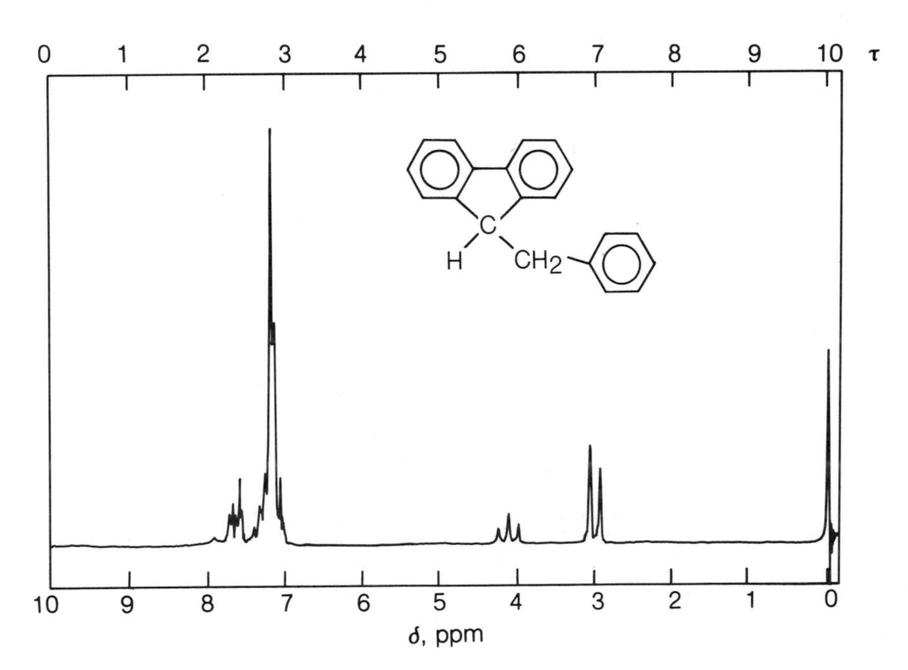

Figure 15.5
The nmr spectrum of 9-benzylfluorene.

sium hydroxide. Swirl the flask to dissolve the KOH and then heat on a steam bath for 1.5 h.

After the indicated heating period, allow the flask to cool and then slowly add 250 mL water while swirling. Allow the flask to stand for 20 to 30 min while the product crystallizes. Collect the crude yellow product on a Buchner funnel and wash it several times with water. Allow the material to suck dry on the Buchner funnel (to remove most of the water) and then air dry on a dry filter paper. Recrystallize the air-dried material from a minimum amount of heptane. The pale yellow product should have mp 75 to 76°C.

The nmr spectra of fluorene, 9-benzalfluorene, and 9-benzylfluorene are shown in Figs. 15.3, 15.4, and 15.5, respectively.

EXPERIMENT 15.2B

REDUCTION OF 9-BENZALFLUORENE TO 9-BENZYLFLUORENE BY HYDRIDE TRANSFER

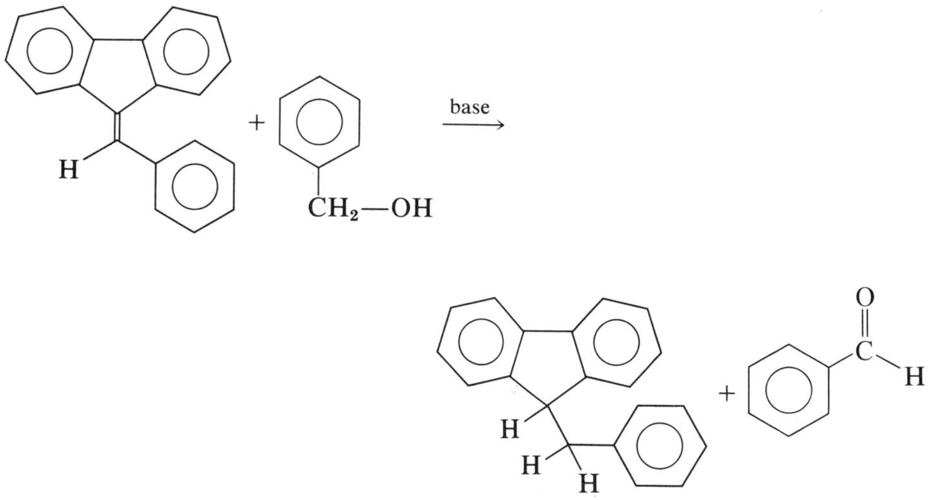

H + CH_2—OH $\xrightarrow{\text{base}}$

H H H + C(=O)H

Time 2 h

Materials 9-Benzalfluorene, 2.5 g (MW 254, mp 75 to 76°C)
Benzyl alcohol, 10 mL (MW 108, bp 205°C, d 1 g/mL)
Potassium hydroxide, 0.75 g (MW 56)

Precautions Refluxing benzyl alcohol boils at 205°C. Be careful to avoid burns when handling the reaction flask.

Hazards Potassium hydroxide is a caustic material. Avoid contact with skin and eyes.

Experimental Procedure

Place 2.5 g 9-benzalfluorene (from Exp. 15.2A) in a 100-mL round-bottom flask. Add 10 mL benzyl alcohol, followed by 0.75 g potassium hydroxide. Add a few boiling chips and place a reflux condenser with lightly greased joints in the top of the flask (see Fig. 9.1).

Heat the reaction at reflux (flame) for 15 to 20 min; then allow it to cool. The yellow 9-benzalfluorene color should fade during this time. After the reaction mixture has cooled, add 25 mL water through the reflux condenser, swirl the reaction mixture, and suction filter the solid. Rinse the crude product several times with cold water.

The solid material may, if necessary, be recrystallized from heptane. The product is a white crystalline material, mp 132 to 134°C.

The nmr spectra of fluorene, 9-benzalfluorene, and 9-benzylfluorene are shown in Figs. 15.3, 15.4, and 15.5, respectively.

EXPERIMENT 15.2C

DIRECT SYNTHESIS OF 9-BENZYLFLUORENE FROM FLUORENE

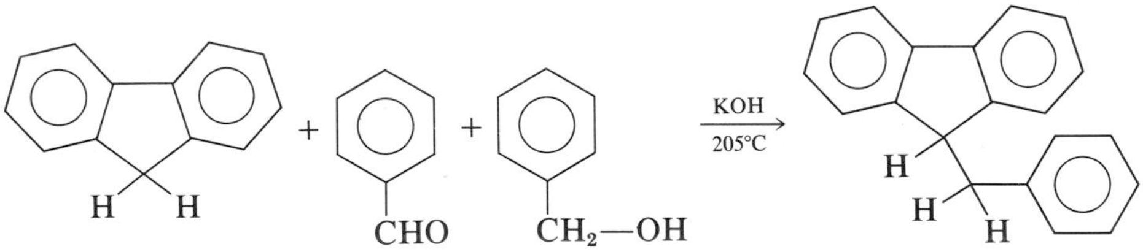

H H + CHO + CH_2—OH $\xrightarrow[205°C]{KOH}$ H H H

Time 3.0 h

Materials Fluorene, 5 g (MW 166, mp 112 to 115°C)
Benzyl alcohol, 15 mL (MW 108, bp 205°C, d 1 g/mL)
Benzaldehyde, 5 drops (MW 106, bp 188°C, d 1 g/mL)
Potassium hydroxide, 2.5 g (MW 56)

Precautions Refluxing benzyl alcohol is very hot; do not touch the flask.

Hazards Potassium hydroxide is a caustic material. Avoid contact with skin and eyes.

Experimental Procedure

Place 5 g (0.030 mol) fluorene in a 100-mL round-bottom flask. Add 15 mL benzyl alcohol, 5 drops benzaldehyde, and finally 2.5 g (0.045 mol) potassium hydroxide. A reflux condenser with lightly greased joints is placed in the top of the flask (Fig. 9.1), followed by several boiling chips. Reflux the reaction mixture (with a flame) for 1 h.

Allow the mixture to cool (**caution: the internal temperature of the flask is near 200°C**), and then add 15 mL water through the reflux condenser. The crude 9-benzylfluorene should crystallize immediately and may be collected by using a Buchner funnel. Rinse the product several times with water and allow it to air dry.

Recrystallize the crude material from heptane. The pure product should be a *white* crystalline material, mp 132 to 134°C.

The nmr spectra of fluorene, 9-benzalfluorene, and 9-benzylfluorene are shown in Figs. 15.3, 15.4, and 15.5, respectively. Very little information can be obtained from an examination of the aromatic region, but the aliphatic region is informative indeed. 9-Benzalfluorene contains no aliphatic protons and no resonance is observed in the nmr spectrum from 0 to 6 ppm. Fluorene itself contains a single type of aliphatic proton, and a large singlet near 3.8 ppm identifies this compound. 9-Benzylfluorene has two kinds of protons (an A_2B system), and the presence of a doublet and triplet in the aliphatic region of the appropriate spectrum clearly indicates that the double bond of 9-benzalfluorene has been saturated.

15.3 PREPARATION OF 3,4-METHYLENEDIOXYCINNAMONITRILE BY ACETONITRILE CONDENSATION

Most of the well-known base-catalyzed condensations which are used in organic chemistry involve relatively strong carbon acids. This is so because the weak bases available in aqueous solution are sufficiently reactive to deprotonate them. An exception to this generalization is the Perkin condensation of acetic anhydride with benzaldehyde to give cinnamic acid. In this case, however, very vigorous conditions are required to drive the reaction to completion. Typical conditions for this reaction are heating with sodium acetate at 180°C for about 8 h. Most of the other base-catalyzed condensations are conducted at much lower temperatures.

$$C_6H_5-CHO + CH_3-\overset{O}{\overset{\|}{C}}-O-\overset{O}{\overset{\|}{C}}-CH_3 \xrightarrow{NaOAc} C_6H_5-CH{=}CH-COOH$$

Perkin condensation

Acetonitrile has been widely used as a solvent. Contrary to this frequent use, however, it has been used relatively little as a reaction partner. Because of its solvent properties, bases such as potassium hydroxide are stronger in acetonitrile than they are in water. The experiment described below involves deprotonation of acetonitrile (in acetonitrile as solvent), followed by condensation with a substituted benzaldehyde.

The intermediate in this reaction is a β hydroxynitrile, which can eliminate water to afford a fully conjugated system. Instead of using benzaldehyde, which would react with acetonitrile to produce cinnamonitrile, the substituted aldehyde 3,4-methylenedioxybenzaldehyde (commonly known as piperonal) has been chosen instead. 3,4-Methylenedioxycinnamonitrile is a solid, whereas the parent compound (cinnamonitrile) is a liquid, and in this reaction both double-bond isomers are formed. Because of the slightly different electronic effect exerted by the methylenedioxy function, only the trans isomer of the solid methylenedioxycinnamonitrile is obtained in this condensation. The latter is of particular advantage in the workup because acetonitrile is toxic in high concentrations. The solid, single-isomer product may be obtained by crystallization, eliminating virtually all the hazards associated with the solvent.

The mechanism of the reaction is essentially the same as that indicated for an aldol condensation except that spontaneous elimination of water occurs. Conceptually, all these reactions are closely related. The differences arise by variation of the base and changes in the nucleophile.

EXPERIMENT 15.3

SYNTHESIS OF 3,4-METHYLENEDIOXYCINNAMONITRILE

Time 3 h

Materials Piperonal, 7.5 g (MW 150, mp 35 to 37°C)
Potassium hydroxide, 3 g (MW 56)
Acetonitrile, 30 mL (MW 42, bp 82°C, d 0.78 g/mL)
Alumina, 15 g
Dichloromethane, 70 mL
Ether, 30 mL

Precautions Carry out all reactions in the hood. Wear gloves when transferring acetonitrile. Use no flames near the reaction apparatus.

Hazards Acetonitrile is flammable and its vapors are toxic. Avoid breathing or contact with skin and eyes. Potassium hydroxide is a caustic compound. Avoid contact with skin and eyes.

Experimental Procedure

Place 3 g solid potassium hydroxide in a 100-mL round-bottom single-neck flask. Stopper the flask with a cork (to prevent excess moisture from being absorbed on the surface of the potassium hydroxide) and attach a reflux condenser. A magnetic stirrer and oil bath are useful for this experiment. Add to the flask a magnetic stirring bar, followed by 25 mL acetonitrile. The magnetic stirrer is turned on and the acetonitrile–potassium hydroxide suspension heated to reflux.

Weigh 7.5 g (0.05 mol) piperonal (3,4-methylenedioxybenzaldehyde) in a 50-mL beaker. Add 5 mL acetonitrile. As the piperonal dissolves, the solution gets cold. During the dissolution of the piperonal, the beaker should be heated briefly on a small hot plate to keep the temperature of the solution near room temperature.

Once the piperonal has dissolved in the acetonitrile and the acetonitrile–potassium hydroxide solution is refluxing vigorously, the piperonal solution is added slowly through the top of the reflux condenser. This addition should be performed by adding approximately 1-mL portions of the solution at about 15-s intervals. After all the solution has been added, 1 to 2 mL fresh acetonitrile is used to wash the contents of the beaker and the reflux condenser into the round-bottom flask.

The reaction solution is now stirred vigorously and refluxed for *exactly* 10 min. As soon as the piperonal solution is added, the potassium hydroxide begins to disintegrate and the clear, colorless acetonitrile solution turns yellow to light orange. The reflux should be maintained at a vigorous level during the 10-min period. [*Note:* The reaction is complete in 8 to 10 min under these conditions. If heating is continued for more than 10 min, the product begins to degrade (e.g., by hydrolysis) and workup becomes difficult.]

After the reflux period is over, pour the reaction mixture into a 600-mL beaker containing 150 mL cold distilled water. Rinse the flask with one 25-mL portion of water and add it to the beaker. Swirl the beaker vigorously (or stir with a magnetic stirrer) until the aqueous solution is intimately mixed. At this point, the oily yellow reaction mixture should coagulate into a purple solid. After several minutes of swirling (to ensure complete coagulation), filter the reaction mixture (Buchner funnel). Collect the solid material and wash it with one 25-mL portion of cold distilled water. The dark mother liquor should be flushed down a sink drain with copious amounts of water.

Transfer the solid filtrate to a 100-mL beaker and add 20 mL dichloromethane. Dissolve the solid by swirling the solution; it will appear purple. Transfer the liquid to a separatory funnel and wash in any remaining solid with 30 mL dichloromethane. Wash the organic layer with three 25-mL portions of saturated aqueous sodium chloride solution (brine). Transfer the purple organic phase to a 125-mL Erlenmeyer flask and dry the solution with anhydrous sodium or magnesium sulfate. After drying, decant the solution into a 250-mL Erlenmeyer flask and evaporate the solvent (**steam bath, hood**). After all the solvent has evaporated and the Erlenmeyer has been removed from the heat source, the heavy purple oil should solidify.

Prepare a small chromatography column (about 22 × 300 mm) as shown in Fig. 15.6. Place a small plug of glass wool in the bottom of the tube, followed by a 1-cm layer of sand and 15 g alumina (Baker, suitable for chromatography).

Dissolve the purple residue described above in 10 mL dichloromethane

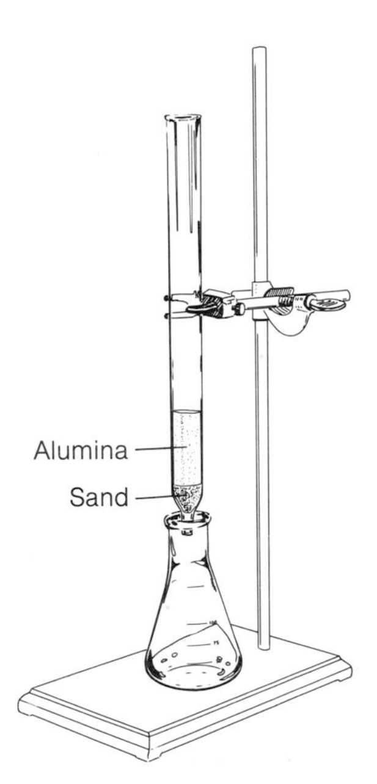

Figure 15.6
A small chromatography column for Exp. 15.3.

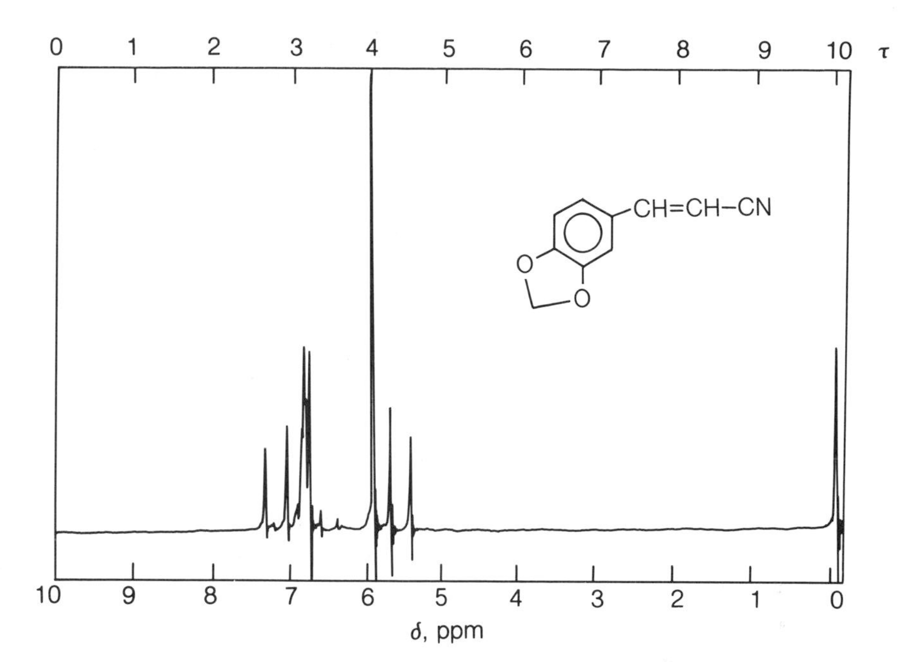

Figure 15.7
The nmr spectrum of 3,4-methylenedioxy-cinnamonitrile.

and add it to the top of the alumina column described above. As the solution percolates through the alumina, the purple polymeric material which contaminates the product will be adsorbed tightly at the top of the column. After all the material has been added to the column and percolated through, another 20 to 25 mL dichloromethane is added at the top of the column to wash the product through the column into a 125-mL Erlenmeyer flask. This entire operation should take approximately 15 to 20 min.

Evaporate the dichloromethane from the yellow solution (**steam bath, hood**). Cover the yellow-tan solid with 15 mL cold methanol and grind the solid with a glass rod (this process is known as trituration). Filter the solid nitrile (Buchner funnel) and wash the filtrate with 5 mL cold methanol. Recrystallize the off-white solid from 90% methanol-water (10 mL/g). The product should be obtained after filtration as small, white, needlelike crystals (mp 90 to 92°C).

An nmr spectrum of the nitrile is shown in Fig. 15.7. Note the trans coupling of the alkene protons (16 Hz) and the tall CH_2 resonance near 6 ppm. The ir spectrum (not shown) of this material reveals the presence of a nitrile band at 2210 cm^{-1} (4.52 μ) and a double-bond vibration at 1626 cm^{-1} (6.15 μ).

15.4 THE BENZOIN CONDENSATION

The benzoin condensation is an anionic condensation whose mechanism is somewhat different from the normal base-catalyzed condensations such as the aldol, Perkin, or Claisen-Schmidt. The benzoin condensation involves the dimerization of an aldehyde *not* via the carbon atom alpha to the carbonyl, but via the carbonyl carbon itself. Thus the dimer produced is an α hydroxyketone instead of the β-hydroxycarbonyl compound formed in the aldol reaction. The two separate reaction paths are shown below for acetaldehyde:

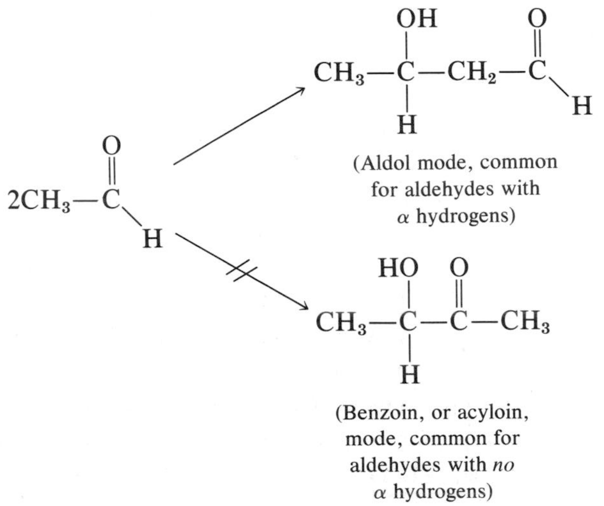

$$2CH_3-\overset{O}{\overset{\|}{C}}-H \longrightarrow CH_3-\underset{H}{\underset{|}{\overset{OH}{\overset{|}{C}}}}-CH_2-\overset{O}{\overset{\|}{C}}-H$$

(Aldol mode, common for aldehydes with α hydrogens)

$$2CH_3-\overset{O}{\overset{\|}{C}}-H \not\longrightarrow CH_3-\underset{H}{\underset{|}{\overset{HO}{\overset{|}{C}}}}-\overset{O}{\overset{\|}{C}}-CH_3$$

(Benzoin, or acyloin, mode, common for aldehydes with *no* α hydrogens)

As one can see from the above, although a dimer is produced in both cases, the benzoin condensation is another reaction type in which carbon-carbon bonds are formed. However, this mode of reaction does not occur spontaneously in basic solution as does the aldol reaction. A catalyst is needed to activate the carbonyl carbon so that the dimerization will proceed. In the synthesis of benzoin from benzaldehyde, cyanide ion functions as the catalyst for the dimerization.

Cyanide ion is not a very basic ion (it is the conjugate base of the weak acid hydrocyanic acid, HCN). It is effective as a catalyst for this reaction because at an intermediate stage an anion is generated which is stabilized both by cyanide and by another functional group within the molecule. The mechanism leading to benzoin is shown in the following equations:

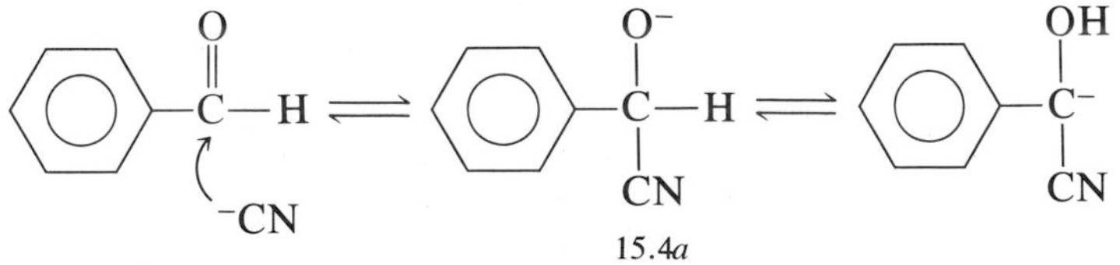

$$C_6H_5-\overset{O}{\overset{\|}{C}}-H + {}^{-}CN \rightleftharpoons \underset{15.4a}{C_6H_5-\underset{CN}{\underset{|}{\overset{O^-}{\overset{|}{C}}}}-H} \rightleftharpoons C_6H_5-\underset{CN}{\underset{|}{\overset{OH}{\overset{|}{C^-}}}}$$

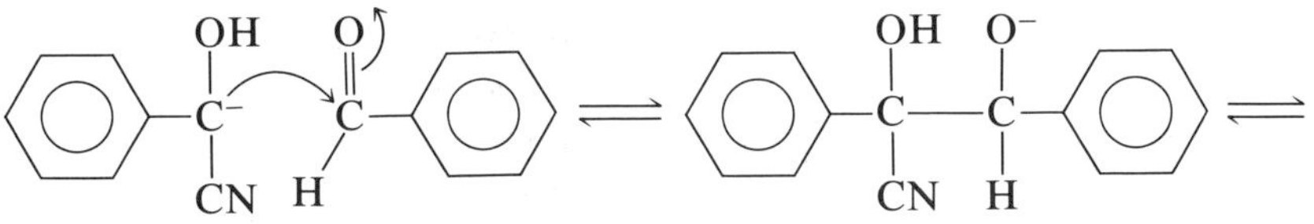

15.4*b*

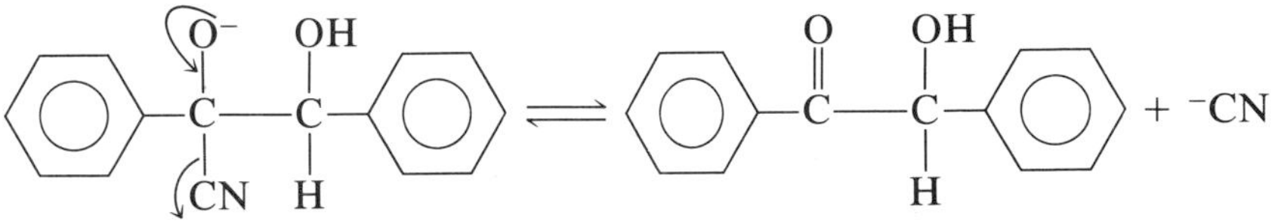

15.4*c*

In the presence of cyanide, benzaldehyde will form a cyanohydrin by the nucleophilic addition of cyanide to the carbonyl. In this intermediate (15.4*a*), the cyanide ion may now stabilize a negative charge on what was originally the carbonyl carbon, thus creating a site of carbon nucleophilicity. The carbon anion may now add to another molecule of benzaldehyde (or to any other carbonyl in solution) in a normal way to form an alkylated cyanohydrin (15.4*b*). Since the formation of the cyanohydrin is reversible, elimination of HCN from the alkylated cyanohydrin will form the α hydroxyketone (15.4*c*), in this case benzoin, which is a dimer of benzaldehyde.

Notice that in the sequence of events shown above cyanide initiates the reaction, plays a central role in stabilizing the anion which leads to the product, and finally functions as a leaving group. Cyanide is an excellent nucleophile, an excellent anion stabilizer, and also an excellent leaving group. It is almost unique in its ability to perform all these functions. Note also that for every cyanide ion which initiates the reaction, a cyanide ion is ultimately lost at the end of the reaction sequence. The reaction is therefore said to be cyanide-catalyzed. The reaction will not occur in the absence of cyanide (or other agent which acts in the same way as cyanide), yet no cyanide ion is ever consumed.

The benzoin (or acyloin) condensation can, in principle, be used to dimerize any aldehyde. Cyanide, aside from being a good nucleophile, is also a base. If an aldehyde has acidic hydrogen atoms next to the carbonyl group (as does acetaldehyde in the example below),

$$CH_3\text{—}CHO + {}^-CN \rightleftharpoons CH_3\text{—}C(OH)(H)\text{—}CN \qquad ({}^-CN\ \text{nucleophilic})$$

$$CH_3\text{—}CHO + {}^-CN \rightleftharpoons {}^-CH_2\text{—}CHO + HCN \qquad ({}^-CN\ \text{basic})$$

the cyanide can (and will) act as a base, the aldol reaction mode will predominate, and no benzoin mode will be observed under normal reaction conditions. Again, in principle any aldehyde which does not contain acidic hydrogen atoms next to the carbonyl group can be dimerized by cyanide. In the case of an aromatic aldehyde, the aromatic ring, along with the cyanide, can stabilize the negative charge, thus additionally stabilizing the intermediate.

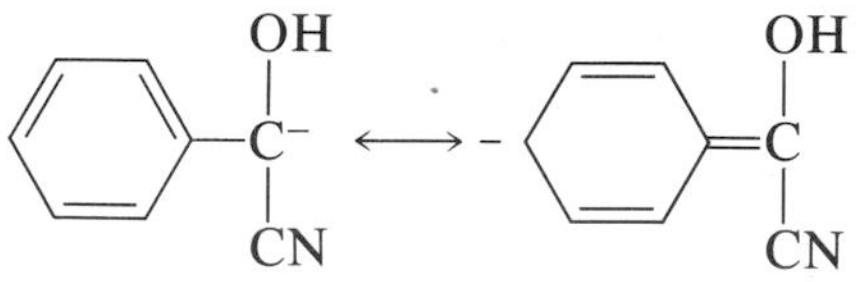

$$C_6H_5\text{-}\overset{OH}{\underset{CN}{C^-}} \longleftrightarrow {}^-C_6H_5{=}\overset{OH}{\underset{CN}{C}}$$

Because of these two constraints, the benzoin condensation is considered a synthetic procedure for the dimerization of aromatic aldehydes and can not, under normal circumstances, be considered a general method for the dimerization of all aldehydes.

EXPERIMENT 15.4

SYNTHESIS OF BENZOIN

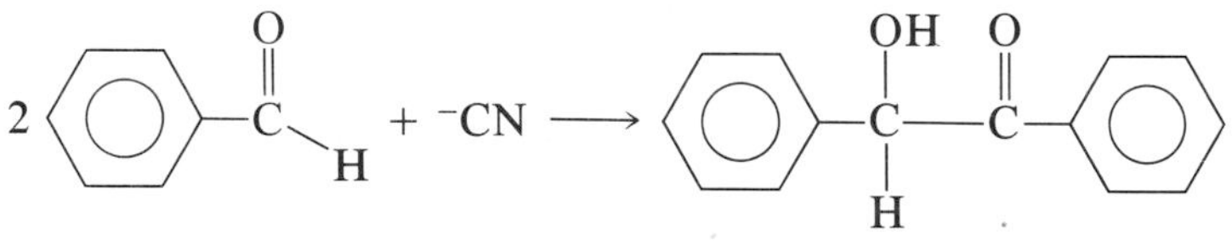

$$2\ C_6H_5\text{-}\overset{O}{\overset{\|}{C}}\text{-}H + {}^-CN \longrightarrow C_6H_5\text{-}\overset{OH}{\underset{H}{C}}\text{-}\overset{O}{\overset{\|}{C}}\text{-}C_6H_5$$

Time 3.0 h

Materials Benzaldehyde, 10 mL (MW 106, bp 182°C, d 1 g/mL)
Sodium cyanide, 1 g (mW 49)
95% Ethanol, 25 to 100 mL

Precautions All operations involving cyanide should be performed in a hood. Wear gloves when weighing sodium cyanide. Do not breath the cyanide dust and clean up spills **immediately.** Small amounts of cyanide waste should be flushed down the drain in a hood with lots of water. Do not allow any NaCN to contact acid (**HCN gas produced**).

Hazards Sodium cyanide is an extremely toxic and hazardous material. Sodium cyanide or its solutions may be absorbed through the skin. Exposure of cyanide to acid produces HCN, a toxic gas. Any dust on the skin should be washed off immediately with copious amounts of water.

Experimental Procedure

Place in a 100-mL round-bottom flask 1 g sodium cyanide and add 10 mL distilled water. Swirl the flask to dissolve the solid. Add 10 mL benzaldehyde to the flask, followed by 25 mL 95% ethanol. Fit the flask with a reflux condenser (lightly greased joints) and reflux the mixture gently for 30 min (with either a heating mantle or free flame). After the heating period is over, remove the heat source and allow the flask to cool to room temperature.

As the reaction mixture cools, the crude benzoin should separate as a yellow solid. Collect this product [**being careful to wear gloves and avoid contact**

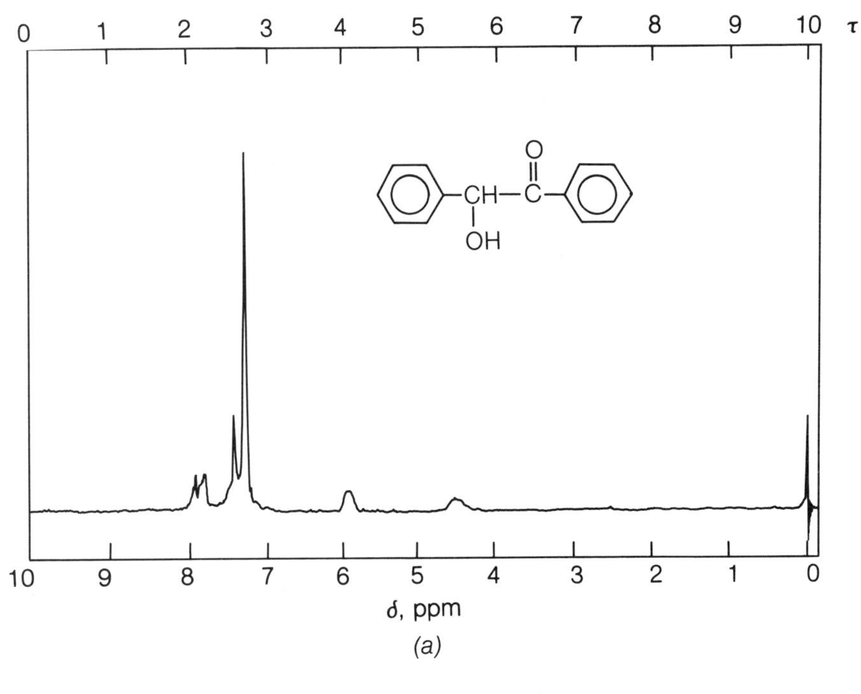

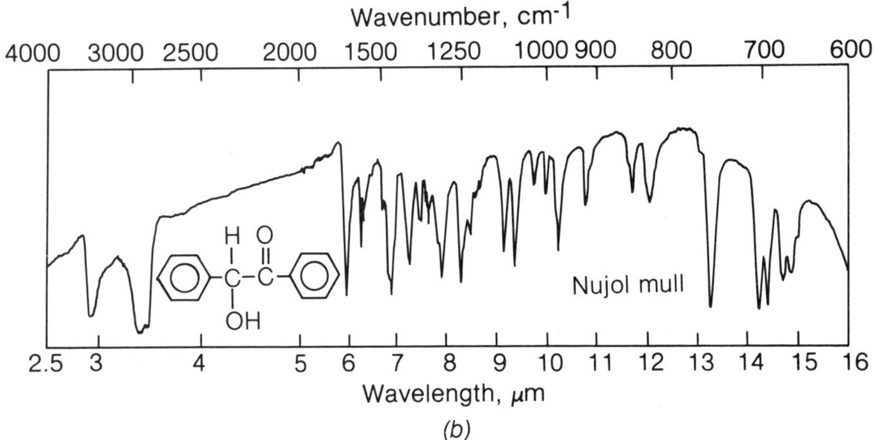

Figure 15.8
The (*a*) nmr and (*b*) ir spectra of benzoin.

with your skin (cyanide)], and wash the solid with 50 mL cold 50% ethanol-water (or 50% methanol-water) to remove any cyanide waste. *Immediately* flush the mother liquors down the hood drain with copious amounts of water.

The crude benzoin should now have a white to pale yellow appearance. It may be recrystallized from either methanol (about 13 mL/g) or ethanol (about 8 mL/g). Benzoin is obtained as small, white, needlelike crystals, mp 134 to 135°C. The yield in this reaction is generally between 70 and 85%.

The nmr and ir spectra of benzoin are shown in Fig. 15.8. Note that one of the aromatic rings is nearly a singlet whereas the other is quite complex. Phenyl rings adjacent to sp^2 carbons often show this type of splitting. It appears in this sample that the hydroxyl hydrogen atom is not exchanging very rapidly because it is coupled to the methine proton. The carbonyl and hydroxyl functions are clearly visible in the ir spectrum.

15.5 THE CANNIZZARO REACTION

In 1835 the famous German chemist Liebig discovered that benzaldehyde was oxidized in the presence of hydroxide to benzoic acid. He did not understand the reaction, however. Benzoic acid was formed only up to a maximum yield of 50%, but Liebig did not recognize that another product was also formed. In 1853 the Italian chemist Stanislao Cannizzaro recognized that the reaction was actually an oxidation-reduction reaction in which half the benzaldehyde was converted to benzoic acid and the other half was reduced to benzyl alcohol. The reaction carries Cannizzaro's name because of his greater understanding of the reaction.

The Cannizzaro reaction begins much the same way the benzoin condensation begins, i.e., with the addition of a nucleophile to benzaldehyde. In this case the nucleophile is hydroxide rather than cyanide. Hydroxide adds to benzaldehyde to give a geminal diol anion designated as 15.5*a* in the mechanism below and corresponding directly to 15.4*a* in the mechanism of the benzoin condensation.

$$C_6H_5CHO + \overset{+}{M}OH^- \rightleftharpoons C_6H_5CH(OH)O^-M^+ + C_6H_5CHO \rightleftharpoons$$

15.5*a*

$$\longrightarrow C_6H_5COOH + C_6H_5CH_2\text{—}OH$$

This anionic material coordinates with the oxygen atom of another benzaldehyde molecule through a metal ion, usually sodium or potassium if sodium hydroxide or potassium hydroxide has been used as the base. At this point a six-membered transition state involving the transfer of a hydride from one molecule of aldehyde to the other is formed. The net effect of this transfer is that one molecule of aldehyde is reduced (to benzyl alcohol) and the other molecule of aldehyde is oxidized (to benzoic acid). Thus, two molecules of benzaldehyde have disproportionated (or dismutated) under the influence of base to an alcohol and acid.

The Cannizzaro reaction is more general than the benzoin condensation. Although most Cannizzaro reactions reported in the literature involve aromatic aldehydes, the reaction works just as well with aliphatic aldehydes with no α-hydrogen atoms (the aldol reaction takes precedence over the Cannizzaro when α hydrogens are present). In fact, mixed Cannizzaro reactions can be observed. If two different aldehydes are subjected to Cannizzaro reaction conditions, a mixture of alcohols and acids will be produced. Although this observation was important in the study of the mechanism, it was not of much synthetic utility until it was observed that formaldehyde (an aldehyde with no α hydrogens) was very prone to act as a reducing agent in this reaction (and thus ended up as formic acid). The other aldehyde present would then act as the oxidizing agent (and end up as an alcohol). This reaction has been used extensively in the past as a reducing process for aromatic aldehydes, since formaldehyde is cheap and formic acid is easy to remove. Today, it has largely been supplanted by hydride reducing agents. This modification of the Cannizzaro reaction, usually referred to as the cross-Cannizzaro reaction, is shown below for benzaldehyde and formaldehyde.

$$C_6H_5{-}\overset{O}{\overset{\|}{C}}{-}H + H{-}\overset{O}{\overset{\|}{C}}{-}H + NaOH \longrightarrow$$

$$C_6H_5{-}CH_2{-}OH + H{-}\overset{O}{\overset{\|}{C}}{-}O^-Na^+$$

There are several industrial uses of the Cannizzaro reaction. One of the largest uses is in the production of furfuryl alcohol from furfural (obtained by the acid distillation of cellulose waste). The furfuryl alcohol is then used in wood adhesives and resins, especially those needed for the production of plywood, now a major structural material in the housing industry. Another important application of the Cannizzaro reaction is the production of pentaerythritol, a synthetic sugar used in the explosives industry to produce PETN by the reaction shown below.

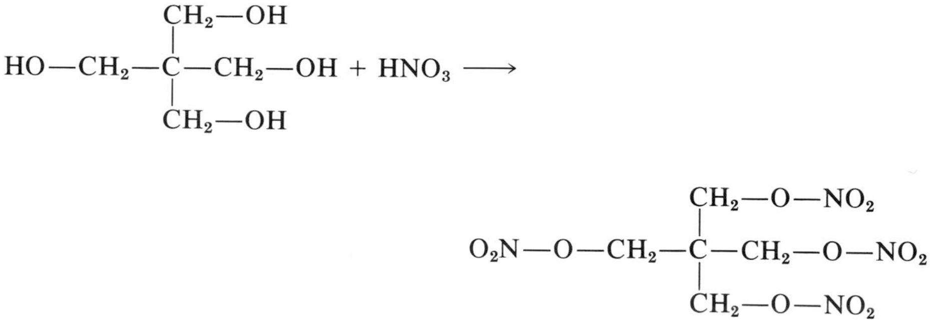

$$\begin{array}{c} CH_2{-}OH \\ | \\ HO{-}CH_2{-}C{-}CH_2{-}OH + HNO_3 \longrightarrow \\ | \\ CH_2{-}OH \end{array}$$

$$\begin{array}{c} CH_2{-}O{-}NO_2 \\ | \\ O_2N{-}O{-}CH_2{-}C{-}CH_2{-}O{-}NO_2 \\ | \\ CH_2{-}O{-}NO_2 \end{array}$$

Pentaerythritol tetranitrate
(PETN, explosive)

In this application the cross-Cannizzaro reaction is used in combination with the aldol reaction. As shown below, acetaldehyde reacts with formaldehyde (in this case acting as the electrophile) to form a molecule which contains three hydroxymethyl groups.

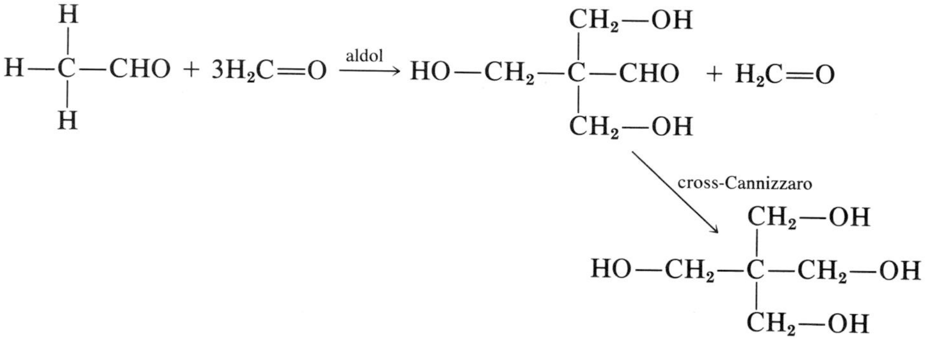

$$\begin{array}{c} H \\ | \\ H{-}C{-}CHO \\ | \\ H \end{array} + 3H_2C{=}O \xrightarrow{\text{aldol}} \begin{array}{c} CH_2{-}OH \\ | \\ HO{-}CH_2{-}C{-}CHO \\ | \\ CH_2{-}OH \end{array} + H_2C{=}O$$

$$\xrightarrow{\text{cross-Cannizzaro}} \begin{array}{c} CH_2{-}OH \\ | \\ HO{-}CH_2{-}C{-}CH_2{-}OH \\ | \\ CH_2{-}OH \end{array}$$

Pentaerythritol

At this point, all the α-hydrogen atoms of the acetaldehyde have been replaced, and the product aldehyde may undergo the cross-Cannizzaro reaction with excess formaldehyde to form pentaerythritol. In this example the formaldehyde is acting as an electrophile and as a reducing agent.

In the experimental procedure below you will subject 4-chlorobenzaldehyde to the Cannizzaro reaction. Although any aromatic aldehyde will work about as well, this aldehyde was chosen because its products, 4-chlorobenzyl alcohol and 4-chlorobenzoic acid, are used as starting materials in other procedures in this book. 4-Chlorobenzyl alcohol also has the advantage that it may be purified by crystallization, whereas many alcohols of this type may be purified only by vacuum distillation.

EXPERIMENT 15.5 CANNIZZARO REACTION OF 4-CHLOROBENZALDEHYDE

$$2\ \text{Cl-C}_6\text{H}_4\text{-CHO} + \text{KOH} \longrightarrow \text{Cl-C}_6\text{H}_4\text{-CH}_2\text{OH} + \text{Cl-C}_6\text{H}_4\text{-COOK}$$

Time 2.5 h (plus 0.2 h in subsequent lab period)
Materials 4-Chlorobenzaldehyde, 15 g (MW 140.57, mp 44 to 47°C)
Potassium hydroxide, 28 g (85% pellets).
Methanol, 40 mL (solvent grade)

Precautions Pour concentrated HCl in the hood.

Hazards Methanol vapors are toxic in high concentrations.

Experimental Procedure

A 250-mL Erlenmeyer flask is charged with potassium hydroxide (KOH, 28 g), followed by distilled water (40 mL). The flask is swirled until all the solid KOH dissolves (hold the neck of the flask; the solution becomes very warm). The aqueous KOH solution is allowed to cool nearly to room temperature.

While the KOH solution is cooling, a 125-mL Erlenmeyer flask is charged with 4-chlorobenzaldehyde (15 g, 107 mmol), followed by methanol (40 mL). The flask is swirled until all the aldehyde has dissolved (brief warming on a steam bath helps accelerate this process). The aldehyde-methanol solution is then poured all at once into the aqueous KOH solution. The 250-mL Erlenmeyer flask is swirled vigorously to mix the two solutions. The reaction is exothermic, so the solution will warm during the swirling. After several minutes the potassium salt of 4-chlorobenzoic acid should begin to precipitate. Keep the mixture at 55 to 65°C (internal temperature, check occasionally with a thermometer) by intermittent heating on the steam bath during a 1-h period. The flask should be swirled regularly when on the steam bath to prevent superheating.

After the heating period, allow the mixture to cool to room temperature and then pour it into a 500-mL separatory funnel containing 200 mL cold distilled water. Any material remaining in the Erlenmeyer flask should be washed into the separatory funnel with dichloromethane (25 mL). Shake the mixture, separate the layers, and then draw off the organic portion. Extract the aqueous solution twice with 25-mL portions of dichloromethane. Reserve the aqueous layer for workup at a later time.

Wash the combined organic fractions with saturated $NaHCO_3$ (two 25-mL portions); then dry over anhydrous sodium sulfate for several minutes.

Gently filter the organic layer into a 250-mL Erlenmeyer flask and remove the solvent on a steam bath. Dissolve the residual oil in 4% acetone-hexane (25 mL),[1] heat to reflux, and then allow the solution to cool to room temperature. Collect the crystals on a Buchner funnel and wash them with ice-cold hexane (10 to 15 mL). After air drying for several minutes, the crystal-

[1] 1 mL acetone + 24 mL hexane.

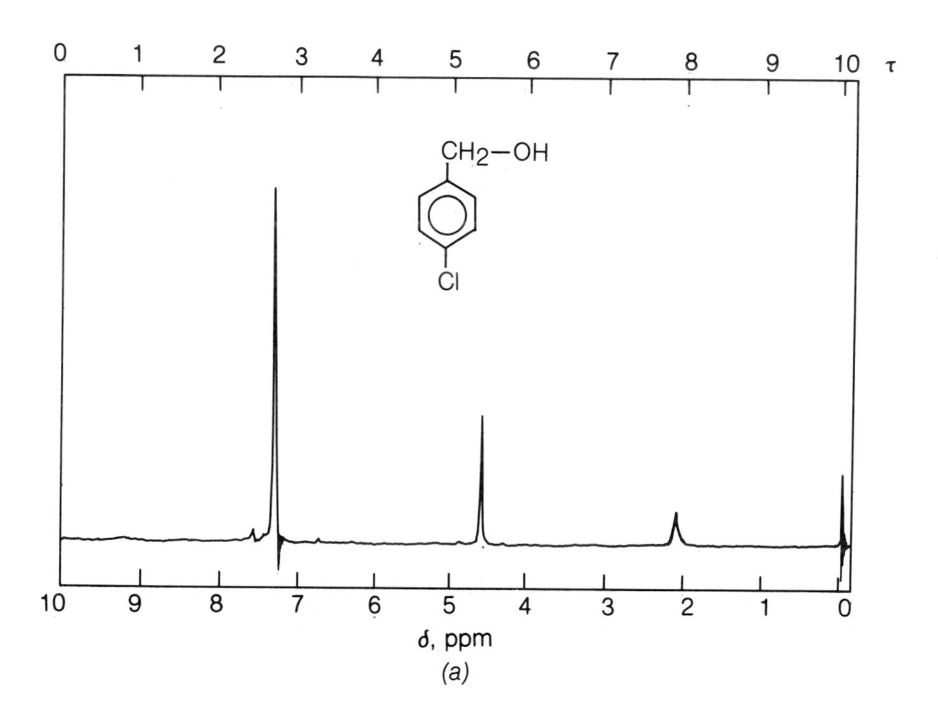

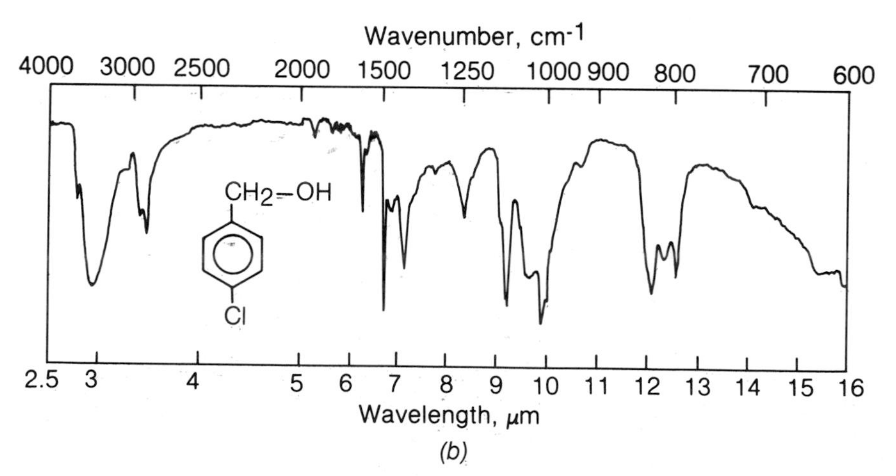

Figure 15.9
The (*a*) nmr and (*b*) ir spectra of 4-chlorobenzyl alcohol.

line 4-chlorobenzyl alcohol (approximately 5.5 g, 70% yield) should have mp 70 to 72°C.

The basic aqueous layer previously set aside is now transferred to a 500-mL Erlenmeyer flask, and charcoal (1 to 2 g) is added. Heat the mixture for several minutes (steam bath) while swirling continuously. Cool the mixture (ice bath) and then filter through a pad of Celite (filter aid). The aqueous mixture should now be clear and water-white. A second filtration is sometimes required.

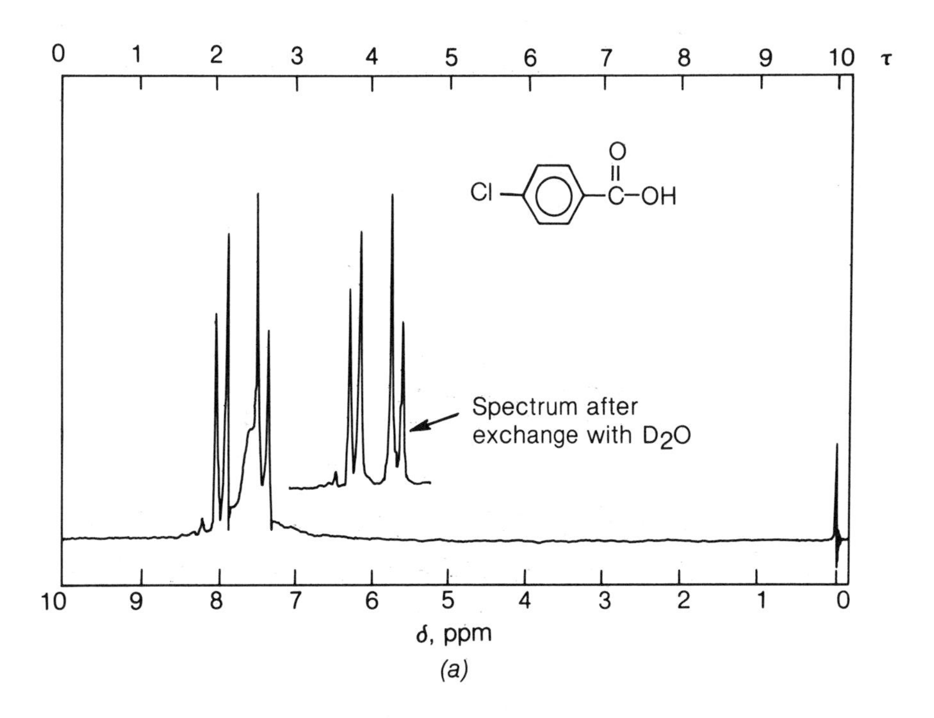

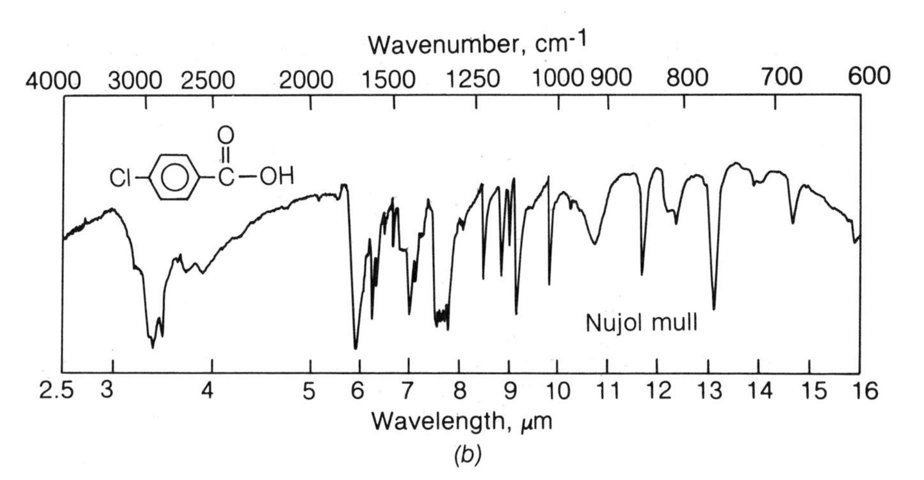

Figure 15.10
The (*a*) nmr and (*b*) ir spectra of 4-chlorobenzoic acid.

Add concentrated hydrochloric acid (12 *N*, 40 mL) with swirling to the filtered aqueous solution. As the acid is added, a white material separates. The aqueous suspension should now be strongly acidic (pH paper) and should be cooled in an ice bath. Suction filter (large Buchner funnel) the white material and wash the residue with cold distilled water (200 mL). Place a second piece of filter paper on top of the filter cake and press lightly with a glass rod to remove as much water as possible. Transfer the damp solid to a watch glass and place it in your desk in such a way that it will neither collect dust nor be upset. Allow your product to air dry until the next laboratory period. When you return, the 4-chlorobenzoic acid (7 g, 85% yield) should be a white powder of mp 237 to 239°C. (*Note:* The melting point should be determined on an apparatus which can be used safely at temperatures as high as 240°C.)

The nmr and ir spectra of 4-chlorobenzyl alcohol are shown in Fig. 15.9 and those of 4-chlorobenzoic acid in Fig. 15.10. Note the different characteristic shapes of the hydroxyl bands in the ir spectra of the alcohol and acid. Note also that the AB pattern characteristic of para substitution is obscured in the nmr spectra of both compounds. In the alcohol spectrum, the aromatic protons coincidentally have the same chemical shifts and appear as a singlet. In the spectrum of the carboxylic acid, the AB pattern overlaps the alcohol peak. The coupling pattern becomes more recognizable when D_2O is added to the solution and the hydroxyl proton exchanges with deuterium too rapidly to be detected. (See discussion in Sec. 5.3.)

QUESTIONS AND EXERCISES

15.1 A careless student neglected to add acetone in Exp. 15.1A, but obtained a white product of mp 121°C. What do you suppose this material is and how might it have been formed?

15.2 A careless student neglected to add benzaldehyde to the reaction mixture (Exp. 15.1A) and obtained a liquid of bp 129°C as the product. This material formed a 2,4-dinitrophenylhydrazine derivative which melted at 203°C. What do you suppose the product was and how can you account for its formation?

15.3 Another student made a similar error in Exp. 15.1A but, in addition to neglecting to add benzaldehyde, heated the reaction mixture. On workup a liquid boiling at 214°C was obtained. This material reacted with 2,4-dinitrophenylhydrazine to give a derivative melting at 130°C. Identify this material and suggest the mechanism by which it must have formed.

15.4 One very careful student isolated a low-melting (mp 39°C) solid in addition to dibenzalacetone in Exp. 15.1A. This material formed an oxime derivative (Sec. 25.7C) of mp 115°C and was found to give dibenzalacetone in the presence of base and benzaldehyde. Identify this substance and suggest how you might prepare it if you desired it rather than the material prepared in this experiment.

15.5 One careless student attempted to carry out the preparation of chalcone as described in Exp. 15.1B, but neglected to add the benzaldehyde. A product was nevertheless obtained. What do you suspect the substance might be?

15.6 Another careless student made a mistake similar to that described in Question 15.5, but instead of the aldehyde neglected to add the ketone. On workup, the student isolated a white solid of mp 121°C. What do you suppose this compound is?

15.7 Ethanol is used as cosolvent in Exp. 15.1B. It is possible that ethanol is playing a role other than that of a solvent. Suggest what this role might be.

15.8 Fluorene and 9-benzylfluorene are both colorless (white) solids but 9-benzalfluorene is yellow. What, if anything, does this tell you? Was a color change observed in the synthesis of fluorenone from fluorene in Exp. 14.1? Is this related?

15.9 Under special conditions, it was found that fluorene could react with 9-benzalfluorene to give an addition compound. The product appeared to derive from two molecules of fluorene and one of benzaldehyde. Two aliphatic resonances were observed in the nmr spectrum in a ratio of 2:1. What might this new compound be and how is it formed?

15.10 If the condensation of acetonitrile with 3,4-methylenedioxybenzaldehyde is conducted for longer than the prescribed period, hydrolysis begins to occur. Suggest structures for two possible by-products resulting from hydrolysis.

15.11 When the condensation described in Sec. 15.3 was attempted on 3-nitrobenzaldehyde, a yellowish solid was isolated whose melting point was found to be 140°C. This water-insoluble substance was found to dissolve in aqueous sodium bicarbonate solution. Suggest a structure for this compound.

15.12 Either ethanol or methanol can be used to recrystallize benzoin. What advantages or disadvantages can you think of for each solvent?

15.13 Do you think that methoxide ion could serve the same purpose as cyanide ion in the benzoin condensation? Why or why not?

15.14 Furfural (furan-2-carboxaldehyde) is an aromatic aldehyde just as benzaldehyde is. It reacts with cyanide ion to give a product known as furoin. Suggest a structure for furoin. Do you think that it would be formed more or less readily than benzoin?

15.15 If the Cannizzaro reaction were conducted on a 1:1 mixture of 4-chlorobenzaldehyde and benzaldehyde but only half the required amount of base was used, which aldehyde would undergo reaction more readily? Why?

15.16 Suggest at least two methods, other than the Cannizzaro reaction, for the preparation of 4-chlorobenzyl alcohol from 4-chlorobenzaldehyde. What starting material would you choose for the synthesis of 4-chlorobenzoic acid by a Grignard carbonation reaction (Sec. 12.3)?

15.17 In the Cannizzaro reaction the alcohol and acid are separated by extraction. What properties of the hydroxyl groups, present in both molecules, allow such easy separation of these two compounds?

15.18 In principle it is possible to isolate the related alcohol, aldehyde, and acid from a Cannizzaro reaction. If a mixture which might contain any or all of these were obtained, what bands in the ir spectrum of the mixture would give a clue to its composition?

XVI
THE FRIEDEL-CRAFTS REACTION

In 1877 a Frenchman named Charles Friedel and an American named James Crafts, working together in Paris in Friedel's laboratory, reported that toluene and ethyl bromide react in the presence of anhydrous aluminum chloride ($AlCl_3$) to yield ethyltoluene. This reaction, which has become known as the Friedel-Crafts alkylation reaction, is known to occur for a large number of aromatic and some nonaromatic hydrocarbons. Later work indicated that not only would alkylating agents undergo this reaction, but acylating agents such as acetyl chloride and acetic anhydride would work as well.

Anhydrous aluminum chloride is a powerful electrophilic catalyst (Lewis acid) and associates with the halogen atom of either an acyl halide or an alkyl halide. In so doing it generates a cation. Depending on its origin, the cation is called a carbonium ion (from an alkyl halide) or an acylium ion (from an acyl halide). The formation of these species is illustrated below.

$$\text{R—Cl} + AlCl_3 \rightleftharpoons R^+ + AlCl_4^-$$

$$\text{R—CO—Cl} + AlCl_3 \rightleftharpoons \text{R—}\overset{+}{C}\text{=O} + AlCl_4^-$$

Neither species is particularly stable and reaction with the aromatic hydrocarbon is rapid.

An important difference in behavior between these two species is that the acylium ion will usually react as formed and the aromatic ring will become attached at the same point from which the halide ion was lost, but this is not necessarily the case for the carbonium ion. The carbonium ion will form at the site from which halide was lost, but rearrangement may occur to afford the more stable cation. In the two reactions which are shown below, benzene is treated with aluminum chloride and either propanoyl chloride or *n*-propyl chloride. The acylation reaction (with propanoyl chloride) occurs by addition of the cation to the aromatic ring, followed by loss of a proton to restore the aromaticity of the ring. In the alkylation reaction, the same general sequence of events occurs, but the cation which forms at first equilibrates to the more stable secondary cation before attack on the aromatic ring can occur. As a result two products, *n*-propyl- and isopropylbenzene, are produced. The amount of each product formed will approximately reflect the stability of each cation.

$$CH_3{-}CH_2{-}CO{-}Cl + AlCl_3 \rightleftharpoons CH_3{-}CH_2{-}\overset{+}{C}{=}O + AlCl_4^-$$

$$CH_3{-}CH_2{-}\overset{+}{C}{=}O + C_6H_6 \longrightarrow CH_3{-}CH_2{-}\overset{O}{\overset{\|}{C}}{-}C_6H_6^+ \xrightarrow{-H^+} CH_3{-}CH_2{-}\overset{O}{\overset{\|}{C}}{-}C_6H_5$$

$$CH_3{-}CH_2{-}CH_2{-}Cl + AlCl_3 \rightleftharpoons CH_3{-}CH_2{-}\overset{+}{C}H_2 + AlCl_4^-$$

$$CH_3{-}CH_2{-}CH_2^+ \longrightarrow CH_3{-}\overset{+}{C}H{-}CH_3$$

$$CH_3{-}CH_2{-}CH_2^+ + C_6H_6 \longrightarrow CH_3{-}CH_2{-}CH_2{-}C_6H_6^+ \xrightarrow{-H^+} CH_3{-}CH_2{-}CH_2{-}C_6H_5$$

$$CH_3{-}\overset{+}{C}H{-}CH_3 + C_6H_6 \longrightarrow (H_3C)_2CH{-}C_6H_6^+ \xrightarrow{-H^+} (H_3C)_2CH{-}C_6H_5$$

There is another important difference between the alkylation and acylation reactions. Alkylated aromatics are electronically more activated than the corresponding starting materials. As a result di- and polyalkylation reactions often occur. This is not the case in acylation reactions. The carbonyl group in the product can complex with aluminum chloride, resulting in deactivation of the ring toward further electrophilic aromatic substitution reactions. Overall, then, the alkylation reaction favors di- and polysubstitution, whereas monosubstituted products usually result from the acylation reaction.

16.1 THE FRIEDEL-CRAFTS ACYLATION REACTION

In the absence of a strong Lewis acid such as aluminum chloride, benzoyl chloride and chlorobenzene, or acetic anhydride and bromobenzene, could be boiled together for weeks to no avail. The driving force for the reaction is the initial formation of the strong bond between aluminum chloride and the chlorine of benzoyl chloride or between aluminum chloride and the oxygen of acetic anhydride. Aluminum is an electropositive element and chlorine (or oxygen, if the anhydride is used) is an electronegative element, and the bond strength between these two species is quite large. The formation of this very strong bond allows the acylium ion to form, but it is itself very unstable and initiates the reaction with the electron-rich aromatic species. A very wide variety of aromatic ketones can be prepared by the Friedel-Crafts acylation. The acylating agent may likewise be any of a wide variety of acid derivatives. Aliphatic as well as aromatic acid chlorides are excellent reaction partners for aromatic hydrocarbons in the Friedel-Crafts reaction, which allows much structural variation in the ketones synthesized.

Two acylation reactions of benzene derivatives are presented in this section. In the first of these, 4-chlorobenzophenone is prepared from chlorobenzene and benzoyl chloride under more or less classical conditions. In the second preparation, acetic anhydride is used as the acylium cation source. The product is 4-bromoacetophenone, which may be used directly in the enol bromination procedure described in Chap. 17.

EXPERIMENT 16.1A SYNTHESIS OF 4-CHLOROBENZOPHENONE

$$C_6H_5COCl + C_6H_5Cl + AlCl_3 \longrightarrow C_6H_5COC_6H_4Cl + HCl$$

Time 3 h

Materials Benzoyl chloride, 6 mL (MW 140.57, bp 198°C, d 1.211 g/mL)
Chlorobenzene, 25 mL (MW 112.56, bp 132°C, d 1.106 g/mL)
Anhydrous aluminum chloride, 7.5 g (MW 133.34, mp 190°C)
Concentrated hydrochloric acid, 25 mL
Dichloromethane, 25 mL
Methanol, 100 mL

Precautions Perform all transfers in a good hood and wear gloves. Make sure the apparatus can dispose of acidic gases (HCl produced during reaction).

Hazards Benzoyl chloride is a skin irritant and lachrymator. Avoid breathing vapors and contact with skin. Aluminum chloride is an acidic solid. Avoid contact with skin, eyes, nose, or any moisture. Aluminum chloride reacts with moisture to produce hydrogen chloride gas. Wash any exposed area with water. Chlorobenzene is toxic in high concentrations. Avoid breathing vapors.

Experimental Procedure

Assemble the apparatus shown in Fig. 9.1 using a 250-mL round-bottom flask. Attach a device for removing gases, such as that shown in Fig. 2.39.

Rapidly weigh (preferably from a freshly opened bottle) 7.5 g (0.056 mol) aluminum chloride ($AlCl_3$) in a 250-mL round-bottom flask. Add to the aluminum chloride in the flask 25 mL chlorobenzene which has been measured from a *dry* graduated cylinder. (*Note:* If the chlorobenzene is not added immediately after weighing the aluminum chloride, stopper the flask to prevent moisture from entering.) After addition of the chlorobenzene measure 6 mL (0.052 mol) benzoyl chloride in a *dry* 10-mL graduated cylinder. Immediately add the benzoyl chloride to the mixture of chlorobenzene and aluminum chloride above. Add several boiling chips to the round-bottom flask and insert the reflux condenser as before. Swirl the entire apparatus to thoroughly mix all the reagents.

Slowly heat (using flame) the round-bottom flask until reflux is obtained. As heat is applied, the aluminum chloride will dissolve in the reaction mixture, hydrogen chloride gas will start to evolve, and the mixture will start turning dark. The heating source should be adjusted to maintain the gentlest possible reflux in the round-bottom flask and the mixture should be refluxed for 45 min.

At the end of the reflux period make a hydrolysis solution by mixing 50 g ice with 25 mL concentrated hydrochloric acid in a 600-mL beaker. Allow the reaction mixture to cool after the reflux period and pour the dark reaction product (**hood, gloves**) into the hydrolysis mixture in a thin stream. After the addition of the reaction mixture, swirl the beaker vigorously to hydrolyze the solution (**heat evolved**). The reaction mixture will lose some of its dark color during this process.

Transfer the hydrolyzed solution to a separatory funnel. Add 25 mL dichloromethane to the round-bottom flask, swirl to dissolve any residue, transfer the organic solvent to the 600-mL beaker, swirl to dissolve any residue, and transfer this solution to the separatory funnel. Shake the layers in the separatory funnel, allow the organic phase to separate, and draw off the *lower,* organic material. Discard the aqueous wash. Add the organic material back to the separatory funnel and wash with two 25-mL portions of water, followed by one 25-mL portion of saturated sodium bicarbonate ($NaHCO_3$). Test the bicarbonate wash with pH paper. If the bicarbonate wash is neutral or acidic by pH paper, wash the organic material with one further 25-mL portion of saturated bicarbonate.

After the aqueous washes transfer the organic layer back into the 250-mL round-bottom flask used in the initial part of the experiment. Place the round-bottom flask on a steam bath and evaporate as much of the methylene chloride as possible. Next add 100 mL water to the residue and set up a simple steam-distillation apparatus. Add several boiling *sticks* (not boiling chips) to the round-bottom flask and steam distill the product. This steam distillation will remove the chlorobenzene as the water-chlorobenzene azeotrope, bp 90°C.

When the temperature of the distilling vapor exceeds 96°C, the steam distillation may be stopped.

(*Note:* This steam distillation is quite rapid. One can remove 25 mL chlorobenzene as the water-chlorobenzene azeotrope in approximately 15 min using this technique. The steam distillation removal of chlorobenzene is much milder than attempting to remove the chlorobenzene by a fractional distillation.)

Cool the round-bottom flask nearly to room temperature and decant the excess water left over from the steam distillation. (*Note:* In many cases the product will solidify during this cooling process.) Remove the boiling sticks. Add 60 mL methanol to the residue and heat to boiling on the steam bath. Transfer the organic solution to a 250-mL Erlenmeyer flask, remove the flask from the steam bath and allow it to cool slightly, and add 1 to 3 g decolorizing carbon. The procedure is recommended so that the last traces of colored contaminants may be removed. The organic residue is placed back on the steam bath and heated briefly with vigorous swirling to complete the adsorption of the impurities on the charcoal. Allow the solution to cool again to room temperature and filter through a 5-mm bed of Celite (filter aid) to remove the charcoal. You should obtain from this treatment a clear methanol solution of 4-chlorobenzophenone. (*Note:* If at any point during this purification the product starts crystallizing out, add more methanol to keep the material in solution.)

The clear methanol solution should be warmed on the steam bath and 15 mL water added all at once. Swirl the flask to ensure good mixing and allow

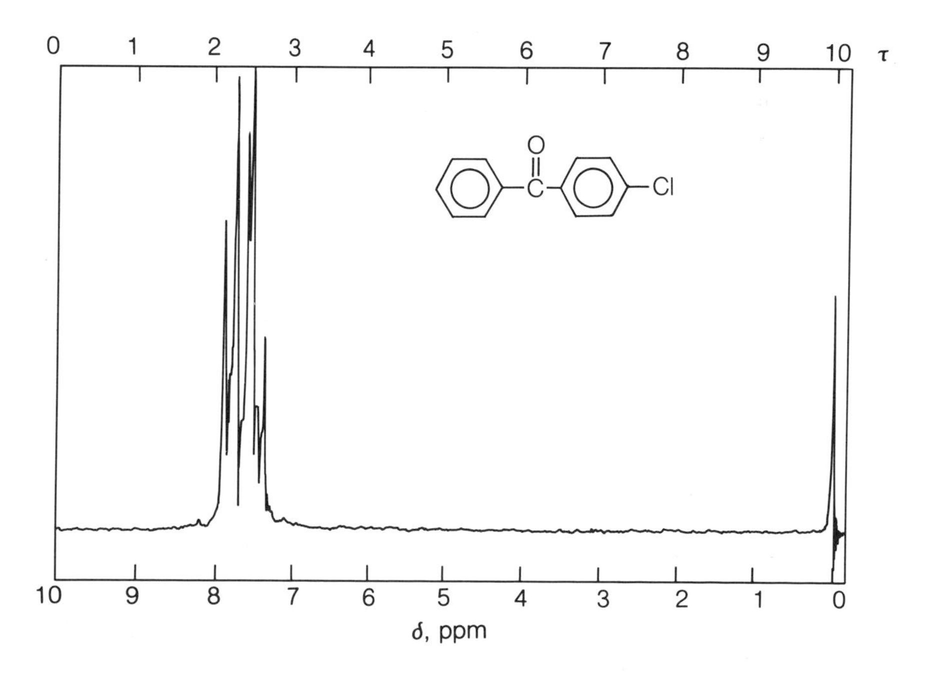

Figure 16.1
The nmr spectrum of 4-chlorobenzeophenone.

the solution to cool to room temperature, followed by cooling in an ice-water bath. Filter the white solid which you obtain using a Buchner funnel and allow the material to air dry briefly. (*Note:* 4-Chlorobenzophenone crystallizes from methanol-water as very small crystalline needles. If an attempt is made to compact the crystalline mass in the Buchner funnel, the overall result is usually that the filter paper becomes clogged and the solvent will not be drawn through. It is therefore recommended that the filtration be done rapidly on a large Buchner funnel and that air drying be used as much as possible to purify the crystalline mass.)

A white crystalline material (4 to 7 g) is obtained at this point, mp 74 to 76°C. Recrystallization, if necessary, may be done in 80% methanol-water as above (this is the preferred solvent) or in cyclohexane. An nmr spectrum of chlorobenzophenone is shown in Fig. 16.1. Compare this with the nmr spectrum of benzophenone (Fig. 14.4) and try to interpret the effect of chlorine on the spectrum.

EXPERIMENT 16.1B SYNTHESIS OF 4-BROMOACETOPHENONE

$$\text{Br-C}_6\text{H}_5 + CH_3\text{—}\overset{O}{\overset{\|}{C}}\text{—O—}\overset{O}{\overset{\|}{C}}\text{—}CH_3 + AlCl_3 \longrightarrow \text{Br-C}_6\text{H}_4\text{—}\overset{O}{\overset{\|}{C}}\text{—}CH_3 + HOAc$$

Time 4 h (may be done in two segments)

Materials Bromobenzene, 10 mL (MW 157, bp 156°C, d 1.49 g/mL)
Anhydrous aluminum chloride, 30 g (MW 133.5)
Acetic anhydride, 8 mL (MW 102, bp 138 to 140°C, d 1.08 g/mL)
Dichloromethane, 50 to 70 mL
Concentrated hydrochloric acid, 30 mL

Precautions Wear gloves and carry out all transfers in a good hood. Avoid breathing aluminum chloride dust.

Hazards Aluminum chloride causes burns and is irritating to the skin, eyes, and respiratory system. It reacts with moisture to form hydrogen chloride. Wash any exposed area with water. Acetic anhydride is flammable and causes skin burns. Bromobenzene is toxic in high concentrations.

Experimental Procedure

Assemble the apparatus shown in Fig. 12.1 using a 250-mL flask. To the top of the reflux condenser attach a device for removal of acidic gases, as shown in Fig. 2.39.[1] Place 30 g aluminum chloride (**hood, gloves**) in the *dry* 250-mL round-bottom flask.

Add 40 mL *dry* dichloromethane to the 250-mL round-bottom flask, and then add 10 mL bromobenzene (0.1 mol) to the suspension (the solution may turn orange). Place 8 mL (0.1 mol) acetic anhydride in the addition funnel. Add the acetic anhydride to the reaction flask dropwise over 15 min. As the acetic anhydride reacts, the aluminum chloride dissolves, hydrogen chloride gas is evolved, and the solution usually turns deep red. After the addition, reflux the mixture for 45 min by heating on a steam bath.

Toward the end of the reflux period make a slurry of 100 g ice and 30 mL concentrated HCl in a 600-mL beaker. At the end of the reflux period slowly add (**hood, gloves: HCl gas and heat produced**) the red dichloromethane solution to the ice-acid mixture. Considerable heat is produced. After the solution has been added, swirl the mixture to hydrolyze the aluminum salts. If necessary, add another 25 to 30 mL dichloromethane. Transfer the two-phase system to a separatory funnel and allow the layers to separate. Remove the lower, dichloromethane layer and discard the upper, milky, aqueous layer. Wash the organic layer with two 25-mL portions of water, one 25-mL portion of sodium hydroxide solution, and two 25-mL portions of half-saturated sodium chloride solution. After the last wash, dry the organic layer with either calcium chloride or calcium sulfate for several minutes. Gravity filter the organic layer into a 250-mL Erlenmeyer flask and remove the dichloromethane by gentle heating on a steam bath. After evaporation of the solvent, the product remains as a heavy yellow liquid.

Note: The procedure may be stopped at this stage. Be certain to stopper the Erlenmeyer flask tightly until the distillation can be performed.

The best method for purification is vacuum distillation (see Sec. 2.1). Set up a simple vacuum distillation apparatus using a 50-mL round-bottom flask and Claisen head fitted to an air-cooled condenser. (Do *not* connect the condenser to a water supply.) Prepare a capillary tube and distill the heavy oil under vacuum using either a mechanical vacuum pump or a water aspirator. The product should distill as a clear liquid, which solidifies in the receiver flask. (*Note:* A heating lamp or heat gun should be used during this distillation to keep the end of the condenser and vacuum adapter at a temperature of about 50°C. This will prevent the ketone from crystallizing and blocking the path of the distillate.) The ketone distills at 100°C at 1 torr, 117°C at 7 torr, 130°C at 15 torr,

[1] The HCl which is generated may be removed by attaching a piece of rubber tubing (preferably old) to the top of the reflux condenser. This is usually accomplished by use of a thermometer adapter. The tube should then be run into the hood sink or into the draft port in the hood. An alternative is to use the inexpensive gas-takeoff adapter described in Sec. 2.8.

and 145°C at 30 torr. The distillation should be stopped when the distillate appears colored.

The solid ketone should be warmed in a water bath or with a heating lamp (to melt it) after distillation and poured into a preweighed, heavy-walled, 125-mL Erlenmeyer flask. After solidification the solid should be scraped out and stored. The yield of solid material is 6 to 10 g. The yield may be increased if two students combine their products for distillation. The ketone is usually white to yellow and melts between 46 and 50°C. Highly colored samples may

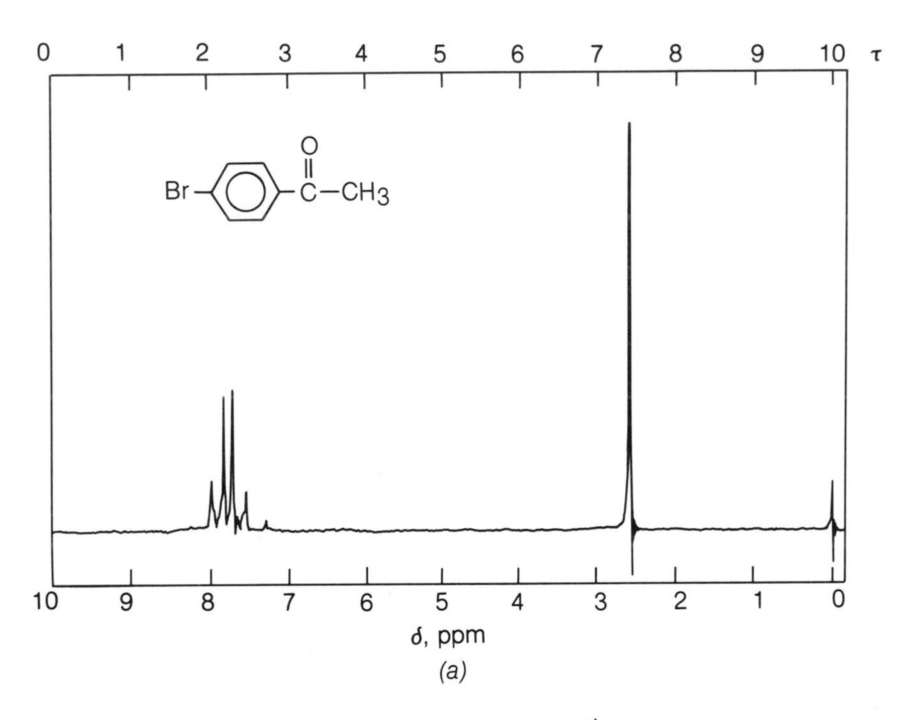

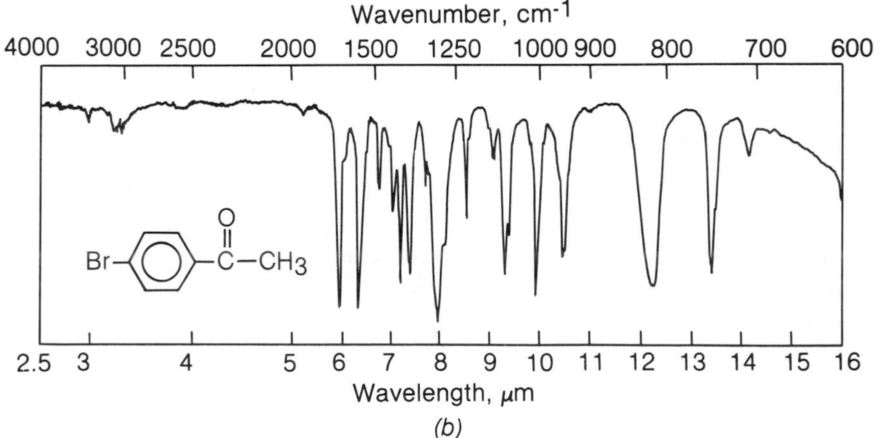

Figure 16.2
The (*a*) nmr and (*b*) ir spectra of 4-bromoacetophenone.

be redistilled to give a pure product *but a heat lamp should always be used during a redistillation.*

(*Note:* One may also distill the ketone at atmospheric pressure (bp 255°C) but the yield is generally lower (5 to 7 g) and the product is more colored. The best method is to distill with a mechanical vacuum pump at the lowest possible pressure.)

The nmr and ir spectra of 4-bromoacetophenone are shown in Fig. 16.2. Note the A_2B_2 pattern in the aromatic protons.

This ketone may be used, even when impure, as starting material in the enol bromination experiment (Exp. 17.1A or 17.1B).

16.2 THE SYNTHESIS OF ACETYLFERROCENE

Ferrocene, the prototype of the so-called sandwich complexes, was discovered accidentally by two different research groups trying to do entirely different things. One group was trying to prepare cyclopentadienylidenecyclopentadiene ($C_{10}H_8$, structure 16.2*a*) using an iron-catalyzed coupling reaction of a Grignard reagent.

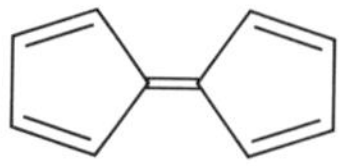

$^{-}$ Fe^{2+} $^{-}$

Fe

16.2*a* 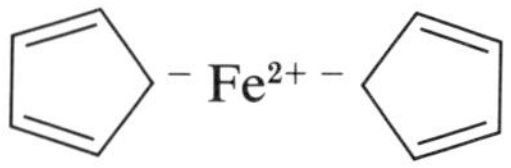 16.2*b* 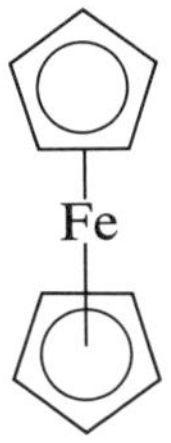 16.2*c*

The other group was looking for a high-temperature catalyst. Both groups isolated an orange solid which melted between 174 and 176°C. The structure of the compound was not clear to either of the groups, although structural possibilities were suggested. One group suggested that the compound was the iron salt (Fe^{2+}) of two cyclopentadienide ions. This was an interesting suggestion but seemed unlikely even then because the compound had too low a melting point (174 to 176°C) and boiling point (250°C) to be a salt.

A compound with the salt structure 16.2*b* would probably hydrolyze very rapidly to cyclopentadiene and ferric hydroxide. This behavior was not observed for the compound. Likewise, one would anticipate that a Friedel-Crafts reaction with the cyclopentadiene rings would lead to at least *two* different monoacylation products. In fact, as you will see in this experiment, this reaction yields only one product when performed on ferrocene. The single-product observation was instrumental in the assignment of the correct structure of ferrocene as the first member of what is now a large family of sandwich complexes.

The sandwich complex structure is illustrated (see p. 411) as 16.2*c*. The structure is harder to draw than it is to visualize. If you imagine two pieces of bread with an orange between the two, you have the approximate structure of ferrocene. You can easily see that the molecule has fivefold symmetry, i.e., all five positions of each aromatic ring are equivalent. Monoacylation of this molecule must therefore lead to a single product. Diacylation, as we will see later, leads to a single acetyl residue in each of the two rings.

The cyclopentadienyl rings of ferrocene can formally be considered as having six π electrons, and these electron-rich aromatics are coordinated to an electron-deficient iron atom. Ferrocene then behaves as an aromatic molecule and readily undergoes electrophilic acylation of the Friedel-Crafts type. There is an important difference in reactivity between ferrocene and benzene, however, because ferrocene is very much more electron-rich. If we think simply of a five-atom ring system containing six electrons, it is clear that the relative electron density at any position is higher than that we would observe in benzene, which has six π electrons distributed over six carbon nuclei. We can ignore, for the purpose of this argument, the presence of the iron atom (this can be done with some justification). Considering that the ferrocene molecule is much more electron-rich, or nucleophilic, than benzene, its reactions with an electrophile should occur much more easily than the same reaction with benzene. The acetylation of benzene with acetyl chloride usually requires the presence of aluminum chloride as a catalyst and over 1 h at reflux temperature (81°C). The reaction of ferrocene with acetyl chloride in the presence of aluminum chloride can be accomplished in less than 1 h at 0°C.

Two experimental procedures are given below for the acylation of ferrocene. One involves aluminum chloride and the other involves the much weaker catalyst phosphoric acid. The latter is given first.

EXPERIMENT 16.2A SYNTHESIS OF ACETYLFERROCENE: THE PHOSPHORIC ACID–CATALYZED METHOD

$$\text{Fe}(C_5H_5)_2 + CH_3{-}\overset{\displaystyle O}{\overset{\|}{C}}{-}O{-}\overset{\displaystyle O}{\overset{\|}{C}}{-}CH_3 \xrightarrow{H_3PO_4} (C_5H_5)\text{Fe}(C_5H_4{-}\overset{\displaystyle O}{\overset{\|}{C}}{-}CH_3) + CH_3COOH$$

Time 2.5 h

Materials Ferrocene, 1 g (MW 186, mp 174 to 176°C)
Acetic anhydride, 10 mL (MW 102, bp 138 to 140°C)
85% Phosphoric acid, 25 drops
Sodium bicarbonate, 10 g

Precautions Wear gloves and perform all transfers in a good hood. Use no flame. When neutralizing acid with $NaHCO_3$, add small portions of the base cautiously to reduce the risk of effervescence and spattering that may occur during this procedure.

Hazards Acetic anhydride and phosphoric acid are corrosive liquids. Make sure flames are kept away from acetic anhydride and avoid breathing its vapors.

Experimental Procedure

Charge a 50-mL, round-bottom flask with 1 g ferrocene and 10 mL acetic anhydride. Slowly add 85% phosphoric acid dropwise, until about 25 drops have been added. Clamp the flask on top of a steam bath, attach a reflux condenser (with lightly greased joints) having a drying tube on top of it (see Fig. 9.1). Heat the mixture on a steam bath for 15 min. The mixture will not reflux during this time because the boiling points of all the reactants are above the steam-bath temperature.

After 15 min of heating on the steam bath, pour the mixture in the round-bottom flask into a 250-mL beaker containing 50 g craked ice. Rinse the reaction flask with about 10 mL cold water and add it to the ice-water solution in the beaker.

Neutralize the acetic anhydride and excess phosphoric acid by the slow and careful addition of 10 g sodium bicarbonate. When bicarbonate reacts with acid, CO_2 is released and vigorous effervescence ensues. Slow addition of bicarbonate ensures that only a small amount of CO_2 will be lost from the mixture during each addition, thus reducing the danger of the acidic solution spattering out of the flask. After neutralization, the mixture should be allowed to stand in an ice-water bath for about 10 min. During this time, the product will precipitate. Suction filter the product and wash it twice with 50-mL portions of ice-cold distilled water. Place a second piece of filter paper over the crude solid and, using a cork, press dry or set the product aside and allow it to air dry until the next laboratory period. The material can easily be recrystallized from boiling heptane (bp 100°C) or petroleum ether to yield an orange solid, mp 85 to 86°C. The ir spectrum of the product ferrocene is shown in Fig. 16.3*a*.

EXPERIMENT 16.2B

SYNTHESIS OF ACETYLFERROCENE: THE ALUMINUM CHLORIDE–CATALYZED METHOD

$$\text{Ferrocene} + CH_3{-}C(=O){-}Cl \xrightarrow{AlCl_3} \text{Acetylferrocene } (C_5H_4{-}C(=O){-}CH_3)\,Fe\,(C_5H_5) + HCl$$

Time 2.5 h

Materials Ferrocene, 9.3 g (MW 186, mp 174 to 176°C)
Acetyl chloride, 4 mL (MW 78.5, bp 52°C, d 1.1 g/mL)
Anhydrous aluminum chloride, 7 g (MW 133.5)
Dichloromethane, 50 mL
10% Sodium bisulfite solution, 1 to 5 mL

Precautions Carry out all transfers in a good hood. Wear gloves when weighing $AlCl_3$ and transferring acetyl chloride. Use no flame.

Hazards Acetyl chloride is a highly flammable and corrosive liquid. It is a potent lachrymator. Avoid breathing of or contact with either vapor or liquid and avoid flames. Anhydrous aluminum chloride is a corrosive solid. It is also a lachrymator. Avoid contact with dust.

Experimental Procedure (hood)

Add acetyl chloride (4.0 mL, 4.4 g, 0.056 mol) to ferrocene (9.3 g, 0.05 mol) and 40 mL dry dichloromethane in a 125-mL Erlenmeyer flask equipped with a magnetic stirring bar and fitted with either a drying tube (if the experiment is conducted in the hood) or a gas trap (Fig. 2.39) (if the reaction is carried out on a bench top). Immerse the flask in an ice-water bath at a temperature between 0 and 5°C. Weigh anhydrous aluminum chloride (7 g, 0.05 mol) in a 50-mL beaker and cover it with a watch glass. Add the aluminum chloride in approximately five portions. Be careful to replace the drying tube after each addition. A short period of time (2 to 5 min) should elapse between successive additions to allow for heat exchange. Reaction begins almost immediately after the first portion of aluminum chloride is added, as evidenced by a color change from red-brown to deep wine red. After the addition is complete, rinse the beaker with 5 mL dichloromethane to transfer all the aluminum chloride to the reaction mixture. Stir or swirl the mixture for 1 h as the ice bath gradually comes to room temperature.

At the end of the reaction period, once again cool the flask by placing it in an ice bath. Hydrolyze the reaction mixture by gradual addition of 2-mL portions of cold water until 10 mL water has been added. Hydrogen chloride gas will be evolved during the hydrolysis, so the flask should be kept in the hood. An additional 50 mL water may be added more rapidly. Next, remove the ice bath and add 10% aqueous sodium bisulfite (prepared and used the same day) dropwise with stirring from a separatory funnel, with an eye dropper, or with Pasteur pipet until the upper layer turns from brown to cream yellow. Stir the solution for 10 min (the sharp odor of sulfur dioxide should no longer be detectable).

Place the reaction mixture in a separatory funnel. Separate the lower, organic phase. Extract the aqueous layer three times with 15-mL portions of dichloromethane. Combine the organic extracts and wash with 10 mL 5% so-

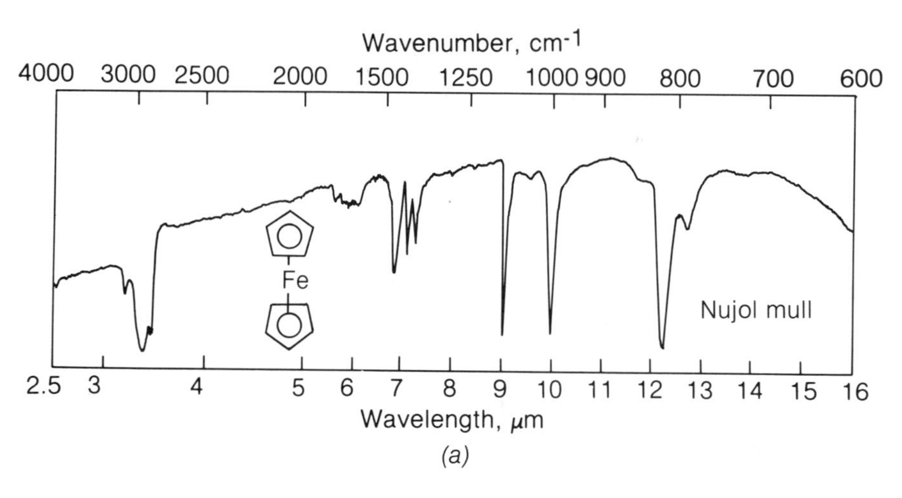

(a)

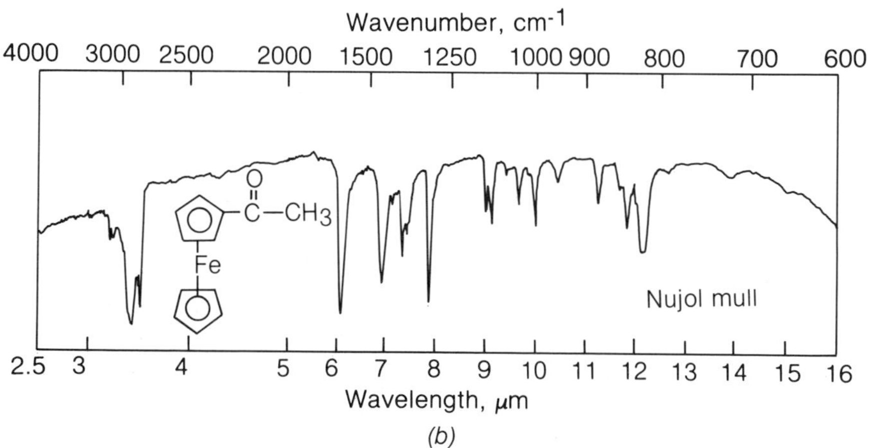

(b)

Figure 16.3
The ir spectra of (*a*) ferrocene and (*b*) acetylferrocene.

dium hydroxide solution and 10 mL saturated aqueous sodium chloride solution. Dry the solution over anhydrous potassium carbonate, sodium sulfate, or magnesium sulfate; filter; decant from the drying agent; and evaporate the solvent by heating on the steam bath to give 10 to 11 g (85 to 95%) of an orange solid, mp 85 to 86°C. The product may be crystallized from heptane.

The ir spectra of ferrocene and acetylferrocene are shown in Fig. 16.3. Note that the carbonyl vibration is observed at 1650 cm^{-1}. This band position is much lower than the carbonyl vibration of acetylbenzene (acetophenone, 1686 cm^{-1}), which indicates the difference in electron-releasing ability of phenyl and ferrocenyl.

16.3 THE FRIEDEL-CRAFTS ALKYLATION

In the examples of Friedel-Crafts acylation described above, the attacking electrophile has been an acylium ion. In the experiment described in this section the electrophile is a carbonium ion. It is generated by treating a tertiary alcohol with a strong acid as the dehydrating agent. The reaction is illustrated below. The tertiary butyl cation is stable for a charged species and does not rearrange, as might be expected for the *n*-butyl cation.

$$(CH_3)_3C{-}OH + H_2SO_4 \rightleftharpoons CH_3{-}C^{+}(CH_3)_2 + H_2O + HSO_4^{-}$$

After the electrophilic cation has been formed, it may attack the benzene ring. Addition to the aromatic compound occurs essentially as illustrated at the beginning of this chapter and is shown below for the example discussed here.

$$p\text{-}(CH_3O)_2C_6H_4 + {}^{+}C(CH_3)_3 \rightleftharpoons [\text{arenium ion: } H, C(CH_3)_3 \text{ on ring}, +] \rightleftharpoons 2\text{-}C(CH_3)_3\text{-}1,4\text{-}(CH_3O)_2C_6H_3 + H^{+}$$

As discussed earlier, alkyl-substituted aromatics are usually more reactive than the starting materials from which they were formed and tend to undergo further reaction. In the example presented here, the second *tert*-butyl group adds (for steric as well as electronic reasons) as far away from the first *tert*-butyl group as possible. No further reaction occurs because the two remaining positions are too sterically hindered.

This sort of reaction is commercially important. If the reaction described in this section is carried out on 4-methylphenol (4-hydroxytoluene), 2,6-di-*tert*-butyl-4-methylphenol results. This compound is referred to industrially as BHT (for butylated hydroxytoluene) and is a food additive which prevents oxidative spoilage.

EXPERIMENT 16.3 SYNTHESIS OF 1,4-DI-tert-BUTYL-2,5-DIMETHOXYBENZENE

$$1,4\text{-}(OCH_3)_2C_6H_4 + (CH_3)_3COH \xrightarrow{H_2SO_4} 1,4\text{-}(OCH_3)_2\text{-}2,5\text{-}[C(CH_3)_3]_2C_6H_2$$

Time 2.5 h

Materials 1,4-Dimethoxybenzene, 6 g (MW 138, mp 56 to 60°C)
tert-Butyl alcohol, 10 mL (MW 74, bp 83°C, d 0.786 g/mL)
Acetic acid, 20 mL (MW 60, bp 116°C, d 1.05 g/mL)
Concentrated (98%) sulfuric acid, 10 mL
30% Fuming sulfuric acid, 10 mL
Dichloromethane, 25 mL

Precautions Wear gloves when handling either concentrated or fuming sulfuric acid. The latter should be poured only in a hood. Acetic acid is not as strongly acidic as sulfuric acid but should be handled very carefully. Make certain that a supply of solid sodium carbonate or sodium bicarbonate is available to be spread on acid spills.

Hazards Acetic and sulfuric acids are dehydrating agents which can cause severe burns. Fuming sulfuric acid poses the same danger and contains the irritating gas SO_3. Avoid contact or inhalation.

Experimental Procedure

Charge a 125-mL Erlenmeyer flask with 6 g 1,4-dimethoxybenzene, 10 mL *tert*-butyl alcohol, and 20 mL glacial acetic acid. Swirl briefly to mix the contents and then clamp the neck of the flask and immerse it in an ice-water bath and let it cool.

Measure 10 mL concentrated sulfuric acid in a graduated cylinder (**gloves**) and pour it into a 125-mL flask. Measure 10 mL fuming sulfuric acid in a graduated cylinder (**gloves, hood**) and add it to the sulfuric acid; then place this flask in the ice bath. Now, holding the first flask by the clamp, swirl the flask in the ice bath until the internal temperature (thermometer) is about 0°C. Leaving the flask in the ice bath, clamp it to a ring stand. Using a second clamp fix a small separatory funnel so the delivery tube is in the neck of the clamped flask. Now transfer the cold sulfuric acid to the separatory funnel and add the acid dropwise during about 5 min. While adding the acid loosen the clamp where it attaches to the ring stand just enough so the flask can be gently rocked from side to side. Agitate the flask gently during the acid addition, being careful not to splash any acid onto your hand. After the addition is complete, stir the slurry briefly with a thermometer. Note the temperature, which should be near 25°C, and swirl the flask for an additional 5 to 10 min.

Add about 25 g cracked ice and then add enough water so that the flask is nearly full but may still be swirled. Filter the reaction mixture (Buchner funnel) using gentle suction. Wash the filter cake with three 25-mL portions of cold distilled water. Turn on the suction full force and place a second piece of filter paper on the solid. Press down on the filter paper with the top of a cork to press out residual water. Remove the filter paper and wash the product with three 15-mL portions of ice-cold methanol.

Gently scrape the solid into a dry 125-mL Erlenmeyer flask and dissolve the solid in 25 mL dichloromethane. Add sodium or magnesium sulfate drying agent and let stand for about 10 min. Gravity filter into a second 125-mL Erlenmeyer flask to remove the drying agent and evaporate the dichloromethane by heating on the steam bath (**hood**). Recrystallization of the crude product from 20 to 30 mL ethanol yields 5 to 7 g 1,4-di-*tert*-butyl-2,5-dimethoxybenzene as white plates, mp 104 to 105°C.

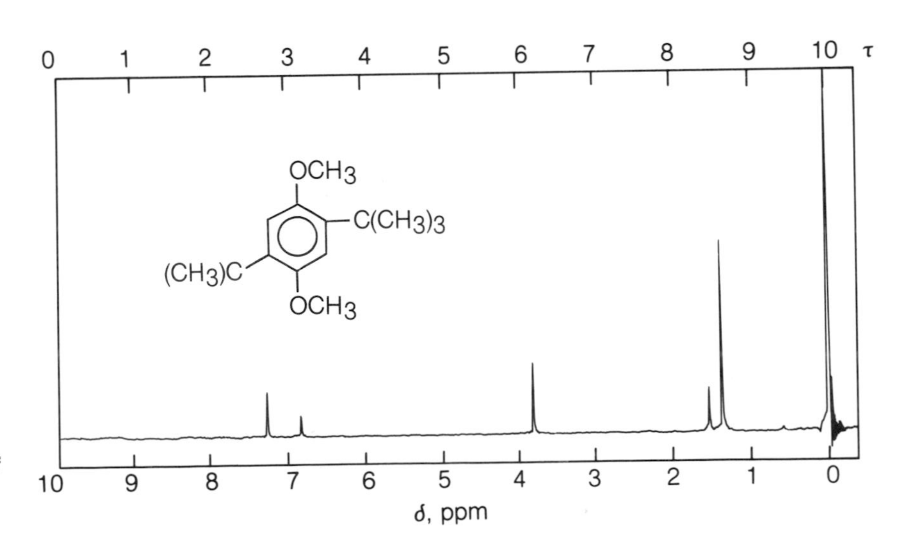

Figure 16.4
The nmr spectrum of 1,4-di-*tert*-butyl-2,5-dimethoxybenzene.

The nmr spectrum of the pure diether is shown in Fig. 16.4. Note that only three singlets are observed although there are 26 hydrogen atoms in this molecule.

QUESTIONS AND EXERCISES

16.1 Would Friedel-Crafts acylation or alkylation be preferred if only one substituent were desired on methoxybenzene?

16.2 The acetylation of ferrocene is described in Sec. 16.2. It is noted there that diacylation sometimes occurs (or can be effected deliberately). When diacylation is observed, the product is always the one in which each ring is monoacylated. Why is no product formed which has both substituents in the same ring? What would be the result if dialkylation had been attempted? Why?

16.3 In Exp. 16.3, *tert*-butyl alcohol is used in the alkylation of 1,4-dimethoxybenzene. What product(s) would be formed if *n*-butyl alcohol were used instead?

16.4 Would it be possible to form the product of Exp. 16.3 by treating 1,4-dimethoxybenzene with 2-methyl-2-propene and sulfuric acid? Why?

ENOL BROMINATION

When a double bond is treated with bromine, the electrons in the double bond react with the electrophilic reagent in such a way that a dibromide forms. This reaction is very general and of great utility in the chemical industry.

Another important reaction between reactive organic groups and bromine (or for that matter any halogen) is the reaction of a methyl ketone with halogen under basic conditions. For instance, if acetone reacts with bromine in the presence of sodium hydroxide, a rapid reaction occurs and the product isolated is sodium acetate and tribromomethane (bromoform). The reaction is formulated below.

$$CH_3-\overset{\overset{\displaystyle O}{\|}}{C}-CH_3 + Br_2 + NaOH \longrightarrow CH_3-\overset{\overset{\displaystyle O}{\|}}{C}-O^-Na^+ + HCBr_3$$

Notice that as the reaction proceeds and the halogen is consumed, a fragmentation reaction occurs which splits a three-carbon ketone into a two-carbon acid and the corresponding one-carbon halogenated methane. This reaction goes under the name of the *haloform* reaction.

If the same reaction is performed either in acid solution or in the absence of excess halide, a somewhat different product is obtained. If only one equivalent of bromine is added to acetone under acidic conditions, the product is a bromoketone. If two molecules of bromine are added to acetone, then the product is a dibromoketone. It is characteristic of the halogenation reaction that a monobromination product is obtained when it is performed in acid solution, whereas in basic solution the bromination often proceeds to the fully brominated product, which is then cleaved.

The bromination of acetone and that of a simple double bond actually occur by a similar mechanism. For a ketone or aldehyde, the first step is enolization to give the enol form of the carbonyl compound. Notice that the enol is an electron-rich, hydroxyl-substituted double bond, which would be expected to add bromine very rapidly. This indeed happens as shown below:

$$R{-}C({=}O){-}CH_3 \overset{H^+}{\rightleftharpoons} R{-}C(OH){=}CH_2 + Br_2 \longrightarrow R{-}C(OH)(Br){-}CH_2{-}Br \longrightarrow R{-}C({=}O){-}CH_2{-}Br + HBr$$

An alternative and equivalent description of this process is shown in the mechanism below. In this case, the enol attacks bromine, with re-formation of the carbon-oxygen double bond and loss of bromide anion. The protonated bromoketone quickly deprotonates to form HBr as a by-product. This latter mechanism is currently favored, especially for reactions which occur with enolate anions (i.e., under basic conditions).

$$R{-}C(\ddot{O}{-}H){=}CH_2 + Br{-}Br \longrightarrow R{-}C({=}\overset{+}{O}H){-}CH_2{-}Br + Br^- \longrightarrow R{-}C({=}O){-}CH_2{-}Br + HBr$$

Under acidic conditions, enol formation is more difficult for the α-bromoketone than it is for the original ketone. Thus, the α-bromoketone will not react further until all the original ketone has been consumed. Once one equivalent of bromine has reacted with the original ketone, dibromination may begin and again proceed in a stepwise fashion. Acid-catalyzed brominations of aldehydes and ketones are easier to control than the base-catalyzed counterparts, and the products are easily purified. The explanation for the difference in reactivity is that under basic conditions the enolization of the halogenated ketone is faster than the enolization of the original starting material. Introduction of halogen into the system speeds up further reaction, and the reaction does not stop at an intermediate stage.

17.1 SYNTHESIS OF 4-BROMOPHENACYL BROMIDE

The two procedures below describe the bromination of 4-bromoacetophenone. The only real difference between the two procedures is the source of bromine. In each case the bromination is conducted under acidic conditions. At the start of the reaction the catalyst is glacial acetic acid. As the reaction proceeds, however, hydrogen bromide is produced and the reaction becomes autocatalytic. The source of bromine can be either molecular bromine (first procedure) or a bromine complex, pyridinium bromide perbromide (second procedure). Both reagents supply reactive bromine to the system. The major difference between the two procedures is that the complexed bromine is safer to handle and weigh, as it is a crystalline solid. Molecular bromine is a viscous, corrosive liquid which has a low boiling point. The course of the reaction is not markedly affected by the difference in bromine sources (except that there is a slightly higher yield with the latter). A practical consideration is that the bromine complex is more costly than molecular bromine.

The choice of 4-bromoacetophenone for this experiment has several advantages. The first advantage is that the crystalline starting material may be synthesized according to the directions in Exp. 16.1B. The second advantage is that the product, α,4-dibromoacetophenone, is a highly crystalline molecule (mp 108 to 110°C). Thus, the crystallinity facilitates workup after the reaction, and purification of the product is very simple. A third advantage is that the product is an important reagent used to characterize organic acids and phenols.

It should also be noted that the crystalline material produced in this reaction is in the family of organic compounds which have been used extensively as tear-gas agents. This particular derivative is not as obnoxious as other members of the family, but it is a skin irritant and a lachrymator. All due precautions should be taken when handling either bromine, complexed bromine reagents, or any product which contains an α-haloketone functional group.

EXPERIMENT 17.1A SYNTHESIS OF 4-BROMOPHENACYL BROMIDE FROM 4-BROMOACETOPHENONE

$$4\text{-}BrC_6H_4COCH_3 + Br_2 \xrightarrow{H^+} 4\text{-}BrC_6H_4COCH_2Br + HBr$$

Time 2.5 h

Materials 4-Bromoacetophenone (Exp. 16.1B), 3 g (MW 199, mp 50°C)
Bromine, 2.4 g (MW 160, d 3.1 g/mL) or stock solution
Glacial acetic acid, 0.5 mL (MW 60, d 1 g/mL)
Dichloromethane, 35 mL
95% Ethanol, 15 mL

Precautions Wear gloves and carry out all transfers in a good hood.

Hazards Bromine is a severe poison, both in liquid and vapor form. It will cause burns when it comes into contact with skin. Avoid breathing vapors and any contact with eyes or skin. Any spill on skin or eyes should be washed immediately with sodium bicarbonate solution. The product, phenacyl bromide, is a mild lachrymator and skin irritant. Avoid contact with skin.

Experimental Procedure

Measure 5 mL 20 vol % bromine in dichloromethane (approximately 1 mL bromine plus 4 mL dichloromethane) stock solution (**hood, gloves**) into a dry 10-mL graduated cylinder. (*Note:* Your instructor may dispense the bromine solution from the stock room.)

Place 3 g (0.015 mol) 4-bromoacetophenone (prepared in Exp. 16.2) in a dry 125-mL Erlenmeyer flask, followed by 25 mL dichloromethane. Swirl to dissolve the ketone. After the ketone dissolves, place 0.5 mL (10 drops) glacial acetic acid in the flask. Add 0.5 mL of the bromine-dichloromethane solution (**hood, gloves**) to the flask with a disposable pipet and swirl. After a short induction period (2 to 5 min), the bromine color discharges and hydrogen bromide gas (**hood, gloves**) is produced. Cool the flask in an ice-water bath until the internal temperature is approximately 15°C (thermometer) and continue to add 0.5-mL portions of bromine solution to the flask. Swirl and allow the bromine color to discharge after each addition. Cool in an ice-water bath, as needed, to keep the solution below 20°C. (*Note:* Toward the end of the addition of one equivalent of bromine, the color discharge will be slower than during the first part of the addition.)

Transfer the mixture to a separatory funnel after the bromine addition is complete (the solution should be colorless to yellow). Wash the mixture with one 25-mL portion of distilled water and drain the *lower,* dichloromethane layer into the 125-mL Erlenmeyer reaction flask. Discard the aqueous wash (drain). Place the Erlenmeyer flask on a steam bath, add several boiling chips, and allow the solvent to evaporate.

After removal of the solvent, add 10 mL 95% ethanol to the residue. The oily yellow product will usually solidify during this addition. Heat the ethanol to boiling and transfer the solution to a 50-mL Erlenmeyer flask. Add another 5-mL portion of ethanol to the 125-mL Erlenmeyer flask, heat to boiling and transfer to the 50-mL Erlenmeyer flask. Bring the entire mixture back to boiling, remove it from the steam bath, and allow it to cool to room temperature. Filter the crystalline mass with the aid of a Buchner funnel and wash with 5 mL *cold* ethanol. The yield is 2.7 to 3.0 g (65 to 70%) of white crystals of mp 108 to 110°C.

If the melting point is low, a single recrystallization from 95% ethanol (5 mL/g) usually gives pure product. The product may be analyzed by tlc (silica gel, 50% dichloromethane-hexane) for purity.

The nmr and ir spectra of 4-bromoacetophenone and 4-bromophenacyl bromide are shown in Figs. 17.1 and 17.2. Note the aromatic A_2B_2 pattern in the nmr spectra of both the starting material and the product. This is a characteristic pattern for this type of compound and is quite useful in identifying derivatives of this compound. The signals of the hydrogen atoms on the carbon next

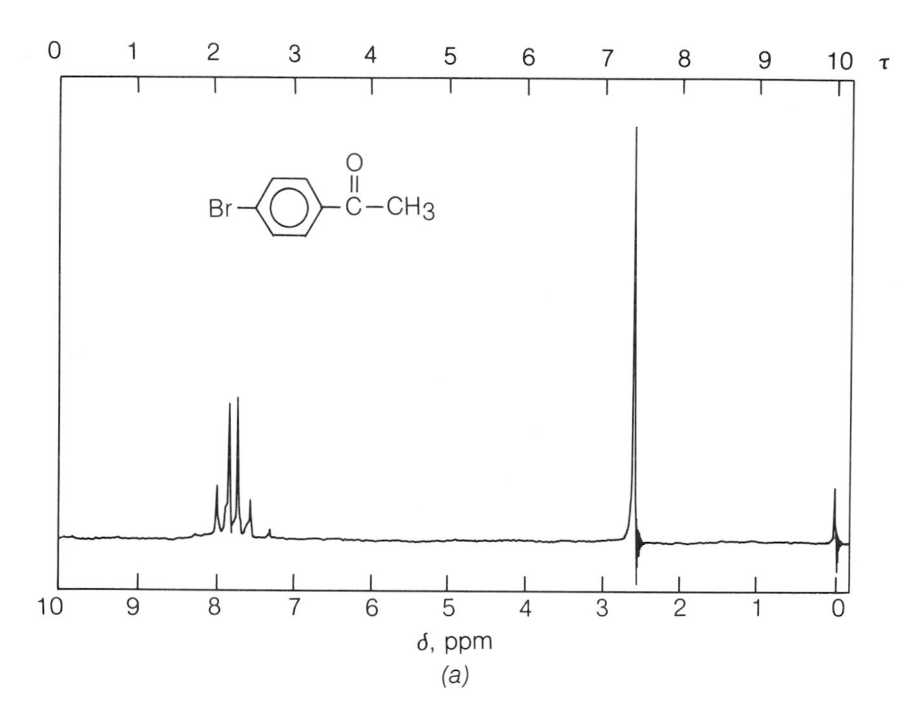

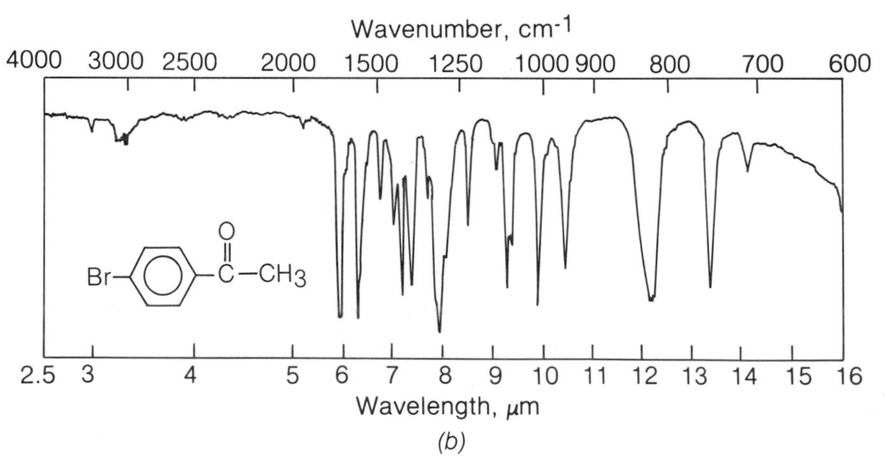

Figure 17.1
The (*a*) nmr and (*b*) ir spectra of 4-bromoacetophenone.

to the ketone move downfield from 2.6 to 4.4 ppm after substitution by bromine. Note also the decrease in size of the proton signal compared with that of the aromatic protons (from 3 to 2). The ir spectra also show evidence of electronegative substitution on the α-carbon atom as the carbonyl frequency shifts from 1670 to 1695 cm^{-1}.

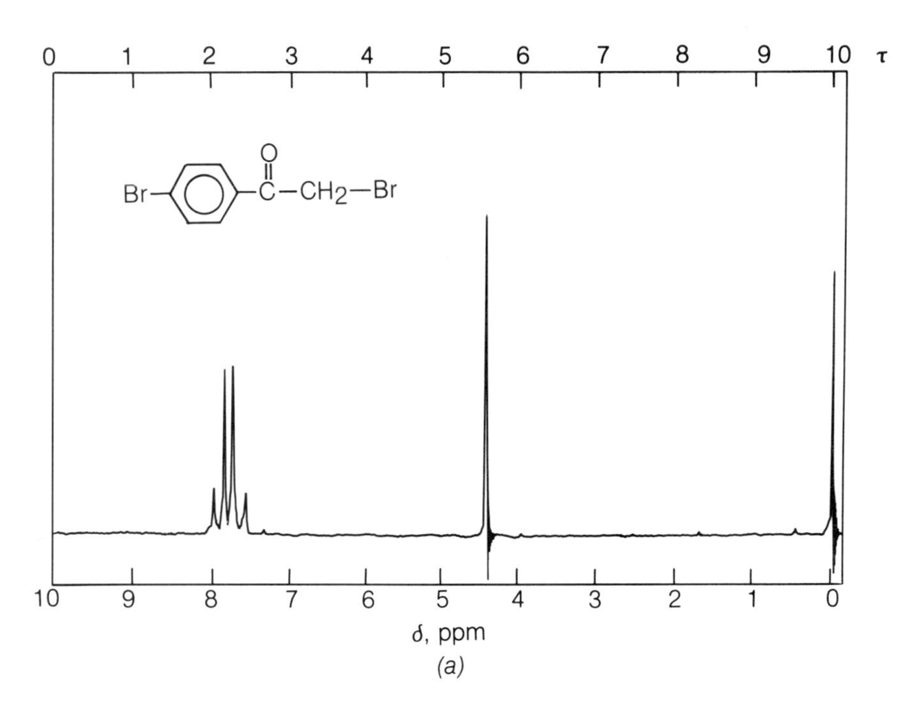

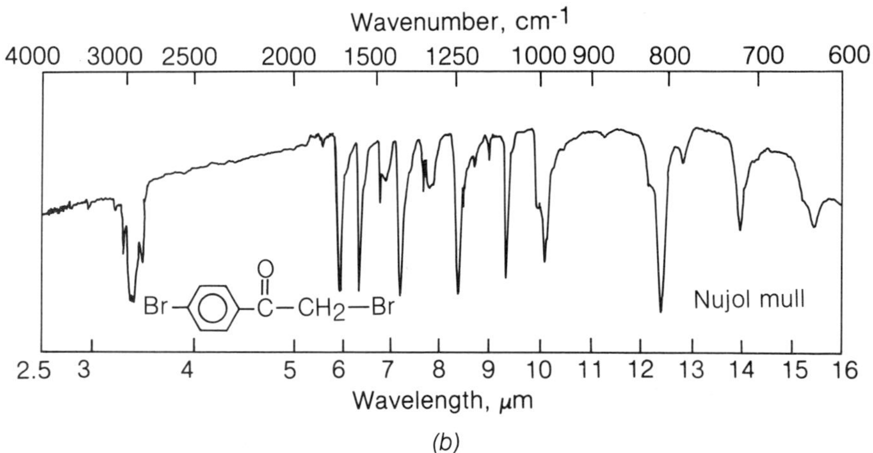

Figure 17.2
The (*a*) nmr and (*b*) ir spectra of 4-bromophenacyl bromide.

EXPERIMENT 17.1B

SYNTHESIS OF 4-BROMOPHENACYL BROMIDE USING PYRIDINIUM BROMIDE PERBROMIDE

O
CH_3 + Br_3^- $\xrightarrow{H^+}$ O Br CH_2 + HBr
Br ^+N Br
H

Time 2.5 h

Materials 4-Bromoacetophenone (Exp. 16.1B), 3 g (MW 199, mp 50°C)
Pyridinium bromide perbromide, 5 g (MW 319)
Glacial acetic acid, 0.5 mL (MW 60, d 1 g/mL)
Cyclohexene, 0.5 mL (MW 86)
Dichloromethane, 30 mL
95% Ethanol, 15 mL

Precautions Wear gloves and carry out all transfers in a good hood.

Hazards Pyridinium bromide perbromide is a lachrymator. It is, however, much safer to handle than liquid bromine. Pyridinium bromide perbromide is a reactive solid. Avoid breathing its dust and avoid contact with skin. Hydrogen bromide is an acidic gas. Avoid contact with skin and eyes. The product, 4-bromophenacyl bromide, is a mild lachrymator and skin irritant. Avoid contact with skin.

Experimental Procedure

Place 5 g (0.016 mol) pyridinium bromide perbromide in a 100-mL beaker (**gloves**) and cover with a watch glass. Place 3 g (0.015 mol) 4-bromoacetophenone (see Exp. 16.1B) in a dry 125 mL Erlenmeyer flask, followed by 30 mL dichloromethane. Swirl to dissolve the ketone. After the ketone dissolves, place 0.5 mL (10 drops) glacial acetic acid in the flask. Add 0.5 g pyridinium bromide perbromide to the flask and swirl vigorously while warming the solution on a steam bath. Heat until the dichloromethane just begins to boil. Remove the flask from the heat source and swirl. As the reaction proceeds, the pyridinium bromide perbromide dissolves and hydrobromic acid is produced. After several minutes, another 0.5-g portion should be added and the procedure repeated. The solution should be heated intermittently to keep the internal temperature close to the boiling point of the solvent.

After addition of all the brominating agent, swirl the solution with heating (**hood, gloves**) for 10 min (the entire procedure should take about 30 min), and then add several drops (0.5 mL) cyclohexene. Swirl the solution and warm briefly on a steam bath. (*Note:* Cyclohexene reacts with any excess bromine present in the reaction mixture. Nevertheless, the color of the solution usually remains yellow to orange because of colored by-products.)

Transfer the mixture to a separatory funnel after the bromination procedure is complete. Wash the mixture with one 25-mL portion of distilled water and return the *lower,* dichloromethane layer, which should now be colorless, to the 125-mL flask. (If any color remains in the dichloromethane, several more drops of cyclohexene should be added and the aqueous wash repeated.) Discard the aqueous wash (drain). Place the Erlenmeyer flask on a steam bath, add several boiling chips, and heat to remove the solvent.

After removal of the solvent, add 10 mL 95% ethanol to the residue. The oily yellow product will usually solidify during this addition. Heat the ethanol to boiling and transfer the solution to a 50-mL Erlenmeyer flask. Add another 5-mL portion of ethanol to the 125-mL Erlenmeyer flask, heat to boiling, and transfer to the 50-mL Erlenmeyer flask. Bring the entire mixture back to boiling, remove it from the steam bath, and allow it to cool to room temperature. Filter the crystalline mass with the aid of a Buchner funnel and wash with 3 to 5 mL *cold* ethanol. The yield is 2.7 to 3.0 g (65 to 70%) of white crystals of mp 108 to 110°C.

If the melting point is low, a single recrystallization from 95% ethanol (5 mL/g) usually gives pure product. The product may be analyzed for purity by tlc (silica gel, 50% dichloromethane-hexane).

QUESTIONS AND EXERCISES

17.1 When 4-bromoacetophenone is treated with bromine, it undergoes bromination only of the methyl group. Why is there no reaction in the aromatic ring to produce, for example, 2,4-dibromoacetophenone?

17.2 Chalcone, prepared in Exp. 15.1B, forms a dibromide rather than a monobromide when treated with bromine. How would you account for this difference in reactivity between chalcone and 4-bromoacetophenone?

17.3 Notice that the heterocyclic compound furan has an enol-type structure. When brominated, it yields the product indicated. How does this reaction occur?

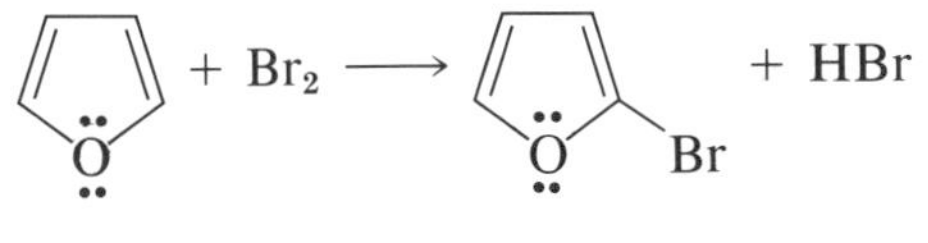

$$\text{furan} + Br_2 \longrightarrow \text{2-bromofuran} + HBr$$

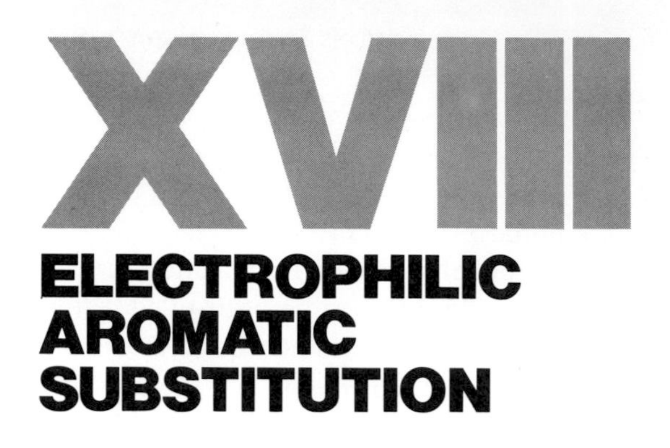

ELECTROPHILIC AROMATIC SUBSTITUTION

A carbon-carbon double bond is an electron-rich species. This electron density is the dominant characteristic of the double bond in the various reactions which it undergoes. Recall from the examples given in Chap. 7 that the electron density in the double-bond species gives this functional group a nucleophilic character. Double bonds therefore react readily with many electrophilic substances, such as bromine.

Aromatic systems, on the other hand, pose a somewhat different problem. Aromatic compounds clearly contain double bonds. However, when one subjects an aromatic substance to the same conditions under which an isolated double bond is reactive, little or no reaction occurs. The reason for this lack of reactivity (or stability) is that there is a driving force for the double bonds in an aromatic system to be retained and not destroyed, as they would be if they were in an isolated situation. For instance, in an earlier experiment (Exp. 7.2), you observed that bromine (Br_2) readily adds to cyclohexene to give 1,2-dibromocyclohexane. Toluene, on the other hand, was observed to be inert to the addition of bromine under the same conditions. It is clear from these experimental observations that the double bond in toluene (being an aromatic double bond) is somehow much less reactive than the double bond in cyclohexene.

The lack of double-bond reactivity in an aromatic system does not mean that there are no aromatic substitution reactions. The aromatic double bond is less reactive than an isolated double bond and reaction will usually occur in such a way that the aromatic system is retained in the product. Bromine requires much more vigorous conditions to react with toluene than with cyclohexene, and a mono-rather than a dibromination product results. The reaction of bromine with cyclohexene produces no by-product. In the toluene reaction, hydrobromic acid is produced as the aromatic system is reestablished. This difference is shown schematically on the next page.

$$\text{cyclohexene} + Br_2 \longrightarrow \text{(2-bromocyclohexyl cation)}\ \ Br^- \longrightarrow \text{1,2-dibromocyclohexane}$$

$$\text{toluene} + Br_2 \longrightarrow \text{(ortho }\sigma\text{-complex)} + Br^- \longrightarrow \text{o-bromotoluene} + HBr$$

$$\text{toluene} + Br_2 \longrightarrow \text{(para }\sigma\text{-complex)} + Br^- \longrightarrow \text{p-bromotoluene} + HBr$$

In most cases, the electrophilic species (such as bromine) will not react with an aromatic double bond unless a catalyst is present. The catalysts required for aromatic substitution reactions are generally acidic. The bromination of toluene, for example, is catalyzed by anhydrous ferric bromide ($FeBr_3$), a Lewis acid. The ferric bromide catalyst is usually generated in the reaction by bromine oxidation of metallic iron. Thus in Exp. 18.2A on the bromination of xylene, an iron nail is added to the reaction mixture. It is not the iron nail but ferric bromide which catalyzes this reaction.

In the nitration experiments described in Sec. 18.1, a strong-acid catalyst is incorporated into the reaction mixture. The catalyst in these reactions is concentrated sulfuric acid. The nitration reaction is therefore initiated by the acid-catalyzed dehydroxylation of nitric acid to form the nitronium ion (NO_2^+). It is the nitronium ion which is the active species in the reaction and the one responsible for the incorporation of the nitro group into the aromatic ring.

$$2H_2SO_4 + HNO_3 \rightleftharpoons H_3O^+ + NO_2^+ + 2HSO_4^-$$

$$NO_2^+ + \text{benzene} \rightleftharpoons \text{(}\sigma\text{-complex with H and } NO_2\text{)}$$

$$HSO_4^- + \text{(}\sigma\text{-complex with H and } NO_2\text{)} \rightleftharpoons \text{nitrobenzene} + H_2SO_4$$

In the bromination of acetanilide (Exp. 18.2B), the catalyst is acetic acid (the solvent). In this case the acid catalyst can be relatively weak because the aromatic nucleus is so reactive.

From the discussion above, it should be obvious that whatever the specific identity of the reaction partners might be, the overall electrophilic substitution reaction can be generalized as shown in the equation below. Note that LA stands for Lewis acid and E for electrophile.

$$\mathrm{LA} + \mathrm{Y{-}E} \rightleftharpoons \mathrm{LA{-}Y} + \mathrm{E^+}$$

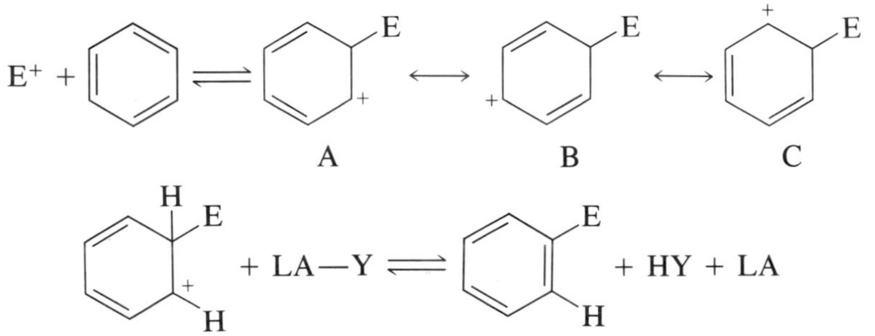

E^+ + (benzene) ⇌ A ⟷ B ⟷ C

A + LA—Y ⇌ (substituted benzene, E) + HY + LA

An aromatic compound such as benzene, plus the reaction partner (herein designated E^+ for the electrophile), goes to an intermediate delocalized carbonium ion (species A above). The ion which is produced is stabilized by several resonance forms throughout the benzene ring (B, C), but it is destabilized by the fact that a previously aromatic system has lost the extensive resonance delocalization which is associated with aromaticity. Loss of a proton in the last step will allow the aromatic system to re-form and when this happens, the resonance energy will be regained. The reaction is essentially the same in every case regardless of the particular electrophile or the particular aromatic compound. The details which are different are the identity of the electrophile E^+, how readily it forms, the rate at which it attacks the aromatic compound, and the stability of the intermediate cation (A). These variations preclude obtaining a uniform yield in all the reactions discussed below. Nevertheless, these are only details. The overall reaction sequence is similar for a large number of reactions.

18.1 ELECTROPHILIC AROMATIC NITRATION

In the experiments below the nitrations of chlorobenzene and bromobenzene are described. In each case two alternate procedures are given. One procedure describes the nitration reaction in the absence of a supporting solvent, and the other procedure describes it in the presence of a moderating solvent. In these examples dichloromethane is the cosolvent. Notice that in the presence of a solvent nitration requires a longer reaction time. The major advantage of the solvent-moderated reaction is that by-products are minimized. Either procedure affords approximately the same yield of product.

The preparations of 1-chloro-2,4-dinitrobenzene and 1-bromo-2,4-dinitrobenzene are both described in this chapter. These two compounds are important in several areas of chemistry in that they are intermediates in nucleophilic aromatic substitution reactions (see Chap. 19).

The preparation of both chlorine- and bromine-substituted aromatics may seem redundant. Both are included in recognition of certain differences of opinion. Some instructors prefer to begin with chlorobenzene, which is less expensive than the bromo compound. On the other hand, the bromine-containing compounds are somewhat more crystalline and a little easier to purify.

The final nitration experiment in this chapter is the synthesis of 3-nitrobenzoic acid. In this procedure, nitration is performed on a derivative of the benzoic acid rather than on benzoic acid itself. This is done for practical reasons. Although nitration of either benzoic acid or methyl benzoate affords the respective 3-nitro derivatives in comparable yield, it is difficult to purify the 3-nitrobenzoic acid obtained in the direct reaction. It is more expedient to nitrate and purify the ester. This nitration procedure is a good example of altering the conditions and starting materials in a reaction in order to enhance the ease of purification.

EXPERIMENT 18.1A

NITRATION OF CHLOROBENZENE

$$C_6H_5Cl + HNO_3 + H_2SO_4 \longrightarrow p\text{-}ClC_6H_4NO_2 + o\text{-}ClC_6H_4NO_2 + 2,4\text{-}Cl C_6H_3(NO_2)_2$$

Time 2.5 h

Materials Chlorobenzene, 10 mL (MW 112.5, bp 132°C, d 1.1 g/mL)
Concentrated (70%) nitric acid, 15 mL
Concentrated (98%) sulfuric acid, 15 mL
Dichloromethane, 40 mL
95% Ethanol, 50 mL

Precautions Wear gloves and carry out all transfers in a good hood. Take care to add sulfuric acid to the nitric acid. Make certain that a supply of solid sodium bicarbonate or sodium carbonate is available in the laboratory to be spread on acid spills.

Hazards Nitric acid is a strong oxidizing acid. Avoid breathing vapor and prevent contact with eyes and skin. Sulfuric acid is a strong dehydrating agent which can cause severe burns. It reacts very exothermically with water. Prevent contact with skin and eyes. Chlorobenzene is a flammable liquid, which is toxic in high concentrations. Avoid breathing of vapor and contact with skin, eyes, and clothing.

Experimental Procedure

Charge a 250-mL Erlenmeyer flask with 15 mL 70% nitric acid, followed by 15 mL concentrated (98%) sulfuric acid. On mixing the internal temperature rises to 55 to 60°C. Cool the mixture in an ice-water bath until the internal temperature falls to 25 to 30°C. (**Caution: Do not reverse the order of addition of the acids.**)

Using a disposable pipet, add 10 mL chlorobenzene (0.097 mol) in 1-mL portions at 1-min intervals to the cool mixture. After each addition of chlorobenzene, vigorously swirl the flask in an ice-water bath in order to maintain an internal temperature of less than 30°C. Cooling in the ice bath may not be needed, but care should be taken that the temperature does not rise above 35°C, as this would allow 1-chloro-2,4-dinitrobenzene, an unwanted by-product, to form.

After the addition is complete, cool the oily mixture in an ice-water bath and add 40 mL dichloromethane, with swirling. After 2 min of swirling, transfer the two-phase mixture into a separatory funnel. The lower, acid layer should be colorless; the dichloromethane layer should appear yellow. Remove and discard the lower, acid layer. Wash the dichloromethane layer with a 20-mL portion of water, three 15-mL portions of saturated $NaHCO_3$, and finally one 25-mL portion of half-saturated NaCl solution. The upper, $NaHCO_3$ layer should turn bright yellow and most of the color in the dichloromethane layer should transfer to the bicarbonate layer.[1]

[1] Maintaining a temperature below 35°C is critical; dinitration and/or sulfonation products are formed at temperatures above 35°C. The longer the temperature remains above 35°C, the greater the yield of the dinitro by-products. The formation of by-product may be monitored by tlc (R_f 0.4); mononitrochlorobenzenes have R_f 0.75.

Dichloromethane is used as a solvent because all the organic products are soluble in it, while the mixed acid mixture is not. When the dichloromethane is washed with saturated $NaHCO_3$, however, the following reactions are observed:

(1-chloro-2,4-dinitrobenzene: Cl, NO_2, NO_2) + $NaHCO_3$ ⟶ (2,4-dinitrophenol: OH, NO_2, NO_2)

(2,4-dinitrophenol: OH, NO_2, NO_2) + $NaHCO_3$ ⟶ (O^-Na^+, NO_2, NO_2)

Therefore the $NaHCO_3$ wash removes the 1-chloro-2,4-dinitrobenzene by-product, and only the 1-chloro-2-nitrobenzene and 1-chloro-4-nitrobenzene are left in the dichloromethane solution. Confirmation of this may again be obtained by tlc analysis before and after the bicarbonate extraction. It is important that 1-chloro-2,4-dinitrobenzene can be removed so easily because it is a severe irritant to some people. The fact that it can be readily removed makes this preparation much safer than it might otherwise be.

[*Note:* The water wash is *less* dense, but the mixed acid is *more* dense than dichloromethane. The aqueous washes (including the bicarbonate solution) are *upper* layers and dichloromethane is the *lower* layer.]

Transfer the *lower,* dichloromethane layer to a 125-mL Erlenmeyer flask, dry over anhydrous granular sodium sulfate, and filter the organic solution. Remove the dichloromethane by heating on a steam bath and add 40 mL 95% ethanol to the bright yellow, oily residue. Briefly heat the solution to boiling (1 to 2 min), cool to room temperature, and then cool in an ice-water bath. Beautiful, needlelike crystals should separate from solution.

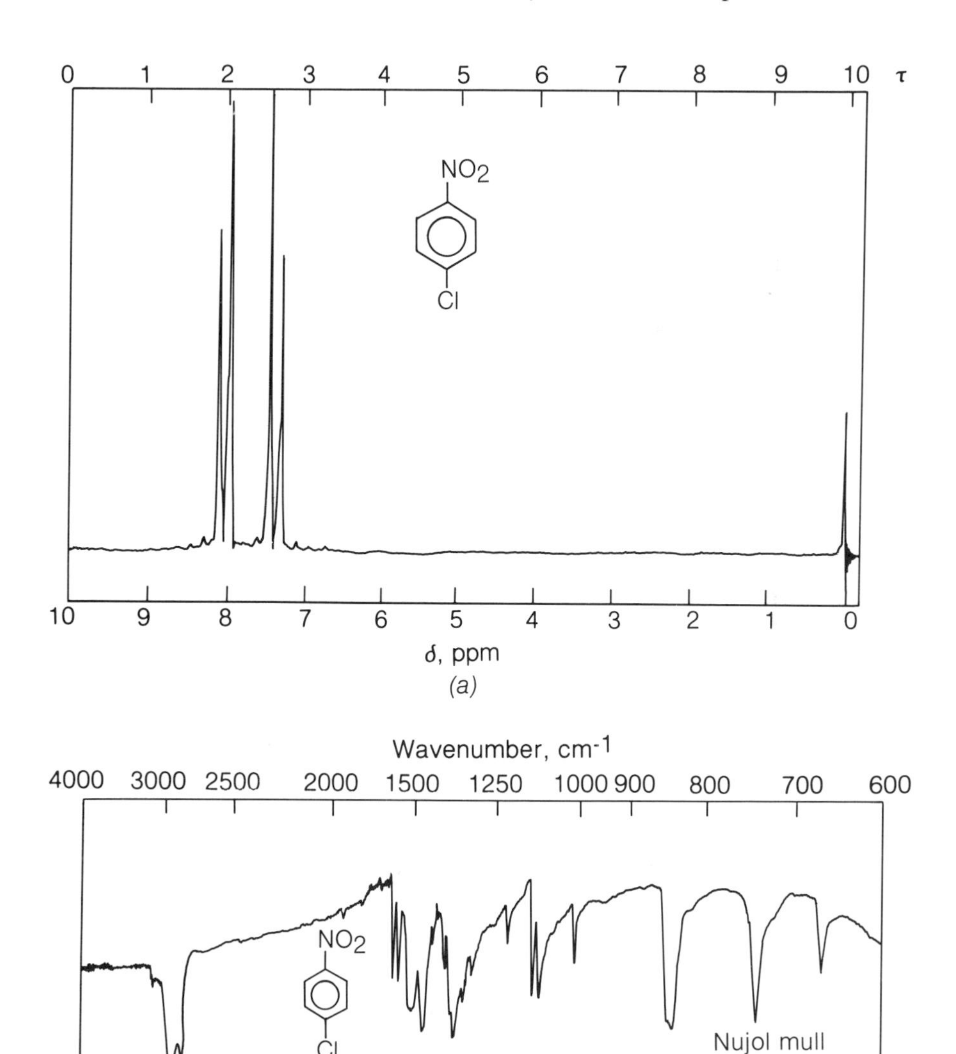

Figure 18.1
The (*a*) nmr and (*b*) ir spectra of 1-chloro-4-nitrobenzene.

Filter the cooled ethanol solution on a Hirsch funnel, wash with 10 mL ice-cold 95% ethanol, and air dry. Between 5 and 6 g of white crystals, mp 80.5 to 82.5°C, should be obtained.

A second crop of slightly less pure crystals may be obtained by reducing the volume of the mother liquors to 25 mL and cooling in the ice-water bath. The crystals obtained in the second fraction are mostly 1-chloro-4-nitrobenzene contaminated by a small amount of 1-chloro-2-nitrobenzene. Yield: 1.5 to 2.2 g of off-white crystals, mp 75 to 79°C.

Thin-layer chromatography can be used to verify the purity of the products obtained. When silica gel is used as an adsorbent and 15% dichloromethane in hexane as an eluting solvent, the first crop shows only one spot on uv visualization. No evidence of 1-chloro-2,4-dinitrobenzene is observed. The two mononitrochlorobenzenes migrate with R_f approximately equal to 0.8. 1-Chloro-2,4-dinitrobenzene migrates with R_f 0.4 and is easy to identify. The product may be recrystallized, if necessary, from 95% ethanol.

The nmr and ir spectra of the 1-chloro-4-nitrobenzene are shown in Fig. 18.1. Note the A_2B_2 pattern in the aromatic protons.

EXPERIMENT 18.1B

ALTERNATE PROCEDURE FOR THE MONONITRATION OF CHLOROBENZENE

Cl | $\xrightarrow[H_2SO_4]{HNO_3}$ | Cl, NO_2 + Cl, NO_2

Time 3.5 h

Materials Chlorobenzene, 10 mL (MW 112.5, bp 132°C, d 1.1 g/mL)
Concentrated (70%) nitric acid, 13 mL
Concentrated (98%) sulfuric acid, 13 mL
Dichloromethane, 60 mL
95% Ethanol, 50 mL

Precautions Wear gloves and carry out all transfers in a good hood. Take care to add sulfuric acid to the nitric acid. Make certain that a supply of solid sodium bicarbonate or sodium carbonate is available in the laboratory to spread on acid spills.

Hazards Nitric acid is a strong oxidizing acid. Avoid breathing vapor and prevent contact with eyes and skin. Sulfuric acid is a strong dehydrating agent, which can cause severe burns. It reacts very exothermically with water. Avoid contact with skin and eyes. Chlorobenzene is a flammable liquid, which is toxic in high concentrations. Avoid breathing vapor and contact with skin, eyes, and clothing.

Note This procedure takes longer to carry out, but the product mixture is free of 1-chloro-2,4-dinitrobenzene, a potential irritant.

Experimental Procedure

Charge a 250-mL single-neck, round-bottom flask with 13.0 mL 70% nitric acid, 13 mL concentrated sulfuric acid, and a magnetic stirring bar in the order noted. Use the same graduated cylinder to measure both acids. A slight exotherm is usually observed when sulfuric acid is added to nitric acid.

Add 60 mL dichloromethane to the acid mixture followed by one 10-mL portion (0.1 mol) of chlorobenzene. Start the magnetic stirrer and adjust to the highest possible stirring rate so that the layers are intimately mixed (see Fig. 18.2). Be careful that the stirrer is not spinning so rapidly that it breaks magnetic contact with the stirrer motor and is thrown through the glass. Spilled acid should be doused quickly with sodium bicarbonate or sodium carbonate. The dichloromethane will turn slightly yellow, and a small amount of brown gas may be observed in the reflux condenser. Reflux the mixture for 2.5 h.

After the reflux period, cool the mixture and transfer the solution to a separatory funnel. Remove the acid mixture from the funnel. Wash the upper, organic layer with one 25-mL portion of distilled water, followed by three 15-mL portions of saturated $NaHCO_3$ solution and finally with one 25-mL portion of half-saturated NaCl solution. (*Note:* The water wash is *less* dense, the acid mixture is *more* dense, than the dichloromethane layer.) Transfer the organic layer to a 125-mL Erlenmeyer flask and dry over granular anhydrous sodium sulfate for 5 to 10 min. Filter the solution from the sodium sulfate, and remove the dichloromethane by evaporation on a steam bath in the hood. Approximately 15 g of a yellow oil should remain.

The oil obtained in this way is a mixture of 1-chloro-4-nitrobenzene, 1-chloro-2-nitrobenzene, and some unreacted starting material. Dissolve the oil in 50 mL 95% ethanol by heating briefly on a steam bath. Cool the ethanol solution to room temperature and then in an ice-water bath (cool until the internal temperature is approximately 10°C). The product (5 to 6 g) should separate as needlelike crystals. Filter and collect the crystals on a Hirsch funnel and air dry briefly. The melting point of this product should be 78 to 81°C.

Thin-layer chromatography can be used to verify the purity of the product. When silica gel is used as adsorbent and dichloromethane in hexane as

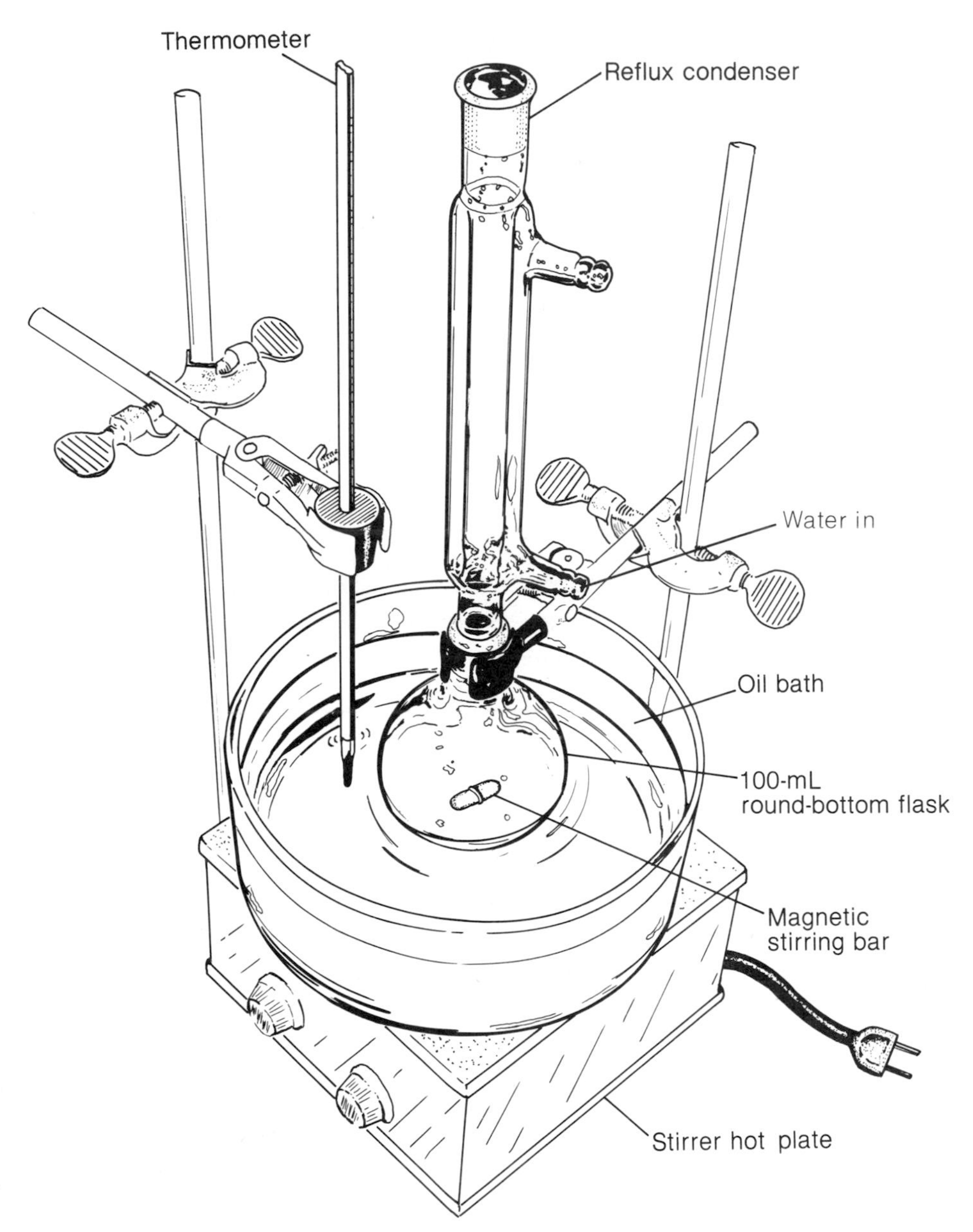

Figure 18.2 Apparatus for Experiment 17.1B.

solvent, the crystalline material shows only one spot under uv visualization, with no evidence of 1-chloro-2,4-dinitrobenzene. The product can be recrystallized from 95% ethanol, if necessary, to obtain material of mp 81 to 83°C.

The nmr and ir spectra of 1-chloro-4-nitrobenzene are shown in Fig. 18.1. Note the A_2B_2 pattern of the aromatic protons. The protons adjacent to chlorine are quite far downfield compared with most other aromatic systems. Compare this spectrum with that of 4-chlorobenzyl acetate (Exp. 13.2).

EXPERIMENT 18.1C

NITRATION OF BROMOBENZENE

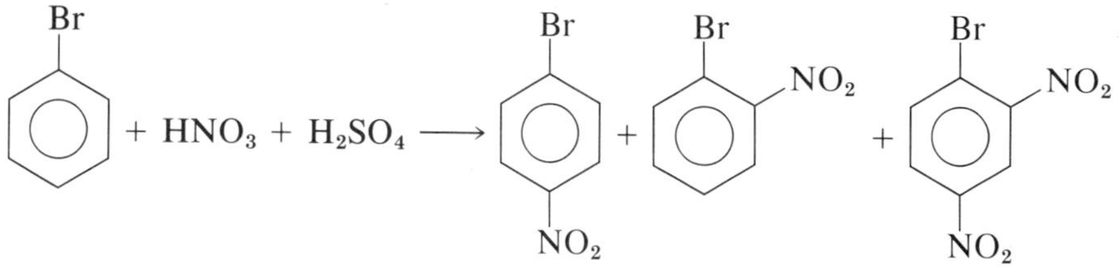

$$C_6H_5Br + HNO_3 + H_2SO_4 \longrightarrow \text{4-}BrC_6H_4NO_2 + \text{2-}BrC_6H_4NO_2 + \text{2,4-}BrC_6H_3(NO_2)_2$$

Time 2 h

Materials Bromobenzene, 10 mL (MW 157, bp 156°C, d 1.49 g/mL)
Concentrated (70%) nitric acid, 15 mL
Concentrated (98%) sulfuric acid, 15 mL
Dichloromethane, 40 mL
95% Ethanol, 90 mL

Precautions Wear gloves and carry out all transfers in a good hood. Take care to add sulfuric acid to the nitric acid. Make certain that a supply of solid sodium bicarbonate or sodium carbonate is available in the laboratory to be spread on acid spills.

Hazards Nitric acid is a strong oxidizing acid. Avoid breathing vapor and prevent contact with eyes and skin. Sulfuric acid is a strong dehydrating agent, which can cause severe burns. It reacts very exothermically with water. Avoid contact with skin and eyes. Bromobenzene is a flammable liquid, which is toxic in high concentrations. Avoid breathing vapor and contact with skin, eyes, and clothing.

Experimental Procedure

Charge a 250-mL Erlenmeyer flask with 15 mL 70% nitric acid followed by 15 mL concentrated (98%) sulfuric acid. On mixing, the internal temperature rises to 55 to 60°C. Cool the mixture in an ice-water bath until the internal temperature falls to 25 to 30°C. (**Caution: Do not reverse the order of addition of the acids.**)

Using a disposable pipet, add 10 mL bromobenzene (0.095 mol) in 1-mL portions at 1-min intervals to the cool mixture. After each addition of bromobenzene, vigorously swirl the flask in an ice-water bath in order to maintain an internal temperature of less than 30°C. Cooling in the ice bath may not be needed, but care should be taken that the temperature does not rise above 35°C, as this would allow 1-bromo-2,4-dinitrobenzene, an unwanted by-product, to form. After adding the final milliliter of bromobenzene, swirl the flask vigorously for 10 to 15 min.

Cool the oily, semisolid mixture in an ice-water bath and add 40 mL dichloromethane, with swirling (solid will deposit after the first milliliter of bromobenzene is added). As the reaction proceeds, this material forms small, pebblelike nodules as a result of the swirling). After 2 min of swirling, transfer the two-phase mixture into a separatory funnel. The lower, acid layer should be colorless; the dichloromethane layer should appear yellow. Remove and discard the lower, acid layer. Wash the dichloromethane layer with a 20-mL portion of water, three 15-mL portions of saturated $NaHCO_3$, and finally one 25-mL portion of half-saturated NaCl solution. The upper, $NaHCO_3$ layer should turn bright yellow and most of the color in the dichloromethane layer will transfer to the bicarbonate layer.[2]

(*Note:* The water wash is *less* dense, but the mixed acid is *more* dense, than dichloromethane. The aqueous washes (including the bicarbonate solution) are *upper* layers and dichloromethane is the *lower* layer.)

[2] Maintaining a temperature below 35°C is critical; dinitration and/or sulfonation products are formed at temperatures above 35°C. The longer the temperature remains above 35°C, the greater the yield of the dinitro by-products. The formation of by-product may be monitored by tlc (R_f 0.4); mononitrobromobenzenes have R_f 0.75.

Dichloromethane is used as a solvent because all the organic products are soluble in it while the mixed acid mixture is not. When the dichloromethane is washed with saturated sodium bicarbonate, however, the following reactions are observed:

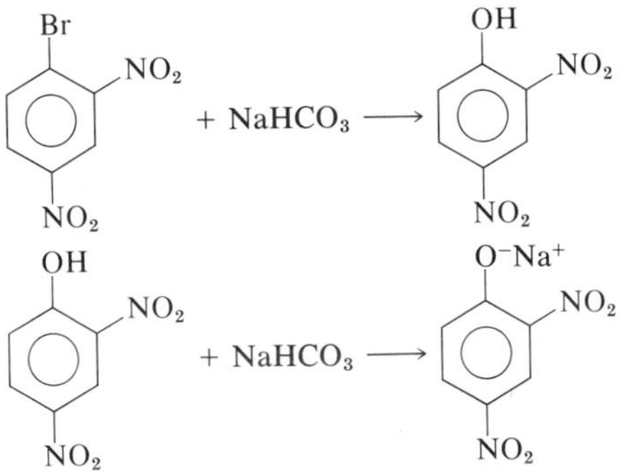

Br, NO_2, NO_2 + $NaHCO_3$ ⟶ OH, NO_2, NO_2

OH, NO_2, NO_2 + $NaHCO_3$ ⟶ O^-Na^+, NO_2, NO_2

Therefore the $NaHCO_3$ wash removes the 1-bromo-2,4-dinitrobenzene by-product, and only the 1-bromo-2-nitrobenzene and 1-bromo-4-nitrobenzene are left in the dichloromethane solution. Confirmation of this may again be obtained by tlc analysis before and after the bicarbonate extraction. It is important that 1-bromo-2,4-dinitrobenzene can be removed so easily, because it is a severe irritant to some people. The fact that it can be readily removed makes this preparation much safer than it might otherwise be.

Transfer the *lower,* dichloromethane layer to a 125-mL Erlenmeyer flask, dry over anhydrous granular sodium sulfate, and filter the organic solution. Remove the dichloromethane by heating on a steam bath and add 80 mL 95% ethanol to the bright yellow, oily, semisolid residue (a solid mass of crystals usually appears on addition of the ethanol). Briefly heat the solution to reflux (1 to 2 min) and then cool to room temperature. Beautiful needlelike crystals should separate from solution. Filter the ethanol solution, collect the product on a Buchner funnel, wash with 10 mL ice-cold 95% ethanol, and air dry. Yield: 8.7 to 10.2 g 1-bromo-4-nitrobenzene, mp 123 to 125°C.

Reduce the mother liquors to 40 to 45 mL and cool in an ice-water bath. Filter the second crop of crystals. This material is a mixture of 1-bromo-2-nitrobenzene and 1-bromo-4-nitrobenzene, 2.1 to 3.2 g, melting between 70 and 110°C.

Thin-layer chromatography can be used to verify the purity of the products. When silica gel is used as adsorbent and 15% dichloromethane in hexane as eluting solvent, the first crop shows only one spot on uv visualization. No evidence of 1-bromo-2,4-dinitrobenzene is observed. The two mononitrobromobenzenes migrate with R_f approximately equal to 0.8. 1-Bromo-2,4-dinitrobenzene migrates with R_f 0.4 and is easy to identify. The product may be recrystallized, if needed, from 95% ethanol.

The nmr and ir spectra of 1-chloro-4-nitrobenzene are shown in Fig. 18.1. The spectra for the bromo compound are similar.

EXPERIMENT 18.1D SYNTHESIS OF 1-CHLORO-2,4-DINITROBENZENE

$$\text{1-chloro-4-nitrobenzene (Cl, NO}_2\text{)} \xrightarrow[H_2SO_4]{HNO_3} \text{1-chloro-2,4-dinitrobenzene (Cl, NO}_2\text{, NO}_2\text{)}$$

Time 3 h

Materials 1-Chloro-4-nitrobenzene, 3 g (MW 157.56, mp 83 to 84°C)
Concentrated (70%) nitric acid, 4 mL
Concentrated (98%) sulfuric acid, 20 mL
Dichloromethane, 60 mL

Precautions Wear gloves and carry out all transfers in a good hood. Take care to add sulfuric acid to the nitric acid. Make certain that a supply of solid sodium bicarbonate or sodium carbonate is available in the laboratory to be spread on acid spills.

Hazards Nitric acid is a strong oxidizing acid. Avoid breathing vapor and prevent contact with eyes and skin. Sulfuric acid is a strong dehydrating agent, which can cause severe burns. It reacts very exothermically with water. Prevent contact with skin and eyes. 1-Chloro-4-nitrobenzene and 1-chloro-2,4-dinitrobenzene are irritants. Some people are very sensitive to the presence of the latter. Avoid contact of any chemicals with eyes, skin, face, and clothes (**gloves, hood**).

Experimental Procedure

Charge a 250-mL single-neck, round-bottom flask with 4.0 mL 70% nitric acid, 20 mL concentrated sulfuric acid, and a magnetic stirring bar in the order noted. Use the same graduated cylinder to measure both acids. A slight exotherm is usually observed when sulfuric acid is added to nitric acid.

Add 60 mL dichloromethane to the acid mixture, followed by 3 g (0.019 mol) 1-chloro-4-nitrobenzene. Start the magnetic stirrer and adjust to the highest possible stirring rate so that the layers are intimately mixed (see Fig. 18.2). Be careful that the stirrer is not spinning so rapidly that it breaks magnetic contact with the stirrer motor and is thrown through the glass. Spilled acid should be doused quickly with sodium bicarbonate or sodium carbonate. The dichloromethane will turn a deep yellow and a small amount of brown gas may be observed in the reflux condenser. Reflux the mixture for 2 h.

After the reflux period, cool the mixture and transfer the solution to a separatory funnel. Remove the acid mixture from the funnel. Wash the upper, organic layer with one 25-mL portion of distilled water and then with one 25-mL portion of half-saturated NaCl solution. (*Note:* The water wash is *less* dense, while the acid mixture is *more* dense than the dichloromethane layer.) Transfer the organic layer to a 125-mL Erlenmeyer flask and dry over granular anhydrous sodium sulfate for 5 to 10 min. Filter the solution from the sodium sulfate and remove the dichloromethane by evaporation on a steam bath in the hood. Approximately 3.5 to 4 g of a yellow oil is obtained.

Subject a small portion of the oil to thin-layer chromatographic analysis using silica gel as the adsorbent and 15% dichloromethane-hexane as the eluting solvent. This analysis should show that no 1-chloro-4-nitrobenzene is present and only the 1-chloro-2,4-dinitrobenzene appears in the oil. The oil, after several hours, solidifies into a low-melting solid, mp 49 to 52°C. This material may be used as either an oil or solid in the synthesis of 2,4-dinitrophenylhydrazine.

The nmr spectrum of the product is shown in Fig. 18.3*a* and is particularly informative. The most downfield resonance is a doublet with a small coupling constant. This peak is the resonance for the proton flanked by both nitro groups. The small coupling is due to the meta proton. The two upfield lines are

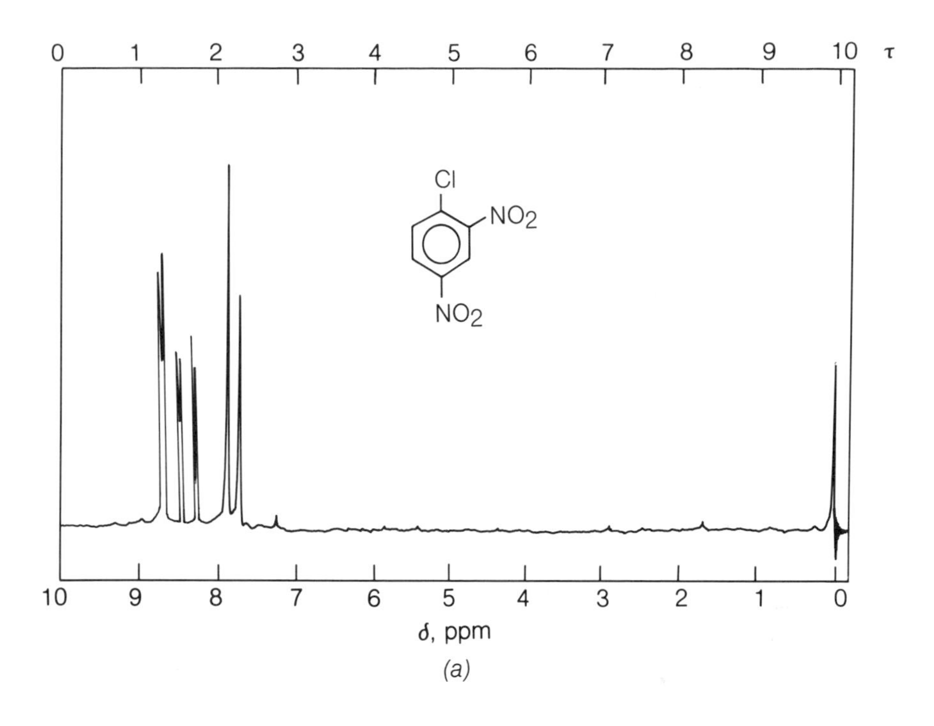

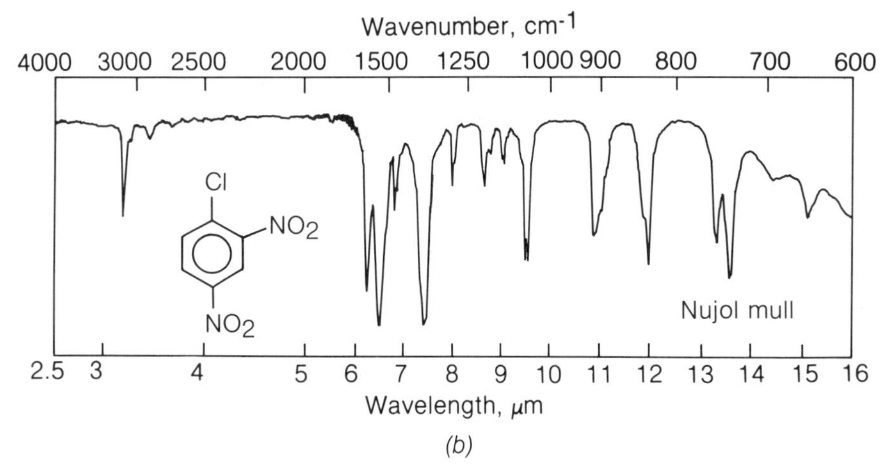

Figure 18.3
The (*a*) nmr and (*b*) ir spectra of 1-chloro-2,4-dinitrobenzene.

due to the proton adjacent to chlorine. It is the upfield half of an AB pattern. This proton is meta to both nitro groups and is therefore not further coupled. The remaining four-line pattern reflects the coupling of the proton adjacent to a single nitro group with the other two protons.

Warning: Remember that 1-chloro-2,4-dinitrobenzene is a skin irritant and a sensitizing agent, which may induce allergic contact dermatitis. All routine precautions (hood, gloves) should be taken to avoid skin contact with this compound.

EXPERIMENT 18.1E

SYNTHESIS OF 1-BROMO-2,4-DINITROBENZENE

$$\text{1-bromo-4-nitrobenzene} \xrightarrow[H_2SO_4]{HNO_3} \text{1-bromo-2,4-dinitrobenzene}$$

Time 3 h

Materials 1-Bromo-4-nitrobenzene, 4 g (MW 202, mp 125 to 126°C)
Concentrated (70%) nitric acid, 4 mL
Concentrated (98%) sulfuric acid, 20 mL
Dichloromethane, 60 mL

Precautions Wear gloves and carry out all transfers in a good hood. Take care to add sulfuric acid to the nitric acid. Make certain that a supply of solid sodium bicarbonate or sodium carbonate is available in the laboratory to be spread on acid spills.

Hazards Nitric acid is a strong oxidizing acid. Avoid breathing vapor and prevent contact with eyes and skin. Sulfuric acid is a strong dehydrating agent, which can cause severe burns. It reacts very exothermically with water. Prevent contact with skin and eyes. 1-Bromo-4-nitrobenzene and 1-bromo-2,4-dinitrobenzene are irritants. Some people are very sensitive to the presence of the latter. Avoid contact of any reactants with eyes, skin, face, and clothes (**gloves, hood**).

Experimental Procedure

Charge a 250-mL single-neck, round-bottom flask with 4.0 mL 70% nitric acid, 20 mL concentrated sulfuric acid, and a magnetic stirring bar in the order noted. Use the same graduated cylinder to measure both acids. A slight exotherm is usually observed when sulfuric acid is added to nitric acid.

Add 60 mL dichloromethane to the acid mixture, followed by 4 g (0.02 mol) 1-bromo-4-nitrobenzene. Start the magnetic stirrer and adjust to the highest possible stirring rate so that the layers are intimately mixed (see Fig. 18.2). Be careful that the stirrer is not spinning so rapidly that it breaks magnetic contact with the stirrer motor and is thrown through the glass. Spilled acid should be doused quickly with sodium bicarbonate or sodium carbonate. The dichloromethane will turn a deep yellow, and a small amount of the brown gas may be observed in the reflux condenser. Reflux the mixture for 2 h.

After the reflux period, cool the mixture and transfer the solution to a separatory funnel. Remove the acid mixture from the funnel. Wash the upper, organic layer with one 25-mL portion of distilled water and then with one 25-

mL portion of half-saturated NaCl solution. (*Note:* The water wash is *less* dense, while the acid mixture is *more* dense than the dichloromethane layer.) Transfer the organic layer to a 125-mL Erlenmeyer flask and dry over granular anhydrous sodium sulfate for 5 to 10 min. Filter the solution from the sodium sulfate, and remove the dichloromethane by evaporation on a steam bath in the hood. Approximately 4.7 to 4.9 g crude 1-bromo-2,4-dinitrobenzene is obtained.

Dissolve this crude material in 15 mL 95% ethanol by heating to boiling on a steam bath. Cool the ethanol solution until crystallization begins (it is useful to save several crystals from the crude material to use them as seed crystals in this process). After crystallization is initiated (either by scratching or by seed crystals) leave the flask undisturbed for at least 1 h (longer if necessary) at room temperature to obtain a good crop of 1-bromo-2,4-dinitrobenzene. Filter this material on a Hirsch funnel and wash with 5 mL ice-cold 95% ethanol. Air dry the crystalline material. About 4 g (83%) of product, mp 70 to 72°C (literature mp 75°C), is obtained after recrystallization.

Perform a tlc analysis (silica gel) on the crystals using 15% dichloromethane-hexane as the eluting solvent (visualization under uv light). In this system 1-bromo-4-nitrobenzene has R_f 0.6 while 1-bromo-2,4-dinitrobenzene has R_f 0.25. (*Note:* This tlc system can be used to monitor the progress of the reaction during the reflux period.)

The nmr spectrum is essentially the same as that of the chloro compound. See the discussion at the end of Exp. 18.1D for the spectral pattern characteristic of this compound.

Warning: Remember that 1-bromo-2,4-dinitrobenzene is a skin irritant and a sensitizing agent, which may induce allergic contact dermatitis. All routine precautions (hood, gloves) should be taken to avoid skin contact with this compound.

EXPERIMENT 18.1F

SYNTHESIS OF 3-NITROBENZOIC ACID BY NITRATION AND HYDROLYSIS

$$C_6H_5\text{–CO–OMe} + HNO_3 \longrightarrow 3\text{-}NO_2C_6H_4\text{–CO–OMe}$$

$$3\text{-}NO_2C_6H_4\text{–CO–OMe} + NaOH \longrightarrow 3\text{-}NO_2C_6H_4\text{–CO–OH}$$

Time 3.0 h

Materials Methyl benzoate, 10 mL (MW 136, bp 198 to 199°C, d 1.1 g/mL)
Concentrated (70%) nitric acid, 8 mL
Concentrated (98%) sulfuric acid, 28 mL
Methanol, 10 mL
Sodium hydroxide, 4 g (MW 40)
Concentrated (36%) hydrochloric acid, 15 mL

Precautions Wear gloves and do all transfers in a good hood. Take care to add the sulfuric acid to the nitric acid. Make certain that a supply of solid sodium bicarbonate or sodium carbonate is available in the laboratory to be spread on acid spills.

Hazards Nitric acid is a strong oxidizing acid. Avoid breathing vapor and prevent contact with eyes and skin. Sulfuric acid is a strong dehydrating agent, which can cause severe burns. It reacts very exothermically with water. Prevent contact with skin and eyes. Sodium hydroxide is caustic.

Experimental Procedure (hood)

Place 8 mL concentrated (70%) nitric acid in a 125-mL Erlenmeyer flask. Add 8 mL concentrated (98%) sulfuric acid. Swirl the flask to thoroughly mix the acids and cool in an ice bath until the internal temperature of the mixture is 5°C.

Place 20 mL concentrated sulfuric acid in a 250-mL Erlenmeyer flask. Add all at once 10 mL (11 g, 0.081 mol) methyl benzoate to the sulfuric acid. Swirl the 250-mL Erlenmeyer flask during the addition of the methyl benzoate and immediately cool the entire mixture in an ice bath until the internal temperature falls to 0 to 5°C.

When the internal temperature of both acid mixtures is below 5°C, add the nitric acid–sulfuric acid mixture (125-mL Erlenmeyer) to the methyl benzoate solution dropwise during 5 to 10 min. Swirl the 250-mL Erlenmeyer flask vigorously during this addition and keep the mixture as close to 5°C as possible. After the addition of the nitric acid–sulfuric acid mixture, swirl the 250-mL Erlenmeyer flask for several minutes in the ice bath and then allow the solution to warm to room temperature (about 10 to 15 min). Swirl the flask occasionally during the warming period. After the mixture has come to room temperature, slowly pour it, with stirring, over 100 g ice (**caution, exotherm**) in a 600-mL beaker. Swirl the material in the beaker and collect the resulting solid by suction filtration using a Buchner funnel. Wash the solid thoroughly with distilled water to remove excess acid. Finally wash the solid with 10 mL ice-cold methanol. Air dry the resulting solid. Methyl 3-nitrobenzoate (9 to 11 g) should be isolated as a practically colorless solid, mp 75 to 76°C. This material is sufficiently pure for the hydrolysis described next.

If material of higher purity is required, the solid may be recrystallized from 10 mL methanol.

Place 4 g sodium hydroxide (NaOH) in a 50-mL Erlenmeyer flask and add 16 mL water. Swirl the Erlenmeyer flask until the sodium hydroxide dissolves (**caution, exotherm**). Add methyl 3-nitrobenzoate (9 to 11 g, as obtained above) to a 50-mL round-bottom boiling flask. Add the aqueous solution to the round-bottom flask through a funnel (be careful not to allow the caustic solution to come in contact with the ground glass joints). Fit the round-bottom flask with a

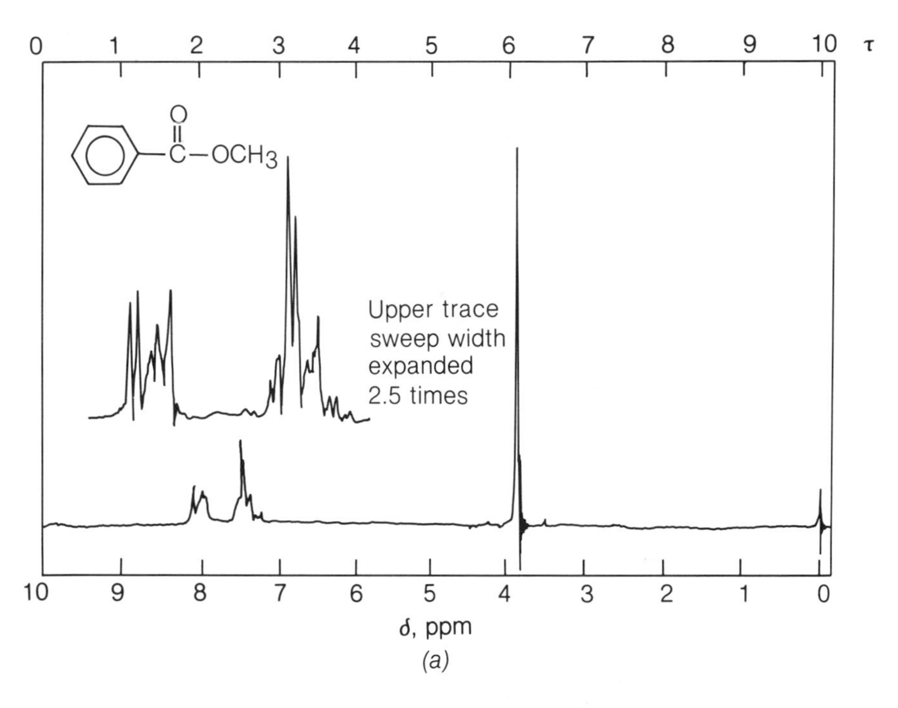

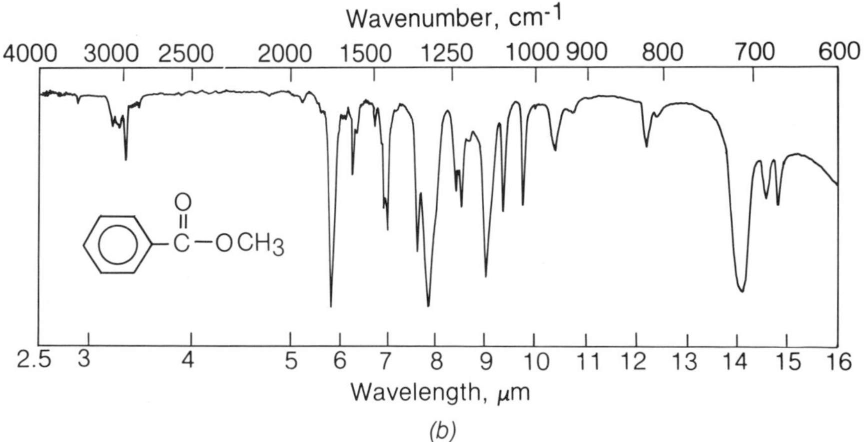

Figure 18.4
The (*a*) nmr and (*b*) ir spectra of methyl benzoate.

reflux condenser (with lightly greased joints) and add several boiling chips to the mixture. Heat the aqueous mixture (with a free flame or oil bath) until the solution boils and continue to heat under reflux for 15 min. The methyl ester should dissolve during this reflux period. After the reflux period, cool the reaction mixture to room temperature and dilute it with 20 mL distilled water. If the solution is colored, treat the mixture with charcoal and filter hot before the next step.

Add, with vigorous swirling, the dilute aqueous base mixture to a 250-mL beaker containing 15 mL hydrochloric acid. After the addition, swirl the flask

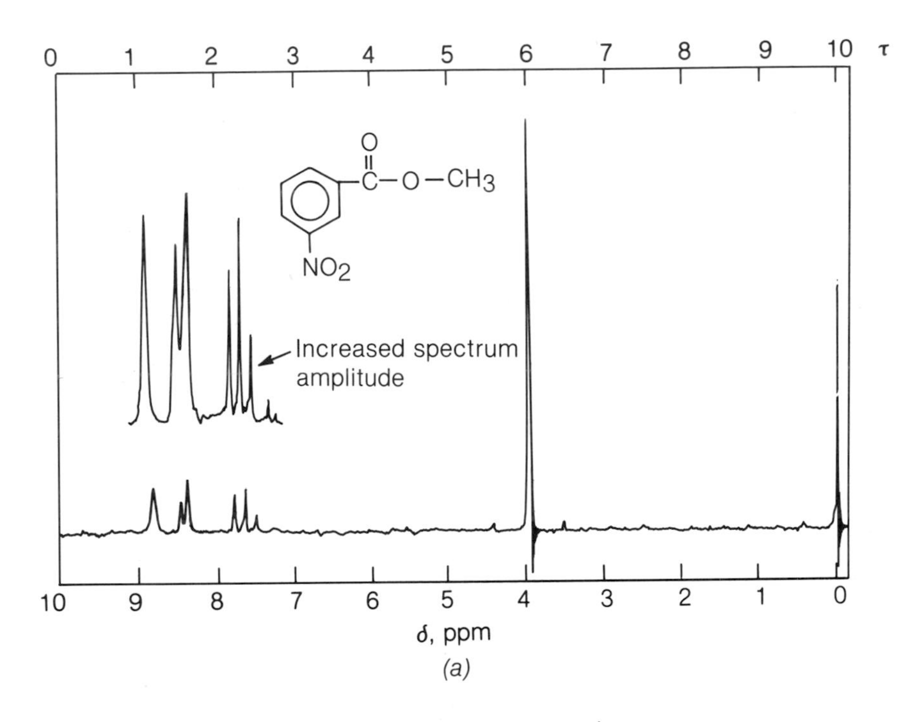

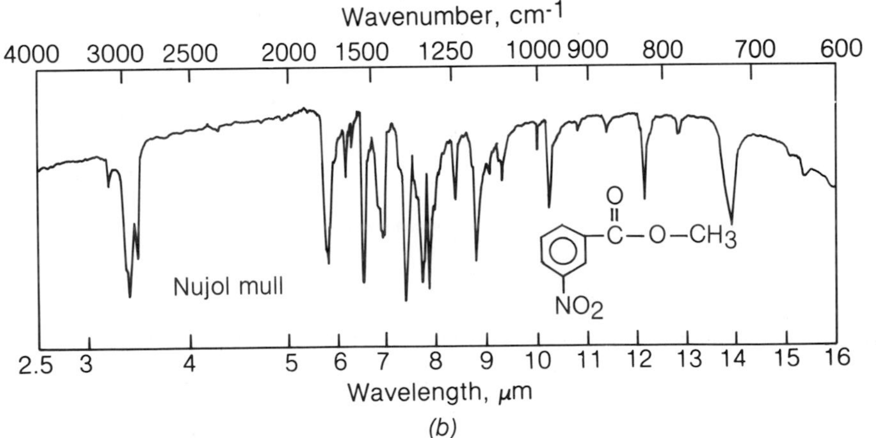

Figure 18.5
The (*a*) nmr and (*b*) ir spectra of methyl 3-nitrobenzoate.

vigorously while cooling in an ice bath until the aqueous solution returns to room temperature. Collect the crude 3-nitrobenzoic acid with the aid of a Buchner funnel and wash the solid with a small amount of distilled water. After air drying, 7 to 9 g crude white to tan acid should be obtained (mp 138 to 140°C).

The crude 3-nitrobenzoic acid may be recrystallized from 1% aqueous hydrochloric acid (1 mL HCl to 99 mL distilled water). The pure product is recovered in 90 to 95% yield and has mp 141°C.

The nmr and ir spectra of methyl benzoate, methyl 3-nitrobenzoate, and

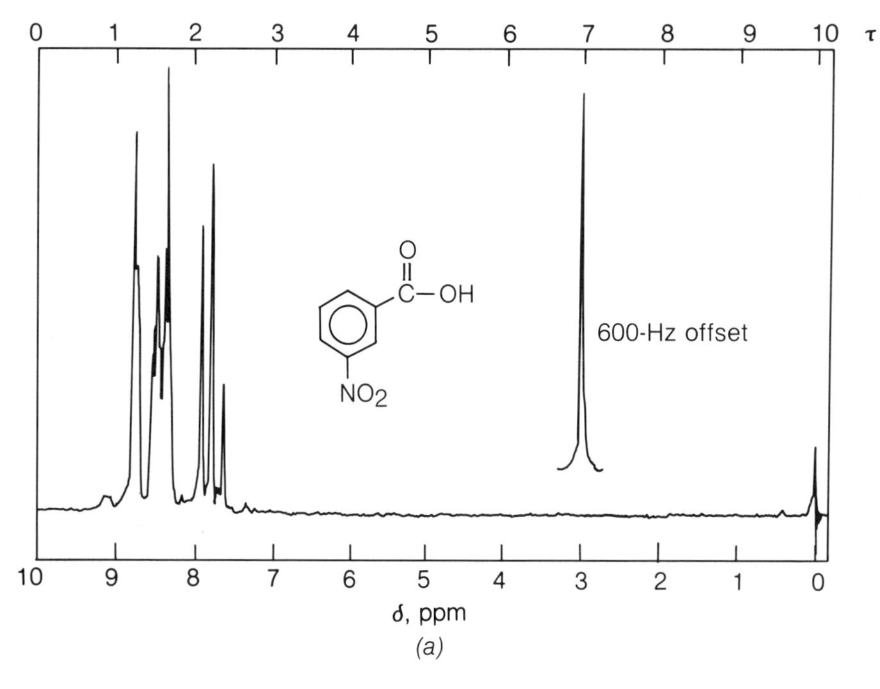

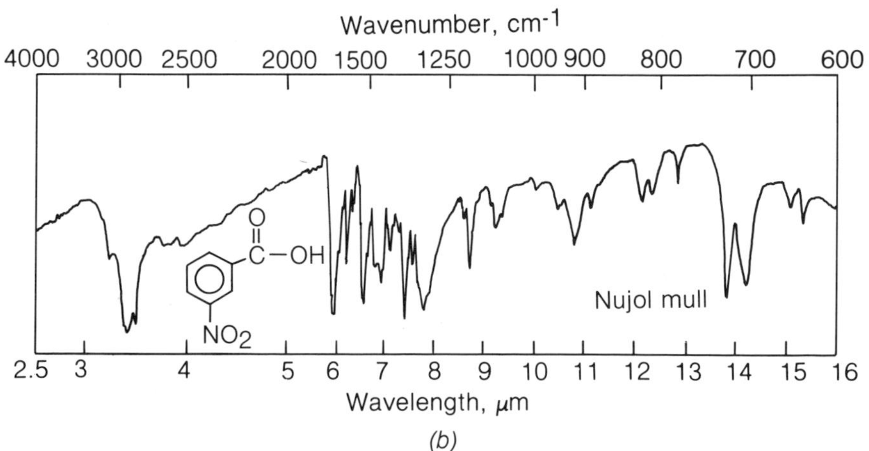

Figure 18.6
The (*a*) nmr and (*b*) ir spectra of 3-nitrobenzoic acid.

3-nitrobenzoic acid are shown in Figs. 18.4, 18.5, and 18.6, respectively. Notice that the introduction of the nitro group in the 3 position of the benzene ring changes the splitting pattern of the aromatic protons. Using the discussion at the end of Exp. 18.1D, attempt to analyze the nmr spectrum of the nitrated product.

18.2 ELECTROPHILIC AROMATIC BROMINATION

The electrophilic bromination of aromatic compounds occurs by the general mechanism discussed at the beginning of this chapter. In both the experiments described in this section, bromination of the aromatic nucleus occurs to give a monosubstitution product. The two preparations differ, however, in that xylene is much less nucleophilic than acetanilide.

The bromination of *p*-xylene occurs to give a single product, 2-bromo-1,4-dimethylbenzene (suitable for use in Exp. 12.3B), because monosubstitution is favored under these conditions and all four substitution sites are equivalent. An electrophilic catalyst, in this case $FeBr_3$, is required to effect the aromatic bromination reaction. The catalyst is generated from an iron nail by bromine oxidation. The reaction works best when a freshly cut nail is used, because bromine oxidation requires a clean surface and is slowed by the presence of rust (oxidation products).

In the second preparation 4-bromoacetanilide is formed either from molecular bromine or pyridinium bromide perbromide, a somewhat safer brominating agent. The most important difference between this and the previous reaction is that no Lewis acid catalyst is required. The presence of amine nitrogen directly attached to the aromatic ring so enhances the reactivity that no catalyst need be used. Aniline itself is so reactive that it forms a tribromide in the absence of an electrophilic catalyst.

4-Bromoacetanilide is a mildly effective analgesic (painkiller). Its relatives, the 4-hydroxy and 4-methoxy compounds, are the important commercial analgesics acetaminophen and phenacetin. These compounds are as effective as aspirin but are noninflammatory.

EXPERIMENT 18.2A BROMINATION OF p-XYLENE

$$p\text{-}CH_3C_6H_4CH_3 + Br_2 \xrightarrow{FeBr_3} 2\text{-}Br\text{-}1,4\text{-}(CH_3)_2C_6H_3 + HBr$$

Time 3 h (may be done in two periods)

Materials *p*-Xylene (1,4-dimethylbenzene), 30 mL (MW 106, bp 138°C, d 0.866 g/mL)
Bromine, 10 mL (MW 160, d 3.1 g/mL)
Iron, 1 to 3 g (iron nail)

Precautions Make all measurements and transfers in a good hood. Wear gloves when handling bromine.

Hazards Bromine is a severe poison, either in liquid or vapor form. It will cause severe and long-lasting burns. Avoid breathing of vapor or any contact with the eyes or skin. The liquid and vapor are heavier than air and will settle. If spilled on skin or eyes, wash *immediately* with 5% sodium bicarbonate solution and inform your instructor. Both *p*-xylene and 2-bromo-*p*-xylene are toxic in high concentrations. Avoid breathing vapors.

Experimental Procedure

Set up a 500-mL three-neck, round-bottom flask in the hood and fit it with a condenser (center neck, lightly greased joints) and an addition funnel. Stopper the third neck with a glass stopper. Clamp the entire assembly securely to a ring stand (center neck). Attach an apparatus to the top of the condenser for the removal of acidic gases (see Sec. 2.8 and consult your instructor). The entire apparatus should be *dry*.

Place 30 mL *p*-xylene in the flask, followed by 1 to 3 g of a freshly cut nail or iron tack[3]. Stopper the flask. Measure 10 mL liquid bromine (**hood, gloves**) in a 25-mL graduated cylinder and transfer the bromine immediately to the addition funnel.

Add 2 mL bromine to the flask. The reaction usually starts in 3 to 10 min (as evidenced by the evolution of HBr at the surface of the iron nail[3]). After the initial reaction subsides, add bromine in 1- to 2-mL portions over a 30-min period, allowing reaction to occur between additions. Keep an ice-water bath available, as heat is evolved and intermittent cooling may be necessary immediately after the addition of each portion of bromine. After the addition of all the bromine, the reaction is allowed to stand at room temperature for 30 min. By the end of this time, there should be no bromine vapors above the liquid and the color of the liquid should be orange-red.

Add 50 mL distilled water to the reaction mixture, swirl, and transfer the mixture to a separatory funnel. Separate the aqueous layer (upper or lower?) and discard. Wash the organic material with two 25-mL portions of distilled water, adding a minimal amount of sodium bisulfite in the first wash to dis-

[3] If the iron nail or tack is cut with a bolt cutter immediately before the start of the reaction, a fresh surface is exposed, the induction time is shortened, and the yield is usually higher.

charge any yellow color in the water. Without drying, transfer the organic material back to the 500-mL round-bottom flask for steam distillation.

Assemble an apparatus for steam distillation as directed by your instructor (see discussion in Sec. 2.2). If live steam is available, arrange a steam-distillation apparatus like the one in Fig. 2-10 (or 2-11), using glass tubing 6 mm in diameter. The length of tubing exposed to the air outside the flask should be as short as possible. The glass tubing must not be flattened or constricted in bending.

Add several boiling sticks (to the round-bottom flask) and start the steam distillation, using a 500-mL Erlenmeyer flask as a collection flask. Collect the milky distillate until about 250-mL of liquid has been collected or until solid appears in the condenser.

If internal steam generation is used, place several boiling sticks (very important to prevent bumping and superheating of the solution) in the flask, along with 250 mL water. Heat the solution with a flame, adding water through the addition funnel to keep the internal volume close to 250 mL. Collect the milky distillate until about 250 mL of liquid is collected or until solid appears in the condenser.

Transfer the distillate to a separatory funnel. Add 30 mL dichloromethane to the 500-mL flask, swirl, and transfer the organic solvent to the separatory funnel. Extract the aqueous layer once and remove the organic phase. Discard the aqueous phase. Wash the organic layer once with 25 mL half-saturated salt solution, dry over anhydrous sodium sulfate, and filter into a dry 125-mL Erlenmeyer flask. Remove the solvent by heating on a steam bath (**hood**).

(*Note:* If the final distillation is to be done in a subsequent period, the procedure may be stopped here. If so, tightly stopper the flask and store the liquid until the next period.)

Assemble a simple distillation apparatus and transfer the oil to a 50-mL distillation flask. Distill the oil at atmospheric pressure and collect a fraction boiling up to 160°C. This fraction is almost pure unreacted xylene. Collect an intermediate fraction, bp 160 to 190°C, which is a mixture of starting material and product. Finally, collect the pure 2-bromo-*p*-xylene (2-bromo-1,4-dimethylbenzene), bp 190 to 200°C. The yield of product ranges between 18 and 22 g (55 to 60%). This material is sufficiently pure to be used directly in Exp. 12.3.

If a vapor-phase chromatograph is available, all three fractions should be analyzed. In some cases, the intermediate fraction accounts for more than 70% of the product. This material can also be used efficiently in the Grignard reaction (the xylene does not interfere). Combining the intermediate fractions from several students and redistilling them through a packed column will provide more material. Again, if available, a vapor-phase chromatograph should be used to monitor the purity of the various fractions.

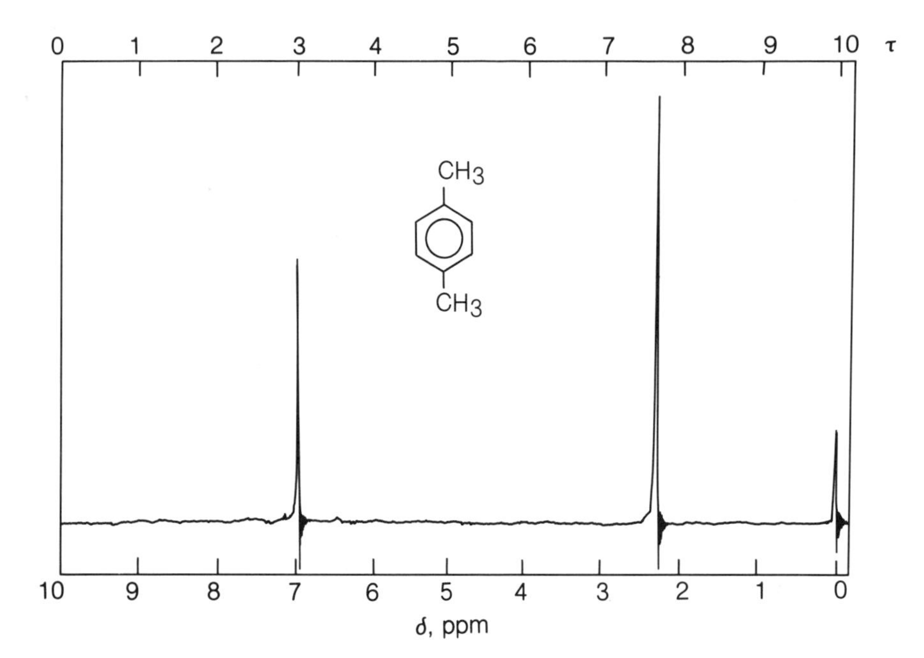

Figure 18.7
The nmr spectrum of *p*-xylene.

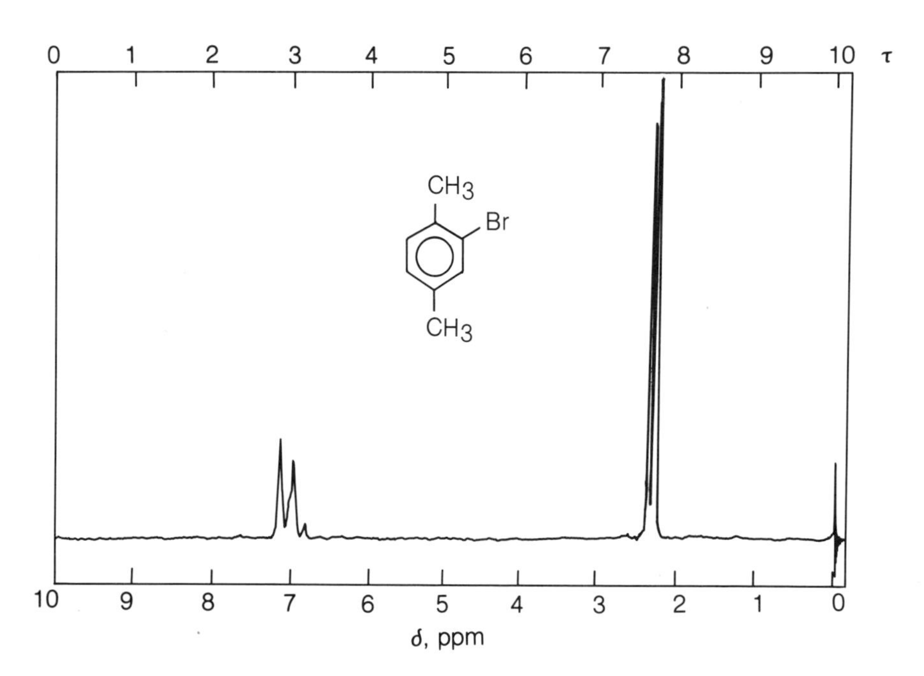

Figure 18.8
The nmr spectrum of 2-bromo-*p*-xylene.

The nmr spectra of *p*-xylene and 2-bromo-*p*-xylene are shown in Figs. 18.7 and 18.8. Notice that the methyl groups in the product are no longer equivalent, as they were in the starting material.

EXPERIMENT 18.2B

SYNTHESIS OF p-BROMOACETANILIDE USING MOLECULAR BROMINE

$$C_6H_5NHC(=O)CH_3 + Br_2 \longrightarrow p\text{-}BrC_6H_4NHC(=O)CH_3$$

Time 2.0 h

Materials Acetanilide, 5.2 g [MW 135, mp 113 to 115°C (from Exp. 11.5A)]
Bromine, 2 mL (MW 160, d 3.1 g/mL) or stock solution
Glacial acetic acid, 30 mL (MW 60, d 1 g/mL)

Precautions Wear gloves during every step of this experiment. Carry out all transfers in a good hood.

Hazards Bromine is a severe poison, both in liquid and in vapor form. It will cause severe and long-lasting burns. Avoid breathing vapor or any contact with the eyes or skin. Bromine liquid and vapor are heavier than air and will settle. If spilled on skin or eyes, wash *immediately* with 5% aqueous sodium bicarbonate solution and inform your instructor. Hydrogen bromide is an acidic gas. Avoid contact with skin and eyes. Glacial acetic acid is flammable. Avoid breathing vapor and contact with eyes and skin.

Experimental Procedure

Place 5.2 g (0.038 mol) acetanilide in a 250-mL Erlenmeyer flask. Add to the acetanilide 20 mL glacial acetic acid (**hood, gloves**) and swirl the mixture, with slight warming on a steam bath if necessary, until all the acetanilide dissolves. After the acetanilide dissolves, allow the flask to cool to room temperature (if necessary).

Pour 2 mL bromine (6.2 g, 0.039 mol) into a dry, 10-mL graduated cylinder (**hood, gloves**). (Inform your instructor before you start this part of the experiment.) Place enough glacial acetic acid in the graduated cylinder to bring the total volume up to 10 mL. [This will give a 0.6 g/mL bromine solution. In some laboratories, this solution will be available from the stockroom. If this is the case in yours (consult your instructor), pour exactly 10 mL of the solution from the storage flask into the graduated cylinder. **As above, all transfers are to be made in the hood.**]

Add the bromine solution to the acetanilide (**hood, gloves**) in three equal portions at room temperature. Swirl the flask for 2 to 3 min between each addition. Note the rate at which the bromine color discharges. During the addition of the third portion, a yellow-white precipitate may start to deposit from solution.

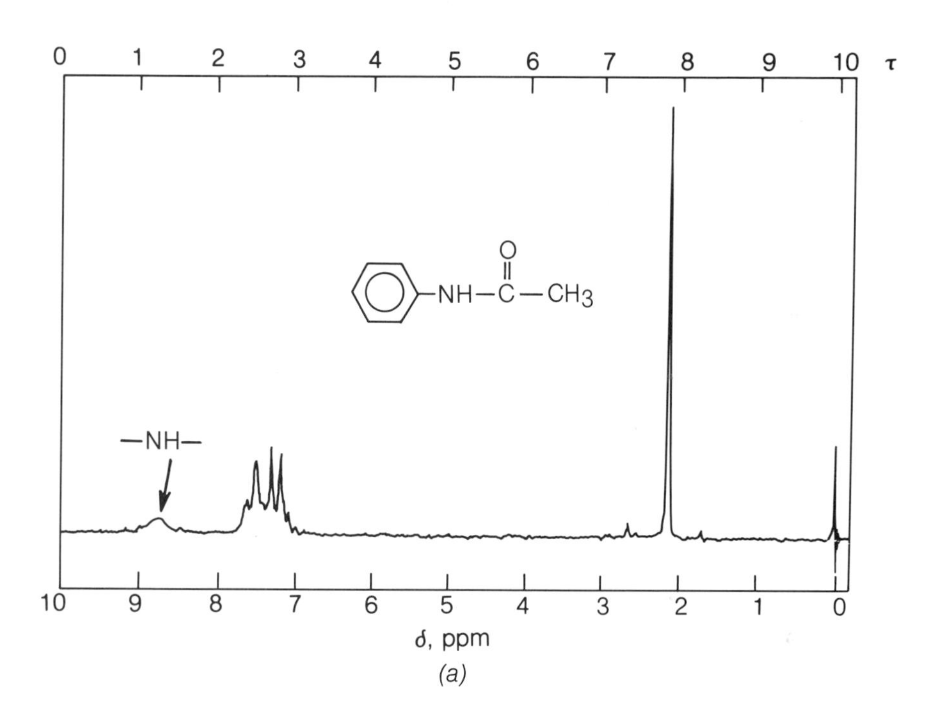

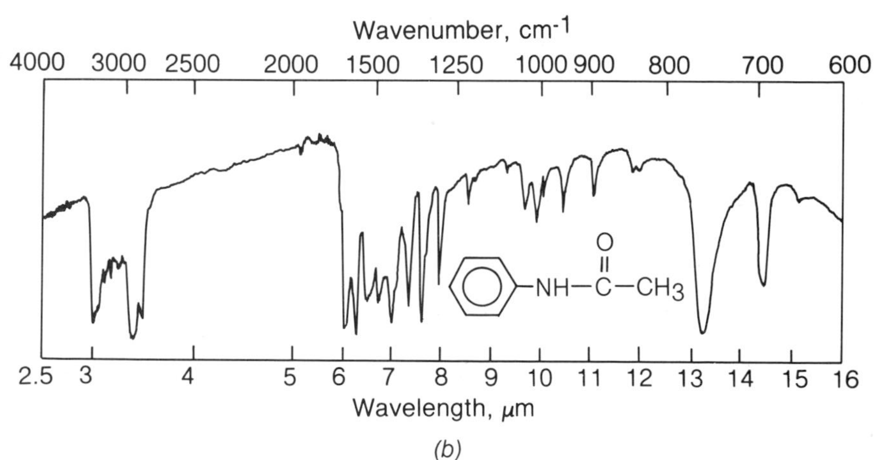

Figure 18.9
The (*a*) nmr and (*b*) ir spectra of acetanilide.

After all the bromine has been added, swirl the flask for five additional minutes. At the end of the reaction period, add 150 mL cold water slowly, while swirling the flask. A white precipitate should deposit. Stir the solution with a glass stirring rod to break up any clumps of crystal. The aqueous solution is usually yellow. Add 0.5 g solid sodium bisulfite ($NaHSO_3$) to the aqueous solution, swirl, and observe the color. Keep adding 0.5-g portions of solid sodium bisulfite until the yellow color disappears (usually a single portion of sodium bisulfite is sufficient).

Filter the white, crystalline product with the aid of a Buchner funnel. Wash

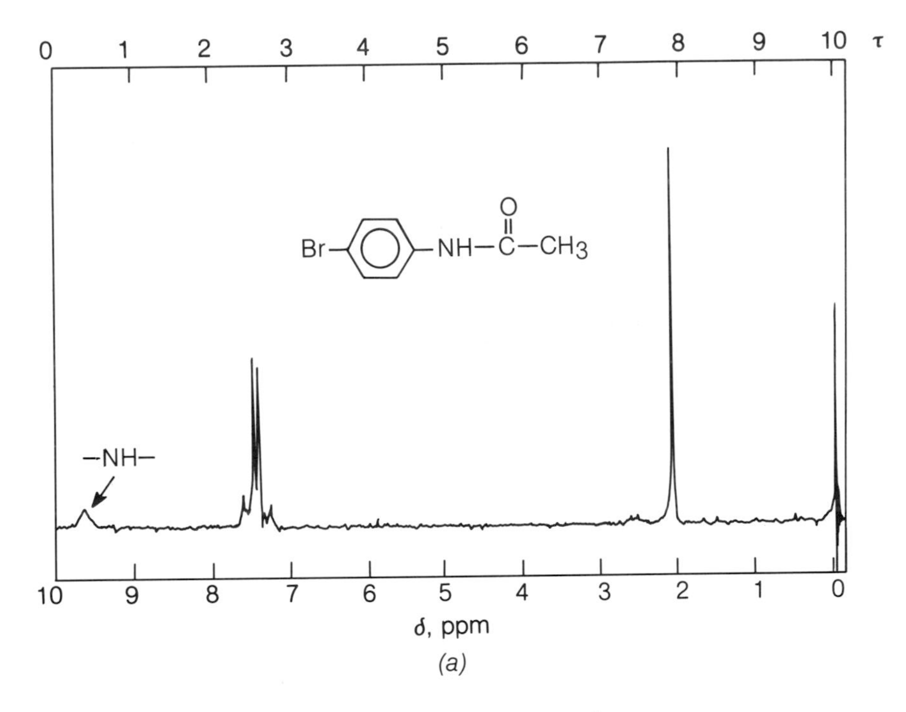

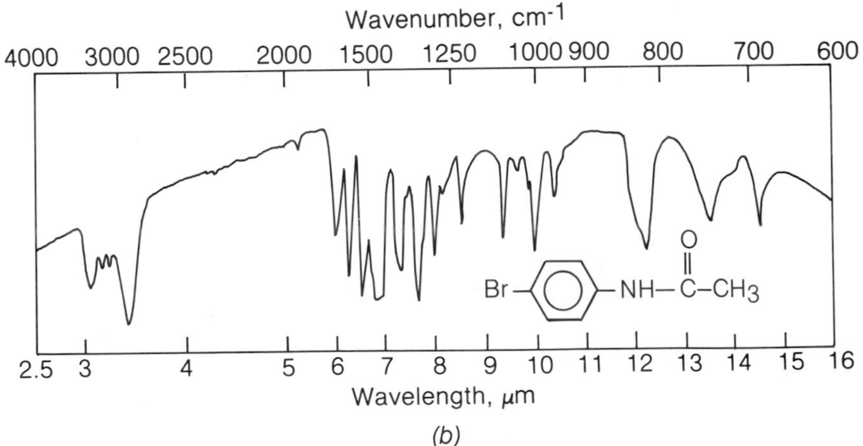

Figure 18.10
The (*a*) nmr and (*b*) ir spectra of 4-bromoacetanilide. The nmr spectrum was prepared using a Nujol mull.

the solid with two 25-mL portions of cold, distilled water and press most of the solvent out of the wet filter cake using a piece of dry filter paper.

Transfer the precipitate to a 125-mL Erlenmeyer flask and dissolve it in a minimum amount of boiling methanol. Once the material has gone into solution, remove the flask from the steam bath and allow it to cool slowly to room temperature. After the crystallization has started, cool the methanol solution in an ice bath for several minutes and collect the crystalline solid on a Buchner funnel. Air dry this material, mp 165 to 167°C. The yield should be approximately 75%.

The mother liquors from the above crystallization can be reduced to a volume of approximately 10 mL on a steam bath, and an additional 1 to 1.5 g of product may be obtained. The overall yield, including the second crop of crystals, is approximately 85 to 90%.

The nmr and ir spectra of acetanilide and 4-bromoacetanilide are shown in Figs. 18.9 and 18.10, respectively.

EXPERIMENT 18.2C ALTERNATE BROMINATION OF ACETANILIDE USING A BROMINE COMPLEX

$$C_6H_5\text{—NH—C(=O)—}CH_3 + \text{Br complex} \longrightarrow 4\text{-Br—}C_6H_4\text{—NH—C(=O)—}CH_3$$

Time 2.0 h

Materials Acetanilide, 4 g (MW 135, mp 113 to 115°C)
Pyridinium bromide perbromide, 10 g (MW 319)
Glacial acetic acid, 20 mL (MW 60, d 1 g/mL)

Precautions Wear gloves and carry out all transfers in a good hood.

Hazards Pyridinium bromide perbromide is a lachrymator and a source of bromine. It is, however, much safer to handle than liquid bromine. Pyridinium bromide perbromide is a crystalline solid, which may be weighed easily. Avoid breathing dust and contact with skin. Hydrogen bromide is an acidic gas. Avoid contact with skin and eyes. Glacial acetic acid is flammable. Avoid breathing vapor and contact with eyes and skin.

Experimental Procedure

Add 4 g (0.030 mol) acetanilide to a 250-mL Erlenmeyer flask. Add to the acetanilide 20 mL glacial acetic acid (**hood, gloves**) and swirl the mixture, with slight warming on a steam bath if necessary, until all the acetanilide dissolves. After the acetanilide dissolves, add 10 g (0.031 mol) pyridinium bromide perbromide to the glacial acetic acid mixture all at once. (*Note:* The pyridinium bromide perbromide is rather insoluble in glacial acetic acid but will slowly go into solution as the reaction proceeds.)

Warm the mixture on a steam bath until the internal temperature is about 60°C (3 to 4 min on the steam bath). Swirl the mixture vigorously for approximately 20 to 30 min while the temperature slowly returns to room temperature. The pyridinium bromide perbromide slowly goes into solution, giving a red color and, after about 30 min a yellowish white precipitate starts coming out of the solvent. Swirl the flask for 10 min after the material starts coming out of solution. (Total time should not be greater than 45 min.)

At the end of the reaction period slowly add 150 mL cold water while swirling the flask. A white precipitate will deposit. After the water has been added, stir the solution with a glass stirring rod to break up any clumps of solid. (The aqueous solution is usually yellow at this point.) Add 0.5 g solid sodium bisulfite, $NaHSO_3$, to the aqueous solution, swirl, and observe the color. Keep adding 0.5-g portions of solid sodium bisulfite until the yellow color disappears (usually a single portion is sufficient).

Filter the white, crystalline product with the aid of a Buchner funnel. Wash the white solid with two 25-mL portions of cold distilled water and press most of the solvent out of the wet filter cake using a piece of dry filter paper.

Transfer the precipitate to a 125-mL Erlenmeyer flask and dissolve it in a minimum amount of boiling methanol. Once the material has gone into solution, remove the flask from the steam bath and allow it to cool slowly to room temperature. After the crystallization has started, cool the methanol solution in an ice bath for several minutes and collect the crystalline solid on a Buchner funnel. Air dry this material, which should have mp 165 to 167°C. The yield should be approximately 75%.

The mother liquors from the above crystallization can be reduced to a volume of approximately 8 to 10 mL on a steam bath, and an additional 0.5 to 1.0 g of product may be obtained. The overall yield, including the second crop of crystals, is approximately 85 to 90%.

The nmr and ir spectra of acetanilide and 4-bromoacetanilide are shown in Figs. 18.9 and 18.10, respectively. Note the influence of the para bromine atom on the spectra. Does this surprise you?

QUESTIONS AND EXERCISES

18.1 Using resonance structures, show why acetanilide can be brominated without using a catalyst whereas most other aromatic compounds require one.

18.2 2,4-Dinitro-1-chlorobenzene is so reactive that it is converted to 2,4-dinitrophenol by the action of sodium bicarbonate. Why is no 2,2′,4,4′-tetranitrodiphenyl ether ever isolated in this hydrolysis?

18.3 Why is a freshly cut nail or tack a better catalyst for aromatic bromination than an old nail?

18.4 What conditions would be required to brominate one of the methyl groups in *p*-xylene rather than the aromatic ring?

18.5 Why does sodium bisulfite discharge the color from the acetanilide bromination mixture?

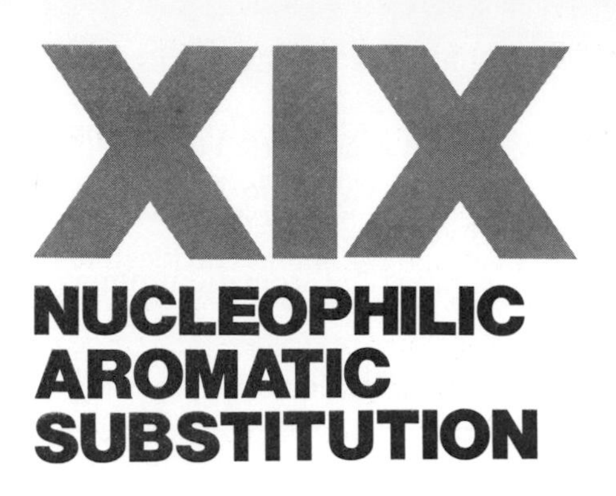

XIX NUCLEOPHILIC AROMATIC SUBSTITUTION

Substitutions in aromatic rings fall basically into three categories: electrophilic aromatic substitution, reactions involving elimination-addition (a benzyne mechanism), and direct nucleophilic aromatic substitution.

Because benzene is an electron-rich system, the addition of nucleophiles is not generally favored. However just as electrophilic substitution is facilitated by the presence of electron-releasing groups, nucleophilic aromatic substitution is facilitated by electron-attracting groups. In order for a nucleophilic aromatic substitution to occur, two things are required. First, functional groups must be present on the aromatic ring which can stabilize the negative charge brought to it by the nucleophile, and second, there must be some leaving group present in the aromatic system which can be lost and can take the negative charge with it. In electrophilic substitution a proton functions as the leaving group and is readily lost. A proton cannot function as a leaving group in nucleophilic substitution reactions because it would have to leave as a hydride ion. Hydride is one of the poorest leaving groups known.

An example of a reagent specifically designed for nucleophilic aromatic substitution is the reagent developed by Sanger for protein analysis. Sanger demonstrated that 1-fluoro-2,4-dinitrobenzene has all the requisites described above. The nitro groups have the ability to stabilize the negative charge brought by some incoming nucleophile, and fluoride ion is probably the best of all leaving groups in nucleophilic aromatic substitution. Sanger showed that if he treated a peptide chain with 1-fluoro-2,4-dinitrobenzene and then hydrolyzed the product, only one amino acid could be isolated which contained the dinitroaromatic residue. By this method Sanger was able to gradually degrade and work out the structure of the protein insulin. This method of peptide sequencing and the elucidation of the structure of insulin led to the award of the Nobel prize to Sanger.

The mechanism of nucleophilic aromatic substitution was first elucidated by Miesenheimer. Miesenheimer found that if he treated 2,4,6-trinitroanisole with ethoxide ion or treated 2,4,6-trinitrophenetole with methoxide ion, he obtained the same intermediate in both cases. This intermediate contained three nitro groups and both ether functions. The structure is illustrated below and is called a Miesenheimer complex. This is the prototype for complexes in nucleophilic aromatic substitution.

$$\text{2,4,6-}(O_2N)_3C_6H_2\text{—O—}CH_3 + CH_3CH_2\text{—}O^- \longrightarrow [CH_3CH_2\text{—O, O—}CH_3 \text{ complex}]^- \longleftarrow {}^-O\text{—}CH_3 + \text{2,4,6-}(O_2N)_3C_6H_2\text{—O—}CH_2CH_3$$

Loss of either methoxide or ethoxide will lead to starting material or a new product.

19.1 SYNTHESIS OF 2,4-DINITROPHENYLHYDRAZINE BY NUCLEOPHILIC AROMATIC SUBSTITUTION

In the experiment described here, the nucleophile is hydrazine and the substrate is the nitration product made in Exp. 18.1D or 18.1E, i.e., 1-chloro-2,4-dinitrobenzene or 1-bromo-2,4-dinitrobenzene, respectively. Both compounds have the requisite functionality to undergo nucleophilic aromatic substitution. Both have halide leaving groups and both have two nitro functions which can effectively stabilize the negative charge generated in the ring during reaction. Hydrazine readily displaces halide under the conditions described below and leads to 2,4-dinitrophenylhydrazine. This material is of considerable utility in organic chemistry because it is used to form derivatives of aldehydes and ketones, as discussed in Sec. 25.5A.

We wish to add a special note of caution regarding this experiment. This experiment was chosen because it is one of the few nucleophilic aromatic substitutions which can be performed in an undergraduate organic laboratory, i.e., the conditions are sufficiently mild that no special apparatus is needed. It should be noted, however, that hydrazine is a toxic material and can cause skin irritation. Likewise, 1-chloro-2,4-dinitrobenzene and 1-bromo-2,4-dinitrobenzene can cause skin irritation. All manipulations should be carried out wearing gloves. The product, 2,4-dinitrophenylhydrazine, is a highly colored amine and in fact is often seen on the hands of students doing qualitative organic analysis. In our experience we have

not observed any ill effects in the performance of this experiment, but there is always danger with compounds which are potential skin irritants. This experiment should be done only by a class with a reasonable amount of laboratory experience and then only under close supervision.

EXPERIMENT 19.1

SYNTHESIS OF 2,4-DINITROPHENYLHYDRAZINE BY NUCLEOPHILIC AROMATIC SUBSTITUTION

$$\text{X-C}_6\text{H}_3\text{-2,4-(NO}_2)_2 + \text{NH}_2\text{—NH}_2 \longrightarrow \text{NH—NH}_2\text{-C}_6\text{H}_3\text{-2,4-(NO}_2)_2 + \text{HX}$$

Time 1 h

Materials 1-Chloro-2,4-dinitrobenzene, 1 g (MW 202.5, mp 49 to 52°C); or 1-bromo-2,4-dinitrobenzene, 1 g (MW 247, mp 70 to 72°C)
64% Hydrazine, 1 mL
Ethanol, 30 mL

Precautions Wear gloves and carry out all reactions in a good hood.

Hazards All the starting materials are toxic and are skin irritants. Avoid skin and eye contact and avoid breathing any vapors. The product, 2,4-dinitrophenylhydrazine, is also a skin irritant. Avoid contact with skin.

Experimental Procedure

Dissolve 1 g 1-chloro-2,4-dinitrobenzene, Exp. 18.1D (or 1-bromo-2,4-dinitrobenzene, Exp. 18.1E) in approximately 15 to 20 mL ethanol in a 125-mL Erlenmeyer flask and warm this mixture on the steam bath. To the warm mixture add 1 mL 64% hydrazine all at once. Continue to warm the solution on the steam bath for several minutes and then cool the darkly colored solution slowly to room temperature over 25 to 30 min. Collect the crystals which are formed on a Hirsch funnel and wash them with 7 to 10 mL ice-cold ethanol. Recrystallize the resulting colored crystals from ethyl acetate. The product should be a light orange solid of mp 198 to 200°C.

QUESTIONS AND EXERCISES

19.1 A student conducting this experiment was very careful to rinse out the Erlenmeyer flask with acetone to remove impurities just before recrystallizing the 2,4-dinitrophenylhydrazine. After recrystallization the student isolated a small amount of a yellow impurity, mp 126°C. What might this impurity be?

19.2 4-Chloro-4′-nitrobenzophenone was treated with a strongly nucleophilic reagent. Analysis of the product showed that a displacement reaction had occurred, but a strong absorption was still observed at 1660 cm^{-1} in the ir spectrum and the Beilstein test proved positive. What must have happened in this reaction?

XX THE CHEMISTRY OF NATURAL PRODUCTS

The study of organic chemistry did not start with simple molecules. Since no structures and no functional groups were known, chemists had to begin by analyzing the molecules nature made available. Since these compounds were not produced synthetically, substances obtained from natural sources became known as *natural products*.

Only a very few compounds are found in nature in pure form. Silicon dioxide (sand) is one of the few examples. Certain salts such as sodium chloride are found in large deposits in pure form. This is true also of some minerals such as zircon (zirconium silicate) which have more complicated structures. Naturally occurring organic compounds are rarely found in pure form. Even readily available sugar must be "mined," just as salt is. The organic chemist's method of "mining" is usually extraction.

In the early days of chemistry the only solvents which were known were water and fermentation products. Ethanol and vinegar were known, as were aqueous acid and base solutions. The latter were particularly useful because naturally occurring compounds having either acidic or basic properties could be obtained by extraction and then removed from solution by adjusting the pH.

A large group of organic bases was known in the early days of chemistry. These compounds generally contained basic nitrogen which could be protonated and made water-soluble. These basic compounds were given the general name *alkaloids*. After treatment with base, the alkaloid was generally deposited from aqueous solution and could often be recrystallized to purity. (For a detailed discussion of extraction, see Sec. 2.5.)

Much of the early study in organic chemistry centered around the isolation, structural elucidation, and synthesis of natural products. Even today, the search for natural compounds with novel structures and novel properties, especially compounds useful in medicinal applications, continues unabated.

Naturally occurring compounds have found wide application in modern society. The most widely used of all natural products is ethanol. Simple aqueous solutions of it are marketed as vodka. The stimulant most widely used in modern society is an alkaloid called *caffeine* which is extracted by hot water from ground coffee beans or tea leaves. Caffeine is the constituent which gives coffee and tea beverages their ability to provide a "lift." Isolation of this alkaloid is described in Exp. 20.1.

A beverage made from the bark of the cinchona tree was widely used by South American natives to combat malaria long before Europeans controlled this disease. Chemical investigations revealed that it was the alkaloid quinine present in the broth which eventually saved uncounted millions of lives.

Not all natural products have beneficial effects or even a single physiological effect. The curare alkaloids, for example, are used by certain South American Indians as arrow poisons but have found medicinal application as muscle relaxants. The excitement of the recent discovery of vernolepin palled when this anticancer compound was synthesized and found to be one of the most potent poisons known.

Many natural products or mixture extracts are prized for their fragrance or taste. The compound pinene may be obtained from many sources and has the odor which most people associate with pine trees. It is used as a perfume base and in a number of other applications. The isolation of eugenol (oil of cloves), from cloves is described in Exp. 20.2. Eugenol is used in perfumery and also as a dental analgesic (pain killer). Other natural products may be used in other chemical transformations, and this is illustrated by the resolution of phenethylamine using tartaric acid (Exp. 20.3).

20.1 ISOLATION OF THE NATURALLY OCCURRING STIMULANT CAFFEINE

Although many stimulants are prescribed for medical purposes, the most widely used of all stimulants, caffeine, may be obtained without a prescription from a variety of sources. Caffeine is found in the customary breakfast beverages coffee and tea. Coffee contains between 1 and 2% caffeine; tea contains 2 to 3% along with a smaller amount of the structurally related stimulant theobromine. Although cocoa (Aztec chocolate) contains relatively little caffeine, it does contain a substantial amount of theobromine. One of the richest sources of caffeine is the African kola nut (*Cola nitida*), which contains about 3% caffeine.

Coffee, tea, and cola are all very refreshing drinks. Part of this is undoubtedly due to their thirst-quenching properties, but a major contributor is the presence of caffeine. The fundamental mechanism by which caffeine operates is not understood. It is known that this compound stimulates the central nervous system and makes the user more alert and often more loquacious. Because of these effects, caffeine is often added to headache medicine so that the user will not only feel pain relief but will also be stimulated. Caffeine may also be purchased in almost pure form in preparations designed to keep the user awake and alert.

The experiment described here is typical of the way natural products are isolated from natural sources. It is atypical in the sense that the extraction is so easy and there is so much of the desired compound in the source. It is common for much more laborious extraction techniques, involving acid, base, and organic solvents, to yield only a small fraction of a percent of the desired compound.

EXPERIMENT 20.1 ISOLATION OF CAFFEINE FROM TEA LEAVES

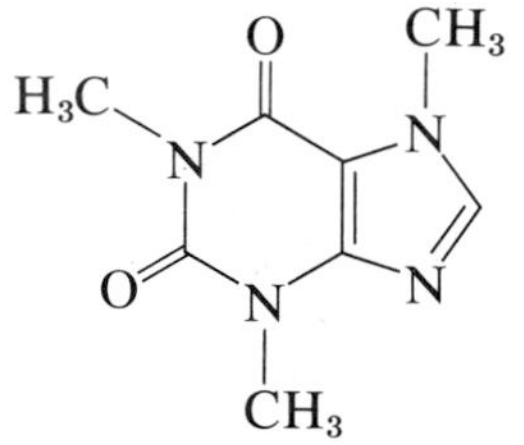

Time 3 h

Materials Tea bags, 10 to 12 (about 25 g—any popular brand may be purchased from a supermarket; loose tea may also be used)
Sodium carbonate, 20 g
Dichloromethane, 200 mL
Toluene, 5 to 7 mL
Heptane, 7 to 10 mL

Precautions Perform extractions in the hood.

Hazards Do not breath organic solvent vapors. Do not burn yourself on hot beakers.

Experimental Procedure

Place 20 g sodium carbonate in a 600-mL or 1-L beaker. Add 275 mL distilled water to the beaker and heat the solution on a ring stand with a flame until the sodium carbonate dissolves.

Add the tea bags (10 to 12, 25 g) to the basic aqueous solution (remove tags from tea bags before adding) and boil the contents of the beaker for 20 to

30 min, using a flame (be careful not to heat too strongly, as bumping will occur). Remove the beaker from the heat and allow to cool somewhat (around 50°C). Decant the dark aqueous layer (**hot apparatus**) from the tea bags into a 600-mL beaker, being careful to obtain as much liquid as possible from the residue. Cool the aqueous solution to room temperature.

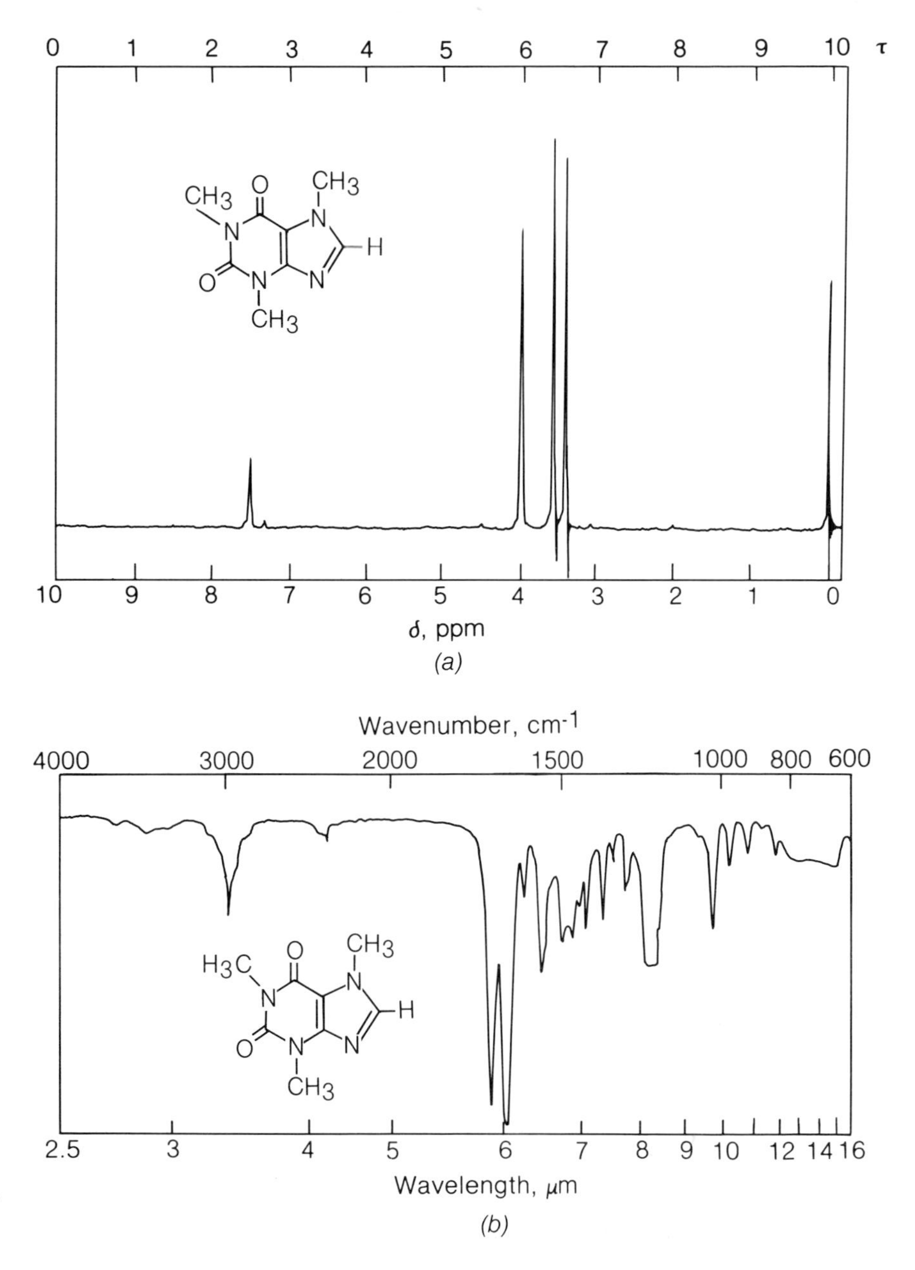

Figure 20.1
The (*a*) nmr and (*b*) ir spectra of caffeine.

Add 30 mL dichloromethane to the cooled beaker and *gently* swirl the beaker for 3 to 5 min. An even better procedure is to use a magnetic stirrer and *slowly* stir the two-phase mixture. [*Note:* If the extraction is performed with vigorous swirling or stirring, an emulsion will form that is very hard to break. The best strategy is to prevent formation of the emulsion rather than to try to deal with it after it has formed; however, if an emulsion forms anyway, filter the mixture through Celite (as described in Sec. 2.4) for 3 to 5 min.] Slowly transfer the two-phase system to a separatory funnel (this is best done through a long-stemmed funnel) and remove the lower organic layer. Return the aqueous solution to the 600-mL beaker and repeat the extraction procedure four more times. Again, *take care to prevent emulsion formation.*

Combine the organic layers and dry with anhydrous granular sodium sulfate. Decant the dichloromethane layer into a *dry* 600-mL beaker and remove the solvent on a steam bath. As the final traces of solvent are removed, the caffeine will crystallize (the caffeine is usually more than 90% pure at this point) as an off-white or cream-colored solid.

The crude caffeine is crystallized from a minimal amount of toluene-hexane to give small needles of a white crystalline solid. This material may be sublimed (Sec. 2.3) to give pure material of mp 233 to 235°C.

The nmr and ir spectra of caffeine are shown in Fig. 20.1. Notice that each methyl group is in a different environment and three separate nmr signals are observed. The small resonance observed in the aromatic region (approximately 7.5 ppm) is due to the single proton in the five-membered ring.

20.2 ISOLATION OF AN ESSENTIAL OIL FROM THE SPICE CLOVE

Many herbs, spices, and so-called essential oils were known millennia before organic chemistry became sophisticated enough for the structures of the active compounds to be elucidated. To this day the precise compositions and structures of some of the constituent oils in perfumes are not established. This is in part by design.

The substances which have been used in cooking and perfumery have always had a special place in society. The use of chemicals to make the human body attractive in one way or another antedates by many years the use of internal medicines. It was often the aromas or essences of oils for which they were prized. These substances and mixtures became known as *essential oils.*

In the experiment described here, the compound eugenol is isolated from the spice clove. The clove is actually the bud of an East Indian evergreen tree, *Eugenia aromatica.* The odor is often characterized as pungent and spicy. The essential oil has found application in perfumery and is of some utility as a topical dental analgesic.

EXPERIMENT 20.2

ISOLATION OF EUGENOL FROM CLOVES

$$HO-C_6H_3(OCH_3)-CH_2-CH{=}CH_2$$

Time 3 h

Materials Whole cloves, 50 g (about one 2-oz can. These may be obtained from a bakery or supermarket.)
Dichloromethane, 200 mL
5% Potassium hydroxide solution
5% Hydrochloric acid solution

Hazards Do not breathe organic solvent vapors.

Experimental Procedure

Set up a steam-distillation apparatus (see Fig. 2.11) for internally generated steam using a 500-mL round-bottom three-neck flask. Add 50 g whole cloves to the 500-mL round-bottom flask, followed by 250 mL water. Add several *boiling sticks* (not boiling chips) and steam distill the mixture over a flame, keeping the internal volume at about 250 mL with appropriate additions of water from the separatory funnel. Steam distill the mixture rapidly until no oily material can be seen in the condenser (usually about 50 to 75 min).

Transfer the distillate to a separatory funnel and extract with 50 mL dichloromethane. Remove the lower organic layer and extract the aqueous phase with an additional 50-mL portion of dichloromethane. Combine the organic layers and discard the aqueous layer.

Transfer the organic layer back to the separatory funnel and extract with three 50-mL portions of 5% potassium hydroxide solution (**heat evolved**). Combine the basic aqueous layers and wash once with one 25-mL portion of dichloromethane.

Transfer the aqueous basic layer to a 600-mL beaker and *slowly* acidify (**heat evolved**) the aqueous layer to pH 1 with 5% hydrochloric acid (test solution with pH paper). After acidification, transfer the solution back to the separatory funnel and extract the aqueous layer with two 40-mL portions of dichloromethane. Discard the aqueous wash. Wash the combined organic layers with one 25-mL portion of distilled water, followed by one 25-mL portion of half-saturated sodium chloride solution. Dry the organic layer over anhydrous granular sodium sulfate, decant the organic material from the drying agent, and remove the dichloromethane on a steam bath. The practically pure (98%) eugenol is obtained as a pale yellow oil with an overpowering odor of cloves. Weigh the oil and calculate the yield of eugenol from whole cloves.

The nmr and ir spectra of eugenol are shown in Fig. 20.2. Notice the strong OH band and double-bond absorption (1600 cm^{-1}) in the ir spectrum. Note also the complex double-bond splitting pattern in the aromatic and double-bond region of the nmr spectrum.

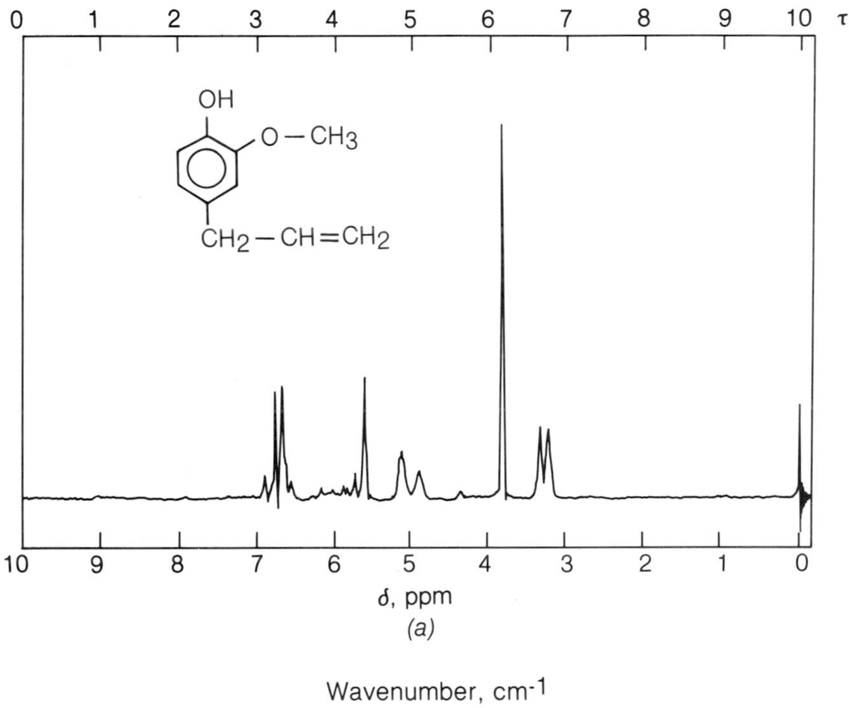

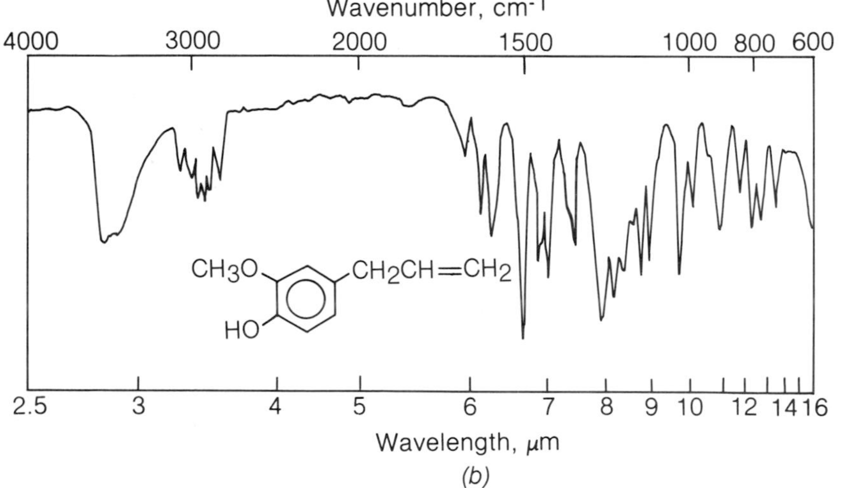

Figure 20.2
The (*a*) nmr and (*b*) ir spectra of eugenol.

20.3 OPTICAL RESOLUTION USING NATURALLY OCCURRING, OPTICALLY ACTIVE TARTARIC ACID

It is often found that only one enantiomer of a chiral structure has the desired biological activity. The other isomer may have unwanted activity or simply be inactive. The need for optically pure products has made the search for new methods of resolution and new resolving agents an important area of investigation over the years.

It is obvious from the outset that resolution requires an optically active resolving agent. Since natural products are synthesized enzymatically, only one of the two or more possible isomers is usually found in the natural source. Since nature has performed the resolution, these optically pure materials are often an ideal source of resolving agents for synthetic materials. Such a compound is naturally occurring tartaric acid (2,3-dihydroxybutanedioic acid). The Germans call this compound *Weinsäure* ("wine acid"), and the name suggests the origin. Tartaric acid is widely distributed in fruits such as the grapes which are fermented to wine. In fact, the pure potassium salt was isolated from fermentation broths and was known in Roman times. It was a tartaric acid salt which Louis Pasteur used to first demonstrate optical activity.

The resolution of 1-aminophenylethane (phenethylamine) is described in Exp. 20.3. It is by procedures such as the one presented here that virtually all modern resolutions are carried out. Phenethylamine, resolved in this way, may serve as a resolving agent for other substances.

In the experiment described in this section, racemic phenethylamine and optically active tartaric acid are combined to form diastereomeric salts. Unlike enantiomers, whose physical properties are all identical, diastereomers may have very different properties. The salts are separated in this experiment by virtue of their differing solubilities.

EXPERIMENT 20.3

RESOLUTION OF RACEMIC PHENETHYLAMINE USING TARTARIC ACID

$$C_6H_5\text{—}CH(CH_3)\text{—}NH_2 + HOOC\text{—}CH(OH)\text{—}CH(OH)\text{—}COOH \longrightarrow C_6H_5\text{—}CH(CH_3)\text{—}\overset{+}{N}H_3\ {}^{-}OOC\text{—}CH(OH)\text{—}CH(OH)\text{—}COOH$$

Time 3 h (over 3 days)

Materials *d*-Tartaric acid, 15.6 g (MW 150, $[\alpha]_D^{20}$ +12.3[c = 20, H_2O], mp 171 to 174°C)
Phenethylamine, 12.5 g [α-methylbenzylamine, 13.1 mL, MW 121, d 0.94 g/mL, bp 185°C]
Methanol, 250 mL
10% Sodium hydroxide solution

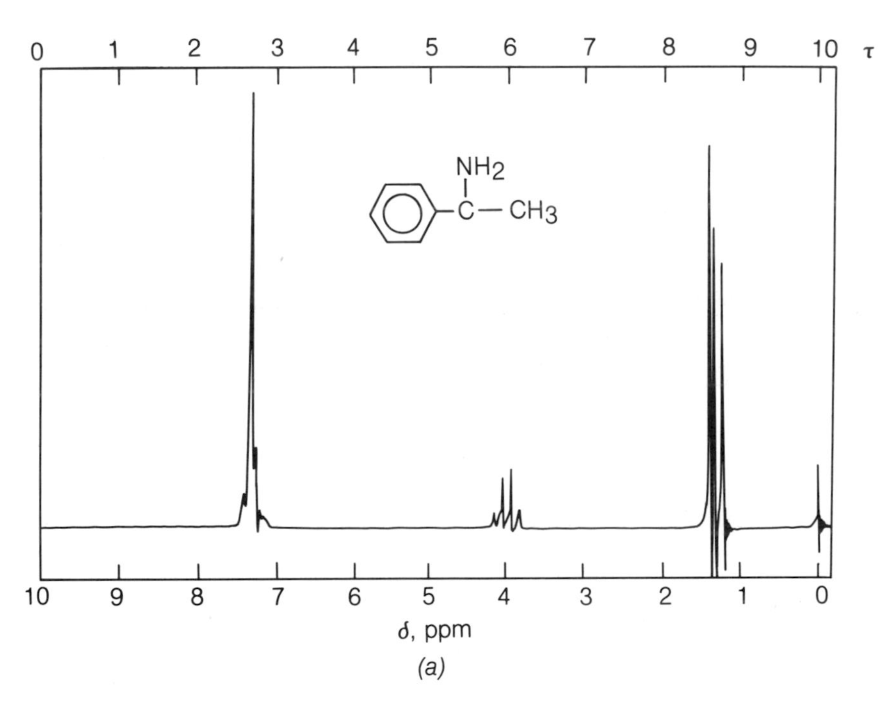

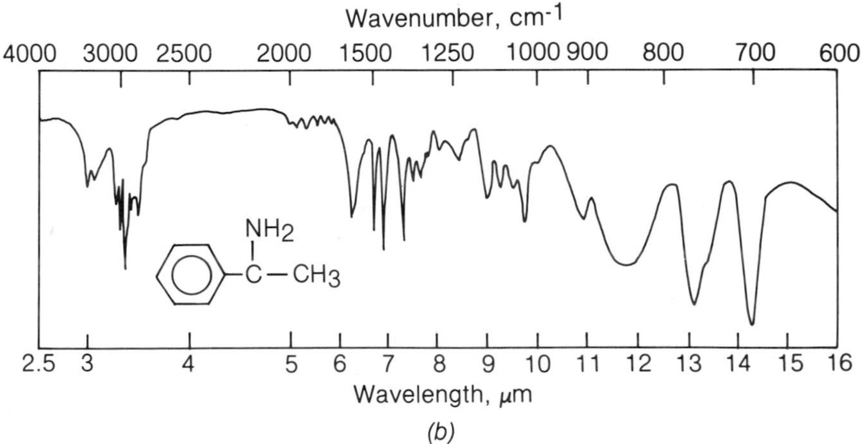

Figure 20.3
The (*a*) nmr and (*b*) ir spectra of racemic phenethylamine.

Hazards Methanol may be dangerous in high concentrations. Distill or evaporate it in the hood to remove it.

Experimental Procedure

Place 15.6 g (0.104 mol) *d*-tartaric acid in a 500-mL Erlenmeyer flask and add 225 mL methanol. Heat the mixture on a steam bath until the internal temperature (thermometer) is about 60°C. While the solution is hot, very carefully add 12.5 g (13.1 mL, 0.103 mol) racemic α-phenethylamine. If the amine is added too quickly, the solution may foam and overflow the flask. Loosely stopper the flask with a cork and set it in a place where it may stand undisturbed overnight. The tartrate salt of the levorotatory amine is relatively insoluble in cold methanol and may be isolated by suction filtration. Wash the white crystals with a small amount of cold methanol. The methanol solution may then be reduced in volume to about 90 mL by heating on the steam bath (**hood**) or by distillation. The smaller-volume solution may now be set aside as before and an additional gram of levorotatory amine salt may be isolated.

After the second crop of levorotatory amine salt has been collected, the methanol solution should be reduced to minimum volume by warming on a steam bath. The residue is predominantly the dextrorotatory amine tartrate.

The free amine may be obtained from the tartrate salt by placing the salt (10 g) in a 125-mL Erlenmeyer flask and adding 50 mL 10% aqueous sodium hydroxide solution. After swirling for a few minutes, an oily layer should be visible. Transfer the aqueous solution to a separatory funnel and extract the free amine with two 25-mL portions of ether. (Dichloromethane may be used for the extraction, but it sometimes forms emulsions when shaken with sodium hydroxide solution.)

The pure amine boils at 185°C and each pure enantiomer has a rotation of $[\alpha]_D^{25}$ ± 40.6 (neat). See Sec. 1.5 for a discussion of polarimetry.

The ir and nmr spectra of the pure amine are shown in Fig. 20.3. The spectra of the enantiomers are, as expected, identical.

QUESTIONS AND EXERCISES

20.1 Based on the discussion at the beginning of this chapter, do you think that those people who drink tea "because coffee makes me nervous" are realistic? Why or why not?

20.2 Alkaloids are often obtained from plants by acid extraction. Why do you suppose the procedure described in Exp. 20.1 does not call for an acid extraction?

20.3 Theobromine is a stimulant closely related to caffeine. In fact, it is identical to caffeine except that there is a methyl group present in the five-membered ring (on the carbon atom flanked by two nitrogen atoms). How would you expect the nmr spectrum of theobromine to differ from that of caffeine?

20.4 Eugenol is isolated from cloves by steam distillation in Exp. 20.2. If extraction had been chosen, would 10% aqueous HCl, NaOH, or Na_2CO_3 have been used? Why?

20.5 *meso*-Tartaric acid is about the same price as *d*-tartaric acid (used in Exp. 20.3). Why was the latter chosen for this experiment?

20.6 Quinine is a naturally occurring, optically active amine whose structure is shown in Sec. 24.2. Would this have been a good choice for the resolution of phenethylamine? Why?

THE WITTIG REACTION

The Wittig reaction is used for the conversion of a carbonyl compound (usually an aldehyde or ketone) to an olefin. Although the Wittig reaction is a relatively new one, its great synthetic utility has made it a very important reaction in synthetic organic chemistry.

The reaction involves a phosphonium salt as a starting material. The phosphonium salt is usually prepared in the first step of the reaction sequence. For example, when triphenylphosphine is heated with methyl iodide, an S_N2 reaction occurs. Trivalent phosphorus has a lone pair of electrons, which can attack an electrophile. The product is a quaternary phosphonium salt. Amines undergo the same sort of reaction; an example may be found in Exp. 13.8. The reaction of triphenylphosphine with methyl iodide is formulated below:

$$(C_6H_5)_3P\!: + CH_3{-}I \longrightarrow (C_6H_5)_3\overset{+}{P}{-}CH_3 \quad I^-$$

The chemistry of the methyl group attached to positively charged phosphorus differs from its chemistry when attached to neutral iodine. Iodomethane reacts with bases primarily by substitution. Methoxide ion, for example, reacts with it to give dimethyl ether. Triphenylmethylphosphonium iodide reacts with strong bases by undergoing deprotonation. Loss of a proton gives a product which is at once positively and negatively charged:

$$(C_6H_5)_3\overset{+}{P}{-}CH_3\ I^- + :B \longrightarrow (C_6H_5)_3\overset{+}{P}{-}\overset{-}{C}H_2 + H\overset{+}{B} + I^-$$

Phosphonium salt — Ylide

These substances are called *ylides* and are said to be *zwitterions* (from the German meaning "hybrid ions"). The negative charge on carbon is stabilized by the positive charge of phosphorus. Nevertheless, ylides are quite reactive nucleophiles.

Once the ylide has formed in a solution containing a carbonyl compound, it reacts to form a carbon-carbon bond. The intermediate in this reaction contains a positively charged phosphorus and a negatively charged oxygen atom. This intermediate undergoes an electronic reorganization, which is driven to completion by formation of the very strong P=O bond. The net reaction is exchange of $=CH_2$ for $=O$. The Wittig reaction is, in fact, a general method for the conversion of an aldehyde or ketone to a substituted vinyl group.

$$(C_6H_5)_3\overset{+}{P}{-}\overset{-}{C}H_2 + \text{cyclohexanone} \longrightarrow (C_6H_5)_3P{-}O \text{ / } CH_2{-}C_6H_{10} \longrightarrow \text{methylenecyclohexane } (=CH_2) + (C_6H_5)_3P{=}O$$

Triphenylphosphine is not the only starting material which is useful in the Wittig reaction but it is one of the most convenient. Another convenient reagent for this application is triethyl phosphite, $(EtO)_3P$. The olefin formation involving a phosphite instead of a phosphine is similar to, but not identical with, the Wittig reaction. This modification was discovered independently by Emmons in the United States and by Horner in Germany. The modification is generally referred to as the Emmons reaction in the United States and the Horner-Wittig reaction in Germany. Fortunately, the reaction occurs by the same mechanism no matter what it is called.

Triethyl phosphite reacts with iodomethane just as triphenylphosphine does, affording a positively charged tetravalent phosphorus atom. The halide ion, which is inert in the phosphonium salt reaction, attacks the salt. This reaction (called the *Arbuzov reaction*) is possible because the leaving group is phosphate rather than phosphine. In any event, a neutral phosphonate is formed. The protons on the carbon atom adjacent to the phosphorus atom are acidified by the P=O group, and reaction can proceed as before. The overall reaction is shown below.

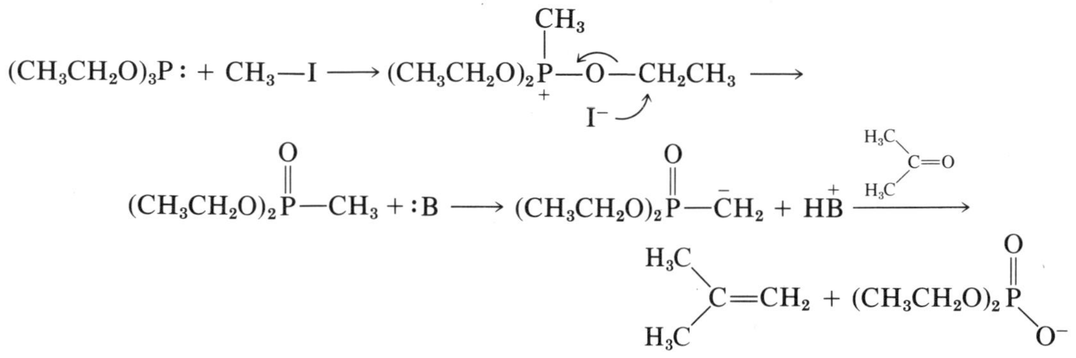

$$(CH_3CH_2O)_3P: + CH_3{-}I \longrightarrow (CH_3CH_2O)_2\overset{+}{P}(CH_3){-}O{-}CH_2CH_3 \;\; I^- \longrightarrow$$

$$(CH_3CH_2O)_2P(=O){-}CH_3 + :B \longrightarrow (CH_3CH_2O)_2P(=O){-}\overset{-}{C}H_2 + H\overset{+}{B} \xrightarrow{(H_3C)_2C=O}$$

$$(H_3C)_2C{=}CH_2 + (CH_3CH_2O)_2P(=O){-}O^-$$

In the reaction described below, triethyl phosphite is treated with benzyl chloride in the first step of a reaction sequence in which $>C{=}O$ is eventually converted to $>C{=}CH{-}C_6H_5$. The carbonyl compound used in this last step can be either benzaldehyde (Exp. 21.2), or cinnamaldehyde (Exp. 21.3). If benzaldehyde is used as the carbonyl compound, the resulting product is stilbene (1,2-diphenylethylene). If cinnamaldehyde is the source of the carbonyl group, then the product is 1,4-diphenyl-1,3-butadiene. In both cases the reaction is the same; only the starting carbonyl component is different. The overall conversion is shown at the base of p. 477.

EXPERIMENT 21.1 SYNTHESIS OF DIETHYL BENZYLPHOSPHONATE BY THE ARBUZOV REACTION

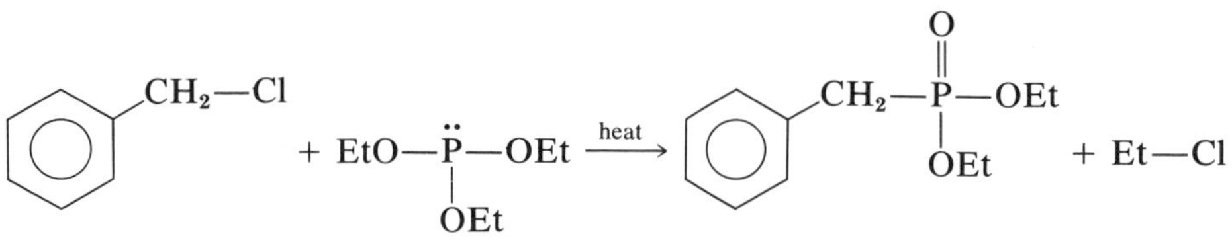

$$C_6H_5{-}CH_2{-}Cl + EtO{-}\ddot{P}(OEt){-}OEt \xrightarrow{\text{heat}} C_6H_5{-}CH_2{-}P(=O)(OEt){-}OEt + Et{-}Cl$$

Time 1.5 h

Materials Benzyl chloride, 3.5 mL (MW 126.5, bp 177 to 181°C, d 1.1 g/mL)
Triethyl phosphite, 5.5 mL (MW 166, bp 156°C, d 0.97 g/mL)

Precautions Carry out all transfers in a good hood. Wear gloves when handling benzyl chloride, triethyl phosphite, dimethylformamide (DMF), or any solution containing phosphorus compounds.

Hazards Benzyl chloride and triethyl phosphite are skin irritants. Benzyl chloride is a lachrymator. Avoid breathing vapors and skin and eye contact with any of the above materials. Be especially careful to avoid skin contact with any organic phosphorus compound.

Experimental Procedure

Using a *dry* 10-mL graduated cylinder, measure 3.5 mL (0.030 mol) benzyl chloride (**hood, gloves**) and add it to a 50-mL round-bottom flask. Using another *dry* 10-mL graduated cylinder (if an organic solvent is used, rinse with methanol, *not* acetone), measure 5.5 mL (0.033 mol) triethyl phosphite (**hood, gloves**) and transfer to the 50-mL round-bottom flask. Add several boiling chips and affix a condenser (with lightly greased joints) to the top of the boiling flask. Attach a drying tube to the top of the reflux condenser and, using a small flame or oil bath, reflux the liquid gently for 1 h. At the end of the reflux period, the only

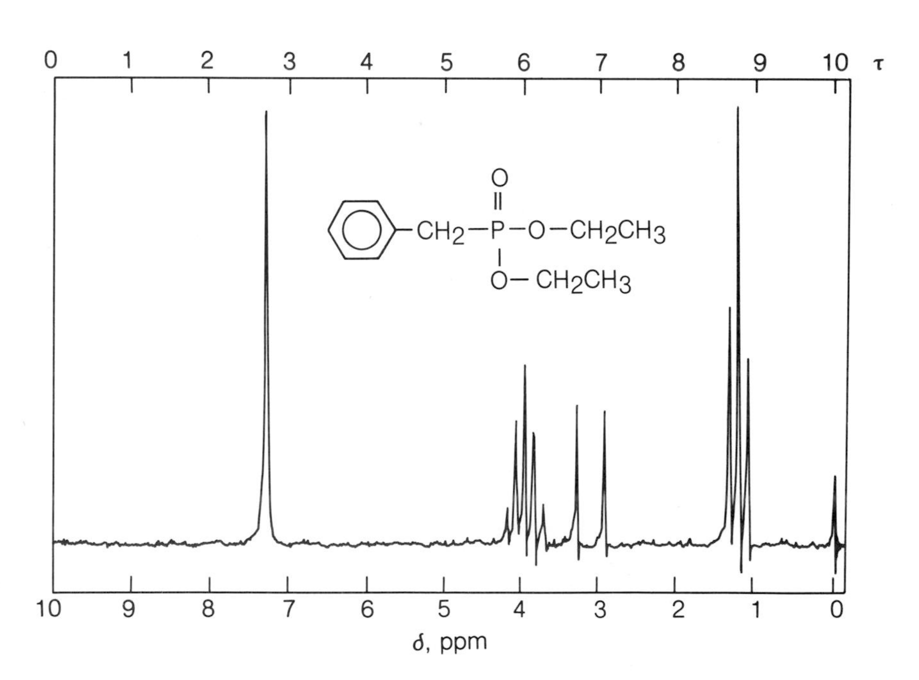

Figure 21.1
The nmr spectrum of diethyl benzylphosphonate.

material in the flask should be yellow. (Elimination of ethyl chloride starts at about 130°C. During the specified time period the temperature of the liquid will rise to approximately 190 to 200°C as the reaction proceeds.) Allow the phosphonate ester to cool to room temperature and *immediately* proceed with either Exp. 21.2 or Exp. 21.3.

The nmr spectrum of diethyl benzylphosphonate is shown in Fig. 21.1. Note the two peaks at approximately 3.0 and 3.3 ppm. This pattern is a doublet because of coupling between the benzyl $—CH_2—$ group and the ^{31}P nucleus.

The Wittig Reaction

$C_6H_5CH_2—Cl + EtO—\ddot{P}(OEt)—OEt \longrightarrow C_6H_5CH_2—\overset{+}{P}(OEt)_2—O—Et \; (Cl^-) \longrightarrow C_6H_5CH_2—P(=O)(OEt)—OEt + Et—Cl$

$C_6H_5CH_2—P(=O)(OEt)—OEt \xrightarrow{CH_3ONa} Na^+ \; C_6H_5\bar{C}H—P(=O)(OEt)—OEt + CH_3OH$

$\xrightarrow{C_6H_5CHO} Na^{+-}O—P(=O)(OEt)—OEt + C_6H_5CH=CHC_6H_5$

$\xrightarrow{C_6H_5CH=CHCHO} C_6H_5CH=CH—CH=CHC_6H_5 + Na^{+}\,^{-}O—P(=O)(OEt)—OEt$

EXPERIMENT 21.2 SYNTHESIS OF *trans*-STILBENE BY THE WITTIG REACTION

$$C_6H_5-CH_2-\overset{\overset{\displaystyle O}{\|}}{\underset{\underset{\displaystyle OEt}{|}}{P}}-OEt + C_6H_5-\overset{\overset{\displaystyle O}{\|}}{C}-H \xrightarrow[\text{DMF}]{NaOCH_3} C_6H_5-CH{=}CH-C_6H_5$$

Time 1.5 h

Materials Benzaldehyde, 3 mL (MW 106, bp 178 to 185°C, d 1.044 g/mL)
Sodium methoxide, 1.8 g (MW 54)
Dimethylformamide (DMF), 30 mL

Precautions Carry out all transfers in a good hood. Wear gloves when handling benzyl chloride, triethyl phosphite, and dimethylformamide (DMF) or any solution containing phosphorus compounds.

Hazards Benzyl chloride, triethyl phosphite, and DMF are skin irritants. Benzyl chloride is a lachrymator. DMF is absorbed through the skin. Avoid breathing vapors and skin or eye contact with any of the above materials. Be especially careful to avoid skin contact with any organic phosphorus compound.

Experimental Procedure

Allow the phosphonate ester (Exp. 21.1) to cool to room temperature. Add 1.8 g (0.033 mol) sodium methoxide (*Note:* Sodium methoxide is *hygroscopic;* weigh rapidly) into a *dry,* 125-mL Erlenmeyer flask. Stopper the flask (with a cork or rubber stopper) to prevent moisture from contaminating the sodium methoxide. After the phosphonate ester has cooled to room temperature, carefully pour the entire reaction mixture into the 125-mL Erlenmeyer flask which contains the sodium methoxide. Rapidly add 30 mL dimethylformamide (DMF) to the Erlenmeyer flask (**hood, gloves**). Swirl the 125-mL Erlenmeyer flask vigorously in an ice-water bath to thoroughly chill the contents (the flask should be stoppered) and quickly add 3 mL (0.030 mol) benzaldehyde by pipet to the cooled reaction mixture. Swirl the flask for 5 min in the ice bath (**exotherm**), remove the flask from the cooling bath, and let it stand at room temperature (with occasional swirling) for 20 to 25 min. The solution turns slightly yellow when the aldehyde is added, and after several minutes the hydrocarbon starts to precipitate from solution.

After the reaction period (not longer than 30 min) add 30 mL distilled water to the reaction mixture, swirl vigorously to dislodge any crystals which form, and collect the product on a Buchner funnel. Wash the product with several

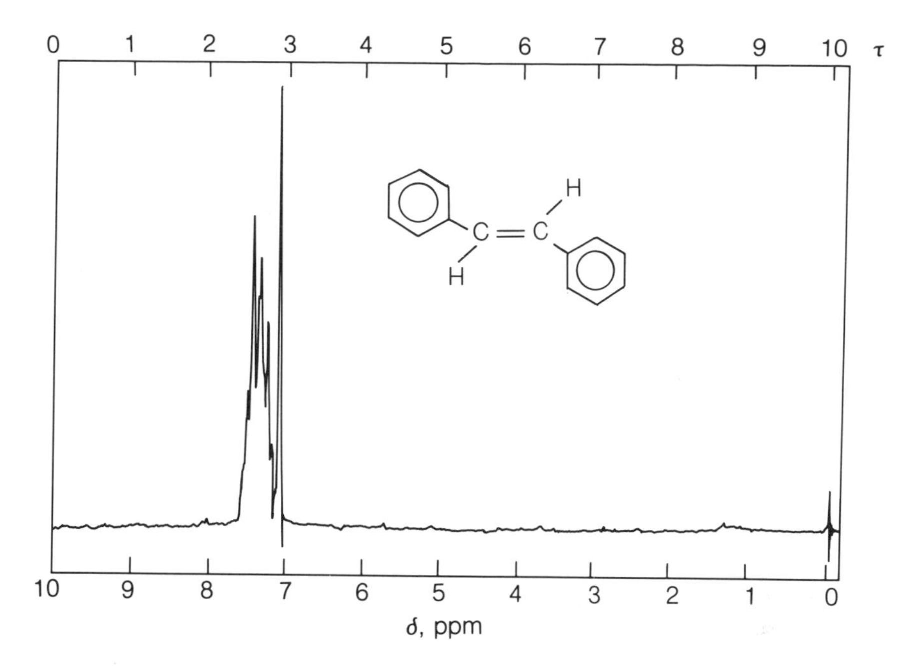

Figure 21.2
The nmr spectrum of *trans*-stilbene.

10-mL portions of distilled water. Air dry the solid material. The yield of *trans*-stilbene, mp 123 to 125°C, is 3.8 to 4.5 g.

The product may be recrystallized (almost never needed for purity, but recommended because of the beauty of the crystals) from 30% toluene-methanol (approximately 10 mL/g product) to give iridescent plates. The recrystallization yield is 90%. The product may also be recrystallized from 95% ethanol.

The nmr spectrum of *trans*-stilbene is shown in Fig. 21.2.

EXPERIMENT 21.3 SYNTHESIS OF 1,4-DIPHENYL-1,3-BUTADIENE

$$C_6H_5CH_2-\overset{O}{\overset{\|}{P}}(OEt)-OEt \; + \; C_6H_5CH{=}CH-CHO \xrightarrow[DMF]{NaOCH_3} C_6H_5CH{=}CH-CH{=}CHC_6H_5$$

Time 1 h

Materials Cinnamaldehyde, 3.5 mL (MW 166, bp 156°C, d 0.97 g/mL)
Sodium methoxide, 1.8 g (MW 54)
Dimethylformamide (DMF), 30 mL
Methanol, 50 mL

Precautions Carry out all transfers in a good hood. Wear gloves when handling benzyl chloride, triethyl phosphite, and dimethylformamide (DMF), or any solution containing phosphorus compounds.

Hazards Benzyl chloride, triethyl phosphite, and DMF are skin irritants. Benzyl chloride is a lachrymator. DMF is absorbed through the skin. Avoid breathing vapors and skin or eye contact with any of the above materials. Be especially careful to avoid skin contact with any organic phosphorus compound.

Experimental Procedure

Allow the phosphonate ester (Exp. 21.1) to cool to room temperature. Add 1.8 g (0.033 mol) sodium methoxide (*Note:* Sodium methoxide is *hygroscopic;* weigh rapidly) into a *dry,* 125-mL Erlenmeyer flask. Stopper the flask (with a cork or rubber stopper) to prevent moisture from contaminating the sodium methoxide. After the phosphonate ester has cooled to room temperature, carefully pour the entire reaction mixture (**hood, gloves**) into the 125-mL Erlenmeyer flask which contains the sodium methoxide. Rapidly add 30 mL dimethylformamide (DMF) to the Erlenmeyer flask (**hood, gloves**). Swirl the 125-mL Erlenmeyer flask vigorously in an ice-water bath to thoroughly chill the contents (the flask should be stoppered). Quickly add 3.5 mL (0.030 mol) cinnamaldehyde by pipet to the cooled reaction mixture. The reaction mixture rapidly turns deep red and the hydrocarbon 1,4-diphenyl-1,3-butadiene starts to crystallize from the solution. Swirl the flask for 5 min in the ice bath (**exotherm**), remove the flask from the cooling bath, and let it stand at room temperature (with occasional swirling) for 20 to 25 min.

Add 15 mL water and 15 mL methanol to the reaction mixture, swirl vigorously to dislodge any crystals which form, and collect the crude product on a Buchner funnel. Wash the product with distilled water until the red color is completely removed from the crystalline mass. Following the water treatment, wash the crude crystals with several 10-mL portions of *cold* methanol to remove a yellow impurity. Continue washing with cold methanol until the methanol wash solution is colorless. The product is a faintly yellow hydrocarbon, mp 150 to 151°C. The yield is usually 60 to 70%.

The reaction product may be recrystallized from 30% toluene-methanol or from a minimum amount of cyclohexane (about 10 mL/g). In most cases this is not necessary.

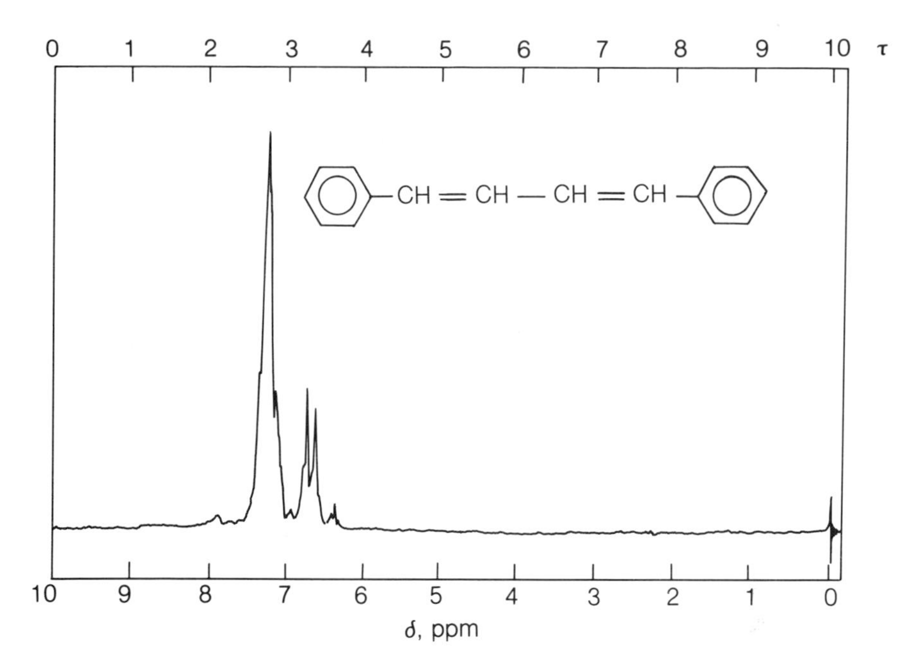

Figure 21.3
The nmr spectrum of 1,4-diphenyl-1,3-butadiene.

The nmr spectrum of the substituted butadiene is shown in Fig. 21.3. Note that those double-bond protons not adjacent to an aromatic ring are upfield of the aromatic resonances. Compare this with the situation in stilbene (Exp. 21.2).

QUESTIONS AND EXERCISES

21.1 What disadvantage(s) can you think of concerning the use of triphenylphosphine instead of triethyl phosphite?

21.2 Under certain conditions, nitrogen ylides can be formed just as is observed for phosphorus compounds. What would be the structure of the most stable ylide which could be formed from benzyltriethylammonium chloride (Sec. 13.6)?

21.3 Would sodium carbonate be a suitable base for the Wittig reaction? Why?

21.4 The nmr spectrum of triphenylmethylphosphonium iodide shows a broad and complex absorption in the aromatic region and two sharp lines upfield. The upfield resonances are separated by about 30 Hz. To what are these lines due?

QUALITATIVE ORGANIC ANALYSIS

TACTICS OF INVESTIGATION

22.1	Introduction	**486**
22.2	Preliminary Examination	**488**
	A Color	
	B Odor	
22.3	Purification	**490**
22.4	Boiling Points	**491**
22.5	Distillation	**493**
22.6	Melting Behavior	**493**
22.7	Flame Test	**494**
22.8	Beilstein Test	**494**
22.9	Specific Gravity	**496**
22.10	Refractive Index	**497**
22.11	Solubility	**498**
	A Solubility in aqueous base	
	B Solubility in aqueous acid	
	C Neutral substances	
	D Solubility in sulfuric acid	
22.12	Carrying On	**504**

22.1 INTRODUCTION

Qualitative organic analysis has played an important role in organic chemistry. For many years the only way organic compounds could effectively be characterized was by elemental analysis, solubility properties, and the transformation of the substance into a compound of known structure and melting point. The conversion of a compound of presumed structure by a known reaction into a compound whose structure and melting point were already known allowed early chemists to characterize a very large number of compounds. Nowadays, ultraviolet, infrared, nuclear magnetic resonance, and mass spectroscopy have largely supplanted traditional qualitative organic analysis in this application. Spectroscopy has become an exceedingly important tool for structure elucidation and characterization.

Nevertheless, qualitative organic analysis is still important for several reasons. For one thing, qualitative organic analysis requires the simultaneous consideration of a variety of properties for the first time in the laboratory program. Instead of having a preparation outlined, more or less in cookbook style, a compound of unknown structure is presented and the problem is to identify and characterize it. Because of this, solubility properties, acid-base behavior, and the potential functional group transformations of the substrate must all be considered. Because of the variety of these considerations, qualitative organic analysis is an important intellectual exercise. It causes one to conceptually synthesize many of the things which have been learned about organic chemistry in different chapters and at different times during the year. Furthermore, a fair amount of imagination and detective work is required for the solution of certain organic qualitative analysis problems. Any exercise which considers a broad and diverse set of facts and requires the investigator to summon forth all his or her knowledge is an important pedagogical tool. Such is the inescapable character of qualitative organic analysis.

A third important aspect of qualitative organic analysis is that it allows a student to mature considerably in laboratory technique. This is because independent work is required in the solution of a qualitative organic analysis problem and the student will probably run certain reactions which are not carried out by anyone else in the laboratory at the same time. The student is therefore responsible for doing all the innovation and development associated with the particular problem. Obviously, this leads to maturing of the scientific technique.

Another point which is often overlooked but which is extremely important is that spectroscopy is not the structural panacea it is believed by some to be. Not every compound which has been prepared is freely soluble in the solvents useful for spectroscopy. For example, if one wishes to detect the presence of a hydroxyl group and the compound is soluble only in water, ir spectroscopy may be difficult and nmr spectroscopy will likewise usually prove fruitless. Also, compounds are occasionally isolated which seem to be insoluble in everything. Sometimes these peculiar materials must be oxidized or reduced (usually the former) to obtain a derivative which can be characterized.

A point often made is that no one in the last quarter of the twentieth century should bother with qualitative organic analysis when ir and nmr spectroscopy are so readily available and prevalent. This is not strictly true because many laboratory workers avail themselves of qualitative organic analysis routinely. Although one may often learn a considerable amount about a structure by examining the spectrum of a crude material, sometimes the spectrum is so complicated that no information can be gained at all. It is often the case that carrying out a quick dinitrophenylhydrazine spot test is faster than evaporating the solution to obtain a sample which is suitable for spectroscopy. It would be foolish to argue that it is faster in all cases to do a simple qualitative organic test, just as it would be foolish to argue that it is always better to run a spectrum. Which technique is applicable and useful depends on how much time and what facilities are available. The most important point is really the nature of the sample and its state of purity. The chemist best able to cope with a structural problem is the one whose background is strong both in functional group transformation and in application of spectroscopic techniques.

Another often overlooked characteristic of qualitative organic analysis is that it teaches the student semimicro technique. Most of the preparations which are carried out in ordinary organic courses involve fairly large scales because larger manipulations are simpler for the beginning student. After most of the year's course in organic chemistry, students are qualified to work on the smaller scales which are associated with research chemistry. The small amounts of material which are available for qualitative organic analysis require the student to become proficient in this important aspect of technique.

The student who does a qualitative organic analysis of structure develops an appreciation of how changes in functional groups must be accommodated by changes in solvents, changes in procedures, changes in reagents involved in those procedures, and so on. Very many more compounds are available for qualitative organic analysis than can be used for the basic experiments presented in undergraduate laboratory manuals. As a result, students become accustomed to a much greater diversity of compounds by doing qualitative organic analysis than they would otherwise.

Regarding the selection of compounds for inclusion in this qualitative organic analysis section, a few points seem worth mentioning. In general, we have tried to include in the tables (and therefore to influence the instructor's choice of compounds) those compounds which give distinct qualitative tests and which do not decompose or undergo unusual rearrangements. Not all compounds which are sometimes given as unknowns in qualitative organic analysis courses are included. There are a variety of reasons for this. In some cases the compounds have recently been shown to pose a health hazard. In other cases the compound undergoes the kinds of unexpected reactions which would be challenging to an advanced student and therefore of relatively little value in educating the beginning student of qualitative organic analysis.

In addition, some functional groups have been avoided in this discussion. The value of qualitative organic analysis lies primarily in the exposure to a variety of functional group transformations and the identification of a compound by formation of suitable derivatives. Alkanes are called *paraffins* because this name connotes their lack of reactivity. To identify an alkane by qualitative organic analysis would be to identify it largely by a process of elimination. In so doing, a student would probably become more frustrated than educated. For variations of this particular reason we have consciously avoided alkanes, haloalkanes, alkenes, and alkynes. Several examples of alkene reactions are included in this textbook, as are certain reactions of haloalkanes. The utility of alkynes in organic synthesis is substantial, but this is primarily as nucleophiles (salts of terminal alkynes) and as cis or trans olefin precursors. Since either type of reaction would be difficult for a typical undergraduate to carry out on a small scale in a limited amount of time, we felt that it served no practical purpose to include these compounds among the functional groups to be analyzed.

One final note: What we hope to accomplish in qualitative organic analysis can be accomplished without having to resort to the use of vile-smelling thiols, sulfoxides, or sulfones, which undergo a relatively limited number of simple reactions, and without having to treat certain other functional groups whose range of simple reactivity is also relatively narrow or too complicated.

22.2 PRELIMINARY EXAMINATION

The identification of any new material necessarily begins with the recognition that there are three classes of compounds: gases, liquids, and solids. The gases are rarely encountered as unknowns in the qualitative organic analysis laboratory because they are difficult to handle. Occasionally, formaldehyde (bp −21°C) is encountered, but it is generally available as a 37% aqueous formalin solution. Formaldehyde-formalin solution poses a problem, but the physical state of most compounds can be determined by inspection. Occasionally there may be some confusion on this point. Some substances have melting points which are close to room temperature. During the winter when the laboratory temperature is 20°C, a low-melting compound will be solid, but on a fine spring afternoon the same material may be liquid. If the laboratory temperature is essentially the same as the compound's melting point, the material may appear to be a mixture of two substances.

The appearance of a liquid compound may indicate whether or not it is a homogeneous substance. The presence of visible impurities may often be obvious. The impurities may appear in the substance as alien solids, specks, or flecks, or one may observe two or more layers. Any of these observations generally implies a mixture. It is also good practice to carefully examine a solid to see if the material is homogeneous, i.e., to see if all the crystals seem to have the same color and crystal habit (the same general shape and constitution).

A Color

Once the physical state of the material has been determined, the color of the substance can be a real clue to the nature of the material. It is a good but qualitative measure of the material's purity. Most pure organic liquids are clear (i.e., they can be seen through easily) and colorless, that is to say that they have the same color as pure water (water-white). The presence of color does not necessarily imply the absence of purity. In fact, pure compounds which are yellow, orange, green, and even blue are all known. Nevertheless, very small amounts of colored impurities may change the color of a liquid quite dramatically. For example, if an unknown has been sitting in a laboratory desk for several weeks while the investigator is working on other things, it may turn from colorless to brown. Such oxidative decomposition is particularly characteristic of such aromatic amines as aniline. These materials, ghastly looking on first examination, will often distill readily and leave behind only a slight residue.

Solids often oxidize at various exposed surfaces, so there may be different kinds of crystals within the material. Often, if the material is clear and colorless or if it is white, it will be of reasonable purity. Although this can be assessed quantitatively only by other methods, usually some yellow color indicates the presence of conjugation and/or aromatic rings. Sometimes, if the solution is held up to the light, a sort of rainbow effect may be apparent. From this the presence of an aromatic ring can often be inferred. Recall again that these are all qualitative observations. Oftentimes a foul-smelling, high-boiling liquid which is brown or black is an aromatic amine (aniline) derivative. Likewise, solid aromatic amines will often have both light and very dark crystals interspersed. Sometimes the impurities dominate, and the substance has the overall appearance of bituminous coal. Several recrystallizations, however, may transform these ugly lumps into glistening white needles.

A particularly valuable observation which can be made on a liquid unknown is the presence of solid. If the solid has been deposited at the top of the solution, nature may have provided a derivative. If the unknown appears to be an aldehyde, the solid may be the corresponding acid. Aromatic aldehydes oxidize in air with particular ease. As the oxidation occurs, the amount of aldehyde is reduced, and as the volume decreases, the acid is deposited on the walls of the vessel. If an acid has been formed, it should be isolated, purified, and characterized.

B Odor

Having determined the physical state and the color of the compound, one should then note very carefully the odor of the sample. Odor is obviously determined by smelling, but there are two techniques for detecting odor without being overwhelmed. First, if the vessel is stoppered, remove the cork or plug and wave it cautiously in front of your nose. Alternatively, remove the stopper completely and wave your hand across the top of the vessel so that some of the vapors reach your nose. Smelling a compound should always be done with considerable care because it may be a noxious substance.

The information available from odor is of the most qualitative sort but sometimes can be valuable in determining the identity of a compound. It is always a good idea to bear in mind daily experience and intuition. For example, if the compound smells strongly of cinnamon and you know there is a group of related compounds, including cinnamic acid, cinnamaldehyde, and cinnamonitrile, all of which have cinnamon odors, you may be able to use this information in identifying your compound. Some classes of compounds have distinct odors but some are nondescript. Where there is no identifiable odor, no information is gained, but this observation should be recorded.

Many low-molecular-weight esters have very distinctive odors. For example, ethyl butyrate, which is used in the manufacture of artificial rum, is called pineapple oil when it is in ethyl alcohol solution. Ethyl heptanoate (also known as ethyl enanthate) is called *synthetic cognac oil, oil of grapes,* and *oleum vitis viniferae*. No matter what it is called, it has an essence with which you will likely be familiar. Methyl benzoate is used in perfumery and is called *protosponia*. Ethyl benzoate, likewise used in perfumery, is called *essence de niobe*. Methyl salicylate, which will be recognizable to you by its wintergreen odor, is called *oil of wintergreen*.

Low-molecular-weight carboxylic acids also have very distinctive odors; acetic acid smells like vinegar and would be readily identifiable as such. Butyric acid is so named because it is found in butter (Latin *butyrum*) and is a constituent of, and therefore contributes an odor to, rancid butter. Butyric acid gives butter the characteristic foul odor it has after it has spoiled.

These are more or less random examples, but will serve to indicate that much information can be drawn from everyday experience. Remember that any identification of a compound or hypothesis of structure must fit all the variables —all the possible indicators that are available to you. If you identify a compound as cinnamaldehyde and you do not return home smelling as if you had visited a doughnut factory, you have probably misidentified the compound. Also beware of very odoriferous impurities; a small amount of impurity may completely mask the true odor of the sample.

22.3 PURIFICATION

The most convenient and very often the best method available for the purification of a liquid is the process known as distillation. This technique is probably familiar from earlier experiments, and it should be easy to assemble a distillation apparatus. However, some preliminary things should be kept in mind before beginning a distillation. It is usually a good idea to have some notion of the compound's boiling point. This can be obtained either qualitatively or quantitatively. The qualitative method is simple: remove the cork or stopper from the vessel containing the material and see if the odor becomes stronger or weaker. If the material is quite volatile, the odor will often be stronger at first than if it is less volatile. Likewise, a sample may be swirled or shaken. In general, the more viscous the substance appears to be, the higher will be its boiling point. Again, be certain to smell organic compounds very cautiously.

22.4 BOILING POINTS

A useful method for determining the exact boiling point is to use a micro-boiling-point determination. Two techniques are commonly employed for this purpose. One technique is the so-called capillary method and the other is a microreflux boiling-point determination. Operationally the microreflux is the simpler of the two and will be dealt with first.

PROCEDURE

BOILING-POINT DETERMINATION

A. Microreflux method

The easiest way to do a microreflux boiling point determination is to clamp a small test tube (7.5 × 100 mm) to a ring stand, place about 0.5 mL of the liquid in it, and suspend a thermometer in the test tube with the bulb about 2 cm above the top of the liquid (Fig. 22.1*a*). Heat the sample, at first with a steam bath and, if insufficient heat is obtained from this source, then with a free flame. Determine the boiling point by watching the vapors rise and just begin to reflux, i.e., condense and fall off the bulb of the thermometer. Note that for the most accurate measurement the entire thermometer bulb should be bathed in vapor. This will give a good idea of the approximate boiling point. The exact boiling point is best determined by distillation of a sample.

B. Capillary method

A somewhat more accurate, if less convenient, method for rapidly determining the approximate boiling point is the capillary method. In the capillary technique, one takes a closed capillary (see Fig. 22.1*b*) and immerses it with the sealed end up in a test tube containing a small amount of the liquid sample. (The best results are usually obtained when the capillary is sealed about 2 cm above the bottom.) The test tube is placed in a Thiele tube or melting-point

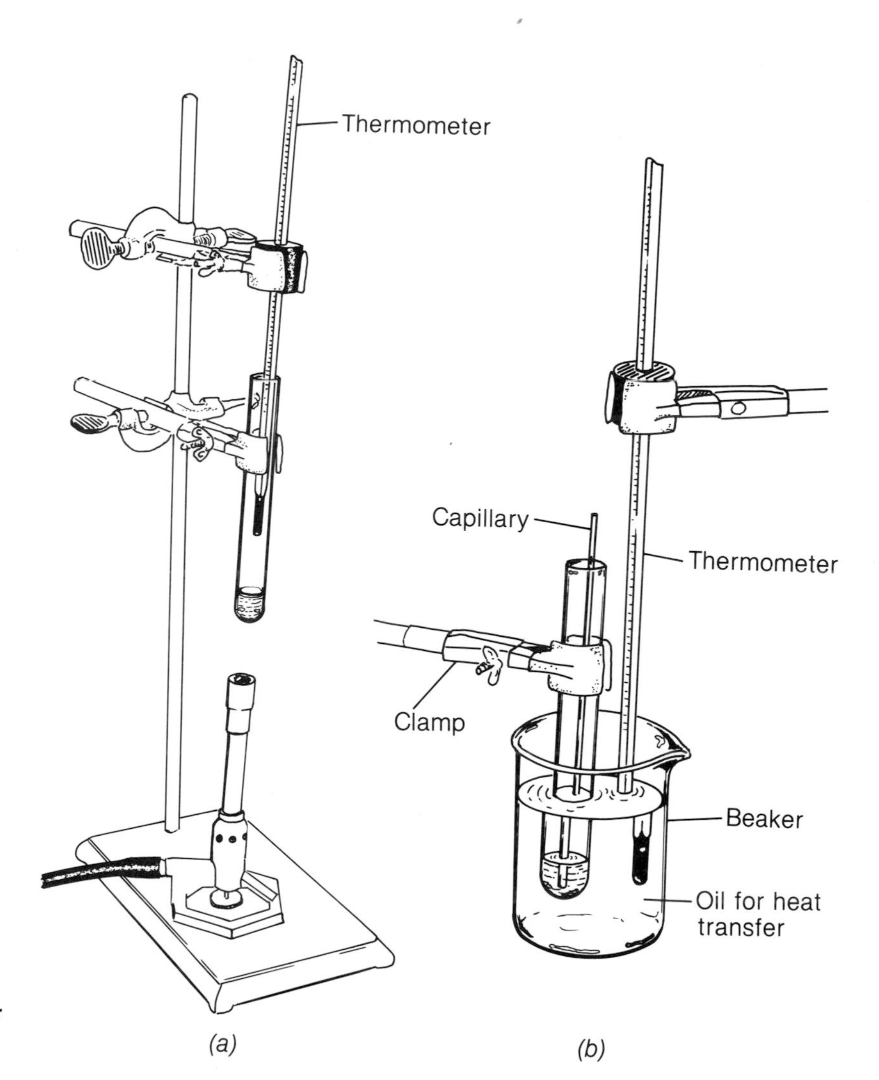

Figure 22.1 Boiling point determination. (*a*) Setup for microreflux method. (*b*) Setup for capillary method.

apparatus designed to indicate the temperature of the surrounding medium. The unknown liquid is slowly heated in the apparatus chosen. As the liquid heats, air is driven out of the sealed capillary. When the boiling point is passed, this slow bubbling stops as vapor from the unknown substance fills the capillary, and a rapid stream of bubbles can be observed exiting from the capillary. The heat is removed and gradually the surrounding medium begins to cool, and by heat transfer the unknown cools as well. At exactly the point at which the level of the liquid in the capillary is at the level of the liquid surrounding it, the vapor pressure both inside and outside the capillary equals the atmospheric pressure; this is the boiling point. The temperature showing on the thermometer is the approximate boiling point of the substance. This procedure may be repeated with the same apparatus simply by allowing the system to cool. Reheat the sample past the boiling point and then allow it to cool slowly.

22.5 DISTILLATION

Once the boiling point has been determined, all but approximately 10% of the substance should be distilled at normal atmospheric pressure in a distillation apparatus. The purpose of the micro-boiling-point procedure is to determine the approximate boiling point of the substance. From this a decision as to whether or not to distill the material on a steam bath can be made. Materials whose boiling points are below 90°C should be distilled by steam bath (or hot water bath) to avoid the possibility of fire or explosion. The materials which are less volatile may be distilled by using an electric mantle, an oil bath, a hot plate, or a free flame (check with your instructor). Materials which boil above about 230°C often decompose at these high temperatures and should be distilled in vacuo or with steam and only at the direction of the laboratory instructor. Alternatives such as aspirator or vacuum-pump distillation should be attempted only when the appropriate techniques have already been added to your experimental repertoire. In any event, be certain that the boiling point is reported under the conditions you choose for the distillation.

22.6 MELTING BEHAVIOR

If the unknown is a solid, its melting point should be determined regardless of its appearance. Let this be a piece of general advice: *Always* determine the melting point of any solid material before further manipulation is attempted. The reason for this is simple. If a substance having a melting point of 110°C is recrystallized from water and its melting point is 113°C, it seems that a slight purification has been achieved. If, on a second recrystallization, no further improvement in the melting point occurs, it can usually be assumed that the original sample was relatively pure. If the melting point decreases on recrystallization, change solvents.

Sometimes relatively peculiar things transpire. One is that the material may occlude or form a complex with a solvent and will be rendered less pure by the process of recrystallization. Perhaps the material will still exhibit a sharp melting point, but if a preliminary melting point of 110°C had not been obtained prior to recrystallization, then the hypothetical new melting point of 70 to 70.5°C might look excellent, when actually there is a 40°C discrepancy. One should not be misled by the fact that a melting point is sharp. There are a number of relatively impure compounds which melt over a 1 to 2°C range, so the experimentalist must be wary. In general, however, a melting point is a good indication of purity, although it is no ultimate criterion.

When recrystallizing the unknown substance always withhold a small amount. This is done for two reasons: the first is so that you may determine the melting point or mixture melting point of the original sample again if need be; and the second is that even the somewhat impure material may still afford excellent seed crystals for the recrystallization of the substance from the solvent.

There is an important difference between dealing with liquids and with solids. In general, 90% of the liquid will be subjected to distillation regardless of its purity. The bulk of a solid sample, on the other hand, may be left as it is if recrystallization of a small sample indicates that the original material is essentially pure. Very often an impurity so small that it is detectable only by a 1 or 2°C decline in the melting point can be ignored for the purposes of solubility classification and derivatization. If the material when obtained has a melting point which is far below the melting point of the material after recrystallization, then the entire sample should be recrystallized so that purity is assured.

22.7 FLAME TEST

There are several other fast and simple tests which can give considerable information about the compound in hand. One of these is a flame test. A flame test is usually conducted by putting a small amount of the sample on a porcelain spoon or clean metal spatula and holding it directly in a flame. The color of the flame and the presence or absence of smoke, ash, and residue should all be noted. If the compound burns with a bright blue, smokeless flame, one can generally assume that the compound is saturated and/or has a high oxygen/carbon ratio. Examples of compounds which burn with bright blue flames are ether, acetone, butyric acid, nitromethane, and propanol. These compounds generally burn with ash-free flames. If the compound burns with a yellow, sooty flame, the inference that one can draw is that the carbon/hydrogen ratio is higher than in the cases in which a blue flame is observed. For example, the carbon/hydrogen ratio is higher in benzene (i.e., there are more carbon atoms per hydrogen atom) than it is in either ethane or butane. A molecule which is highly unsaturated will usually burn with such a yellow, sooty flame. A yellow flame is cooler than a blue flame and implies that the compound is not effectively feeding its own combustion. Soot or ash in the flame (a dark residue is often left) is another manifestation of incomplete combustion. It is reasonable to infer from these observations that the compound being burned is unsaturated. Note also that a gray or white ash often implies the presence of an inorganic substance.

22.8 BEILSTEIN TEST

Another kind of flame test which can be valuable is the so-called Beilstein test. The Beilstein test consists in burning the material on a copper wire. This is done by looping the end of the copper wire and placing it directly in a bunsen burner flame for a minute or two to completely free it of any substances which could contaminate the material to be tested. The copper wire loop is then dipped into the unknown. The copper wire is used as a spatula to place some of the unknown gently into the side of the flame. A fleeting or fugitive green flame occasionally accompanies the burning of the material. This is characteristic of diacids and occasionally monoacids. A very bright, billowing, long-lasting (several seconds) green flame indicates the presence of halogen. It is extremely important to try this test on a known halogen-containing material before drawing any conclusions.

The Beilstein test is a very reliable one. Relatively few compounds afford great difficulty in interpretation. As noted above, these are often the dicarboxylic acids, and of these malonic and succinic seem, in the authors' experience, to be the worst offenders. There are confirmatory methods, however, which will allow one to test for the presence of halogens and, if a neutralization equivalent indicates that the compound is a monoacid, the likelihood that a long-lasting green flame could have resulted from other than a Beilstein-type reaction is very small. Note that the Beilstein test will indicate the presence of halogen, but not which halogen is present. In general it is a rare case in which two compounds which have the same functional group, the same melting or boiling point, and the same derivative will contain different halogens. Other factors will indicate which halogen must be present in the compound so that, although the precise identity of the halogen will not be known directly from the observation of the green flame, it can usually be inferred with confidence.

PROCEDURE

BEILSTEIN'S FLAME TEST FOR HALOGENS

A 20-cm length of copper wire is bent into a 5-mm diameter loop at one end, and the other end is looped tightly about a cork. Holding the cork, place the small loop in a bunsen burner flame and heat to glowing. A faint green coloration may initially appear during this process but should disappear quickly. The copper wire which has been burned free of contaminants is removed from the flame and allowed to cool to room temperature. The cooled copper wire is then dipped into the unknown, so that a small amount of the substance is deposited on the small loop. (See Fig. 3.1.)

The end of the copper wire with the unknown on it is then edged into the flame. The color of the flame is observed. If a vivid green coloration lasting several seconds is observed, the unknown probably contains halogen. If a normal combustion is observed and there is either no or only a very transient green flame, it is safe to assume that the compound is halogen-free. Always conduct this test twice and compare the results with those obtained on a known halogen-containing compound.

Note: This is a very sensitive test for the presence of halogen. A very small amount of halogen-containing compound will give a positive test. Once the copper wire has been flame-cleaned, it should not be touched, as there is usually enough salt on fingers to give a faintly positive test. The wire should be flame-cleaned immediately before each test to ensure that the wire is free of contaminants.

22.9 SPECIFIC GRAVITY

If the unknown is a liquid, determination of its specific gravity can be of considerable help in identifying its chemical structure. The specific gravity is the ratio of the weight of 1 cm^3 of a substance to that of 1 cm^3 of water at 4°C. (Recall that water achieves its greatest density at 4°C.) The specific gravity of a small amount of material can be determined in an approximate fashion as follows.

PROCEDURE

DETERMINATION OF SPECIFIC GRAVITY

A. Approximate method

A 1-mL volumetric flask is weighed, filled to the mark with the liquid, and again weighed. The difference between the two weights will be the weight of a known volume of liquid. The specific gravity is approximately equal to the density, which approximately equals the weight of substance per milliliter. Values obtained by this method are usually sufficiently accurate for unknown identification.

B. Precise method

Weigh a clean, dry 5-mL or 10-mL volumetric flask. Fill the flask to the mark with the desired liquid and weigh the filled flask. Empty and rinse the volumetric flask. When it is clean and dry, fill it to the mark with distilled water. Determine the specific gravity of the substance by calculating the ratio of weights of equal volumes. The specific gravity will not have been determined at 4°C, but it will still be reasonably accurate. Specific gravity values determined in this fashion generally compare favorably with literature values.

It should be obvious that the specific gravity of a solid can also be obtained, but this is a less useful piece of data because the literature values are less available, and in any event this determination is much more cumbersome for a solid than for a liquid.

22.10 REFRACTIVE INDEX

Another property which is often very useful in identifying a liquid substance is the property known as *refractive index*. This is the ratio of the speed of light through a vacuum to the speed of light through the substance. Typical values for the refractive index are 1.33 for water, 1.358 for acetone, 1.386 for propanoic acid, 1.446 for chloroform, and 1.501 for benzene. In general, the refractive index is accurate *at least* to two decimal places, and often to three or four, and may therefore be an excellent means for discriminating among several possible compounds. It is important to note that refractive index is sensitive to the presence of impurities. The device which is used for determining the refractive index has cleverly been called a *refractometer* and consists of a sodium lamp, a constant-temperature bath, and an optical piece. Its use is described below.

PROCEDURE

DETERMINATION OF THE REFRACTIVE INDEX

The operation of most refractometers is simple. Since the refractometer is a prism instrument, first check the prism surface for residue from the previous determination. As a general rule the surface should be cleaned with 95% ethanol and allowed to dry before use. (*Note:* Since the index of refraction is sensitive to small amounts of contaminants, be certain all the alcohol has evaporated). The liquid sample is then placed on the lower prism so that the entire *width* of the prism plate is covered. A dropper should be used for this, and care should be taken not to bring the end of the dropper into direct contact with the prism (scratches). The upper prism is then brought down into contact with the lower prism. The liquid should now form an unbroken layer between the two prisms. At this point the controls are manipulated to bring the light and dark fields into focus with the cross hairs, and at that point the reading is made (see appendix to this chapter for specific instructions).

As implied above, the index of refraction is temperature-dependent. For every 1°C difference in temperature between the operator's refractometer reading and that recorded in the literature (assumed to be 25°C, unless otherwise specified), a correction of 0.0004 units is made. As the temperature goes down, the refractive index goes up, and vice versa. Therefore, if one obtains a reading of 1.5263 at 25°C (written as n_D^{25} 1.5263) and the literature records the refractive index at 20°C, the correction is as follows:

$$0.0004 \times 5 = 0.0020 \qquad n_D^{20} = 1.5263 + 0.0020 = 1.5283$$

Having now characterized the compound in terms of physical constants, appearance, odor, melting- or boiling-point behavior, ignition properties, specific gravity, and refractive index, the investigator knows a good deal. These are the physical constants of the compound. The most important chemical information to obtain about a compound is its reactivity. Much can be learned about the functional groups which are present in a molecule by considering the compound's solubility properties. The solubility properties will allow a classification of the compound into one of several classes.

22.11 SOLUBILITY

For the purposes of most unknown identifications there are three solubility classes: the acids, the bases, and the neutral substances. There are six solubility classification tests that are commonly used. These classification designations are made by observing the solubility of the unknown substance in water, aqueous sodium hydroxide, aqueous sodium bicarbonate, aqueous hydrogen chloride, ether, and concentrated sulfuric acid. If the substance is soluble in water, one can infer from this the presence of one or more polar groups and/or low molecular weight. It is important, if the material is freely soluble in water, to check the solution with either pH or indicator paper. If the material is soluble in water and gives an indication that it is acidic, one can generally presume the presence of either a carboxyl function or a phenolic hydroxyl group (always be sure to check the pH of the distilled water first). If the material affords a basic solution, it is likely an amine. If the substance is water-soluble, it will also be soluble in aqueous sodium hydroxide solution, aqueous sodium bicarbonate solution, and aqueous hydrogen chloride solution. The solubility in these solutions is due to the compound's solubility in water, and no reasonable inference can be drawn as to its reactivity.

Further clarifications are possible. If the material is water-soluble, it should also be checked for ether solubility. Bifunctional molecules (such as amino acids) and low-molecular-weight organic acids (e.g., malonic acid) are insoluble in ether, although most organic compounds readily dissolve in this solvent. Note also that if a substance is insoluble in a particular solvent, the mixture should be heated gently on a steam bath to see if solution occurs.

Alas, there are always borderline cases. If the material is only slightly soluble in water, one can continue with the sodium hydroxide and hydrochloric acid tests anyway to see if the material is more soluble in acid or base than in water. If, however, the material is freely soluble in water, little reliable information can be gleaned from examining the aqueous acid and base solutions. Determining the pH of the aqueous solution will be particularly valuable in the latter case.

A Solubility in Aqueous Base

A material which is insoluble in water will be soluble in an aqueous 5 or 10% solution of sodium hydroxide only by virtue of the formation of a new substance, i.e., because of a reaction. For example, benzoic acid is not soluble in water at room temperature; it is soluble in water at elevated temperature and recrystallizes nicely from this solvent. If solid benzoic acid is shaken with an aqueous solution of sodium hydroxide, the solid gradually begins to dissolve. This is because a proton is transferred from the carboxyl group to the hydroxide ion.

$$C_6H_5COOH + NaOH \rightleftharpoons C_6H_5COO^-Na^+ + H_2O$$

This chemical reaction produces water and sodium benzoate. Sodium benzoate has an ionic bond between the carboxyl oxygen and the sodium ion and is a salt in the same sense that sodium chloride is a salt. Like most salts, it is soluble in water. The solubilization phenomenon can be described very simply as a chemical reaction between a base and an insoluble acid (benzoic acid) to give a soluble salt (sodium benzoate) and water.

Aqueous sodium bicarbonate may be used in much the same way aqueous sodium hydroxide is used. This reagent, however, is valuable because it is a weaker base than is sodium hydroxide and will react only with the stronger acids to release CO_2 bubbles from bicarbonate ion. The most common organic acids are the carboxylic acids, which generally have pK_a values in the range of 1 to 7, and phenols, whose pK_a values are generally higher (often around 10).

The pK_a of a compound depends very much on the substituents present. For example, as shown in the illustration below, phenols can have a broad range of pK_a values, from 10 or above down to nearly 0. The examples that are illustrated below are all phenols. Phenol itself has a pK_a of 9.89. 4-Nitrophenol, with an electron-withdrawing substituent on the benzene ring, has a pK_a of 7.15. 2,4-Dichlorophenol has two electronegative substituents and its pK_a is 7.44. 2,4-Dinitrophenol has a pK_a of 3.96, and 2,4,6-trinitrophenol has a pK_a of 0.38. The latter substance is so strongly acidic that its trivial name is picric acid.

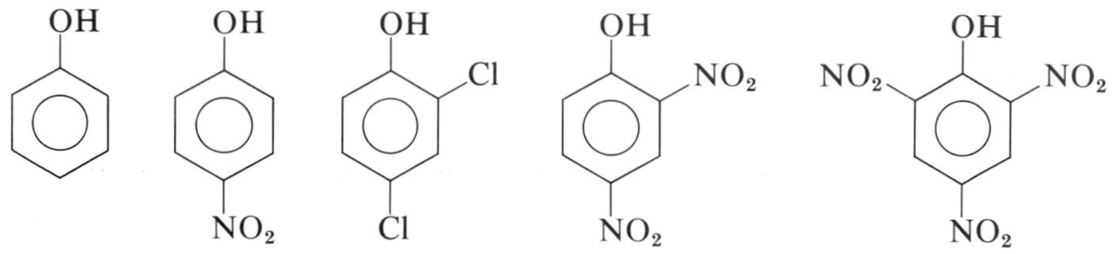

OH	OH, NO_2	OH, Cl, Cl	OH, NO_2, NO_2	OH, NO_2, NO_2, NO_2
9.89	7.15	7.44	3.96	0.38

In considering acidic compounds, one must be aware of the possibility that the pK_a of a substance can be reduced dramatically by electronegative substitution. Two of those which are illustrated have pK_a's which are lower than that of acetic acid. For calibration, note that acetic acid has a pK_a of 4.75 and benzoic acid has a pK_a of 4.19. In general, phenols will be only weakly acidic and will transfer a proton to sodium hydroxide, but they will not react with bicarbonate ion to release CO_2 even though they may be soluble in the basic solution. This is a simple but important means for distinguishing strong from weak acids.

In all this it is important to remember that the solubility of these substances results from chemical reactions which alter the nature of the substance. In other words, 4-nitrophenol, which is only sparingly soluble in water, can be coaxed into a basic solution because it ionizes to give a proton and a phenoxide salt. The salt is then well solvated by water and readily dissolves. This is the same principle, of course, which is illustrated above for the carboxylic acids. The reaction of phenol with sodium hydroxide is illustrated below.

$$NaOH + C_6H_5OH \longrightarrow H_2O + Na^{+-}OC_6H_5$$

$$NaHCO_3 + C_6H_5OH \longrightarrow \text{no reaction}$$

It is always an alteration in the nature of a compound which alters its solubility properties. Solubility is an intrinsic property of any substance, and only a change in its chemical nature can lead to high solubility where only slight solubility was observed before.

B Solubility in Aqueous Acid

A substance which is insoluble in water and which is insoluble in aqueous base solution should be tested next to see if it is soluble in dilute acid. Ordinarily the dilute acid which is used for this examination is 5 to 10% aqueous hydrochloric acid.

One must consider what kinds of organic substances would be anticipated to have reasonable solubility in acid. Any compound which can have a proton transferred to it, i.e., which can be protonated, will be soluble in water if the proton is stable in its new environment.

There are a number of substances which have lone pairs of electrons but whose protonated forms are not stable in aqueous acidic solution. Ethers, for example, usually have pK_a values of -4 or -5. In aqueous solution, a proton would not be stable on an ether oxygen because a solution of a compound more acidic than the proton source is energetically unfavorable.

The most common example of compounds which can readily accept a proton and form water-soluble salts is provided by the amines. Amines fall into three classes, primary, secondary, and tertiary. In each of these classes there is a lone pair of electrons on nitrogen which can accept a porton from the acidic medium. This proton transfer reaction leads to a water-soluble ammonium salt. The equation formulated below illustrates this process.

$$R_3N + HX \longrightarrow R_3\overset{+}{N}H + X^-$$

Most primary, secondary, and tertiary amines react with acids to form ammonium salts. The principal difference among these three classes is that the number of alkyl or aryl groups varies, and as a consequence the lipophilicity or hydrophilicity [see Chap. 4 (Solubility and Reactivity)] of the amine salt may vary from example to example as well. In general, the salts of primary amines will be more soluble than the salts of tertiary amines, although this generalization is not a hard and fast rule. Protonation of an amine which has 10 or more carbon atoms per amine function will often yield a water-insoluble salt. This is so because there is too much hydrocarbon "skin" or organic substitution about the salt. The single charge in this case is not sufficient for the purposes of polarization and solubilization in a polar medium like water to allow these potential solutes to dissolve. In general, there must be one charge or functional group for each five or six carbon atoms for a substance to exhibit water solubility. In short, some amines are insoluble in aqueous acid despite the fact that they can accept a proton. Two notorious examples of this difficulty are α-phenylethylamine and *N*-benzylaniline (illustrated below). Note that the phenylethylamine poses a problem despite the presence of only eight carbon atoms.

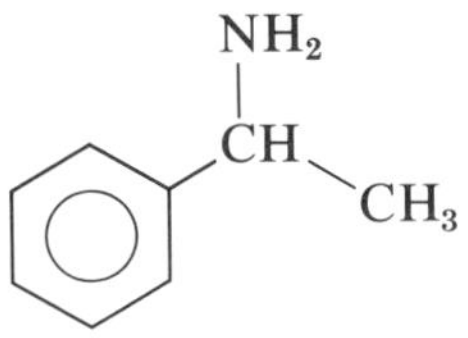

$C_6H_5CH(NH_2)CH_3$ $\qquad$ $C_6H_5CH_2NHC_6H_5$

α-Phenylethylamine 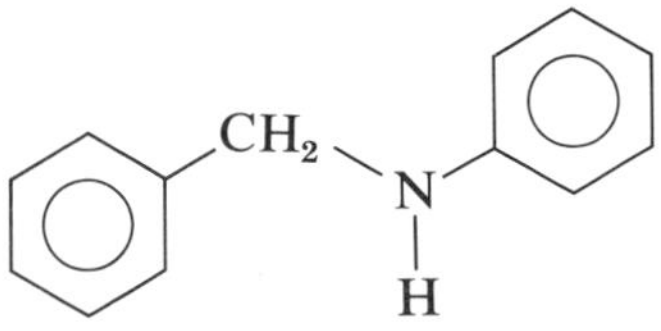 $\qquad$ *N*-Benzylaniline

A difficulty which is often overlooked is that the ability of the lone-pair electrons on nitrogen to be protonated is affected by electronic and steric interactions. The classic example of this is diphenylamine. This amine does not afford reliable results in the normal acid-solubility test because the two phenyl groups are electron-withdrawing both by induction and by resonance. Delocalization of the lone pair reduces the basicity of nitrogen a great deal. The pK_a of the protonated amine is very low; indeed, the protonated substance qualifies as a reasonably strong acid. Protonation does not occur readily and, considering the number of carbon atoms present, the salt probably would not be soluble anyway.

C Neutral Substances

From the discussion above it should be clear that just because a substance is insoluble in aqueous hydrochloric acid, aqueous sodium bicarbonate, or aqueous sodium hydroxide, it is not necessarily a neutral compound. In general, however, those compounds which are soluble in neither aqueous acid nor aqueous base fall into the neutral category. Included in the group of neutral compounds are alcohols, aldehydes, amides, esters, ethers, hydrocarbons, ketones, nitriles, and ureas. If the compound is a neutral substance by the operational criteria offered above and not a compound exhibiting anomalous solubility behavior, there are several approaches one can take. A general approach is given in the next section, and more detailed information can be found in later chapters.

D Solubility in Sulfuric Acid

One final note: The solubility of some substances in concentrated sulfuric acid serves as a further indication of the presence of functional groups. If a substance shows no appreciable solubility in water, dilute acid, or base but does show solubility in organic substances, one can sometimes learn whether one has a pure alkane or has rather an olefin or some other very weak base. In general, alkanes will be insoluble in concentrated sulfuric acid; olefins will be soluble because of reaction; aromatic hydrocarbons will be soluble by a combination of protonation and sulfonation; and certain amines which were not readily soluble in dilute acid will be soluble in concentrated sulfuric acid. Occasionally otherwise insoluble amines give a voluminous white precipitate, which is usually the bisulfate salt, directly from the medium. Sometimes one also will see a black color developing and, although no direct information can be inferred from this, it can often provide confirmation of a structure after the compound has been identified by other means.

PROCEDURE

DETERMINATION OF SOLUBILITY IN 5% AQUEOUS BASE

In order to determine the solubility of a substance in base, start with a clean 10 × 75 mm test tube or some equivalent (glass vials are also acceptable). Put about 0.1 g of the substance into the test tube. If the substance is a solid, 100 mg can be weighed. Often 50 mg is a small lump about 3 to 4 mm across that just about covers, in a circular fashion, the tip of the average spatula. (We offer here a semiofficial definition: An average spatula blade is one which will fit inside a 10 × 75 mm test tube.) If the material to be examined is liquid, then it can be added by dropper. (Usually there are somewhere between 10 and 20 drops/mL. A milliliter corresponds to a gram, so 0.1 g is about 0.1 mL, which is somewhere between 1 and 2 drops.) On top of the substance, whether added as a liquid or a solid, is added about 1 mL of the solubilizing solution. This is not crucial: it may be 1 mL or it may be 2 mL. In fact, if solubilization is either slow or nonexistent, add a little more of the solubilizing solution for confirmation. Once covered with this 1 to 2 mL of solution, the substance should be vigorously stirred or shaken. In the authors' opinion, the best way to perform this operation is to insert a flat-bladed spatula and spin it in the tube vigorously. This allows for agitation and it also allows for a certain amount of grinding, which is advantageous if the material is a solid and also keeps the solution from splashing out.

A technique we have seen applied quite a number of times is to stopper a test tube with the thumb and shake well. *We do not recommend this technique*, particularly if you are trying to dissolve something in sulfuric acid. If the material seems to dissolve slowly, give it some time. Sometimes gentle heating on the steam bath can be very helpful. The material may be slow to dissolve and need heating, stirring, and a little patience. Set it aside for a short time and see if solution does occur. Often, the higher the melting point, the more slowly substances dissolve. This is because the higher lattice energy associated with higher melting points requires more solvent interaction to break down the solute.

The procedure described above for testing the solubility of an acid in dilute aqueous base can be carried out with 5 to 10% aqueous sodium hydroxide or sodium bicarbonate. For determining whether a substance is base-soluble, we recommend first treating it with 5% aqueous sodium hydroxide solution. If the material is soluble, treat it with 5% aqueous sodium bicarbonate solution. Solubility in the latter medium will indicate the presence of a strong acid (usually a carboxylic acid), whereas solubility in hydroxide but not in bicarbonate will usually indicate the presence of a phenol.

PROCEDURE

DETERMINATION OF SOLUBILITY IN ACID

A. Solubility in 5% aqueous hydrochloric acid

In order to determine if the substance is soluble in aqueous HCl, first determine if the material is soluble in water. If it is water-insoluble, use the procedure described above for determining solubility in aqueous base, but use 5% aqueous HCl as the test medium. Again, be sure to stir the solvent well with the unknown, and if necessary warm gently on the steam bath. If solution fails to occur and it appears that vigorous shaking might be helpful, stopper the test tube with a cork or rubber stopper and then shake the tube in such a way that if the cork comes off, no one will be sprayed by the solution.

B. Solubility in concentrated sulfuric acid

This procedure is exactly the same as that described above, except that concentrated (96 to 98%) H_2SO_4 rather than 5% HCl is used.

22.12 CARRYING ON

Once the solubility class of the material is known, other tests must be considered to ascertain the sample's exact structure. The rest of this section is intended to serve as a general guide to solubility classification. In the chapters which deal with acids, amines, alcohols, phenols, etc., specific directions are given for the classification and derivatization of most compounds.

If a substance is soluble in both aqueous sodium hydroxide and sodium bicarbonate, it is usually safe to presume that the compound is a carboxylic acid. Knowing this and the boiling or melting point, consider the various possible carboxylic acids to see if one among them has a property which obviously distinguishes it from the other possibilities. If the compound is bicarbonate-soluble, melts at 133°C, and has the odor of cinnamon, it might be cinnamic acid. Considering the odor, it is more likely to be cinnamic acid than acetylsalicylic acid (aspirin), whose melting point (135°C) is similar.

If the compound is believed to be an acid, a neutralization equivalent is often definitive. This test involves titrating a known weight of acid with a base of known normality. When the neutralization point is reached, the acidic protons have all been transferred to the base, and all the base has been converted to a salt. The amount of base used corresponds directly to the number of acidic protons per gram of sample and therefore to the equivalent weight. The equivalent weight of a compound, when combined with such obvious physical properties as the boiling and melting points, can often allow an unequivocal determination of the structure. At the very least, the neutralization equivalent will narrow the range of possibilities. Basic titration will also work for phenols (of course the more acidic the phenol, the more accurate this technique will be) and can be applied to any substance which has a dissociable proton.

If the substance is soluble in dilute aqueous hydrochloric acid, the most reliable classification test to attempt is the Hinsberg test. This test helps to determine whether an amine is primary, secondary, or tertiary on the basis of its reactions with benzenesulfonyl chloride. The Hinsberg test is especially useful because, if the amine is either primary or secondary, the sulfonamide which forms in this test may be isolated and used as a derivative.

Solubility Classification Chart

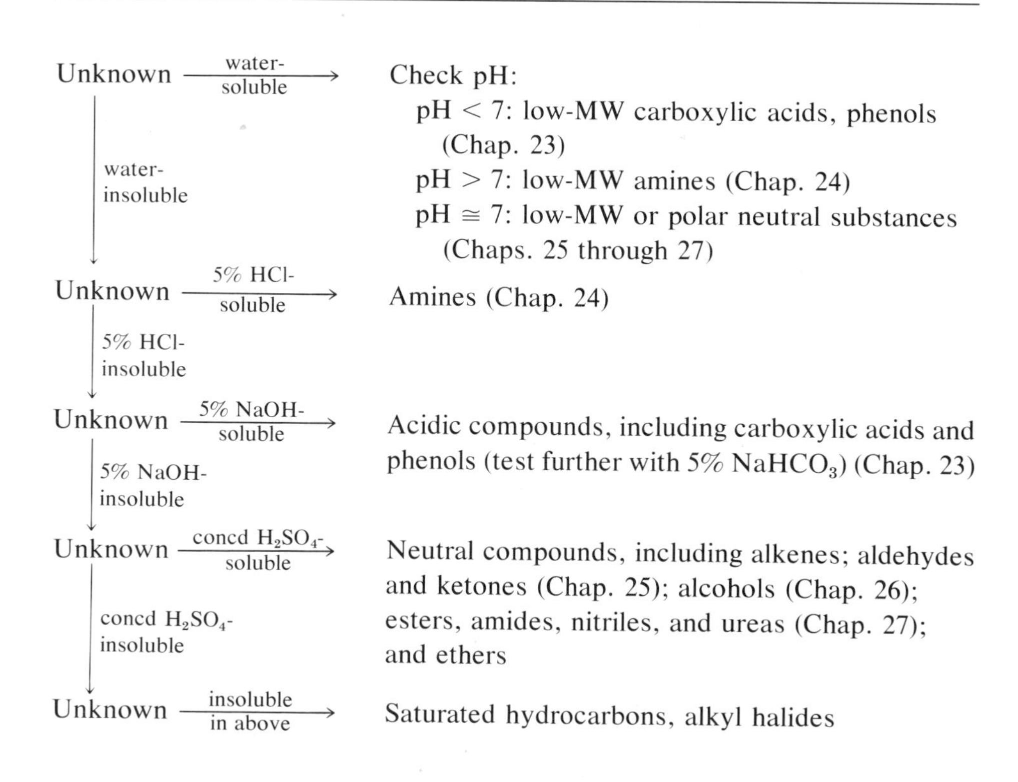

If no indication of acidic or basic properties has been gleaned from preliminary examination, it is best to take a small amount of the material and subject it to the dinitrophenylhydrazine test. This is a fast and reliable way to check for the presence of a carbonyl function (aldehyde or ketone). If this test is negative, i.e., if no orange precipitate forms, most reactive ketones and aldehydes can be ruled out.

Basic hydrolysis is a good way to characterize an ester which exhibits neutral solubility behavior. The Lucas test, which utilizes zinc chloride and hydrochloric acid, is a good test to distinguish among primary, secondary, and tertiary alcohols. Nitro groups can be detected quite readily by reduction techniques, although these tests are not reliable enough to make them a quick means of determining the presence of a functional group.

The information above is summarized in the Solubility Classification Chart. Use it only as a guide. When you are more informed about the nature of your unknown, refer to the appropriate chapter(s).

APPENDIX: SPECIFIC INSTRUCTIONS FOR INDEX OF REFRACTION

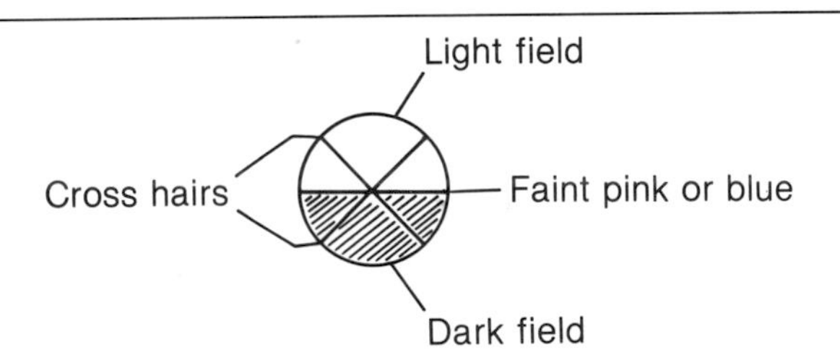

Control for Taking Reading

1 To the right of the refractometer are found coarse and fine adjustment controls for movement of the prisms. These are adjusted so that a light and a dark semicircle appear in the viewer (see above). When these two fields meet at the cross hairs, any necessary corrections involving color aberration are then made.

2 The aberration corrector is found facing the operator and is covered by a clear plate held to the refractometer by two magnets. This is manipulated while looking through the viewer until a faint pink or blue line appears between the light and dark fields just at the cross hairs.

3 A reading is taken by depressing the switch found on the left side of the refractometer. The reading is made to four places beyond the decimal point, and the temperature is noted.

Solvents for Cleaning

The usual solvents for cleaning the refractometer prisms are ethanol, methanol, and toluene, followed by hexane (or petroleum ether) as needed. Under *no* circumstances is *acetone* to be used, as this will dissolve the adhesive holding the prisms in place. Cotton or a soft tissue (Kimwipes) should be used to remove the excess unknown sample from the prisms before cleaning with the above solvents.

CARBOXYLIC ACIDS AND PHENOLS

23.1 Introduction **509**
23.2 Historical **509**
23.3 Traditional acids **510**
- **A** Definitions
- **B** Dissociation constants and pK_a

23.4 Operational Distinctions **511**
- **A** Solubility of acids
- **B** Distinguishing acids from phenols

23.5 Typical Acids **512**
23.6 Derivatization and Reactivity **513**
- **A** Neutralization equivalents
- **B** Nucleophilic addition to the carbonyl group
  - **1** Formation of acid chlorides
  - **2** Amides from acid chlorides
  - **3** Methyl and ethyl esters of carboxylic acids
- **C** Derivatives obtained by alkylation
  - Phenyacyl ester formation

23.7 Phenols: The Other Acidic Class **525**
- **A** Classification of phenols
- **B** Characterization and derivatization of phenols
  - **1** Aryloxyacetic acid derivatives
  - **2** Neutralization equivalent of the derivative
  - **3** Bromination of phenols
  - **4** Benzoate ester derivatives
  - **5** Urethane derivatives of phenols

23.8 Spectroscopic Confirmation of Structure **530**

23.1 INTRODUCTION

The acids are the first major class of compounds which can be distinguished by solubility behavior using the qualitative analysis approach. These substances are distinguished from other compounds by their solubility properties in two basic solutions, sodium bicarbonate and sodium hydroxide. Solubility properties will be dealt with in detail later in this chapter, but before going on let us consider what an acid is and some of its history.

23.2 HISTORICAL

Gay-Lussac was one of the earliest chemists to recognize that acids and bases are related in a complementary way: "They are related interchangeably to one another." Lavoisier recognized acid properties but felt that the culprit was oxygen. Oxygen, he said, is an "acidifying principle" because it is a common feature in nitric, sulfuric, and carbonic acids, i.e., it is the action of oxygen on the elemental substances nitrogen, sulfur, and carbon that produces acids. His observation that oxygen is involved in all these substances was correct, but his presumption that oxygen is a fundamental feature of all acids was not.

Sir Humphry Davy, who worked in England in the early 1800s, tested many of Lavoisier's postulates and ideas. Davy was apparently influenced to become a chemist after reading Lavoisier's work, *Traité Élémentaire*. It is ironic that Davy disputed many of the principles set forth therein after working in his own laboratory and recording his own observations. At the end of the nineteenth century Ostwald and Arrhenius associated acidity specifically with protons and basicity specifically with hydroxide ions. The Arrhenius definition was a very important one because it allowed for the recognition of acidic properties in substances by virtue of structural theory rather than simple empiricism.

The great Danish chemist Johannes Brönsted defined acidity in broader terms when he said that an acid is a substance which is a proton donor. His concept of a base, defined as any substance which is a proton acceptor, was somewhat more general than that postulated by Arrhenius. Built around the proton, the Brönsted acid-base theory has, with few modifications, held up very well over the years. The principal alternative to the Brönsted definition of acids is that offered in 1923 by Gilbert Lewis, who defined acidity in terms of electron-pair acceptors and basicity in terms of electron-pair donors. Of course, a proton is the simplest example of an electron-pair acceptor, and in this sense Lewis acidity is similar to Brönsted acidity. On the other hand, the Lewis definition of an acid is considerably broader than the Brönsted concept because any electron-deficient species which can form even a transient complex with a base, a nucleophile, or a lone-pair donor can be considered a Lewis acid. In fact, such substances as BF_3, aluminum chloride, and PCl_5 all function as catalysts in reactions in which protons may also be catalysts. This course will deal primarily with protonic acids because most of the general Lewis acidic substances also have other characteristic reactivities which make them fairly difficult to handle. Acetic acid dissolves in water to give a solution containing acetic acid, protons, and acetate ions. When aluminum chloride dissolves in water it produces aluminum hydroxide and hydrogen chloride because of hydrolysis (reaction with the solvent water). Therefore, aluminum chloride is a Lewis acid but, as it does not produce a hydrogen ion upon dissolution in water until after it is itself hydrolyzed, it is not a Brönsted acid.

23.3 TRADITIONAL ACIDS

A Definitions

The formality of definition is a very important starting point for any subject. Beyond the definition of what is or may be an acid, the range of commonly encountered acidities must be considered. Many substances can be considered acids only within the very broadest concept of acidity. Compounds whose dissociation constants are many powers of ten smaller than water are not, for all practical purposes, considered acids.

B Dissociation Constants and pK_a

Acidity can be defined in quantitative terms: A generalized acid HA dissociates reversibly to give H^+ and an anion A^-, as shown below.

$$HA \overset{K_a}{\rightleftharpoons} H^+ + A^-$$

This dissociation process is most often observed in aqueous solution. The aquated acid should be shown dissociating, not to simple protons but to hydronium ion, H_3O^+, or to some aggregate of hydronium ion like $H_9O_4^+$, and an anion A^-. (Note that A^- should also be shown as an aggregate.) Since water is pervasive, however, its role is not considered in detail. Moreover, since the mole ratio of water to acid will often be 100 or more, and since the dissociation constant is relatively small, the change in water concentration is inconsequential compared with other considerations. Since the concentration of water is effectively constant, it is neglected in all but the most sophisticated treatments.

Ignoring water and rearranging, the dissociation constant may be expressed as follows:

$$K_a = \frac{[H^+][A]}{[HA]}$$

The equilibrium constant K_a is the dissociation constant of HA. For organic acids K_a is usually small. For carboxylic acids K_a is ordinarily in the range of 10^{-4} to 10^{-6}, K_a for phenol is 10^{-10}, and many other substances are even less acidic.

It would be cumbersome to always express small dissociation constants, as for example 5.2×10^{-6}, so a familiar shorthand is used. The constant is converted to its common logarithm [$\log_{10}(5.2 \times 10^{-6}) = -5.284$]. Although this number is more convenient, the fact that it is a negative number makes it unattractive. The same trick used to make discussion of hydrogen ion concentrations more convenient is applied. In analogy with our use of the pH, we use the pK_a to express the dissociation constant in a convenient form. An acid whose dissociation constant is 5.2×10^{-6} has a pK_a, after rounding off, of 5.3. This method makes it possible to compare the acidities of compounds by using convenient numerical values. Keep in mind, however, that the larger the number, the weaker the acid.

The aqueous pH range is 1 to 14. The substances which are acidic in this range include carboxylic acids, whose pK_a's are about 4 to 6, and the phenols, whose pK_a's are about 10. Ammonium salts have pK_a's in the range of 5 to about 12; β-dicarbonyl compounds, nitro compounds, and cyano compounds all have pK_a values in the range of 10 to 20. Many substances can be considered acidic within a certain frame of reference, but do not dissociate enough in water at room temperature to turn pH paper red. Common examples include water, whose pK_a is 15.7; alcohols (e.g., methanol, ethanol, and propanol), whose pK_a's are 16 to 20; methyl ketones (like acetone), whose pK_a's range from 20 to 25; and certain nitroalkanes, hydrocarbons, and heterocyclic compounds.

23.4 OPERATIONAL DISTINCTIONS

A Solubility of Acids

Qualitative analysis is concerned with the discrimination among compounds on the basis of simple observations or straightforward chemical tests. The carboxylic acids, for example, are readily discernible because they will quantitatively transfer a proton to hydroxide ion in 10% aqueous sodium hydroxide solution. An acid which is insoluble in water will often dissolve when it transfers a proton to hydroxide because the salt is water-soluble. The same is generally true of the less acidic phenols.

B Distinguishing Acids from Phenols

How can we distinguish between carboxylic acids and phenols if both are acidic? Is there, in fact, a way to distinguish so subtle a difference as 4 to 5 pK_a units? If a relatively weak base such as sodium bicarbonate ($NaHCO_3$) is utilized, it will react with a carboxylic acid (see below), but not with a weakly acidic phenol.

$$R\text{—}\overset{\overset{\displaystyle O}{\|}}{C}\text{—}OH + HO\text{—}\overset{\overset{\displaystyle O}{\|}}{C}\text{—}O^-Na^+ \longrightarrow R\text{—}\overset{\overset{\displaystyle O}{\|}}{C}\text{—}O^-Na^+ + H_2CO_3$$

$$H_2CO_3 \rightleftharpoons CO_2 + H_2O$$

A carboxylic acid will transfer a proton to the bicarbonate anion to generate carbonic acid, which is in equilibrium with CO_2 and water. The sodium carboxylate salt, which is soluble in aqueous solution, results. The solution effervesces because, in the presence of a strong acid, the bicarbonate ion releases CO_2. This process is commonly referred to as the decarboxylation of bicarbonate. Most phenols are too weakly acidic to decarboxylate bicarbonate ion (i.e., their anions are not as stable as the bicarbonate anion), so the proton tends to reside on the phenol which remains insoluble in bicarbonate solution.

In this test there are two visual indications of reaction, effervescence and the dissolution of a previously insoluble substance. Certain electronegatively substituted phenols [see Chap. 22 (Tactics of Investigation)] have relatively low pK_a's and will decarboxylate bicarbonate ion. Somewhat later in this chapter we will deal with the mechanisms by which phenols may be distinguished from carboxylic acids.

Classification Scheme for Carboxylic Acids and Phenols

Solubility	If water-soluble, pH $<$ 7 implies an acid or phenol If water-insoluble: Solubility in 5% NaOH implies carboxylic acid or phenol Solubility in 5% $NaHCO_3$ implies carboxylic acid or *very* acidic phenol
Classification tests	Ferric chloride test for phenols Neutralization equivalent for carboxylic acids Neutralization equivalent for aryloxyacetic acid derivative of a phenol
Derivatives of carboxylic acids	Amides, anilides and toluidides, phenacyl and *p*-bromophenacyl esters
Derivatives of phenols	Aryloxyacetic acids, bromophenols, benzoate esters, urethanes

23.5 TYPICAL ACIDS

Clues regarding the structures of unknown compounds come from a variety of sources. Even common names can provide useful information.

The English word *acid* arises from the Latin *acidus,* meaning "sharp" or "sour." In the German language the word for acid is *Säure,* related to the adjective *sauer,* which means "sour." From the origins of these words we can readily anticipate that acidic substances probably have sharp odors and sour tastes, facts readily confirmed by everyday experience with vinegar.

Further clues can be gained from the origins of the names of the common carboxylic acids. Sometimes a consideration of the name will provide a clue to the properties of the substance and assist in its identification.

In the nomenclature of organic chemistry, despite the best efforts of the International Union of Pure and Applied Chemistry (IUPAC), common names frequently prevail over systematic names. For example, HCO_2H is rarely called methanoic acid but is known rather as formic acid. The word formic derives from the Latin *formica*, meaning "ant." Formic acid is so named because it was first isolated by the destructive distillation of ants. As you might expect, it is one of the principal irritants in an ant or bee sting and it should be obvious why the old folk remedy of putting baking soda on a bee sting works. Acetic acid (ethanoic acid) is so called because of its aqueous solution (vinegar), for which the Latin word is *acetum*. There is a slight change when we come to the third compound in the aliphatic carboxylic acid series. The three-carbon acid (systematically propanoic acid) is commonly called propionic acid, a word derived not from the Latin but from the Greek. The Greek words *protos* meaning "first" and *pion* meaning "fat" have been combined into propionic meaning "first fat." Propionic acid is of course the simplest of the fatty acids. This is in a sense a misnomer, because the fatty acids arise from a biochemical buildup sequence involving two-carbon fragments and usually have even numbers of carbon atoms. While this is interesting, it is not of very much help in identifying a substance because its name does not connote any recognizable property. Fortunately, the fourth member of the series does. Butanoic acid is commonly called butyric acid from the Latin word *butyrum*, which means "butter." Butyric acid contributes to the characteristic odor of rancid butter.

Although many other acids have names which provide clues to their structures or to their properties, many do not. Benzoic acid (C_6H_5—COOH) is so named because it was isolated in sequence from compounds arising from gum benzoin. Cinnamic acid (C_6H_5—CH=CH—COOH), the vinylog of benzoic acid, can be identified by its faint odor of cinnamon.

23.6 DERIVATIZATION AND REACTIVITY

In qualitative organic analysis we try to understand chemistry by examining the reactivities of classes of compounds. This is accomplished by the formation of derivatives which are unique to each of the functional groups being identified. The most important criterion for selection of a derivative is the crystallinity of the product. In choosing a derivative, check the tables to be sure your possibilities yield distinguishable crystalline derivatives. Acids and many of their derivatives will be dealt with together, because all are related through the carbonyl group. The acidic property of a carboxylic acid is the principal focus, but it should be recognized that the carboxylate group is bifunctional, i.e., it has two potential electrophilic sites. The proton is one site; the electrophilic carbonyl group is the other. When derivatizing a carboxylic acid, it is sometimes difficult to discriminate between these two positions.

Let us focus first on the acidity of the carboxyl group, i.e., on its ability to donate a proton to a basic molecule. The most common neutralization reaction involves formation of an acid salt by the action of sodium hydroxide, as shown in the equation below.

$$R{-}C(=O){-}OH + HO^- \longrightarrow R{-}C(=O){-}O^- + H_2O$$

Since only one proton is lost per carboxyl group, the number of carboxyl groups in a molecule may be determined if the molecular weight of the compound is known. Likewise, the equivalent weight of a compound may be determined by titration [see Sec. 23.6A (Neutralization Equivalents)].

In addition, an amine may serve as a base for proton transfer from a carboxyl group. The reaction of a carboxylic acid and an amine yields an ammonium carboxylate salt as shown in the equation below.

$$R{-}C(=O){-}OH + :NR'_3 \longrightarrow R{-}C(=O){-}O^- \quad H\overset{+}{N}R'_3$$

The ammonium carboxylate salt may be a sharp-melting solid and may serve as both a derivative and means of characterizing the compound.

A Neutralization Equivalents

The reaction of a carboxylic acid with base may be utilized to determine the equivalent weight of the compound. The amount of base required to neutralize a given weight of a carboxylic acid depends on two things, the molecular weight of the acid and the number of carboxyl groups present. For example, 1 mol hydroxide is required to neutralize 60 g acetic acid because acetic acid (MW 60) has a single carboxyl group.

On the other hand, succinic acid (MW 118) has two carboxyl groups. In order to neutralize the two acidic protons, 2 mol hydroxide would be required. As only 1 mol base is required per carboxyl group, succinic acid, which has approximately twice the molecular weight of acetic acid, will require (within experimental error) the same amount of base to neutralize the same weight of acid. As a consequence of this equivalence, the numerical value which results from such titrations is called a *neutralization equivalent*.

$$CH_3COOH + \overset{+}{Na}OH^- \longrightarrow CH_3COO^-Na^+ + H_2O$$

$$HOOCCH_2CH_2COOH + 2\ \overset{+}{Na}OH^- \longrightarrow Na^{+-}OOCCH_2CH_2COO^{-+}Na + 2\ H_2O$$

In the neutralization procedure, a standard amount of the unknown acid (usually 200 to 500 mg) is carefully weighed, dissolved in a suitable solvent, and titrated with standard base; then the ratio of the molecular weight to the number of carboxyl groups (the neutralization equivalent) can be calculated. For acetic acid, the neutralization equivalent is 60 and since there is one carboxyl group, the molecular weight is also 60. The neutralization equivalent for succinic acid is also approximately 60 (actually 59), but because two carboxyl groups are present, the molecular weight is determined by multiplying the neutralization equivalent by 2.

The question which now arises is: If more than one compound can have the same neutralization equivalent, how is it possible to discriminate among the various possibilities? All the available information should be considered. A compound with a neutralization equivalent of 60 and a distinct odor of vinegar will almost certainly be acetic acid. Dicarboxylic and tricarboxylic acids, because of their polar functional groups, are relatively high-melting. Therefore if a compound has a melting point of 100, 150, or 200°C and a relatively low neutralization equivalent, it probably has more than one carboxyl group.

An important point about the determination of a neutralization equivalent is that it involves a critical weight and a critical volume. The precise weight of acid used must be known, although the amount started with is not crucial. If a 200- to 500-mg sample is suggested, it makes no difference except in relative terms whether 200 mg, 500 mg, or some amount in between is used, but the exact amount must be known because it is used in the calculation. It is also essential to use the purest and driest sample of carboxylic acid available. When titrating, be certain not to pass the endpoint, and measure carefully the amount of base that is used. The normality and volume of the solution must be known exactly.

The approximate amount of acid required for a neutralization equivalent will be determined by two things, the ratio of carboxyl groups to molecular weight and the solubility of the acid in the specified medium. If the neutralization equivalent is being determined in a 125-mL Erlenmeyer flask, you obviously cannot use so much carboxylic acid that it will take 200 mL of base to neutralize it. The exact amount of base required probably cannot be determined in advance, but you should certainly be able to make an educated guess from among the various possibilities.

In general, the easiest way to carry out a neutralization equivalent is to dissolve the carboxylic acid in water, use a few drops of phenolphthalein as indicator, and titrate with 0.1 *N* aqueous sodium hydroxide. The carboxylic acid must be reasonably soluble in water if this technique is to be successful. As the neutralization procedure is carried out, the salt of the acid will be formed and, as it forms, will dissolve in water. Even if a small amount of residual acid remains undissolved at the beginning of the procedure, it can often be titrated anyway. This expedient almost always results in a sacrifice in the accuracy of the procedure. It is usually better to use a homogeneous solution of acid in the titration if possible. This may sometimes require the use of rather substantial amounts of ethanol. The pH at which the indicator changes color may vary to some extent with the solvent in which it is used. Thus, bromthymol blue should be used as an indicator if a great deal of alcohol has been used to dissolve the carboxylic acid.

When an indicator other than phenolphthalein is required, add a small amount to an alcohol solution and add some base. From this quick test, you will know exactly what color transition to expect. While you are probably experienced with phenolphthalein from introductory and analytical chemistry, such indicators as methyl orange, methyl red, and bromcresol green may be relatively unfamiliar. In this context, the use of multiple determinations (at least two) cannot be stressed enough. *At least two* and often three runs should be made to ensure that you know the precision of your data.

PROCEDURE

NEUTRALIZATION EQUIVALENT OF ACIDS

Place about 50 mg or 1 drop of the acid in a small test tube. Add a few drops of water to see if the acid is soluble. If so, accurately weigh out about 0.5 g of the unknown acid and place it in a 125-mL Erlenmeyer flask. Add about 25 mL water and swirl the flask to dissolve the sample. Should the acid be insoluble in water, add alcohol. (If a large amount of alcohol is required, start over, omitting the water.) Add 2 to 3 drops phenolphthalein indicator solution (bromthymol blue if more than 50% alcohol is used) and titrate with 0.5 *N* sodium hydroxide solution. A transition from colorless to faint pink should occur at the endpoint. Record the volume of base used to titrate the acid and calculate the neutralization equivalent using the formula below.

$$\text{Neutralization equivalent} = \frac{1000W}{VN}$$

where W = weight of sample, g
V = volume of base used in titration, mL
N = normality of base solution

Note: This determination should be conducted at least twice in order to ensure accurate results. All data and a sample calculation should be recorded in the laboratory report. If smaller amounts of acid (approximately 100 mg) are used, the normality of the base should be 0.1 to 0.2 *N*.

The sodium hydroxide solution should always be fresh. Old solutions may be considerably weaker than the stated normality because sodium hydroxide absorbs carbon dioxide from the air. If you suspect your base solution is inaccurate, titrate it with a sample of benzoic acid. If the base solution is weak, make up an approximately 0.5 *N* solution using reagent grade sodium hydroxide and distilled water. Titrate three 25-mL samples of standard acid solution (0.5 *N* HCl will do). Calculate the normality of your solution using the following relationships:

$$N_A V_A = N_B V_B$$

$$N_B = \frac{N_A V_A}{V_A}$$

where N_A = normality of acid
N_B = normality of base
V_A = volume of acid
V_B = volume of base

You may use the base solution before you do the calculations if you wish.

B Nucleophilic Addition to the Carbonyl Group

Although the amide derivative of a carboxylic acid may often be prepared by strongly heating the corresponding ammonium carboxylate salt, substitution

$$R{-}COO^{-}NH_4^{+} \xrightarrow{\Delta} R{-}CO{-}NH_2 + H_2O$$

will rarely occur at a simple carboxyl group. The principal reason for this is that hydroxyl is a poor leaving group and proton transfer dominates. Activation of the carbonyl involves essentially two processes: (1) exchange of the hydroxyl group for some other group which can be readily lost; and (2) utilization of a group adjacent to the carbonyl which increases, rather than decreases, the electrophilicity of the carbonyl.

The most common method for activating a carboxyl is to exchange hydroxyl for chloride. The chlorine atom is a particularly favorable activating group for carbonyl because it, like oxygen, has a relatively high electronegativity (3 on the Pauling scale), which makes the carbonyl group electrophilic, and it is also a good leaving group. Because of these two factors, nucleophilic addition to acid chlorides is the major means by which carboxylic acids are converted into derivatives.

$$R{-}\overset{\delta^+}{C}(=\overset{\delta^-}{O}){-}Cl + Nu^- \longrightarrow R{-}C(=O){-}Nu + Cl^-$$

1 Formation of acid chlorides

Before considering all the variations, let us begin with the procedure for converting acids to acid chlorides. The most common reagent used to achieve this conversion is thionyl chloride ($SOCl_2$), which reacts readily with active carboxylic acids to give the acid chloride, with SO_2 and HCl as by-products.

$$R{-}COOH + SOCl_2 \longrightarrow R{-}CO{-}Cl + SO_2\uparrow + HCl\uparrow$$

In cases where this procedure works well it is indeed convenient, because the two by-products formed are readily lost as gases and the crude acid chloride may be used without further purification. A difficulty arises with this procedure when the carboxylic acid is not particularly reactive. For a variety of reasons, the addition of dimethylformamide (DMF) will generally cause a more rapid reaction to occur. Dimethylformamide reacts with thionyl chloride to form an intermediate formamidinium chloride salt, which accelerates the reaction of thionyl chloride with the carboxylic acid. The procedure for forming an acid chloride in the presence of DMF is described below.

Note that only a very small amount of DMF is required in this procedure. Although it takes part in the reaction, it is regenerated and is therefore a catalyst, as shown in the equations below.

$$(CH_3)_2N{-}CHO + SOCl_2 \longrightarrow (CH_3)_2\overset{+}{N}{=}C(Cl)(H)\ Cl^- + SO_2$$

$$R{-}CO{-}OH + (CH_3)_2\overset{+}{N}{=}C(Cl)(H) \longrightarrow R{-}CO{-}Cl + HCl + (CH_3)_2N{-}CHO$$

Recall that a catalyst is a substance which helps the reaction reach equilibrium more rapidly but is not itself consumed in the reaction. In this particular case, because SO_2 and HCl are irreversibly lost, the equilibrium lies far to the right.

Because DMF fulfills its catalytic function very effectively, it should not be added after the reaction has been heated. Although this is noted in the procedure, we call attention to it here also because failure to observe this precaution can lead to a very dangerous reaction.

2 Amides from acid chlorides

Once the acid has been converted to its chloride, most of the other common derivatives can be formed using it. From the equations below, it can be seen that an ester is produced if the acid chloride reacts with alcohol; an amide is produced if the acid chloride reacts with ammonia; and a substituted amide results if a primary or secondary amine is used.

$$R{-}CO{-}Cl + R'{-}OH \longrightarrow R{-}CO{-}OR' + HCl$$

$$R{-}CO{-}Cl + R'{-}NH_2 \longrightarrow R{-}CO{-}NH{-}R' + HCl$$

Although the reactants differ, all these reactions are related examples of nucleophilic substitution. In each reaction, a nucleophile attacks the activated carbonyl to form a tetrahedral intermediate bearing oxygen, chlorine, and the nucleophile.

$$R{-}\overset{\overset{\displaystyle O}{\|}}{C}{-}Cl + Nu^- \longrightarrow R{-}\underset{\underset{\displaystyle Cl}{|}}{\overset{\overset{\displaystyle O^-}{|}}{C}}{-}Nu$$

When a tetrahedral intermediate undergoes electronic reorganization, thus allowing the carbonyl group to re-form, either the chloride or the nucleophile is expelled. When the chloride is lost, a derivative has formed. If the incoming nucleophile is ethanol, an ethyl ester results. The by-product in this reaction, hydrogen chloride, is lost by evaporation or evolution of the gas. Hydrogen chloride will also be a by-product in the case of an amine nucleophile, but the problem here is that one mole of the nucleophile will be deactivated for every mole of HCl produced.

Recall that the nucleophilicity of ammonia is a consequence of the nitrogen lone pair. If HCl is generated in a reaction, the ammonia electrons will be protonated and ammonium chloride will result. Because ammonium chloride has no nucleophilic electron pair, it cannot attack an acyl chloride. At least two equivalents of the derivative-forming amine should therefore be used in the formation of amide derivatives.

$$R{-}CO{-}Cl + 2NH_3 \longrightarrow R{-}CO{-}NH_2 + NH_4Cl$$

$$R{-}CO{-}Cl + NH_4Cl \longrightarrow \text{no reaction}$$

In certain cases hydroxide rather than excess amine is used as base. This procedure, called the *Schotten-Baumann reaction,* refers to the reaction in which an acid chloride is treated with an amine in the presence of aqueous sodium hydroxide solution. Sodium hydroxide reacts with the hydrogen chloride generated in the nucleophilic substitution. This is a relatively easy procedure but it is most commonly used to form benzamide derivatives of amines rather than to derivatize acids.

PROCEDURE (in hood)

AMIDE DERIVATIVES OF CARBOXYLIC ACIDS

$$R{-}\overset{O}{\overset{\|}{C}}{-}OH + SOCl_2 \xrightarrow{DMF} R{-}\overset{O}{\overset{\|}{C}}{-}Cl + HCl + SO_2$$

$$R{-}\overset{O}{\overset{\|}{C}}{-}Cl + NH_3/H_2O \longrightarrow R{-}\overset{O}{\overset{\|}{C}}{-}NH_2$$

Weigh out 0.5 g of the carboxylic acid in a round-bottom flask. Add 1 drop dimethylformaide (DMF). **Warning: Do not add DMF to the hot solution.** Attach a reflux condenser and add 3 to 4 mL thionyl chloride (through the condenser) and warm the mixture carefully on a steam bath for 15 to 20 min.

At the end of this time, cool the solution to room temperature and add it dropwise, *cautiously,* to 10 mL ice-cold concentrated ammonium hydroxide. The ammonium hydroxide solution should be swirled vigorously as the acid chloride is added. The crude amide should be collected and recrystallized from alcohol, aqueous alcohol, water, or acetone–petroleum ether.

Note: Most low-molecular-weight amides are water-soluble. The preferred derivatives of low-molecular-weight carboxylic acids are therefore toluidine or aniline derivatives, i.e., the toluidides and anilides.

PROCEDURE (in hood)

ANILIDES AND p-TOLUIDIDES OF CARBOXYLIC ACIDS

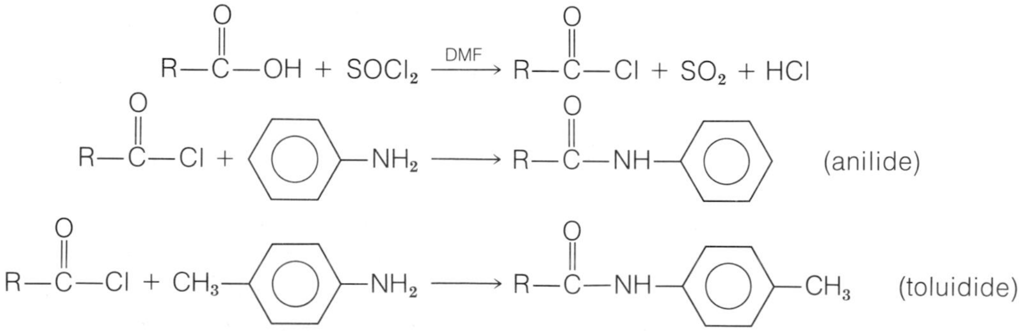

$$R{-}\overset{O}{\overset{\|}{C}}{-}OH + SOCl_2 \xrightarrow{DMF} R{-}\overset{O}{\overset{\|}{C}}{-}Cl + SO_2 + HCl$$

$$R{-}\overset{O}{\overset{\|}{C}}{-}Cl + C_6H_5{-}NH_2 \longrightarrow R{-}\overset{O}{\overset{\|}{C}}{-}NH{-}C_6H_5 \quad \text{(anilide)}$$

$$R{-}\overset{O}{\overset{\|}{C}}{-}Cl + CH_3{-}C_6H_4{-}NH_2 \longrightarrow R{-}\overset{O}{\overset{\|}{C}}{-}NH{-}C_6H_4{-}CH_3 \quad \text{(toluidide)}$$

Weigh out 0.5 g of the carboxylic acid in a round-bottom flask. Add 1 drop dimethylformamide (DMF). **Warning: Do not add DMF to the hot solution.**

Attach a reflux condenser and pour in 4 mL thionyl chloride (through the condenser). Warm gently for 15 to 20 min on a steam bath and allow the solution to cool to room temperature.

In a separate 125-mL Erlenmeyer flask dissolve 1 g *p*-toluidine or 1 mL aniline in 30 mL dichloromethane. Add 20 mL 25% NaOH solution to this flask and cool the two-phase mixture briefly in an ice bath. When acid chloride formation is complete, the oil is cooled to room temperature and diluted with 10 mL dichloromethane. The acid chloride solution is then added to the Erlenmeyer flask in several portions (1 to 2 mL each), with vigorous swirling and cooling after each addition. The upper, aqueous base layer is checked (pH paper) after each addition to ensure that the pH of the aqueous phase remains near 10. A vigorous reaction is observed on addition of each portion of acid chloride solution to the reaction flask. In some cases the heat generated will be sufficient to briefly boil the dichloromethane solution. If a significant amount of dichloromethane is lost in this manner, more solvent should be added.

After the addition is complete swirl the flask vigorously (room temperature) for 10 min. A yellow color is often observed in the dichloromethane layer at this stage.

The mixture is transferred to a separatory funnel and the dichloromethane layer (lower) is removed. The basic aqueous phase is discarded. The organic layer is washed with 15 mL 5% HCl solution and then with 15 mL water (drying the organic phase is not necessary). The dichloromethane is evaporated to dryness on a steam bath (**hood**), and the solid residue is recrystallized from aqueous alcohol. The solid anilide or toluidide may be recrystallized from acetone–petroleum ether if necessary.

3 Methyl and ethyl esters of carboxylic acids

The primary criterion for the selection of a derivative is the crystallinity of the product. It is for this reason that 4-bromophenacyl derivatives of carboxylic acids are usually more desirable than the simple phenacyl esters. A cursory examination of the tables will reveal, however, that the derivative which is usually best is not *always* superior. In selecting a derivative, one should always keep in mind that the product should have a melting point as high as possible and as different as possible from that of the other compounds under consideration.

In some cases, the identity of the compound is known and a confirmatory derivative is desired. In these cases, the reaction selected need not yield a product whose melting point is in the comparative tables. An example of such a derivative might be benzoic acid isolated from benzaldehyde. In the context of carboxylic acids, a useful derivative is sometimes the methyl or ethyl ester.

The methyl and ethyl esters of carboxylic acids are usually easy to prepare and values for their melting points can be obtained readily by examining Table

28.13 (Solid Esters) and looking in the "acid produced" column for the unknown. If the unknown forms both solid methyl and ethyl esters, the former is usually preferable. A procedure for forming esters of carboxylic acids is described below.

PROCEDURE

FORMATION OF METHYL AND ETHYL ESTERS OF CARBOXYLIC ACIDS

$$R{-}CO{-}OH + CH_3OH \xrightarrow{H^+} R{-}CO{-}OCH_3 + H_2O$$

One gram of the carboxylic acid is placed in a 50-mL round-bottom boiling flask, and 20 mL dry methanol (or ethanol), 2 drops concentrated H_2SO_4, and a boiling chip are added. The flask is equipped for reflux and a drying tube is attached to the top of the condenser. The mixture is heated at reflux for 1 to 2 h and then allowed to stand (stoppered) until the next laboratory period. If the ester has separated, collect by filtration. If the ester has not separated, cool the flask in ice. If the ester still does not separate, distill off half the solvent and repeat the above.

In the event that the ester remains soluble, dilute the alcohol solution with 50 mL 5% Na_2CO_3 solution and extract with ether. Evaporation of the ether solution should leave the crude ester, which can be recrystallized from petroleum ether.

C Derivatives Obtained by Alkylation

In discussing the reactivity of the carboxyl group, we have directed attention to the fact that it is an ambident ("both teeth") system, i.e., it has two reactive positions, the electrophilic carbonyl group and the electrophilic proton. Carboxylic acids can be derivatized by removing the hydroxyl group and exchanging it for some other leaving group. The transposition of hydroxyl for chloride has been described in some detail above, and the use of alkoxy as a leaving group is a simple extension of this discussion. The reactivity of an ester differs from the reactivity of an acid chloride primarily in that reactions of esters occur more slowly than the corresponding reactions of acid chlorides. Another possibility is to eliminate not the hydroxyl group but the proton on it. In this case the carbonyl group will remain electrophilic, although less so, because the full charge on oxygen is now a nucleophile for other substrates.

The carboxylate anion has long been known as a poor nucleophile in aqueous solutions. Carboxylate ion can be a useful nucleophile under the appropriate conditions. The use of modern solvents makes it possible to have carboxylate anion present in solution in the absence of those solvent forces which have hitherto kept its nucleophilicity and its basicity very low. The carboxylate anion in such solvents as dimethylformamide (DMF), dimethyl sulfoxide (DMSO), and hexamethylphosphoramide (HMPA) is much more nucleophilic than it is in water or alcohol (the solvents traditionally used). Problems occur, however, because such solvents as HMPA are generally high-boiling and difficult to remove after the reaction, and often have certain toxicities associated with them. For these reasons none of the procedures utilizing such solvents are described in this textbook.

Anion activators, such as quaternary ammonium compounds or crown ethers (see Sec. 4.7), make it possible to work in solvents which are inexpensive and easy to remove. Procedures have been developed for use in a number of derivative-forming reactions, and these are suggested below.

Phenacyl ester formation

In the traditional procedure for preparing a phenacyl ester, a carboxylic acid is neutralized with sodium or potassium hydroxide. The carboxylate salt is then alkylated with a very reactive α-bromoketone, namely phenacyl bromide. The traditional procedure, which involves water and alcohol as solvents, can be effective if the acid is of low enough molecular weight that it is readily soluble in water. The traditional procedure is given below.

PROCEDURE (in hood)

PHENACYL ESTER FORMATION

$$R-\overset{O}{\overset{\|}{C}}-O^-K^+ + C_6H_5-\overset{O}{\overset{\|}{C}}-CH_2-Br \longrightarrow R-\overset{O}{\overset{\|}{C}}-O-CH_2-\overset{O}{\overset{\|}{C}}-C_6H_5$$

$$R-\overset{O}{\overset{\|}{C}}-O^-K^+ + Br-C_6H_4-\overset{O}{\overset{\|}{C}}-CH_2-Br \longrightarrow R-\overset{O}{\overset{\|}{C}}-O-CH_2-\overset{O}{\overset{\|}{C}}-C_6H_4-Br$$

About 0.5 g of the acid is dissolved in 2 mL distilled water and carefully neutralized to a phenolphthalein endpoint with 10% aqueous potassium hydroxide solution. Once the acid has been neutralized, the water is evaporated by heating gently with a free flame. When most of the water has been removed, 5 to 10 mL 95% ethyl alcohol is added, along with about 0.5 g phenacyl bromide (or 2,4′-dibromoacetophenone if the 4-bromophenacyl ester is desired).

The resulting solution is boiled for 1 to 2 h to ensure that all the carboxylate anion has been alkylated. the 4-bromophenacyl ester is usually fairly crystalline and can be isolated by partially evaporating the alcohol. This ordinarily results in a usable derivative. The difficulty, which is substantial, is that only the low-molecular-weight carboxylate salts are soluble enough in this mixture of water and alcohol to react effectively, and it is these compounds which give the poorest (i.e., the lowest-melting) phenacyl ester derivatives.

Alternative procedures which are of somewhat broader applicability are given below.

PROCEDURE (in hood)

QUATERNARY ION–MEDIATED FORMATION OF PHENACYL ESTERS

$$R{-}\overset{O}{\overset{\|}{C}}{-}O^{-}K^{+} + C_6H_5{-}\overset{O}{\overset{\|}{C}}{-}CH_2{-}Br \longrightarrow R{-}\overset{O}{\overset{\|}{C}}{-}O{-}CH_2{-}\overset{O}{\overset{\|}{C}}{-}C_6H_5$$

$$R{-}\overset{O}{\overset{\|}{C}}{-}O^{-}K^{+} + Br{-}C_6H_4{-}\overset{O}{\overset{\|}{C}}{-}CH_2{-}Br \longrightarrow R{-}\overset{O}{\overset{\|}{C}}{-}O{-}CH_2{-}\overset{O}{\overset{\|}{C}}{-}C_6H_4{-}Br$$

Dissolve 0.5 g (or 0.5 mL) of the carboxylic acid in approximately 20 mL methanol. To this solution add 2 drops 1% phenolphthalein in methanol and neutralize (pink endpoint) with 10% methanolic potassium hydroxide. The methanol is then evaporated, either under reduced pressure on a rotary evaporator or by boiling the solution to dryness in a beaker on a small hot plate, steam bath, etc.

The crude potassium salt of the unknown acid (the material may appear pink) is placed in a 100-mL round-bottom flask, along with 0.5 g phenacyl bromide (or 2,4′-dibromoacetophenone if the bromophenacyl ester is desired), and the mixture is diluted with 20 mL toluene. To this suspension is added 5 mL standard catalyst solution (see Sec. 4.7). This suspension is refluxed for 1 h and then cooled to room temperature.

The solution is filtered (Hirsch funnel) to remove precipitated salt. The toluene is evaporated and the residue crystallized from toluene–ethyl acetate, toluene-cyclohexane, or acetone–petroleum ether to generate the crystalline derivative.

In carrying out this procedure, the potassium salt should be present in an excess of the alkylating agent. If all the alkylating agent is not consumed, it will contaminate the phenacyl ester derivative and produce a mixture which is very difficult to separate and purify. (*Note:* This reaction is quite rapid with monocarboxylate salts. Dicarboxylate salts react more slowly, and their derivatives are usually insoluble in the toluene solution. As a general rule, one will have an idea of whether or not the unknown is a monocarboxylic acid from the neutralization equivalent.)

In most cases, the catalyst is removed from the solution by recrystallization of the crude derivative. Another workup procedure which is very efficient is to take the toluene solution of the derivative and percolate it through 10 g dry column silica gel. This percolation is done after the precipitated solids have been filtered. The purpose of the silica gel treatment is to remove any dissolved catalyst, while the derivative passes through unhindered. Subsequent to this step, the toluene is evaporated to obtain the purified derivative.

23.7 PHENOLS: THE OTHER ACIDIC CLASS

A Classification of Phenols

The fact that carboxylic acids have pK_a values close to 5, phenols have pK_a values close to 10, and alcohols have pK_a values close to 16 has already been discussed. The question that arises now is: Why are phenols so much more acidic than alcohols although the hydroxyl functional group appears to be the same in both cases? The very important difference between phenols and alcohols is the stabilization of the anion once the proton is lost. The comparative acidities of any two compounds is determined by the difference in stability between the two resulting anions. This must be so because dissociation of a proton is the same in both cases. The phenoxide anion is stabilized to a greater extent by resonance than is the alkoxide anion, as shown below:

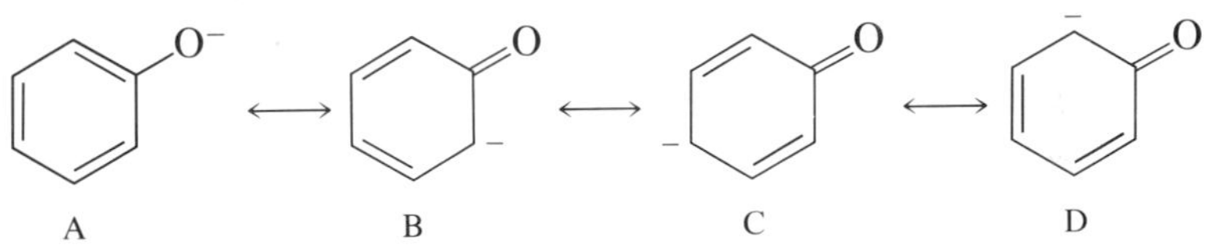

A B C D

When there is a ring substituent in a position to stabilize the negative charge (ortho or para), the acidity of a phenol will be further enhanced because the stability of the conjugate base will be increased. For example, a nitro group attached to the aromatic ring in the 4 position will stabilize the negative charge in structure C above. For this reason nitrophenols are typically 2 to 3 pK_a units more acidic than the parent compound.

The phenols commonly encountered in qualitative organic chemistry are those which have alkyl, alkoxy, nitro, or halogen substitution. Because all but the nitro compounds have pK_a values similar to that of phenol, they are soluble in aqueous sodium hydroxide but not in aqueous sodium bicarbonate solution. Water-soluble phenols will be soluble in bicarbonate solution but CO_2 will not be given off. As a consequence it is possible to distinguish most phenols from carboxylic acids without undue difficulty. Note also that nitrophenols give a deep yellow color in aqueous sodium hydroxide solution. Confirmation is usually obtained by application of the ferric chloride enol test, described in the following procedure.

CLASSIFICATION PROCEDURE

FERRIC CHLORIDE ENOL TEST

Dissolve 1 drop of the liquid sample or approximately 50 mg of the solid sample in 1 mL reagent-grade chloroform. Add 1 mL of a 1% ferric chloride–chloroform solution. After the two solutions have been thoroughly mixed, add 1 drop pyridine. Rapid color development indicates the presence of a phenol or an enol.

This is an extremely sensitive test. The colors produced range from green to blue to deep purple to red. One should always perform this test in the presence of a blank to see the color change (light yellow to dark yellow) in the absence of an enol. The colors observed for several representative enols are recorded below.

2-Naphthol	Green
2,6-Dimethylphenol	Blue-green
1,3-Dihydroxybenzene	Deep blue
4-Methoxyphenol	Blue
4-Allyl-2-methoxyphenol	Blue
4-Ethylphenol	Blue
5-Chlorosalicylic acid	Purple
4-Nitrophenol	Red

It is fortunate that the most common electron-withdrawing groups impart some color to phenols when they are attached to the aromatic ring. For example, phenol, *p*-cresol, and catechol are all colorless solids, but *p*-nitrophenol, which is much more acidic than any of these three, is yellow. This fact can be used to determine in a qualitative way whether a strong acid is a phenol or a carboxyl compound. The very acidic phenols whose pK_a's are similar to those of carboxyl-containing compounds are almost always yellow. Thus, it is easy to distinguish between benzoic acid and picric acid by virtue of color, although both are strong acids which melt at 121°C.

B Characterization and Derivatization of Phenols

Because phenols are acids, they can be titrated with standard base in exactly the same way one would titrate a carboxylic acid. For a variety of reasons, however, titration is not as useful for phenols as for acids. This is due in part to the hygroscopicity of many phenols. A direct molecular weight determination for a phenol may be carried out by a simple extension of a derivative procedure described below. Before you choose a derivative, check the tables to be certain your possibilities give crystalline and distinguishable derivatives.

1 Aryloxyacetic acid derivatives

Phenols are quite nucleophilic and react readily with chloroacetic acid in the presence of aqueous base to form crystalline phenoxyacetic acid derivatives, as shown below:

$$\text{Ar—OH} + \text{Cl—CH}_2\text{—COOH} \xrightarrow{\text{KOH}} \text{Ar—O—CH}_2\text{—COOH} + \text{KCl}$$

The aryloxyacetic acid derivatives serve two purposes. First, they provide a crystalline derivative whose melting point is characteristic of the starting phenol; and second, the resulting carboxylic acid can be titrated to determine its neutralization equivalent. Because most of the phenoxyacetic acids are monocarboxylic acids, this is a direct molecular weight determination for the unknown phenol.

PROCEDURE (in hood)

ARYLOXYACETIC ACID DERIVATIVES

$$\text{Ar—OH} + \text{Cl—CH}_2\text{—COOH} \xrightarrow{\text{KOH}} \text{Ar—O—CH}_2\text{—COOH} + \text{KCl}$$

One to two grams (or milliters) of the compound is added to 5 to 10 mL 30% aqueous potassium hydroxide solution. This is swirled and 2 to 5 mL 50% aqueous chloroacetic acid solution is added. A small amount of water must be added when the salt of the phenol is not completely soluble. The test tube is placed in a steam or hot-water bath and left there for about 30 min (occasionally, a slightly longer period of time is required). The solution should be gently agitated from time to time during this period. If the warm solution is homogeneous, it is cooled, diluted with an approximately equal volume of water, acidified with concentrated hydrochloric acid until the pH is about 3, and allowed to stand. If the warm solution contains solid, add water and wait for the salts to dissolve before acidifying. If crystals do not readily deposit, it may be necessary to transfer this aqueous solution to a separatory funnel and extract it twice with two 25-mL portions of ether. The ether solution is dried over sodium sulfate, placed on a steam bath, and the ether evaporated. The resulting solid should be recrystallized from ethyl alcohol.

2 Neutralization equivalent of the derivative

The aryloxyacetic acid derivative obtained from the procedure described above should be separated into two parts. One portion should be dried or recrystallized, subjected to a melting-point determination, and then utilized as a derivative for the compound in question. The other half of the material should be titrated (at least twice) with standard base and its molecular weight determined (see Sec. 23.6A). A neutralization equivalent is often acceptable as a derivative in qualitative analysis courses (ask your instructor). Therefore, in this single sequence two derivatives can actually be prepared, the usual minimum required for identification of a compound.

3 Bromination of phenols

Bromination is another technique commonly used to derivatize aromatic hydroxy compounds. Because phenols are aromatic alcohols, the aromatic ring is much more electron-rich than it is in benzene. The nucleophilic ring reacts readily with bromine. The reaction of phenol with an aqueous solution of bromine is rapid and, if excess bromine is present, polybromination will generally result. If there are groups blocking the ortho or para position, exposure to bromine water will result in a monobromo or a dibromo derivative. These derivatives are usually yellow solids with sharp melting points. A procedure is described below for the formation of the bromide drivative.

PROCEDURE (in hood)

BROMINATION OF PHENOLS

OH, Br, OH, Br, Br

$$C_6H_5OH + 3Br_2 \longrightarrow C_6H_2Br_3OH + 3HBr$$

Make a solution which consists of 1.6 mL bromine (**gloves**) and 7.5 g potassium bromide in 50 mL water. Swirl this solution until the bromine completely dissolves. Add this solution in small portions to 0.5 g of an unknown phenol dissolved in water-dioxane until the yellow color persists. The product is isolated by dilution with cold water and collection of the precipitated solid by filtration. The solid is recrystallized from ethanol or ethanol-water.

4 Benzoate ester derivatives

Benzoate esters may also be of use in derivatizing phenols. A benzoylation procedure is described below.

PROCEDURE (in hood)

SCHOTTEN-BAUMANN BENZOYLATION OF PHENOLS

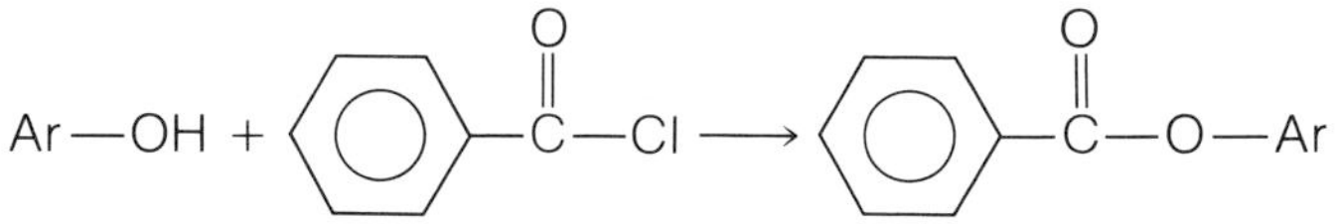

$$Ar{-}OH + C_6H_5{-}\overset{O}{\overset{\|}{C}}{-}Cl \longrightarrow C_6H_5{-}\overset{O}{\overset{\|}{C}}{-}O{-}Ar$$

Place 1 g or 1 mL of the phenol in 10 to 15 mL 10% aqueous NaOH (it should dissolve) in a 50-mL Erlenmeyer flask or glass-stoppered flask. To this solution add 2 mL benzoyl chloride (**hood**), stopper or cork securely, and shake vigorously for 10 to 15 min; the pungent odor of benzoyl chloride should have abated. Be sure that the solution is basic (litmus) and isolate the benzoate ester as described for benzamides in Sec. 24.7A.

5 Urethane derivatives of phenols

In the previous discussion, we have focused on the acidity of phenols. The similarity of phenols to alcohols should not be overlooked. Although the classification of and derivative formation from phenols generally involves chemistry which is not common to alcohols, urethane derivatives of phenols may often be prepared by the method suggested for alcohols (see Sec. 26.6A). The urethane derivative is formed by a simple addition reaction, as shown in the equation below.

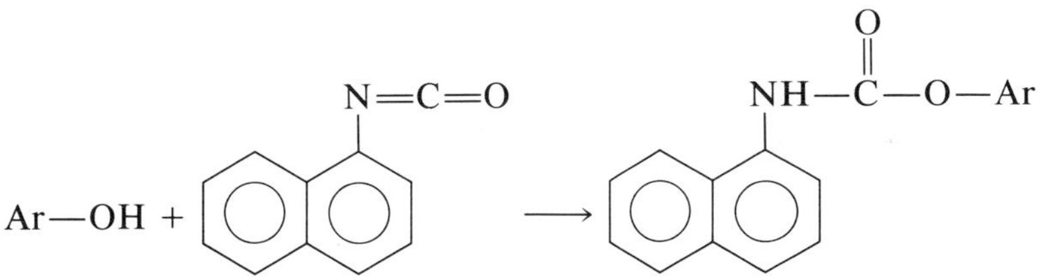

$$Ar{-}OH + C_{10}H_7{-}N{=}C{=}O \longrightarrow C_{10}H_7{-}NH{-}\overset{O}{\overset{\|}{C}}{-}O{-}Ar$$

PROCEDURE (in hood)

URETHANE DERIVATIVES OF PHENOLS

A. Hydrocarbon-soluble phenols

Dissolve 0.5 g of the phenol in 5 to 10 mL petroleum ether (or ligroin) and add to this solution a mixture containing 0.5 to 0.7 mL α-naphthyl isocyanate (**Caution: lachrymator**) in approximately 10 mL petroleum ether. Heat the mixture gently on a steam bath for 5 min (exclude moisture) and filter hot. The filtered petroleum ether solution should then be allowed to cool to room temperature and the crystallized product collected by filtration. The crude product should be recrystallized from petroleum ether or petroleum ether–ethyl acetate.

B. Hydrocarbon-insoluble phenols

Place 1 mL of the phenol and 0.5 to 0.7 mL of the isocyanate in a small test tube (**hood**). Shake the tube. If no reaction is apparent, heat the tube on a steam bath for 5 to 10 min. Purify the product as described above.

23.8 SPECTROSCOPIC CONFIRMATION OF STRUCTURE

Carboxylic acids and phenols may be distinguished from other classes of compounds by their solubility in aqueous base. Phenols may generally be distinguished from carboxylic acids by their different reactions with bicarbonate ion (see Sec. 23.4B). It is often useful, however, to confirm the presence of aromatic hydroxyl (phenol) or carboxyl (acid) by spectroscopic means.

Although both phenols and carboxylic acids contain hydroxyl functions, they appear quite different in the ir spectrum. Phenols show a characteristic —OH vibration between 3600 and 3300 cm^{-1}. Since most phenols are benzene derivatives, it is common also to see a benzene-ring (double-bond) vibration near 1600 cm^{-1}. Although the latter bond is fairly constant for a variety of phenols, the hydroxyl vibration will be observed at different positions depending on the solvent, concentration, etc., used to record the spectrum.

Carboxylic acids will exhibit broad, strong ir bands anywhere from 3600 to 2500 cm^{-1}, depending again on solvent, concentration, etc. The important signal implying the presence of a carboxyl function is the carbonyl vibration, which is usually observed as a strong, sharp band in the 1700 to 1750 cm^{-1} range. This band will be completely absent in the ir spectrum of a phenol.

It may be useful to refer to Chap. 5, which shows spectra of a variety of compounds. In addition, the spectra of acids may be found in a number of sections, including Secs. 12.3 and 12.4. The ir spectrum of phenol is shown in Fig. 23.1. The nmr and uv spectroscopic methods are of relatively less utility than the ir method for discriminating between these two functional groups.

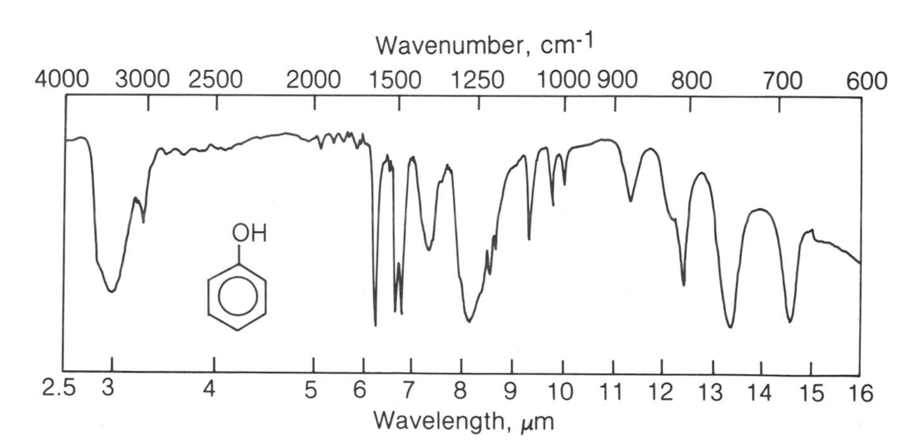

Figure 23.1
The ir spectrum of phenol.

AMINES

24.1	Introduction	**532**
24.2	Historical	**532**
24.3	Classes of Amines	**533**
24.4	Acidity and Basicity	**534**
	Hydrolysis	
24.5	Operational Distinctions	**538**
	A Flow chart for the Hinsberg test	
	B Spectroscopic information	
24.6	Reactivity	**544**
24.7	Derivatives of Primary and Secondary Amines	**545**
	A Amides	
	B Hydrochloride salts	
	C Phenylthioureas	
24.8	Derivatives of Tertiary Amines	**548**
	A Hydrochloride salts	
	B Picrate salts	
	C Methiodide salts	
	D Methyl tosylate salts	

24.1 INTRODUCTION

The amines are organic bases and are generally distinguished qualitatively by their solubility in dilute aqueous acid. Although not all amines have strongly basic properties, the electron pair on the nitrogen atom, which can often be protonated quite readily, serves to distinguish this class of compounds. The enormous group of natural products known as the *alkaloids* consists of amines which are obtained from a variety of natural sources by extraction with aqueous acid. The alkaloids are readily separable from a variety of other naturally occurring materials because most of the latter materials are neutral and are soluble neither in water nor in aqueous acid. The amines, or nitrogenous bases, are ubiquitous in living systems and comprise a major class of compounds within the framework of organic chemistry.

24.2 HISTORICAL

As is true of all organic compounds, the history of the amines begins long before humans were around to recognize their properties. One thing that the earliest people must have recognized, however, was the odor of the ptomaines. These aliphatic diamines (given the trivial names *putrescine* and *cadaverine*) are 1,4-diaminobutane and 1,5-diaminopentane. As their names imply, these are foul-smelling substances which are produced in the decomposition of all mammals, fish, and fowl.

Humans recognized the properties of amines in a variety of ways. Although the chemical structures of amines were unknown to them, early people recognized the effects of a variety of these nitrogenous bases. The amine coniine, for example (2-*n*-propylpiperidine, illustrated below), is the active principle of hemlock, which ended Socrates' search for truth in this life.

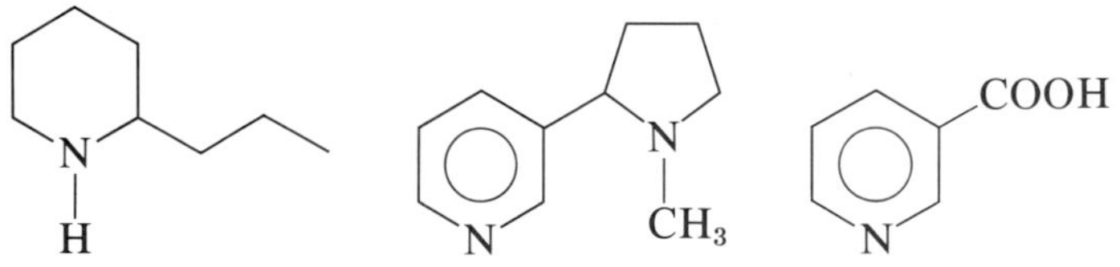

Coniine Nicotine Nicotinic acid

The substance we know as nicotine is also an alkaloid, although it has two nitrogen-containing rings, a pyridine ring and a pyrrolidine ring. The American Indians used nicotine as a stimulant long before Europeans arrived with their culture, and it was in fact from the Indian culture that tobacco was introduced into Europe. We now know that nicotine is poisonous in large concentrations, but it can be a valuable substance nevertheless. Oxidation of nicotine with nitric acid yields 3-carboxypyridine, commonly called nicotinic acid or niacin, the antipellagra vitamin.

Another Indian contribution to European culture was the discovery in 1633 by Father Calancha that the South American natives knew of a cure for malaria. The bark of the so-called fever tree was ground up to produce a drink which mitigated the effects of this dreaded disease. The introduction of this substance to Europe was of inestimable value. In fact, it converted Rome from the hotbed of malaria it was at that time to a stable community. This substance, which we now know as *quinine*, is important in the history of chemistry because it marks the first time that Europeans began a systematic search for chemicals, natural products, or other preparations which might be valuable in the cure of disease. Up to this time, medicine had changed little from the traditional methods (bleeding, leeching) espoused hundreds of years earlier by Galen of Pergamon. The structure of quinine is shown below.

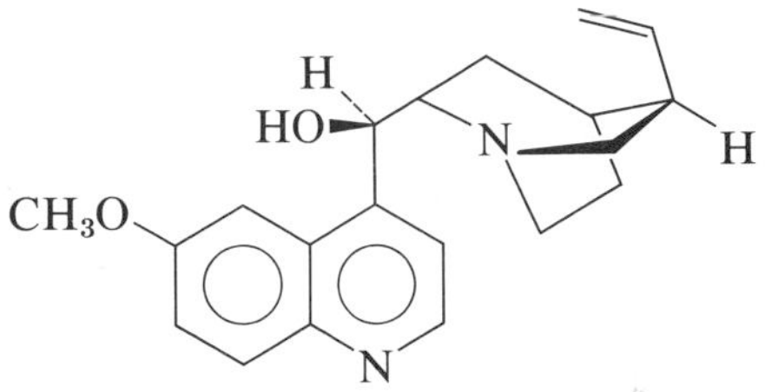

Quinine

It is also interesting to note that the death of the vital force theory of organic chemistry was brought about because of nitrogen-containing compounds. Wöhler in 1828 notified Berzelius by letter that he had isomerized ammonium isocyanate to urea. Ammonium isocyanate is clearly an inorganic substance and urea is an amine, although not a basic one, and a very common natural product. It was previously believed that no substance isolated from a living source could be prepared from inorganic precursors.

24.3 CLASSES OF AMINES

We have already said that the principal factor which distinguishes amines from the variety of other organic compounds is the presence of basic lone pairs of electrons on nitrogen atoms, but it should be recognized that a compound may still be classified as an amine even though there are other functional groups present in the molecule. This is true because the presence of nitrogen usually dominates the solubility characteristics of the whole molecule. We shall see in a later section why this is so and how it occurs. For the present, consider the three principal classes of amines: primary, secondary, and tertiary. These are defined by the substitution pattern about nitrogen. The three classes are often distinguished by using the symbols 1°, 2°, and 3°, which have a somewhat different meaning than when applied to alcohols.

A primary amine is an amine in which one of the three normal valences of nitrogen is satisfied by an alkyl or an aryl group; a secondary amine is one in which two of the three valences of nitrogen are satisfied by aryl or alkyl groups; and a tertiary amine is one in which all three valences are satisfied by aryl or alkyl groups. The consequence of this is that the nitrogen atom in a primary amine bears two hydrogen atoms, that in a secondary amine bears one hydrogen atom, and in a tertiary amine there is no hydrogen bound to nitrogen. Some examples of primary amines are methylamine (CH_3NH_2), *tert*-butylamine [$(CH_3)_3CNH_2$], and aniline ($C_6H_5NH_2$). Notice that all three of these compounds are primary amines because only one valence on nitrogen is satisfied by an organic residue and two hydrogen atoms remain. *tert*-Butylamine is a primary amine despite the fact that it bears a *tert*-butyl group. *tert*-Butyl alcohol, on the other hand, is a tertiary alcohol because in alcohols the designation refers to the carbon bonded to the heteroatom rather than to the heteroatom itself.

Diethylamine, dipropylamine, and methylethylamine are all examples of dialkyl secondary amines. Methylphenylamine, commonly referred to as *N*-methylaniline, is also a secondary amine, as is diphenylamine.

$$H_5C_2—NH—C_2H_5 \qquad C_6H_5—NH—CH_3 \qquad C_6H_5—NH—C_6H_5$$

Diethylamine — *N*-Methylaniline — Diphenylamine

Note that in each of the compounds shown above a single hydrogen atom is bonded to nitrogen, although the organic residues may be aryl, alkyl, or both. The same is true for tertiary amines, which are distinguished by the presence of three organic residues (alkyl or aryl) bonded to nitrogen. Triphenylamine and triethylamine are both examples of tertiary amines. It is also interesting to note that a secondary or tertiary amino function may be part of a ring system. Coniine, illustrated in Sec. 24.2, is an example of this. Note that coniine is a secondary amine because the nitrogen atom bears a single hydrogen atom and that both alkyl residues actually form part of the same ring. It is important to be conversant with the nomenclature of amines and, in particular, not to be confused by the difference in nomenclature between amines and alcohols.

Because nitrogen bears a lone pair of electrons which can accept a proton, nitrogen can, in fact, be tetravalent. Those compounds in which nitrogen is tetravalent are said to be *quaternary ammonium compounds*. The formation of such compounds and the consequences of their structure are discussed in the next section.

24.4 ACIDITY AND BASICITY

An amine which contains a basic nitrogen atom can be protonated by strong organic acids. An amine which is ordinarily insoluble in water dissolves when protonated because an ammonium salt has formed. Recall that in the discussion of carboxylic acids and phenols the pK_a was utilized as a measure of acidity. If we think about the equation that defines K_a (see below), the HA term clearly does not correspond to our amine.

$$HA \rightleftharpoons H^+ + A^-$$

$$K_a = \frac{[H^+][A^-]}{[HA]}$$

$$pK_a = -\log K_a$$

On the other hand, the A^- on the right side of the equation does correspond to the amine because it has a lone pair of electrons (although the amine is not negatively charged). In the equation describing the dissociation of HA to proton (H^+) and anion (A^-), let us put the amine on the right- instead of the left-hand side of the equation. If a neutral amine like ammonia is substituted for A^-, we can consider the equilibrium process of association because it is simply the reverse of the dissociation process. In this case, HA in the equation above corresponds to NH_4^+, the ammonium or quaternary ammonium ion. The overall relationship is expressed in the equations below:

$$NH_4^+ \rightleftharpoons H^+ + :NH_3$$
$$HA \rightleftharpoons H^+ + :A^-$$

The basicity of the amine (A:) can be assessed by considering the dissociation of the ammonium salt. If we consider all ammonium ions as acids (since they all dissociate to a proton plus free amine) we are actually considering the affinity of the amine base for the proton in the opposite direction. By doing this, we are using the ammonium salt acidity to assess the amine basicity. It is important to understand that the pK_a of the ammonium salt is by no means the same as the pK_a of the corresponding amine. In the first place, only primary and secondary amines can dissociate to give amide anions ($R_2\ddot{N}H \rightarrow H^+ + R_2\ddot{\underset{\cdot\cdot}{N}}^-$) and when they do so there are two electron pairs on nitrogen. This process is a much higher-energy process than the dissociation of an ammonium salt. The pK_a's for primary and secondary amines are very high (30 to 40), and amide anions are therefore generated only under rather special conditions. The protonated amines (ammonium salts) usually have pK_a values in the vicinity of 0 to 12, within the normal aqueous pH range.

The simplest amine is ammonia, $\ddot{N}H_3$. The pK_a of ammonia is 36, far beyond the values we deal with in normal laboratory courses. Its corresponding ammonium salt (ammonium ion) has a pK_a of about 9.

$$NH_3 \rightleftharpoons H^+ + :\ddot{N}H_2^- \qquad (pK_a = 36)$$
$$NH_4^+ \rightleftharpoons H^+ + :NH_3 \qquad (pK_a = 9)$$

Those factors which increase the basicity of the amine will increase the value of the pK_a. For example, substitution of one of the hydrogens in ammonia by an ethyl group gives the primary amine ethylamine. The electron-donating property of an alkyl group relative to hydrogen would be expected to increase the electron density about nitrogen, i.e., to increase the availability of the lone pair of electrons. The more basic the amine, the more it will tend to resist dissociation when it exists as its ammonium salt. Ethylamine therefore should be more basic and its ammonium salt should be less acidic. The pK_a of ethylammonium ion is almost 11. The butyl and cyclohexyl substituents show a similar effect when they are substituted in the ammonia molecule to give primary amines, and their ammonium salt pK_a's are similar to each other and to that of ethylammonium ion. The pK_a values of the protonated forms of common amines are listed in Table 24.1.

The secondary ammonium salt of diethylamine has a pK_a of just over 11. The decreased acidity or enhanced basicity of the amine is due largely to the fact that there are now two alkyl groups donating electron density to the nitrogen atom.

An additional factor which plays an important role in the acidity of amines is whether or not the group bonded to nitrogen is capable of resonance. Recall that aniline is far more reactive in electrophilic substitution than is benzene. This is so because the lone pair of electrons on the nitrogen atom is delocalized by resonance with the ring. The consequence of this is increased nucleophilicity of the ring. The corresponding effect on the acidity of the quaternary ammonium compound or the basicity of the free amine is that aniline holds a proton less strongly than does an aliphatic amine. In fact, the anilinium ion is

TABLE 24.1
pK_a Values of ammonium ions

$$\mathrm{H{-}\overset{+}{N}R_3 \rightleftharpoons H^+ + \ddot{N}R_3}$$

Protonated amine	pK_a
Ammonia	9.247
Ethylamine	10.807
Butylamine	10.777
Cyclohexylamine	10.660
Diethylamine	11.090
Benzylamine	9.330
Aniline	4.630
3-Nitroaniline	2.466
4-Nitroaniline	1.000
4-Methoxyaniline (anisidine)	5.340
Diphenylamine	0.790

rather a strong acid being almost as acidic as acetic acid. Obviously, if there is an electron-withdrawing group on the aromatic ring, the lone-pair electron density should be even further depleted and, as expected, 4-nitroaniline has a pK_a of about 2.5. Conversely, an electron-donating substituent on an aromatic amine tends to increase the pK_a of the corresponding ammonium salt. The resonance effect of a benzene ring is additive in diphenylamine, which is so weakly basic that it is not extracted into aqueous solution when treated with 10% hydrochloric acid. That is, diphenylammonium ion dissociates a proton so readily (forming the free amine, which is water-insoluble) that the normal solubility classification test fails for this molecule.

The solubility behavior observed for diphenylamine illustrates an important principle. As emphasized earlier, a molecule which is not water-soluble in its uncharged state may assume a water-soluble form when it is either protonated or deprotonated. Recall that benzoic acid is not very soluble in water at room temperature, but when treated with a base, it forms a water-soluble benzoate salt. The same general principle applies to amines. Amines which are insoluble in water may react with acid to form a water-soluble salt.

When protonation of an amine leads to a substance which is still water-insoluble, the classification test will fail. For example, *N*-benzylaniline (illustrated below) forms *N*-benzylanilinium chloride when protonated by hydrochloric acid.

$$\underset{\text{Water-insoluble}}{C_6H_5\text{—}\ddot{N}H\text{—}CH_2\text{—}C_6H_5} + H^+ \longrightarrow \underset{\textit{Still}\text{ water-insoluble}}{C_6H_5\text{—}\overset{+}{N}H_2\text{—}CH_2\text{—}C_6H_5}$$

N-Benzylanilinium chloride is insoluble in water because there is only one charge in the molecule and there are more than 10 carbon atoms present. The protonated molecule is too lipophilic or hydrophobic (see Chap. 4) to dissolve in water despite the presence of the proton. The same situation is encountered with phenethylamine. Phenethylamine and *N*-benzylaniline are both basic. These amines differ from diphenylamine in the following respect. Diarylamines like diphenylamine are generally much less basic than dialkyl or arylalkyl amines. In aqueous solution, the protonation equilibrium lies on the side of the free amine. *N*-Benzylaniline protonates, but its ammonium salt is not water-soluble. It is likely that diphenylamine, a 12-carbon compound, would be insoluble in water even if it were readily protonated, simply because the carbon number/charge ratio is relatively high.

Hydrolysis

If a compound contains nitrogen and does not dissolve in 5 to 10% aqueous hydrochloric acid, it is probably not an amine. When attempting to classify nitrogen-containing substances, the next logical step is to see if the substance hydrolyzes. Heat 1 drop or 50 mg of the compound in a test tube containing 1 mL 20% NaOH. Evolution of ammonia (test with litmus or pH paper) implies the presence of an amide or a urea.

$$H_2N{-}\overset{\overset{\displaystyle O}{\|}}{C}{-}NH_2 \qquad R{-}NH{-}\overset{\overset{\displaystyle O}{\|}}{C}{-}NH_2 \qquad R{-}\overset{\overset{\displaystyle O}{\|}}{C}{-}NH_2$$

Urea — A substituted urea — An amide

If the substance does not hydrolyze and does not give a test for a nitro group but is known to contain nitrogen, a very careful check should be made to see that the substance is not an amine which simply does not dissolve in acid.

24.5 OPERATIONAL DISTINCTIONS

The most important means by which amines are distinguished from other classes of organic compounds is the basicity of nitrogen. This basicity is discussed above in some detail, and we have seen how certain structural types can be discriminated from other substances by differences in basicity. Beyond the presence of basic nitrogen, the most important distinction which can be drawn is whether the amine is primary, secondary, or tertiary. The means by which these classes are distinguished is known as the Hinsberg test.

A Flow Chart for the Hinsberg Test

Hinsberg found that benzenesulfonyl chloride reacted with primary and secondary amines but not with tertiary amines. The three equations which describe the reactivity of benzenesulfonyl chloride with the different classes of amines are presented below.

$$1^\circ \qquad RNH_2 + C_6H_5SO_2Cl \longrightarrow RNH{-}SO_2C_6H_5 + HCl$$

$$2^\circ \qquad R_2NH + C_6H_5SO_2Cl \longrightarrow R_2N{-}SO_2C_6H_5 + HCl$$

$$3^\circ \qquad R_3N + C_6H_5SO_2Cl \longrightarrow \text{usually no reaction}$$

The simple fact that primary and secondary amines react readily with benzenesulfonyl chloride distinguishes them from tertiary amines, which do not react. (Addition compounds of tertiary amines and benzenesulfonyl chloride do exist, but these are generally slow to form and in any event are beyond the scope of this discussion.)

It is clear, then, that reaction with benzenesulfonyl chloride permits one to readily distinguish primary and secondary from tertiary amines. All three classes of amines can form an addition compound with benzenesulfonyl chloride, but only the primary and secondary amines have protons which can be lost to form product. The tertiary amine–benzenesulfonyl chloride adduct cannot lose a proton to form a product and simply reverts to the starting material. Let us now consider the specific products of the Hinsberg reaction with primary and secondary amines, knowing that the tertiary amines do not afford stable products on reaction with benzenesulfonyl chloride. Formation of the primary amine product is shown below.

$$RNH_2 + Cl{-}SO_2C_6H_5 \longrightarrow [R\overset{+}{N}H_2{-}SO_2C_6H_5]Cl^- \xrightarrow{-HCl} RNH{-}SO_2C_6H_5$$

When Hinsberg's reagent reacts with a primary amine, it produces a benzenesulfonamide in which the nitrogen atom bears a hydrogen atom. Because the sulfonyl group is strongly electron-withdrawing, the proton on the nitrogen can be dissociated much more readily (pK_a 11) than it can in a free amine. When such a primary benzenesulfonamide is treated with dilute aqueous base, the insoluble sulfonamide transfers a proton to the base and forms a salt, which often is soluble in water.

$$RNH{-}SO_2{-}C_6H_5 + HO^- \longrightarrow \underset{\text{Water-soluble}}{R\overline{N}{-}SO_2{-}C_6H_5} + H_2O$$

This is the direct analog of the solubilization of carboxylic acids in aqueous base. A secondary amine reacts with benzenesulfonyl chloride to yield a secondary sulfonamide, which has no dissociable hydrogen bonded to nitrogen and therefore will not dissolve when treated with aqueous base. We can distinguish between a primary and a secondary amine then, not by their initial reaction, but by the reaction of the product with dilute aqueous base. Primary amines, which yield primary sulfonamides, give products most of which are soluble in base, whereas the products of secondary amines are insoluble in base.

Considering the reaction overall, the following sequence of events defines these three possibilities. The amine is treated with benzenesulfonyl chloride and base. There may or may not be a reaction evident (heat generated) after the components are shaken together. If no reaction has occurred, the amine was probably a tertiary amine. It should, therefore, be reisolated from the solution and checked for solubility in acid (unreacted free amine will be acid-soluble). If, after shaking, an obvious reaction has taken place but the product is not visible, the amine was probably primary. Recall that the reaction mixture is basic and the primary sulfonamide should be soluble. If a reaction has obviously occurred and a product has deposited, this product is likely the secondary sulfonamide, which should be insoluble in base. These steps are summarized in the following flow chart.

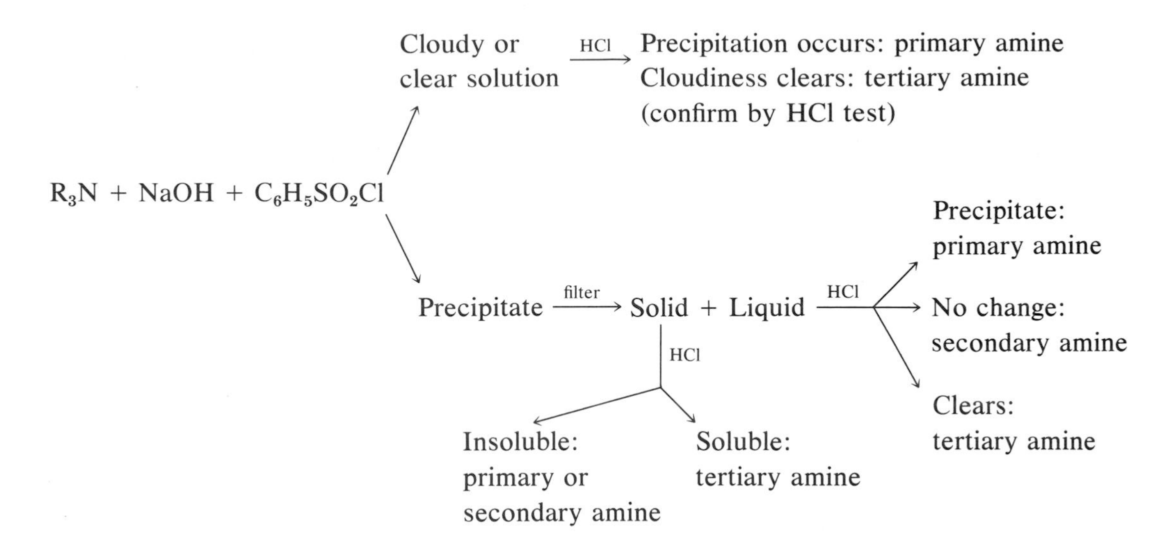

The Hinsberg test must be interpreted very carefully because the sulfonamide derivative has six additional carbon atoms (from the benzene ring). A primary sulfonamide derivative of an amine which had marginal solubility in acid in the first place may well be insoluble in base because its salt is insoluble in water. Great care and good judgment must be exercised in interpreting the results. If the salt does not seem to be soluble, it should be checked to see if perhaps it is unreacted amine. A procedure for the Hinsberg test is given below.

PROCEDURE (in hood)

HINSBERG'S TEST: CLASSIFICATION AND DERIVATIZATION

$$RNH_2 + C_6H_5SO_2Cl + NaOH \longrightarrow RNH{-}SO_2C_6H_5 + NaCl$$

$$RNH{-}SO_2C_6H_5 + NaOH \longrightarrow [RN{-}SO_2C_6H_5]^-Na^+$$

$$R_2NH + C_6H_5SO_2Cl + NaOH \longrightarrow R_2N{-}SO_2C_6H_5 + NaCl$$

$$R_3N + C_6H_5SO_2Cl \longrightarrow \text{no reaction}$$

Place 0.5 mL (or 500 mg) of the amine to be tested in a 13 × 150 mm test tube. Add 1 mL benzenesulfonyl chloride (**hood**) and 10 mL 10% sodium hydroxide solution and then seal with either a rubber or cork stopper. Shake the tube vigorously. After a minute or so of shaking, remove the stopper (there may be a pressure buildup) and sniff **very cautiously** to determine if the odor of benzenesulfonyl chloride persists. Continue shaking and checking for the disappearance of benzenesulfonyl chloride. When the odor of the chloride is no longer detected, check the solution with litmus or pH paper. If the solution is acidic, add sufficient 10% aqueous sodium hydroxide so that the solution is distinctly basic to litmus.

Cool the reaction mixture by immersing the test tube in an ice-water bath. Filter the solution and collect any solid which is present. If a solid is obtained at this stage, it may either be the sulfonamide of a secondary amine, recovered tertiary amine (if the latter was a solid), or the insoluble salt of a primary sulfonamide derivative. Residual tertiary amine will be soluble in 10% aqueous hydrochloric acid. The secondary sulfonamide is no longer basic nor does it have a dissociable proton; as a result, it will be insoluble in both dilute acid and base.

If there is no oily tertiary amine residue and no solid present, acidify the solution by addition of 10% aqueous HCl. If the amine is primary, the sulfonamide will precipitate and should be collected. Either the primary or secondary sulfonamide should be retained and recrystallized (from ethanol) and may be used as derivatives.

Try the Hinsberg test on aniline, morpholine, and triethylamine. Note that certain tertiary amines like pyridine give purple solutions under Hinsberg conditions, and note also that for amines with more than about 10 carbon atoms the test may be unreliable (see text).

Remember: Any time a classification test is conducted, it should be tried first on a compound which is known to give a reliably positive test and on a compound which is known to give a reliably negative test. You cannot know what to look for if you have never before seen the test performed.

How can we deal with the distinction between primary and secondary amines if the Hinsberg test does not always give a reliable result? Fortunately, there are two other tests which *are* reliable, and one of these is very commonly used. Primary amines react readily with nitrous acid (generated from sodium nitrite and hydrochloric acid) as shown below.

$$R\text{—}NH_2 + HONO \longrightarrow R\text{—}N_2^+ \longrightarrow N_2\uparrow + R^+$$

The result is the formation of a diazonium salt. Diazonium compounds are quite reactive and, when maintained at room temperature, quickly lose nitrogen gas. If the amine is readily converted to a diazonium salt and loses nitrogen, bubbles will be visible in a very short period of time. If no bubbles of nitrogen appear, you can assume that the amine is secondary or tertiary rather than primary.

PROCEDURE (in hood)

DIAZOTIZATION OF PRIMARY AMINES

$$R{-}NH_2 + NaNO_2 + HCl \longrightarrow R{-}N_2^+Cl^-$$

$$R{-}N_2^+Cl^- + H_2O \xrightarrow{\text{warm}} R{-}OH + HCl + N_2$$

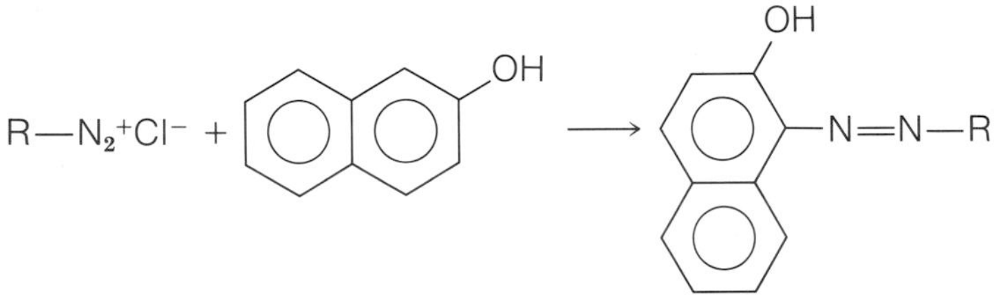

$$R{-}N_2^+Cl^- + \beta\text{-naphthol} \longrightarrow \text{1-(R-N=N)-2-naphthol}$$

Red azo compound

Place 0.5 g (or 0.5 mL) of the amine to be diazotized in a test tube and add 3 mL 6 *N* HCl solution (2 mL concentrated HCl plus 2 mL distilled water). Stir the solution or suspension while cooling in an ice-water or ice-salt bath. Dissolve 0.5 g $NaNO_2$ in 2 mL distilled water and add this solution dropwise with stirring or shaking to the cold amine hydrochloride. The endpoint can be determined by putting a drop of the solution on starch-KI paper. A blue color is observed when excess nitrite is present.

Remove about 2 mL of the diazonium chloride solution and put it in a 10 × 75 mm test tube. Allow the solution to warm to room temperature. If the amine is primary, the unstable diazonium compound will lose nitrogen as an alcohol or phenol is formed in the aqueous medium. Primary aromatic amines form less reactive diazonium ions and brief heating on a steam bath may be helpful.

To 0.1 g β-naphthol in 2 mL 5% aqueous NaOH solution, add 1 mL of the cold diazonium solution. Formation of the red azo compound indicates a primary *aromatic* amine. Dispose of the red azo compound and reaction mixture by very carefully washing down the drain and **do not contact your skin with either.** Some azo compounds and by-products of this reaction may pose a health hazard, but should not be a problem on the scale suggested here.

The Hofmann carbylamine reaction has been used for years as a diagnostic test for primary amines. The test is based on the reaction of primary amines with chloroform and base. Chloroform and base react to form dichlorocarbene, which in turn reacts with a primary amine to form an isonitrile.

$$NaOH + CHCl_3 \longrightarrow :CCl_2 + H_2O + NaCl$$
$$:CCl_2 + R{-}NH_2 \longrightarrow R{-}N{=}C: + 2HCl$$

The basis of this test is that when an isonitrile is present, in even a small concentration, its odor is very, very distinctive and often unpleasant. For example, when a small amount of cyclohexylamine reacts with dichlorocarbene, producing only milligram amounts of cyclohexyl isocyanide, the strong odor can be readily detected even by one stricken with severe nasal congestion. The advantage of the test, therefore, is that it is quite sensitive because of the strong isonitrile odor. This is also the disadvantage of the test.

PROCEDURE (in hood)

THE PTC-HOFMANN CARBYLAMINE TEST FOR PRIMARY AMINES

$$R{-}NH_2 + CHCl_3 + 3NaOH \xrightarrow[PTC]{(:CCl_2)} R{-}NC + 3NaCl$$

Put 1 mL 25% aqueous sodium hydroxide solution in a small test tube. Add 1 mL of chloroform, 2 drops standard catalyst solution, and 1 drop (or about 50 mg) of the unknown primary amine (**hood**). If solid amine is used, it may be dissolved in a minimum of chloroform. Insert in the solution a flat-bladed spatula and spin it to stir the solution. After 30 to 60 s cautiously sniff the top of the tube. The strong odor of isonitrile is an unmistakable indication that a primary amine was present. (*Note:* Although isonitriles are virtually nontoxic, their odors can be obnoxious, and vigorous inhalation should be avoided.)

After the test is complete, pour the reaction mixture into a solution containing 10 mL 10% aqueous HCl solution and 5 mL methanol. Let the solution stand for 10 min before pouring it down the drain. The acid hydrolyzes the isonitrile to the odorless formamide.

Try the test on aniline, *n*-butylamine, and triethylamine so that you are familiar with the odor.

Note that the small amount of chloroform used in this relatively rare test should pose no health problem.

B Spectroscopic Information

The proton nmr spectra of amines will be characteristic of the alkyl or aryl groups attached to nitrogen. If these groups can be identified and the number of them present determined, the amine can be identified. The presence of amine protons is often difficult to detect because they tend to give broad signals. Tertiary amines will not exhibit any proton resonance in any event.

Infrared analysis is the more useful tool for distinguishing the classes of amines. Primary (—NH_2) and secondary ($>$NH) amines show N—H stretching vibrations in the 3000 to 3600 cm^{-1} region (depending on solvent, concentration, etc.) The presence of a band in this region will make it fairly easy to distinguish primary and secondary from tertiary amines, but it will not be as easy to tell the difference between primary and secondary. The value of the Hinsberg test should be apparent in this connection.

Most alkyl amines exhibit a C—N stretching vibration in the 1030 to 1230 cm^{-1} region, and in tertiary amines this often appears as a doublet. This may often be used as additional spectroscopic confirmation of trisubstituted nitrogen.

Infrared spectral analysis is particularly useful in distinguishing between nonbasic amines and nitriles or amides. Nitriles exhibit a characteristic —C≡N vibration near 2000 cm^{-1}. Amides and ureas show carbonyl absorption in the 1650 to 1750 cm^{-1} region. These bands are absent in simple amines.

24.6 REACTIVITY

The chemistry of amines is dominated by the nitrogen lone pair of electrons, which imparts both basic and nucleophilic properties to these compounds. Steric effects notwithstanding, primary, secondary, and tertiary amines all have similar nucleophilic properties. The important difference is that, unlike tertiary amines, primary and secondary amines have a proton which can be lost after the nucleophilic addition has taken place. There is therefore a distinction between the reactivity of primary and secondary versus tertiary amines. This reactivity can be distinguished by noting that primary and secondary amines undergo addition-elimination reactions (nucleophilic addition followed by proton loss), whereas tertiary amines undergo only addition reactions.

The reactivity of the primary and secondary amines can be summarized by the following equations:

$$R'X + R_2NH \longrightarrow R'\overset{+}{N}HR_2\ X^-$$

$$R'\overset{+}{N}HR_2\ X^- \longrightarrow HX + R'NR_2$$

The first step in the reaction represents nucleophilic addition of the amine to some electrophilic species. This intermediate species then loses the proton and the associated leaving group, thus forming a stable derivative. If RX in the first equation above is acetyl chloride, the derivative produced is the acetamide derivative. If RX is benzoyl chloride, the derivative produced is the benzamide derivative. The equations for the formation of both the acetamide and the benzamide derivatives are given below.

$$CH_3—CO—Cl + R_2NH \longrightarrow R_2N—CO—CH_3 + HCl$$

$$C_6H_5—CO—Cl + R_2NH \longrightarrow R_2N—CO—C_6H_5 + HCl$$

24.7 DERIVATIVES OF PRIMARY AND SECONDARY AMINES

A Amides

Benzamides and acetamides are both useful derivatives for primary and secondary amines. Both form quite rapidly and are very inexpensive to prepare. Because the acetic acid derivatives are frequently lower-melting than benzamides and occasionally oily, they are much more difficult to purify than the more crystalline benzamide derivatives. Thus, the benzamides are preferred over the acetamides and are more common derivatives for primary and secondary amines than are acetamides. Note that the reaction of benzoyl chloride, an amine, and sodium hydroxide is the Schotten-Baumann reaction. This reaction was described earlier for ester formation from an alcohol and benzoyl chloride. This Schotten-Baumann reaction is the so-called amine variant; the one discussed in Chap. 26 is the alcohol variant. Note, though, that the reactions are quite similar.

It is easy to recognize that benzoyl chloride and acetyl chloride are both acid chlorides, but it may be less obvious that benzenesulfonyl chloride is also an acid chloride. Whereas carboxylic acid chlorides react with amines to form carboxamides, benzenesulfonyl chlorides form benzenesulfonamides. The principles involved in the reactions are quite similar. Benzenesulfonyl chloride is attacked at the electrophilic sulfur atom by the lone pair of electrons on nitrogen. This intermediate then expels chloride ion and a proton to yield the benzenesulfonamide derivative.

$$C_6H_5SO_2Cl + RNH_2 \longrightarrow (C_6H_5SO_2NH_2R)^+Cl^-$$

$$(C_6H_5SO_2NH_2R)^+Cl^- \longrightarrow C_6H_5SO_2NHR + HCl$$

Not only are the benzenesulfonamides of value in distinguishing among primary, secondary, and tertiary amines, but the sulfonamides formed from the primary and secondary amines can be utilized as derivatives. The benzenesulfonamides are generally quite crystalline and form very rapidly; otherwise the Hinsberg test would not be so valuable. The benzenesulfonamides are also excellent derivatives for amines. Sulfonyl groups often confer crystallinity on a molecule. A point which is often overlooked is that if the Hinsberg test is carefully carried out and the resulting sulfonamide is isolated, not only has the amine been characterized but a derivative has been prepared. If the residue is saved, it may be recrystallized to purity and used as a derivative.

Just as there is very little difference in principle between benzoyl chloride and benzenesulfonyl chloride, the difference between the toluenesulfonamide derivative and the benzenesulfonamide derivative is likewise small (about 14 amu). Toluenesulfonyl chloride is similar to benzenesulfonyl chloride except for the *p*-methyl group on the aromatic ring. The chemistry described above for benzenesulfonyl chloride is the same for toluenesulfonyl chloride; only the melting points of the derivatives differ. Because it is cumbersome to repeatedly say toluenesulfonyl, the abbreviation *tosyl* is frequently used for this group.

To prepare a tosyl derivative of an amine, substitute the appropriate amount of toluenesulfonyl chloride for benzenesulfonyl chloride in the procedure given.

PROCEDURE (in hood)

SCHOTTEN-BAUMANN BENZOYLATION OF AMINES

Place 1 g or 1 mL of the amine and 10 mL 10% aqueous NaOH in a 50-mL glass- or rubber-stoppered flask. Add (**hood**) 2 mL benzoyl chloride, stopper, and shake vigorously (**heat may be evolved**). Be sure the stopper is tight. (If the reaction gets hot, vent the reaction by briefly removing the stopper from time to time.) After 10 to 15 min of shaking, the pungent odor of benzoyl chloride should be gone, indicating that the reaction is complete. Test the solution and add 10% aqueous NaOH solution if it is not basic to litmus. Pour the solution into an equal volume of cold water. The benzamide may separate as a solid or an oil. If a solid is obtained, filter and recrystallize from methanol, ethanol, or acetone-hexane. If an oil separates, extract with dichloromethane, concentrate the solution on the steam bath, and attempt to crystallize.

B Hydrochloride Salts

A derivatization of amines which is commonly employed is formation of the hydrogen chloride addition compound. (This is shown below for tertiary amines.)

$$R_3N + HCl \longrightarrow R_3NH^+Cl^-$$

The advantage of forming hydrochloride salts is that they, like many salts, are highly crystalline and have distinct melting points. Moreover, these derivatives are very inexpensive to prepare. On the other hand, there are several distinct disadvantages inherent in the formation of hydrochloride salts. In general, hydrochlorides are quite hygroscopic, i.e., they readily pick up water, often leading to formation of a gummy or oily residue. These hygroscopic hydrochloride salts cannot be recrystallized to purity in the water-laden atmosphere commonly encountered in undergraduate organic laboratories. Occasionally, these salts are also extremely high-melting, and therefore the melting points cannot readily be determined in the oil-type melting devices most commonly used in undergraduate laboratories. For selected cases, however, hydrochloride salts can be very useful derivatives.

PROCEDURE (in hood)

HYDROCHLORIDE SALTS OF AMINES

$$R_3N + HCl \longrightarrow R_3\overset{+}{N}HCl^-$$

Place 0.5 g (or 0.5 mL) of the amine in a 13 × 100 mm test tube and add 2 mL diethyl ether. Swirl until the amine dissolves. To this solution add 2 drops of concentrated hydrochloric acid. A copious white precipitate should appear immediately. Decant as much of the mother liquor as possible and add fresh ether (2 mL). Stir briefly and then filter.

Wash the solid thus obtained with an additional small portion of diethyl ether. It may be advantageous in some cases to transfer the solid to another small test tube, cover it with 2 mL ether, and grind the solid with a glass rod. After this treatment and a second filtration, the solid should be pure enough for a melting point to be recorded. The operations described above should be done with all deliberate speed, as many hydrochloride salts are hygroscopic.

C Phenylthioureas

Phenylthiourea derivatives are occasionally used as amine derivatives. The reagent known as phenylisothiocyanate has an electrophilic carbon which can add an amine. Only the primary and secondary amines have a dissociable proton, so only primary and secondary amines yield substituted phenylthioureas. The equation for this reaction is given below.

$$R_2NH + C_6H_5—N{=}C{=}S \longrightarrow C_6H_5—NH—CS—NR_2$$

In the first stage of the derivatization reaction, an addition compound forms in which amine nitrogen bonds to the isothiocyanate carbon. At this stage, the amine is protonated and the sulfur exists as an anion. Proton transfer yields the tautomer of the substituted thiourea, which quickly rearranges. Phenylthiourea derivatives are occasionally useful, but phenylisothiocyanate is relatively expensive and water-sensitive. A procedure for the preparation of these derivatives is described below.

PROCEDURE (in hood)

FORMATION OF PHENYLTHIOUREA DERIVATIVES

$$C_6H_5—N{=}C{=}S + R_2NH \longrightarrow C_6H_5—NH—CS—NR_2$$

Place 0.5 g (or 0.5 mL) of the amine in a 16 × 150 mm test tube and add 7 to 10 mL petroleum ether (bp 60 to 70°C). Swirl the mixture to dissolve the amine in the hydrocarbon solvent and then add approximately 0.6 mL phenyl isothiocyanate. Rapidly swirl the mixture so that the reactant is thoroughly dispersed in the solution. In cases in which the reaction is slow, slight warming of the mixture combined with swirling may be required.

For low-molecular-weight amines which are not very soluble in petroleum ether, this reaction can be carried out in ethanol. The crude crystals are collected and recrystallized from ethanol.

24.8 DERIVATIVES OF TERTIARY AMINES

A Hydrochloride Salts

Recall the distinction drawn above between primary, secondary, and tertiary amines. While all three types will form addition compounds, only the primary and secondary amines can lose a proton and form a stable derivative. Tertiary amines form only addition compounds, but some of these are stable and can be used as derivatives. The most common addition compounds formed are those in which the electrophiles are protons and methyl groups. In the proton case an ammonium salt is actually formed. Hydrochloride salt formation is usually more satisfactory for tertiary amines than it ordinarily is for primary and secondary amines. This is because the presence of more organic substituents in the tertiary amine make the salt somewhat lower-melting and also somewhat less hygroscopic. The formation of a hydrochloride salt of a tertiary amine should be effected by the method described above for primary or secondary amine hydrochloride salts.

B Picrate Salts

An organic acid which is commonly used to derivatize tertiary amines is 2,4,6-trinitrophenol (picric acid). Picric acid is a bright yellow solid, whose melting point is 121°C. It is important to remember this melting point because, on many occasions, what *appears* to be the picrate derivative is isolated but, after extensive purification, proves to be only picric acid. The reaction of picric acid with amines occurs by the reaction described above for hydrochloride salts: a proton is transferred from picric acid to the amine in order to form an ammonium salt.

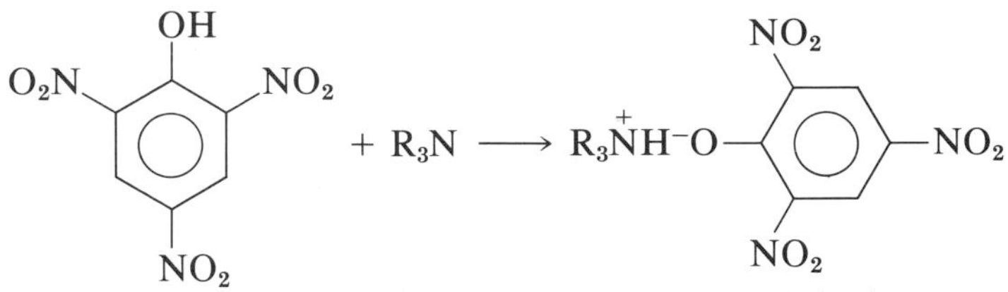

$$\text{Picric acid (2,4,6-}(O_2N)_3C_6H_2OH) + R_3N \longrightarrow R_3\overset{+}{N}H\ {}^{-}O\text{-}C_6H_2(NO_2)_3 \text{ (Picrate derivative)}$$

Picric acid

Picrate derivative

The anion is trinitrophenoxide ion, and these complexes are frequently crystalline and well-behaved. Moreover, these complexes often afford beautiful yellow crystals from aqueous ethanol, which facilitates their handling under laboratory conditions.

The disadvantages of forming a picric acid derivative must also be taken into account. Picric acid itself is an explosive compound; it is, as noted above, 2,4,6-trinitrophenol. If, instead of a hydroxyl function, a methyl group is present in the 1 position, the compound would be 2,4,6-trinitrotoluene. This compound, commonly abbreviated TNT, is quite a potent explosive. Picric acid is not as shock-sensitive as is TNT, but there is a certain danger.

One should, therefore, prepare a picrate derivative only if the laboratory instructor is aware that this derivatization is being conducted, and then only on a small scale.

Another disadvantage, although a minor one, is the fact that picric acid is a staining agent. If not handled carefully, it will yield a lasting yellow stain on whatever it touches. This problem can be overcome by working with particular care when handling this reagent. The procedure written below takes into account the requirements both of safety and of small-scale operation.

PROCEDURE

FORMATION OF PICRATE DERIVATIVES

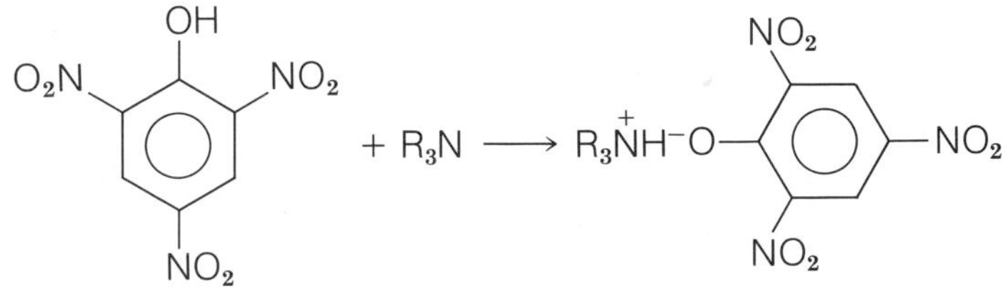

$$\text{2,4,6-}(O_2N)_3C_6H_2OH + R_3N \longrightarrow R_3\overset{+}{N}H\ {}^{-}O\text{-}C_6H_2(NO_2)_3$$

Place 0.5 g (or 0.5 mL) of the amine to be derivatized in a 13 × 100 mm test tube and add 5 mL 95% ethanol. Place 0.75 g picric acid in a small Erlenmeyer flask and add 15 to 20 mL ethanol. Swirl until the picric acid is dissolved. Combine the two solutions and heat briefly on a steam bath (heat to boiling). Allow the yellow solution to cool slowly to room temperature. Collect the salt which crystallizes and recrystallize it from 95% ethanol.

Note: It is important to use approximately equivalent amounts of picric acid and the amine (the weights can be calculated based on the presumed structure of the amine). Excess picric acid tends to coprecipitate with the derivative, complicating purification.

C Methiodide Salts

The third common derivative of tertiary amines is the adduct which forms with methyl iodide. The quaternary ammonium compound thus formed bears a positive charge, and the ammonium cation is associated with the iodide anion.

$$R_3N + CH_3I \longrightarrow R_3\overset{+}{N}{-}CH_3\ I^-$$

The methiodide salts thus formed are frequently crystalline solids and are useful derivatives. They generally form quite readily and often can be recrystallized with ease. The difficulty with methiodide salts is that these compounds, like the hydrochloride salts, are sometimes hygroscopic and, although highly crystalline, may pick up water.

PROCEDURE (in hood)

FORMATION OF TERTIARY AMINE METHIODIDE SALTS

$$R_3N + CH_3I \longrightarrow R_3\overset{+}{N}{-}CH_3\ I^-$$

Caution: Methyl iodide is a volatile alkylating agent and may be dangerous. Inhalation should be avoided.

Place 0.5 g (or 0.5 mL) of the amine to be derivatized in a 16 × 130 mm test tube and add 5 to 10 mL ethyl acetate. Swirl to dissolve the amine. To this solution, add 1 mL methyl iodide, cork the tube and allow to stand undisturbed for 1 to 2 h. In most cases, the methiodide salt crystallizes from the ethyl acetate solution. Recrystallization, if necessary, should be effected with ethyl acetate–ether or, in limited cases, with ethyl alcohol.

Note: Methiodide salts are often hygroscopic. Prolonged exposure to atmospheric moisture should be avoided.

D Methyl Tosylate Salts

You will note in the tables describing tertiary amine derivatives that occasionally a methyl tosylate is reported. Methyl-*p*-toluenesulfonate is conceptually related to methyl iodide. In this case, instead of iodide, the anion is *p*-toluenesulfonate. Reactions are conducted in the same way and the products which form are also quaternary methylammonium salts, but the anion is toluenesulfonate rather than iodide.

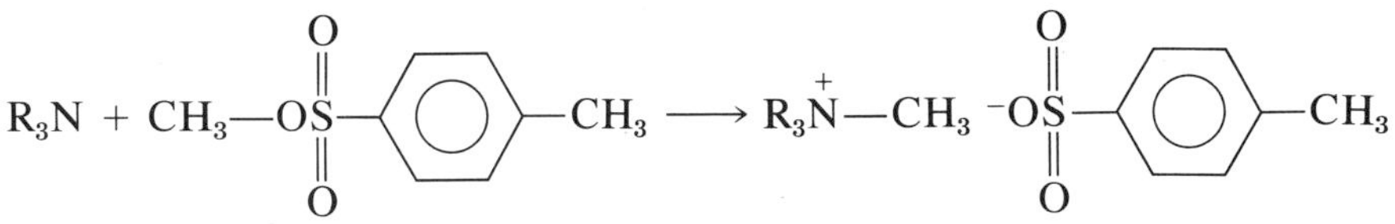

$$R_3N + CH_3{-}OS(=O)_2{-}C_6H_4{-}CH_3 \longrightarrow R_3\overset{+}{N}{-}CH_3 \; {}^{-}OS(=O)_2{-}C_6H_4{-}CH_3$$

PROCEDURE (in hood)

FORMATION OF p-TOLUENESULFONATE SALTS

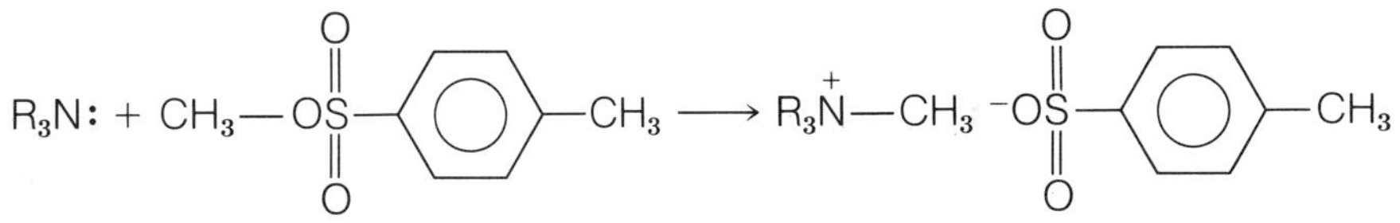

$$R_3N\!: + CH_3{-}OS(=O)_2{-}C_6H_4{-}CH_3 \longrightarrow R_3\overset{+}{N}{-}CH_3 \; {}^{-}OS(=O)_2{-}C_6H_4{-}CH_3$$

Place 0.5 g (or 0.5 mL) of the tertiary amine in a 13 × 100 mm test tube and add 2 to 5 mL ethyl acetate, acetonitrile (**hood**), or absolute alcohol. Swirl to dissolve the amine in the chosen solvent. To this solution add 0.5 g methyl *p*-toluenesulfonate. Swirl the mixture to effect solution and allow it to stand for 1 to 2 h. In most cases the salt will form during this time period and crystallize. Isolate by suction filtration.

If crystallization is not evident at the end of 2 h, the mixture is heated for 15 to 20 min at reflux, cooled to room temperature, and then allowed to stand undisturbed until crystals are deposited. The quaternary ammonium salt should be crystallized from ethyl acetate or ethyl acetate–ether.

THE CARBONYL GROUP

25.1	General Tendencies	**553**
25.2	Odor	**554**
25.3	Structural Variety	**555**
25.4	Other Structural Variations	**558**
25.5	Classification	**558**
	A The 2,4-dinitrophenylhydrazine (2,4-DNP) test	
	B Classification scheme for carbonyl compounds	
	C Reactivity: What a difference a proton makes	
	D The Tollens test	
	E The Fehling and Benedict tests	
	F The Baeyer permanganate test	
	G The Purpald test	
	H The fuchsin aldehyde test (Schiff's test)	
	I The iodoform test for methyl ketones	
25.6	Spectroscopic Confirmation of Structure	**569**
25.7	Derivatives of Aldehydes and Ketones	**571**
	A 2,4-Dinitrophenylhydrazone formation	
	B Semicarbazone derivatives	
	C Oximes	
	D Dimedone derivatives	
	E Oxidation to the corresponding carboxylic acid (aldehydes)	
	1 Potassium permanganate method	
	2 Oxidation by the Cannizzaro reaction	
	F Reduction of aldehydes and ketones	

25.1 GENERAL TENDENCIES

The carbonyl group can be characterized as that functional group containing carbon and oxygen with a double bond between the elements. Many organic molecules are possible which contain the C═O group, and the exact nature of each molecule will depend on what groups are attached to the two other bonding sites on carbon. The differences in properties caused by the bonding of different groups to carbon are discussed below, but for the moment let us focus on the nature of the carbonyl group.

The electronegativity of carbon (on the Pauling scale) is 2.5 and that of oxygen is 3.5. Because of this rather large difference in electronegativity, the carbonyl group is distinctly polarized. The carbon end of this dipolar species is less electronegative and as a consequence is Lewis acidic. (If you are not already familiar with it, you should check Sec. 23.2 in order to understand Lewis acidity.) Lewis acidity or electrophilicity at a carbon atom means that nucleophiles or other electron-bearing species will attack the carbon and often attach themselves to it. The double bond is negatively polarized because oxygen is more electronegative and also bears two electron pairs. The oxygen end of this dipolar bond can behave as a nucleophile, although in most reactions it behaves simply as a proton acceptor. In general, addition of a proton to the oxygen end of the carbonyl group imparts an even greater electropositive character to the carbon atom. In fact, many nucleophilic additions to carbonyl are catalyzed either by protons or by boron trifluoride etherate ($BF_3 \cdot Et_2O$). In these particular cases, the proton or Lewis acid attaches itself to one of the lone pairs on oxygen and increases the overall polarity of the species. After addition of the nucleophile to the carbon end of the carbonyl group, either boron trifluoride or a proton is usually lost as the new product forms.

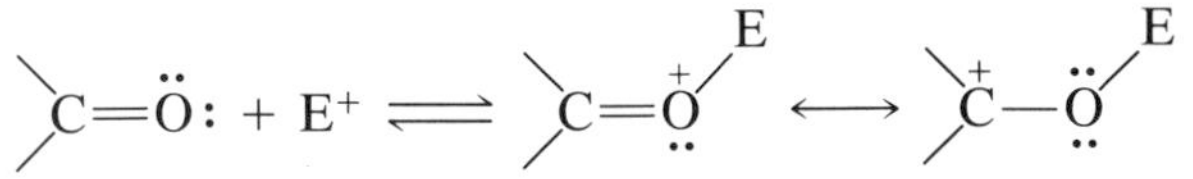

$$\text{>C=}\ddot{\text{O}}\text{:} + \text{E}^+ \rightleftharpoons \text{>C=}\overset{+}{\ddot{\text{O}}}\text{—E} \longleftrightarrow \text{>}\overset{+}{\text{C}}\text{—}\ddot{\underset{\cdot\cdot}{\text{O}}}\text{—E}$$

The polarity of the carbonyl group influences the solubility properties of certain carbonyl-containing molecules. If both the available bonding sites on the carbon atom are occupied by medium-size or large alkyl groups, the carbonyl-containing compound will likely be soluble in hydrocarbon solvents. Obviously, it is not uncommon for an organic compound to be soluble in organic solvents. The interesting property of the group is that because of its polarity (i.e., its dipole) its negative end can form hydrogen bonds with water and with alcohol-type proton-donor solvents, as shown below.

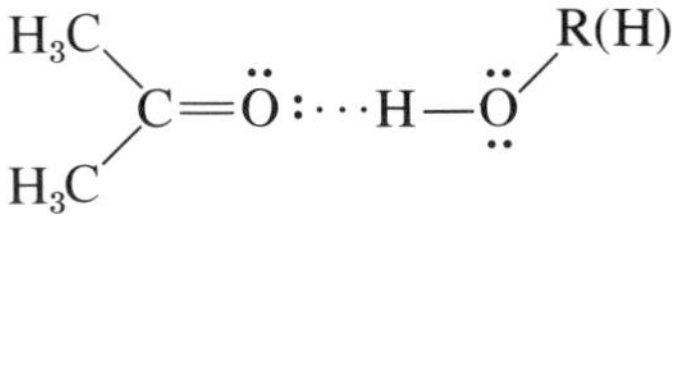

$$(H_3C)_2\text{C=}\ddot{\text{O}}\text{:}\cdots\text{H—}\ddot{\underset{\cdot\cdot}{\text{O}}}\text{—R(H)}$$

If the alkyl groups are not too large, the molecule containing the carbonyl may also be soluble in water. This is generally true for the low-molecular-weight aldehydes and ketones. As one might expect, butane is virtually insoluble in water. Acetone (MW 58), which has the same molecular weight as butane, is freely soluble in water. As the alkyl groups or aryl groups (i.e., the organic residues attached to the carbonyl carbon atom) become larger and larger, the organic portion of the molecule begins to dominate the molecule's solubility behavior. Di-*n*-hexyl ketone, for example, is insoluble in water but is freely soluble in hexane.

25.2 ODOR

We noted above that the solubility properties of aldehydes, ketones, and other carbon-containing compounds as well are dominated either by the carbonyl group or by the attached hydrocarbon residues. In intermediate ranges, there is slight solubility in both kinds of solvents, but at either extreme the solubility properties of the molecule are dominated by either one portion of the molecule or the other. The odor of a molecule, on the other hand, is always the result of the total structure. It is generally believed that odor receptor sites have distinct shapes. Molecules which have quite different functional groups and reactivities may have similar odors because the molecules have similar shapes. While it is difficult to smell a methyl group at C-7 in dodecane or an ethyl group in a steroid, somewhat simpler generalizations can be made and certain common odors can be learned. In other words, common sense can be applied at almost any level. We have noted in Chap. 23 on acids that a consideration of names can often lead to an inference about the odor of a substance. Odor is a less useful property in the case of aldehydes and ketones, but it is certainly good to consider it. Two common examples are anisaldehyde, which has no color but has a distinct licorice odor, and benzaldehyde, which has the odor of almond oil. Another molecule, whose odor you probably recall from molecular-weight determinations in introductory chemistry, is camphor.

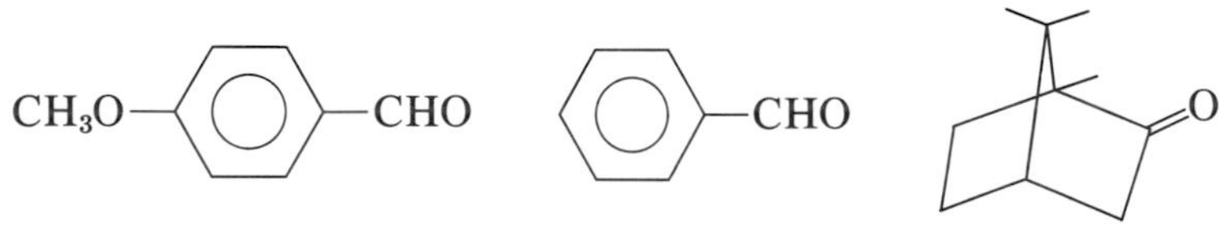

$CH_3O-C_6H_4-CHO$ $\qquad$ C_6H_5-CHO

Anisaldehyde (4-methoxybenzaldehyde) $\qquad$ Benzaldehyde $\qquad$ Camphor

Camphor is a so-called terpene ketone which is roughly spherical in shape. Its odor, in fact, is the standard for spherically shaped molecules. Most spherical molecules are said to have *camphoraceous* odors and smell more or less like mothballs. Carvone (called oil of caraway) is also a terpene ketone, and you may recall its pleasant odor if you did the experiment in Sec. 2.2.

Two aldehydes whose names imply their odors are cinnamaldehyde and vanillin. Although the odor of cinnamon is faint in cinnamaldehyde, it is nevertheless distinct. Vanillin, 4-hydroxy-3-methoxybenzaldehyde, has an odor that is almost sickeningly intense despite the fact that it is a solid.

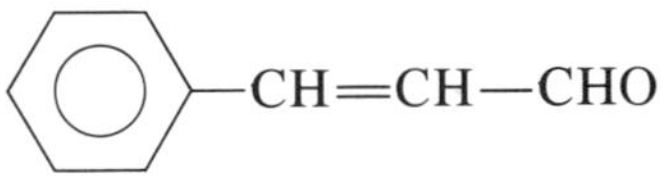

C_6H_5—CH=CH—CHO HO—C_6H_3(OCH$_3$)—CHO

Cinnamaldehyde 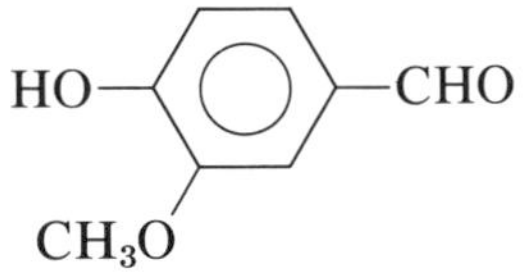 Vanillin

It is hard to conjure up an odor for a compound discussed only in terms of a systematic name such as 2-methylbutyl 3-hexyl ketone. Some compounds which are frequently referred to by their systematic names have common names which do give a clue to their odors. It is always advantageous to look up the names and properties of several possibilities in your identification attempt to see if there is anything unusual about the odor of any of them. This can often lead you, if not to a definite identification, at least to a strong presumption.

25.3 STRUCTURAL VARIETY

We have characterized the carbonyl group as a functionality or substructure containing a carbon-oxygen double bond and have said that the carbon atom is bonded to two other groups. The general structure by which an enormous variety of compounds is defined is written below.

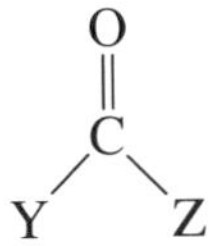

$$\mathrm{Y{-}C({=}O){-}Z}$$

Compounds of this general type are listed in Table 25.1. If Y is an organic group, i.e., an alkyl or aryl group, and Z is hydrogen, then the structure is an aldehyde. (The —CH=O group is called the formyl or carboxaldehyde function.) If Z is also alkyl or aryl, then the compound is a ketone. The distinction, therefore, between aldehyde and ketone is whether or not one of the two groups (Y or Z) is hydrogen. The simplest possible aldehyde (called formaldehyde) violates this rule because it has two hydrogen atoms; all other aldehydes have a carbonyl function bearing a single hydrogen atom. The carbonyl group in ketones is bonded to two organic *residues,* the so-called R groups.

If Y is a typical alkyl or aryl organic residue (methyl, ethyl, phenyl, etc.) and Z is a hydroxyl group, the total function R—CO—Z is now R—CO—OH (a carboxylic acid). If Z is OR or OAr, the substance is either an alkyl or aromatic ester. When Z is NH_2, NHR, or NR_2, the substance is a primary, second-

TABLE 25.1
Compounds of the type Y—CO—Z

Y	Z	Example	Common name	Compound class
H	H	H—CO—H	Formaldehyde	Aldehyde
R, Ar	H	CH_3—CO—H	Acetaldehyde	Aldehyde
R, Ar	R, Ar	CH_3—CO—CH_3	Acetone	Ketone
R, Ar	OH	C_6H_5—CO—OH	Benzoic acid	Carboxylic acid
R, Ar	OR, OAr	C_6H_5—CO—OCH_3	Methyl benzoate	Ester
X	X	Cl—CO—Cl	Phosgene	Acyl halide
R, Ar	X	CH_3—CO—Cl	Acetyl chloride	Acyl halide
H_2N	H_2N	H_2N—CO—NH_2	Urea	Amide (urea)
R, Ar	$NR_2(H_2)$	CH_3—CO—NH_2	Acetamide	Amide

ary, or tertiary amide, respectively. If Z is a halogen (sometimes abbreviated X), i.e., fluorine, chlorine, bromine, or iodine, the compound is an acyl halide. You are probably already familiar with acetyl chloride from Exp. 16.2B. Another obvious variation occurs where Y and Z are both chlorine. That molecule, Cl—CO—Cl, is phosgene, the deadly gas used in World War I. Where Y and Z are both OH, the compound is the hypothetical carbonic acid.

The reactions that any carbonyl group will undergo are essentially similar and are ordinarily dominated by the two reaction types discussed earlier, electrophilic addition at oxygen or nucleophilic addition at carbon. Which product results from these reactions is determined by the nature of Y and Z. In the case of aldehydes and ketones, reactions with a nucleophile most commonly occur by addition of the nucleophile to the carbonyl carbon. The lone pair of electrons on the nucleophile attacks the carbon atom, causing electron flow towards the oxygen atom, which becomes an $—O^-$ function. This is formulated in the equation below

$$Nu^- + R—\overset{\overset{O}{\|}}{C}—R' \longrightarrow R—\underset{\underset{Nu}{|}}{\overset{\overset{O^-}{|}}{C}}—R' \xrightarrow{H^+} R—\underset{\underset{Nu}{|}}{\overset{\overset{OH}{|}}{C}}—R'$$

$$R—\underset{\underset{Nu}{|}}{\overset{\overset{O^-}{|}}{C}}—R' \xrightarrow{-R^-} Nu—\overset{\overset{O}{\|}}{C}—R' \qquad\qquad R—\underset{\underset{Nu}{|}}{\overset{\overset{OH}{|}}{C}}—R' \xrightarrow{-H_2O} \begin{matrix} R \\ & C{=}Nu \\ R' \end{matrix}$$

The intermediate alkoxide can become protonated or, if the nucleophile which added has a proton available, elimination of water can occur. Many reagents which are used for derivatizing aldehydes and ketones are substituted amines. These substances are hydrogen-bearing nucleophiles, which attack the carbon atom and then eliminate hydroxide or water, which results in the formation of a double bond between the carbonyl carbon and the nucleophile. For example, hydroxylamine (NH_2—OH), phenylhydrazine (H_2N—NH—C_6H_5), 2,4-dinitrophenylhydrazine [H_2N—NH—$C_6H_3(NO_2)_2$], and semicarbazide (H_2N—NH—CO—NH_2) are all common derivatizing agents. The general reaction of these nucleophiles with acetone is shown in the equation below.

$$R{-}NH_2 + Me_2C{=}O \longrightarrow R{-}NH{-}\underset{Me}{\overset{OH}{\underset{|}{\overset{|}{C}}}}{-}Me \longrightarrow R{-}N{=}C\begin{matrix} ^{\diagup Me} \\ _{\diagdown Me} \end{matrix} + H_2O$$

The intermediate structure, which contains both an OH group and a nitrogen bonded to it, may look strange at first but really is not unusual. These compounds are the nitrogen analogs of hemiacetals and are called either *carbinolamines* or *aminals*.

Consider the structural variation possible by changes in R in the equation above. If R is OH, the compound is the oxime derivative of acetone. (Note that the melting point of acetone oxime is 59°C.) If R is a urea group, then the derivative is acetone semicarbazone (mp 190°C). When the organic residue is 2,4-dinitrophenylamino, the compound formed is the 2,4-dinitrophenylhydrazone of acetone (mp 126°C). Acetone is mentioned in particular here because it is a simple ketone and, more importantly, because these acetone derivatives may be encountered more frequently than hoped for. If glassware has been washed with acetone immediately before it is used to carry out classification or derivatization reactions, the acetone derivative may form as well as the derivative of the unknown. It is good, therefore, to be familiar with the melting points of these derivatives, as this can save confusion and distress in the future.

The reactivities of carbonyl compounds will be discussed in more detail when we deal with classification tests, but at this point a few details should be mentioned. We noted above that because there is a hydrogen atom bonded to the carbonyl group in an aldehyde, aldehydes are less sterically hindered than ketones and tend to react more rapidly with the same reagents. In addition, aldehydes are slightly more reactive because the hydrogen atom is not as electron-releasing as an alkyl group; the carbon atom is therefore somewhat more electrophilic than it is when alkyl-substituted. If there is an aromatic ring bonded to carbon, the carbonyl group will tend to conjugate with it. This will alter the electron density about carbon as well. In general, aromatic aldehydes and ketones are slightly less reactive than their aliphatic counterparts, but such substances as benzaldehyde are clearly more reactive than diethyl or dibutyl ketone.

25.4 OTHER STRUCTURAL VARIATIONS

If Y (in Y—CO—Z) is an alkyl or an aryl group and if Z is a noncarbon functional group, reactions other than addition followed by elimination of water become more common. For example, if Z is halogen or alkoxy, the coupound is an acyl halide or an ester, respectively. In either case, nucleophilic addition to the carbonyl begins in the same way as it does for an aldehyde or ketone. A tetrahedral intermediate in which the oxygen atom bears a charge is formed but, instead of losing water (with loss of a proton from the nucleophile), the Z group, i.e., chloro, bromo, or alkoxy, is lost instead because it is usually a better leaving group than R^- or Ar^-. The result is a new carbonyl-containing compound, and Z now corresponds to the incoming nucleophile. This reaction is formulated in the equation below.

$$R—CO—Z + Nu^- \longrightarrow R—CO—Nu + Z^-$$

A simple example is the reaction of an acid chloride with ammonia. In this case the incoming nucleophile, NH_3, adds to the carbonyl, but instead of eliminating water expels chloride and a proton (HCl). The result is the formation of an amide. Thus, benzoyl chloride plus ammonia yields benzamide and HCl.

$$R—CO—Cl + NH_3 \longrightarrow R—CO—NH_2 + HCl$$

This technique is also used for the derivatization of esters. The ester is heated with phenylhydrazine to form a hydrazinoamide, which frequently has highly crystalline properties.

25.5 CLASSIFICATION

A The 2,4-Dinitrophenylhydrazine (2,4-DNP) Test

It is presumed at this stage that preliminary classification tests indicate that the unknown compound is neither an acid nor a base. Since a neutral compound may fall into one of several different classes, the simplest step to take at this stage is to treat a small sample of the material with 2,4-dinitrophenylhydrazine (DNP) reagent. (This reagent is a solution of 2,4-DNP in alcohol and phosphoric acid or in diethylene glycol and HCl.) A small amount of the unknown compound can be treated with a drop or so of DNP reagent. If a carbonyl group is present in the form of either an aldehyde or ketone, a yellow to red precipitate will usually appear.

The reaction taking place here is discussed above for the general case. Here, phosphoric acid protonates the carbonyl oxygen atom, making carbon more electrophilic, and the hydrazine nucleophile attacks the carbon atom. Loss of water gives the carbon-nitrogen double bond of the hydrazone derivative. The presence of the nitro groups in the aromatic ring confers color and high crystallinity on these derivatives.

$$O_2N-C_6H_3(NO_2)-NH-NH_2 + R-\overset{O}{\overset{\|}{C}}-R' \xrightarrow{H^+} O_2N-C_6H_3(NO_2)-NH-N=C\begin{matrix} R \\ R' \end{matrix} + H_2O$$

There are several very distinct advantages to using the DNP test as the first classification test after neutral solubility behavior has been demonstrated. The DNP test is rapid and usually definitive. Precipitation of a yellow, orange, or red solid quickly indicates the presence of a ketone or an aldehyde. This positive reaction is almost unmistakable. A further advantage is that, when the test is done on a slightly larger scale, the precipitate can be recovered, recrystallized, and used as a solid derivative.

There are certain difficulties which attend the use of DNP reagent. Some carbonyl compounds (usually ketones) react quite slowly. Examples of this include the ketones benzophenone and benzil. A few, though not many, compounds react with 2,4-dinitrophenylhydrazine to yield an oil rather than a crystalline derivative. We emphasize that examples of this kind, which include 3-phenyl-2-propanone and *o*-methoxyacetophenone, are rare. In general, however, even low-molecular-weight substances like acetone and methyl ethyl ketone give crystalline 2,4-dinitrophenylhydrazone derivatives. The low-molecular-weight dicarbonyl compound biacetyl may react at either or both carbonyl groups. The bis(dinitrophenyl)hydrazone of biacetyl melts at well over 300°C.

$$C_6H_5-\overset{O}{\overset{\|}{C}}-C_6H_5$$

Benzophenone

$$C_6H_5-\overset{O}{\overset{\|}{C}}-\overset{O}{\overset{\|}{C}}-C_6H_5$$

Benzil

$$o\text{-}CH_3O-C_6H_4-\overset{O}{\overset{\|}{C}}-CH_3$$

o-Methoxyacetophenone

$$CH_3-\overset{O}{\overset{\|}{C}}-\overset{O}{\overset{\|}{C}}-CH_3$$

Biacetyl

A somewhat more common difficulty is encountered with less reactive carbonyl compounds, especially if they are also of higher molecular weight. Not only do these compounds form dinitrophenylhydrazone derivatives slowly, but they may cause the DNP reagent to precipitate. A red solid is obtained in these cases but it is reagent rather than product. This is certainly a difficulty, but one which can easily be recognized. 2,4-Dinitrophenylhydrazine is a bright red solid which melts at 198°C with decomposition. A final difficulty is that dinitrophenylhydrazine can form what is known as a *charge-transfer complex*. The DNP reagent contains an aromatic ring which is quite electron-deficient. Some π electron density tends to be withdrawn from the benzene ring by the nitro groups. As a result, the aromatic compound is said to be a π acid. An electron-rich aromatic ring (a π base) can react with it to form a charge-transfer complex. Situations of this sort, however, are relatively uncommon. When they do occur, a simple test is to substitute *p*-nitrophenylhydrazine, which is not nearly as reactive in the charge-transfer sense, for 2,4-dinitrophenylhydrazine. If a derivative is still obtained, complex formation can generally be ruled out.

PROCEDURE

2,4-DINITROPHENYLHYDRAZINE CLASSIFICATION TEST FOR ALDEHYDES AND KETONES

$$O_2N\text{-}C_6H_3(NO_2)\text{-}NH\text{—}NH_2 + O\text{=}CR_2 \longrightarrow O_2N\text{-}C_6H_3(NO_2)\text{-}NH\text{—}N\text{=}CR_2 + H_2O$$

Place 1 mL 2,4-dinitrophenylhydrazine–diethylene glycol reagent in a small test tube. To this dark red-orange solution is added 1 drop of liquid sample or approximately 50 mg of solid sample. The test tube is agitated or swirled until the sample is dispersed or dissolved in the reagent. To this solution is added 2 to 3 drops of concentrated hydrochloric acid. The solution is swirled to completely mix the reagents.

On addition of the hydrochloric acid, an immediate color change from dark red-orange to light yellow should be observed. Very soon after that (1 or 2 min) a precipitate should appear if the material is either an aldehyde or ketone. If no solid appears quickly, agitation is continued at room temperature for 3 to 5 min. Aldehydes and ketones, even deactivated or sterically hindered ones, will yield a derivative within 5 min. In very difficult cases warm the mixture on the steam bath.

If the sample is relatively insoluble in diethylene glycol, it should first be dissolved in a minimum amount of diglyme (the dimethyl ether of diethylene glycol), and this sample-containing solution should be added to the test solution as described above.

[The 2,4-dinitrophenylhydrazine reagent is made by dissolving 1 g 2,4-dinitrophenylhydrazine in 60 mL diethylene glycol (with warming). The solution is allowed to stand at room temperature for 15 to 20 min, and the clear, red-orange solution is decanted from a small amount of residue. This solution is stable for extended periods but should be kept in a closed amber bottle until needed. One milliliter of this reagent is used in the test above.]

It has been noted that certain long-chain aliphatic ketones (4-heptanone, 2-octanone), tend to form an oil rather than a precipitate in the test. The oil in most cases crystallizes in 5 to 10 min. It is obvious, however, even when the derivative is an oil that a positive reaction has taken place.

The simplest way to tell if a precipitate is 2,4-DNP itself or the expected derivative is to isolate it and determine its melting point.

B Classification Scheme for Carbonyl Compounds

Solubility	Unknown is:
	insoluble in dilute acid
	insoluble in dilute base
	soluble in concentrated H_2SO_4
Classification test	
Aldehydes	DNP test positive
	Baeyer test positive
	Tollens test positive
	Fehling and Benedict tests positive
	Purpald test positive
	Fuchsin (Schiff's) test positive
Methyl ketones	Iodoform test positive
Derivatives	DNP derivative
	Semicarbazones
	Oximes
	Dimedone
	Acids from aldehydes
	Permanganate procedure
	Cannizzaro procedure
	Reduction to the corresponding alcohols

C Reactivity: What a Difference a Proton Makes

We have mentioned the two important reactivity differences which distinguish aldehydes from ketones. There are several differences in reactivity besides rate of reaction, and these are discussed below.

Because the carbonyl group can stabilize an adjacent negative charge, and because the aldehydes are generally reactive electrophiles, the self-condensation of aldehydes is occasionally a problem, particularly with low-molecular-weight aldehydes. The simplest example of self-condensation is the reaction of acetaldehyde with itself. Two molecules of acetaldehyde which condense under basic catalysis yield 3-hydroxybutanal, a compound given the trivial name aldol.

$$\underset{\text{Acetaldehyde}}{2CH_3{-}CHO} \longrightarrow \underset{\text{Aldol}}{CH_3{-}\overset{\overset{\displaystyle OH}{|}}{C}H{-}CH_2{-}CHO}$$

This molecule contains both hydroxyl and aldehyde functions, the presence of which is reflected in the name. Because of the aldol condensation, it is not uncommon to find old samples of isobutyraldehyde which distill from 120 to 150°C instead of at 66°C. The high-boiling material is not an isomer; rather it is the so-called aldol dimer.

$$2(CH_3)_2CH{-}CHO \longrightarrow (CH_3)_2CH{-}\overset{\overset{\displaystyle OH}{|}}{C}H{-}\underset{\underset{\displaystyle CHO}{|}}{C}(CH_3)_2$$

One must always be on the lookout for this behavior. If attempted distillation of an aldehyde yields a forerun at one temperature and then gives another fraction at twice the boiling point or higher, it is likely that aldol dimer formation has occurred. There are philosophical differences on how to deal with this problem, but in the authors' opinion the best approach is to discard the contaminated sample and obtain a fresh one. It is occasionally claimed that contaminants in the glass cause aldolization of fresh samples. In the authors' experience, a fresh, clean, uncontaminated sample of isobutyraldehyde or propanal distills with no appreciable aldol dimer formation. In general, aldol dimer formation results from long storage under unfavorable conditions.

D The Tollens Test

An important difference in reactivity between aldehydes and ketones is their potential for oxidation. Ketones do not oxidize as readily as aldehydes because carbon-carbon bond cleavage would be required. Aldehydes can often be oxidized quite rapidly to carboxylic acids, which then resist further oxidation.

$$\underset{\text{Aldehyde}}{R{-}\overset{\overset{\displaystyle O}{\|}}{C}{-}H} \xrightarrow{\text{oxidation}} \underset{\text{Carboxylic acid}}{R{-}\overset{\overset{\displaystyle O}{\|}}{C}{-}O{-}H}$$

Several classical characterization or classification tests have been used which rely on the oxidizability of aldehydes. A carbonyl compound, if it is an aldehyde, will be transformed into a carboxylic acid by an oxidizing metal. Oxidation of the aldehyde is accompanied by reduction of the metal ion. If the oxidizing agent is the silver cation, silver metal will be deposited as a mirror on the surface of the reaction vessel. This is the basis of the Tollens test, in which silver(I) is reduced to a silver(0) mirror. It is interesting to note that this reaction is the very one used commercially for many years in the manufacture of looking glasses. The carbonyl compound used as the reducing agent is formaldehyde, which oxidizes to formic acid and then to CO_2. The silver oxidizing agent is originally in solution and, when reduced, deposits on the glass plate.

PROCEDURE

CLASSIFICATION TEST FOR ALDEHYDES: THE TOLLENS TEST

$$R{-}CHO + [Ag(NH_3)_2]^+ + HO^- \longrightarrow R{-}COO^-NH_4^+ + Ag^\circ \downarrow + H_2O$$

Add 1 drop of the aldehyde or 3 drops of an ethanol solution of the aldehyde to 2 mL ammoniacal silver (Tollens) solution contained in a 10 × 75 mm test tube. The test tube must be very clean or silver will not deposit. Swirl for 2 min. If no silver deposit is visible, warm the solution on a steam bath for about 5 min. If no silver mirror is observed and the test tube was clean, the substance was probably not an aldehyde. As soon as a conclusion is drawn, discard the solution.

(*Note:* If the test tube is not clean but an aldehyde is present, the silver will deposit as a dark brown to black colloidal material. This positive test is much less satisfactory than the observation of the silver mirror above.)

Try the Tollens test on formaldehyde (aqueous formalin solution), benzaldehyde, and diethyl ketone. Note that the presence of a reactive halogen compound (e.g., benzoyl chloride) will result in precipitation of the silver halide salt and may lead to difficulties. Note also that if the sample requires heating on a steam bath and if the substance has a boiling point below that of water, it may be lost when the test solution is heated.

Preparation of the Tollens reagent One milliliter of 10% aqueous silver nitrate solution (1 g $AgNO_3$ in 10 mL water) and 1 mL aqueous sodium hydroxide solution (1 g NaOH in 10 mL water) are mixed in a clean test tube. Dilute aqueous ammonia solution (approximately 2%) is added dropwise until the silver oxide is just dissolved. Add a small amount of the aldehyde to this solution as indicated above.

E The Fehling and Benedict Tests

In the Tollens test, an oxidizable aldehyde is distinguished from a nonoxidizable ketone by the precipitation of silver. The silver-ion reactant is maintained in solution as a complex with ammonia. Related conceptually to the Tollens test are two other important tests, Benedict's test and Fehling's test. Each uses cupric ion [Cu(II) ion] which oxidizes an aldehyde to a carboxylic acid. The cupric [Cu(II)] ion is reduced to cuprous [Cu(I)] ion, which precipitates as its brick red oxide.

$$\text{R—CHO} + \underset{\substack{\text{(Citrate or}\\ \text{tartrate}\\ \text{complex)}}}{Cu^{+}} \longrightarrow \text{R—CO—OH} + \underset{\text{(Brick red)}}{Cu_2O\downarrow}$$

The principal difference between Benedict's and Fehling's solutions is the substance which complexes the cupric ion and keeps it in solution. In Benedict's solution the complexing substance is citric acid; in Fehling's solution it is tartaric acid. The reaction in each of these cases is the same, however, and the appearance of a cupric oxide precipitate is indicative of an aldehyde rather than a ketone.

The Fehling and Benedict tests generally work best for α-hydroxy aldehydes and sugars. Procedures for these tests are not included because in the authors' opinion the ones presented in this chapter are more generally reliable. If it seems necessary to conduct this test, check with your instructor, who can help you locate a procedure.

F The Baeyer Permanganate Test

Although the Fehling, Benedict, and Tollens tests are often referred to, the latter generally is more useful and reliable. Recent advances have led to other newer tests, which are described here. The first of these is an oxidative process based on the phase-transfer method (see Sec. 4.7). In this classification test, the oxidizing agent potassium permanganate is solubilized in toluene by addition of a catalyst solution. Permanganate dissolved in toluene will react readily with an aldehyde but not with a ketone, which resists oxidation. The permanganate ion (MnO_4^-) will be reduced to manganese dioxide (MnO_2). The permanganate reagent is a transparent purple solution, whereas manganese dioxide is a brown, scummy material, insoluble in toluene. It is therefore easy to tell when a test is positive (permanganate has been reduced). The detailed procedure for this classification test follows.

PROCEDURE

BAEYER TEST FOR UNSATURATION (PHASE-TRANSFER METHOD)

$$R\text{—CHO} + MnO_4^- \longrightarrow R\text{—COO}^- + MnO_2\downarrow$$

One milliliter of standard catalyst solution is added to a clean 13 × 100 mm test tube. A small sample (1 drop or 50 mg gives a mound about 5 mm in diameter) of the material to be tested is added, and the test tube is swirled until solution occurs. To this mixture add 5 drops 2% aqueous $KMnO_4$ (potassium permanganate) solution. Agitate again gently for about 15 s; the purple color of permanganate ion should now appear in the toluene layer. Continue to observe the test solution for 2 to 5 min.

If the compound being tested is unsaturated (aldehyde, enolizable ketone, olefin, alkyne, etc.), the purple permanganate color in the toluene layer will fade rapidly (within 2 min) to a scummy brown color. If the unknown compound contains no sites of unsaturation, the toluene layer will remain purple. As in all classification tests, a blank should be run and known compounds, such as benzaldehyde and cyclohexene, should also be tested to provide a basis for comparison.

In this test, color change generally occurs within 2 min for aldehydes, alkenes, and alkynes as a result of the oxidation reaction between permanganate ion and double bonds. The reaction is illustrated below for cyclohexene.

$$\text{cyclohexene} + KMnO_4 + H_2O \longrightarrow \text{cyclohexane-1,2-diol (OH, OH)} + MnO_2\downarrow$$

The reaction works best with electron-rich olefins. If an electron-withdrawing group is conjugated with the olefin (as in an α,β-unsaturated ketone), reaction will be slower and color discharge will also be slower. Overall, this test is quite sensitive and does not suffer from the solubility problems or solvent reactivity problems associated with the classical procedure for the Baeyer test. (Refer to Sec. 26.4B for additional discussion.)

G The Purpald Test

The second of these new classification tests utilizes a heterocyclic reagent called 4-amino-3-hydrazino-5-mercapto-1,2,4-triazole. It is fairly easy to see why the company which markets this substance refers to it as Purpald rather than by its systematic name. This heterocyclic molecule will react to form a purple complex with an aldehyde as shown below, but not with a ketone.

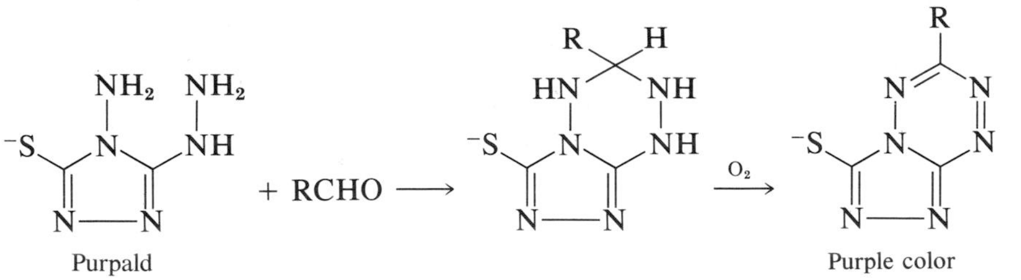

$$\text{Purpald} + \text{RCHO} \longrightarrow \text{(intermediate)} \xrightarrow{O_2} \text{Purple color}$$

Purpald   Purple color

This reagent's utility has been limited to some extent by its insolubility in the media used for the solution of higher-molecular-weight aldehydes. This difficulty has been overcome by the use of phase-transfer catalysis. The phase-transfer Purpald test is definitive for aldehydes. A procedure for this test is given below. When this test is applied, a colored solution is observed only when a reactive aldehyde is present. Ketones generally do not give such a test.

PROCEDURE

PURPALD CLASSIFICATION TEST FOR ALDEHYDES

In each of two test tubes (20 × 150 mm) is placed 50 mg (a small mound about 5 mm in diameter) of Purpald, followed by 1 mL of a toluene solution containing 100 mg/mL of tri-*n*-caprylylmethylammonium chloride (standard catalyst solution). To each tube is added 5 mL toluene (total volume 6 mL). The sample (aldehyde or ketone) is added to one of the test tubes while the other tube is kept as a blank. One milliliter of 10% aqueous sodium hydroxide solution is added to each test tube and the tube is swirled vigorously by hand. As soon as the base solution is added, a yellow color will appear in each tube, and this color will diffuse into the toluene layer. If the sample is an aldehyde, the color will begin to change immediately from yellow (with a slight greenish tinge) to orange to reddish and eventually to a deep rust color. *The appearance of this deep rust color is a positive test.* If no coloration is observed in the toluene layer at the end of 5 min, both test tubes should be placed in a hot water bath and held at 70°C for 2 min. A deactivated aldehyde, which will not react at room temperature, will usually react under these conditions.

Try this classification test on benzaldehyde, acetophenone, and acetone.

H The Fuchsin Aldehyde Test (Schiff's Test)

Another aldehyde classification test which is occasionally applied is Schiff's test. In this classification test, fuchsin (*p*-rosaniline hydrochloride) in solution with sulfurous acid reacts with an aldehyde to form a purple complex. The purple color characteristic of the quinoid dye complex is quite unmistakable and therefore an excellent indication of aldehydic carbonyl. Unfortunately, the test is not always as reliable as one might hope. Most aromatic aldehydes give distinct colors despite their limited solubility (probably due in part to their aromatic character). The low-molecular-weight aliphatic aldehydes generally do not give as strongly colored solutions, although their greater solubility in aqueous solution makes the test successful. Those aliphatic aldehydes whose higher molecular weight limits their solubility in the aqueous reagent often give questionable results.

Schiff's reagent is colorless to faintly pink. When an aldehyde reacts with it, a deep purple solution results. Ketones, on the other hand, tend to give only a pinkish color with Schiff's reagent. Occasionally when a compound believed to be an aldehyde does not give a positive test, this may be due to insolubility of the carbonyl compound. Addition of a small amount of acetone as cosolvent may allow one to observe the characteristic color. If this technique is applied, it is important to add acetone first and observe the color before adding unknown. *Warning:* This variant is not strictly reliable and should be heeded only if it is clearly positive.

PROCEDURE

SCHIFF'S TEST

$NHSO_2H$

$Cl^-H_3\overset{+}{N}$—C_6H_4—C(SO_3H)($C_6H_4NHSO_2H$)—C_6H_4—$NHSO_2H$ $\xrightarrow{RCHO}$ $H_2\overset{+}{N}$=C_6H_4=C(C_6H_4—$NHSO_2$—CHR—OH)$_2$ Cl^-

Add 1 drop or about 50 mg of the aldehyde to 2 mL water or aqueous ethanol (if necessary). To this solution, add 2 mL Schiff's reagent and shake. A purple color will be visible if an aldehyde is present.

Try this test on formalin solution, benzaldehyde, and acetone.

[*Note:* Schiff's reagent may be prepared by dissolving 200 mg *p*-rosaniline hydrochloride in 200 mL distilled water and then adding 8 mL saturated aqueous sodium bisulfite solution. After the mixture has been allowed to stand (react) for 60 to 90 min, 4 mL concentrated HCl solution should be added.]

We have discussed a number of tests which serve to distinguish aldehydes from ketones, but we have presented no tests specifically intended to identify ketones. The reason for this is simple: there are no such tests. Classification of an aldehyde or ketone is done by first eliminating acidic and basic possibilities and then determining if a reactive carbonyl group is present (DNP test). Once this is determined, a further process of elimination is used. If all the aldehyde tests prove negative, the carbonyl compound must be a ketone. One further piece of information can be obtained about a ketone, however, namely, whether or not it contains the acetyl (CH_3—CO—) group.

I The Iodoform Test for Methyl Ketones

It is often of value in identifying an unknown carbonyl compound to know if the material contains a methyl ketone group. A carbonyl group acidifies an adjacent methyl group, and this property can be used in identification. A methyl ketone will often react with sodium hydroxide to give a resonance-stabilized anion, which can attack iodine if it is present in the same solution. Eventually, all three hydrogen atoms of the methyl group are replaced by iodine atoms, and now hydroxide can add to the carbonyl group with loss of triiodomethyl carbanion. When this reaction occurs, a carboxylic acid is formed from the methyl ketone and iodoform (a yellow solid, CHI_3) is formed. Caution should be exercised in interpreting this test as compounds containing the CH_3—CH(OH)—group and acetaldehyde often give false positive tests.

PROCEDURE

THE IODOFORM TEST

$$R—CO—CH_3 + NaOH + I_2 \longrightarrow R—CO_2Na + I_3CH$$

Place a test tube containing 100 mg of the unknown compound and 5 mL dioxane in the steam bath and warm it to 50 to 60°C. Add 1 mL 10% aqueous sodium hydroxide solution and then add iodine-iodide solution dropwise until a slight iodine (brown) color persists. (Iodine-iodide solution is prepared from 10 g KI, 5 g iodine, and 50 mL water.) Continue warming the tube, adding iodine-iodide solution as necessary to maintain a slight excess of iodine. After about 5 min of additional heating, add enough 10% aqueous sodium hydroxide solution to destroy the excess iodine. Add 15 mL water, stopper the tube, and allow it to stand briefly. The appearance of yellow iodoform (mp 120 to 122°C) as a precipitate is a positive test. Try this test on acetone first to be sure you know how to do it properly.

25.6 SPECTROSCOPIC CONFIRMATION OF STRUCTURE

By qualitative organic analytical methods it is possible to distinguish among three groups of carbonyl compounds. The first group is the carboxylic acids. The acids are easily distinguished by the presence of the acidic proton and the resulting solubility in aqueous base. It is relatively more difficult to distinguish neutral carbonyl compounds such as the ketones and esters from amides. This is usually done by elemental analysis.

In modern chemical laboratories pure samples are often distinguished spectroscopically rather than by qualitative organic analysis. For quick distinction among the various types of carbonyl-containing compounds, ir spectroscopy is probably the most useful tool. In earlier times when uv spectroscopy was the only spectroscopic tool available to the chemist, quite sophisticated correlations allowed distinctions to be drawn among the several classes of compounds. This is still possible today, but it is not the most convenient method because the uv spectrum of the carbonyl group is affected less by the structural changes associated with aldehyde, ketone, ester, and other functional groups than is the ir spectrum.

Likewise, the influence of a carbonyl group on the nmr spectrum is really quite similar regardless of whether the protons are near an ester, an aldehyde, a ketone, or an amide carbonyl. There are other features of the spectrum which allow these distinctions to be drawn, but they cannot usually be done as simply as by the use of ir spectroscopy. Recording the ir spectrum of a pure neutral carbonyl-containing compound is probably the easiest way to determine whether the carbonyl is present in an aldehyde, ketone, ester, or other functional group. In many laboratories even today, the presence of a carbonyl group is checked by a 2,4-dinitrophenylhydrazine spot test. Even though the ir spectroscopic technique is a powerful one, a relatively pure sample is required for it. A wet sample placed in a salt cell for ir analysis could have unfortunate consequences. A wet sample tested by a DNP reagent could give useful results on relatively impure material.

Nevertheless, ir spectroscopy can be an extremely useful method for confirming the presence of a carbonyl group and for confirming the results of the classical tests which are emphasized here.

Simple ketones (unstrained systems) such as acetone and cyclohexanone absorb in the ir at 1715 cm^{-1}. The carbonyl affords an intense absorption in the ir region. This same intense absorption is observed whether the carbonyl group is part of an aldehyde or a ketone. The principal difference between the vibrations for these two functional groups is that, while ketone absorbs at 1715 cm^{-1}, aldehydes generally absorb about 10 cm^{-1} higher (approximately 1725 cm^{-1}). The same factors, such as ring strain and conjugation, affect the position of both carbonyls, so it is relatively difficult to determine which is present simply by examining the carbonyl absorption. The way aldehydes and ketones are usually distinguished is by examining the 2700 to 2800 cm^{-1} region in the ir spectrum. Aldehydes exhibit a characteristic doublet (two peaks) at 2720 cm^{-1} and 2820 cm^{-1}. The higher-energy band (2820 cm^{-1}) may often be obscured by vibrations arising from carbon-hydrogen bonds. As a consequence, it is usually the 2720 cm^{-1} band which is searched for and considered diagnostic for aldehydes.

Esters absorb at higher energy (shorter wavelength) than do either aldehydes or ketones. Ethyl propionate, for example, absorbs at 1740 cm^{-1}. Compared with acetone, the standard ketone, esters generally absorb about 25 cm^{-1} higher. Again, the factors which affect changes in vibrational positions for ketones and aldehydes apply also to esters (including cyclic esters, i.e., lactones). In all cases conjugation tends to lower vibrational frequency and ring strain tends to raise it. Ring strain will be of consequence only with the cyclic esters, the lactones.

The presence of an ester carbonyl can usually be corroborated by looking for the very strong carbon-oxygen single bond vibration near 1200 cm^{-1}.

The other neutral carbonyl-containing compounds should be mentioned here. These are the anhydrides and the acid halides. These can readily be distinguished by wet chemical methods from aldehydes, ketones, and esters because when they are exposed to aqueous base, they will rapidly hydrolyze and dissolve therein. The compounds will therefore appear to be carboxylic acids. Because of this deceptive reactivity, they are rarely given as unknowns in qualitative analysis. If they are given, however, they may often be distinguished by the following ir criterion: Anhydrides generally show two strong carbonyl vibrations in the vicinity of 1750 and 1825 cm^{-1}.

It would be particularly useful to refer to Chap. 5 for a more detailed discussion of the functional groups. Also, a check of Secs. 11.1, 11.2, 11.3, and 11.4 will give an idea of typical nmr and ir spectra for esters. Sample spectra of aldehydes and ketones can be observed throughout Chaps. 14, 15, and 16.

25.7 DERIVATIVES OF ALDEHYDES AND KETONES

Most derivatives of aldehydes and ketones obey the reaction principles discussed above, i.e., nucleophilic addition of a reagent to the carbonyl group, followed by water loss, generates an imine or other double-bonded species. The derivatives discussed here are usually solids and can be obtained in a pure state by recrystallization without having to resort to difficult manipulations.

Because ketones and aldehydes differ only by the presence or absence of a hydrogen atom bonded to the carbonyl function, the derivatives suggested here will generally apply equally well to aldehydes or ketones. In the procedures described, aldehydes and ketones may ordinarily be used interchangeably.

A 2,4-Dinitrophenylhydrazone Formation

For the reasons discussed above, either the 4-nitrophenylhydrazone or the 2,4-dinitrophenylhydrazone of an aldehyde, formation of which is shown below, is usually the simplest derivative to prepare.

$$R{-}CO{-}R' + O_2N{-}C_6H_3(N_2O){-}NH{-}NH_2 \longrightarrow \begin{matrix} R \\ \quad \backslash \\ \quad C{=}N{-}NH{-}C_6H_3(O_2N){-}NO_2 \\ \quad / \\ R' \end{matrix}$$

The difficulties which apply to the classification test also apply to derivative formation. The presence of unreacted 2,4-dinitrophenylhydrazine can readily be detected by its characteristic color and its melting point. The derivatives form readily and are usually the best derivatives for an aldehyde or ketone. The procedure for their formation is given below.

A difficulty encountered with aldehydes is that self-condensation is occasionally a problem when using this procedure. The acid-catalyzed aldol condensation, shown below, is especially prevalent with low-molecular-weight aliphatic aldehydes.

$$2\,R{-}CH_2{-}\overset{\overset{\displaystyle O}{\|}}{C}{-}H \longrightarrow R{-}CH_2{-}\underset{\underset{\displaystyle H}{|}}{\overset{\overset{\displaystyle OH}{|}}{C}}{-}CH(R)(CHO) \xrightarrow{-H_2O} R{-}CH_2{-}CH{=}C(R)(CHO)$$

In general, the 2,4-dinitrophenylhydrazone derivatives of aliphatic aldehydes are yellow in color. A brick-red DNP derivative of a presumed aliphatic aldehyde probably indicates that the compound has undergone aldol formation and dehydration to form an α,β-unsaturated aldehyde. When this aldehyde is treated with DNP reagent, it gives a conjugated derivative which is red in color.

If the aldehyde contains either an α-halogen atom or an α-hydroxyl group, other difficulties such as dehalogenation and elimination of water may be encountered.

PROCEDURE

2,4-DINITROPHENYLHYDRAZONES OF KETONES AND ALDEHYDES

A. Diethylene glycol procedure

$$O_2N\text{-}C_6H_3(NO_2)\text{-}NH\text{-}NH_2 + R\text{-}\overset{O}{\overset{\|}{C}}\text{-}R' \xrightarrow{H^+} O_2N\text{-}C_6H_3(NO_2)\text{-}NH\text{-}N{=}C(R)(R') + H_2O$$

In a 125-mL Erlenmeyer flask is placed 1.0 g 2,4-dinitrophenylhydrazine, followed by 35 mL diethylene glycol. This solution is swirled and heated briefly on a steam bath to dissolve the hydrazine. To this red-orange solution (ignore small amounts of precipitated solid) is added 0.5 g of the unknown carbonyl compound. The mixture is swirled to dissolve the carbonyl compound in the solvent. To this solution is added 1.5 mL concentrated HCl and the solution is swirled to thoroughly mix the acid in the solution. The color of the solution should change from red-orange to light yellow. The solution is allowed to stand at room temperature for 15 to 30 min to allow crystallization of the derivative.

At the end of the reaction period, 10 mL water is added to the solution portionwise with swirling. The residue is suction filtered and the solid collected and then washed with 10 to 15 mL 50% aqueous alcohol. The residue is recrystallized from ethanol, ethyl acetate, or ethanol-water.

B. Ethanol procedure

Dissolve 1 mL concentrated hydrochloric acid in enough ethanol to make 10 mL of solution. Add 0.5 g 2,4-dinitrophenylhydrazine and warm the solution until the reagent dissolves. Add 0.25 g of the solid or 5 drops of the liquid carbonyl compound and heat to boiling. The 2,4-DNP derivative should separate on cooling.

This procedure is somewhat simpler than that described in part A above and works well for lower-molecular-weight carbonyl compounds. For more difficult cases, procedure A is superior.

One recrystallization of a DNP derivative is usually sufficient to give a good melting point. Low-melting derivatives of long-chain ketones and aldehydes (4-heptanone or 2-octanone) tend to first form an oil and then crystallize. They also tend to form an oil again when recrystallized. Certain high-melting DNP derivatives are difficult to recrystallize. Often, boiling these materials with 95% ethanol, followed by suction filtration, will afford purer material.

Ketones and aldehydes which are insoluble in diethylene glycol should be dissolved in a minimum amount of dioxane (use dioxane only with your instructor's consent) or tetrahydrofuran (THF) and added to the reagent. The procedure is the same from that point on.

If a *p*-nitrophenylhydrazone (NPH) or a phenylhydrazone derivative is desired (the latter derivatives are not listed in the tables), the procedure given above should be used, except that the following amounts of reagent and solvent should be used:

Dinitrophenylhydrazine (0.5 g) in 35 mL diethylene glycol
p-Nitrophenylhydrazine (0.5 g) in 20 mL diethylene glycol
Phenylhydrazine (liquid) (0.5 g) in 10 mL diethylene glycol

Note that with the less substituted phenylhydrazines the color change when acid is added will be less pronounced.

B Semicarbazone Derivatives

Semicarbazones ordinarily form quite readily and are highly crystalline substances.

$$\underset{}{R_2C{=}O} + \underset{\text{Semicarbazide}}{H_2N{-}NH{-}CO{-}NH_2} \longrightarrow \underset{\text{Semicarbazone}}{R_2C{=}N{-}NH{-}CO{-}NH_2} + H_2O$$

The semicarbazones are usually solid and many form sparkling white crystals. Difficulty with semicarbazone formation is usually encountered in dealing with low-molecular-weight aldehydes. Semicarbazone derivatives of small aldehyde molecules often are water-soluble and are lost in the aqueous wash. This is a frustrating experience, but should be anticipated when using any aliphatic aldehyde with less than five carbon atoms. In such cases, it is often wise to use the thiosemicarbazide reagent to yield a thiosemicarbazone. Use of this reagent is similar to the application of the oxygen-containing material but should be restricted to difficult cases because the material is far more expensive than the oxygen compound.

PROCEDURE

SEMICARBAZONE DERIVATIVES

A. Usual procedure

To 0.5 g of an unknown carbonyl compound in a round-bottom flask is added 1 g semicarbazide hydrochloride, along with 10 mL methanol. To this solution is added 1 g sodium acetate, along with 1 to 2 mL water. The solution is refluxed for 30 min on a steam bath. The solution is then diluted with water to cloudiness, cooled, and the crystals collected. The semicarbazone is recrystallized from methanol, ethanol, methanol-water, or ethanol-water.

B. Modification for low-molecular-weight aldehydes and ketones

Place 1 g semicarbazide hydrochloride and 1.5 g sodium acetate in a 50-mL Erlenmeyer flask. Add 10 mL water and swirl to dissolve the solids. Add 0.5 to 1.0 mL of the carbonyl compound, cork the flask, and shake. If the solution is turbid, add methanol until a clear solution is obtained. Do not add more than 10 mL methanol. Remove the cork and warm on a steam bath for 30 min, cool, and filter to obtain the crystals. Crystallization may be aided by the addition of a *little* water. Wash the crystals with cold water and recrystallize from ethanol, methanol, ethanol-water, or methanol-water.

Note: Some of the semicarbazones of low-molecular-weight carbonyl compounds are quite soluble. Too much methanol will prevent crystallization however much water is added. The water-methanol ratio is not critical, but try to arrange for the carbonyl compound to be *just* in solution.

C Oximes

Oximes are formed from hydroxylamine hydrochloride and a carbonyl compound, as shown below.

$$R_2C{=}O + NH_2{-}OH{\cdot}HCl \longrightarrow R_2C{=}N{-}OH + H_2O$$

The difficulty with this procedure is not in the formation but in the crystallinity of the resulting compound. Formation of the oximes does not appreciably enhance the molecular weight, and as a consequence many oximes are oils. In addition, oximes may form either syn or anti isomers or both. A further complication is that in some cases a syn oxime may be isolated, when it is the melting point of the anti compound which is recorded in the literature. It may be useful in this situation to consult other tables. The syn and anti oxime isomers are not usually as much of a problem with aldehydes as they are with unsymmetrical ketones. This is because the hydroxyl group tends to lie on the same side as the aldehydic hydrogen in aldehyde derivatives. With such ketones as methyl ethyl ketone, the steric requirement of an ethyl group is similar to that of a methyl group, with the result that both syn and anti isomers may be formed. Mixtures of syn and anti oximes may be formed to such an extent that the derivative is obtained as an oily liquid which will not crystallize.

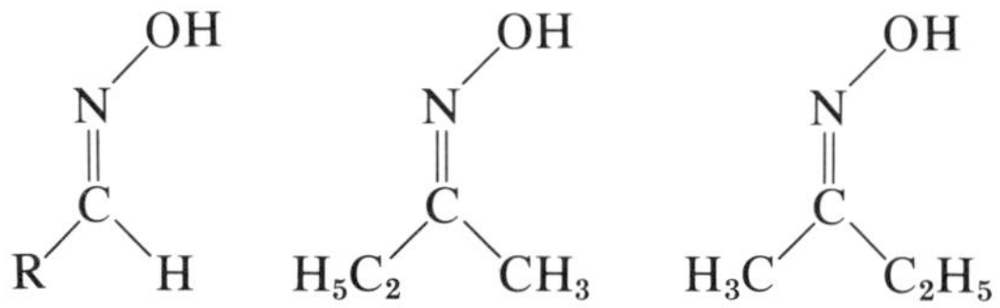

OH OH OH
N N N
C C C
R H H_5C_2 CH_3 H_3C C_2H_5

PROCEDURE

OXIME DERIVATIVES

To 0.5 g of a carbonyl compound in a 25-mL round-bottom flask is added 0.5 g hydroxylamine hydrochloride, along with 3 to 4 mL pyridine and 5 mL ethanol. This solution is heated to reflux on a steam bath for 1.5 to 2 h and most of the solvent evaporated. The residue may be recrystallized from ethanol, methanol, ethanol-water, or methanol-water.

D Dimedone Derivatives

The reaction of 4,4-dimethylcyclohexane-2,6-dione with aldehydes yields a derivative called the dimedone derivative, as shown below.

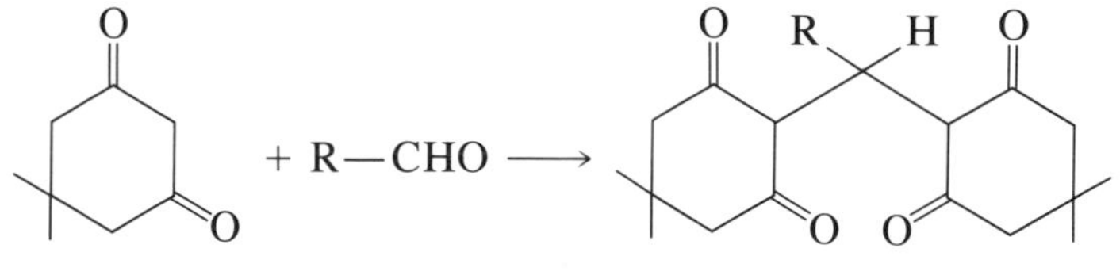

O O R H O

+ R—CHO ⟶

O O O

In many cases the dimedone derivatives form readily and are highly crystalline. In some cases, however, their preparation requires several hours. Since most organic laboratories operate within a 3- to 4-h time constraint, we have chosen not to include a general procedure for this derivative. If a dimedone must be prepared, ask your instructor to refer you to a more specialized text.

E Oxidation to the Corresponding Carboxylic Acid (Aldehydes)

Recall from the classification tests that one means for discriminating between aldehydes and ketones is oxidation. Whereas aldehydes oxidize readily to carboxylic acids, ketones resist such oxidation. In the discussion above, attention has been focused on the reduction of the oxidizing agent, i.e., the transformation of cupric ion to cuprous oxide or of silver ion to a silver mirror. Obviously, if an aldehyde is oxidized to a carboxylic acid which is itself crystalline, it will have a characteristic melting point, and could serve as a derivative.

$$R-\overset{\overset{\displaystyle O}{\|}}{C}-H + O_2 \longrightarrow R-\overset{\overset{\displaystyle O}{\|}}{C}-O-H$$

This is particularly true of aromatic aldehydes. For example, benzaldehyde oxidizes so easily that a white solid around the mouth of the reagent bottle is often encountered. This solid is benzoic acid, which has been formed by the air oxidation of the aldehyde. A number of aldehydes can be derivatized in this way. Consult Table 28.2 (Solid Carboxylic Acids) to see where this process may be useful.

1 Potassium permanganate method

PROCEDURE

OXIDATION OF ALDEHYDES TO THE CORRESPONDING ACIDS

$$R-CHO + KMnO_4 \longrightarrow R-CO-OH + MnO_2 \downarrow$$

Place 0.5 g potassium permanganate in a 50-mL Erlenmeyer flask, followed by 10 mL distilled water. Swirl the mixture to effect solution.

Place 0.5 g (or 0.5 mL if liquid) of the sample in a 125-mL Erlenmeyer flask, followed by 15 mL dichloromethane. After the sample has dissolved, add 1 mL standard catalyst solution (see Sec. 4.7) with swirling. To this dichloromethane solution add all at once the aqueous potassium permanganate solution above. Swirl the flask vigorously to ensure good contact between the phases. (*Note:* If a magnetic stirrer is available, its use in this procedure will greatly increase the oxidation rate and ease of manipulation.)

After swirling for 30 to 40 min, small amounts (approximately 750 mg) of solid $NaHSO_3$ are added with vigorous swirling after each addition. This procedure should discharge almost all the scummy brown precipitate (MnO_2). After the color discharge, the solution is made strongly acidic ($pH < 2$, pH paper) by adding 12 *N* HCl dropwise. The solution is transferred to a separatory funnel and the dichloromethane layer removed. The aqueous layer is discarded.

The dichloromethane layer is extracted with 10 mL saturated $NaHCO_3$. The separated $NaHCO_3$ layer is acidified to pH 2 by cautious dropwise addition of 12 *N* HCl. The acid should precipitate at this point and should be filtered (Buchner funnel), washed with *cold* water, and air dried. The acid may be recrystallized from water, aqueous alcohol, or acetone–petroleum ether.

2 Oxidation by the Cannizzaro reaction

For a detailed discussion of the Cannizzaro reaction, refer to Sec. 15.5.

PROCEDURE

OXIDATION OF ALDEHYDES TO THE CORRESPONDING ACIDS

$$2R\text{—}CHO + KOH \longrightarrow R\text{—}CO\text{—}OH + R\text{—}CH_2\text{—}OH$$

In a large test tube is placed 2 g KOH, followed by 3 mL water. This tube is swirled until all the KOH has dissolved. To this solution is added 0.5 mL of a liquid or 0.5 g of a solid aromatic aldehyde in 3 mL methanol. The mixture is swirled to effect solution and then heated on a steam bath with intermittent swirling for 1 h.

After the reaction period, the solution is diluted with 10 mL water, transferred to a separatory funnel, and extracted with three 10-mL portions of ether. The ether layer is then discarded.

The basic aqueous layer is acidified by cautious addition of 12 *N* HCl. The precipitated acid is filtered, washed with cold distilled water, and air dried. The acid may be recrystallized from hot water, aqueous alcohol, or acetone–petroleum ether.

F Reduction of Aldehydes and Ketones

Another method for derivatizing carbonyl functions is to convert the carbonyl compound into the corresponding alcohol. We have focused above principally on the oxidation of aldehydes to carboxylic acids, but we note that both aldehydes and ketones can be reduced to the corresponding alcohols by the reagent known as sodium borohydride. It is generally the case that alcohols do not form as well-mannered derivatives as do the corresponding aldehydes or ketones. Nevertheless, there are certain cases which are exceptions to this rule. In these cases an alternative to the formation of a simple oxime or DNP derivative is to reduce the carbonyl compound to an alcohol, which is then treated with phenyl isocyanate to form the phenylurethane. A procedure for the reduction of an aldehyde or ketone in alcohol solution using sodium borohydride is given below. The derivatization either with naphthyl isocyanate or phenyl isocyanate can be found in the Sec. 26.6A.

PROCEDURE

BOROHYDRIDE REDUCTION

$$R_2C{=}O \xrightarrow{NaBH_4} R_2CH{-}OH \xrightarrow{Ar{-}N{=}C{=}O} R_2CH{-}O{-}CO{-}NH{-}Ar$$

In a 50-mL Erlenmeyer flask is placed 0.5 g or 0.5 mL of a carbonyl compound, along with 15 mL methanol. To this solution is added 100 mg sodium borohydride. The solution is swirled and warmed on a steam bath for 15 min. Five milliliters of water is added and the solution is heated at reflux for 5 min on the steam bath. In many cases, the product will crystallize on cooling this solution. If crystals do not appear, most of the methanol is removed by evaporation and the remaining solution extracted with two 10-mL portions of dichloromethane. The dichloromethane solution is dried (Na_2SO_4) and the solvent removed on a steam bath. The residue is recrystallized from acetone–petroleum ether or cyclohexane.

If the alcohol cannot be induced to crystallize or if the residue is an oil, dissolve it in 5 to 10 mL petroleum ether and then prepare a phenyl- or α-naphthylurethane, as described in Sec. 26.6A.

XXVI
ALCOHOLS

26.1 Historical and General **580**
26.2 Classes of Alcohols **580**
26.3 Properties of Alcohols **582**
26.4 Operational Distinctions **584**
- **A** Solubility and the 2,4-dinitrophenylhydrazine test
- **B** The Baeyer test
- **C** Oxidation tests for alcohols
- **D** The Lucas test
- **E** The oxidation-aldehyde test sequence
- **F** The periodate reaction

26.5 Spectroscopic Confirmation of Structure **592**
26.6 Derivatives of Alcohols **593**
- **A** Phenyl- and α-naphthylurethane derivatives
- **B** Ester formation
- **C** Low-melting ester derivatives
- **D** Benzoate derivatives of alcohols

26.1 HISTORICAL AND GENERAL

Alcohol and alcohols have been used and recognized from the very earliest of times. The book of Genesis, for example, records the drinking of wine by Noah the Ark-builder. Wine or aqua vitae was and is widely used in almost all modern cultures. Even today alcohol is probably the most widely used of all known drugs.

The isolation of alcohol as a discrete substance owes much to the work of the alchemist Geber, who early in the ninth century refined the crude kerotakis into a usable alembic or distillation apparatus. Medieval alchemists then used this apparatus and further refinements of it to distill the substance now known as alcohol from a variety of wines and spirits. Alcohol has been used over the years as an analgesic, a tranquilizer, a sterilizer, and even as a truth serum. One of the truly important drugs of history, it has been so used for longer than any other substance. In most civilized cultures it is still important and useful, although it contributes to a variety of social problems. It is an unfortunate circumstance that the alcoholism rate in the civilized world is remarkably high and shows no sign of abatement.

The term *alcohol* is synonomous with ethanol or ethyl alcohol. Ethanol is often called grain alcohol to distinguish it from the other common alcohol, methanol, which is obtained from wood. Wood alcohol is much more toxic than ethanol. It is probably the methanol contaminating some moonshine preparations which leads to the reputation of such substances for causing one to become "blind drunk." In fact, ingestion of substantial quantities of methanol will lead to permanent blindness and ultimately to death. Ingestion of sufficient quantities of ethyl alcohol will likewise lead to death, but the body's mechanism for rejecting large quantities of ethanol seems to be more favorable than for methanol. Some other common and familiar alcohols are isopropyl alcohol, commonly known as rubbing alcohol, and 1,2-dihydroxyethane or ethylene glycol, which is the most common constituent of antifreeze. Ethylene glycol is very effective as an antifreeze because it boils at almost 200°C (much higher than water) and freezes well below 0°C; even water solutions of it freeze below 0°C.

26.2 CLASSES OF ALCOHOLS

Alcohols may be divided into four principal classes: primary, secondary, and tertiary aliphatic alcohols, and aromatic alcohols, i.e., the phenols. Primary alcohols are those compounds in which the carbon atom bonded to the hydroxyl group is bonded to one other carbon atom. In a secondary alcohol, the hydroxyl-bearing carbon atom has two carbon atoms attached. A tertiary alcohol can be recognized by the fully substituted carbon atom bonded to the oxygen atom.

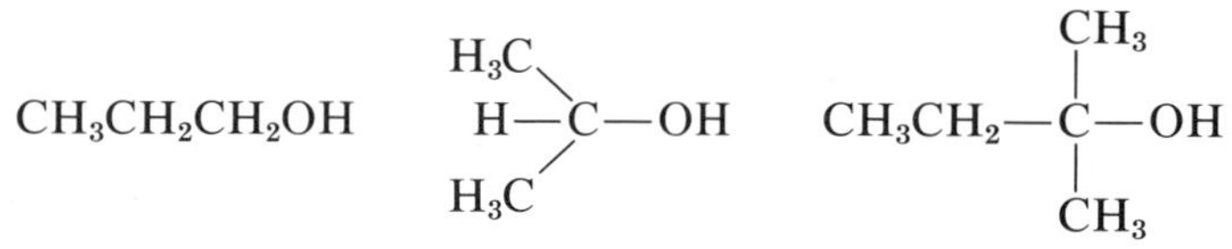

$CH_3CH_2CH_2OH$ — 1-Propanol (primary)

$(H_3C)_2CH—OH$ — 2-Propanol (secondary)

$CH_3CH_2—C(CH_3)_2—OH$ — *tert*-Amyl alcohol (tertiary)

In addition to primary, secondary, and tertiary saturated alcohols, one occasionally encounters vinyl alcohols. In compounds of this class, one of the carbon atoms involved in the double bond is also attached to a hydroxyl group. This is an example of both an alcohol (ol) and an alkene (ene), and vinyl alcohols are therefore referred to as *enols*. Enols generally arise from and are in equilibrium with either aldehydes or ketones. The enol content of a carbonyl compound depends (in any given solvent) upon the overall structure of the compound and the consequent stability of the enol system. In β-diketone systems, for example, an appreciable amount of enol is present because formation of the vinyl alcohol residue leads to a conjugated, six-membered, hydrogen-bonded system, as illustrated below.

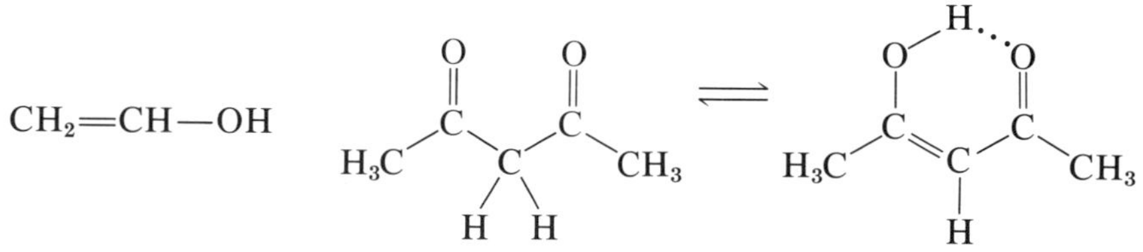

$CH_2{=}CH—OH$ — Enol

$H_3C—C(=O)—CH_2—C(=O)—CH_3$ — Keto form $\rightleftharpoons$ $H_3C—C(OH)=CH—C(=O)—CH_3$ (with $O—H\cdots O$ hydrogen bond) — Enol form

The relative amount of each form which is present depends on temperature, solvent, and other factors and may be appreciable.

The β-dicarbonyl system is an extreme example because the enol side of the equilibrium is so favored. The phenol system is another extreme example.

Phenol is actually the enol form of cyclohexadienone. When cyclohexadienone enolizes, the carbon-carbon double bond is in the ring, which results in a fully conjugated system containing six π electrons. As a result, the enol form of cyclohexadienone is aromatic. This stability afforded by aromaticity causes the enol equilibrium to lie far to the side of the vinyl alcohol, so much so in fact that ketone is not detected in these systems.

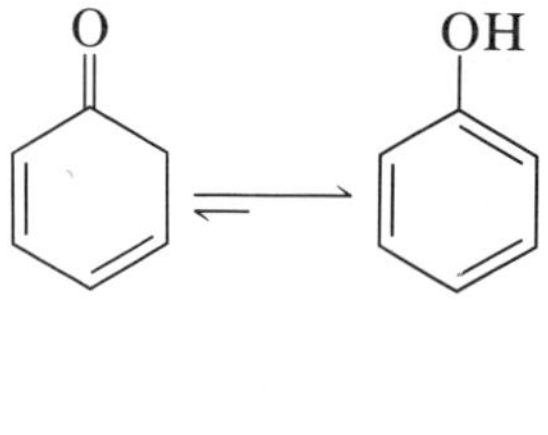

Cyclohexadienone (O) $\rightleftharpoons$ Phenol (OH)

TABLE 26.1
Approximate pK_a's of some acids

Compound	Approximate pK_a
R—COOH	5
Ar—OH	11
R—OH	17

Because the benzene ring is more electron-withdrawing than many substituents, and because the anion formed by the deprotonation of a phenol is stabilized by benzene-ring resonance, phenols are far more acidic than are alcohols. (Note that phenols are discussed as a separate category in Chap. 23.) Phenols are less acidic than carboxylic acids but far more so than simple saturated alcohols. Most carboxylic acids have pK_a values which fall in the range of approximately 2 to 7. Simple phenols generally have pK_a values in the range of 10 to 12 and saturated alcohols have still higher values. Compared with water (pK_a 15.7), methanol has a pK_a of 16, ethanol a pK_a of about 17, and *tert*-butyl alcohol a pK_a of about 19. Saturated alcohols are not readily deprotonated by aqueous sodium hydroxide solution, and are considered neutral compounds. Table 26.1 summarizes these values.

26.3 PROPERTIES OF ALCOHOLS

It has already been noted that alcohols do not react with aqueous sodium hydroxide, and this distinguishes alcohols from the carboxylic acids and the phenols. Water solubility is another useful property and can assist in the identification of alcohols. Many of the lower-molecular-weight alcohols are quite soluble in water. For example, methanol, ethanol, 1-propanol, 2-propanol, *tert*-butyl alcohol, ethylene glycol, propylene glycol, glycerol, and a variety of other alcohols are miscible with water in all proportions.

Consider the forces which act on molecules and aid in solvating them [see Chap. 4 (Solubility and Reactivity)]. Aliphatic alcohols can interact with hydrocarbonlike solvents if hydrocarbon side chains are present. The solubility of low-molecular-weight alcohols in water can be attributed largely to the presence of the hydroxyl group. The hydroxyl oxygen bears two electron pairs, which can function as Lewis base donors in hydrogen-bond formation both with other alcohols and with water. In addition, the hydrogen in the hydroxyl group can form hydrogen bonds with either another molecule of the alcohol or with water. The cooperative donation and acceptance of protons in hydrogen-bond formation by alcohols and water leads to the considerable solubility of many alcohols in aqueous solution.

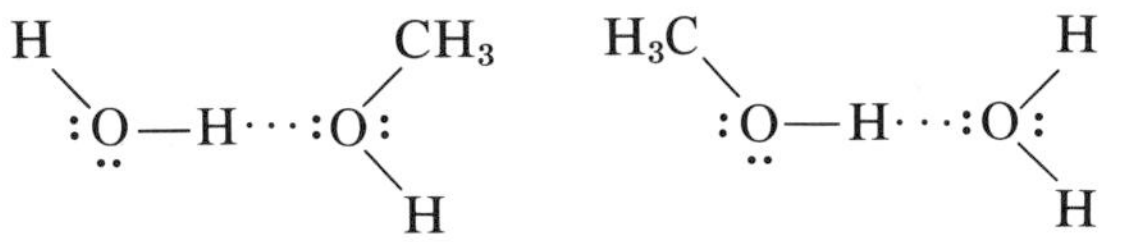

H ─ :Ö─H···:Ö: (CH$_3$, H)   H$_3$C ─ :Ö─H···:Ö: (H, H)

In considering the hydroxyl function it is important to remember that this functional group, while polar, dominates the solubility behavior of the substance only as long as the molecule does not become too large. Alcohols which contain fewer than about six carbon atoms are relatively soluble in water, and even if several more carbons than that are present, the alcohol will have at least some water solubility. On the other hand, if the alkyl group becomes quite long (on the order of 15 or more carbon atoms), the compound must be considered essentially as a hydrocarbon which happens to bear the hydroxyl substituent. In other words, when the hydroxyl group constitutes a major portion of the molecule, it will probably dominate the solubility of the substance, but when the alkyl group is large relative to it, its presence will not be nearly so influential. Note, however that a long-chain alcohol having two or more hydroxyl groups arranged along its length might show appreciable water solubility.

The ability of alcohols to form hydrogen bonds manifests itself in the relative boiling points of a series of related molecules. Consider for example the two-carbon system consisting of ethanol, acetaldehyde, and acetic acid. Each compound in this series contains two carbon atoms, but the oxidation state at the first carbon atom is successively higher in each, and the boiling points differ quite dramatically. Ethanol boils at 78°C, acetaldehyde at 22°C, and acetic acid at 117°C. The difference in the boiling behavior of these substances is due to the presence of hydrogen bonds. Because ethanol can hydrogen bond to other ethanol molecules, it acts as both a hydrogen-bond donor and acceptor. The formation of many hydrogen bonds throughout the solution (polymeric hydrogen bonding) leads to a much higher boiling point for the alcohol than for the aldehyde, which is not capable of hydrogen-bond formation.

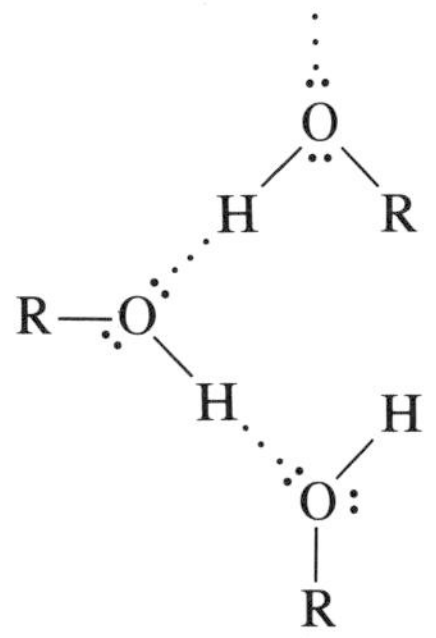

O (H, R) ··· R─O(H) ··· O (H, R)

The carboxylic acid, which has an even higher boiling point, has the unique ability to form dimeric hydrogen-bonded species. Such hydrogen bond dimers are very stable, and rupture of both hydrogen bonds requires more energy than does rupture of polymeric hydrogen bonds.

$$2\ R{-}COOH \rightleftharpoons R{-}C\begin{matrix} \diagup\!\!\!\diagup\ \ddot{O}\!:\cdots H{-}\ddot{O}\!:\ \diagdown \\ \diagdown\ \ddot{O}{-}H\cdots:\!\ddot{O}\ \diagup\!\!\!\diagup \end{matrix}C{-}R$$

The strength of the hydrogen bonds will be reflected in the energy needed to vaporize a substance in a distillation process. The strength of the intermolecular interaction roughly parallels the increases in boiling points. There are intermolecular dipole-dipole interactions in acetaldehyde, but these are far weaker than the hydrogen bonding in ethanol. The formation of carboxylic acid dimers in a substance such as acetic acid is even more important and causes the boiling point to be further elevated.

26.4 OPERATIONAL DISTINCTIONS

Operational distinctions for identifying alcohols are summarized in the following list and are discussed in more detail below.

Operational Distinctions for Alcohols

Unknown solubility (Sec. 26.4A)	Unknown is: insoluble in dilute acid insoluble in dilute base soluble in concentrated H_2SO_4
Preliminary (Secs. 26.4A and B)	DNP test: positive if aldehyde or ketone Baeyer test: positive if olefin, aldehyde, or enolizable ketone
Classification (Secs. 26.4C to 26.4F)	Pyridinium chlorochromate Chromic anhydride The Lucas test Oxidation-aldehyde sequence Periodate test
Derivatives (Sec. 26.6)	Phenyl- and α-naphthyl urethanes Benzoate esters 4-Nitrobenzoate esters 3,5-Dinitrobenzoate esters

A Solubility and the 2,4-Dinitrophenylhydrazine Test

Acid and base solubility tests are negative for alcohols as long as the compound is water-insoluble. There are many neutral compounds, however, from which alcohols must be distinguished. When it appears that a sample is a neutral compound, it is wise to observe the reaction of the substance with 2,4-dinitrophenylhydrazine (DNP) solution. If the material does not react with DNP to give the characteristic orange or red precipitate, it is probably not an aldehyde or a ketone. (Sometimes, sterically hindered ketones react quite slowly with DNP reagent, but this is a relatively uncommon situation. Just as uncommon is the situation in which the alcohol is oxidized during the test to the ketone or aldehyde.) Generally, if the alcohol is pure, it will not react with DNP solution at a rate appreciable enough to cause confusion, although old samples of primary or secondary alcohols may contain enough aldehyde or ketone impurity to cause a problem. The first step after solubility tests, then, is to use the DNP test to determine if the neutral compound is an aldehyde or ketone.

B The Baeyer Test

Next, the Baeyer test for unsaturation should be applied to the neutral substance. This test will readily distinguish a neutral olefin from an alcohol. The latter will not oxidize as rapidly under the conditions described here as will a substance containing an isolated double bond. Note that a tertiary alcohol will not oxidize in any event. This test will be nearly useless unless a known alkene and a known alcohol are compared before the unknown is tested.

$$R{-}CH{=}CH{-}R + KMnO_4 \longrightarrow R{-}\underset{}{\overset{OH}{\overset{|}{C}}}H{-}\overset{OH}{\overset{|}{C}}H{-}R + MnO_2 \qquad \text{(fast)}$$

$$R{-}CH_2{-}OH + KMnO_4 \longrightarrow R{-}CO{-}OH + MnO_2 \qquad \text{(slow)}$$

PROCEDURE

PRELIMINARY CLASSIFICATION OF ALCOHOLS

The Baeyer test for unsaturation

One milliliter standard catalyst solution (see Sec. 4.7) is added to a clean (13 × 100 mm) test tube. A small amount (1 drop or about 50 mg) of the test sample is added and the test tube is swirled until solution occurs. To this mixture is added 5 drops of 2% aqueous potassium permanganate ($KMnO_4$) solution. Agitate again gently for about 15 s. The purple permanganate color should appear in the toluene layer. Continue to observe the test solution for 2 to 5 min. A positive test is indicated by the dissipation of the purple color and the appearance of a scummy brown precipitate (MnO_2). It is imperative that trial tests be run on an olefin, such as cyclohexene, and an alcohol, such as ethanol or 2-propanol.

C Oxidation Tests for Alcohols

Once the nonalcoholic alternatives have been dismissed, the class of alcohol in hand can be distinguished by its oxidation behavior. Recall that primary, secondary, and tertiary alcohols differ by the extent of substitution at the hydroxyl carbon atom. It should be clear that tertiary alcohols cannot readily oxidize because further oxidation would involve cleavage of a carbon-carbon bond. A secondary alcohol can oxidize to a ketone. A primary alcohol will oxidize to yield an aldehyde, which in turn is oxidized to a carboxylic acid.

$$1^\circ \quad RCH_2OH \xrightarrow{[O]} RCHO \xrightarrow{[O]} RCOOH$$

$$2^\circ \quad R_2CHOH \xrightarrow{[O]} R_2C{=}O$$

$$3^\circ \quad R_3COH \xrightarrow{[O]} \text{no reaction}$$

In general, aldehydes are oxidized more rapidly than are the alcohols from which they arise. Once a small amount of aldehyde has been generated in an oxidation reaction, it will usually be oxidized to the carboxylic acid more rapidly than it can be isolated. As a consequence, the product observed in the oxidation of a primary alcohol will be the carboxylic acid.

There are several common reagents which are used to oxidize primary and secondary alcohols. Several of these utilize chromium as an oxidizing agent. Chromium trioxide, when it is used in a mixture of H_2SO_4 and acetone, is called Jones reagent; when complexed with pyridine, it is called Collins reagent. The latter reagent is particularly mild and can be utilized to generate an aldehyde from a primary alcohol without further oxidation. Pyridinium chlorochromate is another common oxidizing reagent. Also based on chromium, this reagent is quite easy to handle and is commercially available at a modest price. Be certain to check with your instructor before preparing any chromium reagents or chromium-containing test solutions because of possible toxicity and disposal problems.

CrO_3 — Chromium trioxide

$O_3Cr \leftarrow N(C_5H_5)$ — Collins reagent

$C_5H_5\overset{+}{N}H \; ClCrO_3^-$ — Pyridinium chlorochromate

The use of oxidizing agents provides a means for distinguishing primary and secondary from tertiary alcohols, since the latter do not oxidize. In the presence of tertiary alcohols, the oxidizing solution will remain yellow-orange and the chromium salt will not be reduced to a lower oxidation state (green color). Secondary alcohols will react but the product of the reaction will be neutral. Finally, primary alcohols which are oxidized to acids show the yellow-orange to green transformation [chromic (+6) to chromous (+3)], and the product of this reaction is acidic. Thus, the three kinds of alcohols can be readily distinguished by the combination of their reactivities and the products produced.

PROCEDURE

CLASSIFICATION OF ALCOHOLS

A. Pyridinium chlorochromate test reagent

Place approximately 30 mg pyridinium chlorochromate in a small test tube (13 × 100 mm). To this tube is added 2 mL dichloromethane. The test tube is swirled to thoroughly wet and mix the solid pyridinium chlorochromate with the solvent. At this point the tube should contain a light yellow dichloromethane solution over solid pyridinium chlorochromate.

Add 1 to 2 drops of liquid or 50 to 100 mg of solid sample. Swirl the solution rapidly to thoroughly mix the components. Within 1 min primary and secondary alcohols will turn the solution a dark green-brown, and within 2 min a very deep brownish, scummy precipitate will appear. (The appearance of this precipitate is very similar to the manganese dioxide precipitate observed in the Baeyer test for unsaturation). The test should be compared with a blank (no unknown added) for about 5 min and run first on a known.

Note: No color change is observed for tertiary alcohols. The only exception to this observation occurs in the case of a tertiary allylic alcohol such as linalool.

$$(CH_3)_2C{=}CHCH_2CH_2\overset{\overset{\displaystyle OH}{|}}{\underset{\underset{\displaystyle CH_3}{|}}{C}}CH{=}CH_2$$

Linalool

Here, the reagent first catalyzes an allylic rearrangement to the primary alcohol and subsequently oxidizes it. This allylic rearrangement–oxidation sequence, however, is slower than the oxidation of a primary or secondary alcohol. Alcohols which dissolve slowly in dichloromethane (such as octadecanol) should be predissolved in 1 to 2 mL of this solvent (warming if necessary) and added to the reagent. This test is then exactly the same as indicated above.

A word of caution: Although the pyridinium chlorochromate reagent is of considerable utility in distinguishing alcohols from a variety of other substances, its use has been limited recently by the suspicion that it, like certain other chromium-containing reagents, might be carcinogenic (cancer-causing). Although there is, at this writing, no evidence on this subject, the possibility is thought to exist because some other chromium compounds are dangerous in this regard.

In addition to the pyridinium chlorochromate method, two traditional methods which have been used for the qualitative identification of alcohols also rely on oxidation reactions. In one case (chromic anhydride), the metal involved in the redox system is chromium (as with the chlorochromate), but this reagent has been used for many years without apparent incident. In the second of the two tests, ceric ion is the oxidizing agent. In each case nontertiary alcohols will be oxidized to carbonyl compounds, and a primary alcohol will undergo further oxidation of the intermediate aldehyde to a carboxylic acid. Tertiary alcohols resist oxidation under these conditions.

PROCEDURE

CLASSIFICATION OF ALCOHOLS

B. Chromic anhydride reagent

$$1° \quad R{-}CH_2OH + Cr^{6+} \xrightarrow{H^+} R{-}CO{-}OH + Cr^{3+}$$

$$2° \quad R_2CHOH + Cr^{6+} \xrightarrow{H^+} R_2C{=}O + Cr^{3+}$$

To a small test tube (13 × 100 mm) add 1 to 2 mL acetone and 50 to 100 mg of unknown (1 to 2 drops if it is a liquid) and swirl the tube to effect solution. Add 1 drop chromium trioxide–sulfuric acid solution and watch the test solution closely. A positive test for a primary or secondary alcohol is indicated by an almost instantaneous precipitate and a color change from orange to blue-green. In the presence of tertiary alcohols the solution will remain orange. *Ignore* any change in the solution *after the first 2 s.* Try the test on ethanol, 2-propanol, and *tert*-butanol.

Chromic anhydride reagent may be prepared by slowly adding a suspension of CrO_3 (25 g) in concentrated H_2SO_4 (25 mL) to 75 mL distilled water and allowing the mixture to cool to room temperature before use.

C. Ceric ammonium nitrate reagent

$$1° \quad R—CH_2OH + Ce^{4+} \xrightarrow{H^+} R—CO—OH + Ce^{3+}$$

$$2° \quad R_2CHOH + Ce^{4+} \xrightarrow{H^+} R_2C{=}O + Ce^{3+}$$

To a small test tube containing 3 mL distilled water add 0.5 mL ceric ammonium nitrate reagent. Swirl the tube to effect solution. Add about 5 drops of the alcohol (a small spatula full if a solid) and stir with the spatula blade. With low-molecular-weight alcohols (less than about 10 carbon atoms) a positive test is indicated by a color change from yellow to red. Phenols tend to give a green to brown precipitate, and, where dioxane has been added to help dissolve the alcohol, a deep red-brown color is usually observed.

Ceric ammonium nitrate reagent may be prepared by dissolving (warming is usually required) 20 g ceric ammonium nitrate in 50 mL 2 *N* HNO_3 solution.

D The Lucas Test

Primary, secondary, and tertiary alcohols may also be distinguished by use of the Lucas test. The Lucas reagent is a combination of hydrochloric acid and zinc chloride. Zinc chloride functions as a catalyst to accelerate the reaction of hydrogen chloride with the hydroxyl group of an alcohol. Tertiary alcohols react quite readily with hydrogen chloride, and in the presence of the zinc chloride catalyst this reaction is frequently instantaneous. When aqueous zinc chloride–HCl solution is shaken with a water-soluble tertiary alcohol, the hydroxyl function will be protonated by the acid and water will be lost. The remaining carbonium ion reacts with chloride anion. The alkyl chloride which results is usually insoluble in the aqueous solution and separates as an insoluble layer. These reactions are shown in the equations below.

$$R_3C—OH \xrightarrow[ZnCl_2]{HCl} R_3C—\overset{+}{O}H_2 \longrightarrow R_3C^+ + H_2O$$

$$R_3C^+ + Cl^- \longrightarrow R_3C—Cl$$

Primary alcohols react slowly with hydrogen chloride, even in the presence of catalytic zinc chloride. The inference can be drawn that an alcohol is indeed primary if no reaction is observed with Lucas reagent even after standing for 10 to 15 min.

The secondary system is intermediate in reactivity between tertiary and primary, i.e., it is more reactive than the primary but less reactive than the tertiary. Application of the Lucas test to a secondary alcohol is therefore quite subjective. The first requirement for a successful Lucas test with a secondary alcohol is the same as for any alcohol: the compound must be soluble in the reagent. If the compound is insoluble, no reaction will occur and no information can be gained. If the substance is soluble, the solution should be shaken and observed for 5 to 10 min. A layer of the secondary alkyl chloride should gradually appear. If no layer forms, the substance is probably primary, and if it appears slowly, it is probably not tertiary. Special care should be used in interpreting this test for another reason: Some insoluble secondary alcohols form emulsions with the Lucas reagent and then gradually separate, giving the appearance of chloride formation.

As in every other test, certain special problems connected with the Lucas test must be recognized. In this test a carbonium ion is formed. Tertiary alcohols react rapidly to form alkyl chlorides because they yield tertiary carbonium ions as intermediates. Primary alcohols, which form unstable primary cations, do not react readily. If reaction of a substance leads to a stable carbonium ion which is *not* tertiary, the results of the Lucas test may easily be misinterpreted. For example, allylic alcohols yield allylic carbonium ions, and benzylic alcohols yield benzylic carbonium ions, as shown below.

$$CH_2{=}CH{-}CH_2{-}Cl \longrightarrow CH_2{=}CH{-}CH_2^+ \longleftrightarrow {}^+CH_2{-}CH{=}CH_2$$

$$C_6H_5{-}CH_2{-}OH \xrightarrow[ZnCl_2]{HCl} C_6H_5{-}\overset{+}{C}H_2 \longleftrightarrow \text{(ring}^+\text{)}{=}CH_2 \longleftrightarrow \text{etc.}$$

Because these cations are stabilized by electron delocalization, they are more stable than simple primary carbonium ions. Since a nontertiary alcohol which yields a stable cation may be mistaken for a tertiary alcohol, observations in the Lucas test should always be verified by an oxidation test.

PROCEDURE (in hood)

CLASSIFICATION OF ALCOHOLS

D. The Lucas alcohol test

$$R_3COH + HCl \xrightarrow{ZnCl_2} R_3CCl + H_2O$$

Add approximately 3 drops of liquid or 100 mg of solid sample to a test tube. Pipet in 2 to 3 mL of the Lucas reagent. Immediately swirl the test tube to dissolve the unknown in the Lucas reagent and time the reaction until positive results are observed.

If the compound is a tertiary alcohol (such as *tert*-butyl alcohol), there will be a virtually instantaneous reaction, and an oily material (alkyl halide) will be obvious as an emulsion or a layer at the bottom of the test tube. This result is also obtained with allylic and benzylic alcohols. If a reaction is observed after 3 to 5 min at room temperature, a secondary alcohol is indicated. If no reaction is observed after 5 min, the sample under study is either a primary alcohol or some other material which does not react with the Lucas reagent.

It should be emphasized again that this test is only conclusive for those alcohols which are relatively water-soluble.

The Lucas reagent is made by dissolving 65 g zinc chloride in 45 mL concentrated hydrochloric acid. During the preparation of this reagent, the entire solution should be cooled.

E The Oxidation-Aldehyde Test Sequence

A useful means for distinguishing a primary from a secondary alcohol when the Lucas test yields little information involves oxidation. Recall that oxidation of either a primary or secondary alcohol will yield the corresponding aldehyde or ketone when pyridinium chlorochromate is used. The product from this reaction can be treated with the nitrogen-containing heterocyclic compound commonly referred to as Purpald (see Sec. 25.5G). If the alcohol was primary, the oxidation product will give the characteristic purple color but the oxidized secondary alcohol will not. Lack of reactivity in the Lucas test coupled with reactivity in the sequential oxidation–Purpald test leads to a definitive identification procedure for primary alcohols.

$$R{-}CH_2OH \longrightarrow R{-}CHO \longrightarrow \text{purple complex}$$

F The Periodate Reaction

Previous discussion has centered largely on distinguishing primary, secondary, and tertiary alcohols from each other. In addition to these, one often encounters 1,2 diols, usually referred to as *glycols*. The 1,2-dihydroxy or vicinal diol system can be distinguished from other functions by treatment with periodic acid. Periodic acid does not generally react with tertiary alcohols, although it will react with α-hydroxyaldehydes and also with acyloins (α-hydroxyketones). The actual reaction involves the addition of periodate to a glycol to form a cyclic periodate, which then undergoes carbon-carbon bond cleavage to form two aldehyde or ketone fragments. A by-product of this oxidative cleavage is iodate ion. If a small amount of silver nitrate is added to the reaction, the formation of a copious white silver iodate precipitate indicates periodate reduction to iodate. The oxidation can be confirmed by the DNP test (see Sec. 25.5A).

$$\begin{array}{l} RCH{-}OH \\ | \\ R'CH{-}OH \end{array} + \begin{array}{c} OH \\ HO\diagdown\ |\ \diagup OH \\ I \\ HO\diagup\ \|\ \diagdown O^- \\ O \end{array} \xrightarrow{-2H_2O} \begin{array}{c} OH \\ RCHO\diagdown\ |\ \diagup OH \\ |\qquad I \\ R'CHO\diagup\ \|\ \diagdown O^- \\ O \end{array} \xrightarrow{-H_2O}$$

$$\begin{array}{c} RCH{-}O\quad O \\ |\qquad \diagdown\ \| \\ \qquad I{-}O^- \\ |\qquad \diagup\ \| \\ R'CH{-}O\quad O \end{array} \longrightarrow \begin{array}{c} R{-}CHO \\ \\ R'{-}CHO \end{array} + IO_3^-$$

The periodic acid test is rapid and definitive, especially for sugars. On the other hand, a positive test may also be observed for a substrate like an α-hydroxy carbonyl compound, which is rapidly oxidized to a diketone or a ketoaldehyde and then cleaved. If the unknown has previously given a 2,4-dinitrophenylhydrazone derivative, the periodic acid may be ambiguous.

26.5 SPECTROSCOPIC CONFIRMATION OF STRUCTURE

The most reliable spectroscopic method for the identification of an alcohol is to examine the ir spectrum. A compound suspected of being an alcohol will have shown neutral solubility properties, so the chance of identifying either a carboxylic or a phenolic hydroxyl as an alcoholic hydroxyl group is minimal.

Primary, secondary, and tertiary alcohols all absorb in the 3600 to 3650 cm^{-1} region, although hydrogen-bonded species may be observed in the 3500 to 3600 cm^{-1} range and occasionally as low as 2500 cm^{-1}. A characteristic oxygen-carbon stretching vibration will also be observed, and with skill may be used to distinguish among primary, secondary, and tertiary alcohols. All things being equal, this vibration will be observed for primary alcohols near 1060 cm^{-1}, for secondary near 1100 cm^{-1}, and for tertiary near 1150 cm^{-1}. The positions at which these vibrations are observed depend to some extent on substitution and should therefore be used with caution.

An alcoholic hydroxyl signal is often observed in the nmr spectrum at 3 to 5 ppm downfield from tetramethylsilane. The exact position of a hydroxyl proton signal depends appreciably on temperature, solvent, and the presence of other proton-bearing functional groups (COOH, NH_2) or water, as well as on other factors. Since this position is often hard to anticipate, interpretation of the nmr spectrum of an alcohol should be done cautiously.

The presence of a suspected hydroxyl group may often be confirmed by exchange with D_2O. The nmr spectrum is recorded and then the sample is removed. A drop or two of D_2O is added to the solution, which is then shaken. If an exchangeable proton (in —OH, NH, etc.) is present, rapid chemical exchange with deuterium may cause the signal to disappear when the spectrum is recorded a second time. This process is not unique to alcoholic hydroxyl groups, but if considered in concert with solubility properties, it can be informative.

26.6 DERIVATIVES OF ALCOHOLS

The important feature of alcohol reactivity in derivative formation is that oxygen behaves as a nucleophile and can add to an electrophilic species. The adduct can then lose a proton with formation of a stable oxygen-carbon bond. The two common kinds of derivatives formed from alcohols are distinguished on the basis of this reactivity. Addition derivatives are formed in those cases in which the alcohol adds to some electrophilic function and the proton then transfers to some other part of the molecule; addition-elimination derivatives are formed when nucleophilic oxygen adds and a proton is then lost as the counterion of some anionic leaving group. Let us consider the addition derivatives first.

A Phenyl- and α-Naphthylurethane Derivatives

The phenyl- and α-naphthylurethanes are particularly useful derivatives of alcohols because they form rapidly and are usually solids. As such, they are easy to manipulate and often easy to crystallize and purify. The reaction for their formation is illustrated in the equation below.

$$\mathrm{R{-}O{-}H + Ar{-}N{=}C{=}O \longrightarrow R{-}O{-}\overset{\overset{\displaystyle O}{\|}}{C}{-}NH{-}Ar}$$

where Ar is [phenyl] or [1-naphthyl]

Note that in the formation of this derivative the alkoxyl function has added to the electrophilic carbon of the isocyanate group and a proton has ultimately found residence on nitrogen. The product is a urethane or, as it is sometimes called, a carbamate. Note that there is an aryl group appended to the isocyanate function in the reaction illustrated above. While the isocyanate function dictates the chemistry of the system, the crystallinity of the derivative formed is determined by the nature of the aryl group. The α-naphthylurethanes are usually more crystalline than the corresponding phenyl compounds, although this tendency is reversed in some cases. A procedure for the preparation of α-naphthyl- and phenylurethanes is described later in this section.

There are several potential disadvantages to this procedure. The most common difficulty is the fact that both α-naphthyl and phenyl isocyanate react with water as well as with alcohols. If water is present in the air, the solvent, or the sample, it may compete with the alcohol as a nucleophile. When this happens, the sequence illustrated below occurs.

$$\mathrm{Ar{-}N{=}C{=}O} + \mathrm{H_2O} \longrightarrow \underset{\text{A carbamic acid}}{\mathrm{Ar{-}NH{-}CO{-}OH}} \longrightarrow \mathrm{Ar{-}NH_2} + \mathrm{CO_2}$$

$$\mathrm{Ar{-}NH_2} + \mathrm{Ar{-}N{=}C{=}O} \longrightarrow \underset{\text{A urea}}{\mathrm{Ar{-}NH{-}CO{-}NH{-}Ar}}$$

In the first step, the aryl isocyanate reacts with water. The carbamic acid then loses CO_2 to form a nucleophilic amine, which reacts with another molecule of aryl isocyanate. The product in this sequence is the symmetrical diarylurea rather than the expected urethane derivative of the alcohol. Bottles of α-naphthyl isocyanate and phenyl isocyanate which have been opened and exposed to moist air sometimes contain appreciable amounts of the corresponding ureas as a contaminant. The unwary student may isolate the urea and observe a melting point totally unrelated to that anticipated for the alcohol derivative.

The isocyanates most often used for derivative formation are α-naphthyl, phenyl, and *p*-tolyl isocyanates. The melting points of the corresponding ureas are noted here, so that if one is encountered it can be identified. The melting points are: di-α-naphthylurea 297°C; diphenylurea 240°C; and di-*p*-tolylurea 268°C. Generally, the diarylureas are insoluble in boiling petroleum ether. If petroleum ether is used as the recrystallization solvent, the hot solution should be filtered to remove contaminants. If this operation is successful, the product which crystallizes out will generally be the urethane. Several extractions of the crude derivative with boiling petroleum ether may be required to obtain the desired result.

Two other disadvantages of using phenyl- or α-naphthylurethanes as derivatives cannot be circumvented as readily as the one just described. First, phenylurethanes react with low-molecular-weight alcohols to give derivatives whose melting points are near or below room temperature. For this reason, the α-naphthylurethane is listed as the first and primary derivative in Tables 28.3 (Liquid Alcohols) and 28.4 (Solid Alcohols). Second, the attempted formation of a urethane from a tertiary alcohol is encumbered by two problems: (1) because the hydroxyl group in a tertiary alcohol is sterically hindered, the reaction between it and an isocyanate is frequently slow; and (2) attempts to force the reaction to completion frequently lead to dehydration of the alcohol. These cases are relatively rare however, and the derivative is a very useful one.

PROCEDURE (in hood)

DERIVATIVES OF ALCOHOLS

A. Phenylurethanes and α-naphthylurethanes

$$\mathrm{R{-}O{-}H + Ar{-}N{=}C{=}O \longrightarrow R{-}O{-}\overset{\overset{\displaystyle O}{\|}}{C}{-}NH{-}Ar}$$

Dissolve 0.5 g of the alcohol in 5 to 10 mL petroleum ether (or ligroin) and add to this solution a mixture containing 0.5 to 0.7 mL phenyl isocyanate or α-naphthyl isocyanate in approximately 10 mL petroleum ether. Heat the mixture gently on a steam bath for 5 min with the exclusion of moisture and *filter hot.* The filtered petroleum ether solution should then be allowed to cool slowly to room temperature and the crystallized product collected by filtration. The crude product should be recrystallized from petroleum ether or petroleum ether–ethyl acetate.

Notes

1. The alcohol should be as dry (water-free) as possible before initiation of this derivatization procedure. If the alcohol is soluble in petroleum ether, use a slight excess of this solvent to dissolve it and add some sodium or magnesium sulfate (**do not use $CaCl_2$**). Let the mixture stand for about 10 min, filter to remove the drying agent, and then proceed as described.
2. For liquid alcohols or alcohols which will not dissolve in petroleum ether, the procedure above may be modified as follows: Place 1 mL of the alcohol and 0.5 mL of the isocyanate in a small test tube (**hood**). Swirl vigorously or stopper and shake the tube. If no reaction is apparent, heat the tube on a steam bath for 5 to 10 min. Purify the product as described above.
3. If the reaction mixture has set to a resin, attempt to scrape out the oil with a spatula. In difficult cases, the test tube must be broken to remove the material.

B Ester Formation As noted above, addition-elimination derivatives are those in which nucleophilic addition of the alcohol to some electrophile is followed by loss of an anionic leaving group paired with the proton of the alcohol. The two most common examples are the *p*-nitrobenzoate and 3,5-dinitrobenzoate derivatives of alcohols. Both these derivatives are prepared from the acid chlorides and form according to the equation formulated below.

$$Ar{-}CO{-}Cl + ROH \longrightarrow Ar{-}CO{-}OR + HCl$$

Formation of either the 4-nitro- or the 3,5-dinitrobenzoate derivative is advantageous from the point of view of speed, cost, and ease of recrystallization. The derivatives form rapidly because the electron-withdrawing substituents in the aromatic ring facilitate nucleophilic addition to the carbonyl. 4-Nitrobenzoic acid and 3,5-dinitrobenzoic acid are both inexpensive and easy to obtain. The esters which result from these two starting materials are usually crystalline, and the dinitro-substituted compounds are often bright yellow. The derivatives are therefore easy to distinguish from the starting materials.

There are certain disadvantages which attend the use of these derivatives. The difficulty is that these derivatives are formed with facility only from the acid chlorides. 3,5-Dinitrobenzoyl chloride is a highly reactive substance, which undergoes hydrolysis with water in the solvent and water in the air and often hydrolyzes even in a closed bottle. One must be forever vigilant in dealing with 3,5-dinitrobenzoyl chloride. A freshly opened bottle of the acid which has been sealed until the time of use may have a melting point near 205°C (the melting point of the pure acid). The acid chloride melts at 70°C and is a lachrymator. There are two options in using the 3,5-dinitrobenzoate as a derivative: the alcohol in question may simply be esterified directly with 3,5-dinitrobenzoic acid, or 3,5-dinitrobenzoyl chloride may be prepared prior to use just as any other acid chloride would be, particularly for the formation of an amide derivative. In most cases in which a 3,5-dinitrobenzoate derivative is formed from what is believed to be dinitrobenzoyl chloride, it is the direct esterification process which is occurring. Derivative formation using freshly prepared acid chloride is usually quite satisfactory. The equations for both reactions are given below.

$$Ar{-}COCl + ROH \longrightarrow Ar{-}CO{-}OR + HCl \qquad \text{(acylation)}$$
$$Ar{-}COOH + ROH \longrightarrow Ar{-}CO{-}OR + H_2O \qquad \text{(esterification)}$$

As expected, tertiary alcohols react with both acids, although they do so rather slowly. The alcohols generally do not dehydrate under the reaction conditions and, despite the fact that the reaction is slow, a useful product eventually results. In fact, dinitro- and nitrobenzoate esters are probably the best available derivatives for tertiary alcohols.

C Low-Melting Ester Derivatives

Students frequently wonder why a derivative of a particular compound is not listed in a table. There are two obvious explanations for this: (1) No one has ever taken the trouble to prepare the derivative; and (2) The derivative melts below room temperature. A survey of the 4-nitrobenzoate esters of several simple alcohols is presented below and in Table 26.2. Notice that only the methyl (mp 96°C) and ethyl (mp 57°C) esters of 4-nitrobenzoic acid are sufficiently high-melting to be easily manipulated in an undergraduate organic laboratory. As the number of carbon atoms in the alcohol increases, the length of the chain attached to the carbonyl group of the ester also increases. The presence of the long, flexible groups tends to reduce the overall crystallinity of the derivative. The melting point of the *n*-propyl ester is 35°C, 22°C lower than that of the ethyl ester; the *n*-butyl ester melts at 17°C, almost another 20°C lower. The branched butyl derivative, *sec*-butyl, has a melting point of 26°C, the *n*-pentyl ester 11°C, and the *n*-hexyl ester 5°C, which appears to be the minimum point. As the number of carbon atoms in the chain of normal alcohols increases beyond 6, the intermolecular forces apparently also increase, and the melting points of the corresponding 4-nitrobenzoate esters gradually rise. The derivative of *n*-heptanol melts at 10°C; the *n*-decyl derivative at 30°C. It is evident that relatively few of the 4-nitrobenzoate esters noted here are actually of value as derivatives. In these particular cases the 3,5-dinitrobenzoate derivatives would be more useful.

D Benzoate Derivatives of Alcohols

PROCEDURE (in hood)

DERIVATIVES OF ALCOHOLS

B. Benzoate esters from the acid chloride

$$C_6H_5-C(=O)-Cl + ROH \longrightarrow C_6H_5-C(=O)-OR + HCl$$

$$O_2N-C_6H_4-C(=O)-Cl + ROH \longrightarrow O_2N-C_6H_4-C(=O)-OR + HCl$$

$$O_2N-C_6H_3(NO_2)-C(=O)-Cl + ROH \longrightarrow O_2N-C_6H_3(NO_2)-C(=O)-OR + HCl$$

Dissolve 0.5 g (or 0.5 mL) of the alcohol in 5 mL toluene. To this toluene solution add 2.5 mL anhydrous pyridine, followed by 2 g benzoyl chloride,

TABLE 26.2
Melting points of certain 4-nitrobenzoates

Alcohol	Melting point of derivative, °C
Methanol	96
Ethanol	57
1-Propanol	35
1-Butanol	17
2-Butanol	26
1-Pentanol	11
1-Hexanol	5
1-Heptanol	10
1-Octanol	12
1-Nonanol	10
1-Decanol	30

4-nitrobenzoyl chloride, or 3,5-dinitrobenzoyl chloride.[1] The solution should be swirled during the addition of the acid chloride. After several minutes, heat the mixture on the steam bath for 15 to 20 min, cool it to room temperature, and pour it into a mixture of ice (15 g) and hydrochloric acid (15 mL of a 10% solution). The suspension is transferred to a separatory funnel and 10 mL toluene is added. The acid layer is then drawn off and the toluene solution washed with water, 10% aqueous sodium carbonate, and saturated sodium chloride solution. The toluene is evaporated and the residue crystallized from alcohol or acetone–petroleum ether.

C. Benzoate esters from the acid

$$3,5\text{-}(O_2N)_2C_6H_3\text{-}C(=O)OH + ROH \longrightarrow 3,5\text{-}(O_2N)_2C_6H_3\text{-}C(=O)OR + H_2O$$

Place 2 mL of the alcohol and 0.5 g 3,5-dinitrobenzoic acid in a 15×170 mm test tube and heat for 10 min in a beaker of hot water (water

[1] If the melting point indicates that the 3,5-dinitrobenzoyl chloride is contaminated by significant amounts of acid, the acid chloride should be prepared as it is in Exp. 11.4.

heated with a steam bath or free flame, depending on the boiling point of the alcohol). Carefully pour the hot solution into a 25-mL beaker containing 10 mL cold water. Collect the precipitated solid and recrystallize from alcohol, alcohol-water, or petroleum ether–ethyl acetate. This procedure is most effective for primary and secondary alcohols. Tertiary alcohols do not yield esters by direct reaction with aromatic acids.

ESTERS, AMIDES, NITRILES, AND UREAS

27.1 General and Historical **601**
27.2 Characterization of the Classes **601**
27.3 Operational Distinctions **602**
- **A** Behavior with 2,4-dinitrophenylhydrazine (DNP) reagent
- **B** Basicity of nitriles
- **C** Reactivity of esters
- **D** Hydrolysis

27.4 Classification Tests **604**
- **A** The hydroxylamine–ferric chloride test
- **B** Hydrolysis of amides and nitriles

27.5 General Classification Scheme for Esters, Amides, Nitriles, and Ureas **607**
27.6 Spectroscopic Confirmation of Structure **607**
27.7 Derivative Formation Reactions **608**
- **A** Derivatives of esters: The neutralization equivalent
- **B** Ester saponification
- **C** Ester exchange: 3,5-Dinitrobenzoate derivatives
- **D** Amides
- **E** Synthesis of an authentic sample
- **F** Reduction of amides
- **G** Nitriles
- **H** Controlled hydrolysis of nitriles
- **I** Ureas

27.1 GENERAL AND HISTORICAL

All four of the main classes of compounds in this chapter are derivatives of carboxylic acids. Esters, amides, and ureas are carbonyl derivatives; nitriles are also in this class because they can be hydrolyzed to amides, or further to the corresponding carboxylic acids, in a single step. The interconvertibility of these systems is alluded to in Chap. 25 (Carbonyl Compounds), and if the general set of distinctions given there is used and the various Y and Z functions are substituted on the carbonyl group, the relationships should be quite obvious.

Of all the compounds in these four classes, the simplest of the ureas is actually the most interesting and important. Urea is diaminocarbonyl ($Y = Z = NH_2$), which is the diamide of carbonic acid. It is an important compound because it is a metabolic sink, i.e., it is the nitrogenous waste product excreted by most mammals. (The notable exception is the dalmatian dog which, like fowl, excretes uric acid.) Urea is also important historically, not because it is the end of the metabolic chain but because it was the first organic compound to be synthesized from an inorganic one. In 1828 the great German chemist Wöhler wrote a letter to Berzelius explaining that he had heated ammonium isocyanate and thereby isomerized it to urea without the intercession of any mammalian organ. Berzelius was the grand master of organic chemistry in his time and one of the foremost proponents of the so-called vital force theory. Wöhler's observation "that ammonium isocyanate is urea" was very important in that it began to dispel this erroneous theory. He showed that inorganic compounds could be converted to organic ones, and therefore that there was no mystical principle involved in organic chemistry (or at least that this was not it). This single discovery did not immediately put an end to the exposition of the vital force theory; it remained for other confirmatory experiments to be done before the theory collapsed beneath the weight of accumulating evidence. Nevertheless, Wöhler's experiment was the beginning.

Esters, amides, nitriles, and ureas are quite common substances. The amides and the ureas are ordinarily solid compounds, but many representatives of the ester and nitrile classes are liquids. The utility of odor in identifying various compounds has been discussed from time to time in this book but is of relatively little use with the often solid amides and ureas. The esters and nitriles, however, frequently have strong and very distinctive odors. If a distinct fragrance can be discerned in the compound, it is often reasonable to assume that it is an ester, although this assumption must be verified chemically. Any distinctive odor may be of value in identifying or confirming the identity of an unknown substance, and this is particularly true for esters.

27.2 CHARACTERIZATION OF THE CLASSES

We have mentioned above that esters, amides, and ureas are all structurally related by virtue of containing a carbonyl group and that because the nitrile is readily transformed into an amide, it is also classed with these compounds. All these substances can and should be considered as derivatives of carboxylic acids. The ureas constitute a special class in this sense because all ureas are derivatives of carbonic acid. For the two-carbon system, examples of the other three classes of related compounds are acetonitrile, acetamide, and ethyl acetate. From the sequence of illustrations below, it can be seen that acetonitrile, which is a colorless liquid (bp 83°C), when hydrolyzed (i.e., when one molecule of water is added) forms the solid compound acetamide (CH_3CONH_2, mp 82°C).

$$CH_3CN \underset{-H_2O}{\overset{H_2O}{\rightleftharpoons}} CH_3CONH_2 \xrightarrow{H_2O} CH_3COOH + NH_3$$

$$CH_3COOH \xrightarrow[CH_3CH_2OH]{H^+} CH_3CO_2C_2H_5 + H_2O$$

Treatment of acetamide with a variety of dehydrating agents, e.g., phosphorus oxychloride or acetic anhydride, reconverts the amide to the corresponding nitrile. Acetamide when further hydrolyzed affords acetic acid (bp 117°C), and if the acid is esterified with ethyl alcohol, ethyl acetate (bp 77°C) is formed. In fact, if nitriles are treated with acid and alcohol, esters can usually be formed directly. It should be clear then that all these compounds are readily interconvertible, and that ureas constitute a special case only because they are diamides.

27.3 OPERATIONAL DISTINCTIONS

While the interrelationship of the chemistry in these systems is a very useful thing to bear in mind, it would not be a good means of grouping these compounds without an important operational distinction which relates them. Obviously, esters are distinct from amides, nitriles, and ureas because they do not contain nitrogen. Nevertheless, the nitrogen found in amides, nitriles, and ureas is nonbasic and all four of these classes of compounds show similar solubility properties. Esters, amides, nitriles, and ureas may all be hydrolyzed under acidic or basic conditions, but they ordinarily do not dissolve in either dilute acid or base at room temperature. This solubility behavior distinguishes them from the carboxylic acids, which dissolve in dilute base, and from the amines, which dissolve in dilute aqueous acid.

A Behavior with 2,4-Dinitrophenylhydrazine (DNP) Reagent

This group of compounds must also be distinguished from the other neutral compounds. Alcohols and carbonyl compounds have already been discussed in this text. The amides, ureas, and esters are clearly carbonyl compounds but they do not give a positive test when treated with the 2,4-dinitrophenylhydrazine reagent (see Sec. 25.5A). In addition, the nitrogen-containing members of this class are nonbasic. The inertness of esters to the DNP reagent and the lack of nitrogen basicity are related. The resonance structures for an amide function are shown below. The amide carbonyl shares its electrons by resonance in the formation of a double bond between carbon and nitrogen. This corresponds to an enolate structure. Because there is partial double-bond character between carbon and nitrogen as well as between carbon and oxygen, the reactivity of the carbonyl group is considerably reduced, as is the basicity of the nitrogen lone pair. The same situation is encountered with ureas, which have two nitrogen atoms sharing electron density with the single carbonyl function. The resonance structures for both amides and ureas are shown below.

$$R{-}C({=}O){-}NH_2 \longleftrightarrow R{-}C(O^-){=}\overset{+}{N}H_2$$

$$H_2N{-}C({=}O){-}NH_2 \longleftrightarrow H_2N{-}C(O^-){=}\overset{+}{N}H_2 \longleftrightarrow H_2\overset{+}{N}{=}C(O^-){-}NH_2$$

B Basicity of Nitriles

Nitriles constitute a somewhat different case because they do not contain the carbonyl. Nitrogen in the carbon-nitrogen triple bond is *sp*-hybridized. One might anticipate that nitriles could function as Lewis bases. These compounds are not strongly basic, though, because the *sp* hybridization of the nitrogen orbital causes the lone pair electrons on the nitrile nitrogen atom to be bound quite tightly. As a result, the electron pair is less available in the Lewis base sense. The lone pair can, however, be used to solvate cations which are Lewis acids, and examples of this are shown in Secs. 4.4 and 15.3.

C Reactivity of Esters

The lack of nitrogen basicity in the amides, ureas, and nitriles is the property which makes these substances insoluble in an aqueous acidic medium. Why is it then that these carbonyl-containing compounds do not react with 2,4-dinitrophenylhydrazine to give the characteristic yellow to red precipitate? A consideration of the amide structure (illustrated in Sec. 27.3A) is informative in this regard. From our earlier discussion of hydrazone formation recall that 2,4-dinitrophenylhydrazine adds its nucleophilic electron pair to the carbon atom of the carbonyl group. Elimination of water then leads to the formation of a carbon-nitrogen double bond. Although in principle this can occur for such compounds as amides, it does not do so readily because the carbonyl is not very electrophilic. The reaction of 2,4-dinitrophenylhydrazine with an ester can begin in the same way it does with a carbonyl compound, but more than one reaction is possible at the stage in which three heteroatoms are bonded to carbon (illustrated below in brackets).

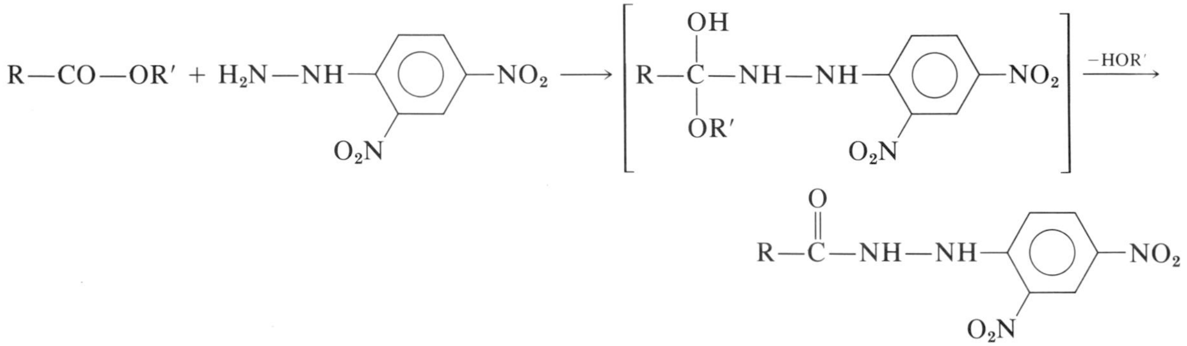

$$R{-}CO{-}OR' + H_2N{-}NH{-}C_6H_3(NO_2)_2 \longrightarrow \left[R{-}\underset{OR'}{\overset{OH}{C}}{-}NH{-}NH{-}C_6H_3(NO_2)_2 \right] \xrightarrow{-HOR'} R{-}\overset{O}{\overset{\|}{C}}{-}NH{-}NH{-}C_6H_3(NO_2)_2$$

The intermediate may lose the entering nucleophile, in which case no reaction will occur; it may eliminate water; or alkoxide may be eliminated from the ester to form the resonance-stabilized amide. Thus, derivatization of these substances does not correspond to normal aldehyde or ketone reactions and should therefore be considered in a different class.

D Hydrolysis

The final point which it is important to note regarding these substances is that they can all undergo either acidic or basic hydrolysis to yield an acid and an alcohol or amine. The esters will hydrolyze to a carboxylic acid and an alcohol, *each of which can be characterized independently*. Primary amides and nitriles will hydrolyze to a carboxylic acid and ammonia. Ammonia will be lost but can often be detected, and the carboxylic acid can be isolated and characterized. Secondary and tertiary amides when hydrolyzed yield a carboxylic acid and a primary or secondary amine, each of which can be characterized separately. The equations below show the products of hydrolysis.

Esters $$R{-}CO{-}OR' + H_2O \xrightarrow[\text{or } OH^-]{H^+} R{-}CO{-}OH + R'OH$$

Amides $$R{-}CO{-}NR'_2 + H_2O \xrightarrow[\text{or } OH^-]{H^+} R{-}CO{-}OH + R'_2NH$$

Nitriles $$R{-}CN + 2H_2O \xrightarrow[\text{or } OH^-]{H^+} R{-}CO{-}OH + NH_3$$

Ureas $$R{-}NH{-}CO{-}NH{-}R' + 2H_2O \xrightarrow[\text{or } OH^-]{H^+} RNH_2 + R'NH_2 + H_2CO_3$$

27.4 CLASSIFICATION TESTS

At the stage at which one is aware of the presence of an ester, an amide, a nitrile, or a urea, the preliminary acid-base solubility sequence has presumably already been established. The presence or absence of nitrogen should be established (see Sec. 3.4), and this will serve to distinguish the esters from the other classes. There is, however, a more important property which discriminates among these classes, namely, reactivity. Because the reactivity of the carbonyl group is relatively low, 2,4-dinitrophenylhydrazine does not react readily at room temperature with these substances. There are, nevertheless, other means by which these classes may be distinguished.

A The Hydroxylamine–Ferric Chloride Test

A very good test for distinguishing an ester from other neutral species is to treat the compound with hydroxylamine and ferric chloride. Hydroxylamine reacts with the ester as an amine would to form a new compound, the hydroxylated amide. This amide forms a very highly colored transition-metal complex with ferric chloride. When the distinct purple color associated with this complex is seen, it can only have arisen from an ester.

$$\mathrm{R{-}CO{-}OR' + NH_2OH \longrightarrow R{-}CO{-}NH{-}OH + R'OH}$$

$$\mathrm{R{-}CO{-}NH{-}OH \xrightarrow{FeCl_3} \text{colored complex}}$$

In principle an amide could have exchanged to give such a derivative, but amides are too unreactive for such an interconversion to occur under the conditions used in the test. Both amides and nitriles can undergo this reaction, but only under much more drastic conditions. The classification test procedure is described below.

PROCEDURE (in hood)

FERRIC CHLORIDE TEST FOR ESTERS

$$\mathrm{R{-}\overset{\overset{\displaystyle O}{\|}}{C}{-}OR' + NH_2{-}OH \longrightarrow R{-}\overset{\overset{\displaystyle O}{\|}}{C}{-}NH{-}OH + R'OH}$$

$$\mathrm{R{-}\overset{\overset{\displaystyle O}{\|}}{C}{-}NH{-}OH + FeCl_3 \longrightarrow \text{colored complex}}$$

To 50 mg of a solid or 1 drop of a liquid sample in a test tube is added 0.5 to 1.0 mL of 1 *M* hydroxylamine hydrochloride in ethanol. To this solution is added dropwise a solution of 10% potassium hydroxide in methanol until the pH (pH paper) of the solution is about 10. The mixture is then heated to reflux for several minutes, cooled to room temperature, and acidified to pH 3 to 4 with 5% HCl. To this acidified solution is added 2 drops 5% aqueous ferric chloride.

If an ester is present, a red to purple color should develop very rapidly. This test is quite sensitive and most esters give a very deep color. A blank and a known (ethyl acetate) should always be run for comparison.

B Hydrolysis of Amides and Nitriles

We have mentioned the difference in reactivity between the compounds considered here (esters, amides, nitriles, ureas) and other carbonyl compounds, but it is well to remember that there is also a significant reactivity difference among the four classes themselves. The esters often hydrolyze readily because alkoxide (RO^-) may be lost as a stable anion. The corresponding amide (NH_2^-) or anilide ($^-NHC_6H_5$) anions are less readily lost, and as a result amides, nitriles, and ureas hydrolyze more slowly than do esters. For example, it is not uncommon to find an ester which hydrolyzes completely in 1 h in 10% aqueous sodium hydroxide. The related amide often requires boiling ethylene glycol (200°C) for 24 h to achieve a comparable yield of the corresponding acid. The reason for this is simple: The resonance stabilization of the amide is much greater than the resonance stabilization of the ester. The reactivity of the carbonyl is, therefore, reduced less in the ester than it is in the amide.

Amides and nitriles can often be detected by hydrolysis. If a small amount of amide or nitrile is added to 10% aqueous sodium hydroxide solution and refluxed, continuous sampling of the effluent at the top of the reflux condenser often allows one to determine the presence of an amide or a nitrile. As the hydrolysis proceeds, a carboxylic acid is produced, which gradually dissolves in the basic solution. At the same time an amine or ammonia is produced, which is insoluble in aqueous base. If the amine released in this procedure is also volatile, it is driven from the solution and may be detected at the top of the condenser. A piece of moist litmus paper will change from red to blue as the gas passes across it. (If pH paper is used, a color corresponding to pH >7 should be observed.) This test often provides a good indication that the sample is an amide or nitrile. A typical procedure for the test is given below.

PROCEDURE

DIAGNOSTIC TEST FOR NITRILES AND AMIDES

Amides $$R{-}CO{-}NR_2 \xrightarrow{NaOH} R{-}CO{-}ONa + R_2NH$$

Nitriles $$R{-}CN \longrightarrow R{-}CO{-}ONa + NH_3$$

One drop of a liquid or 50 to 100 mg of a solid sample is placed in a small test tube. To this is added 3 mL 15% sodium hydroxide solution and a boiling chip. The mixture is heated to reflux with a very small flame. Care should be taken not to reflux this solution too vigorously.

As the solution is heated, a strip of pH paper should be moistened with *distilled* water and placed on the top of the tube. If the unknown is a primary amide or nitrile, ammonia will be liberated and pass over the moist pH or litmus paper. The pH paper should turn dark green to blue; litmus will go from red to blue if ammonia is present.

If the solution is boiled vigorously, a small amount of the sodium hydroxide solution may be spattered onto the pH paper, giving a false positive test. One way to eliminate this spattering is to put a loose plug of glass wool in the top of the test tube.

Ammonia from nitriles or primary amides and low-molecular-weight amines from secondary and tertiary amides will usually give a positive test.

27.5 GENERAL CLASSIFICATION SCHEME

Solubility	Unknown is: insoluble in dilute acid insoluble in dilute base soluble in concentrated H_2SO_4
Classification tests	2,4-DNP test negative (Sec. 25.5A) Baeyer test negative (Sec. 25.5F) Chromic anhydride (or equivalent) test negative (Sec. 26.4C2)
Indications	Not acid, phenol, amine, hydrocarbon, halide, aldehyde, ketone, or alcohol Ester likely if N absent; perform $FeCl_3$–NH_2OH test (Sec. 27.4A) Amide, nitrile, or urea likely if N present; perform diagnostic hydrolysis (Sec. 27.4B)
Derivative formation	Hydrolyze Isolate solid fragment if possible Perform neutralization equivalent on acid if obtained

27.6 SPECTROSCOPIC CONFIRMATION OF STRUCTURE

The —CO—O—, —CN, and —CO—N< functional groups may or may not bear protons, depending on how they are substituted. If protons are present on the amide group, they may be detected by proton nmr spectroscopy, and useful information may result. Often, however, protons attached to amide (—CO—NH—) exhibit broad or nearly undetectable resonances, so detection of the functional group per se is hard to do by the nmr method. On the other hand, the organic residue attached to the functional group may be identified by this technique.

Infrared spectroscopic analysis of these samples will usually produce the most useful information. The cyanide group may be detected very readily by a usually strong absorption in the 2200 to 2250 cm^{-1} range. Few other organic functional groups show strong absorption in this region, so nitriles are usually readily identified.

The ester, amide, and urea functions all contain the carbonyl group ($>C=O$), and the appearance of a strong band in the 1650 to 1750 cm^{-1} region usually signals its presence. Aldehydes and ketones can be ruled out on the basis of a 2,4-DNP test. The other carbonyl-containing groups can be identified as follows.

Amides usually show strong absorption in the 1650 to 1700 cm^{-1} region, and if the amide has an aromatic ring or double bond adjacent to the carbonyl group, this frequency will be lowered by 20 to 30 cm^{-1}. This band is sometimes referred to as the *amide I band*. If the amide is either primary or secondary, the N—H vibration gives rise to a so-called *amide II band* between about 1530 and 1640 cm^{-1}. If the compound is a tertiary amide, the amide II band will obviously be absent.

Ureas are structurally related to amides and exhibit the same sorts of absorptions. They are generally observed at about the same frequencies as amides, i.e., 1550 to 1650 cm^{-1}. The amide II band usually appears in the 1550 to 1575 cm^{-1} range.

27.7 DERIVATIVE FORMATION REACTIONS

Esters, amides, and nitriles are all characterized most easily by converting these substances to one of their constituents. All may be converted hydrolytically to the corresponding carboxylic acid. A cursory glance at the derivative tables for esters, amides, and nitriles (Tables 28.7, 28.12, 28.13, 28.16, and 28.17) will indicate that the corresponding acids mentioned as derivatives provide a means by which these materials may be characterized. In each of the cases below, the best way to form a derivative is simply to cleave the molecule hydrolytically, determine how much acid or base is required for the cleavage, and isolate or further derivatize the acid, alcohol, or amine which is produced.

A Derivatives of Esters: The Neutralization Equivalent

Esters usually hydrolyze readily in dilute sodium hydroxide solution. This technique has been used for centuries to hydrolyze fats (esters of glycerol) and thus produce soap. It is for this reason that the hydrolysis process used in ester cleavage is known as *saponification*. The amount of base required to cleave or saponify an ester corresponds to the molecular weight of the ester. Just as an acid may be neutralized with standardized base to determine its molecular weight, an ester may be saponified for this purpose. For any given ester, this numerical value is called the *saponification equivalent*. The equation for saponification of an ester is written below. Note that one equivalent of hydroxide affords cleavage, and therefore there is a 1:1 correspondence between each ester which is present in the molecule and the amount of base used. A procedure for the saponification equivalent is described below.

PROCEDURE

SAPONIFICATION EQUIVALENT OF ESTERS AND AMIDES

$$R—CO—OR' + H_2O \xrightarrow{KOH} R—CO—OK + R'OH$$

Fit *two* 250-mL conical flasks with efficient reflux condensers by means of rubber stoppers. Accurately weigh about 0.5 g of the ester into one flask. Introduce 25.0 mL 0.5 *N* alcoholic potassium hydroxide from a buret into each flask (one flask acts as a blank or control), and add a boiling chip to each flask. Boil each flask gently under *efficient* reflux for 2 h. Pour 20 to 25 mL water down each condenser, remove the flasks from the respective condensers, and cool them in cold water. Titrate the contents of each flask with standard 0.5 *N* (or 0.25 *N*) hydrochloric acid, using phenolphthalein as indicator. The endpoint should be a faint pink.

Calculation Calculate the saponification equivalent of the ester from the formula:

$$\text{Saponification equivalent} = \frac{1000W}{(V_1 - V_2)N_1}$$

where W = weight of sample, g
V_1 = volume of acid required for blank, mL
V_2 = volume of acid required for sample, mL
N_1 = normality of hydrochloric acid

B Ester Saponification

Before attempting to isolate an acid from an ester, careful consideration should be given to the possibilities which exist for the unknown compound. In a given melting or boiling range there are often several alternatives. These possibilities generally include esters of aliphatic acids, esters of aromatic acids, and even phenolic esters. A simple flame test will often serve to distinguish aliphatic from aromatic esters. For example, ethyl caprylate, 3-cresyl acetate, and ethyl benzoate all have boiling points of 210 ± 5°C. If a yellow, sooty flame is observed for the unknown, ethyl caprylate can be eliminated as a reasonable possibility. This judgment can be confirmed by checking the refractive index. The aliphatic ester has a refractive index of 1.4166, while the other two possibilities have refractive index values which are much higher.

Faced with the possibility of having either ethyl benzoate or 3-cresyl acetate as an unknown, a decision must be made regarding which portion of the ester should be isolated. Saponification of ethyl benzoate will produce tractable solid benzoic acid, while acetic acid from cresyl acetate will be very difficult to obtain in a pure state. A ferric chloride enol test on the solution obtained in the saponification equivalent will tell whether a phenol (cresol) is present, and this will facilitate the decision. If the unknown gives a negative enol test, the carboxylic acid portion should be pursued. If the enol test is positive, the phenol (alcohol portion) should be isolated and characterized. The saponification reactions of the three esters are shown below.

$$CH_3CH_2O{-}CO{-}(CH_2)_4CH_3 \xrightarrow{NaOH} CH_3CH_2OH + CH_3(CH_2)_4COOH$$

$$3\text{-}CH_3C_6H_4{-}O{-}\overset{\overset{O}{\|}}{C}CH_3 \xrightarrow{NaOH} 3\text{-}CH_3C_6H_4OH + CH_3COOH$$

$$C_6H_5{-}\overset{\overset{O}{\|}}{C}{-}OCH_2CH_3 \xrightarrow{NaOH} C_6H_5{-}\overset{\overset{O}{\|}}{C}{-}OH + CH_3CH_2OH$$

In order to isolate the acid it is probably easiest to work in dilute aqueous base solution. If an ester is added to 10% aqueous sodium hydroxide solution and this mixture is heated, there will be some solubilization of the ester in the aqueous solution simply because it is hot. There will also be some surface reaction and the ester will gradually dissolve. As the hydrolysis of the ester proceeds, a water-soluble carboxylic acid salt and an alcohol will be produced. If the alcohol is of reasonably low molecular weight, it will be soluble in the aqueous hydroxide system.

The reaction can therefore be monitored visually by the disappearance of the oil or the solid. When a clear (not necessarily colorless) solution results, the reaction will be very near completion. This process usually takes only 1 h or so of boiling. On workup, the carboxylate salt will remain in the aqueous base.

The alcohol can be separated either by distillation or by extraction and derivatives prepared by the procedures discussed in Secs. 23.6 and 26.6. Once the aqueous base solution is extracted or distilled and freed of any ester and/or alcohol, it may be acidified; the carboxylate ion protonates and often separates from the aqueous solution. If the acid itself is a solid at room temperature, it can usually be obtained by filtration. If the acid is an oil or an oily solid which will not filter readily, it can be extracted. A detailed procedure for saponification and isolation of the fragments is given below.

PROCEDURE

ESTER SAPONIFICATION AND FRAGMENT ISOLATION

$$\text{R—CO—OR}' \xrightarrow[\text{H}_2\text{O}]{\text{NaOH}} \text{R—CO—ONa} + \text{R}'\text{OH}$$

Place 1 g (or 1 mL) of the ester in a 50-mL round-bottom flask and cover it with 25 mL 10% aqueous sodium hydroxide solution. Add a boiling chip, attach a condenser, and reflux the mixture for 1 h or until no more insoluble ester is visible. At this point one of two approaches may be used, depending on the expected molecular weight of the alcohol fragment. If this fragment is expected to contain more than six carbon atoms, allow the solution to cool (immerse the flask in an ice-water bath) and then transfer the liquid to a separatory funnel. Extract the aqueous solution twice with 15-mL portions of ether. The extract will contain the alcohol fragment from the ester, as well as any unreacted ester. If the alcohol fragment is expected to have fewer than six carbon atoms, distill the aqueous base solution. The mixture of alcohol and water which results should be saturated with salt and then extracted with ether. The remaining base solution should contain essentially pure carboxylate salt.

To the aqueous base solution, continuously add 6 *N* HCl until the resulting solution is distinctly red to litmus (pH approximately 2). If the acid separates as a solid, filter it and wash the crystals with water. If the acid separates as an oil, decant the aqueous solution or extract with two 15-mL portions of ether. The oily acid can sometimes be induced to crystallize by trituration with ligroin–ethyl acetate (9:1). If the acid will not crystallize, it may be derivatized as discussed in Sec. 23.6.

C Ester Exchange: 3,5-Dinitrobenzoate Derivatives

Derivatization of esters, amides, nitriles, and ureas is dominated by a single reaction: hydrolysis. In most cases it is expeditious to hydrolytically cleave the compound and then characterize and derivatize the fragments. An exception to this general approach occurs in the case of esters.

The reactivity of the carbonyl group in esters is not diminished as much as it is in amides or ureas, and in many cases one can effect an acid exchange with 3,5-dinitrobenzoic acid. This derivatization is rather a brute force technique but is simple to carry out and is therefore of some value. The reaction itself is an acid-catalyzed exchange in which the crystalline dinitrobenzoate derivative of the ester's alcohol portion is isolated. A procedure is given below. Melting points of the derivatives can be found by referring to Tables 28.3 (Liquid Alcohols) and 28.4 (Solid Alcohols).

PROCEDURE

3,5-DINITROBENZOATE DERIVATIVES OF ESTERS

$$R{-}CO{-}OR' + 3,5\text{-}(O_2N)_2C_6H_3{-}COOH \xrightarrow{H_2SO_4} R{-}CO{-}OH + 3,5\text{-}(O_2N)_2C_6H_3{-}CO{-}OR'$$

Place 2 g each of the ester and 3,5-dinitrobenzoic acid in a large test tube and add 2 to 3 drops concentrated H_2SO_4. Put a plug of glass wool or cotton in the mouth of the test tube and heat. If the boiling point of the ester is lower than about 100°C, heat on the steam bath for about 1 h. If the boiling point of the ester is significantly below 100°C, a boiling flask with reflux condenser may be required. If the boiling point of the ester is high, heat the test tube in an oil bath at about 150°C for 30 to 45 min (longer if the reaction fails the first time).

Allow the reaction mixture to cool and dilute with 25 mL dichloromethane. Wash the dichloromethane solution (separatory funnel) with two 25-mL portions of 5% sodium carbonate and then with a 25-mL portion of distilled water. Evaporate the dichloromethane solution (steam bath). The crude ester should be recrystallized from 95% ethanol or acetone–petroleum ether.

D Amides

The principles discussed for the hydrolysis of esters apply equally well to amides with one important difference. Because of carbonyl-heteroatom resonance, it is much harder for a nucleophile to add to the carbonyl group of an amide than to add to the carbonyl of an ester. If it is difficult for hydroxide to enter and participate in the reaction, the reaction will be correspondingly slower. Likewise, the loss of an amide anion is less favorable than is the loss of an alkoxide ion. These two factors in concert impede the saponification of amides. The logical extension of this is that these difficulties will also be encountered in nitrile hydrolysis to an acid, which must involve an amide as an intermediate. A saponification equivalent can be determined on an amide as well as on an ester, but the values obtained are usually neither reliable nor reproducible for a variety of reasons which do not merit a detailed discussion here. Suffice it to say that many factors are involved and, although saponification may occasionally be achieved quantitatively, there are enough difficulties associated with the technique to make it an erratic one.

Amides may be hydrolyzed by heating either with aqueous acid or with base. Although both work well in certain cases, for practical reasons it is usually best to saponify an amide in basic solution. The ammonia produced if the amide is primary will be lost, and the acid itself will be the only substance remaining. Acidification and filtration or extraction should yield a carboxylic acid, which can then be derivatized by any of the standard procedures. Alternatively, the neutralization equivalent of the material can be determined. Refer to Sec. 23.6 to determine the best approach for derivatization. Also note that the common derivatives of acids are primary amides, anilides, and toluidides. A primary amide can be saponified to the acid, and simply by formation of the amide (characterized by a mixed melting point) one can return to the same substance.

PROCEDURE

SAPONIFICATION OF AMIDES

$$R—CO—NR'_2 \xrightarrow{KOH} R—CO—OH + R'_2NH$$

Dissolve 5 g KOH in 35 mL distilled water and dilute this solution with 10 mL methanol. Place 2 g of the amide in a 100-mL round-bottom flask and add the solution prepared above and a boiling chip. Equip the flask for reflux and boil the mixture for 2 h. If the amide is at first insoluble in the hot solution but gradually dissolves, the reaction may be terminated as soon as all the amide disappears.

When the reaction is over, cool the solution, acidify to pH 2 with 6 *N* HCl, and collect the acid by filtration if it precipitates. If no solid separates or if an oil is observed, extract the mixture with ether or dichloromethane, evaporate the solvent, and recrystallize or derivatize (Sec. 23.6). If the acid fragment is expected to be of low molecular weight, use the procedure in Sec. 27.7H.

If the amide in question is not primary, and if the amine is not volatile enough to be lost in the saponification procedure, it may be isolated by extraction. The amine produced in the hydrolysis procedure will not generally be soluble in the basic medium and can be extracted with ether or dichloromethane. When the amine is separated from the soluble carboxylic acid, it can be derivatized by using phenyl isothiocyanate (formation of the phenylthiourea derivative). The preparation of the phenylthiourea derivative is described below and in Sec. 24.7C.

PROCEDURE (in hood)

PHENYLTHIOUREA DERIVATIVES

$$C_6H_5{-}N{=}C{=}S + R_2NH \longrightarrow C_6H_5{-}NH{-}CS{-}NR_2$$

Place 0.5 g (or 0.5 mL) of the amine in a 13 × 150 mm test tube and add 7 to 10 mL petroleum ether (bp 60 to 70°C). Swirl the mixture to dissolve the amine in the hydrocarbon solvent and then add approximately 0.6 mL phenyl isothiocyanate. Rapidly swirl the mixture so that the reactant is thoroughly dispersed in the solution. After several minutes the phenylthiourea begins to crystallize. In cases in which the reaction is slow, slight warming of the mixture combined with swirling may be required.

For low-molecular-weight amines which are not very soluble in petroleum ether, this reaction can be effected in ethyl alcohol. The crude crystals are collected and recrystallized from ethanol.

E Synthesis of an Authentic Sample

It is useful to note that a variety of common carboxylic acids are available in undergraduate laboratories, as are many primary and secondary amines. If a compound is thought to be a certain amide, the known carboxylic acid may be converted to the chloride and then treated with the desired amine. This approach is called *synthesis of an authentic sample*. If the compound synthesized from known fragments has the same melting point as the unknown compound and if the melting points of the two substances do not depress each other (mixed melting point), the structure and composition of the unknown compound have been confirmed. (See Secs. 11.4, 11.5, 23.6, and 24.7.)

F Reduction of Amides

One other technique should be noted, although its utility in this course will be determined by your instructor; it is based on the fact that secondary amides are readily reduced by lithium aluminum hydride under a variety of conditions to the corresponding amines.

$$R{-}CO{-}NR'_2 \xrightarrow{LiAlH_4} R{-}CH_2{-}NR'_2$$

Treatment of the resulting amine with any of the standard derivatizing reagents will yield useful derivatives. This technique is of more value for structure determination in an advanced qualitative organic analysis course than in a general laboratory course because lithium aluminum hydride is a potentially dangerous reagent. This material is extremely water-sensitive, and when it comes in contact with even small amounts of water, it bursts into flame. Although this reduction technique is often useful, it should be applied only with the advice and consent of your instructor. The technique of analyzing fragments is more generally useful.

G Nitriles

The nitriles and amides have hydrolysis properties in common. In the first stage, water adds to the nitrile to form a primary amide, which reacts further to yield a carboxylic acid and ammonia. As with amides, the saponification equivalents for these systems are usually erratic and not very useful.

The discussion of amide hydrolysis presented above will generally suffice for nitriles. There are certain properties of the nitrile function which allow various different derivatives to be obtained, but these often require the application of sophisticated techniques. The best way to derivatize a nitrile is to saponify it under vigorous conditions to the corresponding carboxylic acid. Ammonia will be lost during the saponification procedure and it can be detected as described for the primary amide case. Occasionally, it may be desirable to prepare the amide from ammonia and the carboxylic acid chloride and then dehydrate to obtain the desired nitrile. Conditions for dehydration are not given in this text, but your laboratory instructor can help you obtain them.

$$\mathrm{R{-}CO{-}Cl + NH_3 \longrightarrow R{-}CO{-}NH_2 + HCl}$$

$$\mathrm{R{-}CO{-}NH_2 + Ac_2O \longrightarrow R{-}CN + 2HOAc}$$

H Controlled Hydrolysis of Nitriles

Another approach to nitrile derivatization is to hydrolyze under controlled instead of vigorous conditions. Dilute acid in contact with nitriles will often effect hydrolysis only to the amide stage, as shown below.

$$\mathrm{R{-}CN + H_2O \xrightarrow{dil.\ HCl} R{-}CO{-}NH_2}$$

This is accomplished by keeping both the acid concentration and the temperature low. Melting point data for the amide can be found listed for the amide derivative of the corresponding carboxylic acid (Tables 28.1 and 28.2).

In addition to hydrolysis achieved by using dilute acid under mild conditions, basic hydrolysis in the presence of 30% hydrogen peroxide is sometimes utilized. In this case, the oxidative hydrolysis usually proceeds quite cleanly from the nitrile to the corresponding amide. Once again, the amide may be further converted or it may itself be identified.

$$R{-}CN + H_2O \xrightarrow[H_2O_2]{NaOH} R{-}CO{-}NH_2$$

Controlled hydrolysis of nitriles is a tricky business. If hydrolysis is not complete, the amide will be contaminated by the starting nitrile. If hydrolysis proceeds too far, the amide will be contaminated by carboxylic acid. The precise conditions required to achieve hydrolysis only to the amide stage depend on the structure of the nitrile. It is therefore even more difficult to present a general procedure in this case than in others in this text.

If controlled hydrolysis seems to be the most reasonable alternative in derivatizing the unknown, discuss the reasons for this assessment with your laboratory instructor. Procedures for acid and peroxide hydrolysis of specific compounds are available in such reference works as *Organic Syntheses*. Performance of one of these hydrolyses, particularly the one involving 30% hydrogen peroxide, should only be done with the instructor's consent.

PROCEDURE

SAPONIFICATION OF AMIDES AND NITRILES

$$R{-}CO{-}NH_2 + KOH \xrightarrow[EtOH]{H_2O} NH_3 + R{-}CO{-}OK \xrightarrow{H^+} R{-}CO{-}OH$$

$$R{-}CN + KOH \xrightarrow[EtOH]{H_2O} NH_3 + R{-}CO{-}OK \xrightarrow{H^+} R{-}CO{-}OH$$

Weigh out 5 g KOH pellets and place them in a 100-mL round-bottom flask (with a ground glass joint). Add 10 mL water and swirl. The solution will heat as the KOH dissolves. After the KOH has dissolved, add 40 mL 95% ethanol, swirl to mix the liquids, and then add 2 g or 2 mL of the unknown amide or nitrile, add a boiling chip, attach a condenser, and reflux for 1 to 2 h. The progress of the reaction can be measured by checking for ammonia or low-molecular-weight amine evolution (see Sec. 27.4B) at the top of the condenser (moist red litmus). If the amide or nitrile was not very soluble in the hot solution, its disappearance will often signal that the reaction is near completion.

After the reaction is over or after 2 h, whichever comes first, add 25 mL water (do not overfill the flask), change from a reflux to a distillation setup, and remove most of the ethanol. The remaining aqueous solution should contain the potassium salt of the acid. Add some chips of ice to cool and dilute the solution and then acidify by careful addition of 1 *N* HCl. If the acid is a very low-molecular-weight acid like acetic or propanoic, attempt to isolate the acid by distillation (beware of azeotropes).

Note: Amides and nitriles which contain more than about 10 carbon atoms will hydrolyze slowly. Higher-molecular-weight nitriles or amides with unreactive carbonyl groups (e.g., benzanilide) may not hydrolyze at all.

I Ureas

To repeat the same series of steps for ureas that we have already discussed for amides and nitriles would be redundant. Suffice it to say that the principles which were discussed for nitriles and amides apply equally well to ureas, except that isolation of the amine (or amines) instead of the acid is the objective because of the transient nature of carbonic acid. Basic hydrolysis (amide-nitrile procedure) will yield an amine as shown below, which can be isolated by acid extraction and then derivatized by reaction with phenyl isothiocyanate, acetyl chloride, or benzoyl chloride.

$$R{-}NH{-}\overset{\overset{\displaystyle O}{\|}}{C}{-}NHR' \xrightarrow{NaOH} R{-}NH_2 + R'{-}NH_2 + CO_2 + H_2O$$

Saponification equivalents for ureas are as erratic as they are for amides and nitriles and are of little utility. There is no prospect of further derivatizing the urea as it exists. As a consequence, the amine must usually be obtained for further study.

DERIVATIVE TABLES

28.1	Liquid Carboxylic Acids	**621**
28.2	Solid Carboxylic Acids	**623**
28.3	Liquid Alcohols	**626**
28.4	Solid Alcohols	**629**
28.5	Liquid Aldehydes	**630**
28.6	Solid Aldehydes	**632**
28.7	Amides	**633**
28.8	Liquid Primary and Secondary Amines	**639**
28.9	Liquid Tertiary Amines	**642**
28.10	Solid Primary and Secondary Amines	**643**
28.11	Solid Tertiary Amines	**645**
28.12	Liquid Esters	**646**
28.13	Solid Esters	**651**
28.14	Liquid Ketones	**654**
28.15	Solid Ketones	**656**
28.16	Liquid Nitriles	**658**
28.17	Solid Nitriles	**659**
28.18	Liquid Phenols	**661**
28.19	Solid Phenols	**662**

Abbreviations Used in Derivative Tables

Carboxylic Acids	TOL	*p*-Toluidide
	NPE	β-Naphthacyl ester
	BPE	*p*-Bromophenacyl ester
	ANI	Anilide
	AMD	Amide
Alcohols	NPU	α-Naphthylurethane
	PHU	Phenylurethane
	NBE	4-Nitrobenzoate
	DNB	3,5-Dinitrobenzoate
Aldehydes and Ketones	DNP	2,4-Dinitrophenylhydrazone
	SCB	Semicarbazone
	OXM	Oxime
	NPH	4-Nitrophenylhydrazone
	PHZ	Phenylhydrazone
Amines	BSA	Benzenesulfonamide
	TSA	Toluenesulfonamide
	PTU	Phenylthiourea
	BZM	Benzamide
	ACM	Acetamide
	HCl	Hydrochloride
	PIC	Picrate
	MeI	Methiodide salt
Phenols	AAA	Aryloxyacetic acid derivative
	ANE	Aryloxyacetic acid neutralization equivalent
	BRD	Bromide
	BZE	Benzoate ester
	NPU	Naphthylurethane

(Continued on next page)

Abbreviations Used in Derivative Tables (*Continued*)

General Abbreviations	A	anti
	d, dec	Decomposes
	DEN	Density
	di	Disubstituted
	m	Monosubstituted
	nr	Not reported
	S	syn
	SE	Saponification equivalent (esters)
	SN	Systematic name
	subl	Sublimes
	t	Trisubstituted
	te	Tetrasubstituted
	TN	Trivial name
	*	See note to right

Note on Table Usage

The derivatives in each table are arranged from left to right in approximate order of utility for characterization of the compound type at hand. This order is based on the authors' experience and instincts and may not be strictly correct for every compound. If used judiciously, however, it can serve as a useful decision-making tool.

TABLE 28.1
Liquid carboxylic acids

Name	bp, °C	mp, °C	MW	RI	DEN, g/mL	TOL	BPE	ANI	AMD	Notes
Formic	100.8	8.5	46	1.3721	1.220	53	138	47	liquid	Methanoic; aqueous azeotrope; **skin irritant**
Acetic	117	16.2	60	1.3720	1.049	152	86	114	80	Ethanoic
Acrylic	139	13	72	1.4202	1.062	141	nr	104	84	Propenoic acid
Propionic	141	−23	74	1.3860	0.993	125	63	104	79	Propanoic
Propiolic	144d	9	70	1.4302	1.138	nr	nr	87	61	Propynoic
Isobutyric	154	−47	88	1.3930	0.950	106	77	105	127	2-Methylpropanoic
Methacrylic	163	16	86	1.4310	1.015	nr	nr	87	109	2-Methylpropenoic
Butyric	163	−8	88	1.3969	0.964	74	63	96	115	Butanoic
Pivalic	163	34	102	—	—	120	76	129	154	Trimethylacetic
Vinylacetic	163	−39	86	1.4249	1.013	nr	nr	58	73	3-Butenoic
Isocrotonic	169	14	86	1.4450	1.026	132	82	101	101	(*Z*)-Crotonic; **skin irritant**
2-Methylbutanoic	174	<−80	102	1.4051	0.941	92	55	110	112	Ethylmethylacetic; values reported for racemate
Isovaleric	176	−37	102	1.4043	0.931	106	68	109	135	3-Methylbutanoic
Crotonic	181	71	86	—	—	132	95	118	160	(*E*)-2-Butenoic
tert-Butylacetic	185	−11	116	1.4100	0.912	134	nr	132	132	3,3-Dimethylbutanoic
2-Chloropropionic	186 (170 to 190)	nr	108	1.4345	1.182	124	nr	92	80	Racemate
Pentanoic	186	−18	102	1.4076	0.939	72	75	63	106	Valeric
2,2-Dimethylbutanoic	186	−13	116	1.4145	0.928	83	nr	92	103	
2,3-Dimethylbutanoic	189	−1.5	116	1.4146	0.927	112	nr	78	129	
Dichloroacetic	190 (194)	7	128	1.4642	1.563	153	99	118	99 (subl)	
2-Ethylbutanoic	195	−15	116	1.4179	0.924	116	nr	124	112	Diethylacetic
2-Methylpentanoic	195	nr	116	1.4112	0.908	81	nr	95	79	Racemate
3-Methylpentanoic	197	−42	116	1.4159	0.930	75	nr	87	125	
4-Methylpentanoic	199	−33	116	1.4144	0.923	63	77	110	119	Isocaproic
Caproic	202	−3	116	1.4161	0.927	74	72	94	100	Hexanoic; hygroscopic
Ethoxyacetic	206	nr	104	1.4190	1.102	32	104	93	80	
α-Bromobutanoic	217d	−4	167	1.4720	1.567	92	nr	98	110	α-Bromobutyric

Abbreviations used: d, decomposes; subl, sublimes; nr, not reported; RI, refractive index; DEN, density; TOL, *p*-toluidide; BPE, *p*-bromophenacyl ester; ANI, anilide; AMD, amide; *, see note to right.

Note: All derivative values are melting points, °C.

TABLE 28.1 (*Continued*)

Name	bp, °C	mp, °C	MW	RI	DEN, g/mL	TOL	BPE	ANI	AMD	Notes
Heptanoic	223	−10	130	1.4221	0.918	80	72	68	96	Enanthic
Octanoic	237	16	144	1.4278	0.910	70	67	57	108	Caprylic
Levulinic	245	33	116	1.4420	1.134	108	84	102	107d	β-Acetylpropionic; oxime mp 96°C
Nonanoic	254	9	158	1.4319	0.906	84	68	57	99	Pelargonic
Decanoic	268	30	172	1.4288	0.878	78	66	63	99	Capric
Undecanoic	275d	28	186	1.4294	0.891	80	68	71	103	Undecylic

Abbreviations used: d, decomposes; subl, sublimes; nr, not reported; RI, refractive index; DEN, density; TOL, *p*-toluidide; BPE, *p*-bromophenacyl ester; ANI, anilide; AMD, amide; *, see note to right.

Note: All derivative values are melting points, °C.

TABLE 28.2
Solid carboxylic acids

Name	mp, °C	bp, °C	MW	TOL	BPE	ANI	AMD	Notes
Pivalic	34	163	102	120	76	129	154	Trimethylacetic
Tridecanoic	41	312	214	88	75	80	100	Tridecylic
Dodecanoic	44	299	200	87	76	76	98	Lauric
Hydrocinnamic	47	280	150	135	104	95	82	β-Phenylpropanoic
Pentadecanoic	52	*	242	nr	77	78	102	Pentadecylic; *bp 257°C/100 torr
Tetradecanoic	54	*	228	93	81	84	102	Myristic; *bp 250°C/100 torr
Trichloroacetic	57	197	163	113	nr	95	141	Hygroscopic
Heptadecanoic	59	*	270	nr	83	nr	106	Margaric; *227°C/100 torr; *p*-nitrobenzyl ester, 48°C
Chloroacetic	61	189	94	162	105	134	120	Reported mp's: 52, 56, 61°C **Caution: Skin irritant, toxic**
Hexadecanoic	61	*	256	98	86	90	106	Palmitic; *271°C/10 torr
(*Z*)-2-Methyl-2-butenoic	63	198	100	70	68	77	75	Tiglic
2,3-Dibromopropanoic	64	*	231	nr	nr	nr	130	*160°C/20 torr
Octadecanoic	67	361	284	102	91	93	108	Stearic
(*E*)-2-Butenoic	71	180	86	132	95	118	160	(*E*)-Crotonic
Phenylacetic	76 (subl)	206	136	135	89	117	154	Musty odor
Eicosanoic	75	*	312	96	89	92	108	Arachidic; *bp 204°C/1 torr
Hydroxyacetic	77	nr	76	143	140	96	120	Glycolic; hygroscopic (anhydride mp 128°C)
α-Hydroxyisobutyric	79	212	104	132	98	136	98	2-Hydroxy-2-methylpropanoic
Methylmaleic	91d	dec	130	170(m)	nr	153(m) 175(di)	185d(di)	Citraconic
Glutaric	98	302	132	218(di)	137(di)	223(di)	174(di)	Pentanedioic
Phenoxyacetic	98	285d	152	nr	148	99	101	
Citric	100*	nr	192	189(t)	148(t)	192(t)	210(t)	2-Carboxyglutaric; anhydrous; *monohydrate
L-Malic	100	nr	134	206(di)	179(di)	197(di)	156(di)	Hydroxysuccinic
2-Methoxybenzoic	100	200	152	nr	113	131	128	*o*-Anisic
o-Toluic	103	258	136	144	57	125	142	2-Methylbenzoic
Pimelic	104	*	160	206(di)	136(di)	108(m) 155(di)	175(di)	Heptanedioic; *bp 212°C/10 torr

Abbreviations used: nr, not reported; subl, sublimes; m, mono; t, tri; TOL, *p*-toluidide; BPE, *p*-bromophenacyl ester; ANI, anilide; AMD, amide; *, see note to right; d, dec, decomposes.

Note: All derivative values are melting points, °C.

TABLE 28.2 (*Continued*)

Name	mp, °C	bp, °C	MW	TOL	BPE	ANI	AMD	Notes
4-Chlorophenylacetic	104	nr	170	190	nr	164	175	
Azelaic	106	>360	182	201(di)	131(di)	107(m) 185(di)	93(m) 175(di)	Nonanedioic
m-Toluic	110	263	136	118	108	125	95	3-Methylbenzoic
dl-Mandelic	118	nr	152	172	nr	151	133	α-Hydroxyphenylacetic
Benzoic	121	249	122	158	119	163	127	
2-Benzoylbenzoic	126	257	226	nr	nr	195	165	Monohydrate, mp 91°C Methyl ester, mp 52°C
Maleic	130	nr	116	142(di)	168	nr	nr	mp range 130 to 137°C; double mp, 54°C (anhydride)
(*E*)-Cinnamic	133	300	148	168	145	153	147	Aniline Michael adduct, mp 109°C
Sebacic	133	*	202	201(di)	147(di)	122(m) 200(di)	170(m) 209(di)	Decanedioic; *294°C/100 torr
2-Furoic	133	230	112	170	138	123	141	
Malonic	135	dec	104	156d(m) 252(di)	nr	132(m) 228(di)	106(m) 168(di)	Decomposes on heating; acetic acid distills
o-Acetylsalicylic	135	140d	180	nr	nr	136	138	Aspirin; *p*-nitrobenzyl ester, mp 90°C
(*E*)-α-Chlorocinnamic	137	nr	182	116	nr	118	121	Methyl ester, mp 33°C
2-Chlorobenzoic	138	nr	156	131	106	116	140	Amide also reported, mp 202°C
3-Nitrobenzoic	140	nr	167	162	132	153	143	
(*E*)-β-Chlorocinnamic	142	nr	182	122	nr	128	118	
Suberic	142	*	174	218(di)	144(di)	128(m) 186(di)	125(m) 216(di)	Octanedioic; *230°C/15 torr
2-Nitrobenzoic	146	nr	167	nr	107	155	174	
Diphenylacetic	148	nr	212	172	nr	180	167	
Diglycolic	148 (142)	nr	150	148(m)	nr	118(m) 152(di)	135(m)	Anhydride, mp 92°C; $HOCOCH_2OCH_2COOH$
2-Bromobenzoic	148	nr	201	nr	102	141	155	
Benzilic	151	nr	228	189	152	174	153	1-Hydroxy-1,1-diphenylacetic
Citric	153	nr	192	189(t)	148(t)	192(t)	210(t)	Monohydrate, mp 100°C
4-Nitrophenylacetic	153	nr	181	nr	207	198	198	
Adipic	153	*	146	241	153	151(m) 240(di)	126(m) 220(di)	Hexanedioic; *265°C/100 torr
3-Bromobenzoic	155	nr	201	nr	120	136	155	Methyl ester, mp 31°C

Abbreviations used: nr, not reported; subl, sublimes; m, mono; t, tri; TOL, *p*-toluidide; BPE, *p*-bromophenacyl ester; ANI, anilide; AMD, amide; *, see note to right; d, dec, decomposes.

Note: All derivative values are melting points, °C.

TABLE 28.2 (*Continued*)

Name	mp, °C	bp, °C	MW	TOL	BPE	ANI	AMD	Notes
4-Chlorophenoxyacetic	157	nr	186	nr	136	125	133	
3-Chlorobenzoic	157	nr	156	nr	116	122	134	
Salicylic	158	nr	138	156	140	135	139	2-Hydroxybenzoic
1-Naphthoic	160	300	172	nr	135	162	203	
2-Iodobenzoic	162	233*	248	nr	110	141	183	***Explodes**
4-Nitrophthalic	164	nr	211	172(m)	nr	192	200d	4-Nitro-1,2-dicarboxybenzene
Haconic	166d	dec	130	nr	117(di)	151(m) 190(di)	191(di)	
d- or *l*-Tartaric	171	nr	150	nr	210(di)	180d(m) 263(di)	171(m) 196(di)	2,3-Dihydroxysuccinic
4-Toluic	179	275	136	162	153	142	159	4-Methylbenzoic
Veratric	181	nr	182	nr	nr	154 (166)	164	3,4-Dimethoxybenzoic
2-Naphthoic	184	nr	172	192	nr	172	193	
4-Anisic	184	nr	152	186	152	168	165	4-Methoxybenzoic
Succinic	188	235d	118	179(m) 255(di)	211(di)	148(m) 228(di)	157 260(di) 242(?)	Anhydride, mp 119°C
Hippuric	188	nr	179	nr	151	208	183	*N*-Benzoylglycine
3-Iodobenzoic	188	nr	248	nr	128	nr	186	
d-Camphoric	188 (183)	nr	200	212 (193)	nr	204(m) 196(di) 226(di)	176(m) 182(m) 192(di)	
3,4-Dihydroxybenzoic	200d	nr	154	nr	nr	166	212	Methyl ester, mp 134°C
3-Hydroxybenzoic	201 (subl)	nr	138	163	176	155	170	
Mesaconic	202	nr	130	196(m) 212(di)	nr	163(m) 202(m) 185(di)	176(m) 222(m) 176(di)	Methyltumaric
3,5-Dinitrobenzoic	204	nr	212	nr	159	234	183	
Phthalic	210 (203d)	subl	166	150(m) 201(di)	153(di)	170(m) 253(di)	149(m) 220(di)	1,2-Dicarboxybenzene; anhydride, mp 131°C
4-Hydroxybenzoic	214	nr	138	203	191	199	162	

Abbreviations used: nr, not reported; subl, sublimes; m, mono; t, tri; TOL, *p*-toluidide; BPE, *p*-bromophenacyl ester; ANI, anilide; AMD, amide; *, see note to right; d, dec, decomposes.

Note: All derivative values are melting points, °C.

TABLE 28.3
Liquid alcohols

Name	bp, °C	mp, °C	RI	DEN, g/mL	NPU	PHU	NBE	DNB	Notes
Methanol	65	−76	1.3306	0.791	124	47	96	107	
Ethanol	78	−117	1.3610	0.789	79	52	57	93	
Isopropyl alcohol	82	−82	1.3770	0.785	106	75 (88)	110	122	2-Propanol
tert-Butanol	83	25	1.3860	0.786	101	136	116	142	
Allyl alcohol	97	−129	1.4119	0.854	108	70	28	49	3-Hydroxypropene
1-Propanol	97	−127	1.3840	0.804	80	57 (51)	35	74	
2-Butanol	98	−115	1.3971	0.808	97	64	25	76	
tert-Amyl alcohol	102	−12	1.4038	0.805	71	42	85	116	2-Methyl-2-butanol
Isobutanol	108	−108	1.3960	0.786	104	86	69	87	2-Methyl-1-propanol
3-Methyl-2-butanol	112	nr	1.4089	0.818	109	68	nr	76	
3-Pentanol	115	nr	1.4096	0.815	95 (71)	48	17	100	
1-Butanol	118	−90	1.3985	0.810	71	61	17	64	
2-Pentanol	119	nr	1.4055	0.812	75	nr	17	61	
3,3-Dimethyl-2-butanol	120	4.8	1.4148	0.812	nr	78	nr	107	
2-Methyl-2-pentanol	121	−103	1.4100	0.835	nr	239	nr	72	
2-Methoxyethanol	124	nr	1.4008	0.965	112	nr	50	nr	
2-Chloroethanol	129	−89	1.4412	1.201	101	51	nr	92	**Caution: Toxic**
2-Methyl-1-butanol	130	nr	1.4100	0.815	82 (97)	31	nr	70 (62)	
3-Methyl-1-butanol	130	−117	1.4061	0.809	67	55	21	61	Isoamyl alcohol
4-Methyl-2-pentanol	132	nr	1.4100	0.802	88	143	nr	65	
2-Ethoxyethanol	135	nr	1.4068	0.930	67	nr	nr	75	
1-Pentanol	137	−78	1.4093	0.811	68	46	11	46	*n*-Amyl alcohol
Cyclopentanol	139	−19	1.4521	0.949	118	132	nr	115	
2-Bromoethanol	149	nr	1.4870	1.763	86	76	nr	nr	**Caution: Irritant**
4-Heptanol	156	−41	1.4190	0.818	78	nr	35	64	
1-Hexanol	156	−52	1.4179	0.814	59	42	5	58	
Cyclohexanol	160	24 (16)	—	—	129	82	50	112	

Abbreviations used: nr, not reported; di, disubstituted; RI, refractive index; DEN, density; NPU, α-naphthylurethane; PHU, phenylurethane; NBE, 4-nitrobenzoic ester; DNB, 3,5-dinitrobenzoate ester; *, see note to right.

Note: All derivative values are melting points, °C.

TABLE 28.3 (*Continued*)

Name	bp, °C	mp, °C	RI	DEN, g/mL	NPU	PHU	NBE	DNB	Notes
2-Methylcyclohexanol	163	−9	1.4625	0.930	nr	92	53	98	*E*,*Z* mixture
3-Methylcyclohexanol	164 (174)	nr	1.4578	0.914	122 (128)	95	58	97	*E*,*Z* mixture
Furfuryl alcohol	170	nr	1.4862	1.135	129	45	76	80	
4-Methylcyclohexanol	173	nr	1.4559	0.914	160	112 (125)	67	127 (139)	*E*,*Z* mixture
3-Methylcyclohexanol	174 (164)	nr	1.4578	0.914	122 (128)	95	58	97	
1-Heptanol	176	−36	1.4232	0.822	62	60 (68)	10	46	
Tetrahydrofurfuryl alcohol	178	−80	1.4512	1.054	90	61	47	83	
2-Octanol	178	nr	1.4234	0.819	63	oil (114)	28	32	
Cyclohexylmethanol	181	nr	1.4621	0.914	108	83	nr	95	
2-Ethyl-1-hexanol	183	nr	1.4308	0.833	60	34	nr	nr	
1-Octanol	196	−15	1.4297	0.827	67	73	12	61	Capryl alcohol
Ethylene glycol	197	−13	1.4310	1.104	176 (di)	157 (di)	140 (di)	169 (di)	1,2-Dihydroxyethane; extremely hygroscopic
2-Nonanol	198 (194)	−36	1.4307	0.827	55	nr	nr	43	
α-Phenylethyl alcohol	204	nr	1.5265	1.013	106	92	43	95	Methylphenylcarbinol
Benzyl alcohol	204	−15	1.5396	1.045	134	77	85	112	
2-Decanol	211	nr	1.4340	0.827	69	nr	nr	44	
1-Nonanol	214	−7	1.4334	0.827	65	62 (69)	10	52	
1,3-Propanediol	214	−30	1.4400	1.053	164 (di)	137 (di)	119 (di)	178 (di)	
β-Phenylethyl alcohol	219	−26	1.5315	1.023	119	79	62	108	
dl-α-Terpineol	219	37	1.4831	0.939	152 (147)	113	139	78	2-Phenylethanol
1,4-Butanediol	230	16	1.4452	1.017	199 (di)	183 (di)	175 (di)	nr	

Abbreviations used: nr, not reported; di, disubstituted; RI, refractive index; DEN, density; NPU, α-naphthylurethane; PHU, phenylurethane; NBE, 4-nitrobenzoic ester; DNB, 3,5-dinitrobenzoate ester; *, see note to right.

Note: All derivative values are melting points, °C.

TABLE 28.3 (*Continued*)

Name	bp, °C	mp, °C	RI	DEN, g/mL	NPU	PHU	NBE	DNB	Notes
1-Decanol	231	7	1.4372	0.829	73	60	30	58	
3-Phenylpropanol	235	nr	1.5257	1.008	nr	45	47	92	
1,5-Pentanediol	242	nr	1.4494	0.994	147 (di)	174 (di)	104 (di)	nr	
1-Undecanol	243	15	1.4400	0.830	73	62	99	55	
Cinnamyl alcohol	250	33	1.5819	1.040	114	90	77	121	Odor of hyacinths

Abbreviations used: nr, not reported; di, disubstituted; RI, refractive index; DEN, density; NPU, α-naphthylurethane; PHU, phenylurethane; NBE, 4-nitrobenzoic ester; DNB, 3,5-dinitrobenzoate ester; *, see note to right.

Note: All derivative values are melting points, °C.

TABLE 28.4
Solid alcohols

Name	mp, °C	bp, °C	NPU	PHU	NBE	DNB	Notes
4-Methoxybenzyl alcohol	24	259	nr	94	nr	nr	Anisic acid, mp 184°C
Cyclohexanol	24 (16)	160	129	82	50	112	
tert-Butanol	25	83	101	136	116	142	
Lauryl alcohol	25	259	80	74	44	60	Dodecanol
Cinnamyl alcohol	33	250	114	90	77	121	
dl-α-Terpineol	37	219	152 (147)	113	139	78	
Myristyl alcohol	38	280	82	73	51	67	Tetradecanol
dl-Fenchyl alcohol	39	201	149	104	109	104	
(−)-Menthol	44	212	126 (119)	102	61	153	
Hexadecanol	49	*	82	73	58	66	Cetyl alcohol; *bp 190°C/18 torr
Neopentyl alcohol	52	113	100	144	nr	nr	
Octadecanol	59	*	89	80	64	77 (66)	Stearyl alcohol; *bp 212°C/15 torr
Benzhydrol	66	297	136	139	131	141	Diphenylcarbinol
Benzoin	134	344	140	165	123	nr	2,4-DNP, mp 245°C; oxime, mp 151°C [see Table 28.15 (Solid Ketones)]
(−)-Cholesterol	148	360	176	168	185 (190)	nr	
(+)-Borneol	208	210	132	138	137	154	

Abbreviations used: nr, not reported; NPU, α-naphthylurethane; PHU, phenylurethane; NBE, 4-nitrobenzoate ester; DNB, 3,5-dinitrobenzoate ester; *, see note to right.

Note: All derivative values are melting points, °C.

TABLE 28.5
Liquid aldehydes

Name	bp, °C	mp, °C	RI	DEN, g/mL	DNP	SCB	OXM	NPH	Notes
Formaldehyde	−21*	−91	*	*	167	169	oil	181	*Available as 37% aq soln (formalin), so RI and DEN are characteristic of water solvent, or as solid trimer; dimedone, mp 192°C
Acetaldehyde	21	−124	1.3316*	0.788	168	163	47	128	Dimedone, mp 139°C; RI at 20°C
Propionaldehyde	46 to 50	−81	1.3650	0.805	149	89	40	124	Propanal
Glyoxal	50	15	*	*	328	270	178	311	*Available as 40% aq soln; see formaldehyde notes
Acrolein	53	−88	1.4050	0.839	165	171	oil	150	Propenal
Isobutyraldehyde	63	−66	1.3723	0.794	182 (187)	125	oil	131	
α-Methylacrolein	71	−81	1.4160	0.847	206	198	oil	nr	1-Methyl-1-formylethene
n-Butyraldehyde	75	−97	1.3790	0.817	122	105	oil	nr	
Pivalaldehyde	75	6	1.3791	0.793	209	190	41	119	2,2-Dimethylpropanal
Isovaleraldehyde	90	−51	1.3882	0.803	123	107	48	110	3-Methylbutanal
Chloral	98	−57	1.4580	1.512	131	90d	56	131	Trichloroacetaldehyde
Pentanal	103	−92	1.3942	0.810	107	oil	52	nr	Valeraldehyde
Crotonaldehyde	104	−69	1.4312	0.858	190	199	119	184	2-Butenal
Diethylacetaldehyde	117	nr	1.4018	0.814	95	99	oil	nr	2-Ethylbutanal
Caproaldehyde	131	nr	1.4035	0.834	104	106	51	80	Hexanal
2-Methyl-3-ethylacrolein	136	nr	1.4488	0.840	159	207	48	nr	2-Methyl-2-pentenal
Heptanal	153	−43	1.4125	0.850	106	109	57	73	Enanthaldehyde
Furfural	162	−36	1.5243	1.160	230 (202)	202 (191)	75, 90*	127	Furan-2-carboxaldehyde; *syn and anti forms
Cyclohexanecarboxaldehyde	162	nr	1.4500	0.926	172	174	90	nr	
1,2,3,6-Tetrahydrobenzaldehyde	163	nr	1.4745	0.940	nr	154	76	163	
Octanal	171	14	1.4183	0.821	106	101	60	80	Caprylaldehyde
2-Ethylhexanal	173	nr	1.4518	nr	124	152	nr	nr	

Abbreviations used: nr, not reported; d, decomposes; A, anti; S, syn; RI, refractive index; DEN, density; DNP, 2,4-dinitrophenylhydrazone; SCB, semicarbazone; OXM, oxime; NPH, 4-nitrophenylhydrazone; *, see note to right.

Note: All derivative values are melting points, °C.

TABLE 28.5 (*Continued*)

Name	bp, °C	mp, °C	RI	DEN, g/mL	DNP	SCB	OXM	NPH	Notes
Benzaldehyde	180	−26	1.5454	1.044	237	222	35	190	
Nonanal	185	nr	1.4240	0.827	100	100	64	nr	Pelargonaldehyde
5-Methylfurfural	187	nr	1.5263	1.107	212	211	112S (52A)	130	
Phenylacetaldehyde	195	−10	1.5293	1.027	121	155	102	151	
Salicylaldehyde	197	1	1.5719	1.146	250d	231	57	227	2-Hydroxybenzaldehyde
o-Tolualdehyde	199	nr	1.5472	1.020	194	208	49	222	2-Methylbenzaldehyde
m-Tolualdehyde	199	nr	1.5411	1.019	212	204	60	157	
p-Tolualdehyde	204	nr	1.5447	1.019	239 (233)	234	80 (110)	200	
2-Phenylpropionaldehyde	204	nr	1.5175	1.009	nr	nr	nr	nr	
Citronellal	207	nr	1.4485	nr	80	84	oil	nr	
Decanal	208	nr	1.4280	0.830	104	102	69	nr	
o-Chlorobenzaldehyde	210	10	1.5658	1.248	208	228	75	249	
m-Chlorobenzaldehyde	213	17	1.5645	1.241	252	228	70	216	
Hydrocinnamaldehyde	223	47	solid	solid	149	127	94	122	3-Phenylpropanal
3-Bromobenzaldehyde	228	nr	1.5935	nr	nr	205	72	220	
m-Methoxybenzaldehyde	230	nr	1.5523	1.119	219	233	40 (112)	171	*m*-Anisaldehyde
4-Isopropylbenzaldehyde	235	nr	1.5301	0.978	244	211	42 (112)	190	Cumyl aldehyde; cuminal
4-Methoxybenzaldehyde	247	−1	1.5713	1.190	253d	210	α 64 β 133	160	*p*-Anisaldehyde
Cinnamaldehyde	250	−7	1.6219	1.050	255d	215	64 (138)	195	3-Phenylpropenal

Abbreviations used: nr, not reported; d, decomposes; A, anti; S, syn; RI, refractive index; DEN, density; DNP, 2,4-dinitrophenylhydrazone; SCB, semicarbazone; OXM, oxime; NPH, 4-nitrophenylhydrazone; *, see note to right.

Note: All derivative values are melting points, °C.

TABLE 28.6
Solid aldehydes

Name	mp, °C	bp, °C	RI	DEN, g/cm^3	DNP	SCB	OXM	NPH	PHZ	Notes
1-Naphthaldehyde	1	292	1.6520	1.150	nr	221	90	224	80	
3-Chlorobenzaldehyde	17	213	1.5645	1.214	252	228	70	216	134	
Myristyl aldehyde	23.5	>250	—	—	108	106	83	95	nr	Tetradecanal
Pentadecanal	24	>250	—	—	106	106	86	94	nr	
Hexadecanal	34	>250	—	—	108	108	88	96	nr	Palmitaldehyde
1-Naphthaldehyde	35	292	1.6520	1.150	nr	221	90	224	80	
5-(Hydroxymethyl)furfural	35	>250	1.5627	1.210	209	192	78A (108S)	185	140	
o-Iodobenzaldehyde	37	206	—	—	nr	206	108	214	79	
Piperonal	37	264	—	—	265d	230	110A (146S)	199	101	3,4-Methylenedioxybenzaldehyde
o-Anisaldehyde	37	238	1.5600	1.127	253	215	99	204	nr	
Veratraldehyde	42	281	—	—	264	177	94	nr	121	
3,4-Dichlorobenzaldehyde	43	247	—	—	nr	nr	114A (120S)	276	nr	
Dodecanal	44	238	—	0.835	106	104	77	90	nr	Lauraldehyde
o-Nitrobenzaldehyde	44	>250	—	1.28*	250d	256	102A (154S)	263	156	*DEN at 50°C
Hydrocinnamaldehyde	47	223	—	—	149	127	94	122	nr	3-Phenylpropanal
p-Chlorobenzaldehyde	48	214	—	—	265	230	110A (146S)	237	127	
Pyrrole-2-carboxaldehyde	48	217	1.5390	—	nr	183	164	182	139	
4-Bromobenzaldehyde	56	nr	—	—	257	228	111 (157)	207	113	
3-Iodobenzaldehyde	57	nr	—	—	nr	226	62	212	155	
m-Nitrobenzaldehyde	58	nr	—	—	290d	246	120	247	122	
2-Naphthaldehyde	59	nr	—	—	270	255	156	230	206d	
4-(Dimethylamino)-benzaldehyde	74	nr	—	—	236	222	185	182	148	HCl salt, mp 109°C
Vanillin	81	285d	—	—	271d	229	117 (122)	223	105	3-Methoxy-4-hydroxybenzaldehyde
m-Hydroxybenzaldehyde	101	nr	—	—	257d	198	88	221	130	
p-Nitrobenzaldehyde	105	nr	—	—	>300	221	133A (182S)	249	159	
p-Hydroxybenzaldehyde	116	nr	—	—	270	224	72	266	177	

Abbreviations used: nr, not reported; d, decomposes; S, syn; A, anti; RI, refractive index; DEN, density; DNP, 2,4-dinitrophenylhydrazone; SCB, semicarbazone; OXM, oxime; NPH, 4-nitrophenylhydrazone; PHZ, phenylhydrazone.

Note: All derivative values are melting points, °C.

TABLE 28.7
Amides

Name	mp, °C	Acid formed	mp of acid, °C	bp of acid, °C
Formamide	nr	Formic	8.5	100.8
Formanilide	47	Formic	8.5	100.8
Nonananilide	57	Nonanoic	9	254
Octananilide	57	Octanoic	16	237
Vinylacetanilide	58	Vinylacetic	−39	163
Pentananilide	63	Pentanoic	−18	186
Heptananilide	68	Heptanoic	−10	223
Undecananilide	71	Undecanoic	28	275d
Vinylacetamide	73	Vinylacetic	−39	163
2-Methyl-2-butenamide	75	2-Methyl-2-butenoic (tiglic)	63	nr
Dodecananilide	76	Dodecanoic	44	nr
2-Methyl-2-butenanilide	77	2-Methyl-2-butenoic (tiglic)	63	nr
2,3-Dimethylbutananilide	78	2,3-Dimethylbutanoic	−1.5	189
Pentadecananilide	78	Pentadecanoic	52	nr
2-Methylpentanamide	79	2-Methylpentanoic	nr	195
Propionamide	79	Propanoic	−23	141
Acetamide	80	Acetic	16.2	117
α-Chloropropionamide	80	α-Chloropropanoic	nr	186
Ethoxyacetamide	80	Ethoxyacetic	nr	206
Tridecananilide	80	Tridecanoic	41	nr
Hydrocinnamamide	82	Hydrocinnamic	47	280
Acrylamide	84	Acrylic	13	139
Tetradecananilide	84	Tetradecanoic (myristic)	54	nr
Methacrylanilide	87	Methacrylic	16	163
3-Methylpentananilide	87	3-Methylpentanoic	−42	197
Propynanilide	87	Propynoic (propiolic)	9	144d
Hexadecananilide	90	Hexadecanoic	64	nr
2-Chloropropionanilide	92	2-Chloropropionic	nr	186
2,2-Dimethylbutananilide	92	2,2-Dimethylbutanoic	−13	186
Eicosananilide	92	Eicosanoic	75	nr
Ethoxyacetanilide	93	Ethoxyacetic	nr	206
Octadecananilide	93	Octadecanoic (stearic)	67	nr
Hexananilide	94	Hexanoic	−3	202

TABLE 28.7 (*Continued*)

Name	mp, °C	Acid formed	mp of acid, °C	bp of acid, °C
Hydrocinnamanilide	95	Hydrocinnamic (3-phenylpropanoic)	47	280
2-Methylpentananilide	95	2-Methylpentanoic	nr	195
m-Toluamide	95	*m*-Toluic (3-methylbenzoic)	110	263
Trichloroacetanilide	95	Trichloroacetic	57	197
Butyranilide	96	Butyric (butanoic)	−8	163
Heptanamide	96	Heptanoic	−10	223
Hydroxyacetanilide	96	Hydroxyacetic	77	nr
α-Bromobutananilide	98	α-Bromobutanoic	−4	217d
Dodecanamide	98	Dodecanoic	44	299
α-Hydroxyisobutyramide	98	α-Hydroxyisobutyric	79	nr
Dichloroacetamide	99	Dichloroacetic	7	190
Nonanamide	99	Nonanoic	9	254
Phenoxyacetanilide	99	Phenoxyacetic	98	nr
Hexanamide	100	Hexanoic	−3	202
Isocrotonamide	101	Isocrotonic	14	169
Isocrotonanilide	101	Isocrotonic	14	169
Phenoxyacetamide	101	Phenoxyacetic	98	nr
N-Methylacetanilide	102	Acetic	16	117
Hydroxyacetamide	102	Hydroxyacetic (glycolic)	77	nr
Levulinanilide	102	Levulinic (β-acetylpropionic)	33	245
Pentadecanamide	102	Pentadecanoic	52	nr
Tetradecanamide	102	Tetradecanoic	54	nr
Undecanamide	103	Undecanoic	28	275d
Acrylanilide	104	Acrylic	13	139
Propionanilide	104	Propanoic	−23	141
Isobutyranilide	105	Isobutyric	−47	154
Hexadecanamide	106	Hexadecanoic	61	nr
Heptadecanamide	106	Heptadecanoic	59	nr
Pentanamide	106	Pentanoic	−18	186
Levulinamide	107	Levulinic (β-acetylpropionic)	33	245
Eicosanamide	108	Eicosanoic	75	nr
Octadecanamide	108	Octadecanoic (stearic)	67	nr
Octanamide	108	Octanoic	16	237
Isovaleranilide	109	Isovaleric	−37	176
Methacrylamide	109	Methacrylic	16	163

TABLE 28.7 (*Continued*)

Name	mp, °C	Acid formed	mp of acid, °C	bp of acid, °C
α-Bromobutyramide	110	α-Bromobutyric	−4	217d
Ethylmethylacetanilide	110	Ethylmethylacetic	−80	174
2-Iodobenzamide	110	2-Iodobenzoic	162	nr
4-Methylpentananilide	110	4-Methylpentanoic	−33	199
Ethylmethylacetamide	112	Ethylmethylacetic	−80	174
2-Ethylbutanamide	112	2-Ethylbutanoic	−15	195
Acetanilide	114	Acetic	16.2	117
Butyramide	115	Butyric (butanoic)	−8	163
2-Chlorobenzanilide	116	2-Chlorobenzoic	138	nr
Phenylacetanilide	117	Phenylacetic	76	nr
2-Butenanilide	118	2-Butenoic	71	nr
α-Chlorocinnamanilide	118	α-Chlorocinnamic	137	nr
β-Chlorocinnamamide	118	β-Chlorocinnamic	142	nr
Dichloroacetanilide	118	Dichloroacetic	7	190
4-Methylpentanamide	119	4-Methylpentanoic	−33	199
Chloroacetamide	120	Chloroacetic	61	nr
α-Chlorocinnamamide	121	α-Chlorocinnamic	137	nr
3-Chlorobenzanilide	122	3-Chlorobenzoic	157	nr
2-Furanilide	123	Furoic	133	nr
Succinimide	123	Succinic	188	235d
Diethylacetanilide	124	Diethylacetic	−15	195
4-Chlorophenoxyacetanilide	125	4-Chlorophenoxyacetic	157	nr
3-Methylpentanamide	125	3-Methylpentanoic	−42	197
o-Toluanilide	125	*o*-Toluic (2-methylbenzoic)	103	nr
m-Toluanilide	125	*m*-Toluic (3-methylbenzoic)	110	nr
Benzamide	127	Benzoic	121	nr
Isobutyramide	127	Isobutyric	−47	154
β-Chlorocinnamanilide	128	β-Chlorocinnamic	142	nr
2-Methoxybenzamide	128	2-Methoxybenzoic	100	nr
2,3-Dimethylbutanamide	129	2,3-Dimethylbutanoic	−1.5	189
Trimethylacetanilide	129	Trimethylacetic (pivalic)	34	163
2,3-Dibromopropionamide	130	2,3-Dibromopropanoic	64	nr
2-Methoxybenzanilide	131	2-Methoxybenzoic	100	nr
tert-Butylacetamide	132	*tert*-Butylacetic	−11	185
tert-Butylacetanilide	132	*tert*-Butylacetic	−11	185
4-Chlorophenoxyacetamide	133	4-Chlorophenoxyacetic	157	nr

TABLE 28.7 *(Continued)*

Name	mp, °C	Acid formed	mp of acid, °C	bp of acid, °C
dl-Mandelamide	133	*dl*-Mandelic	118	nr
Chloroacetanilide	134	Chloroacetic	61	nr
3-Chlorobenzamide	134	3-Chlorobenzoic	157	nr
Isovaleramide	135	Isovaleric	−37	176
Salicylanilide	135	Salicylic (2-hydroxybenzoic)	158	nr
o-Acetylsalicylanilide	136	*o*-Acetylsalicylic (aspirin)	135	nr
3-Bromobenzanilide	136	3-Bromobenzoic	155	nr
α-Hydroxyisobutyranilide	136	α-Hydroxyisobutyric	79	nr
o-Acetylsalicylamide	138	*o*-Acetylsalicylic (aspirin)	135	nr
Salicylamide	139	Salicylic (2-hydroxybenzoic)	158	nr
2-Chlorobenzamide	140	2-Chlorobenzoic	138	nr
2-Bromobenzanilide	141	2-Bromobenzoic	148	nr
2-Furamide	141	2-Furoic	133	nr
2-Iodobenzanilide	141	2-Iodobenzoic	162	nr
Trichloroacetamide	141	Trichloroacetic	57	nr
o-Toluamide	142	*o*-Toluic (2-methylbenzoic)	103	nr
p-Toluanilide	142	*p*-Toluic (4-methylbenzoic)	179	nr
3-Nitrobenzamide	143	3-Nitrobenzoic	140	nr
Cinnamamide	147	Cinnamic	133	nr
dl-Mandelanilide	151	*dl*-Mandelic	118	nr
Diglycolanilide	152(di)	Diglycolic	148	nr
Benzilamide	153	Benzilic	151	nr
Cinnamanilide	153	Cinnamic	133	nr
3-Nitrobenzanilide	153	3-Nitrobenzoic	140	nr
Phenylacetamide	154	Phenylacetic	76(subl)	nr
Pivalamide	154	Pivalic (trimethylacetic)	34	nr
3,4-Dimethoxybenzanilide	154(166)	3,4-Dimethoxybenzoic (veratric)	181	nr
Trimethylacetamide	154	Trimethylacetic (pivalic)	34	163
2-Bromobenzamide	155	2-Bromobenzoic	148	nr
3-Bromobenzamide	155	3-Bromobenzoic	155	nr
3-Hydroxybenzanilide	155	3-Hydroxybenzoic	201(subl)	nr
2-Nitrobenzanilide	155	2-Nitrobenzoic	146	nr
Pimelanilide	155(di)	Pimelic (heptanedioic)	104	nr
l-Malamide	156(di)	*l*-Malic	100	nr
4-Toluamide	159	4-Toluic (4-methylbenzoic)	179	nr

TABLE 28.7 (*Continued*)

Name	mp, °C	Acid formed	mp of acid, °C	bp of acid, °C
Crotonamide	160	Crotonic (2-butenoic)	71	nr
4-Hydroxybenzamide	162	4-Hydroxybenzoic	214	nr
Naphthanilide	162	Naphthoic	160	nr
Benzanilide	163	Benzoic	121	nr
4-Chlorophenylacetanilide	164	4-Chlorophenylacetic	104	nr
3,4-Dimethoxybenzamide	164	3,4-Dimethoxybenzoic (veratric)	181	nr
2-Benzoylbenzamide	165	2-Benzoylbenzoic	126	nr
4-Methoxybenzamide	165	4-Methoxybenzoic (anisic)	184	nr
3,4-Dimethoxybenzanilide	166	3,4-Dimethoxybenzoic (veratric)	181	nr
Diphenylacetamide	167	Diphenylacetic	148	nr
Malonamide	168(di)	Malonic	135	nr
4-Methoxybenzanilide	168	4-Methoxybenzoic (4-anisic)	184	nr
3-Hydroxybenzamide	170	3-Hydroxybenzoic	201	nr
2-Naphthanilide	172	2-Naphthoic	184	nr
Benzilanilide	174	Benzilic	151	nr
Glutaramide	174	Glutaric	98	nr
2-Nitrobenzamide	174	2-Nitrobenzoic	146	nr
Azelaic acid amide	175(di)	Azelaic	106	nr
4-Chlorophenylacetamide	175	4-Chlorophenylacetic	104	nr
Methylmaleanilide	175(di)	Methylmaleic	91d	nr
Pimelamide	175(di)	Pimelic (heptanedioic)	104	nr
Mesaconamide	176(di)	Mesaconic	202	nr
Diphenylacetanilide	180	Diphenylacetic	148	nr
3,5-Dinitrobenzamide	183	3,5-Dinitrobenzoic	204	nr
Hippuramide	183	Hippuric	188	nr
Azelaic acid amide	185(di)	Azelaic	106	nr
Mesaconanilide	185(di)	Mesaconic	202	nr
Methylmaleamide	185d(di)	Methylmaleic	91d	nr
3-Iodobenzamide	186	3-Iodobenzoic	188	nr
Suberanilide	186(di)	Suberic (octanedioic)	142	nr
Haconanilide	190(di)	Haconic	166d	nr
Haconamide	191(di)	Haconic	166d	nr
d-Camphoramide	192(di)	*d*-Camphoric	188	nr
Citranilide	192(t)	Citric (monohydrate)	100	nr
Citranilide	192(t)	Citric	153	nr

TABLE 28.7 (*Continued*)

Name	mp, °C	Acid formed	mp of acid, °C	bp of acid, °C
2-Naphthamide	193	2-Naphthoic	184	nr
2-Benzoylbenzanilide	195	2-Benzoylbenzoic	126	nr
d-Camphoranilide	196(di)	*d*-Camphoric	188	nr
d- or *l*-Tartaramide	196(di)	*d*- or *l*-Tartaric	171	nr
Malanilide	197(di)	*l*-Malic	100	nr
4-Nitrophenylacetamide	198	4-Nitrophenylacetic	153	nr
4-Nitrophenylacetanilide	198	4-Nitrophenylacetic	153	nr
4-Hydroxybenzanilide	199	4-Hydroxybenzoic	214	nr
4-Nitrophthalamide	200d	4-Nitrophthalic	164	nr
Sebacanilide	200(di)	Sebacic (decanedioic)	133	nr
1-Naphthamide	203	1-Naphthoic	160	nr
Hippuranilide	208	Hippuric	188	nr
Sebacamide	209(di)	Sebacic (decanedioic)	133	nr
Citramide	210(t)	Citric (monohydrate)	100	nr
Citramide	210(t)	Citric	153	nr
3,4-Dihydroxybenzamide	212	3,4-Dihydroxybenzoic	200d	nr
Suberamide	216(di)	Suberic (octanedioic)	142	nr
4-Nitrophthalamide	220d	4-Nitrophthalic	164	nr
Adipamide	220(di)	Adipic	153	nr
Phthalamide	220(di)	Phthalic	203	nr
Glutaranilide	223(di)	Glutaric	98	nr
d-Camphoranilide	226(di)	*d*-Camphoric	188	nr
Malonanilide	228(di)	Malonic	135	nr
Succinanilide	228(di)	Succinic	188	nr
3,5-Dinitrobenzanilide	234	3,5-Dinitrobenzoic	204	nr
Phthalimide	238	Phthalic	203	nr
Adipanilide	240(di)	Adipic	153	nr
Phthalanilide	253(di)	Phthalic	203	nr
Succinamide	260(di)	Succinic	188	nr
d- or *l*-Tartaranilide	263(di)	*d*- or *l*-Tartaric	171	nr

TABLE 28.8
Liquid primary and secondary amines

Amine	bp, °C	mp, °C	RI	DEN, g/mL	BSA	TSA	PTU	BZM	ACM	HCl	Notes
Methylamine	−6	nr	nr	nr	30	75	113	80	28	nr	
Dimethylamine	7	nr	nr	nr	47	79	135	41	oil	nr	
Ethylamine	16	nr	nr	nr	58	63	106 (135)	71	oil	109	
Isopropylamine	33	nr	1.3770	0.691	26	51	101	100	oil	nr	
tert-Butylamine	46	nr	1.3774	0.696	nr	nr	120	134	101	270	
n-Propylamine	48	nr	1.3890	0.719	36	52	63	84	nr	nr	
Allylamine	53	−88	1.4203	0.761	39	64	98	nr	nr	nr	3-Aminopropene
Diethylamine	55	−50	1.3861	0.707	42	60	34	42	oil	227	
sec-Butylamine	63	−19	1.3928	0.724	70	55	101	76	nr	nr	
Isobutylamine	69	nr	1.3970	0.736	53	78	82	57	nr	nr	
n-Butylamine	78	nr	1.4015	0.740	nr	65	65 (108)	42	nr	195	
Pyrrolidine	87	2	1.4431	0.852	49	123	149	nr	nr	nr	
Isoamylamine	96	nr	1.4089	0.751	nr	65	102	nr	nr	215	
Piperidine	106	−13	1.4525	0.861	93	96	101	48	nr	245	
Di-*n*-propylamine	110	−63	1.40455	0.738	51	nr	69	nr	oil	nr	
Ethylenediamine	118	8.5	1.4565	0.899	168	160 (360)	102	247	172	nr	
dl-2-Methylpiperidine	118	nr	1.4459	0.844	nr	55	nr	45	oil	207	
1,2-Diaminopropane	119	nr	1.4460	0.870	nr	103	nr	192	139	nr	
n-Hexylamine	129	−19	1.4255	0.763	96	nr	77	40	nr	nr	
Morpholine	129	−7	1.4541	0.999	118	147	136	75	nr	nr	
Cyclohexylamine	134	−17	1.4580	0.867	89	nr	148	149	102	nr	
Diisobutylamine	137	−77	1.4081	0.740	55	nr	113	nr	86	nr	
1,3-Diaminopropane	140	−12	1.4575	0.888	96	148	nr	140 (147)	126 (107)	246	
Di-*n*-butylamine	159	−82	1.4168	0.767	nr	nr	86	nr	nr	nr	
N-Methylbenzylamine	184	nr	1.5224	0.945	nr	95	nr	nr	nr	nr	
Aniline	184	−6	1.5855	1.022	112	103	54	160	114	196	

Abbreviations used: nr, not reported; RI, refractive index; DEN, density; BSA, benzenesulfonamide; TSA, toluenesulfonamide; PTU, phenylthiourea; BZM, benzamide; ACM, acetamide; HCl, hydrochloride salt.

Note: All derivative values are melting points, °C.

TABLE 28.8 *(Continued)*

Amine	bp, °C	mp, °C	RI	DEN, g/mL	BSA	TSA	PTU	BZM	ACM	HCl	Notes
Benzylamine	185	nr	1.5424	0.981	88	116 (185)	156	105	65	253	
dl-α-Phenethylamine	187	nr	1.5253	0.940	nr	nr	nr	120	57	158	1-Phenylethylamine
4-Fluroaniline	187	−1	1.5395	1.173	nr	nr	nr	185	152	nr	
N-Methylaniline	196	nr	1.5704	0.989	79	94	87	63	102	nr	
β-Phenethylamine	199	nr	1.5332	0.965	69	64	135	145 (116)	114 (51)	217	2-Phenylethylamine
o-Toluidine	199	−28	1.5709	1.004	124	185 (108)	136	144	111	218	2-Methylaniline
m-Toluidine	203	nr	1.5669	0.999	95	171 (114)	93 (104)	125	65	nr	
1,6-Hexanediamine	204	39	—	—	154	nr	nr	55	126	nr	
N-Ethylaniline	205	−63	1.5538	0.963	oil	87	89	60	54	nr	
2-Chloroaniline	209	−1	1.5877	1.213	129	105 (193)	156	99	87	nr	
2-Ethylaniline	210	−44	1.5590	0.983	nr	nr	nr	147	111	nr	
4-Ethylaniline	216	−5	1.5542	0.975	nr	nr	104	151	94	nr	
2,5-Diethylaniline	218	11	1.5592	0.973	138	232 (119)	148	140	139	nr	
2,4-Dimethylaniline	218	nr	1.5586	0.980	128	181	152 (133)	192	133	nr	
2,6-Dimethylaniline	218	11	1.5601	0.984	nr	212	204	168	177	nr	
3,5-Dimethylaniline	218	nr	1.5578	0.972	136	nr	153	144	144	nr	
2,3-Dimethylaniline	222	2.5	1.5673	0.993	nr	nr	nr	189	135	254	
o-Anisidine	225	5	1.5730	1.092	89	127	136	60 (84)	86	nr	2-Methoxyaniline
o-Phenetidine	229	nr	nr	nr	102	164	137	104	79	nr	2-Ethoxyaniline
o-Bromoaniline	229	30	1.6113	1.578	nr	90	146 (161)	116	99	nr	
m-Chloroaniline	230	−10	1.5937	1.216	121	138 (210)	124 (116)	119	72 (78)	nr	
1,2,3,4-Tetrahydroisoquinoline	232	−30	1.5668	1.064	154	nr	nr	129	46	nr	

Abbreviations used: nr, not reported; RI, refractive index; DEN, density; BSA, benzenesulfonamide; TSA, toluenesulfonamide; PTU, phenylthiourea; BZM, benzamide; ACM, acetamide; HCl, hydrochloride salt.

Note: All derivative values are melting points, °C.

TABLE 28.8 (*Continued*)

Amine	bp, °C	mp, °C	RI	DEN, g/mL	BSA	TSA	PTU	BZM	ACM	HCl	Notes
2-Bromo-4-methylaniline	240	15 (26)	1.6015	1.500	nr	nr	154	149	117	221	
Phenylhydrazine	240	19	1.6070	1.099	148	151	172	168	128	252	
p-Phenetidine	250	4	1.5609	1.065	143	106	136	173	136	nr	4-Ethoxyaniline
m-Anisidine	251	1	1.5794	1.096	nr	68	nr	nr	81	167	3-Methoxyaniline
m-Bromoaniline	251	18	1.6250	1.579	nr	nr	143	136 (120)	87	nr	
Dicyclohexylamine	256	2	1.4842	0.910	nr	nr	nr	153	103	nr	
Methyl anthranilate	256d	24	1.5820	1.168	107	nr	nr	100	101	nr	Methyl 2-aminobenzoate
Dibenzylamine	300	nr	1.5731	1.026	68	159	nr	112	nr	256	
Diphenylamine	302	52	—	1.160	124	144	152	180	101	nr	
N-Benzylaniline	306	36	—	1.016	119	nr	103	119	58	nr	

Abbreviations used: nr, not reported; RI, refractive index; DEN, density; BSA, benzenesulfonamide; TSA, toluenesulfonamide; PTU, phenylthiourea; BZM, benzamide; ACM, acetamide; HCl, hydrochloride salt.

Note: All derivative values are melting points, °C.

TABLE 28.9
Liquid tertiary amines

Name	bp, °C	mp, °C	RI	DEN, g/mL	PIC	MeI	HCl	Notes
Triethylamine	89	−7	1.4000	0.726	173	280	nr	
Pyridine	115	−42	1.5102	0.978	167	117	nr	Methyl tosylate, mp 139°C
2-Methylpyridine	128	−70	1.5000	0.943	166	230	200	Methyl tosylate, mp 150°C
2-Dimethylaminoethanol	139	nr	1.4294	0.887	96	263	nr	
2,6-Dimethylpyridine	144	−6	1.4976	0.920	168	233	230	2,6-Lutidine
3-Methylpyridine	144	−19	1.5060	0.957	147	92	nr	3-Picoline
4-Methylpyridine	145	2	1.5045	0.957	167	149	nr	4-Picoline
4-Chloropyridine	147	nr	1.5300*	1.194	146	nr	210 (142)	*Estimated
3-Chloropyridine	149	nr	1.5304	1.194	135	nr	160	
N-Propylpiperidine	152	nr	nr	nr	121	181	212	
Tri-*n*-propylamine	156	−93	1.4160	0.753	116	207	nr	
2,4-Dimethylpyridine	159	−60	1.4991	0.927	180	113	nr	2,4-Lutidine
N,*N*-Diethylaminoethanol	161	nr	1.4414	0.884	nr	nr	nr	4-Nitrophenylurethane, mp 59°C
2-Chloropyridine	170	nr	1.5320	1.200	nr	nr	nr	Methyl tosylate, mp 120°C
3-Bromopyridine	173	nr	1.5695	1.640	154	165	nr	Methyl tosylate, mp 156°C
2,4,6-Trimethylpyridine	172	−43	1.4979	0.917	155	nr	nr	Collidine
N,*N*-Dimethyl-*o*-toluidine	185	nr	1.5150	0.929	116	210	nr	*N*,*N*,2-Trimethylaniline
N,*N*-Dimethylbenzylamine	183	−75	1.5011	nr	93	179	nr	
5-Ethyl-2-picoline	178	nr	1.4974	0.919	164	nr	nr	2-Methyl-5-ethylpyridine
N,*N*-Dimethylaniline	193	2	1.5581	0.956	163	228d	nr	Methyl tosylate, mp 161°C
2-Bromopyridine	193	nr	1.5720	1.657	105	nr	nr	Methyl tosylate, mp 127°C
N-Ethyl-*N*-methylaniline	201	nr	nr	0.919	134	125	114	
N,*N*-Diethyl-*o*-toluidine	206	nr	nr	nr	180	224	nr	*N*,*N*-Diethyl-2-methylaniline
N,*N*-Dimethyl-*p*-toludine	211	nr	1.5458	0.937	127	219	nr	Methyl tosylate, mp 85°C
Tri-*n*-butylamine	216	−70	1.4283	0.778	106	180	nr	
N,*N*-Diethylaniline	217	−38	1.5409	0.938	142	102	nr	
Quinoline	237	−15	1.6245	1.095	203	133	nr	Methiodide monohydrate, mp 72°C; methyl tosylate, mp 126°C
Isoquinoline	242	26	1.6230	1.099	222	159	227	Methyl tosylate, mp 163°C
2-Methylquinoline	248	−2	1.6108	1.058	191	195	nr	Methyl tosylate, mp 161°C (134°C)

Abbreviations used: nr, not reported; d, decomposes; RI, refractive index; DEN, density; PIC, picrate; MeI, methiodide; HCl, hydrochloride salt; *, see note to right.

Note: All derivative values are melting points, °C.

TABLE 28.10
Solid primary and secondary amines

Name	mp, °C	bp, °C	BSA	TSA	PTU	BZM	ACM	HCl	Notes
2-Bromo-4-methylaniline	15 (26)	240	nr	nr	154	149	117	221	
Phenylhydrazine	19	240	148	151	172	168	128	252	
Methyl anthranilate	24	256d	107	nr	nr	100	101	nr	
m-Iodoaniline	27	*	nr	128	nr	157	119	nr	*bp, 145°C/50 torr
o-Bromoaniline	30	229	nr	90	146	116	99	nr	
1,1-Diphenylhydrazine	34 (44)	*	nr	nr	nr	192	184	nr	*bp, 220°C/40 torr
N-Benzylaniline	36	306	119	nr	103	119	58	nr	
1,6-Hexanediamine	39	204	154(di)	nr	nr	155(di)	126	nr	
p-Toluidine	43	200	120	118	141	158	147	nr	4-Methylaniline
2-Aminobiphenyl	50	299	*	*	*	*	*	*	***Carcinogen**
2,5-Dichloroaniline	50	251	nr	nr	nr	120	132	191	
1-Naphthylamine	50	300	*	*	*	*	*	*	***Carcinogen**
4-Aminobiphenyl	52	302	*	*	*	*	*	*	***Carcinogen**
Diphenylamine	53	302	124	144	152	180	101	nr	
o-Iodoaniline	55	nr	nr	nr	nr	139	109	153	
p-Anisidine	57	240	95	114	154	154	127	nr	4-Methoxyaniline
p-Iodoaniline	61	nr	nr	nr	153	222	183	nr	
m-Phenylenediamine	63	282	194	172	nr	240(di) 125(m)	191(di) 87(m)	nr	1,3-Diaminobenzene
2,4-Dichloroaniline	63	245	128	126	nr	117	145	nr	
4-Bromoaniline	64	245	134	101	148	204	167	220(subl)	HBr salt, mp 230°C
Pseudocumidine	68	235	136	nr	nr	167	161	nr	1-Amino-2,4,5-trimethylbenzene
p-Chloroaniline	69	232	121	95 (119)	152	192	179	nr	
o-Nitroaniline	70	284	104	142(?)	142(?)	96 (110)	93	nr	
2-Amino-4-nitromesitylene	75	nr	163	nr	nr	169	191	nr	4-Nitromesidine
4-Methyl-3-nitroaniline	77	nr	160	164	171 (145)	172	148	nr	
2,4,6-Trichloroaniline	77	263	152	nr	nr	174	204	nr	
2,4-Dibromoaniline	78	nr	nr	134	171	134	146	nr	

Abbreviations used: nr, not reported; d, decomposes; subl, sublimes; BSA, benzenesulfonamide; TSA, toluenesulfonamide; PTU, phenylthiourea; BZM, benzamide; ACM, acetamide; HCl, hydrochloride salt; *, see note to right.

Note: All derivative values are melting points, °C.

TABLE 28.10 (*Continued*)

Name	mp, °C	bp, °C	BSA	TSA	PTU	BZM	ACM	HCl	Notes
N-Ethyl-*p*-nitroaniline	96	nr	nr	107	nr	98	118	nr	
2-Amino-4-methylpyridine	98	nr	nr	nr	nr	114	102	nr	Picrate mp 227°C
2,4-Diaminotoluene	99	284	178 (192)	192	nr	224	224	nr	
o-Phenylenediamine	100	256	185	260	nr	301	185	nr	1,2-Diaminobenzene
Piperazine*	106 (112)	146	282	173	nr	193	144	nr	*Hexahydrate, mp 44°C
p-Aminoacetophenone	106	295	128	203	nr	205	167	nr	
2-Naphthylamine	112	300	*	*	*	*	*	*	***Carcinogen**
3-Nitroaniline	113	284	136	138	160	155	155	nr	
4-Methyl-2-nitroaniline	115	nr	102	146 (166)	148	97	170	nr	
2,4,6-Tribromoaniline	120	300	nr	157	156	198	232(di) 127(m)	nr	
3-Aminophenol	122	nr	nr	157	156	153	148 (101)	nr	
Benzidine	127	400	*	*	*	*	*	*	***Carcinogen**
2-Methyl-4-nitroaniline	131	nr	158	174	nr	nr	202	nr	
2-Methoxy-4-nitroaniline	140	nr	181	175	nr	150	153	nr	
p-Phenylenediamine	140	267	247	266	nr	300(di) 128(m)	304(di) 162(m)	nr	1,4-Diaminobenzene
4-Nitroaniline	148	*	139	191	nr	200(di)	213	nr	*bp, 260°C/100 torr
4-Nitro-*N*-methylaniline	152	nr	120	nr	nr	111	153	nr	
Picramic acid	168	nr	nr	191	nr	220 (300)	201	nr	**Irritant**
o-Aminophenol	174	nr	141	139 (146)	146	165 (182)	201 (209)	nr	
2,4-Dinitroaniline	180	nr	*	*	*	*	*	*	***Irritant and toxic**
4-Aminophenol	186	nr	125	252	150	216 (234)	150 (168)	nr	
Picramide	190	nr	211	nr	nr	196	203	nr	**Irritant**
1-Amino-4-nitronaphthalene	195	nr	173 (158)	185	nr	224	190	nr	
2,4-Dinitrophenylhydrazine	198d	nr	nr	nr	nr	206	197	nr	

Abbreviations used: nr, not reported; d, decomposes; subl, sublimes; BSA, benzenesulfonamide; TSA, toluenesulfonamide; PTU, phenylthiourea; BZM, benzamide; ACM, acetamide; HCl, hydrochloride salt; *, see note to right.

Note: All derivative values are melting points, °C.

TABLE 28.11
Solid tertiary amines

Name	mp, °C	bp, °C	PIC	MeI	HCl	Notes
Isoquinoline	26	242	222	159	227	Methyl tosylate, mp 163°C
2,3-Dimethyl-5,6,7,8-tetrahydroquinoline	38	nr	169	117	nr	
6-Chloroquinoline	41	262	nr	248	nr	Methyl tosylate, mp 143°C
2,4,8-Trimethylquinoline	43	nr	193	229	238	
2,6,8-Trimethylquinoline	46	nr	187	nr	207	
5-Bromoquinoline	48	280	nr	205	225	
8-Methoxyquinoline	50	283	143	160	nr	
7-Bromoquinoline	52	290	nr	240	213	
2,6-Dimethylquinoline	60	266	178 (186)	236	nr	Methyl tosylate, mp 175°C
N,*N*-Dimethyl-3-nitroaniline	60	285	119	205	nr	
2,4,6-Trimethylquinoline	65	281	200	246	268	
N,*N*-Dibenzylaniline	70	300d	131d	135	nr	
8-Hydroxyquinoline	73	270	204	143d	nr	Benzoate ester, mp 120°C
Tribenzylamine	91	380	190	184	nr	
1-Phenylisoquinoline	95	300	165	242	235	
Acridine	97	346	208	224	nr	10-Azaanthracene
3,5-Dibromopyridine	112	222	nr	274	nr	Methyl tosylate, mp 219°C
4,4′-Bis(dimethylamino)benzophenone	175	nr	156	105	nr	2,4-DNP, mp 273°C; oxime, mp 233°C
Hexamethylenetetramine	280	nr	nr	190	nr	Urotropine; methyl tosylate, mp 205°C

Abbreviations used: nr, not reported; d, decomposes; PIC, picrate; MeI, methiodide; HCl, hydrochloride salt.

TABLE 28.12
Liquid esters

Name	bp, °C	mp, °C	SE	RI	DEN, g/mL	Acid produced	bp of acid, °C	mp of acid, °C
Methyl formate	34	−100	60	1.3434	0.974	Formic	101	8
Ethyl formate	53	−80	74	1.3592	0.917	Formic	101	8
Methyl acetate	57	−98	74	1.3652	0.932	Acetic	117	16
Methyl chloroformate	71	nr	47	1.3865	1.223	(Hydrochloric)	—	—
Ethyl acetate	77	−83	88	1.3720	0.900	Acetic	117	16
Methyl propionate	79	−88	88	1.3770	0.915	Propionic	141	−23
Methyl acrylate	80	−75	86	1.4021	0.956	Acrylic	139	13
n-Propyl formate	81	−92	88	1.3779	0.9071	Formic	101	8
Isopropyl acetate	85	−73	102	1.3770	0.872	Acetic	117	16
Dimethyl carbonate	90	3	45	1.3682	1.069	(Carbonic)	—	—
Methyl isobutyrate	90	−86	102	1.3821	0.892	Isobutyric	154	−47
Ethyl chloroformate	93	−80	54	1.3941	1.135	(Hydrochloric)	—	—
sec-Butyl formate	97	nr	102	1.3840	0.884	Formic	101	8
tert-Butyl acetate	98	nr	116	1.3853	0.862	Acetic	117	16
Isobutyl formate	98	−95	102	1.3854	0.885	Formic	101	8
Methyl methacrylate	99	−48	100	1.4140	0.936	Methacrylic	163	16
Ethyl propionate	99	−73	102	1.3835	0.891	Propionic	141	−23
Ethyl acrylate	99	−71	100	1.4049	0.924	Acrylic	139	13
Methyl trimethylacetate (methyl pivalate)	101	nr	116	1.3900	0.873	Trimethylacetic (pivalic)	163	34
n-Propyl acetate	102	−95	102	1.3840	0.888	Acetic	117	16
Methyl *n*-butyrate	102	−85	102	1.3879	0.898	Butyric	163	−8
Trimethyl orthoformate	102	nr	—	1.3790	0.970	(Carbonic)	—	—
Allyl acetate	104	nr	100	1.4049	0.928	Acetic	117	16
n-Butyl formate	107	−92	102	1.3894	0.888	Formic	101	8
Ethyl isobutyrate	110	−88	116	1.3903	0.869	Isobutyric	154	−47
sec-Butyl acetate	112	nr	116	1.3877	0.876	Acetic	117	16
Methyl isovalerate	117	nr	116	1.3900	0.881	Isovaleric	176	−37
Isobutyl acetate	117	nr	116	1.3901	0.875	Acetic	117	16
Ethyl trimethylacetate (ethyl pivalate)	118	nr	130	1.3906	0.855	Trimethylacetic (pivalic)	163	34
Ethyl methacrylate	118	nr	114	1.4147	0.9106	Methacrylic	163	16

Abbreviations used: nr, not reported; d, decomposes; SE, saponification equivalent; RI, refractive index; DEN, density.

TABLE 28.12 (*Continued*)

Name	bp, °C	mp, °C	SE	RI	DEN, g/mL	Acid produced	bp of acid, °C	mp of acid, °C
Methyl crotonate	119	nr	100	1.4233	0.944	Crotonic	180	71
Ethyl *n*-butyrate	120	−93	116	1.3920	0.878	Butyric	163	−8
n-Propyl propionate	123	−76	116	1.3935	0.881	Propionic	141	−23
tert-Amyl acetate	124	nr	130	1.4010	0.874	Acetic	117	16
Isoamyl formate	124	−94	116	1.3976	0.882	Formic	101	8
n-Butyl acetate	127	−78	116	1.3940	0.882	Acetic	117	16
Diethyl carbonate	127	−43	59	1.3837	0.975	(Carbonic)	—	—
Methyl valerate	128	nr	116	1.3962	0.885	Valeric	186	−18
Isopropyl butyrate	128	nr	130	nr	nr	Butyric	163	−8
Methyl methoxyacetate	130	nr	104	1.3964	1.051	Methoxyacetic	203	nr
Ethyl isovalerate	132	−99	130	1.3964	0.868	Isovaleric	176	−37
sec-Amyl acetate (2-pentyl acetate)	134	nr	130	1.3960	0.869	Acetic	117	16
n-Propyl isobutyrate	135	nr	130	1.3959	0.884	Isobutyric	154	−47
Methyl pyruvate	135	nr	102	1.4065	1.130	Pyruvic	165	12
Isobutyl propionate	137	−71	130	1.3975	0.888	Propionic	141	−23
n-Amyl acetate (1-pentyl acetate)	142 (149)	−100 (−70)	130	1.4019	0.876	Acetic	117	16
Ethyl crotonate	142	nr	114	1.4248	0.918	Crotonic	180	71
Isoamyl acetate	142	−78	130	1.4000	0.876	Acetic	117	16
Ethyl chloroacetate	143	−26	122	1.4205	1.114	Chloroacetic	189	62
n-Propyl *n*-butyrate	144	−95	130	1.4005	0.872	Butyric	163	−8
Ethyl pyruvate	144 (155)	nr	116	1.4056	1.060	Pyruvic	165	12
Methyl bromoacetate (bp 51°C/15 torr)	144d	nr	153	1.4586	1.657	Bromoacetic	206	48
Triethyl orthoformate	146	−76	—	1.3909	0.891	(Carbonic)	—	—
Ethyl valerate	146	−91	130	1.4009	0.874	Valeric	186	−18
n-Butyl propionate	147	−90	130	1.4040	0.870	Propionic	141	−23
Isobutyl isobutyrate	149	−81	144	1.3999	0.875	Isobutyric	154	−47
n-Amyl acetate (1-pentyl acetate)	149 (142)	−70 (−100)	130	1.4019	0.876	Acetic	117	16
Methyl caproate	151	−71	130	1.4050	0.885	Caproic	202	−3

Abbreviations used: nr, not reported; d, decomposes; SE, saponification equivalent; RI, refractive index; DEN, density.

TABLE 28.12 *(Continued)*

Name	bp, °C	mp, °C	SE	RI	DEN, g/mL	Acid produced	bp of acid, °C	mp of acid, °C
Isopropyl valerate	153	nr	144	1.4009	0.858	Valeric	186	−18
Ethyl pyruvate	144	nr	116	1.4056	1.060	Pyruvic	165	12
	(155)							
2-Ethoxyethyl acetate	156	nr	132	1.4040	0.975	Acetic	117	16
Isobutyl butyrate	157	nr	144	1.4030	0.862	Butyric	163	−8
Ethyl 2-bromopropionate	158	nr	181	1.4470	1.394	Bromopropionic	205	24
	(179)							
Ethyl bromoacetate	159	nr	167	1.4510	1.506	Bromoacetic	206	48
	(169)							
n-Butyl butyrate	167	−92	144	1.4060	0.869	Butyric	163	−8
Ethyl caproate	168	−68	144	1.4075	0.873	Caproic	202	−3
Ethyl trichloroacetate	168	nr	191	1.4447	1.378	Trichloroacetic	197	57
Ethyl bromoacetate	169	nr	167	1.4510	1.506	Bromoacetic	206	48
	(159)							
Hexyl acetate	169	−80	144	1.4090	0.876	Acetic	117	16
Methyl acetoacetate	169	−80	116	1.4185	1.076	Acetoacetic	100d	36
Methyl heptanoate (Methyl enanthate)	173	−56	144	1.4108	0.870	Heptanoic	223	−10
Cyclohexyl acetate	175	nr	142	1.4420	0.970	Acetic	117	16
Furfuryl acetate	176	nr	140	1.4618	1.118	Acetic	117	16
Hexyl acetate	178	−80	144	1.4090	0.876	Acetic	117	16
	(169)							
Ethyl 2-bromopropionate	179	nr	181	1.4470	1.394	Bromopropionic	205	24
	(158)							
Ethyl acetoacetate	181	−43	130	1.4190	1.021	Acetoacetic	100d	36
Methyl furoate	181	nr	126	1.4862	1.179	Furoic	230	133
Dimethyl malonate	181	−62	66	1.4135	1.156	Malonic	—	135d
Diethyl oxalate	185	−41	73	1.4096	1.076	Oxalic (dihydrate)	—	101
n-Pentyl butyrate	186	−72	158	1.4120	0.866	Butyric	163	−8
n-Butyl valerate	187	−93	158	1.4123	0.868	Valeric	186	−18
n-Propyl caproate	187	−75	158	1.4170	0.867	Caproic	202	−3
Ethyl heptanoate	187	−66	158	1.4144	0.8685	Heptanoic	223	−10
Isoamyl isovalerate	190	nr	172	1.4130	0.870	Isovaleric	176	−37

Abbreviations used: nr, not reported; d, decomposes; SE, saponification equivalent; RI, refractive index; DEN, density.

TABLE 28.12 (*Continued*)

Name	bp, °C	mp, °C	SE	RI	DEN, g/mL	Acid produced	bp of acid, °C	mp of acid, °C
Methyl caprylate	193	−40	158	1.4160	0.8775	Caprylic	237	16
Phenyl acetate	196	nr	136	1.5030	1.073	Acetic	117	16
Methyl benzoate	198	−12	136	1.5165	1.094	Benzoic	249	121
Diethyl malonate	199	−50	80	1.4135	1.055	Malonic	—	135d
Dimethyl succinate	200	19	73	1.4190	1.117	Succinic	235d	188
Benzyl formate	203	nr	136	1.5160	1.080	Formic	101	8
Dimethyl maleate	204	8	72	1.4416	1.145	Maleic	nr	130
Ethyl levulinate	206	nr	144	1.4222	1.016	Levulinic	250d	33
Benzyl acetate	206 (216)	−51	150	1.5006	1.040	Acetic	117	16
n-Amyl valerate	206 (204)	−79	172	1.4130 (1.4181)	0.858	Valeric	186	−18
n-Butyl caproate	208	−64	172	1.4153	0.8623	Caproic	202	−3
γ-Valerolactone	208	nr	100	1.4576	1.079	γ-Hydroxyvaleric	—	—
n-Propyl heptanoate	208	−65	172	1.4184	0.8656	Heptanoic	223	−10
Ethyl caprylate	208	−47	172	1.4166	0.878	Caprylic	237	16
2-Cresyl acetate	208	nr	150		1.048	Acetic	117	16
Phenyl propionate	211	75	150	1.5003	1.047	Propionic	141	−23
3-Cresyl acetate	212	12	150	1.4978	1.049	Acetic	117	16
4-Cresyl acetate	212	nr	150	1.4991	1.051	Acetic	117	16
Ethyl benzoate	212	−34	150	1.5049	1.051	Benzoic	249	121
Benzyl acetate	216 (206)	−51	150	1.5006	1.040	Acetic	117	16
Diethyl succinate	218	−20	87	1.4200	1.047	Succinic	235d	188
Methyl *p*-toluate	217	33	150	solid	solid	*p*-Toluic	179	275
Isopropyl benzoate	218	nr	164	1.4890	1.012	Benzoic	249	121
Methyl phenylacetate	218 (254)	nr	150	1.5075	1.044	Phenylacetic	206	76
n-Propyl levulinate	221	nr	158	1.4258	0.989	Levulinic	250d	33
Methyl salicylate	222	−8	152	1.5362	1.174	Salicylic	nr	158
Diethyl maleate	225	−10	86	1.4390	1.064	Maleic	nr	130
Menthyl acetate	227	nr	198	1.4469	0.9185	Acetic	117	16
Ethyl phenylacetate	229	nr	164	1.4980	1.031	Phenylacetic	206	76

Abbreviations used: nr, not reported; d, decomposes; SE, saponification equivalent; RI, refractive index; DEN, density.

TABLE 28.12 *(Continued)*

Name	bp, °C	mp, °C	SE	RI	DEN, g/mL	Acid produced	bp of acid, °C	mp of acid, °C
n-Propyl benzoate	231	−52	164	1.5014	1.027	Benzoic	249	121
Diethyl glutarate	234 (237)	−24	94	1.4240	1.022	Glutaric	302	98
Ethyl salicylate	234	3	166	1.5219	1.131	Salicylic	nr	158
Ethyl 3-toluate	234	nr	164	1.5050	1.026	3-Toluic	263	110
Ethyl 4-toluate	235	nr	164	1.5089	1.027	4-Toluic	275	179
Isopropyl salicylate	237 (242)	nr	180	1.5065	1.073	Salicylic	nr	158
n-Propyl salicylate	239 (250)	nr	180	1.5161	1.098	Salicylic	nr	158
Ethyl decanoate	245	−20	200	1.4248	0.862	Decanoic	268	30
Diethyl adipate	245	−21	101	1.4276	1.009	Adipic	265	153

Abbreviations used: nr, not reported; d, decomposes; SE, saponification equivalent; RI, refractive index; DEN, density.

TABLE 28.13
Solid esters

Name	mp, °C	bp, °C	SE	Acid produced	mp of acid, °C	bp of acid, °C
Dimethyl succinate	18	196	73	Succinic	185	nr
Methyl myristate	18	323	242	Myristic	54	nr
Phenyl propionate	20	211	150	Propionic	−23	141
Methyl 3-chlorobenzoate	21	231	171	3-Chlorobenzoic	158	nr
Benzyl benzoate	21	323	212	Benzoic	121	249
3,4-Dimethylphenyl acetate	22	235	164	Acetic	16	117
Methyl anthranilate	24	300	151	Anthranilic	146	nr
Dimethyl sebacate	27	nr	115	Sebacic	133	nr
d-Bornyl acetate	29	224	196	Acetic	16	117
Eugenyl acetate	30	282	206	Acetic	16	117
Methyl palmitate	30	nr	270	Palmitic	63	nr
Ethyl 2-nitrobenzoate	30	275	195	2-Nitrobenzoic	146	nr
n-Octadecyl acetate	30	nr	312	Acetic	16	117
Ethyl 2-naphthoate	32	304	200	2-Naphthoic	185	nr
Methyl 3-bromobenzoate	32	nr	229	3-Bromobenzoic	155	nr
Methyl 4-toluate	33	217	150	4-Toluic	179	275
Thymyl benzoate	33	nr	255	Benzoic	121	249
Ethyl furoate	34	197	140	Furoic	133	230
Diethyl 4-nitrophthalate	34	nr	133	4-Nitrophthalic	165	nr
Ethyl benzilate	34	nr	256	Benzilic	150	nr
Methyl cinnamate	36	261	162	Cinnamic	133	300
Ethyl *dl*-mandelate	37	254	180	Mandelic	118	nr
Dimethyl itaconate	38	208	79	Itaconic	165	nr
Monomethyl sebacate	38	288	216	Sebacic	133	nr
Benzyl cinnamate	39	nr	238	Cinnamic	133	300
Phenyl salicylate	42	nr	214	Salicylic	158	nr
Dibenzyl phthalate	43	nr	173	Phthalic	201	nr
Diethyl terephthalate	44	302	111	Terephthalic	300	nr
Cinnamyl cinnamate	44	nr	264	Cinnamic	133	300
Ethyl 3-nitrobenzoate	47	296	195	3-Nitrobenzoic	140	nr
2-Phenylethyl cinnamate	47	nr	252	Cinnamic	133	300
1-Naphthyl acetate	48	nr	186	Acetic	16	117

Abbreviations used: nr, not reported; SE, saponification equivalent.

TABLE 28.13 (*Continued*)

Name	mp, °C	bp, °C	SE	Acid produced	mp of acid, °C	bp of acid, °C
Methyl 4-methoxybenzoate	49	255	166	4-Methoxybenzoic	185	277
Phenacyl acetate	49	nr	178	Acetic	16	117
Dibenzyl succinate	51	nr	149	Succinic	183	235
Methyl piperonylate	52	271	180	Piperonylic	229	nr
Methyl *dl*-mandelate	53	250	166	Mandelic	118	nr
3-Cresyl benzoate	55	314	212	Benzoic	121	249
Ethyl 4-nitrobenzoate	56	186	195	4-Nitrobenzoic	241	nr
1-Naphthyl benzoate	56	nr	248	Benzoic	121	249
Ethyl diphenylacetate	58	nr	240	Diphenylacetic	148	nr
Ethyl 2-benzoylbenzoate	58	nr	254	2-Benzoylbenzoic	128	nr
Methyl diphenylacetate	60	nr	226	Diphenylacetic	148	nr
Dimethyl 4-nitrophthalate	66	nr	120	4-Nitrophthalic	165	nr
Dicyclohexyl phthalate	66	nr	165	Phthalic	201	nr
Phenyl benzoate	70	314	198	Benzoic	121	249
Dimethyl 3-nitrophthalate	70	nr	134	3-Nitrophthalic	218	nr
Methyl 3-hydroxybenzoate	70	nr	152	3-Hydroxybenzoic	200	nr
2-Naphthyl acetate	71	nr	186	Acetic	16	117
4-Cresyl benzoate	71	316	212	Benzoic	121	249
Glyceryl tribenzoate	72(76)	nr	135	Benzoic	121	249
Phenyl cinnamate	72	nr	238	Cinnamic	133	300
Ethylene glycol dibenzoate	73	nr	135	Benzoic	121	249
Methyl 2-nitrocinnamate	73	nr	207	2-Nitrocinnamic	240	nr
Diphenyl phthalate	74	nr	159	Phthalic	201	nr
Methyl benzilate	75	nr	242	Benzilic	150	nr
Methyl 2-naphthoate	77	290	186	2-Naphthoic	185	nr
Methyl 3-nitrobenzoate	78	279	181	3-Nitrobenzoic	140	nr
Ethyl 3-nitrocinnamate	79	nr	221	3-Nitrocinnamic	199	nr
Methyl 2-benzoylbenzoate	80	350	240	2-Benzoylbenzoic	128	nr
Benzoin acetate	83 (104)	nr	254	Acetic	16	117
Catechol dibenzoate	84	nr	159	Benzoic	121	249
Methyl 4-nitrobenzoate	96	nr	181	4-Nitrobenzoic	241	nr
n-Propyl 4-hydroxybenzoate	96	nr	180	4-Hydroxybenzoic	213	nr
Diphenyl adipate	106	nr	149	Adipic	153	216

Abbreviations used: nr, not reported; SE, saponification equivalent.

TABLE 28.13 (*Continued*)

Name	mp, °C	bp, °C	SE	Acid produced	mp of acid, °C	bp of acid, °C
2-Naphthyl benzoate	107	nr	248	Benzoic	121	249
Methyl 3,5-dinitrobenzoate	108	nr	226	3,5-Dinitrobenzoic	204	nr
Ethyl 4-hydroxybenzoate	116	nr	166	4-Hydroxybenzoic	213	nr
Resorcinol dibenzoate	117	nr	159	Benzoic	121	249
Diphenyl succinate	121	nr	135	Succinic	184	235
Di-4-cresyl succinate	121	nr	149	Succinic	184	235
Methyl 3-nitrocinnamate	124	nr	207	3-Nitrocinnamic	199	nr
Methyl 4-hydroxybenzoate	131	nr	152	4-Hydroxybenzoic	213	nr
Hydroquinone dibenzoate	202	nr	159	Benzoic	121	249

Abbreviations used: nr, not reported; SE, saponification equivalent.

TABLE 28.14
Liquid ketones

Name	bp, °C	mp, °C	RI	DEN, g/mL	DNP	SCB	OXM	NPH	Notes
Acetone	56	−95	1.3582	0.791	126	190	59	148	Propanone
2-Butanone	80	−86	1.3780	0.804	118	146	oil	128	
Biacetyl	88	−2	1.3951	0.981	>300	235(m) 278d(di)	75(m) 245d(di)	230(m)	2,3-Butanedione
3-Methyl-2-butanone	94	nr	1.3879	0.804	120	113	oil	108 (120)	
2-Pentanone	100	nr	1.3897	0.812	144	111	58	117	
3-Pentanone	102	−40	1.3920	0.816	156	138	69	144	
Pinacolone	106	−50	1.3964	0.801	127	157	76	nr	3,3-Dimethyl-2-butanone
4-Methyl-2-pentanone	114	−80	1.3962	0.801	95d (81)	133	58	nr	
3-Methyl-2-pentanone	118	nr	1.4002	0.815	71	94	nr	nr	
2,4-Dimethyl-3-pentanone	124	−80	1.3986	0.806	88	160	34	nr	
3-Hexanone	124	nr	1.4002	0.815	130	113	oil	nr	
2-Hexanone	127	nr	1.4005	0.812	110	123	40	88	
Mesityl oxide	129	−53	1.4445	0.858	203	164 (133)	48	134	4-Methyl-3-penten-2-one
Cyclopentanone	130	−51	1.4359	0.951	144	205	56	154	
Acetylacetone	134	−30	1.4510	0.975	122(m) 209(di)	209	149(di)	nr	2,4-Pentanedione
4-Heptanone	145	−34	1.4070	0.817	75	132	oil	nr	
3-Heptanone	146	nr	1.4085	0.818	nr	nr	oil	nr	
2-Heptanone	149	−35	1.4085	0.820	89 (74)	125	oil	73	
Cyclohexanone	155	−47	1.4500	0.947	162	166	90	146	
2-Methylcyclohexanone	162	−14	1.4478	0.924	137	196	43	142	
Diisobutyl ketone	169	nr	1.4120	0.805	92 (66)	122	nr	nr	
3-Methylcyclohexanone	169	−73	1.4450	0.914	155	180	43	119	
4-Methylcyclohexanone	169	−40	1.4445	0.915	132	201	38	128	
2-Octanone	172	−16	1.4150	0.819	58	123	oil	92	
Methyl cyclohexyl ketone	180	nr	1.4514	nr	140	177	60	154	

Abbreviations used: nr, not reported; d, decomposes; m, monosubstituted; di, disubstituted; RI, refractive index; DEN, density; DNP, 2,4-dinitrophenylhydrazone; SCB, semicarbazone; OXM, oxime; NPH, 4-nitrophenylhydrazone; *, see note to right.

Note: All derivative values are melting points, °C.

TABLE 28.14 (*Continued*)

Name	bp, °C	mp, °C	RI	DEN, g/mL	DNP	SCB	OXM	NPH	Notes
Cycloheptanone	180	nr	1.4611	0.951	148	163	oil	137	
5-Nonanone	186	−6	1.4190	0.826	70 (41)	90	oil	nr	
2,5-Hexanedione	191	−9	1.4260	0.973	255*	220*	137*	210*	*All values are di
Phorone	198	27	nr	0.885	115	186 (221)	48	nr	2,6-Dimethyl-2,5-heptadien-4-one
Acetophenone	202	19	1.5325	1.030	338 (250)	198	60	184	Methyl phenyl ketone
β-Thujone	202	nr	1.4500	0.913	114	174	55	nr	3-Isopropyl-6-methylbicyclo-[3.1.0$^{3.5}$]hexanone
l-Menthone	207	−6	1.4505	0.895	146	187	59	nr	
Isophorone	214	nr	1.4759	0.923	130	191 (199)	77	nr	
Methyl 2-thienyl ketone	214	10	1.5660	1.168	nr	190	81	181	
o-Methylacetophenone	214	nr	1.5302	1.026	160*	206	61	nr	*Often oils
Phenylacetone	216	27	1.5158	1.000	156*	198	70	145	*Often oils
Propiophenone	218	18	1.5258	1.009	191	173	53	nr	Ethyl phenyl ketone
m-Methylacetophenone	218	−10	1.5290	0.986	207	200	56	nr	
Isobutyrophenone	218	nr	1.5172	0.986	163	181	61 (94)	nr	Isopropyl phenyl ketone
n-Butyrophenone	221	11	1.5195	1.021	190	188	50	nr	
Pulegone	224	nr	1.4850	0.937	142	174	119	nr	2-Isopropylidene-5-methylcyclohexanone
Isobutyl phenyl ketone	225 (236)	nr	1.5139	0.9701	240 (124)	210	72	nr	Isovalerophenone
p-Methylacetophenone	226	27	1.5328	1.005	260	205	86	198	
(+)-Carvone	230	nr	1.4989	0.965	187	162 (142)	*d*, 72 *dl*, 92	174	1-Methyl-4-isopropenyl-Δ^6-cyclohexen-2-one
p-Chloroacetophenone	232	20	1.5549	1.192	233	202	95	239	
Benzylacetone	235	nr	1.5122	0.989	127	142	86	nr	
Isobutyl phenyl ketone	236 (225)	nr	1.5139	0.9701	240 (124)	210	72	nr	Isovalerophenone
Valerophenone	245	nr	1.5143	0.988	166	160	52	162	Butyl phenyl ketone
o-Methylacetophenone	245	nr	1.5393	1.090	nr	183	83	nr	

Abbreviations used: nr, not reported; d, decomposes; m, monosubstituted; di, disubstituted; RI, refractive index; DEN, density; DNP, 2,4-dinitrophenylhydrazone; SCB, semicarbazone; OXM, oxime; NPH, 4-nitrophenylhydrazone; *, see note to right.

Note: All derivative values are melting points, °C.

TABLE 28.15
Solid ketones

Name	mp, °C	bp, °C	DNP	SCB	OXM	NPH	PHZ	Notes
Phorone	27	198	115	186 (221)	48	nr	nr	2,6-Dimethyl-2,5-heptadien-4-one
Phenylacetone	27	216	156	198	70	145	87	
p-Methylacetophenone	27	226	260	205	86	198	97	
2-Hydroxyacetophenone	28	218	212	210	117	nr	110	
Levulinic acid	33	245	206	nr	45	174	108	β-Acetylpropionic
1-Acetylnaphthalene	34	302	>300	230	136	nr	146	1-Acetonaphthone
Dibenzyl ketone	34	330	100	145	125	nr	120 (128)	1,3-Diphenylacetone
p-Chloropropiophenone	35	nr	223	175	62	nr	nr	
4-Methoxyacetophenone	38	258	220 (230)	198	87	195	142	
1,2-Cyclohexanedione	39	194	nr	nr	nr	nr	nr	
Benzalacetone	39 (41)	260	223	187	115	166	157	(*E*)-4-Phenyl-3-buten-2-one
1-Indanone	40	242	252	233	145	234	134	
Benzophenone	48	305	238	164	142	154 (144)	137	
p-Bromoacetophenone	51	255	230 (237)	208	128	nr	126	
3,4-Dimethoxyacetophenone	51	287	206	218	140	227	131	
2-Acetonaphthone	53	300	262d	235	147	nr	176	Methyl 2-naphthyl ketone
4-Methylbenzophenone	56	326	200	121	136 (154)	nr	109	
Deoxybenzoin	56	320	204	148	98	161	116	α-Phenylacetophenone
Chalcone	57	345d	244	168 (180)	140 (70)	nr	120	Benzalacetophenone
4-Methoxybenzophenone	60	354	180	nr	138 (115)	198	132 (90)	
4-Chlorobenzophenone	75	nr	185	nr	163	nr	106	
m-Nitroacetophenone	76	202	228	257	132	nr	128 (135)	
p-Nitroacetophenone	78	202	nr	nr	nr	nr	nr	

Abbreviations used: nr, not reported; d, decomposes; m, monosubstituted; d, disubstituted; DNP, 2,4-dinitrophenylhydrazone; SCB, semicarbazone; OXM, oxime; NPH, 4-nitrophenylhydrazone; PHZ, phenylhydrazone; *, see note to right.

Note: All derivative values are melting points, °C.

TABLE 28.15 (*Continued*)

Name	mp, °C	bp, °C	DNP	SCB	OXM	NPH	PHZ	Notes
Fluorenone	79	341	283	234	195	269	151	
Di-*p*-tolyl ketone	92	334	218 (229)	143	163	nr	100	
Benzil	94	347	189(m)	177(m) 243(di)	137(m) 237(di)	193(m) 290(di)	134(m) 225(di) 235(di)	
Dibenzalacetone	112	nr	180	187	144	173	153	Distyryl ketone
4-Acetylbiphenyl	117	nr	241	nr	184	nr	nr	
Benzoin	134	344	245	205d	151	nr	106 (158)	
p-Hydroxybenzophenone	134	nr	242	194	152	nr	144	
Furoin	135	nr	217	nr	161 (102)	nr	80	
2,4-Dihydroxyacetophenone	143	nr	206	216	200	nr	157	
4,4′-Bis(dimethylamino)benzophenone	175	nr	273	nr	233	nr	174	Michler's ketone
dl-Camphor	176	205	164	235 (247)	118	217	233	
d-Camphor	177	205	177	237	118	217	233	

Abbreviations used: nr, not reported; d, decomposes; m, monosubstituted; d, disubstituted; DNP, 2,4-dinitrophenylhydrazone; SCB, semicarbazone; OXM, oxime; NPH, 4-nitrophenylhydrazone; PHZ, phenylhydrazone; *, see note to right.

Note: All derivative values are melting points, °C.

TABLE 28.16
Liquid nitriles

Name	bp, °C	mp, °C	RI	DEN, g/mL	Acid produced	bp of acid, °C	mp of acid, °C	Notes
Acetonitrile	81	−48	1.3440	0.786	Acetic	117	16	
Propionitrile	97	−93	1.3660	0.772	Propanoic	141	−23	**Poison**
Trimethylacetonitrile	105	15	1.3774	0.752	Trimethylacetic (pivalic)	163	34	
Isobutyronitrile	107	−72	1.3720	0.760	Isobutyric	154	−47	
n-Butyronitrile	116	−112	1.3842	0.794	Butyric	163	−8	
Valeronitrile	140	−96	1.3973	0.795	Valeric	186	−18	
2-Furonitrile	147	nr	1.4798	1.064	2-Furoic	nr	133	
Hexanenitrile	163	−80	1.4061	0.809	Hexanoic	202	−3	
Mandelonitrile	170	nr	1.5315	1.117	Mandelic	nr	118	
Benzonitrile	188	−13	1.5280	1.010	Benzoic	249	121	
n-Heptyl cyanide	199	−45	1.4200	0.814	Octanoic	237	16	
2-Toluonitrile	205	13	1.5279	0.989	2-Toluic	258	103	**Severe poison**
2-Methylbenzyl cyanide	212	nr	1.5275	1.056	2-Tolylacetic	nr	89	
3-Toluonitrile	212	−23	1.5256	0.976	3-Toluic	263	110	
4-Toluonitrile	217	26	nr	0.981	4-Toluic	275	179	
n-Octyl cyanide	224	nr	1.4260	0.786	*n*-Nonanoic	254	9	
Benzyl cyanide	233	−24	1.5230	0.972	Phenylacetic	206	76	
4-Fluorophenylacetonitrile	240	nr	1.5002	1.126	4-Fluorophenylacetic	nr	81	
3-Methylbenzyl cyanide	240	nr	1.5200	1.002	3-Tolylacetic	nr	63	
4-Methylbenzyl cyanide	242	18	1.5190	0.992	4-Tolylacetic	265	92	
Cinnamonitrile	255	19	1.6010	1.028	Cinnamic	300	133	
Glutaronitrile	286	−29	1.4345	0.995	Glutaric	302	98	
1,4-Dicyanobutane	295	1	1.4380	0.951	Adipic	nr	153	

Abbreviations used: nr, not reported; RI, refractive index; DEN, density.

TABLE 28.17
Solid nitriles

Name	mp, °C	bp, °C	RI	DEN, g/cm³	Acid produced	mp of acid, °C	bp of acid, °C
4-Methylbenzyl cyanide	18	242	1.5190	0.992	4-Tolylacetic	92	265
Cinnamonitrile	19	255	1.6010	1.028	Cinnamic	133	300
2-Chlorophenylacetonitrile	24	241	1.5440	nr	2-Chlorophenylacetic	95	nr
4-Toluonitrile	26	217	nr	0.981	4-Toluic	179	275
2-Cyanopyridine	27	213	1.5288	nr	2-Carboxypyridine (2-picolinic acid)	137	nr
4-Chlorobenzyl cyanide	30	266	nr	nr	4-Chlorophenylacetic	104	nr
1-Naphthylacetonitrile	34	nr	1.6192	nr	1-Naphthylacetic	130	nr
4-Fluorobenzonitrile	36	188	nr	nr	4-Fluorobenzoic	183	nr
3-Bromobenzonitrile	39	225	nr	nr	3-Bromobenzoic	155	nr
3-Chlorobenzonitrile	40	nr	nr	nr	3-Chlorobenzoic	157	nr
2-Chlorobenzonitrile	44	232	nr	nr	2-Chlorobenzoic	138	nr
Succinonitrile	47	265	nr	0.985	Succinic	188	235d
Anthranilonitrile	48	267	nr	nr	Anthranilic	146	nr
4-Bromophenylacetonitrile	48	nr	nr	nr	4-Bromophenylacetic	118	nr
3-Cyanopyridine	51	243	nr	nr	Nicotinic	236	nr
2-Bromobenzonitrile	55	252	nr	nr	2-Bromobenzoic	148	nr
p-Acetylbenzonitrile	57	nr	nr	nr	4-Acetylbenzoic	nr	208
p-Anisonitrile	58	240	nr	nr	4-Anisic	183	nr
Diphenylacetonitrile	72	nr	nr	nr	Diphenylacetic	148	nr
4-Cyanopyridine	79	nr	nr	nr	Isonicotinic	313	nr
2-Naphthylacetonitrile	83	303	nr	nr	2-Naphthylacetic	142	nr
4-Chlorobenzonitrile	92	223	nr	nr	4-Chlorobenzoic acid	240	nr
Piperonylonitrile	92	nr	nr	nr	Piperonylic acid	230	nr
2-Nitrobenzonitrile	105	nr	nr	nr	2-Nitrobenzoic acid	146	nr
4-Bromobenzonitrile	110	236	nr	nr	4-Bromobenzoic	252	nr
2-Chloro-5-nitrobenzonitrile	106	nr	nr	nr	2-Chloro-5-nitrobenzoic	167	nr
4-Nitrophenylacetonitrile	115	nr	nr	nr	4-Nitrophenylacetic acid	153	nr
3-Nitrobenzonitrile	116	nr	nr	nr	3-Nitrobenzoic acid	140	nr
3,5-Dinitrobenzonitrile	128	nr	nr	nr	3,5-Dinitrobenzoic acid	240	nr
1,2-Dicyanobenzene	140	nr	nr	nr	Phthalic acid	201 (subl)	nr

Abbreviations used: nr, not reported; d, decomposes; subl, sublimes; RI, refractive index; DEN, density.

TABLE 28.17 ***(Continued)***

Name	mp, °C	bp, °C	RI	DEN, g/cm^3	Acid produced	mp of acid, °C	bp of acid, °C
4-Nitrobenzonitrile	147	nr	nr	nr	4-Nitrobenzoic acid	240	nr
1,3-Dicyanobenzene	159	nr	nr	nr	Isophthalic acid	300	nr
Anthracene carbonitrile	175	nr	nr	nr	Anthracene-9-carboxylic acid	214	nr
Adamantane carbonitrile	187	nr	nr	nr	1-Adamantylcarboxylic acid	174	nr
1,4-Dicyanobenzene	225	nr	nr	nr	Terephthalic	300 (subl)	nr

Abbreviations used: nr, not reported; d, decomposes; subl, sublimes; RI, refractive index; DEN, density.

TABLE 28.18
Liquid phenols

Name	bp, °C	mp, °C	RI	DEN, g/mL	AAA	ANE	NPU	BRD	BZE	Notes
2-Chlorophenol	176	8	1.5579	1.241	145	186.5	120	48(m) 76(di)	nr	
Phenol	181	40	—	1.071	99	152	132	95(t)	69	Hydroxybenzene
2-Cresol	191	31	—	1.048	152	166	141	56(di)	nr	2-Methylphenol
2-Bromophenol	195	5	1.5892	1.492	143	231	129	95(t)	nr	
2-Chloro-4-methylphenol	196	nr	1.5200	1.178	108	200.5	nr	nr	71	
2-Ethylphenol	196	−18	1.5372	1.037	141	180	nr	nr	38	
Salicylaldehyde	197	1	1.5719	1.146	132	180	nr	nr	nr	DNP, mp 250°C(d)
4-Cresol	202	33	—	1.034	136	166	146	47(di) 198(te)	70	4-Methylphenol
3-Cresol	203	10	1.5392	1.034	103	166	127	84(t)	55	3-Methylphenol
Guaiacol	205	28	1.5429	1.129	118	182	118	116(t)	57	2-Methoxyphenol
2,4-Dichlorophenol	209	42	—	nr	137	221	nr	68(m)	97	
2,4-Dimethylphenol	212	26	1.5390	1.027	141	180	135	nr	37	
3-Chlorophenol	214	34	1.5632	nr	110	186.5	158	nr	71	
3-Ethylphenol	217	−4	1.5300	1.000	77	180	nr	nr	52	
4-Chlorophenol	220	44	—	1.306	156	186.5	166	90(di)	88	
2-*n*-Propylphenol	225	nr	1.5279	0.989	99	194	nr	nr	nr	
4-Isobutylphenol	236	nr	1.5319	0.979	124	208	nr	nr	nr	
3-Bromophenol	236	31	—	nr	108	231	108	nr	86	
Carvacrol	237	0	1.5240	0.976	151	150	116	46(m)	nr	3-Hydroxy-4-methylcumene
2,4-Dibromophenol	238	36	—	nr	153	310	nr	95(m)	97	
3-Methoxyphenol	243	−17	1.5510	1.131	118	182	128	104(t)	nr	
4-*n*-Butylphenol	248	22	1.5165	0.978	81	208	nr	nr	27	
Eugenol	254	−11	1.5408	1.066	80	222	122	118(te)	70	4-Allyl-2-methoxyphenol
4-*n*-Pentylphenol	254	23	1.5272	0.962	90	222	nr	nr	51	
Isoeugenol	267	nr	1.5782	1.085	94 (116)	222	150	94(di)	103 (*Z*, 68)	2-Methoxy-4-propenylphenol

Abbreviations used: AAA, aryloxyacetic acid derivative; ANE, arloxyacetic acid neutralization equivalent; BRD, bromide; BZE, benzoate ester; NPU, naphthylurethane; m, monosubstituted; di, disubstituted; t, trisubstituted; te, tetrasubstituted; nr, not reported.

Note: All derivative values are melting points, °C.

TABLE 28.19
Solid phenols

Name	mp, °C	bp, °C	AAA	ANE	NPU	BRD	BZE	Notes
4-*n*-Butylphenol	22	248	81	208	nr	nr	27	
4-*n*-Pentylphenol	23	254	90	222	nr	nr	51	
2,4-Dimethylphenol	26	212	141	180	135	nr	37	
Guaiacol	28	205	118	182	118	116	57	2-Methoxyphenol
o-Cresol	31	191	152	166	141	56(di)	nr	2-Methylphenol
2-Bromo-4-chlorophenol	33	nr	139	265.5	nr	nr	99	
p-Cresol	33	202	136	166	146	47(di) 198(te)	70	4-Methylphenol
2,4-Dibromophenol	36	238	153	310	nr	95(m)	97	
3-Iodophenol	40	nr	115	278	nr	nr	72	
Phenol	40	181	99	152	132	95(t)	69	
2,4-Dichlorophenol	42	209	137	221	nr	68(m)	97	
4-Chlorophenol	44	220	156	186.5	166	90(di)	88	
2-Nitrophenol	44	214	158	197	113	117(di)	59	
4-Ethylphenol	44	218	97	180	128	nr	59	
2,6-Dimethylphenol	45	203	nr	180	176	79	nr	
Thymol	50	232	149	208	160	55	32	5-Methyl-2-isopropyl-1-phenol
4-Methoxyphenol	55	243	110	182	nr	nr	87	
2,3-Dichlorophenol	57	206	174	221	nr	90(di)	nr	
Orcinol hydrate	58	290	217	120	160	104(t)	88(di)	3,5-Dihydroxytoluene
4-Isopropylphenol	59	212	nr	194	nr	nr	71	
4-Bromophenol	64	238	157	231	169	95(di)*	58	*di = tribromophenol
2,4,6-Trichlorophenol	64	246	182	255.5	188	nr	70(75)	
2,4,5-Trichlorophenol	64	248	153	255.5	nr	nr	93	
3,4-Dimethylphenol	66	227	162	180	141	171(t)	59	
3,5-Dichlorophenol	67	233	nr	221	nr	189(t) (166)	55	
2,4,6-Trimethylphenol	68	220	142	194	nr	158(di)	62	Mesitol
2,4,5-Trimethylphenol	71	232	132	194	nr	35	63	Pseudocuminol
2,5-Dimethylphenol	71	212	118	180	172	178(t)	61	
2,3-Dimethylphenol	74	217	187	180	nr	nr	nr	5-Methoxy-4-hydroxybenzaldehyde
Vanillin	81	285d	187	210	nr	nr	78	2,4-DNP, mp 271°C(d); semicarbazone, mp 229°C

Abbreviations used: AAA, aryloxyacetic acid derivative; ANE, aryloxyacetic acid neutralization equivalent; BRD, bromide; BZE, benzoate ester; NPU, naphthylurethane; m, monosubstituted; di, disubstituted; t, trisubstituted; te, tetrasubstituted; nr, not reported.

Note: All derivative values are melting points, °C.

TABLE 28.19 (*Continued*)

Name	mp, °C	bp, °C	AAA	ANE	NPU	BRD	BZE	Notes
2-Hydroxybenzyl alcohol	83	nr	120	182	nr	nr	51(di)	Salicyl alcohol
4-Iodophenol	92	nr	156	278	nr	nr	119	
1-Naphthol	95	278	193	202	152	105 (2,4-di)	56	
2,4,6-Tribromophenol	95 (87)	286	200	388	153	120(te)	81	
3-Nitrophenol	97	nr	156	197	167	91(di)	95	
4-*tert*-Butylphenol	100	237	86	208	110	50(m) 67(di)	81	
3-Hydroxybenzaldehyde	101	nr	148	180	nr	nr	48	2,4-DNP, mp 257°C(d); semicarbazone, mp 198°C
Catechol	104	245	136	113	175	192(te)	84(di)	1,2-Dihydroxybenzene
Orcinol (anhydrous)	107	290	217	120	160	104(t)	88(di)	3,5-Dihydroxytoluene; orcinol hydrate, mp 58°C
1,2-Dihydroxynaphthalene	108	nr	*	*	*	*	*	***Potent blistering agent**
2,2-Dihydroxybiphenyl	109	326	nr	151	nr	188(di)	101(di)	
Resorcinol	110	275	175 (195)	113	nr	112(t)	117(di) 135(m)	1,3-Dihydroxybenzene
4-Nitrophenol	113	279	187	197	150	142(di)	142	
2,4-Dinitrophenol	114	nr	*	*	*	*	*	***Toxic**
4-Hydroxybenzaldehyde	116	nr	198	180	nr	181(di)	90	2,4-DNP, mp 242°C; semicarbazone, mp 224°C
2-Naphthol	122	286	95	202	156	84	107	
2,5-Dihydroxytoluene	124	nr	153	120	nr	84	119(di)	
1,2,3-Trihydroxyphenol	113	309	nr	100	nr	158(di)	140(m) 108(di) 90(t)	Pyrogallol
4-Hydroxybenzophenone	134	nr	nr	256	nr	nr	115	2,4-DNP, mp 242°C; semicarbazone, mp 194°C
2-Hydroxybenzoic acid	158	nr	191	98	nr	nr	132	Salicylic acid; toluidide, mp 156°C
2,4,6-Triiodophenol	158	nr	nr	530	nr	nr	137	
4-Phenylphenol	165	305	nr	228	nr	nr	150	

Abbreviations used: AAA, aryloxyacetic acid derivative; ANE, aryloxyacetic acid neutralization equivalent; BRD, bromide; BZE, benzoate ester; NPU, naphthylurethane; m, monosubstituted; di, disubstituted; t, trisubstituted; te, tetrasubstituted; nr, not reported.

Note: All derivative values are melting points, °C.

TABLE 28.19 ***(Continued)***

Name	mp, °C	bp, °C	AAA	ANE	NPU	BRD	BZE	Notes
Hydroquinone	172	285	250	113	247	186(di)	200	
1,4-Dihydroxynaphthalene	176	nr	nr	138	220	nr	186(di)	
2,7-Dihydroxynaphthalene	187	nr	147	138	nr	nr	139(di) 199(m)	
Pentachlorophenol	190	nr	196	324	nr	nr	164	
3-Hydroxybenzoic acid	201	nr	206	98	nr	nr	nr	Toluidide, mp 163°C; anilide, mp 155°C; amide, mp 170°C
4-Hydroxybenzoic acid	214	nr	278	98	nr	nr	221	Toluidide, mp 203°C; anilide, mp 199°C
1,3,5-Trihydroxybenzene	220	nr	nr	100	nr	151(t)	173(t)	

Abbreviations used: AAA, aryloxyacetic acid derivative; ANE, aryloxyacetic acid neutralization equivalent; BRD, bromide; BZE, benzoate ester; NPU, naphthylurethane; m, monosubstituted; di, disubstituted; t, trisubstituted; te, tetrasubstituted; nr, not reported.

Note: All derivative values are melting points, °C.

INDEX

Abbreviations, 619
Accident reporting, 2
Acetanilide:
  bromination of, 452
  ir spectrum of, 263, 453
  nmr spectrum of, 263, 453
  preparation of, 261
Acetoacetic acid thioester, 375
Acetoacetic ester, 322
  (*See also* Ethyl acetoacetate)
Acetoacetic ester condensation, 320
Acetonitrile condensation, 385
Acetophenone, ir spectrum of, 171
Acetylcoenzyme A, 375
Acetylferrocene, 411
  ir spectrum of, 415
  preparation of, 412, 414
Acetylsalicylic acid, 254
  ir spectrum of, 257
  nmr spectrum of, 257
  (*See also* Aspirin)
Acid-base extraction, 88
Acid chloride formation, 518
Acidity, 499
  of carboxylic acids, 499
  of phenols, 499
Acids (*see* Carboxylic acids; Phenols)
Activated hyrdrocarbons, 379
Acyloin, 390
Aggregation pheromone, 293
Air oxidation, 335
Alarm pheromone, 293
Alcohols, 579
  tert-amyl, 581
  Bacycr tcst for, 585
  benzoate esters from, 597
  ceric ammonium nitrate test for, 589
  chromic anhydride reagent, 588
  classes of, 580
  classification of, 584
  3,5-dinitrobenzoate derivatives, 598
  2,4-dinitrophenylhydrazine test
    for, 585
  ester formation from, 595
  general discussion of, 580
  from Grignards, 275
  hydrogen bonding in, 583
  Lucas test for, 589
  α-naphthylurethanes, 593
  4-nitrobenzoate derivatives, 598
  oxidation-aldehyde sequence, 591
  oxidation tests for, 586
  periodate test for, 591
  phenylurethane derivative, 593
  properties of, 582
  spectroscopy of, 592
  table of liquid, 626
  table of solid, 629
Aldehydes:
  Cannizzarro oxidation, 577
  odor of, 554
  oxidation of acids, 576
  permanganate oxidation, 576
  table of liquid, 630

Aldehydes:
  table of solid, 632
  (*See also* Carbonyl compounds)
Aldol condensation, 372, 373
Alkaloids, 462
Alkanes, 195
  preparation of, 197
  solubility of, 198
  table of, 196
Alkenes, 200
  bromination of, 205
  bromine addition to, 204
  formation of, 203
  hardening of oils, 202
  properties of, 200
  reactivity of, 202
Alkyl halides, 232
Alkynes, 217
Amide derivatives, 518
  procedure for forming, 520
Amides, 240
  classification scheme for, 607
  derivatives, 612
  diagnostic hydrolysis of, 606
  hydrolysis of, 604
  preparation of, 243
  in qualitative organic analysis, 600
  reduction of, 614
  resonance in, 242
  saponification of, 612
  spectroscopy of, 607
  table of, 633
Amines, 531
  basicity of, 534
  classes of, 533
  derivatives of, 545
  diazotization of, 542
  Hinsberg test for, 538
  Hofmann carbylamine test for, 543
  hydrochlorides of, 546, 548
  methiodide salts, 550
  methyl tosylate derivatives, 551
  nitrous acid test, 541
  phenylthiourea derivatives, 547
  picrate derivative, 548
  reactivity of, 544
  Schotten-Baumann benzoylation of, 546
  spectroscopy of, 543
  table of liquid primary and
    secondary, 639
  table of liquid tertiary, 642
  table of solid primary and
    secondary, 643
  table of solid tertiary, 645
4-Amino-3-hydrazino-5-mercapto-1,2,4-triazole
  (*see* Purpald test)
tert-Amyl alcohol, 581
Anilide derivatives, 520
Aniline:
  ir spectrum of, 369
  from nitrobenzene, 364
  nmr spectrum of, 369
Anisaldehyde, 554
Anthracene, 126
Arbuzov reaction, 475
Aromatic substitution:
  electrophilic, 428, 429
  nucleophilic, 458
Aryloxyacetic acid derivatives, 527
Aspirin, 253
  preparation of, 255
Azeotropes, 56
  diagrams for, 57
  table of, 57
Azo compounds, 127

Baeyer test, 564
  for alcohols, 585
Base solubility, procedure for
    determining, 499, 503
Beer-Lambert law, 154
Beilstein test, 129, 494
  procedure for, 130, 495
Benedict's test, 564
Benzalacetophenone (*see* Chalcone)
Benzaldehyde:
  ir spectrum of, 170
  reaction with Grignard, 281
9-Benzalfluorene, 380
  hydride reduction of, 383
  nmr spectrum of, 382
  preparation of, 381
Benzamide derivatives, 546
Benzene, 126
  C-H ratio in, 130
  restrictions on use of, 4
  use of, 3
  uv spectrum of, 160
Benzenesulfonamide derivatives, 545
Benzhydrol:
  from benzaldehyde, 281
  hypochlorite oxidation of, 352
  ir spectrum of, 283, 346
  nmr spectrum of, 283, 346
  oxidation of, 342, 352
Benzil, 559
Benzoate ester derivatives, 597
  of phenols, 528
Benzoic acid:
  from Grignard reagent, 287
  ir spectrum of, 171, 289

Benzoic acid:
nmr spectrum of, 289, 393
(*See also* Benzonitrile, hydrolysis of)
Benzoin:
ir spectrum of, 393
nmr spectrum of, 393
preparation of, 392
Benzoin condensation, 389
mechanism for, 391
Benzonitrile:
hydrolysis of, 304
ir spectrum of, 172
Benzophenone:
from benzhydrol, 342
by hypochlorite oxidation of benzhydrol, 352
ir spectrum of, 345
nmr spectrum of, 345
reaction with phenylmagnesium bromide, 276
N-Benzylaniline, solubility in acid, 501
Benzyl cyanide (*see* Phenylacetonitrile)
9-Benzylfluorene, 380
nmr spectrum of, 382
preparation of, 383
preparation from fluorene, 384
Benzyltriethylammonium chloride, (BTEAC), 331
nmr spectrum of, 332
Biacetyl, 559
Biphenyl, from Grignard reagent, 277
Boiling-point (bp) determination, 491
by capillary method, 491
apparatus for, 492
factors affecting, 33
by microreflux method, 37, 491
apparatus for, 38
procedure for, 34
Boiling points (bp), 32
apparatus for measuring, 35
Borohydride reduction of aldehydes and ketones, 578
Bromination:
of acetanilide, 452
of alkenes, 205
of 4-bromoacetophenone, 422, 426
of enols, 420
of phenols, 528
of stilbene, 208
of *p*-xylene, 448
Bromine solution, 205
4-Bromoacetanilide:
ir spectrum of, 454
nmr spectrum of, 454
preparation of, 452
4-Bromoacetophenone:
bromination of, 422, 426
ir spectrum of, 410, 424
4-Bromoacetophenone:
nmr spectrum of, 410, 424
preparation of, 408
Bromobenzene, nitration of, 437
1-Bromobutane:
ir spectrum of, 169, 236
nmr spectrum of, 236
preparation of, 234
2-Bromo-1,4-dimethylbenzene, Grignard carbonation of, 290
1-Bromo-2,4-dinitrobenzene, 442
warning, 443
1-Bromo-4-nitrobenzene:
nitration of, 442
preparation of, 437
4-Bromophenacyl bromide, 421
ir spectrum of, 425
nmr spectrum of, 425
preparation of, 422, 426
2-Bromo-*p*-xylene:
nmr spectrum of, 451
(*See also* 2-bromo-1,4-dimethylbenzene)
BTEAC (*see* Benzyltriethylammonium chloride)
Bubble-cap column, 54
Bunsen burner, 115, 116
(S)-cis-Butadiene, 224
n-Butanol, ir spectrum of, 169
n-Butylacetoacetic ester (*see* Ethyl *n*-butylacetoacetate)
n-Butyl benzoate:
ir spectrum of, 245
nmr spectrum of, 245
by phase-transfer catalysis, 244
preparation of, 243
reaction with Grignard, 279
n-Butyl bromide:
preparation of, 234
(*See also* 1-Bromobutane)
tert-Butyl chloride:
ir spectrum of, 238
nmr spectrum of, 238
preparation of, 237
Butyrophenone, nmr spectrum of, 185

Caffeine, 463
ir spectrum of, 465
isolation from tea leaves, 464
nmr spectrum of, 465
Calcium chloride as drying agent, 91
Calcium sulfate as drying agent, 91
Calculation of yield, 8
Camphor, 44, 67
ir spectrum of, 350
from isoborneol, 342
nmr spectrum of, 350

Camphor:
  preparation of, 347
  sublimation of, 68, 349
Cannizzaro reaction, 394
  of 4-chlorobenzaldehyde, 397
  cross-Cannizzaro as modification of, 395
  oxidation of aldehydes by, 577
Capillary bp determination, 491
  apparatus for, 492
Caraway seeds, 64
Carbon dioxide, vibrational modes of, 165
Carbonyl compounds, 552
  Baeyer test for, 564
  Benedict's test for, 564
  classification of, 558, 561
  dimedone derivatives, 575
  2,4-dinitrophenylhydrazine test for, 558, 560
  2,4-dinitrophenylhydrazone derivatives, 571
  Fehling's test for, 564
  Fuchsin aldehyde test for, 567
  general discussion of, 553
  iodoform test for, 568
  odor of, 554
  oxime derivatives, 574
  Purpald test for, 565
  reduction of, 577
  Schiff's test for, 567
  semicarbazone derivatives, 573
  spectroscopy of, 569
  Tollens test for, 562
  types of, 555
Carboxylic acids, 508
  acidity of, 499
  amide derivatives of, 518
  classification scheme for, 512
  derivatives of, 513
  dissociation constants of, 510
  ethyl esters of, 521
  history of, 509
  hydrogen bonding in, 584
  methyl esters of, 521
  neutralization equivalent of, 514
  phenacyl ester derivatives of, 523
  solubility of, 511
  spectroscopy of, 530
  table of liquid, 621
  table of solid, 623
Carbylamine test for amines, 543
Carvone:
  from caraway seeds, 64
  2,4-DNP derivative of, 66
Caryophyllene, 201
Cedrene, 201
Celite, 78
Ceric ammonium nitrate reagent, 589
Chalcone, 374
  ir spectrum of, 378
  nmr spectrum of, 378
  preparation of, 377
Charcoal, 76
  apparatus for treatment of, 77
Chemical Abstracts, 12, 13
Chemical literature, 9
  locating compounds in, 11
  locating preparations in, 10
  sources of, 13
Chemical shift (*see* Nuclear magnetic resonance)
Chemical spills, 2
4-Chlorobenzaldehyde:
  Cannizzaro reaction of, 397
  hydride reduction of, 358
  ir spectrum of, 360
  nmr spectrum of, 360
  reaction with Grignard reagent, 284
Chlorobenzene, nitration of, 431
4-Chlorobenzhydrol:
  from 4-chlorobenzaldehyde, 284
  hypochlorite oxidation of, 354
  ir spectrum of, 285, 356
  nmr spectrum of, 285, 356
4-Chlorobenzoic acid:
  from Cannizarro reaction, 397
  ir spectrum of, 399
  nmr spectrum of, 399
4-Chlorobenzophenone, 357
  ir spectrum of, 357
  nmr spectrum of, 357, 407
  preparation of, 354
  preparation by Friedel-Crafts reaction, 405
4-Chlorobenzyl acetate:
  hydrolysis of, 309
  ir spectrum of, 311
  nmr spectrum of, 311
  preparation of, 308
4-Chlorobenzyl alcohol:
  from the Cannizzaro reaction, 397
  from 4-chlorobenzaldehyde, 358
  ir spectrum of, 312, 361, 398
  nmr spectrum of, 312, 361, 398
  preparation of, 309, 359
bis-4-Chlorobenzyl ether, 318
  nmr spectrum of, 320
1-Chloro-2,4-dinitrobenzene, 439
  ir spectrum of, 441
  nmr spectrum of, 441
  preparation of, 439
  reaction with hydrazine, 460
  warning, 441
Chloromethanes, polarization in, 142

4-Chloronitrobenzene, 430
  ir spectrum of, 433
  nitration of, 439
  nmr spectrum of, 433
  preparation of, 431, 434
4-Chlorophenyl phenyl carbinol (*see* 4-Chlorobenzhydrol)
Cholesterol, 44
Chromatography, 92
  column (*see* Column chromatography)
  development in, 98, 103
  equipment for, 96
  gas (gc), 97, 98
  gas-liquid (glc) 93, 97
  high-pressure liquid (hplc), 95
  mobile phases in, 98
  paper, 95, 98
  solvents for, 98, 102
  thin-layer (*see* Thin-layer chromatography)
  vapor-phase (vpc), 93, 97
Chromic anhydride reagent, 588
Chromium trioxide, 342
Chromophore, uv spectroscopy for, 157, 159
Cinnamaldehyde, 555
Claisen condensation, 371
Clove (spice), 466
Collins reagent, 586
Color, 126
  in qualitative organic analysis, 489
Column chromatography, 94, 108
  illustration of, 110
  packing the column for, 109
  sample addition in, 112
  sample elution in, 113
  solvent delivery in, 112
  steps in, 111
Coniine, 532
Cross-Cannizzaro reaction, 395
18-Crown-6, 149
Crown ethers, 149
Crystallization, 70
  advantages of, 71
  procedure for, 74, 80
  solvents for, 72, 73
Crystals, seed, 79
Cyanide:
  in aldol reactions, 391
  as catalyst, 390
  (*See also* Sodium cyanide)
Cycloaddition by Diels-Alder reaction, 223, 225
1,3-Cyclohexadiene, 224
Cyclohexanol, dehydration of, 205
Cyclohexanone, ir spectrum of, 170
Cyclohexene:
  ir spectrum of, 168
Cyclohexene:
  nmr spectrum of, 208
  preparation of, 205
  reaction with :CCl_2, 213
4-Cyclohexene-1,2-dicarboxylic anhydride:
  ir spectrum of, 228
  nmr spectrum of, 228
  preparation of, 226
Cyclohexenone, ir spectrum of, 170
Cyclopentadiene, 224
  reaction with maleic anhydride, 229

2,4-D (*see* 2,4-Dichlorophenoxyacetic acid)
DEET (see *N,N*-Diethyl-*m*-toluamide)
Defoliating agent, 313
Density, 39
  procedure for determining, 40, 41
  table of selected densities, 40
Derivative tables, 618
  usage note on, 620
Deuterium oxide, 179
  exchange with alcohols, 592
Diatomaceous earth, 78
Dibenzalacetone, 373
  nmr spectrum of, 376
  preparation of, 375
1,3-Dibromopropane, nmr spectrum of, 184
1,4-Di-*tert*-butyl-2,5-dimethoxybenzene:
  nmr spectrum of, 418
  preparation of, 417
β-Dicarbonyl groups, 321
Dichlorocarbene, 212
7,7-Dichloronorcarane, 212
2,4-Dichlorophenoxyacetic acid (2,4-D), 316
  nmr spectrum of, 317
  preparation of, 312, 317
Dicyclohexano-18-crown-6, 150
Dicyclopentadiene, 224
Diels-Alder reaction, 223, 225
Dienophile (diene component), 223
Diethyl benzylphosphonate, 475
  nmr spectrum of, 476
Diethyl *n*-butylmalonate:
  ir spectrum of, 330
  nmr spectrum of, 330
  preparation of, 327
Diethyl ether (*see* Ethyl ether)
Diethyl malonate, alkylation of, 327
N,N-Diethyl-*m*-toluamide (DEET):
  ir spectrum of, 172, 260
  nmr spectrum of, 260
  preparation of, 258
Dihydroxyacetone phosphate, 375
Diisopropyl ether (*see* Isopropyl ether)

Dimedone derivatives, 575
1,4-Dimethoxybenzene, alkylation of, 416
1,4-Dimethylbenzene (*see* Xylene)
2,5-Dimethylbenzoic acid:
  ir spectrum of, 292
  nmr spectrum of, 292
  preparation of, 290
2,5-Dimethylbromobenzene, 448
3,5-Dinitrobenzoate derivatives, 598
  of esters, 611
2,4-Dinitrophenylhydrazine (2,4-DNP), 459
  preparation of, 460
  test for alcohols, 585
  test for aldehydes and ketones, 560
  test for carbonyl compounds, 558, 560
  warning, 459
2,4-Dinitrophenylhydrazones:
  color of, 127
  preparation of, 571
Diphenylacetylene (tolan):
  ir spectrum of, 220
  nmr spectrum of, 220
  preparation of, 217, 219
1,4-Diphenylbutadiene:
  nmr spectrum of, 481
  preparation of, 479
Diphenyl carbinol (*see* Benzhydrol)
Dipolar aprotic solvents, 143
Dipole moments, 140
Di-*n*-propyl ether (see *n*-Propyl ether)
Distillation, 45
  apparatus for, 36, 51, 59, 207
  azeotrope table, 57
  of azeotropes, 56
  bubble-cap column for, 54
  column efficiency in, 50
  columns for, 49
  fractional, 47
    procedure for, 58
  fractionation apparatus, 52, 53
  of qualitative organic analysis samples, 493
  simple, 46
  spinning-band column, 54
  steam, 62
    apparatus for, 63, 65
  theoretical plates in, 49
  vacuum, 53
    apparatus for, 56
2,4-DNP (*see* 2,4-Dinitrophenylhydrazine)
Dry ice, 286
Drying agents, 90

Electrophilic aromatic bromination, 448
Electrophilic aromatic nitration, 430
Electrophilic aromatic substitution, 428
Electrophilic aromatic substitution:
  mechanism of, 429
Elemental analysis, 132
Eluotropic series, 98
Emmons reaction, 474
Ene component (dienophile), 223
Enol bromination, 420
Enol test, 526
Essential oil, 466
Esters, 240, 521, 600
  benzoate derivatives of, 597
  classification scheme for, 607
  3,5-dinitrobenzoate derivatives of, 611
  ethyl esters of carboxylic acids, 521
  formation from alcohols, 595
  fragment isolation, 611
  hydrolysis of, 604
  hydroxamic acid test for, 605
  malonic ester condensation, 320
  methyl esters of carboxylic acids, 521
  preparation of, 240
  reactivity of, 603
  saponification equivalent, 609
  spectroscopy of, 607
  table of liquid, 646
  table of solid, 651
Ethyl acetate, ir spectrum of, 169
Ethyl acetoacetate:
  alkylation of, 320, 325
  sodium salt of, 324
Ethylbenzene, uv spectrum of, 160
Ethyl *n*-butylacetoacetate, 322
  ir spectrum of, 326
  nmr spectrum of, 326
  preparation of, 322
Ethyl ether, nmr spectrum of, 182
Ethyl phenyl ketone (*see* Propiophenone)
Eugenol:
  ir spectrum of, 468
  isolation of, 467
  nmr spectrum of, 468
Experimental techniques, 17
Extraction, 81
  acid-base 88
  funnel size for, 83
  performance of, 83
  procedure for, 92
  separatory funnels for, 84–86
Extraction volumes, 83

Fehling's test, 564
Fermi contact mechanism, 177
Ferric chloride test, 526
Ferrocene, 67, 411
  acylation of, 412

Ferrocene:
  ir spectrum of, 415
Filter aid, 78
Filter paper, 78
  fluted, 78
Filtration, gravity, 77
    apparatus for, 79
Fischer esterification, 246
Flame test, 129, 494
Fluorene:
  9-benzylfluorene from, 384
  color of, 126
  fluorenone from, 335
  nmr spectrum of, 381
  partial oxidation of, 339
  reaction with benzaldehyde, 381
Fluorene oxidation, 335
  apparatus for, 340, 341
  mechanism of, 336
  partial, 339
  by phase-transfer catalysis, 337
Fluorenol:
  from fluorenone, 358
  ir spectrum of, 363
  nmr spectrum of, 363
  preparation of, 362
Fluorenone:
  color of, 126
  from fluorene, 335
  hydride reduction of, 358
  nmr spectrum of, 342
  preparation of, 337
Formaldehyde, 488
Fractional distillation, 47, 58
Fractionation apparatus, 52, 53
Friedel-Crafts acylation, 404
Friedel-Crafts alkylation, 416
Fructose-1,6-diphosphate, 375
Fuchsin aldehyde test, 567
Fumaric acid, 212
  preparation of, 210

Gas chromatography (gc), 97, 98
Gas-liquid chromatography (glc), 93, 97
Gas removal, 122
Gc (gas chromatography), 97, 98
Glas-Col, 119
Glc (gas-liquid chromatography), 93, 97
Glyceraldehyde phosphate, 375
Gravity filtration, 77, 79
Grignard reagent, 268
  addition to a nitrile, 271
  apparatus for, 273
  benzoic acid from, 287
  biphenyl from, 277
Grignard reagent:
  carbonation of, 286, 294
  reaction with benzaldehyde, 281
  reaction with *n*-butyl benzoate, 279
  reaction with 4-chlorobenzaldehyde, 284
  reaction with 4-methyl-3-heptanol, 296
  synthesis of alcohols from, 275

Haloform reaction, 240
Halogens:
  analysis for, 133
  Beilstein test for, 129
Heating mantles, 119
Heating methods, 114
  burners, 115, 116
  free flame, 115
  oil bath, 117, 119
  steam bath, 116, 117
Heptahelicene, 44
Herbicide, 313
High-pressure liquid chromatography,
    (hplc), 95
Hinsberg test, 538
  procedure for, 540
Hofmann carbylamine test, 542, 543
Horner-Wittig reaction, 474
Hplc (high-pressure liquid chromatography), 95
Hydrates, 20
Hydride reduction:
  of 9-benzalfluorene, 383
  of 4-chlorobenzaldehyde, 358
  of fluorenone, 358
Hydrocarbons, activated, 379
Hydrochloride derivatives of amines, 546, 548
Hydrogen bonding, 136
  in alcohols, 583
  in carboxylic acids, 584
Hydrolysis:
  of amides, 604
  of esters, 604
  of methyl 3-nitrobenzoate, 443
  of nitriles, 604, 606, 615
  of nitrogen-containing compounds, 537
  of phenylacetonitrile, 303
  of ureas, 604
Hydroxamic acid test, 605
Hydroxylamine–ferric chloride test, 605
Hydroxylamine hydrochloride, 574
Hypochlorite oxidation, 352

IAA (indoleacetic acid), 313
Immersion heater, 120
Index of refraction, 37, 497, 506
Indoleacetic acid (IAA), 313

Infrared (ir) spectra:
  acetanilide, 263, 453
  acetophenone, 171
  acetylferrocene, 415
  acetylsalicylic acid, 257
  aniline, 369
  benzaldehyde, 170
  benzhydrol, 283, 346
  benzoic acid, 171, 289
  benzoin, 393
  benzonitrile, 172
  benzophenone, 345
  4-bromoacetanilide, 454
  4-bromoacetophenone, 410, 424
  1-bromobutane, 169, 236
  4-bromophenacyl bromide, 425
  n-butanol, 169
  n-butyl benzoate, 245
  tert-butyl chloride, 238
  caffeine, 465
  camphor, 350
  chalcone, 378
  4-chlorobenzaldehyde, 360
  4-chlorobenzhydrol, 285, 356
  4-chlorobenzoic acid, 399
  4-chlorobenzophenone, 357
  4-chlorobenzyl acetate, 311
  4-chlorobenzyl alcohol, 312, 361, 398
  1-chloro-2,4-dinitrobenzene, 441
  4-chloronitrobenzene, 433
  cyclohexanone, 170
  cyclohexene, 168
  4-cyclohexene-1,2-dicarboxylic anhydride, 228
  cyclohexenone, 170
  diethyl benzylphosphonate, 476
  diethyl *n*-butylmalonate, 330
  N,N-diethyl-*m*-toluamide, 172, 260
  2,5-dimethylbenzoic acid, 292
  diphenylacetylene, 220
  ethyl acetate, 169
  ethyl *n*-butylacetoacetate, 326
  eugenol, 468
  ferrocene, 415
  fluorenol, 363
  isoamyl acetate, 253
  isoborneol, 351
  methyl benzoate, 248, 445
  methyl 4-chlorobenzoate, 250
  4-methyl-3-heptanol, 299
  methyl 3-nitrobenzoate, 446
  nitrobenzene, 368
  3-nitrobenzoic acid, 447
  norbornene-5,6-*endo*-dicarboxylic anhydride, 231

Infrared (ir) spectra:
  pentanoic acid, 171, 295
  phenacetin, 265
  α-phenethylamine, 470
  phenol, 350
  phenylacetic acid, 308
  phenylacetonitrile, 172, 307
  triphenylcarbinol, 278
Infrared (ir) spectroscopy, 161
  functional group absorptions, 166
  instrumentation in, 166
  reduced mass, 164
  references for, 190
  sampling techniques in, 166
  theory of, 161
Iodoethane, nmr spectrum of, 180
Iodoform test, 568
1-Iodopropane, nmr spectrum of, 181
Ir (*see* Infrared spectra; Infrared spectroscopy)
Isoamyl acetate (pear oil):
  ir spectrum of, 253
  nmr spectrum of, 253
  preparation of, 251
Isoborneol:
  ir spectrum of, 351
  nmr spectrum of, 351
  oxidation of, 342
Isomorphs, 22
4-Isopropenylcyclohexanone, uv spectrum of, 158
4-Isopropylcyclohex-2-enone, uv spectrum of, 158
Isopropyl ether, nmr spectrum of, 183

Ketones:
  table of liquid, 654
  table of solid, 656
  (*See also* Carbonyl compounds)
Knoevenagel condensation, 371

Laboratory notebook, 4, 5
Lewis bases, 139
Linalool, 587
Liquid-vapor composition curve, 48
Literature (*see* Chemical literature)
Lithium aluminum hydride, 358
London forces, 140
Lucas test, 589

Macrocyclic polyethers (*see* Crown ethers)
Magnesium sulfate as drying agent, 91

Magnetic stirring apparatus, 353
Maleic acid:
  isomerization of, 210
  nmr spectrum of, 212
Maleic anhydride, 210
  isomerization of, 210
  reaction with cyclopentadiene, 229
  reaction with sulfolene, 225
Malonic acid, 322
Malonic ester condensation, 320
Margarine, 202
Mass spectrometry, 185
  instrumentation in, 186
  molecular weight determination using, 186
  references for, 190
  theory of, 185
Meisenheimer complex, 459
Melting behavior of qualitative organic analysis samples, 493
Melting point, 19
  apparatus for determining, 23
  calibration curve, 31
  Fisher-Johns apparatus, 27
  of glasses, 21
  of isomorphs, 22
  Meltemp apparatus, 28
  mixture, 21, 32
  procedure for determining, 30, 31
  sample preparation, 27
  thermometer calibration, 27, 30
  Thiele tube apparatus, 25
  Thomas-Hoover apparatus, 26
Menschutkin reaction, 331
Methiodide salts, 550
o-Methoxyacetophenone, 559
Methyl benzoate:
  ir spectrum of, 248, 445
  nitration of, 443
  nmr spectrum of, 248, 445
  preparation of, 246
Methyl 4-chlorobenzoate:
  ir spectrum of, 250
  nmr spectrum of, 250
  preparation of, 248
3,4-Methylenedioxycinnamonitrile, 385
  nmr spectrum of, 389
  preparation of, 386
4-Methyl-3-heptanol:
  by Grignard reaction, 296
  ir spectrum of, 299
  nmr spectrum of, 299
Methyl 3-nitrobenzoate:
  hydrolysis of, 443
  ir spectrum of, 446
  nmr spectrum of, 446
4-Methylphenoxyacetic acid:
  nmr spectrum of, 315
  preparation of, 312
Methyl salicylate, 254
Methyl tosylate salts, 551
Microreflux apparatus, 38
Microreflux bp determination, 491
Microreflux method, 37, 491
Mixture melting points, 21, 32
Molecular sieves, 91
Muscalure, 218

Naphthalene, 216
  uv spectrum of, 161
α-Napthylurethane derivative, 593
Natural products, 462
Neutral substances, solubility of, 502
Neutralization equivalent, 514
  of aryloxyacetic acid derivatives, 528
  procedure for, 516
Niacin, 532
Nicotine, 532
Nicontinic acid, 532
Nitration, 430
  apparatus for, 436
  of bromobenzene, 437
  of 1-bromo-4-nitrobenzene, 442
  of chlorobenzene, 431
  of 1-chloro-4-nitrobenzene, 439
  of methyl benzoate, 443
Nitriles:
  basicity of, 603
  classification scheme for, 607
  diagnostic hydrolysis of, 606
  Grignard reagent addition to, 271
  hydrolysis of, 604
    procedure for, 615
  in qualitative organic analysis, 600
  spectroscopy of, 607
  table of liquid, 658
  table of solid, 659
Nitrobenzene:
  ir spectrum of, 368
  nmr spectrum of, 368
  reduction of, 364
    apparatus for, 366
4-Nitrobenzoate derivatives, 598
3-Nitrobenzoic acid:
  ir spectrum of, 447
  nmr spectrum of, 447
  preparation of, 443
Nitrogen, analysis for, 133
Nitrogen-containing compounds, hydrolysis of, 537

Nitromethane condensation, 379
β-Nitrostyrene, 379
Nitrous acid test, 541
Nmr (*see* Nuclear magnetic resonance; Nuclear magnetic resonance spectra)
Norbornene-5,6-*endo*-dicarboxylic anhydride:
  ir spectrum of, 231
  synthesis of, 230
Norite, 76
Notebook:
  format for page, 6
  ink for records, 5, 6
  laboratory, 4, 5
  sample page in, 7
Nuclear magnetic resonance (nmr), 173
  chemical shift in, 175
  chemical shift ranges in, 180
  coupling constant in, 176
  decoupling a resonance in, 179
  instrumentation for, 175
  references for, 191
  shielding cones in, 176
  structure determination using, 179
  theory of, 173
  Zeeman diagram in, 174
Nuclear magnetic resonance (nmr) spectra:
  acetanilide, 263, 453
  acetylsalicylic acid, 257
  aniline, 369
  9-benzalfluorene, 382
  benzhydrol, 283, 346
  benzoic acid, 289, 393
  benzoin, 393
  benzophenone, 345
  9-benzylfluorene, 382
  benzyltriethylammonium chloride, 332
  4-bromoacetanilide, 454
  4-bromoacetophenone, 410, 424
  1-bromobutane, 236
  4-bromophenacyl bromide, 425
  2-bromo-*p*-xylene, 451
  n-butyl benzoate, 245
  tert-butyl chloride, 238
  butyrophenone, 185
  caffeine, 465
  camphor, 350
  chalcone, 378
  4-chlorobenzaldehyde, 360
  4-chlorobenzhydrol, 285, 356
  4-chlorobenzoic acid, 399
  4-chlorobenzophenone, 357, 407
  4-chlorobenzyl acetate, 311
  4-chlorobenzyl alcohol, 312, 361, 398
  bis-4-chlorobenzyl ether, 320
  1-chloro-2,4-dinitrobenzene, 441
  4-chloronitrobenzene, 433
Nuclear magnetic resonance (nmr) spectra:
  cyclohexene, 208
  4-cyclohexene-1,2-dicarboxylic anhydride, 228
  dibenzalacetone, 376
  1,3-dibromopropane, 184
  1,4-di-*tert*-butyl-2,5-dimethoxybenzene, 418
  2,4-dichlorophenoxyacetic acid, 317
  diethyl benzylphosphonate, 476
  diethyl *n*-butylmalonate, 330
  N,N-diethyl *m*-toluamide, 260
  2,5-dimethylbenzoic acid, 292
  2,5-dimethylbromobenzene, 451
  diphenylacetylene, 220
  1,4-diphenylbutadiene, 481
  ethyl *n*-butylacetoacetate, 326
  ethyl ether, 182
  eugenol, 468
  fluorene, 381
  fluorenol, 363
  fluorenone, 342
  iodoethane, 180
  2-iodopropane, 181
  isoamyl acetate, 253
  isoborneol, 351
  isopropyl ether, 183
  maleic acid, 212
  methyl benzoate, 248, 445
  methyl 4-chlorobenzoate, 250
  3,4-methylenedioxycinnamonitrile, 389
  4-methyl-3-heptanol, 299
  methyl 3-nitrobenzoate, 446
  4-methylphenoxyacetic acid, 315
  nitrobenzene, 368
  3-nitrobenzoic acid, 447
  pentanoic acid, 295
  phenacetin, 265
  α-phenethylamine, 470
  phenylacetic acid, 308
  phenylacetonitrile, 307
  propiophenone, 184
  n-propyl ether, 182
  trans-stilbene, 479
  1,1,2-trichloroethane, 177
  triethyl orthoformate, 183
  triphenyl carbinol, 278
  p-xylene, 451
Nucleophilic aromatic substitution, 458

Octane number (RON), 196
Odor, 128
  of aldehydes and ketones, 554
  in qualitative organic analysis, 489
Oil bath, 117, 119
Oil of wintergreen, 254

Optical activity, 42, 470
Optical rotation (*see* Polarimetry)
Oxidation:
  air, 335
  of benzhydrol, 342, 352
  with chromium trioxide, 342
  of fluorene, 335
    partial, 339
  hypochlorite, 352
  of isoborneol, 342
Oxidation reactions, 334
Oxime derivatives, 574

Paper chromatography, 95, 98
Partition coefficients, 82
  procedure for determination of, 87
Pascal's triangle, 178
Pear oil (*see* Isoamyl acetate)
Pentaerythritol, 396
Pentanoic acid:
  ir spectrum of, 171, 295
  nmr spectrum of, 295
  (*See also* Valeric acid)
Periodate test, 591
Perkin condensation, 371, 385
PETN (pentaerythritol tertranitrate), 396
Phase-transfer catalysis (PTC), 145
  Baeyer test for, 565
  n-butyl benzoate by, 244
  4-chlorobenzyl acetate by, 309
  bis-4-chlorobenzyl ether by, 318
  crown ethers in, 149
  cycle diagram for, 148
  dichlorocarbene generation, 212
  hypochlorite oxidation by, 352
  oxidation of fluorene, 337
  phenacyl ester derivatives, 524
  references for, 151
Phase-transfer catalyst, 331
Phenacetin:
  ir spectrum of, 265
  nmr spectrum of, 265
  preparation of, 261
Phenacyl ester derivatives, 523
  equation for formation of, 150
  phase-transfer catalysis, 524
α-Phenethylamine:
  ir spectrum of, 470
  nmr spectrum of, 470
  resolution of, 469
  solubility in acid, 501
Phenols, 508
  acidity of, 499
  aryloxyacetic acid derivatives of, 527
  benzoate ester derivatives of, 528

Phenols:
  benzoylation of, 529
  bromination of, 528
  classification scheme for, 512, 525
  derivatives of, 527
  ferric chloride test for, 526
  ir spectrum of a phenol, 530
  pK_a of, 499
  spectroscopy of, 530
  table of liquid, 661
  table of solid, 662
  urethane derivatives of, 529
Phenylacetic acid:
  ir spectrum of, 308
  nmr spectrum of, 308
  preparation of, 303
Phenylacetonitrile:
  hydrolysis of, 303
  ir spectrum of, 172, 307
  nmr spectrum of, 307
  preparation of, 303
Phenylmagnesium bromide, 271
  (*See also* Grignard reagent)
Phenylthiourea derivatives, 547, 593, 614
Pheromones, 292
Phosphonium salts, 474
Physical measurements, 19
Picrates, 548
Pinene, 201
pK_a:
  of amines, 534
  of ammonium ions, 536
  definition of, 510
  of phenols, 499
Polarimeter, 43
Polarimetry, 41
Polyethers (*see* Crown ethers)
Potassium carbonate as drying agent, 91
Potassium hydroxide as drying agent, 91
Propiophenone, nmr spectrum of, 184
n-Propyl ether, nmr spectrum of, 182
Propyl phenyl ketone (*see* Butyrophenone)
Proton exchange using D_2O, 179
PTC (*see* Phase-transfer catalysis)
Purpald test, 565
Pyridinium bromide perbromide, 426, 455
Pyridinium chlorochromate, 586
Pyridinium chlorochromate test, 587
  warning, 588

Qualitative organic analysis (QOA), 485
  boiling point determination in, 491
  color in, 489
  melting behavior of samples, 493
  nitriles in, 600

Qualitative organic analysis (QOA):
odor in, 489
reasons for study, 486
sample purification, 490
solubility in, 498
tables for, 618
tactics in, 484
Quinine, 533

Raoult's law, 62
Records, maintenance of, 4
Recrystallization (*see* Crystallization)
Reduced mass, ir spectroscopy, 164
Reduction reactions, 334
of aldehydes and ketones, 577
of amides, 614
of nitrobenzene, 364
Reflux:
air-cooled condenser for, 220
with exclusion of moisture, 249
Reflux apparatus, 227
Refractive index, 37, 497, 506
Refractometer, 39
Resolution, 469
R_f (retardation factor) in thin-layer chromatography, 106
p-Rosaniline hydrochloride (*see* Schiff's test)

Safety glasses, 2
Safety information, 1
Safrole, 201
Salting out, 89
Sample purification, 490
Sanger's reagent, 459
Saponification equivalent, 609
Schiff's test, 567
Schotten-Baumann reaction, 519
with phenols, 529
Seed crystals, 79
Semicarbazide, 573
Semicarbazone derivatives, 573
Separatory funnel(s):
draining, 86
illustration of, 84
shaking, 84, 85
types of, 84
venting, 85
Sex pheromone, 293
S-isomer, 224
S_N1 reaction, 233
S_N2 reaction, 233, 302
Sodium borohydride, 358
Sodium cyanide, precautions and warning, 304
Sodium ethoxide, 323
Sodium fusion, 132
Sodium hydroxide as drying agent, 91
Sodium hypochlorite (*see* Hypochlorite oxidation)
Sodium sulfate as drying agent, 90
Solubility, 135
acid, procedure for determining, 500, 504
base, procedure for determining, 499, 503
of *N*-benzylaniline, 501
of carboxylic acids, 511
of neutral substances, 502
of phenethylamine, 501
in qualitative organic analysis, general discussion of, 498
in sulfuric acid, 502
Solubility classification chart, 505
Solvation, 135
Solvent affinity, 81
Solvent properties, table, 146
Specific gravity, 39, 496
procedure for determining, 40, 41
Spectroscopy, 152
of alcohols, 592
of amides, 607
of amines, 543
of carbonyl compounds, 569
of carboxylic acids, 530
of esters, 607
of nitriles, 607
of phenols, 530
references for, 190
of ureas, 607
(*See also* Infrared spectra; Infrared spectroscopy; Nuclear magnetic resonance; Nuclear magnetic resonance spectra; Ultraviolet spectra; Ultraviolet spectroscopy)
Spills, chemical, 2
Spinning-band column, 54
"Standard catalyst" solution, 145
Steam bath, 116, 117
Steam distillation, 62, 63, 65
Stilbene:
bromination of, 208
nmr spectrum of, 479
preparation of, 478
Stilbene dibromide:
diphenylacetylene from, 217, 219
elimination from, 217
preparation of, 208
Sublimation, 67
apparatus for, 69, 70

Sublimation:
  of camphor, 68, 349
Sulfolene, 225
Sulfur, analysis for, 133
Sulfuric acid, solubility in, 502
Swirling a flask, 214

Tables for qualitative organic analysis, 618
Tartaric acid, 469
TEBAC (*see* Benzyltriethylammonium chloride)
Tetracene, 126
Tetrahydroisoquinolines, 379
Thin-layer chromatography (tlc), 94
  capillary applicator for, 100
  developing chambers for, 104
  development of plate, 103
  microscope slides for, 99
  plates for, 99
  procedure for, 107
  R_f in, 106
  solvents for, 98, 102
  steps in, 101
  visualization process in, 104
  (*See also* Chromatography)
Tin, 365
Tlc (*see* Thin-layer chromatography)
Tolan (*see* Diphenylacetylene)
Tollens reagent, 563
Tollens test, 562
Toluenesulfonate derivatives, 551
p-Toluidides, 520
1,1,2-Trichloroethane, nmr spectrum of, 177
Triethyl orthoformate, nmr spectrum of, 183
Triethyl phosphite, 474
Triphenyl carbinol:
  from benzophenone, 276
  from *n*-butyl benzoate, 279
  ir spectrum of, 278
  nmr spectrum of, 278
Triphenylphosphine, 473
Trituration of a solid, 389

Ultraviolet (uv) spectra:
  benzene, 160
  ethylbenzene, 160
  4-isopropenylcyclohexanone, 158
Ultraviolet (uv) spectra:
  4-isopropylcyclohex-2-enone, 158
  naphthalene, 161
Ultraviolet (uv) spectroscopy, 153
  chromophores, 157, 159
  electronic absorption in, 156
  instrumentation of, 155
  references for, 190
  structural analysis using, 157
  Woodward-Fieser rules in, 159
Ureas, 600
  classification scheme for, 607
  derivatives of, 617
  hydrolysis of, 604
  spectroscopy of, 607
Urethane derivatives of phenols, 529
Uv (*see* Ultraviolet spectra; Ultraviolet spectroscopy)

Vacuum distillation, 53, 56
Valeric acid:
  by carbonation of a Grignard, 294
  (*See also* Pentanoic acid)
Vanillin, 555
Vapor-phase chromatography (vpc), 93, 97
Vigreux column, 49, 50
Vpc (vapor-phase chromatography), 93, 97

Williamson ether synthesis, 318
Wittig reaction, 473
  mechanism for, 474
Woodward-Fieser rules, 159

Xylene:
  bromination of, 448
  nmr spectrum of, 451

Yield calculation, 8
Ylides, 474

Zeeman diagram, 174
Zeolites, 91
Zerewittenoff reaction, 270

HTC DOWNS-JONES LIBRARY
Experimental organic chemistry
QD 261 .D87
3 3081 00026645 3